VOLUME ONE HUNDRED AND FOUR

VITAMINS AND HORMONES

Neurotrophins

VOLUME ONE HUNDRED AND FOUR

VITAMINS AND HORMONES

Neurotrophins

Series Editor

GERALD LITWACK, PhD
Toluca Lake, North Hollywood, California

ELSEVIER

ACADEMIC PRESS
An imprint of Elsevier
elsevier.com

Cover photo credit:
Bertrand, T.
Crystal Structures of Neurotrophin Receptors Kinase Domain
Vitamins and Hormones (2017) **104**, pp. 1-18.

Academic Press is an imprint of Elsevier
50 Hampshire Street, 5th Floor, Cambridge, MA 02139, United States
525 B Street, Suite 1800, San Diego, CA 92101-4495, United States
The Boulevard, Langford Lane, Kidlington, Oxford OX5 1GB, United Kingdom
125 London Wall, London, EC2Y 5AS, United Kingdom

First edition 2017

ISBN: 978-0-12-812263-1
ISSN: 0083-6729

For information on all Academic Press publications
visit our website at https://www.elsevier.com/books-and-journals

Publisher: Zoe Kruze
Acquisition Editor: Alex White
Editorial Project Manager: Helene Kabes
Production Project Manager: Vignesh Tamil
Cover Designer: Miles Hitchen

Typeset by SPi Global, India

Former Editors

CONTENTS

Contributors xiii
Preface xvii

1. Crystal Structures of Neurotrophin Receptors Kinase Domain **1**

T. Bertrand

1. Introduction 1
2. Crystal Structure of TrkA, TrkB, and TrkC Kinase Domain 4
3. Pharmacological Approach: The Selectivity Issue 13
4. Conclusion and Future Directions 16

Acknowledgments 16
References 16

2. BDNF Propeptide: A Novel Modulator of Synaptic Plasticity **19**

M. Kojima and T. Mizui

1. Neurotrophins and Their Biological Actions 20
2. NT Receptors and Their Signaling 20
3. Synthesis, Intracellular Trafficking, and Secretion of BDNF Protein 21
4. Modulation of Synaptic Function by BDNF 22
5. BDNF Prodomain: New Roles Beyond a Molecular Chaperone 22
6. BDNF and Its Propeptide Are Localized in Presynaptic Dense Core Vesicles 23
7. The BDNF Propeptide Is a Novel Synaptic Modulator 24
8. Role of the Val66Met Polymorphism in BDNF Propeptide-Dependent LTD 25
9. Discussion 25

References 26

3. Zinc Interactions With Brain-Derived Neurotrophic Factor and Related Peptide Fragments **29**

A. Travaglia and D. La Mendola

1. Introduction 30
2. Brain-Derived Neurotrophic Factor 31
3. Zinc in the Brain 38
4. Zinc(II) Interaction With BDNF and Related Peptides 41
5. Zinc, BDNF, and Diet 47

6. Conclusion 48
Acknowledgments 49
References 49

4. Structural Characterization of the p75 Neurotrophin Receptor: A Stranger in the TNFR Superfamily **57**

M. Vilar

1. TNFR Superfamily 58
2. Structural Characterization of $p75^{NTR}$ 60
3. The Transmembrane Domain (TMD) 66
4. The Intracellular Domain (ICD) 70
5. Conclusions and Future Directions 79
Acknowledgments 79
References 79

5. Releasing Mechanism of Neurotrophic Factors via Polysialic Acid **89**

C. Sato

1. Introduction 90
2. Biochemical Features of BDNF 91
3. Regions and Cellular Distribution of BDNF 95
4. Extracellular Secretion of BDNF 97
5. Conclusion 105
Acknowledgments 105
References 106

6. Emergent Role of Coronin-1a in Neuronal Signaling **113**

M. Martorella, K. Barford, B. Winkler, and C.D. Deppmann

1. Introduction 114
2. The Coronin Family of Proteins 114
3. Type I Coronins as Actin Regulators 118
4. Coronin-1a as a Signaling Effector Protein 121
5. Conclusions and Future Directions 128
Acknowledgments 129
References 129

7. Neuronal Roles of the Bicaudal D Family of Motor Adaptors **133**

M. Budzinska, K.B. Wicher, and M. Terenzio

1. Introduction 134
2. The BICD Family of Cargo Adaptors 136

3. Role of BicD in *Drosophila* Development 138
4. Mammalian BICDs and Their Role in Mitotic Cells 139
5. Role of BICD in Neurons 141
6. BICDs and NT Signaling 143
7. Conclusions 146
Acknowledgments 147
References 147

8. BDNF and Hippocampal Synaptic Plasticity 153

G. Leal, C.R. Bramham, and C.B. Duarte

1. Introduction 154
2. Basic Mechanisms of LTP 157
3. The Neurotrophin Family and Its Receptors 159
4. TrkB Localization and Signaling 160
5. Synthesis and Secretion of BDNF 162
6. BDNF and LTP in the Hippocampus 164
7. $p75^{NTR}$ and LTD 177
8. Interplay Between Glucocorticoids and BDNF to Determine Cellular Response to Stress 180
9. Concluding Remarks 182
References 183

9. Neurotrophins in the Brain: Interaction With Alcohol Exposure During Development 197

K.E. Boschen and A.Y. Klintsova

1. Introduction 198
2. Neurotrophins and the Developing Brain 198
3. Alcohol as a Teratogenic Agent 199
4. Neurotrophins as Targets of Alcohol Exposure 204
5. Interactions Between Developmental Alcohol Exposure and Neurotrophins in the Brain: Molecular and Intracellular Effects 208
6. Interactions Between Developmental Alcohol Exposure and Neurotrophins: Neuroanatomical and Behavioral Effects 215
7. Neurotrophins as Compensatory/Neuroprotective Molecules Following Developmental Alcohol Exposure 225
8. Conclusions 228
Acknowledgments 228
References 228

10. Neurotrophins: Role in Placental Growth and Development **243**

A.S. Sahay, D.P. Sundrani, and S.R. Joshi

1. Introduction 244
2. Role of Neurotrophins in Placental Growth and Development 246
3. Neurotrophins in Different Regions of the Placenta 248
4. Cross Talk of Neurotrophins With Other Growth Factors and Proteases Involved in Placental Development 250
5. Role of Neurotrophins in Pregnancy 252
6. Conclusion and Future Directions 255

Acknowledgments 256
References 256

11. Development of Precision Small-Molecule Proneurotrophic Therapies for Neurodegenerative Diseases **263**

J. Janssens, D. Lu, B. Ni, W. Chadwick, S. Siddiqui, A. Azmi, H. Etienne, A. Jushaj, J. van Gastel, B. Martin, and S. Maudsley

1. Introduction 264
2. Mammalian Neurotrophin Ligand–Receptor System 265
3. Neurotrophin Ligand–Receptor Biological Functionality 270
4. Proneurotrophin-Based Therapy for Age-Related Neurodegeneration 280
5. Future Developments in Small-Molecule Neurotrophic Agents 294

References 295

12. The Role of Neurotrophins in Inflammation and Allergy **313**

S. Manti, P. Brown, M.K. Perez, and G. Piedimonte

1. Introduction 314
2. NTs and Allergic Disease 320
3. NTs and Inflammatory Responses 323
4. Conclusions 335

References 336

13. Neurotrophic Factors and Maternal Nutrition During Pregnancy **343**

M. Dhobale

1. Introduction 344
2. Neurotrophic Factors 345
3. Maternal Nutrition 347
4. Association Between Maternal Nutrition and Neurotrophic Factors Through One Carbon Cycle 355

5. Consequences of Altered Maternal Nutrition in Pregnancy 358
References 359

14. The Role of Neurotrophin Signaling in Gliomagenesis: A Focus on the p75 Neurotrophin Receptor ($p75^{NTR}$/CD271) 367

M.M. Alshehri, S.M. Robbins, and D.L. Senger

1. General Overview 368
2. Malignant Glioma: A Major Clinical Challenge 368
3. Glioma Disease Reservoirs 370
4. Molecular Determinants of Glioma Invasion and Progression 372
5. The Role of the Neurotrophins and Their Cognate Receptors in Glioma Progression 374
6. A Role for $p75^{NTR}$ in Cancer: Extending Beyond Glioma 384
7. Conclusions and Future Directions 385

Acknowledgments 389
References 389

15. Neurotrophic Factors Used to Treat Spinal Cord Injury 405

S.I. Hodgetts and A.R. Harvey

1. Introduction 406
2. Neurotrophins 428
3. Selected Examples of Other Growth Factors 434
4. Combinations of Factors 438
5. Location of Neurotrophic Factor Delivery 439
6. Mode of Neurotrophic Factor Delivery 441
7. Timing of Neurotrophic Factor Delivery 443
8. Conclusion 444

Acknowledgments 444
References 444

16. Neurotrophins and Migraine 459

L.B. Martins, A.L. Teixeira, and R.B. Domingues

1. Introduction 459
2. Neurotrophins and Pain 460
3. Migraine Pathophysiology 463
4. Neurotrophins and Migraine 465
5. Conclusion and Future Directions 469

References 470

17. BDNF During Withdrawal **475**

H. Geoffroy and F. Noble

1. Introduction 476
2. BDNF Alterations Following Chronic Administration of Drugs of Abuse 476
3. Protracted Abstinence, Incubation, and Craving: How Is BDNF Involved? 483
4. Peripheral BDNF Levels 486
5. Polymorphism in BDNF Gene: Predictive Value in Drug Addiction? 489
6. Conclusion and Future Directions 490

Acknowledgments 491

References 491

18. Neurotrophin Receptors and Perineural Invasion: Analyses in Select Lineage-Unrelated Cutaneous Malignancies With a Propensity for Perineural Invasion **497**

N.F. Frydenlund and M. Mahalingam

1. Introduction 498
2. PNI in Cancer 498
3. Definitions 499
4. Incidence of PNI in Cutaneous Malignancies 500
5. Prognostic Impact of PNI in cSCC and DM 502
6. The Biological Underpinnings of PNI: The Role of Neurotrophins 506
7. Clinical Potential of Neurotrophins as Therapeutic Targets 518
8. Personal Observations 519
9. Conclusions 523

References 524

CONTRIBUTORS

M.M. Alshehri
Arnie Charbonneau Cancer Centre, University of Calgary, Calgary, AB, Canada

A. Azmi
Translational Neurobiology Group, University of Antwerp, Antwerpen, Belgium

K. Barford
University of Virginia, Charlottesville, VA, United States

T. Bertrand
Sanofi, Vitry-sur-Seine, France

K.E. Boschen
University of Delaware, Newark, DE, United States

C.R. Bramham
K.G. Jebsen Center for Neuropsychiatric Disorders, University of Bergen, Bergen, Norway

P. Brown
Center for Pediatric Research, Cleveland Clinic Foundation, Cleveland, OH, United States

M. Budzinska
Molecular NeuroPathobiology Laboratory, UCL Institute of Neurology, University College London, London, United Kingdom

W. Chadwick
Receptor Pharmacology Unit, National Institute on Aging, National Institutes of Health, Baltimore, MD, United States

C.D. Deppmann
University of Virginia, Charlottesville, VA, United States

M. Dhobale
BioTRaK Research and Diagnostics Centre, Navi Mumbai, India

R.B. Domingues
Universidade Federal de Minas Gerais, Belo Horizonte, Minas Gerais, Brazil

C.B. Duarte
CNC—Center for Neuroscience and Cell Biology; University of Coimbra, Coimbra, Portugal

H. Etienne
Translational Neurobiology Group, University of Antwerp, Antwerpen, Belgium

N.F. Frydenlund
University of Iowa Carver College of Medicine, Iowa City, IA, United States

H. Geoffroy
Université Paris Descartes, Inserm U1124, CNRS ERL3649, Neuroplasticité et Thérapies des addictions, Paris, France

A.R. Harvey
School of Anatomy, Physiology and Human Biology, The University of Western Australia; Western Australian Neuroscience Research Institute, Perth, WA, Australia

S.I. Hodgetts
School of Anatomy, Physiology and Human Biology, The University of Western Australia; Western Australian Neuroscience Research Institute, Perth, WA, Australia

J. Janssens
Translational Neurobiology Group, University of Antwerp, Antwerpen, Belgium

S.R. Joshi
Interactive Research School for Health Affairs, Bharati Vidyapeeth University, Pune, India

A. Jushaj
Translational Neurobiology Group, University of Antwerp, Antwerpen, Belgium

A.Y. Klintsova
University of Delaware, Newark, DE, United States

M. Kojima
Biomedical Research Institute, Advanced Industrial Science and Technology (AIST), Osaka; Core Research for Evolutional Science and Technology (CREST), Japan Science and Technology Agency (JST), Kawaguchi, Japan

D. La Mendola
Department of Pharmacy, University of Pisa, Pisa, Italy

G. Leal
CNC—Center for Neuroscience and Cell Biology, University of Coimbra, Coimbra, Portugal

D. Lu
Receptor Pharmacology Unit, National Institute on Aging, National Institutes of Health, Baltimore, MD, United States

M. Mahalingam
VA Consolidated Laboratories, West Roxbury, MA, United States

S. Manti
Center for Pediatric Research, Cleveland Clinic Foundation, Cleveland, OH, United States

B. Martin
Metabolism Unit, National Institute on Aging, National Institutes of Health, Baltimore, MD, United States

L.B. Martins
Universidade Federal de Minas Gerais, Belo Horizonte, Minas Gerais, Brazil

M. Martorella
University of Virginia, Charlottesville, VA, United States

S. Maudsley
Translational Neurobiology Group, University of Antwerp, Antwerpen, Belgium; Receptor Pharmacology Unit, National Institute on Aging, National Institutes of Health, Baltimore, MD, United States

T. Mizui
Biomedical Research Institute, Advanced Industrial Science and Technology (AIST), Osaka; Core Research for Evolutional Science and Technology (CREST), Japan Science and Technology Agency (JST), Kawaguchi, Japan

B. Ni
Receptor Pharmacology Unit, National Institute on Aging, National Institutes of Health, Baltimore, MD, United States

F. Noble
Université Paris Descartes, Inserm U1124, CNRS ERL3649, Neuroplasticité et Thérapies des addictions, Paris, France

M.K. Perez
Center for Pediatric Research; Pediatric Institute and Children's Hospital, Cleveland Clinic Foundation, Cleveland, OH, United States

G. Piedimonte
Center for Pediatric Research; Pediatric Institute and Children's Hospital, Cleveland Clinic Foundation, Cleveland, OH, United States

S.M. Robbins
Arnie Charbonneau Cancer Centre, University of Calgary, Calgary, AB, Canada

A.S. Sahay
Interactive Research School for Health Affairs, Bharati Vidyapeeth University, Pune, India

C. Sato
Bioscience and Biotechnology Center, Nagoya University, Nagoya, Japan

D.L. Senger
Arnie Charbonneau Cancer Centre, University of Calgary, Calgary, AB, Canada

S. Siddiqui
Receptor Pharmacology Unit, National Institute on Aging, National Institutes of Health, Baltimore, MD, United States

D.P. Sundrani
Interactive Research School for Health Affairs, Bharati Vidyapeeth University, Pune, India

A.L. Teixeira
Universidade Federal de Minas Gerais, Belo Horizonte, Minas Gerais, Brazil

M. Terenzio
Weizmann Institute of Science, Rehovot, Israel

A. Travaglia
Center for Neural Science, New York University, New York, NY, United States

J. van Gastel
Translational Neurobiology Group, University of Antwerp, Antwerpen, Belgium

M. Vilar
Molecular Basis of Neurodegeneration Unit, Institute of Biomedicine of Valencia (IBV-CSIC), València, Spain

K.B. Wicher
Ossianix, Stevenage Bioscience Catalyst, Stevenage, United Kingdom

B. Winkler
University of Virginia, Charlottesville, VA, United States

PREFACE

Neurotrophins were first discovered to be growth factors involved in neuronal development and functioning. Later on, they were shown to function in the immune system and reproductive system. These factors are first synthesized in the form of proneurotrophin precursors that become cleaved to generate the C-terminal mature neurotrophins. They bind and activate Trk (tropomyosin-related kinase) receptors and a p75 neurotrophin receptor ($p75^{NTR}$). The four family members of neurotrophins are comprised by the nerve growth factor (NGF), the brain-derived nerve growth factor (BDNF), neurotrophin-3 (NT-3), and neurotrophin 4/5 (NT-4/5). NGF and BDNF are prominent and have been most studied. NGF binds and activates the membrane Trk receptor leading to the activation of Ras-mitogen-activated protein kinase, phospholipase C, extracellular signal-regulated kinase, and phosphatidylinositol 3-kinase. BDNF activates the TrkB receptor that is involved in the development of the visual cortex, growth of neurons, and brain development. NT-3 activates TrkC (and to a lesser extent TrkB). NT-3 promotes neurite ganglion outgrowth. NT-4/5 activates TrkB and may be involved in bipolar disorder.

This volume is organized so that crystal structures and basic structural studies on neurotrophins and their receptors appear first. The following section focuses on neurotrophin functions and the final section concentrates on the biological actions of neurotrophins related to clinical conditions and disease.

To open with structural studies, T. Bertrand is the author of "Crystal structures of neurotrophin receptors kinase domain." This is followed by "BDNF pro-peptide: a novel modulator of synaptic plasticity" by M. Kojima and T. Mizui. Then, A. Travaglia and D. La Mendola report on "Zinc interactions with brain-derived neurotrophic factor and related peptide fragments." The "Structural characterization of p75 neurotrophin receptor: a stranger in the TNFR superfamily" is the subject of M. Vilar. Next, C. Sato reviews "Releasing mechanism of neurotrophic factors via polysialic acid." This is followed by M. Martorella, K. Barford, B. Winkler, and C.D. Deppmann who describe the "Emergent role of Coronin-1a in neuronal signaling." Finally, M. Budzinska, K.B. Wicher, and M. Terenzio report on the "Neuronal Roles of the Bicaudal D Family of Motor Adaptors."

Chapters on biological activities of neurotrophins open with "BDNF and hippocampal synaptic plasticity" by G. Leal, C.R. Bramham, and C.B. Duarte. K.E. Boschen and A.Y. Klintsova describe "Neurotrophins in the brain: interaction with alcohol exposure during development" and A.S. Sahay, D.P. Sundrani, and S.R. Joshi report on "Neurotrophins: role in placental growth and development."

In the final section devoted to clinical conditions, J. Janssens, D. Lu, B. Ni, W. Chadwick, S. Siddiqui, A. Azmi, H. Etienne, A. Jushaj, J. van Gastel, B. Martin, and S. Maudsley offer "Development of precision small-molecule pro-neurotrophic therapies for neurodegenerative diseases." "The role of neurotrophins in inflammation and allergy" is described by S. Manti, P. Brown, M.K. Perez, and G. Piedimonte. M. Dhobale reports on "Neurotrophic factors and maternal nutrition during pregnancy." "The role of neurotrophin signaling in gliomagenesis: a focus on the p75 neurotrophin receptor ($p75^{NTR}$/CD271)" is authored by M.M. Alshehri, S.M. Robbins, and D.L. Senger. S.I. Hodgetts and A.R. Harvey write on "Neurotrophic factors used to treat spinal cord injury." "Neurotrophins and migraine" is the topic of L.B. Martins, A.L. Teixeira, and R.B. Domingues. H. Geoffroy and F. Noble focus on "BNDF during withdrawal." In conclusion, N.F. Frydenlund and M. Mahalingam report on "Neurotrophin receptors and perineural invasion—analyses in select lineage-unrelated cutaneous malignancies with a propensity for perineural invasion."

The illustration on the cover is a reproduction of Fig. 1 in Chapter 1 by Thomas Bertrand: "Crystal structures of neurotrophin receptors kinase domain." The legend of this figure is: "Structure of the dimeric nerve growth factor complexed with the extracellular domain of TrkA (pnb code 2IFG). Both immunoglobulin-like domains (d5, d4) and the leucine-rich repeats (d3, d2, and d1) are labeled. The localization of the cell membrane is indicated by a *rectangular box*."

Publication of this volume was facilitated by the efforts of Helene Kabes of Elsevier (Oxford, UK) and by Vignesh Tamilselvvan of Elsevier (S&T Book Production, Chennai, India).

GERALD LITWACK
Toluca Lake, North Hollywood, CA, United States
September 15, 2016

CHAPTER ONE

Crystal Structures of Neurotrophin Receptors Kinase Domain

T. Bertrand[1]
Sanofi, Vitry-sur-Seine, France
[1]Corresponding author: e-mail address: thomas.bertrand@sanofi.com

Contents

1. Introduction 1
   1.1 Trk Receptors Global Structure 2
2. Crystal Structure of TrkA, TrkB, and TrkC Kinase Domain 4
   2.1 Trk Kinase Domain apo Structures 5
   2.2 Trk Kinase Domain Complexed Structures 7
3. Pharmacological Approach: The Selectivity Issue 13
4. Conclusion and Future Directions 16
Acknowledgments 16
References 16

Abstract

Neurotrophins and their receptors (Trk) play key roles in the development of the nervous system and in cell survival. Trk receptors are therefore attractive pharmacological targets for brain disorders as well as for cancers. While the druggability of the extracellular domain of the receptors, that specifically binds neurotrophins, is yet to be proven, the intracellular kinase domains are attractive targets for small-molecule binding. The recent crystal structures of the three isoforms of the Trk family, TrkA, TrkB, and TrkC have been described in their apo forms and in complex with potent and selective pan-Trk inhibitors. The description of the kinase domain of each of the isoforms will be discussed in their apo forms or bound to potent inhibitors of interest in cancer therapy. Non-oncology indications and selectivity issues will also be discussed.

1. INTRODUCTION

The tropomyosin receptor kinase (*trk*) gene was originally identified as oncogene: the 5′ region of the tropomyosin gene was fused with a gene coding for an unknown protein but showing a high degree of homology with the insulin receptor gene. The product of the fused oncogene showed a tyrosine kinase activity (Martin-Zanca, Hughes, & Barbacid, 1986) and

Vitamins and Hormones, Volume 104
ISSN 0083-6729
http://dx.doi.org/10.1016/bs.vh.2016.10.001

the newly discovered gene coding for the tyrosine kinase took the name *trkA*. The proteins TrkA, TrkB, and TrkC are the three homologous isoforms of the Trk family. They are highly expressed in cells of neuronal origin and bind neurotrophins with different specificity. The binding of neurotrophins to the extracellular domain of the receptor Trk induces the kinase domain, located inside the cell, to autophosphorylate at several tyrosines, triggering downstream signals, for instance, inducing transduction pathways such as RAS/ERK and PI3K/AKT.

Neurotrophins are highly homologous dimeric growth factors and include nerve growth factor (NGF), brain-derived neurotrophic factor (BDNF), neurotrophin-3 (NT-3), neurotrophin-4/5 (NT-4/5), neurotrophin-6 (NT-6), and neurotrophin-7 (NT-7). TrkA binds preferentially NGF, NT-7, and, to a lesser extent, NT-6 (Gotz et al., 1994; Kaplan, Hempstead, Martin-Zanca, Chao, & Parada, 1991; Lai, Fu, Ip, & Ip, 1998); TrkB interacts with BDNF, NT-3, and NT-4/5 (Hubbard, Wei, Ellis, & Hendrickson, 1994; Ip et al., 1992; Klein, Lamballe, Bryant, & Barbacid, 1992; Klein et al., 1991), whereas TrkC is specific for NT-3 (Lamballe, Klein, & Barbacid, 1991). This observed ligand specificity suggests different roles of the various Trk proteins.

Implication of Trk kinases in cancer has been reported in prostate (Weeraratna, Arnold, George, DeMarzo, & Isaacs, 2000), pancreas (Miknyoczki et al., 1999), breast (Tognon et al., 2002), lung (Davies et al., 2005), thyroid, and colon (Nakagawara, 2001). Trk kinases are also strongly associated with central and peripheral nervous system processes such as memory (Yamada & Nabeshima, 2003), pain (Indo et al., 1996), depression (Rantamaki & Castren, 2008), neuronal development, plasticity, and protection (Huang & Reichardt, 2001).

1.1 Trk Receptors Global Structure

The structural arrangement of the full-length receptor is conserved between TrkA, TrkB, and TrkC. From N- to C-terminus, the receptor consists of five extracellular domains (domains 1 and 3 are cysteine-rich regions, domain 2 a leucine-rich region, and domains 4 and 5 are immunoglobulin-like domains), a transmembrane region, and the intracellular kinase domain (Schneider & Schweiger, 1991).

Studies on TrkB and TrkC have shown that domain 5 (d5) is sufficient for the binding of ligands and is responsible for their binding specificity (Holden et al., 1997; Perez, Coll, Hempstead, Martin-Zanca, & Chao,

1995; Urfer, Tsoulfas, O'Connell, & Presta, 1997). Structural information in the extracellular domain of Trk can be found in the literature for more than a decade ago: structures of neurotrophin binding domain 5 for TrkA, TrkB, and TrkC (Ultsch et al., 1999), TrkA-d5 in complex with NGF (Wiesmann, Ultsch, Bass, & de Vos, 1999), TrkB-d5 in complex with NT-4/5 (Banfield et al., 2001), or even the full extracellular domain of TrkA in complex with NGF (Wehrman et al., 2007) (Fig. 1).

Studies on TrkB with selective and commercially available polyclonal and monoclonal antibodies showed that biologics can demonstrate distinct pharmacological profiles through the allosteric modulation of the receptor (Cazorla, Arrang, & Premont, 2011). Some of the antibodies recognized the entire extracellular domain of the receptor, while some recognized only subdomains d4 and d5. While the rational design of biologics for a wanted therapeutic effect remains difficult, it is still feasible when targeting the extracellular domain of Trk receptors. However, the druggability of such domain, by a small-molecule approach, is yet to be established compared to the proven high druggability of a kinase domain.

In this chapter, we review our current knowledge of the structures of the kinase domain of each of the three isoforms of the receptor, alone and in

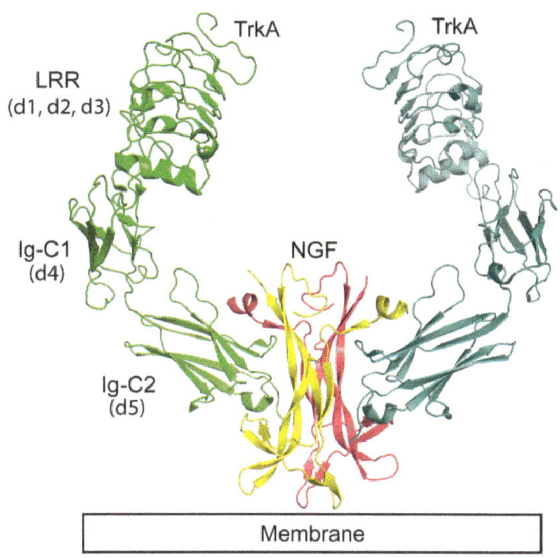

Fig. 1 Structure of dimeric nerve growth factor complexed with the extracellular domain of TrkA (pdb code 2IFG). Both immunoglobulin-like domains (d5, d4) and the leucine-rich repeats (d3, d2, and d1) are labeled. The localization of the cell membrane is indicated by a *rectangular box*.

complex with small-molecule inhibitors, and analyze key regions of these structures to design potent and selective drug molecules.

2. CRYSTAL STRUCTURE OF TRKA, TRKB, AND TRKC KINASE DOMAIN

The kinase domains of TrkA, TrkB, and TrkC share between 71.9% and 78.3% sequence identity, TrkB and TrkC being the closest homologues (Fig. 2).

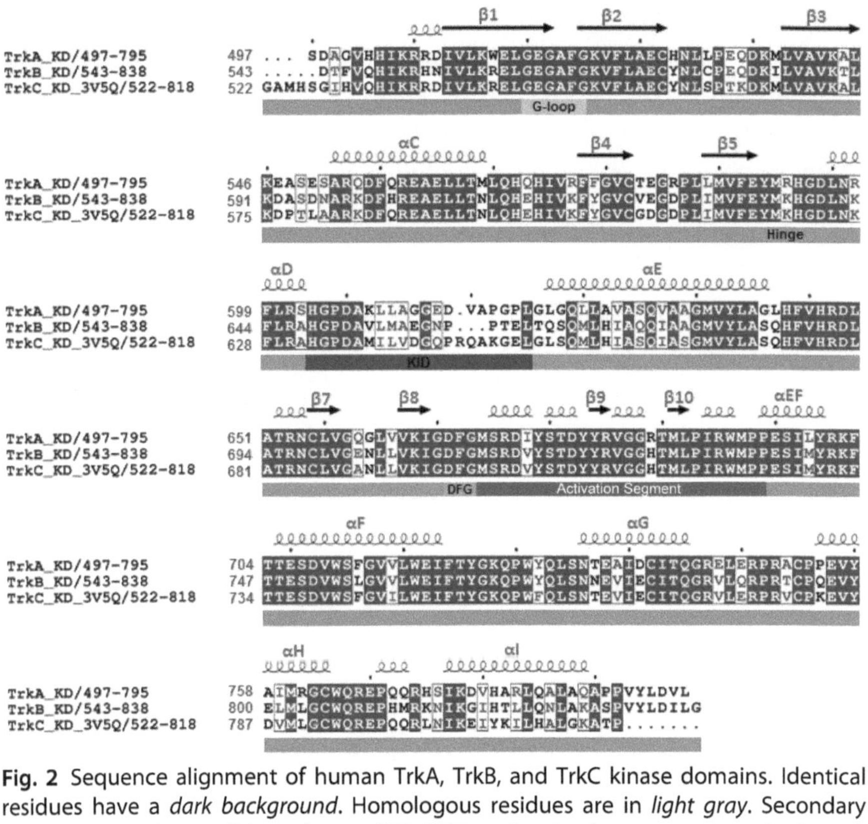

Fig. 2 Sequence alignment of human TrkA, TrkB, and TrkC kinase domains. Identical residues have a *dark background*. Homologous residues are in *light gray*. Secondary structures are shown. G-loop, hinge, KID, DFG, and activation segment are indicated. *Adapted from Bertrand, T., Kothe, M., Liu, J., Dupuy, A., Rak, A., Berne, P. F., et al. (2012). The crystal structures of TrkA and TrkB suggest key regions for achieving selective inhibition.* The Journal of Molecular Biology, 423*(3), 439–453. Copyright (2012) with permission from Elsevier.*

The challenges of obtaining selective kinase inhibitors for oncology applications have been extensively discussed (Dancey & Sausville, 2003); in the case of the Trk kinases, additional complications are introduced as central nervous system (CNS)-related side effects might appear. Anticancer drugs are largely described in reviews showing, for instance, that pan-Trk inhibitors can reduce the expansion of transplanted or xenografted cancer cell lines. However, the observed antitumor effects could also appear to be unspecific to Trk, involving kinases from other pathways (Desmet & Peeper, 2006; Wang, Yu, & Lamb, 2009). Structural information appears crucial to find selectivity against other protein kinases and even more toward the three isoforms of Trk.

Recently, the crystal structures of the kinase domain of TrkA and TrkB, apo, or in complex with potent and selective inhibitors have been recently published (Bertrand et al., 2012). Since then, several costructures of TrkA and TrkC appeared in the literature in complex with compounds, showing a broad degree of biochemical and cellular activity.

2.1 Trk Kinase Domain apo Structures

TrkA, TrkB, and TrkC kinase domains possess the principal architectural features common to protein kinases: one N-terminal lobe mainly constituted of β-strands, an α-helical C-terminal lobe composed of α-helices, and a hinge linker bridging both lobes of the kinase.

Trk kinases also possess two main structural elements involved in the regulation of protein kinases. The first element is the activation segment containing three tyrosine residues that, when phosphorylated, trigger downstream signaling cascades. The activation segment starts with a DFG (Aspartate–Phenylalanine–Glycine) motif which, in a so-called DFG-out conformation, can flip with reference to the active conformation (DFG-in) so that the aspartate no longer coordinates the magnesium ion at the catalytic site, and the phenylalanine points toward the active site preventing ATP binding. A second regulatory system involves the α-C helix moving away from the C-terminal lobe, breaking a conserved salt-bridge interaction between a glutamate residue of the α-C helix and the N-lobe lysine responsible for the stabilization of the α-phosphate of ATP. Both regulatory systems can independently lead to an inactive kinase (Cherry & Williams, 2004; Noble, Endicott, & Johnson, 2004) (Fig. 3).

Trk kinases possess an additional structural element, less common: the kinase insert domain (KID). The KID can be found in the VEGFR protein

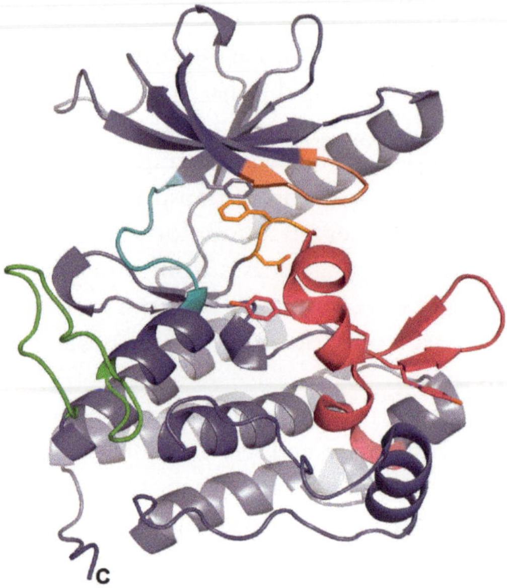

Fig. 3 Global structure of apo TrkA kinase domain. The hinge is in *cyan*, the G-loop in *pink*, the DFG motif in *orange*, the activation segment in *magenta*, the α-C helix in *yellow*, and the KID in *green*. The Gatekeeper Phe588, the DFG motif, and the phosphorylated tyrosines in the activation segment are in *sticks*.

kinase family and consists of an extension of the loop located between helices α-D and α-E of the C-terminal lobe. Studies on deletion or mutations of the KID domain on kinases of the VEGFR family showed that the KID is not important for the intrinsic kinase activity, but it is important for the binding of other proteins involved in signal transduction via autophosphorylation of KID tyrosine residues (McTigue et al., 1999).

The apo crystal structures of TrkA and TrkB depict an inactive conformation: both kinases are in a DFG-out conformation and TrkB is also in an α-C helix out conformation. There is no apo-TrkC described in the literature. The overall structures of apo-TrkA and apo-TrkB are quite similar with an r.m.s.d. of around 1 Å. Major structural differences reside in the KID region, different in lengths and in sequence between both kinases. In TrkA, the KID region lies above the C-terminal end of the kinase, in the close vicinity of the hinge region, whereas in the apo-TrkB structure the KID domain is under the C-terminal and, because it is shorter, the KID is even further from the hinge region of TrkB.

2.2 Trk Kinase Domain Complexed Structures

Molecules binding to the ATP binding site of protein kinases typically interact with the hinge region, the glycine-rich loop, and the DFG region of the activation segment. Some compounds can also exploit a back pocket, mainly hydrophobic, whose shape and size are controlled by the nature of the "gatekeeper" residue side chain. The gatekeeper is the first residue of the hinge connecting the C- and N-lobes of the kinase: when the gatekeeper has a bulky side chain, the back pocket is small, while when the gatekeeper has a small side chain, the back pocket is large. Targeting a kinase with a small gatekeeper by placing a substituent in the larger cavity is an attractive way to gain in potency, by creating extra interactions between the ligand and the protein, as well as gain selectivity against kinases with a large gatekeeper. For an extensive review on protein kinase inhibition, see Zuccotto, Ardini, Casale, and Angiolini (2010).

All of the Trk isoforms possess a bulky phenylalanine gatekeeper, leading to a small and unattractive back pocket. Nevertheless, potent and selective inhibitors are described in the literature and several costructures of TrkA, TrkB, and TrkC in complex with inhibitors are summarized in Table 1.

Some of the compounds described in Table 1 show micromolar IC_{50} values because there were originally discovered by high-throughput screenings and then used as starting points for compound optimization, quickly leading to highly potent pan-Trk inhibitors.

The inhibitors described in Table 1 all bind to the ATP binding site of the Trk kinase domain. They bind in either a DFG-in or a DFG-out conformation, following a classic binding pattern in protein kinases: the main scaffold (usually aromatic) binds, via hydrogen bonds, to residues of the hinge region of the kinase. The scaffold is stacked between residues of the C-terminal lobe of the kinase, and the glycine-rich loop in the N-terminal lobe. Chemical groups then can interact with the site entrance, near the solvent region of the crystal structures, while other moieties can point toward the back pocket of the kinase, either interacting with the DFG motif or pointing toward the α-C helix.

As an example, Cpd5n binds TrkB kinase domain in a DFG-in conformation. The glycine-rich loop closes as a lid on the inhibitor and Cpd5n is bound to the hinge of TrkB via its thiazole carboxamide scaffold and its urea group. The benzocycloheptene moiety is sandwiched between the gatekeeper Phe633 and the glycine-rich loop (Fig. 4). This binding mode is also observed in the other DFG-in TrkA and TrkC kinases.

Table 1 List of Available Liganded Trk Kinase Structures

Kinase	Ligand Name	Ligand Structure	IC_{50} (nM)	DFG	PDB	References
TrkA	GNF-10	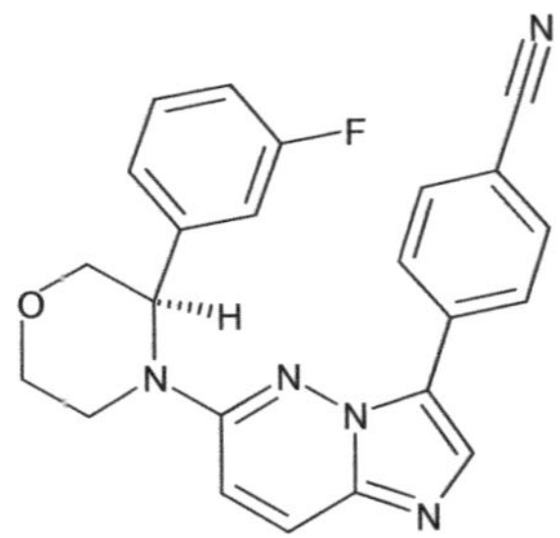	75[a]	In	4YPS	Choi et al. (2015)
TrkA	GNF-14	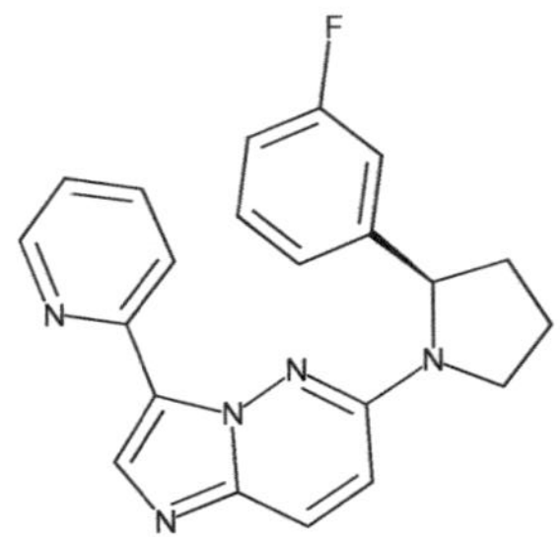	3[a]	In	4YNE	Choi et al. (2015)
TrkA	MK-1	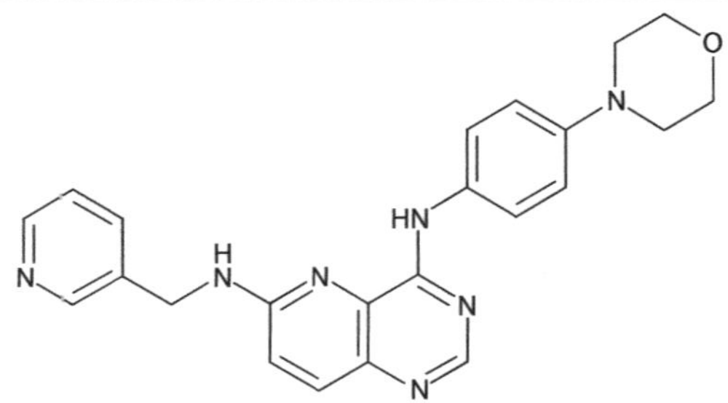	997	In	4PMT	Stachel et al. (2014)

TrkA	MK-2	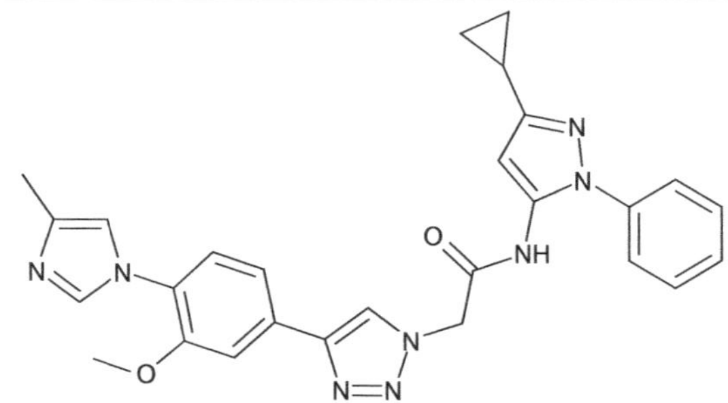	662	Out	4PMM	Stachel et al. (2014)
TrkA	MK-3	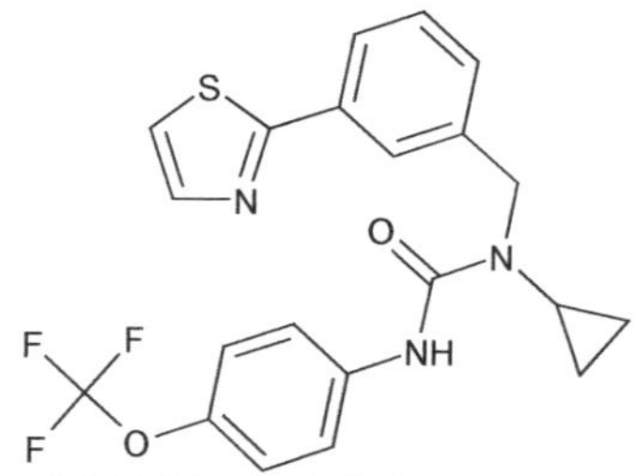	4233	Out	4PMP	Stachel et al. (2014)
TrkA	MK-4	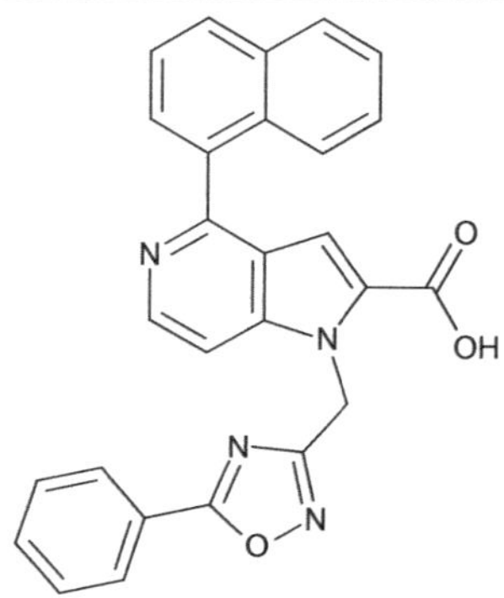	163	Out	4PMS	Stachel et al. (2014)

Continued

Table 1 List of Available Liganded Trk Kinase Structures—cont'd

Kinase	Ligand Name	Ligand Structure	IC_{50} (nM)	DFG	PDB	References
TrkA	AZ-23	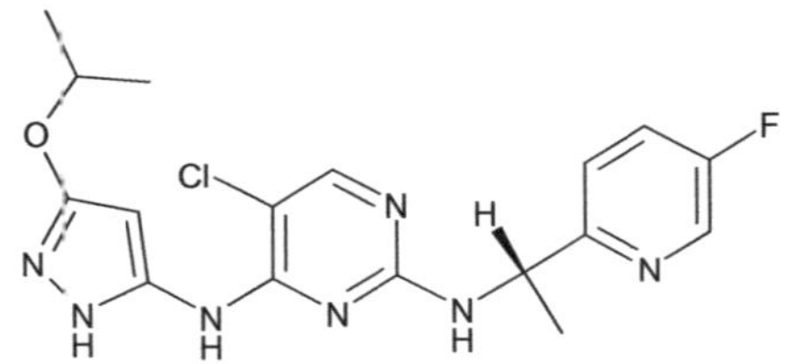	3	In	4AOJ	Wang et al. (2012)
TrkB	Cpd5n	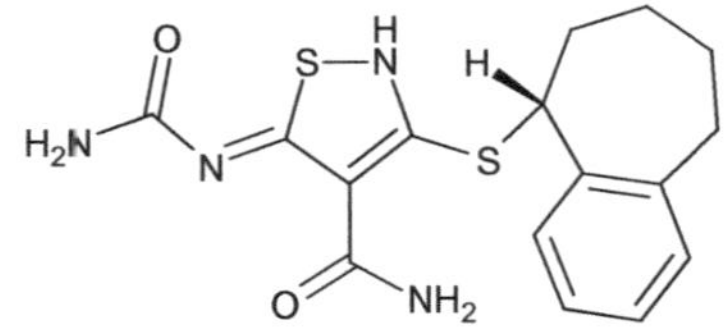	4	In	4AT3	Bertrand et al. (2012)
TrkB	EX429	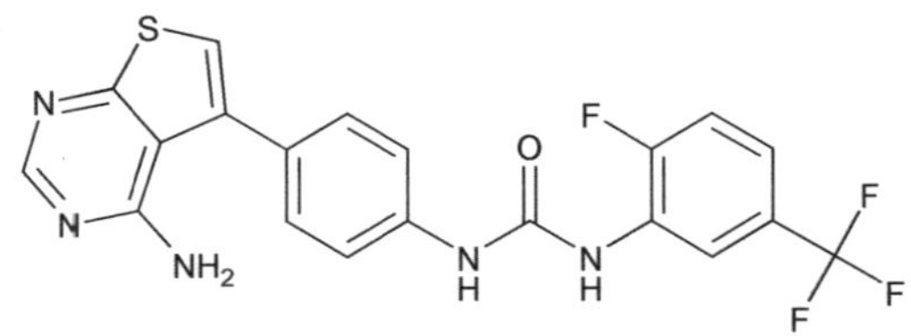	12	Out	4AT4	Bertrand et al. (2012)
TrkB	GW2580	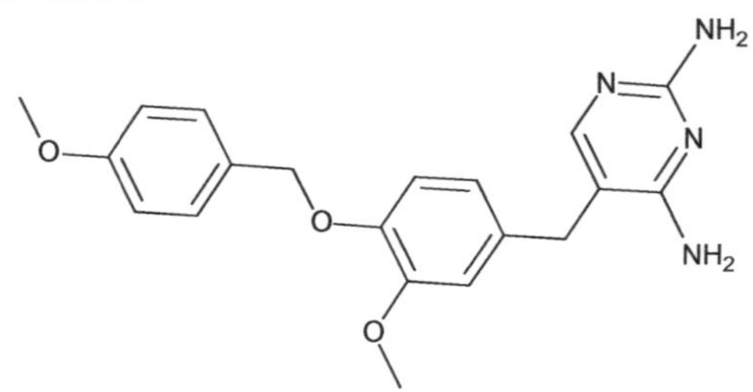	38	Out	4AT5	Bertrand et al. (2012)

TrkC	GNF-20	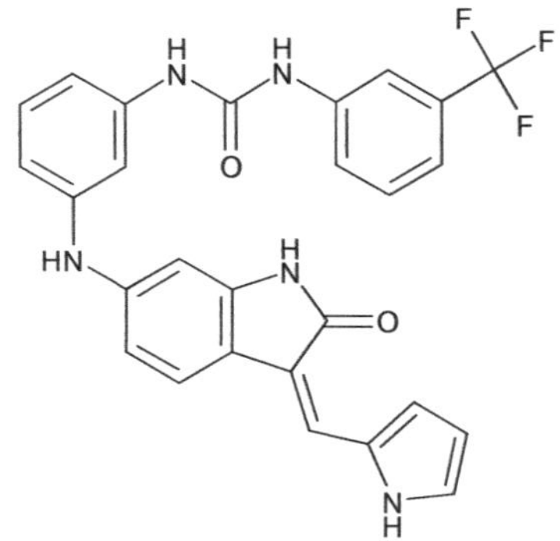	40	Out	3V5Q	Albaugh et al. (2012)
TrkC	GNF-1	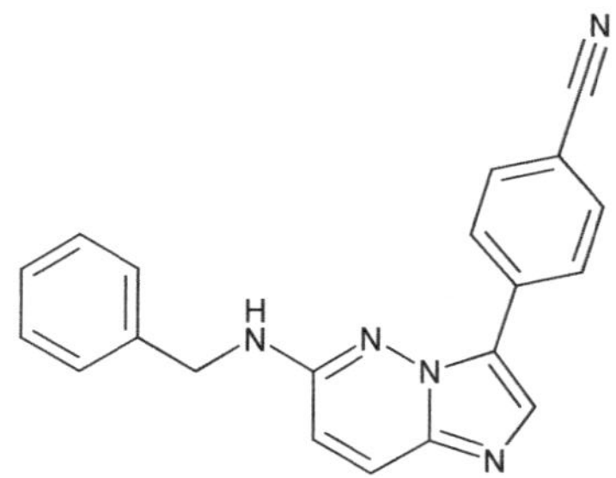	100[a]	Out	4YMJ	Choi et al. (2015)

[a]Cellular IC_{50}, other values are biochemical IC_{50}.

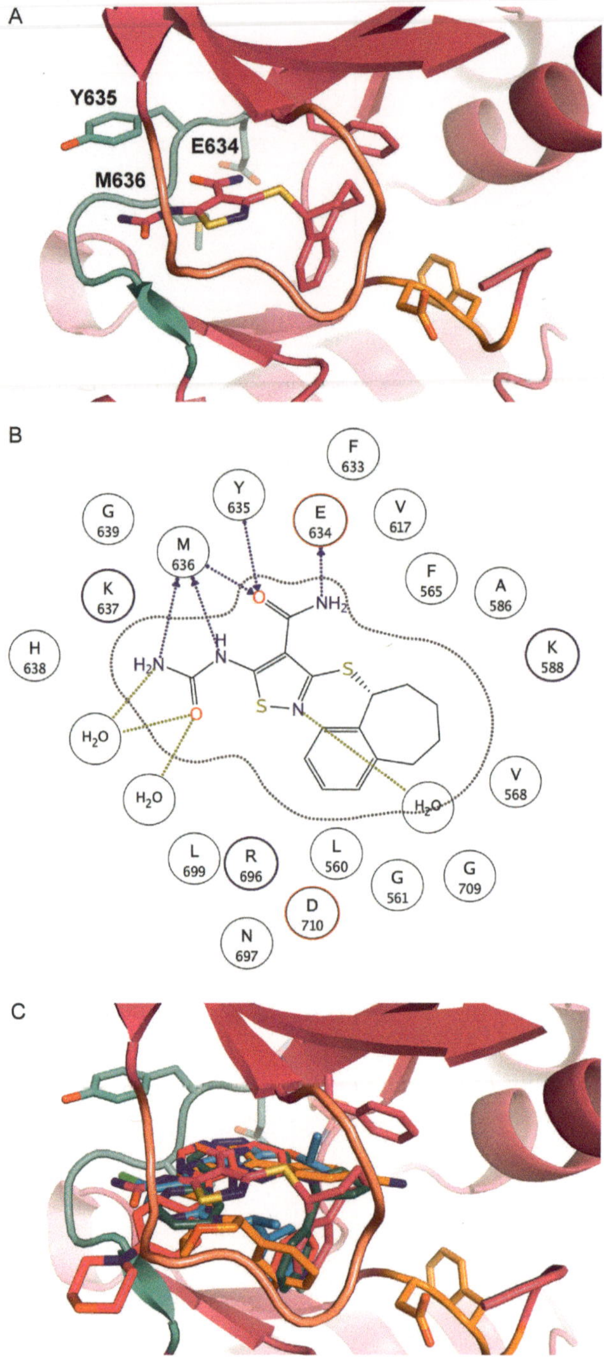

Fig. 4 See legend on opposite page.

EX429, in contrast, binds TrkB in a DFG-out conformation. The inhibitor is bound to the TrkB hinge region via its thienopyrimidine moiety. The phenyl linker of EX429 is sandwiched in a hydrophobic environment made of gatekeeper Phe633 on one side and Phe711 of the DFG motif on the other side. The fluoro-phenyl moiety of EX429 is located in the back pocket of the kinase, embedded in a hydrophobic environment (Fig. 5). Likewise, TrkA and TrkC depict similar interactions with inhibitors binding the DFG-out conformation of the kinases, despite different chemical scaffolds.

3. PHARMACOLOGICAL APPROACH: THE SELECTIVITY ISSUE

Interestingly, Trk kinases have the quite uncommon feature to bind either DFG-in or DFG-out inhibitors without apparent preference. This flexibility has proven useful to design selective Trk inhibitors. However, achieving selectivity between the tree isoforms of Trk is not straightforward and all compounds in Table 1 are pan-Trk inhibitors.

An exhaustive analysis, based on more than 40 kinases complexed with ATP or ATP analogues, has shown that at least 40 residues of the kinase domain in its DFG-in conformation might potentially interact with a ligand in the ATP binding site (Cherry & Williams, 2004). If such comparison is performed between TrkA and TrkB, only 2 out of the 40 residues are different. TrkB Lys637 and Lys643 are arginines in TrkA. Comparing TrkB with TrkC shows that all 40 residues are identical between the two isoforms. These observations suggest that selectivity between the three isoforms of Trk is not easily achievable. A closer look to all interacting residues in the described structures in Table 1 shows that, indeed, all residues interacting with the described inhibitors are conserved among all three isoforms whether the inhibitors bind a DFG-in or a DFG-out conformation of the kinase. Noticeably, in the 4PMT structure, the morpholine moiety of the DFG-in MK-1 inhibitor is in the close vicinity of Arg599 in TrkA. Since MK-1 is described as a pan-Trk inhibitor, targeting this residue is not sufficient for achieving selectivity TrkA toward TrkB.

Fig. 4 TrkB–Cpd5n binding mode. (A) 3D view. Structural elements are color-coded as in Fig. 3. Residues of the hinge H-bonded with Cpd5n are labeled. (B) 2D interaction pattern. Hydrogen bonds are depicted as *dashed lines*. (C) Superimposition of all DFG-in inhibitors from Table 1 on the TrkB–Cpd5n structure.

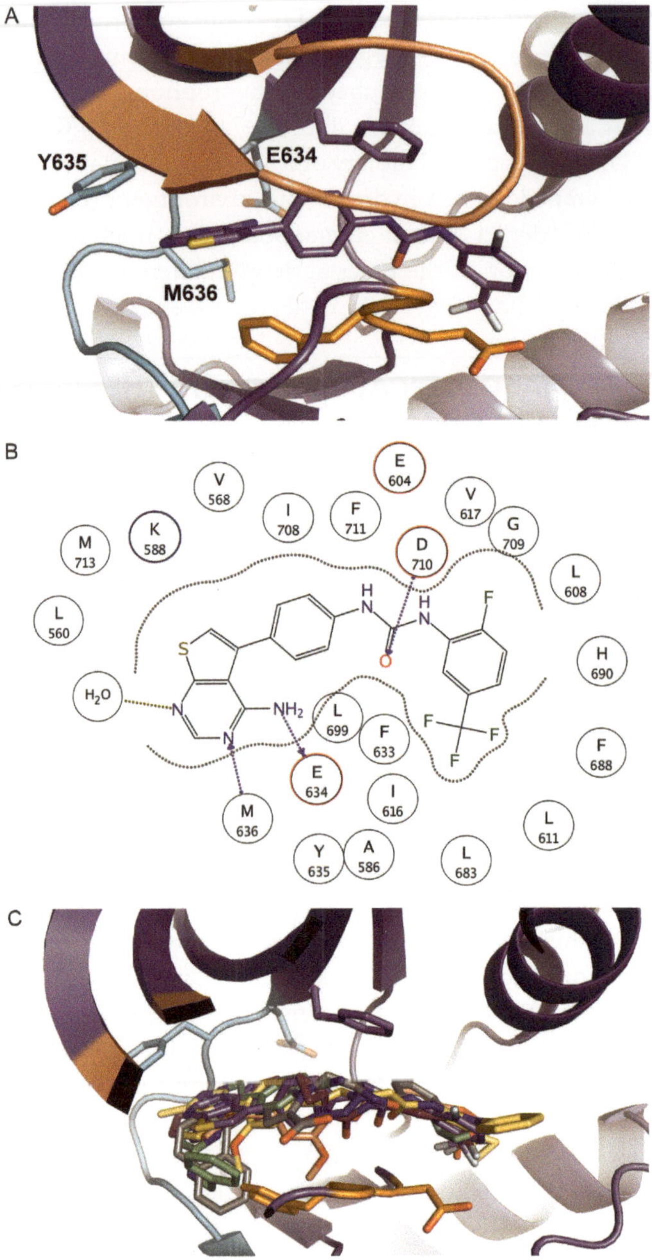

Fig. 5 TrkB–EX429 binding mode. (A) 3D view. Structural elements are color-coded as in Fig. 3. Residues of the hinge H-bonded with EX429 are labeled. (B) 2D interaction pattern. Hydrogen bonds are depicted as *dashed lines*. (C) Superimposition of all DFG-out inhibitors from Table 1 on the TrkB–EX429 structure.

However, it has been hypothesized that selectivity between TrkA, TrkB, and TrkC might be achieved by targeting the KID region of the kinases. All three isoforms possess a KID, located between helices α-D and α-E of the C-terminal region. Residues of this loop are not conserved between TrkA, TrkB, and TrkC, and the loop is different in length depending on the isoform. A closer look in the apo-TrkA structure reveals novel interactions between the tip of the loop and the hinge region in both TrkA monomers present in the asymmetric unit of the published crystal form, despite different packing environments (Bertrand et al., 2012). Interestingly, this conformation of the KID domain is identical to a later published structure of apo-TrkA, in a different packing environment, leading to the conclusion that the observed conformation is not imposed by crystal packing (Fig. 6). In TrkB, the KID is shorter and its conformation does not allow an interaction with the hinge region of the kinase. No conclusion can be drawn on the configuration of the KID in the TrkC isoform as this domain was not visible in the released PDB structures. Selectivity might, therefore, be achieved by designing compounds that are capable of reaching the KID of TrkA to interact or to clash with it. This hypothesis could offer attractive avenues to develop novel-selective compounds within the Trk family. While a similar

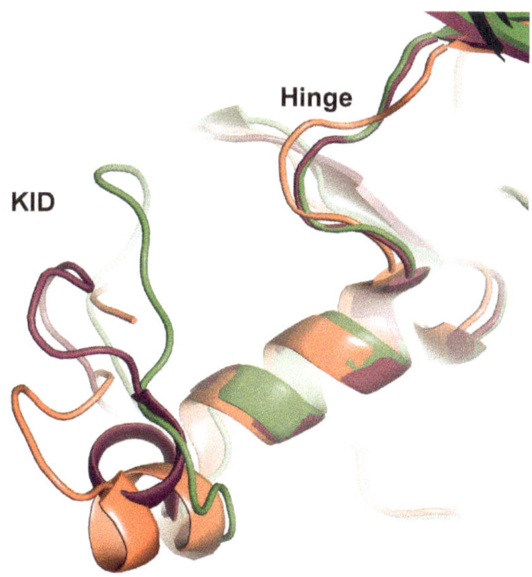

Fig. 6 Examples of conformation configurations of the kinase insert domain found in structures. Apo (pdb code 4F0I) and MK-3 complexed TrkA (4PMP) are in *green* and *purple*, respectively; TrkB in complex with GW2580 (4AT5) is in *pink*.

approach has been described for another serine/threonine kinase (Wang et al., 2006), the costructures of TrkA available in the literature do not make use of this structural element in their binding mode, and this hypothesis still needs to be proven.

4. CONCLUSION AND FUTURE DIRECTIONS

Tropomyosin-related kinases are reported as targets of high interest for the treatment of numerous disorders of the peripheral and CNS including depression, memory loss, and pain. Trk receptors play also key roles in cell survival and have been largely documented to be involved in a wide panel of cancers. Trk proteins are attractive targets for drug therapies, however contrasting effects that may arise whether targeting either the extracellular domain of the receptor with an agonist or the intracellular kinase domain with an inhibitor. For instance, unless kinase inhibitors can be restricted from passing through the blood–brain barrier, treatment of cancers could lead to brain disorders. Moreover, current Trk kinase inhibitors are pan-Trk inhibitors: the lack of selectivity increases the risk of CNS side effects if several Trk kinases are simultaneously targeted. The high-resolution crystal structures of the three isoforms of Trk kinase domain, apo, or in complex with potent and various inhibitors provide, however, valuable information on ligand binding and inform strategies to design potent and selective inhibitors. While selectivity might not be achievable toward the three isoforms through the classical regions of the ATP binding pocket of the kinase, other flexible and heterologous elements such as the KID near the hinge region of the kinases might be of importance in the rational design of inhibitors.

ACKNOWLEDGMENTS

This chapter is dedicated to our present and past colleagues who contributed to this work. The author thanks Dr. Magali Mathieu for advice and critical reading of this manuscript.

REFERENCES

Albaugh, P., Fan, Y., Mi, Y., Sun, F., Adrian, F., Li, N., et al. (2012). Discovery of GNF-5837, a selective TRK inhibitor with efficacy in rodent cancer tumor models. *ACS Medicinal Chemistry Letters*, *3*(2), 140–145.

Banfield, M. J., Naylor, R. L., Robertson, A. G., Allen, S. J., Dawbarn, D., & Brady, R. L. (2001). Specificity in Trk receptor:neurotrophin interactions: The crystal structure of TrkB-d5 in complex with neurotrophin-4/5. *Structure*, *9*(12), 1191–1199.

Bertrand, T., Kothe, M., Liu, J., Dupuy, A., Rak, A., Berne, P. F., et al. (2012). The crystal structures of TrkA and TrkB suggest key regions for achieving selective inhibition. *Journal of Molecular Biology*, *423*(3), 439–453.

Cazorla, M., Arrang, J. M., & Premont, J. (2011). Pharmacological characterization of six trkB antibodies reveals a novel class of functional agents for the study of the BDNF receptor. *British Journal of Pharmacology, 162*(4), 947–960.

Cherry, M., & Williams, D. H. (2004). Recent kinase and kinase inhibitor X-ray structures: Mechanisms of inhibition and selectivity insights. *Current Medicinal Chemistry, 11*(6), 663–673.

Choi, H. S., Rucker, P. V., Wang, Z., Fan, Y., Albaugh, P., Chopiuk, G., et al. (2015). (R)-2-Phenylpyrrolidine substituted imidazopyridazines: A new class of potent and selective Pan-TRK inhibitors. *ACS Medicinal Chemistry Letters, 6*(5), 562–567.

Dancey, J., & Sausville, E. A. (2003). Issues and progress with protein kinase inhibitors for cancer treatment. *Nature Reviews. Drug Discovery, 2*(4), 296–313.

Davies, H., Hunter, C., Smith, R., Stephens, P., Greenman, C., Bignell, G., et al. (2005). Somatic mutations of the protein kinase gene family in human lung cancer. *Cancer Research, 65*(17), 7591–7595.

Desmet, C. J., & Peeper, D. S. (2006). The neurotrophic receptor TrkB: A drug target in anti-cancer therapy? *Cellular and Molecular Life Sciences, 63*(7–8), 755–759.

Gotz, R., Koster, R., Winkler, C., Raulf, F., Lottspeich, F., Schartl, M., et al. (1994). Neurotrophin-6 is a new member of the nerve growth factor family. *Nature, 372*(6503), 266–269.

Holden, P. H., Asopa, V., Robertson, A. G., Clarke, A. R., Tyler, S., Bennett, G. S., et al. (1997). Immunoglobulin-like domains define the nerve growth factor binding site of the TrkA receptor. *Nature Biotechnology, 15*(7), 668–672.

Huang, E. J., & Reichardt, L. F. (2001). Neurotrophins: Roles in neuronal development and function. *Annual Review of Neuroscience, 24*, 677–736.

Hubbard, S. R., Wei, L., Ellis, L., & Hendrickson, W. A. (1994). Crystal structure of the tyrosine kinase domain of the human insulin receptor. *Nature, 372*(6508), 746–754.

Indo, Y., Tsuruta, M., Hayashida, Y., Karim, M. A., Ohta, K., Kawano, T., et al. (1996). Mutations in the TRKA/NGF receptor gene in patients with congenital insensitivity to pain with anhidrosis. *Nature Genetics, 13*(4), 485–488.

Ip, N. Y., Ibanez, C. F., Nye, S. H., McClain, J., Jones, P. F., Gies, D. R., et al. (1992). Mammalian neurotrophin-4: Structure, chromosomal localization, tissue distribution, and receptor specificity. *Proceedings of the National Academy of Sciences of the United States of America, 89*(7), 3060–3064.

Kaplan, D. R., Hempstead, B. L., Martin-Zanca, D., Chao, M. V., & Parada, L. F. (1991). The trk proto-oncogene product: A signal transducing receptor for nerve growth factor. *Science, 252*(5005), 554–558.

Klein, R., Lamballe, F., Bryant, S., & Barbacid, M. (1992). The trkB tyrosine protein kinase is a receptor for neurotrophin-4. *Neuron, 8*(5), 947–956.

Klein, R., Nanduri, V., Jing, S. A., Lamballe, F., Tapley, P., Bryant, S., et al. (1991). The trkB tyrosine protein kinase is a receptor for brain-derived neurotrophic factor and neurotrophin-3. *Cell, 66*(2), 395–403.

Lai, K. O., Fu, W. Y., Ip, F. C., & Ip, N. Y. (1998). Cloning and expression of a novel neurotrophin, NT-7, from carp. *Molecular and Cellular Neurosciences, 11*(1–2), 64–76.

Lamballe, F., Klein, R., & Barbacid, M. (1991). trkC, a new member of the trk family of tyrosine protein kinases, is a receptor for neurotrophin-3. *Cell, 66*(5), 967–979.

Martin-Zanca, D., Hughes, S. H., & Barbacid, M. (1986). A human oncogene formed by the fusion of truncated tropomyosin and protein tyrosine kinase sequences. *Nature, 319*(6056), 743–748.

McTigue, M. A., Wickersham, J. A., Pinko, C., Showalter, R. E., Parast, C. V., Tempczyk-Russell, A., et al. (1999). Crystal structure of the kinase domain of human vascular endothelial growth factor receptor 2: A key enzyme in angiogenesis. *Structure*, 7(3), 319–330.

Miknyoczki, S. J., Lang, D., Huang, L., Klein-Szanto, A. J., Dionne, C. A., & Ruggeri, B. A. (1999). Neurotrophins and Trk receptors in human pancreatic ductal adenocarcinoma: Expression patterns and effects on in vitro invasive behavior. *International Journal of Cancer, 81*(3), 417–427.

Nakagawara, A. (2001). Trk receptor tyrosine kinases: A bridge between cancer and neural development. *Cancer Letters, 169*(2), 107–114.

Noble, M. E., Endicott, J. A., & Johnson, L. N. (2004). Protein kinase inhibitors: Insights into drug design from structure. *Science, 303*(5665), 1800–1805.

Perez, P., Coll, P. M., Hempstead, B. L., Martin-Zanca, D., & Chao, M. V. (1995). NGF binding to the trk tyrosine kinase receptor requires the extracellular immunoglobulin-like domains. *Molecular and Cellular Neurosciences, 6*(2), 97–105.

Rantamaki, T., & Castren, E. (2008). Targeting TrkB neurotrophin receptor to treat depression. *Expert Opinion on Therapeutic Targets, 12*(6), 705–715.

Schneider, R., & Schweiger, M. (1991). A novel modular mosaic of cell adhesion motifs in the extracellular domains of the neurogenic trk and trkB tyrosine kinase receptors. *Oncogene, 6*(10), 1807–1811.

Stachel, S. J., Sanders, J. M., Henze, D. A., Rudd, M. T., Su, H. P., Li, Y., et al. (2014). Maximizing diversity from a kinase screen: Identification of novel and selective pan-Trk inhibitors for chronic pain. *Journal of Medicinal Chemistry, 57*(13), 5800–5816.

Tognon, C., Knezevich, S. R., Huntsman, D., Roskelley, C. D., Melnyk, N., Mathers, J. A., et al. (2002). Expression of the ETV6-NTRK3 gene fusion as a primary event in human secretory breast carcinoma. *Cancer Cell, 2*(5), 367–376.

Ultsch, M. H., Wiesmann, C., Simmons, L. C., Henrich, J., Yang, M., Reilly, D., et al. (1999). Crystal structures of the neurotrophin-binding domain of TrkA, TrkB and TrkC. *Journal of Molecular Biology, 290*(1), 149–159.

Urfer, R., Tsoulfas, P., O'Connell, L., & Presta, L. G. (1997). Specificity determinants in neurotrophin-3 and design of nerve growth factor-based trkC agonists by changing central beta-strand bundle residues to their neurotrophin-3 analogs. *Biochemistry, 36*(16), 4775–4781.

Wang, T., Lamb, M. L., Block, M. H., Davies, A. M., Han, Y., Hoffmann, E., et al. (2012). Discovery of disubstituted imidazo[4,5-b]pyridines and purines as potent TrkA inhibitors. *ACS Medicinal Chemistry Letters, 3*(9), 705–709.

Wang, Z., Liu, J., Sudom, A., Ayres, M., Li, S., Wesche, H., et al. (2006). Crystal structures of IRAK-4 kinase in complex with inhibitors: A serine/threonine kinase with tyrosine as a gatekeeper. *Structure, 14*(12), 1835–1844.

Wang, T., Yu, D., & Lamb, M. L. (2009). Trk kinase inhibitors as new treatments for cancer and pain. *Expert Opinion on Therapeutic Patents, 19*(3), 305–319.

Weeraratna, A. T., Arnold, J. T., George, D. J., DeMarzo, A., & Isaacs, J. T. (2000). Rational basis for Trk inhibition therapy for prostate cancer. *Prostate, 45*(2), 140–148.

Wehrman, T., He, X., Raab, B., Dukipatti, A., Blau, H., & Garcia, K. C. (2007). Structural and mechanistic insights into nerve growth factor interactions with the TrkA and p75 receptors. *Neuron, 53*(1), 25–38.

Wiesmann, C., Ultsch, M. H., Bass, S. H., & de Vos, A. M. (1999). Crystal structure of nerve growth factor in complex with the ligand-binding domain of the TrkA receptor. *Nature, 401*(6749), 184–188.

Yamada, K., & Nabeshima, T. (2003). Brain-derived neurotrophic factor/TrkB signaling in memory processes. *Journal of Pharmacological Sciences, 91*(4), 267–270.

Zuccotto, F., Ardini, E., Casale, E., & Angiolini, M. (2010). Through the "gatekeeper door": Exploiting the active kinase conformation. *Journal of Medicinal Chemistry, 53*(7), 2681–2694.

CHAPTER TWO

BDNF Propeptide: A Novel Modulator of Synaptic Plasticity

M. Kojima*,†,1, T. Mizui*,†

*Biomedical Research Institute, Advanced Industrial Science and Technology (AIST), Osaka, Japan
†Core Research for Evolutional Science and Technology (CREST), Japan Science and Technology Agency (JST), Kawaguchi, Japan
[1]Corresponding author: e-mail address: m-kojima@aist.go.jp

Contents

1. Neurotrophins and Their Biological Actions 20
2. NT Receptors and Their Signaling 20
3. Synthesis, Intracellular Trafficking, and Secretion of BDNF Protein 21
4. Modulation of Synaptic Function by BDNF 22
5. BDNF Prodomain: New Roles Beyond a Molecular Chaperone 22
6. BDNF and Its Propeptide Are Localized in Presynaptic Dense Core Vesicles 23
7. The BDNF Propeptide Is a Novel Synaptic Modulator 24
8. Role of the Val66Met Polymorphism in BDNF Propeptide-Dependent LTD 25
9. Discussion 25
References 26

Abstract

One of the most characterized neurotrophic factors is brain-derived neurotrophic factor (BDNF), which regulates neuronal survival and differentiation, and functions in activity-dependent plasticity processes such as long-term potentiation, long-term depression (LTD), and learning memory. Similar to other growth factors, BDNF protein is produced by transcriptional and translational mechanisms. Nevertheless, a post-translational mechanism, the proteolytic conversion of precursor BDNF into at least two fragments, bioactive BDNF and the prodomain, has not been well elucidated. Recently, we demonstrated that the BDNF prodomain, which is named the BDNF propeptide, was endogenously secreted from neuronal cells and facilitated a cellular mechanism of LTD, suggesting the manner through which this posttranslational mechanism multiplies the biological actions of BDNF. In this chapter, we focus on the BDNF propeptide, especially in synaptic plasticity, and discuss the role of this molecule in the brain. The findings regarding the BDNF propeptide would provide new insights for understanding the mechanisms of action of the propeptides of growth factors as well as the biological roles of neurotrophins.

Vitamins and Hormones, Volume 104
ISSN 0083-6729
http://dx.doi.org/10.1016/bs.vh.2016.11.006

1. NEUROTROPHINS AND THEIR BIOLOGICAL ACTIONS

Neurotrophins (NTs), including nerve growth factor (NGF), brain-derived neurotrophic factor (BDNF), NT-3, and NT-4, exert their biological activities by binding to their cognate tyrosine kinase receptors (NGF to tropomyosin-related kinase (Trk) A, BDNF and NT-4 to TrkB, and NT-3 to TrkC) as well as to a common low-affinity NT receptor ($p75^{NTR}$) (Chao, 2003). The discoveries of NT family proteins provided valuable insights into the formation of the neuronal network during the development of the nervous system and into synaptic plasticity, memory, and learning in the adult brain (Bibel & Barde, 2000; Park & Poo, 2013). Moreover, specific neuronal populations require one or more NTs, suggesting that they function in spatial- and temporal-dependent manners (Bibel & Barde, 2000; Park & Poo, 2013). Interestingly, NTs have long- and short-term actions. The long-term trophic actions depend on gene regulation, whereas the short-term effects, including chemotrophic effects on developing neurons and synaptic events, are controlled by NT activation of intracellular effectors. Because the large majority of the literature and other chapters of this book deal with the biological actions, signaling mechanism, and transcriptional and translational control of NTs, we will focus on the posttranslational mechanism and its products, in particular the effect of the BDNF propeptide on synaptic plasticity in the brain.

2. NT RECEPTORS AND THEIR SIGNALING

The first receptor, $p75^{NTR}$, is a member of the tumor necrosis factor (TNF) receptor family (Chao, 2003). $p75^{NTR}$ was initially discovered as a low-affinity NGF receptor but was subsequently found to bind to BDNF, NT-3, and NT-4/5 with a similar affinity (Chao, 2003). The extracellular domain of $p75^{NTR}$ has cysteine-rich motifs, and the cytoplasmic domain includes a "death" domain, which is also found in the cytoplasmic domains of other TNF receptors (He & Garcia, 2004; Liepinsh, Ilag, Otting, & Ibanez, 1997). While $p75^{NTR}$ does not contain a catalytic kinase motif, it interacts with many proteins that transmit signals important for neuronal survival and differentiation (Chao, 2003; Hempstead, 2014). $p75^{NTR}$ activates three major signaling pathways. First, NF-kappa B activation leads to the transcription of multiple genes, including several that promote neuronal survival. Second, activation of the Jun kinase via $p75^{NTR}$ leads to neuronal apoptosis. Third, ligand engagement of $p75^{NTR}$, through the regulation of Rho activity, can control growth cone motility (Reichardt, 2006).

The second group of NT receptors is the Trk family of receptor tyrosine kinases (TrkA, TrkB, and TrkC) (Reichardt, 2006). Trks have a transmembrane region that spans the plasma membrane and a cytoplasmic domain that has tyrosine kinase activity. Binding of an NT to a specific Trk receptor activates its tyrosine kinase activity for the activation of the phosphatidylinositol 3-kinase, mitogen-activated protein kinase, and phospholipase C-γ (PLC-γ) pathways.

$p75^{NTR}$ and Trk receptors and their signaling mechanisms have been reviewed by others (e.g., see Chao, 2003; Hempstead, 2002; Ibanez & Simi, 2012; Reichardt, 2006; Teng, Felice, Kim, & Hempstead, 2010) and will not be described further here.

3. SYNTHESIS, INTRACELLULAR TRAFFICKING, AND SECRETION OF BDNF PROTEIN

Like most growth factors, BDNF is initially synthesized as a precursor form of approximately 270 amino acids. The protein product of BDNF is composed of a signal sequence, a prodomain, and the mature domain. To produce the mature form, the N-terminal fragment of approximately 120 amino acids or the prodomain is proteolytically cleaved by intracellular and/or extracellular proteases (e.g., furin, prohormone convertase, and plasmin) (Lee, Kermani, Teng, & Hempstead, 2001; Lu, Pang, & Woo, 2005; Pang et al., 2004; Seidah, Benjannet, Pareek, Chretien, & Murphy, 1996). Several studies reported the intracellular mechanisms of BDNF processing (Mowla et al., 1999; Seidah et al., 1996). In nonneuronal cells, BDNF is constitutively secreted. On the other hand, in neurons, the N-terminal prodomain is cleaved in the trans-Golgi and secretory granules, which in turn produce the mature form of BDNF. BDNF is secreted through constitutive and neuronal activity-regulated secretion pathways in neurons, and it has been reported that the secretion of BDNF occurs in neuronal activity- and Ca^{2+}-dependent manners (reviewed by Lessmann & Brigadski, 2009). Thus, the activity-dependent secretion of BDNF is a crucial mechanism in BDNF-dependent regulation of synaptic transmission and long-term synaptic plasticity (Park & Poo, 2013).

4. MODULATION OF SYNAPTIC FUNCTION BY BDNF

BDNF mRNA is highly expressed in the brain, particularly in the hippocampus and cortex (Leibrock et al., 1989). Expression and secretion of BDNF are controlled by neuronal activity and excitatory transmission (Canossa et al., 1997; Castren et al., 1993; Goodman et al., 1996; Isackson, Huntsman, Murray, & Gall, 1991; Patterson, Grover, Schwartzkroin, & Bothwell, 1992; Zafra, Hengerer, Leibrock, Thoenen, & Lindholm, 1990). Acute application of exogenous BDNF increases neuronal activity and synaptic transmission in cultured neurons (Lessmann, Gottmann, & Heumann, 1994; Levine, Dreyfus, Black, & Plummer, 1995). These findings suggest a role of BDNF in synaptic plasticity. Indeed, application of exogenous BDNF facilitated the induction of long-term potentiation (LTP) in neonatal hippocampal slices (Figurov, Pozzo-Miller, Olafsson, Wang, & Lu, 1996). Treatment with TrkB-IgG, a fusion protein that scavenges endogenous BDNF, decreases the magnitude of LTP in adult hippocampal slices (Figurov et al., 1996; Yan et al., 1997). The role of BDNF in hippocampal synaptic plasticity was further studied using BDNF knockout mice (Ernfors, Lee, & Jaenisch, 1994; Jones, Farinas, Backus, & Reichardt, 1994; Patterson et al., 1996). Furthermore, various findings showed that the mature forms of BDNF contribute to the activation of Trk receptors rather than $p75^{NTR}$. First, the inhibition of $p75^{NTR}$ did not block BDNF regulation of presynaptic function or hippocampal LTP (Xu et al., 2000). Second, mutation of TrkB receptors at the PLC-γ-docking site, but not at the Shc-docking site, resulted in a deficit in hippocampal LTP (Minichiello et al., 2002).

5. BDNF PRODOMAIN: NEW ROLES BEYOND A MOLECULAR CHAPERONE

In contrast with the role of BDNF, the role of precursor BDNF (proBDNF) and its prodomain was poorly understood. Recently, Lee et al. demonstrated that the precursor form of NGF (proNGF) promoted neuronal death, while the mature form of NGF elicited neuronal survival (Lee et al., 2001). A similar notion was proposed by the reports of proBDNF; proBDNF promoted cell death, growth cone retraction, spine shrinkage, and long-term depression (LTD), while BDNF promoted spine formation, neuronal survival, and LTP (Deinhardt et al., 2011; Koshimizu et al., 2009; Teng et al., 2005; Woo et al., 2005). The biological actions of proNGF and proBDNF required a receptor complex consisting of p75^{NTR} and sortilin, a member of the Vps10p domain receptor family (Deinhardt et al., 2011; Teng et al., 2005).

Another important finding is from a study of a single nucleotide polymorphism (SNP) in the *bdnf* gene, which changes a valine to a methionine at codon 66 (Val66Met) in the prodomain of human BDNF. Interestingly, the Val66Met mutation significantly affected memory function, neuronal activity-dependent secretion of BDNF, and synapse sorting of BDNF (Egan et al., 2003). An increasing number of reports indicate that the Val66Met genetic variant of the BDNF gene is associated with susceptibility to brain disorders, including Alzheimer's disease (Bath & Lee, 2006).

These recent reports together raise several new hypotheses: (1) precursor and mature BDNF exert opposing biological actions; (2) the BDNF prodomain can play a new role beyond a molecular chaperone to assist the folding of BDNF (Kolbeck, Jungbluth, & Barde, 1994); and (3) as a more general notion, posttranslational mechanisms, including proteolytic cleavage of precursor proteins, may lead to diverse actions of proteins.

6. BDNF AND ITS PROPEPTIDE ARE LOCALIZED IN PRESYNAPTIC DENSE CORE VESICLES

Theoretically, it is expected that BDNF and its propeptide (prodomain) are both produced from their precursor protein (proBDNF) and are present at the same time. It was reported that the BDNF prodomain acts as a molecular chaperone to assist the folding of BDNF (Kolbeck et al., 1994). As mentioned earlier, proBDNF consists of the prodomain and the mature domain of BDNF, and has the opposite biological activity to BDNF (Lu et al., 2005). A SNP that changes a valine to a methionine at codon 66 in the proregion of human BDNF affects memory function as well as the secretion of BDNF (Egan et al., 2003). Together, these reports raised the possibility that the BDNF propeptide (prodomain) may have its own function and influence the biological actions of BDNF.

To understand the modes and actions of secreted proteins, their subcellular localizations need to be determined. To understand the substantial function of the BDNF propeptide (prodomain), the presence and localization of endogenous BDNF were investigated. Recently, Dieni et al. investigated the distribution of BDNF in the adult hippocampus using both light and ultra-structural microscopy, and showed that BDNF and its propeptide both localized in dense core vesicles in excitatory presynaptic terminals of the adult mouse hippocampus (Dieni et al., 2012). This report provided a notable basis for an anterograde mode of action of BDNF in the central nervous system, which is distinct from the retrograde model derived from experiments with NGF in the peripheral nervous system. They further demonstrated that BDNF and its propeptide were approximately 10-fold more abundant than proBDNF in adult brain, suggesting that the BDNF propeptide functions in mature neurons.

7. THE BDNF PROPEPTIDE IS A NOVEL SYNAPTIC MODULATOR

The BDNF propeptide is a portion of proBDNF. Recently, we reported that the BDNF propeptide is a new synaptic modulator in the central nervous system; the BDNF propeptide enhanced hippocampal LTD (Mizui et al., 2015). In this study, several lines of evidence indicated that the BDNF propeptide is a newly discovered facilitator of hippocampal LTD. For LTD induction, a sequence of low-frequency stimulation (1 Hz, 900 pulses, 15 min) was applied to Schaffer collaterals of hippocampal slices prepared from 3- to 4-week-old juvenile mice and then field excitatory postsynaptic potential slopes in the CA1 were measured. First, we showed that 30 min treatment with the BDNF propeptide (10 ng/mL) facilitated LTD without affecting basal synaptic transmission. Second, a subnanomolar concentration of the BDNF propeptide was sufficient for LTD enhancement. Third, application of the BDNF propeptide to *Bdnf*$^{-/-}$ hippocampal slices facilitated LTD, suggesting that the effect of the BDNF propeptide was not via an interaction of the peptide with endogenous BDNF. Furthermore, the BDNF propeptide-dependent facilitation of hippocampal LTD required the activation of GluN2B-containing NMDA receptors and p75^{NTR}. Moreover, the BDNF propeptide facilitated NMDA-induced AMPA receptor endocytosis, a critical mechanism for LTD expression. These findings suggest that the BDNF propeptide induces synaptic plasticity, in particular LTD, and regulates a mechanism critical for promoting synaptic depression.

8. ROLE OF THE Val66Met POLYMORPHISM IN BDNF PROPEPTIDE-DEPENDENT LTD

A growing body of clinical evidence indicates that the Val66Met polymorphism increases susceptibility to a variety of brain disorders (Bath & Lee, 2006). However, the molecular role of this BDNF polymorphism has not been fully investigated. Recently, it was reported that hippocampal slices prepared from mice with the Val66Met mutation are defective in NMDAR-dependent plasticity (Ninan et al., 2010). We also showed that the BDNF propeptide with this mutation unexpectedly attenuated hippocampal LTD (Mizui et al., 2015). The BDNF propeptide-induced endocytosis of AMPA receptor subtype GluA2, a crucial mechanism for the expression of LTD, was reversed by the amino acid substitution from valine to methionine in the BDNF propeptide (Mizui et al., 2015). These two reports provide mechanistic insights into the role of this BDNF polymorphism in synaptic plasticity.

9. DISCUSSION

While the BDNF prodomain was predicted to be disordered (Anastasia et al., 2013), we showed that the BDNF propeptide elicited LTD (Mizui et al., 2015), and Anastasia et al. reported that the Met-type BDNF propeptide promoted acute growth cone retraction (Anastasia et al., 2013). Recently, a report by Sugase et al. examined how disordered proteins fold upon binding to their targets (Sugase, Dyson, & Wright, 2007). Another recent report showed that many eukaryotic proteins exist in a disordered form under physiological conditions and fold into ordered structures upon binding to their cellular targets (Wright & Dyson, 1999). The BDNF propeptide might play important biological roles beyond its traditional role in assisting the folding of BDNF.

The BDNF propeptide and BDNF showed antagonistic functions in LTD, i.e., facilitation and blockade of LTD induction (Mizui et al., 2015). Thus, how these two proteins function under physiological conditions is an interesting question to be addressed. Notably, a recent report demonstrated the role of neuronal activity on the levels of BDNF-induced TrkB activation; field stimulation with TBS (*t*heta-*b*urst *s*timulation) changed BDNF-induced TrkB phosphorylation from a transient to a sustained mode, while another stimulation protocol used to induce the LTD paradigm did not (Kinoshita et al., 1999). Activity-dependent secretion of BDNF is a crucial mechanism for BDNF-dependent synaptic plasticity (Park & Poo, 2013). The BDNF propeptide is located at presynaptic dense core vesicles (Dieni et al., 2012) and is secreted in an activity-dependent manner (Anastasia et al., 2013). Thus, the mechanism underlying secretion of the propeptide should be elucidated in the future. It is expected that the reports about the BDNF propeptide will provide new insights for understanding the mechanisms of action of the prodomains (propeptides) of growth factors.

REFERENCES

Anastasia, A., Deinhardt, K., Chao, M. V., Will, N. E., Irmady, K., Lee, F. S., et al. (2013). Val66Met polymorphism of BDNF alters prodomain structure to induce neuronal growth cone retraction. *Nature Communications*, *4*, 2490.

Bath, K. G., & Lee, F. S. (2006). Variant BDNF (Val66Met) impact on brain structure and function. *Cognitive, Affective, & Behavioral Neuroscience*, *6*, 79–85.

Bibel, M., & Barde, Y. A. (2000). Neurotrophins: Key regulators of cell fate and cell shape in the vertebrate nervous system. *Genes & Development*, *14*, 2919–2937.

Canossa, M., Griesbeck, O., Berninger, B., Campana, G., Kolbeck, R., & Thoenen, H. (1997). Neurotrophin release by neurotrophins: Implications for activity-dependent neuronal plasticity. *Proceedings of the National Academy of Sciences of the United States of America*, *94*, 13279–13286.

Castren, E., Pitkanen, M., Sirvio, J., Parsadanian, A., Lindholm, D., Thoenen, H., et al. (1993). The induction of LTP increases BDNF and NGF mRNA but decreases NT-3 mRNA in the dentate gyrus. *Neuroreport*, *4*, 895–898.

Chao, M. V. (2003). Neurotrophins and their receptors: A convergence point for many signalling pathways. *Nature Reviews. Neuroscience*, *4*, 299–309.

Deinhardt, K., Kim, T., Spellman, D. S., Mains, R. E., Eipper, B. A., Neubert, T. A., et al. (2011). Neuronal growth cone retraction relies on proneurotrophin receptor signaling through Rac. *Science Signaling*, *4*, ra82.

Dieni, S., Matsumoto, T., Dekkers, M., Rauskolb, S., Ionescu, M. S., Deogracias, R., et al. (2012). BDNF and its pro-peptide are stored in presynaptic dense core vesicles in brain neurons. *The Journal of Cell Biology*, *196*, 775–788.

Egan, M. F., Kojima, M., Callicott, J. H., Goldberg, T. E., Kolachana, B. S., Bertolino, A., et al. (2003). The BDNF val66met polymorphism affects activity dependent secretion of BDNF and human memory and hippocampal function. *Cell*, *112*, 257–269.

Ernfors, P., Lee, K. F., & Jaenisch, R. (1994). Mice lacking brain-derived neurotrophic factor develop with sensory deficits. *Nature*, *368*, 147–150.

Figurov, A., Pozzo-Miller, L. D., Olafsson, P., Wang, T., & Lu, B. (1996). Regulation of synaptic responses to high-frequency stimulation and LTP by neurotrophins in the hippocampus. *Nature*, *381*, 706–709.

Goodman, L. J., Valverde, J., Lim, F., Geschwind, M. D., Federoff, H. J., Geller, A. I., et al. (1996). Regulated release and polarized localization of brain-derived neurotrophic factor in hippocampal neurons. *Molecular and Cellular Neurosciences*, 7, 222–238.

He, X. L., & Garcia, K. C. (2004). Structure of nerve growth factor complexed with the shared neurotrophin receptor p75. *Science*, *304*, 870–875.

Hempstead, B. L. (2002). The many faces of p75NTR. *Current Opinion in Neurobiology, 12,* 260–267.

Hempstead, B. L. (2014). Deciphering proneurotrophin actions. *Handbook of Experimental Pharmacology, 220,* 17–32.

Ibanez, C. F., & Simi, A. (2012). p75 neurotrophin receptor signaling in nervous system injury and degeneration: Paradox and opportunity. *Trends in Neurosciences, 35,* 431–440.

Isackson, P. J., Huntsman, M. M., Murray, K. D., & Gall, C. M. (1991). BDNF mRNA expression is increased in adult rat forebrain after limbic seizures: Temporal patterns of induction distinct from NGF. *Neuron, 6,* 937–948.

Jones, K. R., Farinas, I., Backus, C., & Reichardt, L. F. (1994). Targeted disruption of the BDNF gene perturbs brain and sensory neuron development but not motor neuron development. *Cell, 76,* 989–999.

Kinoshita, S., Yasuda, H., Taniguchi, N., Katoh-Semba, R., Hatanaka, H., & Tsumoto, T. (1999). Brain-derived neurotrophic factor prevents low-frequency inputs from inducing long-term depression in the developing visual cortex. *The Journal of Neuroscience, 19,* 2122–2130.

Kolbeck, R., Jungbluth, S., & Barde, Y. A. (1994). Characterisation of neurotrophin dimers and monomers. *European Journal of Biochemistry, 225,* 995–1003.

Koshimizu, H., Kiyosue, K., Hara, T., Hazama, S., Suzuki, S., Uegaki, K., et al. (2009). Multiple functions of precursor BDNF to CNS neurons: Negative regulation of neurite growth, spine formation and cell survival. *Molecular Brain, 2,* 27.

Lee, R., Kermani, P., Teng, K. K., & Hempstead, B. L. (2001). Regulation of cell survival by secreted proneurotrophins. *Science, 294,* 1945–1948.

Leibrock, J., Lottspeich, F., Hohn, A., Hofer, M., Hengerer, B., Masiakowski, P., et al. (1989). Molecular cloning and expression of brain-derived neurotrophic factor. *Nature, 341,* 149–152.

Lessmann, V., & Brigadski, T. (2009). Mechanisms, locations, and kinetics of synaptic BDNF secretion: An update. *Neuroscience Research, 65,* 11–22.

Lessmann, V., Gottmann, K., & Heumann, R. (1994). BDNF and NT-4/5 enhance glutamatergic synaptic transmission in cultured hippocampal neurones. *Neuroreport, 6,* 21–25.

Levine, E. S., Dreyfus, C. F., Black, I. B., & Plummer, M. R. (1995). Brain-derived neurotrophic factor rapidly enhances synaptic transmission in hippocampal neurons via postsynaptic tyrosine kinase receptors. *Proceedings of the National Academy of Sciences of the United States of America, 92,* 8074–8077.

Liepinsh, E., Ilag, L. L., Otting, G., & Ibanez, C. F. (1997). NMR structure of the death domain of the p75 neurotrophin receptor. *The EMBO Journal, 16,* 4999–5005.

Lu, B., Pang, P. T., & Woo, N. H. (2005). The yin and yang of neurotrophin action. *Nature Reviews. Neuroscience, 6,* 603–614.

Minichiello, L., Calella, A. M., Medina, D. L., Bonhoeffer, T., Klein, R., & Korte, M. (2002). Mechanism of TrkB-mediated hippocampal long-term potentiation. *Neuron, 36,* 121–137.

Mizui, T., Ishikawa, Y., Kumanogoh, H., Lume, M., Matsumoto, T., Hara, T., et al. (2015). BDNF pro-peptide actions facilitate hippocampal LTD and are altered by the common BDNF polymorphism Val66Met. *Proceedings of the National Academy of Sciences of the United States of America, 112,* E3067–E3074.

Mowla, S. J., Pareek, S., Farhadi, H. F., Petrecca, K., Fawcett, J. P., Seidah, N. G., et al. (1999). Differential sorting of nerve growth factor and brain-derived neurotrophic factor in hippocampal neurons. *The Journal of Neuroscience, 19,* 2069–2080.

Ninan, I., Bath, K. G., Dagar, K., Perez-Castro, R., Plummer, M. R., Lee, F. S., et al. (2010). The BDNF Val66Met polymorphism impairs NMDA receptor-dependent synaptic plasticity in the hippocampus. *The Journal of Neuroscience, 30,* 8866–8870.

Pang, P. T., Teng, H. K., Zaitsev, E., Woo, N. T., Sakata, K., Zhen, S., et al. (2004). Cleavage of proBDNF by tPA/plasmin is essential for long-term hippocampal plasticity. *Science*, *306*, 487–491.

Park, H., & Poo, M. M. (2013). Neurotrophin regulation of neural circuit development and function. *Nature Reviews. Neuroscience*, *14*, 7–23.

Patterson, S. L., Abel, T., Deuel, T. A., Martin, K. C., Rose, J. C., & Kandel, E. R. (1996). Recombinant BDNF rescues deficits in basal synaptic transmission and hippocampal LTP in BDNF knockout mice. *Neuron*, *16*, 1137–1145.

Patterson, S. L., Grover, L. M., Schwartzkroin, P. A., & Bothwell, M. (1992). Neurotrophin expression in rat hippocampal slices: A stimulus paradigm inducing LTP in CA1 evokes increases in BDNF and NT-3 mRNAs. *Neuron*, *9*, 1081–1088.

Reichardt, L. F. (2006). Neurotrophin-regulated signalling pathways. *Philosophical Transactions of the Royal Society of London. Series B, Biological Sciences*, *361*, 1545–1564.

Seidah, N. G., Benjannet, S., Pareek, S., Chretien, M., & Murphy, R. A. (1996). Cellular processing of the neurotrophin precursors of NT3 and BDNF by the mammalian proprotein convertases. *FEBS Letters*, *379*, 247–250.

Sugase, K., Dyson, H. J., & Wright, P. E. (2007). Mechanism of coupled folding and binding of an intrinsically disordered protein. *Nature*, *447*, 1021–1025.

Teng, K. K., Felice, S., Kim, T., & Hempstead, B. L. (2010). Understanding proneurotrophin actions: Recent advances and challenges. *Developmental Neurobiology*, *70*, 350–359.

Teng, H. K., Teng, K. K., Lee, R., Wright, S., Tevar, S., Almeida, R. D., et al. (2005). ProBDNF induces neuronal apoptosis via activation of a receptor complex of p75NTR and sortilin. *The Journal of Neuroscience*, *25*, 5455–5463.

Woo, N. H., Teng, H. K., Siao, C. J., Chiaruttini, C., Pang, P. T., Milner, T. A., et al. (2005). Activation of p75NTR by proBDNF facilitates hippocampal long-term depression. *Nature Neuroscience*, *8*, 1069–1077.

Wright, P. E., & Dyson, H. J. (1999). Intrinsically unstructured proteins: Re-assessing the protein structure-function paradigm. *Journal of Molecular Biology*, *293*, 321–331.

Xu, B., Gottschalk, W., Chow, A., Wilson, R. I., Schnell, E., Zang, K., et al. (2000). The role of brain-derived neurotrophic factor receptors in the mature hippocampus: Modulation of long-term potentiation through a presynaptic mechanism involving TrkB. *The Journal of Neuroscience*, *20*, 6888–6897.

Yan, Q., Rosenfeld, R. D., Matheson, C. R., Hawkins, N., Lopez, O. T., Bennett, L., et al. (1997). Expression of brain-derived neurotrophic factor protein in the adult rat central nervous system. *Neuroscience*, *78*, 431–448.

Zafra, F., Hengerer, B., Leibrock, J., Thoenen, H., & Lindholm, D. (1990). Activity dependent regulation of BDNF and NGF mRNAs in the rat hippocampus is mediated by non-NMDA glutamate receptors. *The EMBO Journal*, *9*, 3545–3550.

CHAPTER THREE

Zinc Interactions With Brain-Derived Neurotrophic Factor and Related Peptide Fragments

A. Travaglia*, D. La Mendola†,1
*Center for Neural Science, New York University, New York, NY, United States
†Department of Pharmacy, University of Pisa, Pisa, Italy
[1]Corresponding author: e-mail address: lamendola@farm.unipi.it

Contents

1. Introduction 30
2. Brain-Derived Neurotrophic Factor 31
   2.1 Peptides-Mimicking BDNF 34
3. Zinc in the Brain 38
   3.1 Zinc Metallostasis in the Brain 39
4. Zinc(II) Interaction With BDNF and Related Peptides 41
   4.1 Peptide Delivery 43
5. Zinc, BDNF, and Diet 47
6. Conclusion 48
Acknowledgments 49
References 49

Abstract

Brain-derived neurotrophic factor (BDNF) is a neurotrophin essential for neuronal development and survival, synaptic plasticity, and cognitive function. Dysregulation of BDNF signaling is involved in several neurodegenerative disorders, including Alzheimer's disease. Alteration of metal ion homeostasis is observed both in normal aging and in many neurodegenerative diseases. Interestingly, there is a significant overlap between brain areas characterized by metal ion dyshomeostasis and those where BDNF exerts its biological activity. Therefore, it is reasonable to speculate that metal ions, especially zinc, can modulate the activity of BDNF. The synthesis of BDNF peptidomimetic can be helpful both to understand the molecular interaction of BDNF with metal ions and to develop new drugs for neurodegenerative diseases.

Vitamins and Hormones, Volume 104
ISSN 0083-6729
http://dx.doi.org/10.1016/bs.vh.2016.10.005

1. INTRODUCTION

Neurotrophins (NTs) are a family of secreted proteins essential for the development, maintenance, and survival of the central and peripheral nervous system (Huang & Reichardt, 2001).

In the early 1950s, Levi-Montalcini isolated and characterized the first neurotrophin, nerve growth factor (NGF) (Levi-Montalcini & Hamburger, 1951). Since this discovery, not only have other members of the NT family have been described (Chao, 2003; Leibrock et al., 1989) but also compelling and exciting evidence highlighted a crucial role of NTs beyond neuronal development and cell survival (Lewin & Barde, 1996).

Brain-derived neurotrophic factor (BDNF), the second discovered member of the NT family, emerged as key regulator in activity-dependent modification of synaptic transmission, especially in the hippocampus and neocortex (Chao, 2003; Huang & Reichardt, 2001; Minichiello, 2009; Park & Poo, 2013).

Taking into account the essential role that BDNF exerts in the central nervous system in physiological conditions, it is evident that alteration of the levels, distribution, structure, and activity of this protein can result in neuronal dysfunction. BDNF dysregulation has been proposed as being responsible for the onset of neurodegenerative diseases, such as Alzheimer's disease (AD), and neuropsychiatric disorders such as schizophrenia (Greenberg, Xu, Lu, & Hempstead, 2009; Lu et al., 2012).

Noteworthy alteration of metal ion homeostasis is also observed in normal aging and in many neurodegenerative diseases (Bonda et al., 2011). Furthermore, there is a significant overlap between brain areas featuring metal ion dyshomeostasis and those where NTs exert their biological activity (Bonda et al., 2011; Travaglia, Pietropaolo, La Mendola, Nicoletti, & Rizzarelli, 2012; Zatta, Drago, Bolognin, & Sensi, 2009). Therefore, it is reasonable to speculate that metal ions, especially zinc and copper, can modulate the activities of NTs both directly (e.g., through conformational changes) and indirectly (e.g., activating the downstream signaling of NT in an NT-independent mode).

The coordination features of BDNF with zinc are not completely elucidated. The synthesis of peptide fragments that are either a part or a functional domain of specific NTs can be helpful to understand the molecular aspects of metal ion interactions with the protein. Moreover, if they are able to mimic the activity of NT, these peptide fragments may represent potential biological active systems; thus, they could be the basis for the development of new drugs to treat neurodegenerative diseases.

2. BRAIN-DERIVED NEUROTROPHIC FACTOR

BDNF is a secreted protein essential for survival and wiring regulation of the central and peripheral nervous system (Chao, 2003; Huang & Reichardt, 2001; Park & Poo, 2013).

Like other NTs, BDNF is synthesized as a precursor (pro-BDNF) and cleaved by proteases to produce the C-terminal mature and active form, which exerts its biological action as a noncovalent homodimer (Chao & Bothwell, 2002; Huang & Reichardt, 2001).

In general, NTs trigger their biologic activity through interactions with two distinct classes of cell surface receptors: p75 and tropomyosin receptor kinase (Trk). Low-affinity NT receptor p75 is the common NT receptor. In fact, p75 is able to bind all NTs, even though with different kinetic constants. p75 is a member of the tumor necrosis factor receptor family, and its signal pathways (independent of Trk activation) are proapoptotic (Chao, 2003; Huang & Reichardt, 2001; Park & Poo, 2013). The Trk receptor belongs to a family of three different members: TrkA, TrkB, and TrkC, but BDNF only binds and activates TrkB. BDNF binding to TrkB induces TrkB dimerization which, in turn, results in receptor autophosphorylation on multiple tyrosine residues (Chao, 2003; Huang & Reichardt, 2001; Park & Poo, 2013). TrkB phosphorylation is the first step of a signaling cascade, which leads to a variety of intracellular phosphorylation and activations, such as the Ras–MAP (mitogen-activated protein) kinase cascade and the phosphorylation of cyclic AMP response element-binding protein (CREB) (Chao, 2003; Minichiello, 2009; Park & Poo, 2013).

The expression, secretion, and activity of BDNF are neural activity dependent. The BDNF/TrkB system has a key function in the central nervous system, promoting neuronal survival, and also influences cell fate choices (Chao, 2003; Park & Poo, 2013). BDNF is neuroprotective in the hippocampus, particularly against ischemic damage (Larsson, Nanobashvili, Kokaia, & Lindvall, 1999), preventing peroxide accumulation and increasing antioxidant enzymes.

BDNF, through its receptor TrkB, influences differentiation, the morphological development of neurons, as well as their synaptic connectivity. BDNF is implicated in synaptic plasticity and has a role in long-term potentiation (LTP) (Kang & Schuman, 1995; Kang, Welcher, Shelton, & Schuman, 1997; Korte et al., 1995, 1996; Patterson et al., 1996), which is a persistent synaptic strengthening, widely considered one of the major cellular mechanisms that underlies learning and memory (Malenka, 2003; Minichiello, 2009; Nicoll & Malenka, 1995).

For example, hippocampal slices from BDNF knockout animals exhibit impaired LTP induction (Korte et al., 1995) which is rescued by BDNF treatment (Korte et al., 1996; Patterson et al., 1996). Incubation of hippocampal slices with TrkB inhibitors blocks LTP (Figurov, Pozzo-Miller, Olafsson, Wang, & Lu, 1996). BDNF has been shown to mediate the maintenance of memory consolidation in the hippocampus, the brain region that is required for many forms of long-term memory in humans and animals (Bambah-Mukku, Travaglia, Chen, Pollonini, & Alberini, 2014; Beckinschtein et al., 2007). In support of this hypothesis, the expression of hippocampal BDNF mRNA, as well as BDNF protein, is increased by contextual learning and memory consolidation (Bambah-Mukku et al., 2014; Beckinschtein et al., 2007). Supporting the critical role of BDNF in memory consolidation and persistence, memory is impaired by function-blocking antibodies to BDNF (Alonso et al., 2002), BDNF knockout (Linnarsson, Bjorklund, & Ernfors, 1997), and knockout of forebrain TrkB signaling (Minichiello et al., 1999).

BDNF has a critical role in adult neurogenesis, the process by which neurons are generated from neural stem cells and progenitor cells. In fact, although the process is most active during the prenatal period and early development, adult neurogenesis has been shown to occur in multiple brain structures, including the hippocampus and the olfactory bulb (Deng, Aimone, & Gage, 2010; Hoerder-Suabedissen & Molnár, 2015). BDNF has been identified as a strong candidate molecule regulating adult neurogenesis. For example, increasing the levels of BDNF through either intraventricular infusion or adenoviral infection increases neurogenesis (Benraiss, Chmielnicki, Lerner, Roh, & Goldman, 2001; Pencea, Bingaman, Wiegand, & Luskin, 2001; Zigova, Pencea, Wiegand, & Luskin, 1998).

Several neurodegenerative and neuropsychiatric disorders characterized by abnormalities in synaptic plasticity and cognitive function have been associated with deregulation of BDNF (Arancio & Chao, 2007; Greenberg et al., 2009; Greenberg, Xu, Lu, & Hempstead, 2013).

BDNF is involved in lasting structural and functional changes underlying epileptogenesis. In animal models of epilepsy, seizure activity increases the expression of BDNF mRNA and protein in the hippocampus (Ernfors, Bengzon, Kokaia, Persson, & Lindvall, 1991; Isackson, Huntsman,

Murray, & Gall, 1991; Lindvall, Kokaia, Bengzon, Elmér, & Kokaia, 1994). Interfering with BDNF signal transduction inhibits the development of the epileptic state in vivo. Increasing BDNF levels trigger the onset of the epileptic state: direct application of protein induces hyperexcitability in vitro (Scharfman, 1997; Scharfman, Goodman, & Sollas, 1999). Overexpression of BDNF in transgenic mice leads to spontaneous seizures, and its intrahippocampal infusion is sufficient to induce seizure activity in vivo (Scharfman, Goodman, Sollas, & Croll, 2002). On the other hand, infusion of anti-BDNF agents (Binder, Routbort, Ryan, Yancopoulos, & McNamara, 1999) or use of BDNF knockout (Kokaia et al., 1995) in mice inhibits epileptogenesis in animal models.

Converging lines of evidence suggest that BDNF could be implicated in the neurodevelopmental abnormalities found in the schizophrenic brain. BDNF levels are increased in the hippocampus and anterior cingulate cortex of patients with schizophrenia, and rat models of schizophrenia show altered expression of the BDNF gene. Interestingly, lithium, a major drug for the treatment of bipolar disorder, increases BDNF and TrkB activation in cerebral cortical neurons (Hashimoto et al., 2002).

Decreased expression of BDNF characterizes neurodegenerative diseases such as AD and Parkinson's (Murer, Yan, & Raisman-Vozari, 2001), and reduced levels of this NT might contribute to the progressive atrophy and death of neuronal populations in the brains of patients affected by AD (Hock, Heese, Hulette, Rosenberg, & Otten, 2000; Zuccato & Cattaneo, 2009).

Moreover, the reduction of pro-BDNF and mature BDNF levels are related to loss of memory and learning capabilities (Peng, Wuu, Mufson, & Fahnestock, 2005).

An unbalanced activity of the different proteases involved in the maturation/degradation of BDNF can be responsible for deficits and failures of BDNF trophic stimulation, or loss of protection. A failure in the conversion of proneurotrophins to their mature form, which is then exacerbated by an increased NT degradation resulting from a rise in MMP-9 activity, has been associated with progressive atrophy of the basal forebrain cholinergic system and the consequent cholinergic contribution to AD-related learning and memory decline (Hock et al., 2000; Zuccato & Cattaneo, 2009).

It is interesting to note that transition-metal ions are active in neuromodulation and neurotransmission (Sensi et al., 2011; Sindreu & Storm, 2011; Xie & Smart, 1991) and have been implicated in the onset and progression of several neurodegenerative diseases, including AD.

2.1 Peptides-Mimicking BDNF

Observation of the essential role and biological activity of NTs suggested their use as potential drugs in the treatment of many neurodegenerative diseases (Skaper, 2011). However, there are several inherent drawbacks with the use of proteins applied as drugs. Mainly, proteins have poor pharmacokinetic properties such as in vivo instability, limited diffusion, low blood–brain barrier penetrability, unsuccessful tissue targeting, and pleiotropic effects due to multiple receptor binding (with the consequence of activation of multiple pathways not intentionally targeted) (Poduslo & Curran, 1996; Zuccato & Cattaneo, 2009). Therefore, in the last several years, a variety of strategies have been devised to design and develop effective low molecular weight mimicking systems able to: (i) target selective receptors as agonist or antagonist; (ii) penetrate tissue barriers; and (iii) be stable to proteolysis. The pharmacological advantages of mimetics over protein therapeutics are well documented and include lack of immunogenicity, low molecular mass, low manufacturing costs, and good pharmacokinetic profiles (Teixido & Giralt, 2008).

A convenient starting point is the synthesis of peptide fragments that are part of the whole protein. This approach has the advantage of being able to maintain intact the functional domain of a protein, thus mimicking the protein's biological activity. These systems also provide information on protein properties' structure, which is helpful to develop other small molecules (peptide or not) with better pharmacokinetic properties such as increased selectivity and resistance to proteases. It is possible to make a combination of both backbone and side-chain amino acid residues in order to identify ideal agonists or antagonists of biologically relevant proteins. In principle, synthetic peptides can be used to regulate almost all receptor responses. The high specificity and low toxicity of peptide drugs derive from the extremely tight binding they have to their targets (Otvos & Wade, 2014).

A protein domain notably studied for BDNF and for other NTs is the loop 2. It has been shown that insertion of the loop 2 region of BDNF into NGF structures results in a chimeric protein that, unlike native NGF, is able to bind to TrkB and show BDNF-like biological activity (Ibánez, Ilag, Murray-Rust, & Persson, 1993). NT loop regions are responsible for the specificity among different Trk receptors and thus are important regions in receptor binding and in subsequent biological activity of the protein.

A series of conformationally constrained short peptides based on the loop 2 region of BDNF have been synthesized using the three-dimensional structure of BDNF as a template (O'Leary & Hughes, 1998). Peptides have been tested in cultures of embryonic chick sensory neurons, a subpopulation of which requires BDNF for survival, and were shown to act as competitive antagonists of the binding of BDNF to TrkB. The inhibitory activity of these peptides correlates with their ability to mimic the 3-hairpin loop conformation of loop 2. This is further confirmed by the absence of activity shown by the linear peptide (EKVPVSKGQLKQ) that does not assume a loop-like conformation, whereas its cyclic form obtained by a disulfide bridge (CEKVPVSKGQLKQC) inhibits BDNF-mediated neuronal survival. These data support the hypothesis that these peptides mimic the native conformation loop 2 of BDNF (O'Leary & Hughes, 1998).

NTs are biologically active as dimers, and thus bicyclic and tricyclic dimeric forms of the previous mimic model have been synthesized. These peptides have a reduced conformational flexibility around their dimerizing linker and, distinct from their monomeric form, did not act as antagonists. In fact, these peptides were able to promote neuron survival in vitro (O'Leary & Hughes, 2003). In particular, the tricyclic dimer was 10-fold more active than the best of the other bicyclic dimers, and twofold less potent than the protein itself. Despite this activity, both bicyclic and tricyclic peptides behaved as partial agonists with respect to BDNF, being able to support maximally about 35% of those neurons kept alive by BDNF and inhibited the neuronal survival effect of BDNF when added in competition. Such behavior could result from the compound being less efficient than BDNF in bringing about dimerization of proteins, or because of inappropriate conformational flexibility of the dimerizing linkages. An alternative mechanism for the partial agonist behavior of the dimeric peptides is that they act through TrkB only, even as full TrkB agonists, but not through p75 (O'Leary & Hughes, 2003).

Peptides that bind to functional sites on TrkB have been also identified from the NT-4/TrkB crystal structure (Banfield et al., 2001). The SRRGE sequence at the amino terminus of NT-4 makes the main contact at the a1/b1 interface of the NT-4/TrkB crystal. A simple cyclic peptide (N-Ac-CSRRGECNH_2, cyclized by a disulfide bridge formed by cysteine residues) serves as a very effective NT-4 and BDNF antagonist. A constrained cyclic peptide mimetic of this motif (N-Ac-CSRRGEC-NH_2) binds to TrkB and acts as antagonist interfering with the ability of the BDNF to activate the receptor (Williams et al., 2005).

Starting from the same motif, a TrkB agonist was designed to constrain the sequence RRGE tandem repeat within a cycle. The cyclic

N-Ac-CSRRGELAASRRGELC-NH_2 peptide showed binding to TrkB, promoting neurite outgrowth in an inhibitory environment, whereas BDNF failed in the same conditions. A tentative hypothesis is that the peptide might engage and stabilize the receptor dimer in an active conformation (Williams et al., 2005).

Mimicking systems were designed by taking cues from other NT loops. The cyclo(-D-Pro-Ala-Lys-Arg-) peptide was synthesized to mimic a tripeptide motif present in loop 4 of BDNF, responsible to the protein binding to the p75 receptor (Fletcher et al., 2008). The peptide adopts a well-defined backbone conformation in solution, and it is an effective promoter of the survival of embryonic chick dorsal root ganglion sensory neurons without affecting TrkB signaling. Furthermore, the cyclo-pentapeptide was found to be effectively resistant to proteolysis when incubated in a solution of rat plasma in vitro, thus representing a template for further design of neuronal survival-promoting agents with higher diffusion with transmembrane passage characteristics (Fletcher & Hughes, 2009).

Peptides able to selectively target TrkB receptors promoted oligodendrocyte myelination in vitro (Douvaras et al., 2014). The tricyclic dimer peptide named TDP6, a structural mimetic of BDNF that targets TrkB, enhances oligodendrocyte myelination in vitro in a TrkB-dependent manner. In contrast, cyclo-DPAKKR (a structural mimetic of BDNF that targets p75NTR) exerted no effect.

Other peptides derived from loops 3 and 4, named Betrofin 3 (RGIDKRHWNSQ) and Betrofin 4 (SYVRALTMDSKKRIGWR), respectively, were capable of binding to both TrkB and p75 (Fobian et al., 2010). Both peptides were found to be biologically active, stimulating neurite outgrowth and promoting survival when applied to primary cerebellar granule neurons. The neuritogenic peptide effect was partially inhibited by BDNF, which did not induce neurite outgrowth under the experimental conditions, thus indicating partial agonistic properties of the Betrofin peptides.

The N-terminal region is another domain of NTs considered critical for the binding selectivity and activation of their cognate Trk receptors (Ibánez, 1998; Pattarawarapan & Burgess, 2003). N-terminal regions of NTs bind and activate Trk receptors by fitting into a binding pocket formed by a patch of charged amino acids in the fifth subdomain of Trk receptors. The primary sequences of NT N-terminal ends are different here, but also show some interesting similarities (Fig. 1).

	1 10 20 30 40
NGF	SSSHPIFHRG EFSVCDSVSV WVGDKTTATD IKGKEVMVLG
BDNF	HSDPARRGEL SVCDSISEWV TAADKKTAVD MSGGTVTVLE
NT3	YAEHKSHRG EYSVCDSESL WVTDKSSAID IRGHQVTVLG
NT4	GVSETAPASR RGELAVCDAV SGWVTDRRTA VDLRGREVEV

Fig. 1 Sequence alignment of neurotrophins N-terminal domain. In the *red rectangle*, NGF(1–14) and BDNF(1–12) sequences are included.

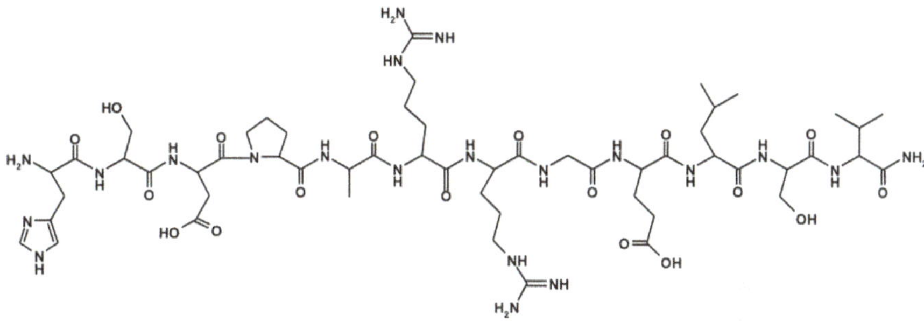

Fig. 2 Schematic view of BDNF(1–12) primary sequence. The peptide is blocked at C-terminus by amidation.

Different peptide fragments of NT N-terminal domains have been synthesized in order to mimic the whole NT activity (Colangelo et al., 2008; Travaglia et al., 2011; Travaglia, La Mendola, et al., 2012, 2013). The linear peptide NGF(1–14), encompassing the first 14 residues of NGF, activates TrkA receptor signaling in PC12. NGF(1–14) exerts a dose- and time-dependent function, an effective and specific NGF-like action on some highly conserved and biologically crucial intermediates of its intracellular targets such as Akt and CREB (Travaglia et al., 2015).

An analogous linear peptide fragment encompassing the N-terminal of BDNF (BDNF(1–12)) has been shown to be biologically active on neuroblastoma SHSY5Y cells (Travaglia, La Mendola, et al., 2012, 2013) (Fig. 2).

Interestingly, both the acetylated form of BDNF(1–12) and the single point mutated (in which aspartate has been substituted by asparagine) showed higher activity. This suggests that the different overall charge of the peptide is able to tune different intramolecular and/or intermolecular interactions with the TrkB receptor, resulting in divergent biochemical signaling. These data indicate how charge and conformational features are essential for peptide activity; potential interaction with metal ions might trigger similar effects.

3. ZINC IN THE BRAIN

Metal ions, especially zinc, iron, and copper, are emerging as key players in biological processes such as respiration, photosynthesis, gene regulation, replication and repair of DNA, antioxidant defense, and neurotransmission (Sensi et al., 2011; Sindreu & Storm, 2011). Such a broad range of activities are due to the fact that metal ions are essential for the function of over one-third of all proteins, enzymes, and transcription factors.

Zinc is an essential element for brain physiology. The first indication of the accumulation of zinc in nerve fibers came from Maske (1955). He found that injection of dithizone, a chelator of heavy metal, results in the staining of the mossy fiber region of the hippocampus. Subsequent histochemical techniques revealed not only the localization of zinc throughout the brain but also an interesting localization into presynaptic synaptic vesicles (Danscher, 1981; Danscher, Howell, Perez-Clausell, & Hertel, 1985; Frederickson, 1989; Timm, 1958). This is conserved across all the vertebrate species analyzed (Frederickson, 1989). Moreover, the presence of presynaptic zinc is greatly restricted to cortical and limbic regions of the brain in a distinct laminar pattern (Birinyi, Parker, Antal, & Shupliakov, 2001; Danscher et al., 2001; Friedman & Price, 1984).

These studies revealed that the majority of zinc in the brain, about 80% of the total, is not detected with histochemical techniques (Cole, Wenzel, Kafer, Schwartzkroin, & Palmiter, 1999). In fact, zinc is mostly tightly bound to proteins, such as metallothioneins (MTs) and zinc-dependent transcription factors (Sensi et al., 2011; Sindreu & Storm, 2011), where it is normally coordinated by four to six ligands, frequently supplied by histidine and cysteine, in conserved binding motifs (Auld, 2001; Vallee & Falchuk, 1993). A fraction of zinc (~10%) is present in its free ionic form (Zn^{2+}) or loosely bound. This free pool is mainly localized within presynaptic vesicles at glutamatergic terminals (Beaulieu, Dyck, & Cynader, 1992; Martinez-Guijarro, Soriano, Del Rio, & Lopez-Garcia, 1991; Sindreu, Varoqui, Erickson, & Perez-Clausell, 2003).

A growing and exciting body of evidence indicates that Zn^{2+} plays a role in synaptic transmission. Synaptic activation triggers the release of vesicular Zn^{2+}, bringing its concentrations in the synaptic cleft to transiently rise. The exact amount of zinc released is still controversial because this measurement is technically challenging. Nevertheless, many laboratories have indicated that increases in transient Zn^{2+} may reach 1–100 μM

(Frederickson et al., 2006; Qian & Noebels, 2005; Vogt, Mellor, Tong, & Nicoll, 2000). From a cognitive perspective, the abundance of vesicular zinc in synapses of the cortex, hippocampus, and amygdala suggests a possible role in learning and memory.

Convincing observation supports the contention that synaptic zinc plays a critical role in neural plasticity, learning, and memory. Alteration of zinc homeostasis at the synaptic level produces an effect on cognitive function. Decreased zinc levels in the hippocampus correlate with spatial memory deficit in aged rats. Zn^{2+} chelation into the hippocampus produces a selective disruption of hippocampus-dependent spatial working memory (Frederickson, Frederickson, & Danscher, 1990). Sindreu, Palmiter, and Storm (2011) showed that ZnT3-KO mice, lacking the synapse-specific vesicular zinc transporter ZnT3, have reduced activation of the Erk1/2 MAPK in hippocampus during hippocampus-dependent learning. The authors also showed that these mice have complete deficits in contextual discrimination and spatial working memory. Contextual memory was also impaired in wild-type mice after locally blocking zinc or MAPK, leading the authors to conclude that ZnT3 is important for zinc homeostasis and is required for hippocampus-dependent memory (Sindreu et al., 2011).

3.1 Zinc Metallostasis in the Brain

The broad and powerful effects of metal ions necessitate that each tissue, cell, and even subcellular compartment has to maintain its metal ion concentration within a narrow range and be able to fine-tune that concentration in response to stimuli such as increased stressors, metabolic activity, or injury. Metal ion homeostasis (metallostasis) is achieved through a complex network of molecular interactions, and so balances intracellular metal uptake, trafficking, storage, speciation, and signaling. Prominent modulators of metallostasis are metal chaperones, metal transporters, metalloproteins, small molecules, and metal transcription factors (Finney & O'Halloran, 2003; Que, Domaille, & Chang, 2008).

Metallostasis regulators can partially correct metal ion dyshomeostasis (e.g., sequestering the metal ion in cytoplasmic vesicles or increasing its efflux/influx) that contributes to a broad range of human diseases, mostly linked to aging (Zatta et al., 2009).

In previous decades, several papers have shown interesting insights into zinc metallostasis through specific transporters. In fact, because of their positive charge, zinc ions are not able to cross cellular membranes by passive diffusion but require the aid of specialized carrier proteins (Liuzzi & Cousins, 2004; Palmiter & Huang, 2004).

ZRT/IRT-related proteins (ZIPs) and zinc transporters (ZnTs) families are multipass transmembrane proteins involved in zinc homeostasis. ZIPs and ZnTs are encoded, respectively, by the Slc39A and Slc30A gene family protein (Liuzzi & Cousins, 2004; Palmiter & Huang, 2004). These carriers are distinguished on the basis of their structural and functional features, having the opposite function in modulating zinc homeostasis. ZIP proteins are influx transporters that mediate Zn uptake from the extracellular environment, or intracellular vesicles, into the cytoplasm. ZnT carriers are efflux transporters able to reduce the cytoplasmic Zn by promoting zinc movement directly out of the cell or into the intracellular compartment (Liuzzi & Cousins, 2004; Palmiter & Huang, 2004).

It is worth noting that both ZnTs and ZIPs exhibit unique tissue- and subcellular-specific expression. For instance, ZnT1 is a cellular zinc exporter and is predominantly expressed on the plasma membrane of cells; it seems to play a key role in early embryonic development and a protective role in the nervous system (Sekler et al., 2002). Moreover, ZnT1 is at least part of the mechanism by which dietary zinc is absorbed, responding to dietary zinc levels (McMahon & Cousins, 1998).

ZnT2 may have a protective role when intracellular zinc concentrations increase rapidly by taking part in metal sequestration into the endosomal/lysosomal vesicles of cells, whereas ZnT5, 6, and 7 are associated with Golgi Zn^{2+} transport (Liuzzi & Cousins, 2004).

ZnT3 transporter is particularly abundant in regions that have high levels of labile, chelatable Zn^{2+}, and it has been suggested that ZnT3 accumulates zinc in synaptic vesicles and has a neuromodulatory role in synaptic transmission. ZnT3 has been linked to degenerative neurological diseases such as AD (Huang et al., 2000).

Intracellular zinc ions can either be shuttled to various subcellular organelles or bound by MTs (Ngu & Stillman, 2009; Vasák, 2005). MTs also have been proposed to play an important role in controlling zinc metabolism. MTs are a family of cysteine-rich proteins able to form metal–thiolate clusters, thus representing a cytosolic buffer for heavy metals in all human cells (Andrews, 2000; Torreggiani & Tinti, 2010).

Under steady-state conditions, a primary function of MTs is to buffer the relatively large zinc content, whereas in nonsteady-state conditions, zinc-binding proteins and transporters act in concert to modulate transient changes in cytosolic zinc ion concentrations (Colvin, Holmes, Fontaine, & Maret, 2010).

In addition, the expression of several genes is induced through direct binding of zinc to the Zn-finger domain of metal response element-binding transcription factor-1 (MTF-1) which, after binding Zn, translocates to the nucleus to bind the metal response element in genes for ZnT1, MT, and gamma glutamylcysteine synthetase (an enzyme that is essential for glutathione synthesis) (Andrews, 2001).

4. ZINC(II) INTERACTION WITH BDNF AND RELATED PEPTIDES

As mentioned earlier, there is a significant overlap between brain areas featured by metal ion dyshomeostasis, and those where the NTs exert their biological activity. Furthermore, some evidences suggest that there is a cross talk between zinc ion homeostasis and BDNF expression. For example, BDNF has been reported to increase the intracellular Zn^{2+} level in retinal pigment epithelium cells, modulating the expression of ZnTs, which in turn increase metal ion uptake (Leung et al., 2008). Moreover, treatments with zinc(II) ions induced the increase of BDNF expression in the cortex and hippocampus (Sowa-Kućma et al., 2008) and in mice models of AD (Corona et al., 2010).

Zinc ions have been reported to modulate the level of BDNF and other NTs. As mentioned earlier, BDNF is synthesized as a precursor, named pro-BDNF, which is then cleaved to give the mature form intracellularly through the action of furin or prohormone convertases, or extracellularly through the action of plasmin and matrix metalloproteinases (MMPs) REF. Hwang et al. showed that zinc(II) ions induce MMP activity and proneurotrophin processing (Hwang, Park, Choi, & Koh, 2005; Hwang, Park, & Koh, 2007). In particular, they showed that zinc(II) ions increased the activity of MMP2 and MMP9 in cortical neurons. This event, in turn, promotes the cleavage of pro-BDNF, thus increasing the secretion of mature BDNF in the media (Hwang et al., 2005, 2007). Addition of MMP inhibitors completely blocked the zinc-induced increases in BDNF, indicating that these enzymes act upstream in pro-BDNF processing (Hwang et al., 2005, 2007).

In addition, high concentrations of Zn^{2+} have been reported to inhibit the biological activity of BDNF as well as other NTs in vitro (Ross et al., 1997). These effects have been attributed to metal-induced conformational changes, which should alter the NT binding to a Trk receptor and the activation of its downstream pathways.

The direct Zn^{2+} and Cu^{2+} binding to the N-terminal domain of NTs may alter interaction with the specific Trk receptors. However, very few studies have been carried out on the effect of metal binding to peptide NT mimicking systems.

The addition of metal ions induces a remarkable difference in the activity of NGF(1–14) and its acetylated analogue AcNGF(1–14) (Travaglia et al., 2011). The effect of NGF(1–14) on SHSY5Y cells is significantly increased in the presence of Zn^{2+} or Cu^{2+} ions, whereas the activity of AcNGF(1–14) is practically unaffected under the same experimental conditions. This is in agreement with the thermodynamic and spectroscopic results, which underline the role played by the N-terminal amino group as the main anchoring site of Cu^{2+} and Zn^{2+}. The metal binding can influence the activity of NGF in different ways: addition of Cu^{2+} induces a synergic effect similar to NGF (1–14) peptide, whereas Zn^{2+} decreases the protein activity (Travaglia et al., 2011).

The linear mimicking systems BDNF(1–12) peptide is able to bind both Cu^{2+} and Zn^{2+} (Travaglia, La Mendola, et al., 2012, 2013). Computational modeling approach revealed that BDNF(1–12), BDNF(1–12)D3N, and Ac-BDNF(1–12) peptides when docked to domain 5 of TrkB receptor (TrkB-D5) show similar binding poses to those of the N-terminal sections of NT 4/5. All the peptides tend to maintain their intramolecular interactions, such as those involving Asp-3 and the N-terminal amine, which can give reasons of the different activity in SHSY5Y cells. The peptide BDNF (1–12) is able to bind copper ions and the complexes formed showed a decrease in the biological activity as observed for the whole protein.

The peptide BDNF(1–12) is able to bind Zn^{2+} and the predominant complex species, at physiological pH, is $[ZnL]^{2+}$, in which the metal ion has a histamine-like coordination mode (Travaglia, La Mendola, et al., 2013). The Zn^{2+} is bound to one amino and one imidazole, so forming a six-membered chelate ring, and two water molecules in a tetrahedral environment (NH_2, N_{Im}, and $2O_{water}$) (Fig. 3).

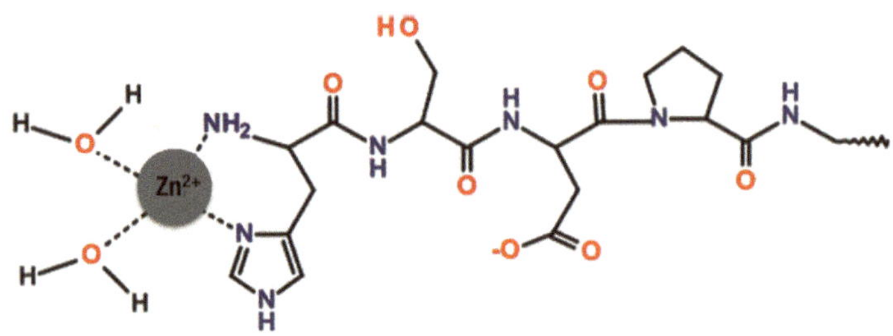

Fig. 3 Scheme of $[ZnL]^{2+}$ complex. The metal ion coordination environment is completed by water molecules.

A comparison with the analogous complex species formed by the acetylated peptide Ac-BDNF(1–12) and the single-point-mutated BDNF (1–12)D3N, in which the aspartate has been substituted by an asparagine residue, clarifies the essential role of the imidazole side chain as the first anchoring site of the metal and rules out the direct involvement of the aspartate residue in metal binding.

The coordination environment is further confirmed by DFT-based geometry optimizations. In fact, comparing the [ZnL] species formed by BDNF(1–12) and BDNF(1–12)D3N, there is a different network of hydrogen bonds. On the whole the structural data have given evidence of different thermodynamic stability constant for the zinc complexes (Travaglia, La Mendola, et al., 2013).

The Zn^{2+} addition induces an increase in the proliferative activity of both the BDNF(1–12) peptide and BDNF. Interestingly, Cu^{2+} has the opposite effect, inducing a decrease in activity of both peptide and protein in this in vitro model (Travaglia, La Mendola, et al., 2012, 2013).

All together, these data indicate that copper and zinc may discriminate and differently modulate the activity of NGF and BDNF.

4.1 Peptide Delivery

Peptides can be designed to target a broad range of biomolecules, giving them almost limitless possibilities in fields such as oncology, immunology, infectious disease, and endocrinology. However, peptides are subject to protease action, so a major topic in new drug formulation is peptide stabilization in delivery devices, as well as the design of appropriate target-specific carriers.

Many efforts have been made for effective delivery of peptides through various routes of administration for successful therapeutic effects (Bruno, Miller, & Lim, 2013; Jain et al., 2013). Chemical modification of peptides can improve their enzymatic stability and/or membrane penetration properties. A direct structural modification such as cyclization may result in an increase of resistance, as for the BDNF-mimicking systems reported earlier. It is also possible to conjugate moieties to a peptide sequence to give more selectivity, solubility, and resistance. Cell surface receptors that can be targeted for internalization of peptides include carbohydrate receptors, lipoprotein receptors, and receptors involved in cell adhesion (Otvos & Wade, 2014). Increasing the hydrophobicity of a peptide (by surface modification using lipophilic moieties) may be of particular benefit for passive or active absorption through cell membranes and may increase peptide stability.

Many drug carrier systems are currently being developed in an attempt to avoid direct chemical modification of peptides. A group of carrier systems is represented by hydrophilic mucoadhesive polymers (polyacrylates, cellulose, chitosan), which can be altered to suit the needs of the peptide being delivered (Asane et al., 2008). Such polymers are available in both cationic and anionic forms and can be also combined with other ligands able to bind metal ions (Bernkop-Schnurch, 2005; Bernkop-Schnurch & Krajicek, 1998).

Nanoemulsions are carrier systems especially used for oral peptide therapeutics. They are defined as oil-in-water (o/w) or water-in-oil (w/o) emulsions, with mean droplet diameters ranging from 50 to 1000 nm. The average droplet size is usually between 100 and 500 nm (Shah, Bhalodia, & Shelat, 2010). Nanoemulsions have a large surface area, making them an effective transport system.

A liposome is defined as a spherical vesicle with an aqueous internal cavity enclosed by a lipid bilayer membrane. Liposome systems have been widely used, especially for oral drug delivery, and there is a concern with the stability of these vesicles under the physiologic conditions of the gastrointestinal tract (Fricker et al., 2010; Werle & Takeuchi, 2009).

Nanoparticles (NPs) are solid particles with sizes in the range of 10–1000 nm (Cerquiera, Lasham, Shelling, & Al-Kassas, 2015; Montalti, Prodi, Rampazzo, & Zaccheroni, 2014). The composition of engineered NPs may vary widely. Source materials may be of biological origin, like phospholipids, lipids, lactic acid, dextran, caprolactone, and chitosan, or have more inorganic characteristics, like various polymers, carbon, silica, and metals. NPs as well as solid lipids have shown great potential in the delivery of peptides. NPs allow for the encapsulation of peptides inside a polymeric matrix, thus protecting them against hydrolysis and enzymatic degradation (Jung et al., 2000).

The NPs are attractive for medical purposes due to their surface-to-mass ratio being higher in comparison with that of other particles, so their ability to bind, to adsorb, and to carry other compounds such as drugs, probes, and proteins is increased (De Jong & Borm, 2008). It is also possible to coat NP surfaces with other compounds or polymers such as polyethylene glycol so as to increase the efficiency of peptide delivery to target cells and tissue (Suk, Xu, Kim, Hanes, & Ensign, 2015).

Solid lipid nanoparticles and nanostructured lipid carriers are two types of lipid-based NPs. They are stabilized with an emulsifying layer in an aqueous dispersion, and their size ranges between 50 and 1000 nm (Naseri, Valizadeh, & Zakeri-Milani, 2015). These systems overcome some of the limitations of other colloidal carriers such as emulsions, liposomes, and polymeric NPs because they combine similar advantages, like a good release profile and targeted drug delivery with excellent physical stability.

A novel approach is represented by a supported lipid bilayer (SLB) formed via fusion of liposomes to a porous silica NP (Irvine, 2011). Such a versatile hybrid nanocarrier (NP–SLB) synergistically combines features of both mesoporous silica particles and liposomes, thus exhibiting many features of an ideal targeted therapeutic delivery platform. The high pore volume and the surface area of the spherical porous silica core allow for an enormous encapsulation capacity for a spectrum of molecules. The fluid-phase lipid bilayer decorating the outer surface of the NP serves as a modular, reconfigurable scaffold, enabling the attachment of a variety of molecules for cell-specific targeting (Ashley et al., 2012). In vitro experiments proved the stability of the SLB acting as a capping system by sealing off colloidal mesoporous silica NPs, thus enabling a tunable release-on-demand of the adsorbed molecules (Cauda et al., 2010).

An advantage of these carrier systems is that lipid molecules can freely diffuse within the SLB so as to increase delivery efficiency and targeting specificity (Liu, Jiang, Ashley, & Brinker, 2009). Furthermore, SLB-based nanoplatforms can be functionalized with biomolecules, either via covalent or via physical binding. The latter approach offers a tunable and reversible adsorption/release by proper matching of the physicochemical properties of both the substrate and the ligand molecules, including charge and hydrophilic character (Satriano, Svedhem, & Kasemo, 2012). The peptide association to the SLB might enable the improvement of its resistance to degradation and bioavailability, due to the synergic combination of materials and biophysical properties of the SLB-based nanoplatform without the need for chemical modification.

The peptide NGF(1–14) has been loaded in SLB–NP. The peptide is hydrophilic and positively charged at physiological pH, so predominantly resides in the vesicle inner volume and on the outer shell of the vesicles, especially for the negatively charged lipids (Fig. 4).

The supported membrane functionalized with NGF(1–14) maintained their fluidic and viscoelastic properties as assessed by fluorescence recovery after photobleaching measurements.

The biological activity of NGF(1–14) associated with the SLB is similar to that observed for peptides alone. This result highlighted the availability of peptide molecules at the outer surface of the supported membrane to interact with cells (Travaglia, Satriano, et al., 2013).

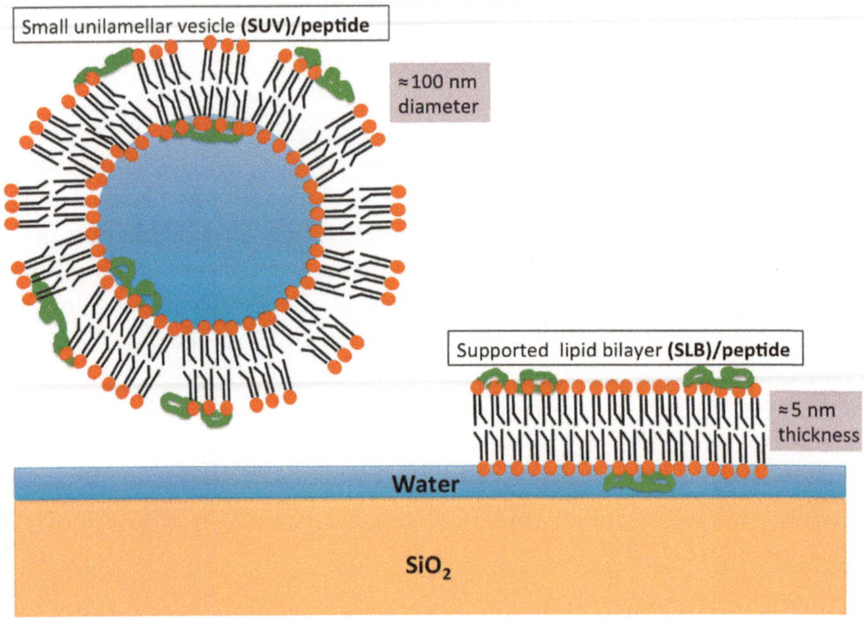

Fig. 4 Graphical representation of peptide–lipid association to the small unilamellar vesicles and the following NGF peptide-supported lipid bilayer.

Another type of carrier is made from gold nanoparticles (GNPs), which are emerging as promising agents for cancer therapy and are being investigated as drug carriers, photothermal agents, contrast agents, and radiosensitizers (Jain, Hirst, & O'Sullivan, 2012).

GNPs can penetrate widely throughout the body and can bind many peptides, proteins, and drugs (Daniel & Astruc, 2004). It is possible to attach covalently different peptides on GNP, but an important issue concerns the maintenance of their properties once linked to gold. The immobilization of biomolecules on the solid surface through controlled physisorption offers several advantages in terms of easy maintenance of conformational degrees of freedom, reversible interaction with the surface, and enhanced stability for the selective cell and nuclear targeting. Peptides NGF(1–14) and BDNF (1–12) were adsorbed on a gold surface under competitive/synergic conditions from the binary equimolecular solution (Forte, Travaglia, Magrì, Satriano, & La Mendola, 2014).

Experimental findings by QCM-D and CD analyses, as well as theoretical MD calculations, highlighted the interactions between each peptide and the gold surface, which affects, in turn, hydrodynamic thickness and related viscoelastic properties of peptide adlayers. The differences found under the two different pH conditions investigated suggest the possibility of fine-tuning and triggering of the viscoelastic properties of the peptide adlayers, including viscosity, effective thickness, and water content (Forte et al., 2014).

At physiological pH, a synergic play of the two peptides in the adsorption on the gold is found, whereas there is a competitive adsorption with a prevalence of BDNF(1–12) at acidic pH.

Cellular assays demonstrated the maintenance of biological properties of the peptides adsorbed, but the cellular response to the NGF(1–14) and BDNF(1–12) peptide fragments immobilized on the gold surfaces is not a direct consequence of the corresponding areal mass coverage; instead, it is affected by a more complex scenario, where the specific peptide–surface and peptide–peptide interactions, and the consequent viscoelastic properties of the hydrodynamic adlayer, modulate overall behavior at the biointerface (Forte et al., 2014). Therefore, other than the development of strategies in peptide design, chemistry, and assembly, an effort is necessary to design and scrutinize the response of peptide-mimicking materials exhibiting complexity, coordinated molecular processes, and dynamic interactions with the cells.

5. ZINC, BDNF, AND DIET

As mentioned, zinc is a very abundant "trace element," function of which is emerging. A large body of evidence suggests that dietary zinc deficiency, both during pregnancy and after birth, impairs the development of the hippocampus and thus impacts learning and memory. For example, mild maternal zinc deficiency during pregnancy/lactation induces deficits in memory formation in offspring. Zinc supplementation of pups after weaning rescues memory impairment.

The effects of zinc supplementation on cognition, particularly in adults, are less clear. Chronic zinc supplementation starting during gestation and lasting throughout adulthood resulted in impaired learning of spatial tasks in rats (Railey, Micheli, Wanschura, & Flinn, 2010). Interestingly, addition of copper to zinc-supplemented water rescued memory impairment, demonstrating the importance of maintaining appropriate homeostasis. Other investigators reported that zinc supplementation (starting with gestation and continuing postnatally through juvenile life) improves spatial memory (Piechal, Blecharz-Klin, Pyrzanowska, & Widy-Tyszkiewicz, 2012). In mice, supplementation with high zinc dosages (increasing zinc levels in drinking water for 3 months) resulted in zinc deficiency in the hippocampus. This was accompanied by a hippocampus-dependent memory deficit and decreased expression of levels of learning and memory-related receptors and synaptic proteins, including NMDA-NR2A, NR2B, AMPA-GluR1, PSD-93, and PSD-95. The same treatment impaired BDNF signaling in the hippocampus. Namely, high-dose zinc supplementation induced a significant decrease in hippocampal BDNF, TrkB, and phosphorylated TrkB levels. Moreover, levels of the downstream target CREB were also reduced. This effect is specific for the hippocampus, since zinc supplementation in low dosages did not interrupt the BDNF–TrkB signaling pathway (Yang et al., 2013).

Contestabile et al. showed that 4-week-long zinc supplementation in adult rats, approximately doubling the normal intake, impairs memory consolidation and facilitates forgetting of stressful experiences in two hippocampal-dependent cognitive tasks (Contestabile, Peña-Altamira, Virgili, & Monti, 2016). The levels of hippocampal BDNF have been reported to increase during memory consolidation (Bambah-Mukku et al., 2014; Beckinschtein et al., 2007). The authors found that the increase of BDNF expression found in the hippocampus of normally fed rats 24 h after IA training did not occur in zinc-supplemented animals (Contestabile et al., 2016). Corona et al. found that Zn^{2+} supplementation (30 ppm) in a transgenic mouse model of AD, the 3xTg-AD model, greatly delays hippocampal-dependent memory deficits and strongly reduces both Aβ and tau pathology in the hippocampus (Corona et al., 2010). Moreover, Zn^{2+} supplementation increased the levels of BDNF in whole-brain extract, possibly through the activation of MMP and consequential maturation of BDNF from inactive pro-BDNF via MMP activation (Corona et al., 2010). These contradictory results can be due to the timing and dosage of zinc deficiency/supplementation, the animal model used, and toxicity of extrasynaptic zinc excess.

The implementation of the zinc may be more effective if administered with peptides capable of binding the metal ion as carnosine (β-alanyl-L-histidine), a dipeptide present in abundance in the muscles (Corona et al., 2011). Therefore, the peptides able to mimic the activity of BDNF and to bind Zn^{2+} may be used together as potential supplements.

6. CONCLUSION

At present, neurodegenerative diseases are pathologies without cures. Thus, it is necessary to revise old players and their links with new molecular entities. Metal ions are emerging as key regulators of many biological processes, including neurotransmission, and their dyshomeostasis may be one of the underlying causes of these diseases. BDNF received the attention of neuroscientist because its role beyond survival and differentiation, that is, activity dependent modification of synaptic transmission.

The interaction of metals and NTs might represent a crossroad for neuronal physiology, and it is therefore strongly rising as one of the most critical step in the study of neurodegenerative diseases. Better understanding of metal ion-driven NTs signal transduction and intercellular signaling could unravel helpful information in the design of new therapeutic interventions. The synthesis of BDNF peptidomimetic revealed to be helpful both to understand the molecular interaction of BDNF with metal ions and be the basis for the development of new drugs and supplements for neurodegenerative diseases.

ACKNOWLEDGMENTS

The Italian Ministry of University and Research is gratefully acknowledged for partial support (PRIN 2010M2JARJ). The authors wish to thank Ian Gardiner for his invaluable contribution in preparing the manuscript.

REFERENCES

Alonso, M., Vianna, M. R., Depino, A. M., Mello e Souza, T., Pereira, P., Szapiro, G., et al. (2002). BDNF-triggered events in the rat hippocampus are required for both short- and long-term memory formation. *Hippocampus*, *12*, 551–560.

Andrews, G. K. (2000). Regulation of metallothionein gene expression by oxidative stress and metal ions. *Biochemical Pharmacology*, *59*, 95–104.

Andrews, G. K. (2001). Cellular zinc sensors: MTF-1 regulation of gene expression. *Biometals*, *14*, 223–237.

Arancio, O., & Chao, M. V. (2007). Neurotrophins, synaptic plasticity and dementia. *Current Opinion in Neurobiology*, *17*, 325–330.

Asane, G. S., Nirmal, S. A., Rasal, K. B., Naik, A. A., Mahadik, M. S., & Rao, Y. M. (2008). Polymers for mucoadhesive drug delivery system: A current status. *Drug Development and Industrial Pharmacy*, *34*, 1246–1266.

Ashley, C. E., Carnes, E. C., Epler, K. E., Padilla, D. P., Phillips, J. K., Castillo, R. E., et al. (2012). Delivery of small interfering RNA by peptide-targeted mesoporous silica nanoparticle-supported lipid bilayers. *ACS Nano*, *6*, 2174–2188.

Auld, D. S. (2001). Zinc coordination sphere in biochemical zinc sites. *Biometals*, *14*, 271–313.

Bambah-Mukku, D., Travaglia, A., Chen, D. Y., Pollonini, G., & Alberini, C. (2014). A positive autoregulatory BDNF feedback loop via C/EBPß mediates hippocampal memory consolidation. *The Journal of Neuroscience*, *34*, 12547–12559.

Banfield, M. J., Naylor, R. L., Robertson, A. G., Allen, S. J., Dawbarn, D., & Brady, R. L. (2001). Specificity in Trk receptor:neurotrophin interactions: The crystal structure of TrkB-d5 in complex with neurotrophin-4/5. *Structure*, *9*, 1191–1199.

Beaulieu, C., Dyck, R., & Cynader, M. (1992). Enrichment of glutamate in zinc-containing terminals of the cat visual cortex. *NeuroReport*, *3*, 861–864.

Beckinschtein, P., Cammarota, M., Igaz, L. M., Bevilaqua, L. R. M., Izquierdo, I., & Medina, J. H. (2007). Persistence of long-term memory storage requires a late protein synthesis- and BDNF-dependent phase in the hippocampus. *Neuron, 53*, 261–277.

Benraiss, A., Chmielnicki, E., Lerner, K., Roh, D., & Goldman, S. A. (2001). Adenoviral brain-derived neurotrophic factor induces both neostriatal and olfactory neuronal recruitment from endogenous progenitor cells in the adult forebrain. *The Journal of Neuroscience, 21*, 6718–6731.

Bernkop-Schnurch, A. (2005). Thiomers: A new generation of mucoadhesive polymers. *Advanced Drug Delivery Reviews, 57*, 1569–1582.

Bernkop-Schnurch, A., & Krajicek, M. E. (1998). Mucoadhesive polymers as platforms for peroral peptide delivery and absorption: Synthesis and evaluation of different chitosan-EDTA conjugates. *Journal of Controlled Release, 50*, 215–223.

Binder, D. K., Routbort, M. J., Ryan, T. E., Yancopoulos, G. D., & McNamara, J. O. (1999). Selective inhibition of kindling development by intraventricular administration of TrkB receptor body. *The Journal of Neuroscience, 19*, 1424–1436.

Birinyi, A., Parker, D., Antal, M., & Shupliakov, O. (2001). Zinc co-localizes with GABA and glycine in synapses in the lamprey spinal cord. *The Journal of Comparative Neurology, 433*, 208–221.

Bonda, D. J., Lee, H. G., Blair, J. A., Zhu, X., Perry, G., & Smith, M. A. (2011). Role of metal dyshomeostasis in Alzheimer's disease. *Metallomics, 3*, 367–370.

Bruno, B. J., Miller, G. D., & Lim, C. S. (2013). Basics and recent advances in peptide and protein drug delivery. *Therapeutic Delivery, 4*, 1443–1467.

Cauda, V., Engelke, H., Sauer, A., Arcizet, D., Bräuchle, C., Rädler, J., et al. (2010). Colchicine-loaded lipid bilayer-coated 50 nm mesoporous nanoparticles efficiently induce microtubule depolymerization upon cell uptake. *Nano Letters, 10*, 2484–2492.

Cerquiera, B. B., Lasham, A., Shelling, A. N., & Al-Kassas, R. (2015). Nanoparticle therapeutics: Technologies and methods for overcoming cancer. *European Journal of Pharmaceutics and Biopharmaceutics, 97*, 140–151.

Chao, M. V. (2003). Neurotrophins and their receptors: A convergence point for many signalling pathways. *Nature Reviews. Neuroscience, 4*, 299–309.

Chao, M. V., & Bothwell, M. (2002). Neurotrophins: To cleave or not to cleave. *Neuron, 33*, 9–12.

Colangelo, A. M., Bianco, M. R., Vitagliano, L., Cavaliere, C., Cirillo, G., De Gioia, L., et al. (2008). A new nerve growth factor-mimetic peptide active on neuropathic pain in rats. *The Journal of Neuroscience, 28*, 2698–2709.

Cole, T. B., Wenzel, H. J., Kafer, K. E., Schwartzkroin, P. A., & Palmiter, R. D. (1999). Elimination of zinc from synaptic vesicles in the intact mouse brain by disruption of the ZnT3 gene. *Proceedings of the National Academy of Sciences of the United States America, 96*, 1716–1721.

Colvin, R. A., Holmes, W. R., Fontaine, C. P., & Maret, W. (2010). Cytosolic zinc buffering and muffling: Their role in intracellular zinc homeostasis. *Metallomics, 2*, 306–317.

Contestabile, A., Peña-Altamira, E., Virgili, M., & Monti, B. (2016). Zinc supplementation in rats impairs hippocampal-dependent memory consolidation and dampens post-traumatic recollection of stressful event. *European Neuropsychopharmacology, 26*, 1070–1082. http://dx.doi.org/10.1016/j.euroneuro.2015.12.041. pii: S0924-977X(15)00431-9 [Epub ahead of print].

Corona, C., Frazzini, V., Silvestri, E., Lattanzio, R., La Sorda, R., Piantelli, M., et al. (2011). Effects of dietary supplementation of carnosine on mitochondrial dysfunction, amyloid pathology, and cognitive deficits in 3xTg-AD-mice. *PLoS One, 6*. e17971.

Corona, C., Masciopinto, F., Silvestri, E., Del Viscovo, A., Lattanzio, R., Sorda, R., et al. (2010). Dietary zinc supplementation of 3xTg-AD mice increases BDNF levels and prevents cognitive deficits as well as mitochondrial dysfunction. *Cell Death & Disease, 1*. e91. http://dx.doi.org/10.1038/cddis.2010.73.

Daniel, M. C., & Astruc, D. (2004). Gold nanoparticles: Assembly, supramolecular chemistry, quantum-size-related properties, and applications toward biology, catalysis, and nanotechnology. *Chemical Reviews*, *104*, 293–346.

Danscher, G. (1981). Histochemical demonstration of heavy metals. A revised version of the sulphide silver method suitable for both light and electron microscopy. *Histochemistry*, *71*, 1–16.

Danscher, G., Howell, G., Perez-Clausell, J., & Hertel, N. (1985). The dithizone, Timm's sulphide silver and the selenium methods demonstrate a chelatable pool of zinc in CNS. A proton activation (PIXE) analysis of carbon tetrachloride extracts from rat brains and spinal cords intravitally treated with dithizone. *Histochemistry*, *83*, 419–422.

Danscher, G., Jo, S. M., Varea, E., Wang, Z., Cole, T. B., & Schroder, H. D. (2001). Inhibitory zinc-enriched terminals in mouse spinal cord. *Neuroscience*, *105*, 941–947.

De Jong, W. H., & Borm, P. J. A. (2008). Drug delivery and nanoparticles: Applications and hazards. *International Journal of Nanomedicine*, *3*, 133–149.

Deng, W., Aimone, J. B., & Gage, F. H. (2010). New neurons and new memories: How does adult hippocampal neurogenesis affect learning and memory? *Nature Reviews. Neuroscience*, *11*, 339–350.

Douvaras, P., Wang, J., Zimmer, M., Hanchuk, S., O'Bara, M. A., Sadiq, S., et al. (2014). Efficient generation of myelinating oligodendrocytes from primary progressive multiple sclerosis patients by induced pluripotent stem cells. *Stem Cell Reports*, *3*, 250–259.

Ernfors, P., Bengzon, J., Kokaia, Z., Persson, H., & Lindvall, O. (1991). Increased levels of messenger RNAs for neurotrophic factors in the brain during kindling epileptogenesis. *Neuron*, *7*, 165–176.

Figurov, A., Pozzo-Miller, L. D., Olafsson, P., Wang, T., & Lu, B. (1996). Regulation of synaptic responses to high-frequency stimulation and LTP by neurotrophins in the hippocampus. *Nature*, *381*, 706–709.

Finney, L. A., & O'Halloran, T. V. (2003). Transition metal speciation in the cell: Insights from the chemistry of metal ion receptors. *Science*, *300*, 931–936.

Fletcher, J. M., & Hughes, R. A. (2009). Modified low molecular weight cyclic peptides as mimetics of BDNF with improved potency, proteolytic stability and transmembrane passage in vitro. *Bioorganic & Medicinal Chemistry*, *17*, 2695–2702.

Fletcher, J. M., Morton, C. J., Zwar, R. A., Murray, S. S., O'Leary, P. D., & Hughes, R. A. (2008). Design of a conformationally defined and proteolytically stable circular mimetic of brain-derived neurotrophic factor. *Journal of Biological Chemistry*, *283*, 33375–33383.

Fobian, K., Owczarek, S., Budtz, C., Bock, E., Berezin, V., & Pedersen, M. V. (2010). Peptides derived from the solvent-exposed loops 3 and 4 of BDNF bind TrkB and p75NTR receptors and stimulate neurite outgrowth and survival. *Journal of Neuroscience Research*, *88*, 1170–1181.

Forte, G., Travaglia, A., Magrì, A., Satriano, C., & La Mendola, D. (2014). Adsorption of NGF and BDNF derived peptides on gold surfaces. *Physical Chemistry Chemical Physics*, *16*, 1536–1544.

Frederickson, C. J. (1989). Neurobiology of zinc and zinc-containing neurons. *International Review of Neurobiology*, *31*, 145–238.

Frederickson, R. E., Frederickson, C. J., & Danscher, G. (1990). In situ binding of bouton zinc reversibly disrupts performance on a spatial memory task. *Behavioural Brain Research*, *38*, 25–33.

Frederickson, C. J., Giblin, L. J., Krezel, A., McAdoo, D. J., Mueller, R. N., Zeng, Y., et al. (2006). Concentrations of extracellular free zinc $(pZn)_e$ in the central nervous system during simple anesthetization, ischemia and reperfusion. *Experimental Neurology*, *198*, 285–293.

Fricker, G., Kromp, T., Wendel, A., Blume, A., Zirkel, J., Rebmann, H., et al. (2010). Phospholipids and lipid-based formulations in oral drug delivery. *Pharmaceutical Research*, *27*, 1469–1486.

Friedman, B., & Price, J. L. (1984). Fiber systems in the olfactory bulb and cortex: A study in adult and developing rats, using the Timm method with the light and electron microscope. *The Journal of Comparative Neurology*, *223*, 88–109.

Greenberg, M. E., Xu, B., Lu, B., & Hempstead, B. L. (2009). New insights in the biology of BDNF synthesis and release: Implications in CNS function. *The Journal of Neuroscience*, *29*, 12764–12767.

Greenberg, M. E., Xu, B., Lu, B., & Hempstead, P. (2013). BDNF-based synaptic repair as a disease-modifying strategy for neurodegenerative diseases. *Nature Reviews. Neuroscience*, *14*, 401–416.

Hashimoto, R., Takei, N., Shimazu, K., Christ, L., Lu, B., & Chuang, D. M. (2002). Lithium induces brain-derived neurotrophic factor and activates TrkB in rodent cortical neurons: An essential step for neuroprotection against glutamate excitotoxicity. *Neuropharmacology*, *43*, 1173–1179.

Hock, C., Heese, K., Hulette, C., Rosenberg, C., & Otten, U. (2000). Region-specific neurotrophin imbalances in Alzheimer disease: Decreased levels of brain-derived neurotrophic factor and increased levels of nerve growth factor in hippocampus and cortical areas. *Archives of Neurology*, *57*, 846–851.

Hoerder-Suabedissen, A., & Molnár, Z. (2015). Development, evolution and pathology of neocortical subplate neurons. *Nature Reviews. Neuroscience*, *16*, 133–146.

Huang, X., Cuajungco, M. P., Atwood, C. S., Moir, R. D., Tanzi, R. E., & Bush, A. I. (2000). Alzheimer's disease, β-amyloid protein and zinc. *Journal of Nutrition*, *130*, 1488S–1492S.

Huang, E. J., & Reichardt, L. F. (2001). Neurotrophins: Roles in neuronal development and function. *Annual Review of Neuroscience*, *24*, 677–736.

Hwang, J. J., Park, M. H., Choi, S. Y., & Koh, J. Y. (2005). Activation of the Trk signaling pathway by extracellular zinc. Role of metalloproteinases. *Journal of Biological Chemistry*, *280*, 11995–12001.

Hwang, J. J., Park, M. H., & Koh, J. Y. (2007). Copper activates TrkB in cortical neurons in a metalloproteinase-dependent manner. *Journal of Neuroscience Research*, *85*, 2160–2166.

Ibánez, C. F. (1998). Emerging themes in structural biology of neeurotrophic factors. *Trends in Neurosciences*, *21*, 438–444.

Ibánez, C. F., Ilag, L. L., Murray-Rust, J., & Persson, H. (1993). An extended surface of binding to Trk tyrosine kinase receptors in NGF and BDNF allows the engineering of a multifunctional pan-neurotrophin. *EMBO Journal*, *12*, 2281–2293.

Irvine, D. J. (2011). Drug delivery: One nanoparticle, one kill. *Nature Materials*, *10*, 342–343.

Isackson, P. J., Huntsman, M. M., Murray, K. D., & Gall, C. M. (1991). BDNF mRNA expression is increased in adult rat forebrain after limbic seizures: Temporal patterns of induction distinct from NGF. *Neuron*, *6*, 937–948.

Jain, S., Hirst, D. G., & O'Sullivan, J. M. (2012). Gold nanoparticles as novel agents for cancer therapy. *The British Journal of Radiology*, *85*, 101–113.

Jain, A., Jain, A., Gulbake, A., Shilpi, S., Hurkat, P., & Jain, S. K. (2013). Peptide and protein delivery using new drug delivery systems. *Critical Reviews in Therapeutic Drug Carrier Systems*, *30*, 293–329.

Jung, T., Kamm, W., Breitenbach, A., Kaiserling, E., Xiao, J. X., & Kissel, T. (2000). Biodegradable nanoparticles for oral delivery of peptides: Is there a role for polymers to affect mucosal uptake? *European Journal of Pharmaceutics and Biopharmaceutics*, *50*, 147–160.

Kang, H., & Schuman, E. M. (1995). Long-lasting neurotrophin-induced enhancement of synaptic transmission in the adult hippocampus. *Science*, *267*, 1658–1662.

Kang, H., Welcher, A. A., Shelton, D., & Schuman, E. M. (1997). Neurotrophins and time: Different roles for TrkB signaling in hippocampal long-term potentiation. *Neuron*, *19*, 653–664.

Kokaia, M., Ernfors, P., Kokaia, Z., Elmér, E., Jaenisch, R., & Lindvall, O. (1995). Suppressed epileptogenesis in BDNF mutant mice. *Experimental Neurology, 133*, 215–224.

Korte, M., Carroll, P., Wolf, E., Brem, G., Thoenen, H., & Bonhoeffer, T. (1995). Hippocampal long-term potentiation is impaired in mice lacking brain-derived neurotrophic factor. *Proceedings of the National Academy of Sciences of the United States of America, 92*, 8856–8860.

Korte, M., Griesbeck, O., Gravel, C., Carroll, P., Staiger, V., Thoenen, H., et al. (1996). Virus-mediated gene transfer into hippocampal CA1 region restores long-term potentiation in brain-derived neurotrophic factor mutant mice. *Proceedings of the National Academy of Sciences of the United States of America, 93*, 12547–12552.

Larsson, E., Nanobashvili, A., Kokaia, Z., & Lindvall, O. (1999). Evidence for neuroprotective effects of endogenous brain-derived neurotrophic factor after global forebrain ischemia in rats. *Journal of Cerebral Blood Flow & Metabolism, 19*, 1220–1228.

Leibrock, J., Lottspeich, F., Hohn, A., Hofer, M., Hengerer, B., Masiakowski, P., et al. (1989). Molecular cloning and expression of brain-derived neurotrophic factor. *Nature, 341*, 149–152.

Leung, K. W., Liu, M., Xu, X., Seiler, M. J., Barnstable, C. J., & Tombran-Tink, J. (2008). Expression of ZnT and ZIP zinc transporters in the human RPE and their regulation by neurotrophic factors. *Investigative Ophthalmology & Visual Science, 49*, 1221–1231.

Levi-Montalcini, R., & Hamburger, V. (1951). Selective growth stimulating effects of mouse sarcoma on the sensory and sympathetic nervous system of the chick embryo. *Journal of Experimental Zoology, 116*, 321–361.

Lewin, G. R., & Barde, Y. A. (1996). Physiology of the neurotrophins. *Annual Review of Neuroscience, 19*, 289–317.

Lindvall, O., Kokaia, Z., Bengzon, J., Elmér, E., & Kokaia, M. (1994). Neurotrophins and brain insults. *Trends in Neurosciences, 17*, 490–496.

Linnarsson, S., Bjorklund, A., & Ernfors, P. (1997). Learning deficit in BDNF mutant mice. *European Journal of Neuroscience, 9*, 2581–2587.

Liu, J. W., Jiang, X. M., Ashley, C., & Brinker, C. J. (2009). Electrostatically mediated liposome fusion and lipid exchange with a nanoparticle-supported bilayer for control of surface charge, drug containment, and delivery. *Journal of the American Chemical Society, 131*, 7567–7569.

Liuzzi, J. P., & Cousins, R. J. (2004). Mammalian zinc transporters. *Annual Review of Nutrition, 24*, 151–172.

Lu, W., Zhang, C., Yi, Z., Li, Z., Wu, Z., & Fang, Y. (2012). Association between BDNF Val66Met polymorphism and cognitive performance in antipsychotic-naïve patients with schizophrenia. *Journal of Molecular Neuroscience, 47*, 505–510.

Malenka, R. C. (2003). The long-term potential of LTP. *Nature Reviews. Neuroscience, 4*, 923–926.

Martinez-Guijarro, F. J., Soriano, E., Del Rio, J. A., & Lopez-Garcia, C. (1991). Zinc-positive boutons in the cerebral cortex of lizards show glutamate immunoreactivity. *Journal of Neurocytology, 20*, 834–843.

Maske, H. (1955). A new method for demonstrating A and B cells in the islands of Langerhans. *Klinische Wochenschrift, 33*, 1058.

McMahon, R. J., & Cousins, R. J. (1998). Regulation of the zinc transporter ZnT-1 by dietary zinc. *Proceedings of the National Academy of Sciences of the United States of America, 95*, 4841–4846.

Minichiello, L. (2009). TrkB signalling pathways in LTP and learning. *Nature Reviews. Neuroscience, 10*, 850–860.

Minichiello, L., Korte, M., Wolfer, D., Kühn, R., Unsicker, K., Cestari, V., et al. (1999). Essential role for TrkB receptors in hippocampus-mediated learning. *Neuron, 24*, 401–414.

Montalti, M., Prodi, L., Rampazzo, E., & Zaccheroni, N. (2014). Dye-doped silica nanoparticles as luminescent organized systems for nanomedicine. *Chemical Society Reviews, 43*, 4243–4268.

Murer, M. G., Yan, Q., & Raisman-Vozari, R. (2001). Brain-derived neurotrophic factor in the control human brain, and in Alzheimer's disease and Parkinson's disease. *Progress in Neurobiology*, *63*, 71–124.
Naseri, N., Valizadeh, H., & Zakeri-Milani, P. (2015). Solid lipid nanoparticles and nanostructured lipid carriers: Structure, preparation and application. *Advanced Pharmaceutical Bulletin*, *5*, 305–313.
Ngu, T. T., & Stillman, M. J. (2009). Metal-binding mechanisms in metallothioneins. *Dalton Transactions*, *28*, 5425–5433.
Nicoll, R. A., & Malenka, R. C. (1995). Contrasting properties of two forms of long-term potentiation in the hippocampus. *Nature*, *377*, 115–118.
O'Leary, P. D., & Hughes, R. A. (1998). Structure-activity relationships of conformationally constrained peptide analogues of loop 2 of brain-derived neurotrophic factor. *Journal of Neurochemistry*, *70*, 1712–1721.
O'Leary, P. D., & Hughes, R. A. (2003). Design of potent peptide mimetics of brain-derived neurotrophic factor. *Journal of Biological Chemistry*, *278*, 25738–25744.
Otvos, L., & Wade, J. D. (2014). Current challenges in peptide-based drug discovery. *Frontiers in Chemistry*, *2*, 62.
Palmiter, R. D., & Huang, L. (2004). Efflux and compartmentalization of zinc by members of the SLC30 family of solute carriers. *Pflügers Archiv: European Journal of Physiology*, *447*, 744–751.
Park, H., & Poo, M. M. (2013). Neurotrophin regulation of neural circuit development and function. *Nature Reviews. Neuroscience*, *14*, 7–23.
Pattarawarapan, M., & Burgess, K. (2003). Molecular basis of neurotrophin-receptor interactions. *Journal of Medicinal Chemistry*, *46*, 5277–5291.
Patterson, S. L., Abel, T., Deuel, T. A., Martin, K. C., Rose, J. C., & Kandel, E. R. (1996). Recombinant BDNF rescues deficits in basal synaptic transmission and hippocampal LTP in BDNF knockout mice. *Neuron*, *6*, 1137–1145.
Pencea, V., Bingaman, K. D., Wiegand, S. J., & Luskin, M. B. (2001). Infusion of brain-derived neurotrophic factor into the lateral ventricle of the adult rat leads to new neurons in the parenchyma of the striatum, septum, thalamus, and hypothalamus. *The Journal of Neuroscience*, *21*, 6706–6717.
Peng, S., Wuu, J., Mufson, E. J., & Fahnestock, M. (2005). Precursor form of brain-derived neurotrophic factor and mature brain-derived neurotrophic factor are decreased in the pre-clinical stages of Alzheimer's disease. *Journal of Neurochemistry*, *6*, 1412–1421.
Piechal, A., Blecharz-Klin, K., Pyrzanowska, J., & Widy-Tyszkiewicz, E. (2012). Maternal zinc supplementation improves spatial memory in rat pups. *Biological Trace Element Research*, *147*, 299–308.
Poduslo, J. F., & Curran, G. L. (1996). Permeability at the blood–brain and blood–nerve barriers of the neurotrophic factors: NGF, CNTF, NT-3 BDNF. *Brain Research. Molecular Brain Research*, *36*, 280–286.
Qian, J., & Noebels, J. L. (2005). Visualization of transmitter release with zinc fluorescence detection at the mouse hippocampal mossy fibre synapse. *The Journal of Physiology*, *566*, 747–758.
Que, E. L., Domaille, D. W., & Chang, C. J. (2008). Metals in neurobiology: Probing their chemistry and biology with molecular imaging. *Chemical Reviews*, *108*, 1517–1549.
Railey, A. M., Micheli, T. L., Wanschura, P. B., & Flinn, J. M. (2010). Alterations in fear response and spatial memory in pre- and post-natal zinc supplemented rats: Remediation by copper. *Physiology & Behavior*, *100*, 95–100.
Ross, G. M., Shamovsky, I. L., Lawrance, G., Solc, M., Dostaler, S. M., Jimmo, S. L., et al. (1997). Zinc alters conformation and inhibits biological activities of nerve growth factor and related neurotrophins. *Nature Medicine*, *3*, 872–878.
Satriano, C., Svedhem, S., & Kasemo, B. (2012). Well-defined lipid interfaces for protein adsorption studies. *Physical Chemistry Chemical Physics*, *14*, 16695–16698.

Scharfman, H. E. (1997). Hyperexcitability in combined entorhinal/hippocampal slices of adult rat after exposure to brain-derived neurotrophic factor. *Journal of Neurophysiology, 78*, 1082–1095.

Scharfman, H. E., Goodman, J. H., & Sollas, A. L. (1999). Actions of brain-derived neurotrophic factor in slices from rats with spontaneous seizures and mossy fiber sprouting in the dentate gyrus. *The Journal of Neuroscience, 19*, 5619–5631.

Scharfman, H. E., Goodman, J. H., Sollas, A. L., & Croll, S. D. (2002). Spontaneous limbic seizures after intrahippocampal infusion of brain-derived neurotrophic factor. *Experimental Neurology, 174*, 201–214.

Sekler, I., Moran, A., Hershfinkel, M., Dori, A., Margulis, A., Birenzweig, N., et al. (2002). Distribution of the zinc transporter ZnT-1 in comparison with chelatable zinc in the mouse brain. *Journal of Comparative Neurology, 447*, 201–209.

Sensi, S. L., Paoletti, P., Koh, J. Y., Aizenman, E., Bush, A. I., & Hershfinkel, M. (2011). The neurophysiology and pathology of brain zinc. *The Journal of Neuroscience, 31*, 16076–16085.

Shah, P., Bhalodia, D., & Shelat, P. (2010). Nanoemulsion: A pharmaceutical review. *Systematic Reviews in Pharmacy, 1*, 24–32.

Sindreu, C., Palmiter, R. D., & Storm, D. R. (2011). Zinc transporter ZnT-3 regulates presynaptic Erk1/2 signaling and hippocampus-dependent memory. *Proceedings of the National Academy of Sciences of the United States of America, 108*, 3366–3370.

Sindreu, C., & Storm, D. R. (2011). Modulation of neuronal signal transduction and memory formation by synaptic zinc. *Frontiers in Behavioral Neuroscience, 5*, 1–14.

Sindreu, C. B., Varoqui, H., Erickson, J. D., & Perez-Clausell, J. (2003). Boutons containing vesicular zinc define a subpopulation of synapses with low AMPAR content in rat hippocampus. *Cerebral Cortex, 13*, 823–829.

Skaper, S. D. (2011). Neurotrophic factors as novel therapeutic targets. In C. G. Shayne (Ed.), *Pharmaceutical sciences encyclopedia: Drug discovery, development, and manufacturing*: 41, John Wiley & Sons; New York, USA.

Sowa-Kućma, M., Legutko, B., Szewczyk, B., Novak, K., Znojek, P., Poleszak, E., et al. (2008). Antidepressant-like activity of zinc: Further behavioral and molecular evidence. *Journal of Neural Transmission (Vienna), 115*, 1621–1628.

Suk, J. S., Xu, Q., Kim, N., Hanes, J., & Ensign, L. M. (2015). PEGylation as a strategy for improving nanoparticle-based drug and gene delivery. *Advanced Drug Delivery Reviews, 99*, 28–51. http://dx.doi.org/10.1016/j.addr.2015.09.012.

Teixido, M., & Giralt, E. (2008). The role of peptides in blood–brain barrier nanotechnology. *Journal of Peptide Science, 14*, 163–173.

Timm, F. (1958). Histochemistry of the region of Ammon's horn. *Zeitschrift für Zellforschung und Mikroskopische Anatomie (Vienna), 48*, 548–555.

Torreggiani, A., & Tinti, A. (2010). Raman spectroscopy a promising technique for investigations of metallothioneins. *Metallomics, 4*, 246–260.

Travaglia, A., Arena, G., Fattorusso, R., Isernia, C., La Mendola, D., Malgieri, G., et al. (2011). The inorganic perspective of nerve growth factor: Interactions of Cu2+ and Zn2+ with the N-terminus fragment of nerve growth factor encompassing the recognition domain of the TrkA receptor. *Chemistry, 17*, 3726–3738.

Travaglia, A., La Mendola, D., Magrì, A., Nicoletti, V. G., Pietropaolo, A., & Rizzarelli, E. (2012). Copper, BDNF and its N-terminal domain: Inorganic features and biological perspectives. *Chemistry, 18*, 15618–15631.

Travaglia, A., La Mendola, D., Magrì, A., Pietropaolo, A., Nicoletti, V. G., Grasso, G., et al. (2013). Zinc(II) interactions with brain-derived neurotrophic factor N-terminal peptide fragments: Inorganic features and biological perspectives. *Inorganic Chemistry, 52*, 11075–11083.

Travaglia, A., Pietropaolo, A., Di Martino, R., Nicoletti, V. G., La Mendola, D., Calissano, P., et al. (2015). A small linear peptide encompassing the NGF N-terminus partly mimics the biological activities of the entire neurotrophin in PC12 cells. *ACS Chemical Neuroscience*, *6*, 1379–1392.

Travaglia, A., Pietropaolo, A., La Mendola, D., Nicoletti, V. G., & Rizzarelli, E. (2012). The inorganic perspectives of neurotrophins and Alzheimer's disease. *Journal of Inorganic Biochemistry*, *111*, 130–137.

Travaglia, A., Satriano, C., Giuffrida, M. L., La Mendola, D., Rampazzo, E., Prodi, L., et al. (2013). Electrostatically driven interaction of silica-supported lipid bilayer nanoplatforms and a nerve growth factor-mimicking peptide. *Soft Matter*, *9*, 4648–4654.

Vallee, B. L., & Falchuk, K. H. (1993). The biochemical basis of zinc physiology. *Physiological Reviews*, *73*, 79–118.

Vasák, M. (2005). Advances in metallothionein structure and functions. *Journal of Trace Elements in Medicine and Biology*, *19*, 13–17.

Vogt, K., Mellor, J., Tong, G., & Nicoll, R. (2000). The actions of synaptically released zinc at hippocampal mossy fiber synapses. *Neuron*, *26*, 187–196.

Werle, M., & Takeuchi, H. (2009). Chitosan-aprotinin coated liposomes for oral peptide delivery: Development, characterisation and in vivo evaluation. *International Journal of Pharmaceutics*, *370*, 26–32.

Williams, G., Williams, E. J., Maison, P., Pangalos, M. N., Walsh, F. S., & Doherty, P. (2005). Overcoming the inhibitors of myelin with a novel neurotrophin strategy. *Journal of Biological Chemistry*, *280*, 5862–5869.

Xie, X. M., & Smart, T. G. (1991). A physiological role for endogenous zinc in rat hippocampal synaptic neurotransmission. *Nature*, *349*, 521–524.

Yang, Y., Jing, X. P., Zhang, S. P., Gu, R. X., Tang, F. X., Wang, X. L., et al. (2013). High dose zinc supplementation induces hippocampal zinc deficiency and memory impairment with inhibition of BDNF signaling. *PLoS One*, *8*. e55384.

Zatta, P., Drago, D., Bolognin, S., & Sensi, S. L. (2009). Alzheimer's disease, metal ions and metal homeostatic therapy. *Trends in Pharmacological Sciences*, 7, 346–355.

Zigova, T., Pencea, V., Wiegand, S. J., & Luskin, M. B. (1998). Intraventricular administration of BDNF increases the number of newly generated neurons in the adult olfactory bulb. *Molecular and Cellular Neuroscience*, *11*, 234–245.

Zuccato, C., & Cattaneo, E. (2009). Brain-derived neurotrophic factor in neurodegenerative diseases. *Nature Reviews. Neurology*, *5*, 311–322.

CHAPTER FOUR

Structural Characterization of the p75 Neurotrophin Receptor: A Stranger in the TNFR Superfamily

M. Vilar[1]

Molecular Basis of Neurodegeneration Unit, Institute of Biomedicine of Valencia (IBV-CSIC), València, Spain
[1]Corresponding author: e-mail address: mvilar@ibv.csic.es

Contents

1. TNFR Superfamily 58
   1.1 The p75 Neurotrophin Receptor 59
2. Structural Characterization of p75^{NTR} 60
   2.1 The Extracellular Domain (ECD) 60
3. The Transmembrane Domain (TMD) 66
4. The Intracellular Domain (ICD) 70
   4.1 p75NTR Juxtamembrane Domain (JTM) 70
   4.2 The p75NTR Death Domain (DD) 72
5. Conclusions and Future Directions 79
Acknowledgments 79
References 79

Abstract

Although p75 neurotrophin receptor (p75^{NTR}) was the founding member of the tumor necrosis factor (TNF) receptor superfamily (TNFRSF), it is an atypical TNFRSF protein. p75^{NTR} like TNF-R1 and Fas-R contain an extracellular domain with four cysteine-rich domains (CRD) and a death domain (DD) in the intracellular region. While TNFRSF proteins are activated by trimeric TNFSF ligands, p75^{NTR} forms dimers activated by dimeric neurotrophins that are structurally unrelated to TNFSF proteins. In addition, although p75^{NTR} shares with other members the interaction with the TNF receptor-associated factors to activate the NF-κB and cell death pathways, p75^{NTR} does not interact with the DD-containing proteins FADD, TRADD, or MyD88. By contrast, the DD of p75^{NTR} is able to recruit several protein interactors via a full catalog of DD interactions not described before in the TNFRSF. p75-DD forms homotypic symmetrical DD–DD complexes with itself and with the related p45-DD; forms heterotypic DD–CARD interactions with the RIP2–CARD domain, and forms a new interaction between a DD and RhoGDI. All these features, in addition to its promiscuous interactions with several ligands and coreceptors, its processing by α- and γ-secretases, the dimeric nature of

Vitamins and Hormones, Volume 104
ISSN 0083-6729
http://dx.doi.org/10.1016/bs.vh.2016.10.007

its transmembrane domain and its "special" juxtamembrane region, make p75^{NTR} a truly stranger in the TNFR superfamily. In this chapter, I will summarize the known structural aspects of p75^{NTR} and I will analyze from a structural point of view, the similitudes and differences between p75^{NTR} and the other members of the TNFRSF.

1. TNFR SUPERFAMILY

The TNFSF/tumor necrosis factor receptor superfamily (TNFRSF) constitutes one of the largest families of ligand/receptor proteins (Croft et al., 2012; Locksley, Killeen, & Lenardo, 2001; Smith, Farrah, & Goodwin, 1994). TNFRSF members participate in several biological responses and depending on the cell type and the context trigger cellular responses as diverse as cell death or survival (Cabal-Hierro & Lazo, 2012) and due to their implication in several diseases, they constitute one of the main pharmaceutical targets (Croft, Benedict, & Ware, 2013). To date, 18 TNFSF ligands and 29 TNFRSF receptors have been described in humans. TNFSF are generally homotrimers type II transmembrane proteins that sometimes are shed from the membrane and act as soluble signaling ligands (Bodmer, Schneider, & Tschopp, 2002). TNFSF bind to receptors of the TNFRSF (Fig. 1). The TNFRSF differ in the number of cysteine-rich domains (CRD) in the extracellular domain (ECD) and in the presence or not of a special domain in the intracellular region, called death domain (DD) (Wu & Hymowitz, 2009) (Fig. 1). Those receptors without a DD are called

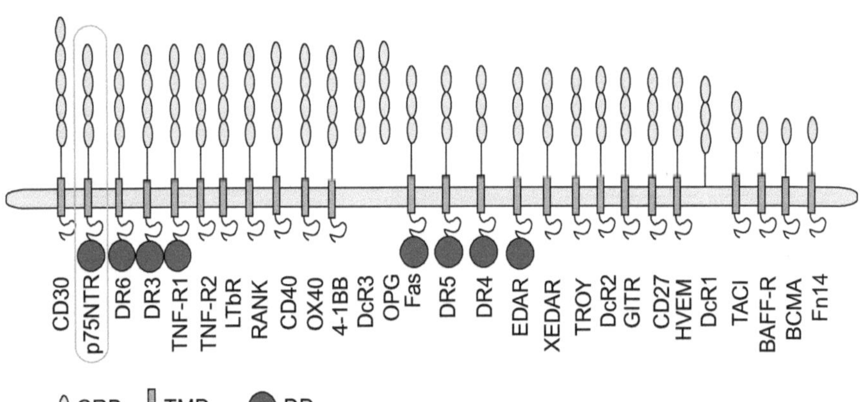

Fig. 1 The TNF receptor superfamily. DD; death domain; TMD, transmembrane domain; CRD, cysteine-rich domain.

survival receptors and include TNFR2, CD40, CD30, OX40, LTβR, RANK, and TACI. DD-containing receptors are called death receptors and include TNFR1, Fas, and $p75^{NTR}$ (Fig. 1).

Receptors of the TNFR superfamily do not contain an enzymatic activity. Instead activation takes place by recruitment of signaling proteins to the intracellular regions (Andera, 2009; Li, Yin, & Wu, 2013; Song & Buchwald, 2015). Two different types of these adaptors are described: (i) adaptor proteins with a DD such as TRADD or FADD, implicated in the death receptor signaling (Chinnaiyan, O'Rourke, Tewari, & Dixit, 1995; Hsu, Shu, Pan, & Goeddel, 1996; Naismith & Sprang, 1998) and (ii) adaptor proteins that do not have a DD, like TNF receptor-associated factors (TRAFs) which interact directly to the receptor or indirectly with other proteins acting as adaptors (Arch, Gedrich, & Thompson, 1998; Wajant, Grell, & Scheurich, 1999). The binding of these adaptor proteins to the TNFRs implies the activation of different pathways promoted by the transcription factors NF-κB or AP-1, as well as the triggering of cell death processes by apoptosis or necroptosis.

1.1 The p75 Neurotrophin Receptor

After an intense effort in 1986, two independent laboratories cloned the p75 as the first NGF receptor (NGFR) (Johnson et al., 1986; Radeke, Misko, Hsu, Herzenberg, & Shooter, 1987). p75 was recognized as the sole NGFR, however later it was found that the Trk-family of receptor tyrosine kinase also bind to neurotrophins (Kaplan, Hempstead, Martin-Zanca, Chao, & Parada, 1991; Klein, Jing, Nanduri, O'Rourke, & Barbacid, 1991; Lamballe, Klein, & Barbacid, 1991). Trks and p75 bind to neurotrophins with a similar affinity (10^{-9} M), however because in the presence of p75, Trks increase their affinity for the neurotrophins 100 times (10^{-11} M), Trk was considered as the high-affinity neurotrophin receptor and p75 was known as the low-affinity NGFR. However, this designation is misleading because p75 and Trk independently bind to neurotrophins with a similar affinity although with a different kinetics (reviewed in Bothwell, 1995). Due that brain-derived neurotrophic factor (Rodriguez-Tebar, Dechant, & Barde, 1990), neurotrophin-3 (NT3) (Rodriguez-Tebar, Dechant, Gotz, & Barde, 1992), and neurotrophin-4 (NT4) (Ryden et al., 1995), bind to p75 with similar affinity, although with different kinetics (Rodriguez-Tebar et al., 1992) p75 was renamed as the p75 neurotrophin receptor ($p75^{NTR}$), rather than the NGFR. Sequence analysis showed that p75 is a type I membrane protein (N-terminal outside of the cell) with an extracellular region very rich in cysteines residues, and an intracellular region without a catalytic activity (Fig. 1). The specific arrangement of cysteine residues in the ECD of $p75^{NTR}$ was the main characteristic of the receptor (Johnson et al., 1986). Later the receptor for TNF was cloned and sequenced (Gray, Barrett, Chantry, Turner, & Feldmann, 1990), followed by other members sharing the same structural characteristics; a cysteine-rich ECD and an intracellular region devoid of catalytic activity. Initially, the family was coined the NGFR/TNFR family, and later just TNFR superfamily.

While Trk receptor signaling is involved in survival and differentiation (Deinhardt & Chao, 2014; Reichardt, 2006), $p75^{NTR}$ participates in several signaling pathways (Kraemer, Yoon, & Carter, 2014). $p75^{NTR}$-mediated signaling is governed by the cell context and the formation of complexes with different coreceptors and ligands, such as sortilin/pro-NGF in cell death (Nykjaer et al., 2004), Nogo/Lingo-1/NgR in axonal growth inhibition (Mi et al., 2004; Wang, Kim, Sivasankaran, Segal, & He, 2002), and TrkA/NGF in survival and differentiation (Hempstead, Martin-Zanca, Kaplan, Parada, & Chao, 1991). $p75^{NTR}$ also undergoes shedding and receptor intramembrane proteolysis (RIP), resulting in the release of its intracellular domain (ICD), which itself possesses signaling capabilities (Jung et al., 2003; Kanning et al., 2003; Skeldal et al., 2012).

2. STRUCTURAL CHARACTERIZATION OF $p75^{NTR}$

2.1 The Extracellular Domain (ECD)

The ECD of $p75^{NTR}$ is characterized by four cysteine-rich domains (CRDs) (Fig. 1). p75ECD also contains putative sites of *O*-linked glycosylation and a single site of *N*-linked glycosylation in the asparagine residue number 32 in the rat protein. Experimental data from several laboratories demonstrated that $p75^{NTR}$ interacts with the neurotrophins through the four cysteine-rich domains in its ECD (Baldwin, Bitler, Welcher, & Shooter, 1992; Yan & Chao, 1991), and does it by forming homodimers (Grob, Berlot, & Bothwell, 1983; Grob, Ross, Koprowski, & Bothwell, 1985), however the molecular details of the p75ECD recognition to a neurotrophin were not revealed until 2004. In 2004, He and Garcia reported the structure of the complex of p75ECD with NGF (He & Garcia, 2004) (Table 1). This structure, which was prepared with tunicamycin to prevent *N*-glycosylation, contains only one $p75^{NTR}$ protomer for each NGF dimer, forming a 2:1 stoichiometry (p75:NT) (He & Garcia, 2004). Based on the

Table 1 p75^{NTR} and Related Proteins Structures Discussed in the Text

Construct	Year	PDB Codes	Technique
p75-DD	1997	1ngr	NMR
RhoGDI/RhoA	1999	1cc0	X-ray
p75ECD/NGF	2004	2sg1	X-ray
p75ECD/NT3	2008	3buk	X-ray
Sortilin/neurotensin	2009	3f6k	X-ray
p75ECD/proNGF	2010	3ij2	X-ray
TNFR2/TNF	2010	3alq	X-ray
p75DD/p75DD	2013	4f42,4f44	X-ray
p45-DD	2014	1ib1	NMR
p75-DD/p75DD	2015	2n97	NMR
p75-DD/Rip2–CARD	2015	2n83	NMR
p75-DD/RhoGDI	2015	2n80	NMR
p75-TM-wt	2016	2mic	NMR
p75-TM-C257A	2016	2mjo	NMR

crosslinking data (Grob et al., 1983, 1985), the homodimeric nature of isolated neurotrophins (Wiesmann & de Vos, 2001), and their 2:2 interaction with the Trks (Jing, Tapley, & Barbacid, 1992; Wiesmann, Ultsch, Bass, & de Vos, 1999), this finding was totally unexpected in the field. Later, however, experiments of the interaction of the isolated p75ECD and NGF in solution (Aurikko et al., 2005), the reported crystal structure of p75ECD with NT-3 (Gong, Cao, Yu, & Jiang, 2008) and the crystal structure of p75ECD with pro-NGF from the same Garcia's laboratory in 2010 (Feng et al., 2010) confirmed a 2:2 stoichiometry for p75ECD:NT. These studies suggested that the 2:1 complex was due to the lack of glycosylation in the protein used in the crystallization with NGF. However, the role of p75ECD glycosylation is still an unsolved issue. Since the same unglycosylated construct was used later in the crystallization of p75ECD/pro-NGF with a 2:2 stoichiometry, together with the finding that the glycans do not make any contact with the neurotrophin in the complex p75ECD/NT-3, and that the structure of p75ECD is the same in a glycosylated or nonglycosylated form, leaves the role of glycosylation in NT recognition a matter of debate. Nowadays, it is admitted that the 2:1 complex may represent a trapped conformation in the process of p75/NGF crystallization and it is an intermediate in the formation of (or from) the native 2:2 complex, indicating the existence of a dynamic equilibrium of stoichiometries (Feng et al., 2010).

The extracellular region of $p75^{NTR}$ is almost identical in all the three crystal structures known (p75ECD/NGF, p75ECD/NT3, and p75ECD/pro-NGF). p75ECD is made up of all four CRDs, termed CRD1 through CRD4, in an elongated and kinked structure (Fig. 2A). Twelve pairs of disulfide bonds are evenly spaced along p75ECD simulating a ladder

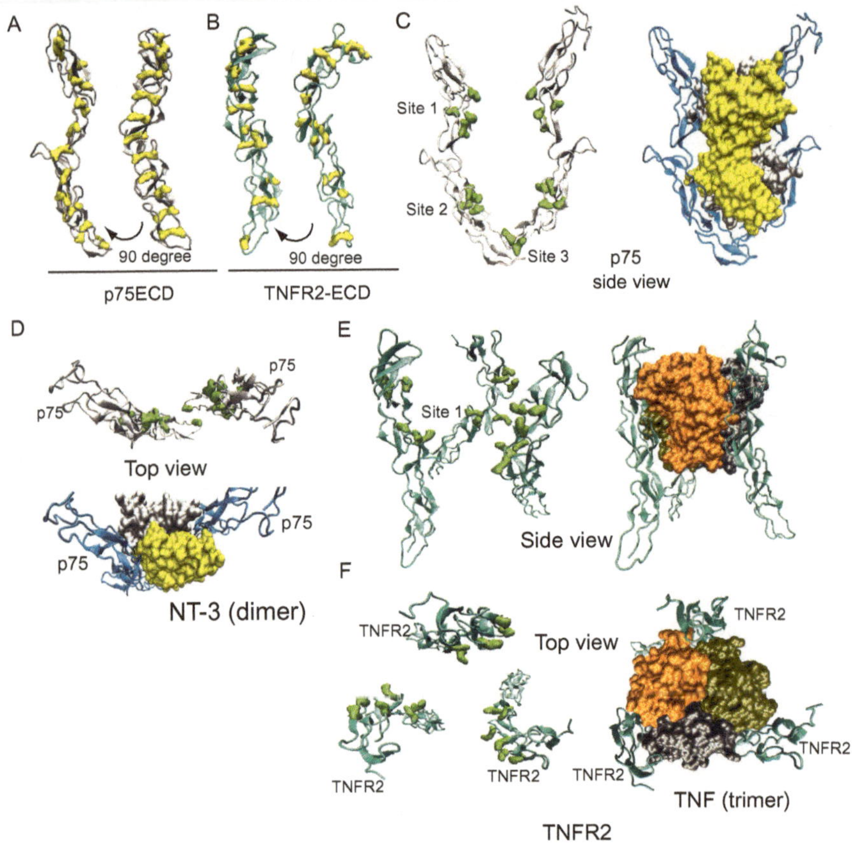

Fig. 2 Structural features of p75ECD. (A) p75ECD structure. In *yellow*, the cysteine disulfide bonds are shown. (B) TNFR2 ECD structure. Side view (C) and top view (D) of p75ECD/NT-3 complex. Contact sites are shown in *green*. Side view (E) and top view (F) of TNFR2/TNF complex. Contact sites are shown in *green*.

(Fig. 2A). The structure of some members of the TNFRSF with four CRDs has been solved by X-ray crystallography, like the TNFR1 (Naismith, Devine, Brandhuber, & Sprang, 1995; Naismith, Devine, Kohno, & Sprang, 1996), TNFR2 (Mukai et al., 2010), and OX40 (Compaan & Hymowitz, 2006). They share the same long stretched folding of the four-tandem CRD repeats as in p75^{NTR} (see as an example the structure of TNFR2 in Fig. 2B). The presence of the 12 pairs of disulfide bonds induces a strong structure in the ECD of this receptor family, rather than the assumption that a long domain should be flexible.

Detailed biochemical analyses have shown that many TNFRs are preclustered as dimers at the cell surface (Chan et al., 2000). This process, known as preligand assembly is mediated by a specific domain located in the CRD1 called PLAD (preligand assembly domain) and is a ligand-independent process. In addition to increasing the local concentration of the receptors, PLADs also contribute to the stability of the CRD2 (Branschadel et al., 2010) which establishes direct contacts with the ligand as evidenced by the ligand–receptor structures (see below). The formation of a precomplex appears to be a common process for TNFRs containing multiple CRDs, but not with TNFRSF with one or two CRDs. It has so far been experimentally validated for several TNF receptors (Chan, 2007) and non-TNF family receptors such as the interleukin receptor (Kramer et al., 2007). Although the protein sequence of the PLAD is not conserved, recently it has been suggested that the p75-CRD1 also contains a partial PLAD because deletion of the p75^{NTR} CRD1 induce less p75^{NTR} multimerization (Anastasia, Barker, Chao, & Hempstead, 2015). However, biochemical analysis by the same authors showed that all four CRDs as well as the transmembrane domain contribute to p75^{NTR} oligomerization, suggesting that in the case of p75^{NTR} the PLAD is not localized in the CRD1 only. Further structural characterization of p75ECD in the absence of ligand is needed to clarify if the p75ECD contains a PLAD region to form preligand oligomers like others TNFRs.

2.1.1 p75ECD Interaction With Neurotrophins

Besides the strikingly different stoichiometry between p75ECD/NGF (He & Garcia, 2004) and p75ECD/NT3 (Gong et al., 2008), the overall structure of p75ECD is very similar. The p75ECD-binding interface to NT3 can be divided into three main contact sites (two in the case of NGF) that are stabilized by hydrophobic interactions, salt bridges, and hydrogen bonds (Fig. 2C and D). The junction regions between CDR1 and CDR2 form the site 1 that contains five hydrogen bonds and one salt bridge (Fig. 2C and D). Site 2 is formed by equal contributions from CDR3 and CRD4 and involves two salt bridges and two hydrogen bonds. Site 3, in the CRD4, includes only one salt bridge. A comparison of the p75ECD/NT-3 complex with TNF/TNF-R2 receptor complexes reveals some differences and similarities (Fig. 2). The first difference is that these two receptors use opposite faces to interact with their respective ligands. p75ECD uses the convex face formed by the curvature of CRDs 1 and 2, whereas TNF-R2 uses the concave face to bind to TNF (Fig. 2C–F). The second difference is that TNF-R2 makes only a single point of contact with TNF, whereas $p75^{NTR}$ has three contact points with NT-3 and two with NGF. The third difference is that, in contrast to other members of the TNFR superfamily, which are activated as preformed trimeric receptor complexes by trimeric ligands, $p75^{NTR}$ is induced to form a dimer by dimeric NT-3 ligand (Fig. 2C–F). However, they share a common binding style in which the receptors bind along the seam of the interfaces between two ligands to form symmetrical complexes.

2.1.2 p75ECD Interaction With Pro-NGF

$p75^{NTR}$ can form a complex with the unprocessed neurotrophins and sortilin to induce cell death (Hempstead, 2014; Nykjaer et al., 2004). Increased pro-NGF is observed in pathological conditions resulting in neuronal cell death, such as seizures, spinal cord injury, and Alzheimer's disease (Harrington et al., 2004). Despite the flexibility of the proregion of pro-NGF in 2010, the laboratory of Dr. Garcia reported the X-ray structure of the complex between p75ECD and pro-NGF to 3.75-Å resolution (Fig. 3) (Feng et al., 2010). Due to its high flexibility the proregion of pro-NGF was not assigned in the complex, although the proregion was proposed to be in a specific position since an empty space was observed in the crystal lattice (ovals in Fig. 3). In the complex, pro-NGF exists as a homodimer, with the dimerization interface mediated by the mature NGF domain of pro-NGF, suggesting that the mature domain is necessary and sufficient to induce pro-NGF dimerization (Fig. 3). The p75ECD/pro-NGF structure has a 2:2 symmetric binding stoichiometry, contrary to the 2:1 crystal structure of p75ECD/NGF reported by the same laboratory and using the same nonglycosylated p75ECD construct (see above). However, in both structures the p75ECD bind to the mature NGF almost identically, along the two seams formed by the edges of the ligand dimerization interface.

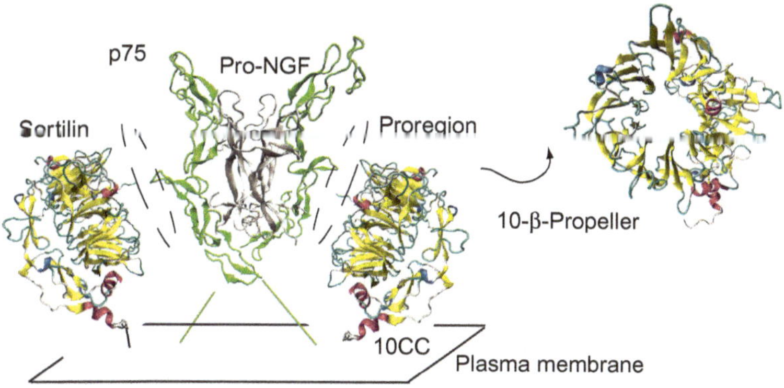

Fig. 3 Structural features of p75ECD/proNGF complex. Model of the ternary complex between p75ECD/proNGF/sortilin. *Ovals* indicate the possible location of the proregion of pro-NGF as described in Feng et al. (2010).

In addition, the contacts between p75ECD are maintained in both mature NGF and pro-NGF complexes, suggesting that $p75^{NTR}$ do not experience any conformational change upon binding to any of these ligands. Only small displacements were described at the two ends of the p75ECD molecules if we compare the 2:2 complex of p75/proNGF to the p75/NT3 complex, reflecting small but significant differences between the mature NGF and NT-3 modes of interaction with p75NTR. If these small differences are able to explain the different affinity between NGF and NT3 to $p75^{NTR}$ (Dechant, Tsoulfas, Parada, & Barde, 1997) needs to be established.

2.1.3 p75ECD Interaction With Sortilin

A ternary complex between $p75^{NTR}$/pro-NGF/sortilin is necessary for cell death (Nykjaer et al., 2004). Although the structure of this ternary complex is still unknown, extensive biochemical studies have provided clues of how the complex may be formed. Surface plasmon resonance experiments suggested that the presence of calcium ions in the solution is needed to form the ternary complex, although it was not possible to crystallize such complex (Feng et al., 2010). In 2012, the laboratories of Dr. Nykjaer and Dr. Coulson reported the mapping between $p75^{NTR}$ and sortilin to a few residues within their ECD of $p75^{NTR}$; specifically to the 23 residues of the extracellular juxtamembrane domain (Skeldal et al., 2012). Mapping of the interaction site between sortilin and $p75^{NTR}$ revealed that the ECD of sortilin, formed by a 10 β-propeller domain and the 10CC domain, is sufficient to mediate an interaction with full-length p75NTR. Previous studies had established that the proregion of pro-NGF binds to sortilin and the mature form to $p75^{NTR}$

(Feng et al., 2010). Neurotensin, the natural ligand of sortilin, binds to the β-propeller domain of sortilin (Quistgaard et al., 2014) and is not competing with p75ECD, suggesting that p75 is probably binding to the 10CC domain of sortilin (Skeldal et al., 2012). Although the crystal structure of sortilin was reported in 2009 (Quistgaard et al., 2009) it is difficult to imagine how a 10-bladed β-propeller may bind to p75NTR and to pro-NGF at the same time, but one possibility is that the β-propeller binds to the proregion of the pro-NGF and the 10CC domain is bound to the 23 residues in the juxtamembrane region of $p75^{NTR}$. A model of this interaction is shown in Fig. 3. Based on the important role of pro-NGF/p75/sortilin in neuronal cell death, to have the molecular details of the ternary complex is really necessary.

To conclude, the structural studies of the p75ECD showed the dimeric interaction between $p75^{NTR}$ and the neurotrophins (mature or immature), contrary to the other members of the TNFRSF where homotrimers of TNFSF ligands induce trimerization of TNF receptors.

3. THE TRANSMEMBRANE DOMAIN (TMD)

Based on the evidence of the dimeric nature of neurotrophins, and the 2:2 complexes observed by X-ray crystallography, crosslinking and biochemical studies in solution, a general model of $p75^{NTR}$ dimerization by all neurotrophins is quite accepted in the field. Supporting that $p75^{NTR}$ dimerization may represent the active receptor, in 2009 the laboratory of Dr. Ibañez reported the design of new $p75^{NTR}$ constitutively active receptors, by way of disulfide-linked dimers between cysteine residues in the extracellular region of $p75^{NTR}$ (Vilar, Charalampopoulos, Kenchappa, Reversi, et al., 2009). However, an important step further was the identification of constitutive disulfide-linked $p75^{NTR}$ dimers. The same laboratory reported that full-length $p75^{NTR}$ forms disulfide-linked dimers in the absence of neurotrophins through the highly conserved Cys257 in the transmembrane domain of $p75^{NTR}$ (Vilar, Charalampopoulos, Kenchappa, Simi, et al., 2009). The formation of disulfide-linked dimers of p75 was already observed in the laboratory of Dr. Bothwell back in 1985 (Grob et al., 1985) and this study reported the identification of the cysteine 257 (rat numbering) as the responsible residue. The presence of the Cys257 in $p75^{NTR}$ protein sequences dates back to almost 600 million years and $p75^{NTR}$ orthologues protein sequences found in some protostomes like the water flea *Daphnia* sp. (Wilson, 2009) or the recently sequenced annelid genome *Platynereis durmelii* (Lauri, Bertucci, & Arendt, 2016) have this same cysteine residue in the TM domain, indicating an important functional role in p75^{NTR}. It will be interesting to see if p75^{NTR} homologues are also present in Cnidarian genomes and if the cysteine residue is present.

In 2016, our laboratory in collaboration with the laboratory of Dr. Arseniev reported the NMR structure of p75 transmembrane domain (Fig. 4) (Nadezhdin et al., 2016). The structure shows that the TM form covalent dimers mediated by the Cys257. This constitutes an important difference between p75NTR and Fas receptor transmembrane domains. The recent NMR structure of Fas receptor in bicelles shows the formation of trimers (Fu et al., 2016) supporting the role of the TMD in the trimerization of FasR upon ligand binding. The formation of Cys disulfide dimers was

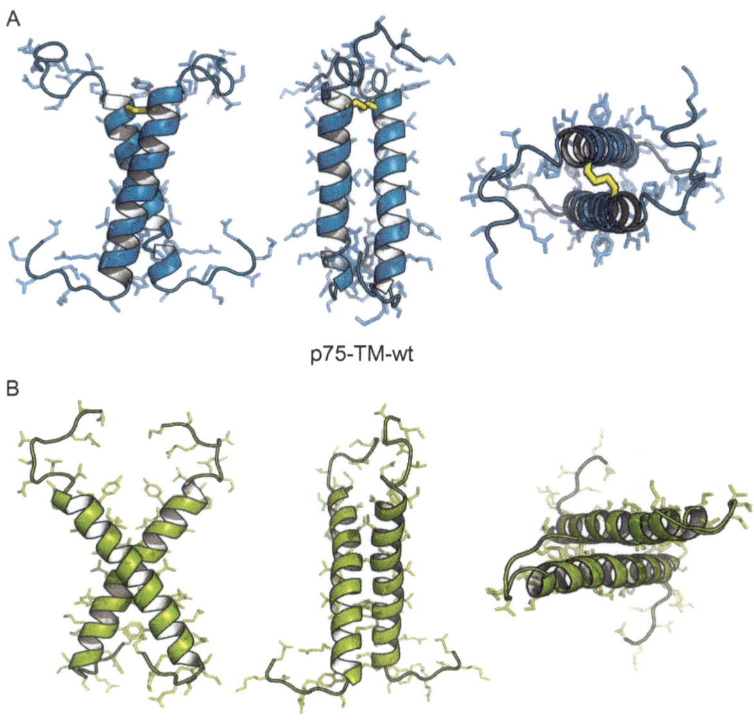

Fig. 4 NMR solution structures of p75-TM domain. Schematic representation of the spatial structure of the p75-TM-wt dimer (A) and p75-TM-C257A (B) from three different angles: front, 90 degree rotation, and top-to-bottom view. TM helices are represented as ribbons and side chains as sticks. The Cys257–Cys257 disulfide bond is shown in *yellow*. The PDB accession code for p75-TM-wt is 2mic and p75-TM-C257A 2mjo. *Adapted from Nadezhdin, K. D., Garcia-Carpio, I., Goncharuk, S. A., Mineev, K. S., Arseniev, A. S., & Vilar, M. (2016). Structural basis of p75 transmembrane domain dimerization.* Journal of Biological Chemistry, 23, *12346–12357. http://dx.doi.org/10.1074/jbc.M116.723585.*

spontaneous in micelles, however because in the plasma membrane not all p75 is in the form of disulfide-linked dimers, this may suggest that other domains of p75 or some components of the membrane may modulate their formation. The same authors solved the structure of the p75-TM-C257A and found a different dimer conformation mediated by the motif $A_{262}xxxG_{266}$ (Nadezhdin et al., 2016). The difference in the dimer conformation may explain why the mutant C257A is not active, although other possibilities could be possible like differences in the cleavage mediated by the γ-secretase or the impossibility in the p75-C257A to transduce the conformational change upon NGF binding.

Functional analysis of the $p75^{NTR}$ mutant lacking Cys257 revealed the key importance of disulfide-mediated receptor dimerization and suggested that predimerization of $p75^{NTR}$ is not enough for its activity and only Cys257-linked dimers respond to NGF (Vilar, Charalampopoulos, Kenchappa, Simi, et al., 2009). Recently the laboratory of Dr. Ibañez reported the generation of a new knock-in mouse with the mutation C256A (mouse numbering), validating and supporting the key role of this residue in the activation of cell death by p75NTR in vivo (Tanaka, Kelly, Goh, Lim, & Ibáñez, 2016). The mutant p75-C257A, although it is still capable to form dimers and bind to NGF, is unable to transduce the signaling triggered by NGF binding in some cell paradigms (Vilar, Charalampopoulos, Kenchappa, Simi, et al., 2009). The laboratory of Dr. Ibañez proposed that the Cys residue acts as the fulcrum of a snail-tong model, where the binding of NGF to the p75ECD induces an opening of the p75ICDs allowing the entrance of intracellular adaptors (Vilar, Charalampopoulos, Kenchappa, Simi, et al., 2009). This model is supported by in vivo FRET studies, mutagenesis and the recent structural studies of p75ICD from the same laboratory (see below) (Lin et al., 2015). However, it is difficult to imagine how a covalent dimer is able to transduce the conformational change upon NGF binding, taking into account, in addition, the flexible regions above and below the TM domain. Another possibility is that NGF would induce the clustering of several p75 molecules joined by disulfide bonds. In this scenario, p75 will form clusters and trigger a signaling event similar to other members of the TNFRSF. Further work is necessary to clarify which of these two, or a combination of both, possibilities are taking place after NGF binding.

In addition to play an importnat role in p75NTR activation, the transmembrane domain of $p75^{NTR}$ has been suggested as the interaction site between $p75^{NTR}$ and TrkA receptors (Esposito et al., 2001). Several lines of evidence implicate functional interactions between TrkA and $p75^{NTR}$ in

NGF-triggered signal transduction (Bothwell, 1995; Chao, 2003; Chao & Hempstead, 1995; Yano & Chao, 2000). TrkA and $p75^{NTR}$ receptors have nanomolar affinities for NGF and cooperate in transducing NGF signals (Bibel, Hoppe, & Barde, 1999; Huang & Reichardt, 2003). The expression patterns of these two receptors overlap extensively (Rifkin, Todd, Anderson, & Lefcort, 2000) and in some instances, such as in the neurons of the dorsal root ganglion (DRG), TrkA is exclusively expressed in conjunction with $p75^{NTR}$ (White et al., 1996). Primary DRG and sympathetic neurons derived from p75-null animals show attenuated survival responses to NGF (Lee et al., 1992), confirming the physiological role of $p75^{NTR}$/TrkA interactions. Moreover, $p75^{NTR}$ has been experimentally demonstrated to enhance the response of TrkA to NGF. Indeed, the concept of the NGF high-affinity-binding site was developed based on the observed increase in NGF-binding affinity when $p75^{NTR}$ and TrkA are coexpressed in the same cell (Hempstead et al., 1991; Mahadeo, Kaplan, Chao, & Hempstead, 1994). While this phenomenon is widely accepted in the field, the mechanism by which $p75^{NTR}$ converts TrkA from a low ($k_d = 10^{-9}$ M) to a high ($k_d = 10^{-11}$ M) affinity-binding site is not completely understood. Experimental evidences of a TrkA/$p75^{NTR}$ complex exist based on coimmunoprecipitation studies (Bibel et al., 1999; Huber & Chao, 1995; Jung et al., 2003) and biophysical methods such as copatching (Ross et al., 1996) and fluorescence recovery after photobleaching (Wolf et al., 1995). The interaction seems not taking place in the ligand-binding domains of the extracellular region (Wehrman et al., 2007), instead, a handful of studies have suggested that the transmembrane and intracellular domains of p75 could be responsible for its interaction with TrkA (Bibel et al., 1999; Esposito et al., 2001; Iacaruso et al., 2011; Matusica et al., 2013; Sykes et al., 2012). However, the structural basis of such direct interaction has not been yet demonstrated.

Finally, the transmembrane domain of $p75^{NTR}$ plays an important role in $p75^{NTR}$ signaling because it is the place where the γ-secretase complex cleaves the receptor, thereby releasing the p75-ICD. The cleavage site has been mapped to the residues $A_{262}VV/VG_{266}$, between Val264 and Val265 (Jung et al., 2003). Although the functional role of the regulated RIP is still under debate (Bronfman, 2007), the p75-ICD released after the cleavage has been implicated in several activities (Skeldal, Matusica, Nykjaer, & Coulson, 2011); like transcription and nuclear localization (Frade, 2005; Parkhurst, Zampieri, & Chao, 2010), cell proliferation (Forsyth et al., 2014), nuclear gating (Schachtrup et al., 2015), and cell death (Majdan et al., 1997; Vicario, Kisiswa, Tann, Kelly, & Ibanez, 2015).

4. THE INTRACELLULAR DOMAIN (ICD)

4.1 p75NTR Juxtamembrane Domain (JTM)

The intracellular domain of p75 presents two different regions. A flexible intrinsically disordered region of 60–70 residues called the juxtamembrane domain (JTM) and a well-folded region that constitutes the DD (Fig. 5). Due to its highly flexible nature, the structural data of the JTM have been scarce. In 1997, the laboratory of Dr. Ibañez in collaboration with Dr. Otting reported the NMR relaxation data from isolated p75ICD where they demonstrated that the JTM is flexible and without any secondary structure (Liepinsh, Ilag, Otting, & Ibanez, 1997). Those observations were confirmed and supported in 2015 by the laboratory of Dr. Arseniev (Mineev, Goncharuk, Kuzmichev, Vilar, & Arseniev, 2015). In that study, the entire TM-ICD of $p75^{NTR}$ was expressed and inserted into lipid bilayers in the form of nanodiscs (Mineev et al., 2015). This was a formidable task as constitutes one of the first studies where the whole intracellular domain of a receptor is studied in nanodiscs. They were able to determine the relaxation kinetics of the entire ICD and found that the JTM is flexible even when linked to the TM and DD domains (Mineev et al., 2015). In addition, they did not found any interaction between the JTM and the DD, and neither the JTM and the lipid bilayer. However, a possibility exists that in vivo the JTM is constitutively bound to intracellular proteins or to a specific component of the plasma membrane absent in the nanodiscs.

A

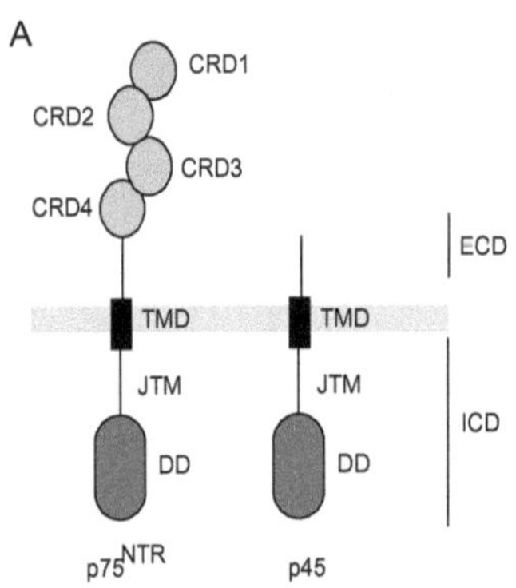

B

hCD40	230-KQE**P**Q**E**IN**F**
mCD40	234-RQD**P**Q**E**ME**D**
hTRANCE-R	341-RQM**P**T**E**DE**Y**
hIRAK-1	539-PPS**P**Q**E**NS**Y**
hRIP2	191-IYM**P**P**E**NY**E**
consensus	**P**x**E**xxAr/Ac
rp75NTR	294- TP**P**P**E**GEK

Fig. 5 Protein domains of p75. (A) Protein domains present in p75 and p45. (B) Alignment of the TRAF6-binding sequences with the putative TRAF6-binding site in the p75-JTM domain. The consensus residues are shown in *bold face.*

Several protein adaptors have been described to bind to the JTM, like NRIF (Casademunt et al., 1999), SC-1 (Chittka & Chao, 1999), TRAF4 (Ye et al., 1999), and TRAF6 (Khursigara, Orlinick, & Chao, 1999), however a detailed structural study of the p75-JTM and any of those proteins is still lacking. One good candidate would be TRAF6, that is a special TRAF member with nonredundant activities to the other TRAFs (Wu & Arron, 2003). Since TRAF6 binds directly to receptors of the TNFSF that do not contain a DD, the case of p75NTR is special between the other members of the TNFRSF and a quantitative demonstration that TRAF6 binds directly to the p75-JTM is still needed. The structure of the TRAF domain of TRAF6 bound to a peptide from CD40 and TRANCE-R showed that TRAF6 uses a specific protein motif that is not shared by other TRAFs, like TRAF1, 2, 3, and 5 (Wu, 2004; Ye et al., 2002). A closer examination using deletion mutants showed that residues 300–315 of the juxtamembrane region of $p75^{NTR}$ were required for the interaction with TRAF6 (Khursigara et al., 1999). In this region of $p75^{NTR}$ a sequence that partially matches the TRAF6 consensus site is present (Fig. 5). Studies in solution by surface plasmon resonance, calorimetry, or structural studies with a peptide derived from this region and the TRAF domain of TRAF6 are required to validate this region as a bona fide TRAF6-binding site.

Although TRAF6 induces activation of the survival pathway after binding to $p75^{NTR}$, the role of the JTM in cell death has received more attention. In 2000, the laboratory of Dr. Coulson reported that a 29-amino acid region, called *Chopper*, inside the p75-JTM was required and sufficient to induce rapid cell death (Coulson et al., 2000). One of the interesting characteristics of the *Chopper* region is that to induce cell death activity it needs to be associated to the plasma membrane by a lipid anchor (Underwood, Reid, May, Bartlett, & Coulson, 2008). $p75^{NTR}$ contains a cysteine in the juxtamembrane domain that is posttranslationally modified by palmytoylation (Barker, Barbee, Misko, & Shooter, 1994) and this cysteine forms part of the chopper domain. It is possible that the *Chopper* may recruit a cell death machinery close to the plasma membrane or in certain lipid microdomains, like the lipid rafts (Underwood et al., 2008).

Protein interactors that bind specifically to the *Chopper* region are unknown for the moment. However in a free state, not bound to the plasma membrane, this same peptide has been recently shown to interact with TrkA intracellular domain and its overexpression potentiates binding of NGF to TrkA leading to increased neurite outgrowth due to enhanced MAPK and Akt signaling (Matusica et al., 2013). However, where and how this peptide binds to TrkA is unknown at the moment and further structural studies should address this interaction.

4.2 The p75NTR Death Domain (DD)

p75^{NTR} contains a DD in the intracellular region (Fig. 6). DDs are found in other death receptors such as Fas and TNF-R1 and death receptor-interacting adapter proteins such as FADD and TRADD. The DD superfamily is one of the largest and most studied domain superfamilies (Park et al., 2007). It is comprised of four subfamilies, the DD subfamily, the death effector domain (DED) subfamily, the caspase recruitment domain (CARD) subfamily, and the pyrin domain (PYD) subfamily (Park et al., 2007; Weber & Vincenz, 2001). These proteins are evolutionarily conserved in many multicellular organisms such as mammals, *Drosophila*, and *C. elegans*.

The unifying structural feature of the DD superfamily is the six-helical bundle structural fold as first revealed by the NMR structures of Fas-DD (Huang, Eberstadt, Olejniczak, Meadows, & Fesik, 1996). DDs are protein–protein interaction domains and they function as modules to bring closer different proteins involved in several signaling cascades (Kersse, Verspurten, Vanden Berghe, & Vandenabeele, 2011). Both caspase

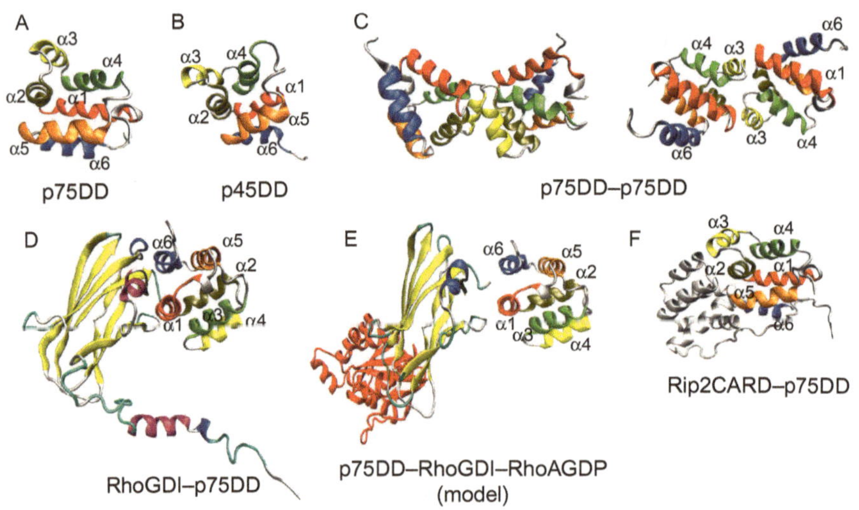

Fig. 6 The p75DD interactome. (A) Structure of p75DD. (B) Structure of p45DD. (C) Structure of the p75DD homodimer complex (lateral and top views). (D) Structure of the p75DD/RhoGDI complex. (E) Model of a ternary complex RhoA/RhoGDI/p75DD. (F) Structure of the p75DD/RIP2CARD complex.

activation and NF-κB activation occur by the assembly of oligomeric signaling complexes in response to internal or external stimuli (Kersse et al., 2011; Riedl & Shi, 2004). For instance Fas, upon ligand activation, recruits the FADD adaptor protein via a homotypic interaction with the C-terminal DD of FADD. FADD also contains an N-terminal DED that interacts homotypically with the tandem DED in the prodomain of caspase-8 or -10. These interactions form the ternary DISC-containing Fas, FADD, and caspase-8 or -10 (Wang et al., 2010). Recruitment of procaspases into the DISC initiates caspase proteolytic autoprocessing. Other high molecular weight complexes based on DD superfamily interactions are the Mydosome and the PIDDosome (see the recent review, Ferrao & Wu, 2012).

Interestingly, the interactions of the DD superfamily are all homotypic, meaning that the DD superfamily binds to the DD of the same subfamily. In addition, they have been reported to be asymmetrical, where the interaction of the two proteins is not using the same interface (Ferrao & Wu, 2012; Weber & Vincenz, 2001). Following this, the homotypic interactions between DDs have been classified as types I, II, and III according to the interfaces involved (Ferrao & Wu, 2012; Park et al., 2007; Weber & Vincenz, 2001). In 2015 in a remarkable work the group of Dr. Ibañez reported three new complexes of p75DD, showing that the DD of $p75^{NTR}$ uses a different repertoire of DD interactions (Lin et al., 2015).

4.2.1 p75DD–p75DD Symmetric Complex

The NMR solution structure of the p75-DD was achieved in 1997 (Fig. 6A) (Liepinsh et al., 1997). After Fas-DD, this was the second DD to be solved confirming that in the DD-containing proteins, the three-dimensional fold of the DD is more conserved than its protein sequence. The solution structure was obtained in pure water and no self-association of the p75-DD was evident in those conditions (Liepinsh et al., 1997). However in 2014, we reported that in a buffered solution containing phosphate ions p75DD is able to self-associate and form dimers (Vilar et al., 2014). We mapped by NMR chemical shifts the residues that mediate dimerization, suggesting a homotypic interaction. However, the definitive binding interface was not revealed until the structure of the complex was obtained. In 2015, the laboratory of Dr. Ibañez confirmed and extended our results further with the NMR structure of the dimer of p75DD (Lin et al., 2015). Interestingly, they confirmed that only when phosphate ions are present in the solution the p75DD dimer is formed and not when other buffered molecules, like

HEPES or TRIS are present (Lin et al., 2015). This suggests that the phosphate ions may stabilize a dimer conformation. Based on the fact that the p75-DD has several putative phosphorylation sites, it could be possible that some phosphate ions may occupy the position of the phosphorylation sites favoring its dimerization, although this possibility is unknown for the moment. The structure of the p75-DD homodimer is symmetric and is held together by low-affinity interactions involving residues in helix α3 and the α1–α2 and α3–α4 loops (Fig. 6C). Unlike the classical type IIIb surface (Ferrao & Wu, 2012), the DD:DD interaction in $p75^{NTR}$ makes extensive use of the residues in the α3 helix, and the same surface in the two interacting DDs is used to form a symmetric dimer, making this complex a new case in the DD superfamily.

In 2013, it was reported that $p75^{NTR}$ may form a symmetrical covalent dimer via cysteine-linked dimers through the Cys379 in the DD (Qu et al., 2013). The NMR and the X-ray structures of p75DD dimers are different, although the position of the Cys379 is close in the NMR dimer interface. In vitro p75-DD form covalent dimers in solution through the Cys379 in the presence of oxidant conditions, like with hydrogen peroxide (Vilar et al., 2014). However, the functional role of Cys379 dimers in vivo is uncertain. Taking into account the highly reduced environment of the cytosol the formation of intracellular disulfide-linked dimers is questionable. Nevertheless, oxidative conditions are present in stressed cells and the possibility that disulfide-linked p75DD dimers are formed in certain circumstances cannot be totally discarded.

4.2.2 p75-DD/p45-DD Interaction

In 2003, the laboratory of Dr. Bothwell described two $p75^{NTR}$ homologues which they designate neurotrophin receptor homologue 1 and 2, NRH1 and NRH2 (Kanning et al., 2003). Alignment of the sequences of these homologs reveals extensive similarity in the transmembrane and cytoplasmic domains to $p75^{NTR}$. However, NRH2 lacks the cysteine-rich repeat domain that constitutes the ligand-binding domain of $p75^{NTR}$ (Fig. 5). NRH1 homologs exist in amphibians, fish, and birds but none is present in mammalian databases, and none can be detected in the human or mouse genomic databases (Kanning et al., 2003). In contrast, NRH2 homologs are present among mammals (mouse, rat, bovine, and porcine), although in humans it is present but it is not expressed (Kanning et al., 2003). NRH2 is also named, neurotrophin receptor alike DD protein (NRADD) (Gowrishankar, Zeidler, & Vincenz, 2004), or p75-like apoptosis inducing

DD protein (PLAIDD) (Frankowski, Castro-Obregon, del Rio, Rao, & Bredesen, 2002) and p45 (Sung et al., 2013; Vilar et al., 2014). In this chapter the name p45 will be used.

p45 and p75^{NTR} share a high degree of amino acid similarity in their TM domain (94%), including the conserved cysteine in the TM domain and the ICD (50%) (Vilar et al., 2014). p45 associates with p75^{NTR} and with TrkA receptors (Frankowski et al., 2002; Murray, Perez, Lee, Hempstead, & Chao, 2004) and participates in the trafficking of sortilin to the plasma membrane (Kim & Hempstead, 2009). In 2014, our laboratory in collaboration with the laboratories of Dr. Lee and Dr. Riek described the NMR structure of p45-DD and the interaction between p75^{NTR} and p45, and how p45 modulates the p75/NgR signaling (Vilar et al., 2014). p75^{NTR} activity is involved in the regulation of axon elongation through the modulation of the activity of the RhoA GTPase (Yamashita, Tucker, & Barde, 1999). RhoA is a small GTPase that regulates the state of actin polymerization. In its active GTP-bound form, RhoA rigidifies the actin cytoskeleton, thereby inhibiting axonal elongation and causing growth cone collapse. Neurotrophin binding to p75^{NTR} inactivates RhoA and myelin-associated proteins Nogo-66, MAG, and OMgp, activate RhoA by a Nogo receptor/p75NTR-dependent mechanism, thus inhibiting neurite outgrowth (Wang et al., 2002; Wong et al., 2002).

The three-dimensional solution structure of mouse p45ICD by NMR spectroscopy showed that p45-ICD contains a flexible domain at the N-terminus and the DD domain at the C-terminus (141–218), in a similar manner as the ICD of p75^{NTR} and other TNFRSF members. The DD of p75 and p45 share many structural features and the same arrangement of all the six α-helices, which are not surprising since p75DD and p45DD are homologues (Fig. 6B). Only the length of the loop between α1 and α2 is longer in p45DD because of an insertion of four amino acid residues in this segment. When compared with p75DD, the longer loop reorients α1 respect to α2 and α3 and brings residue E153 of the loop in close neighborhood to other negative charged residues. Together with some additional amino acid residue differences, this small structural reorientation significantly changes the charge distribution around helix α3 of p45DD. The presence of such a negative patch in p45 explained why p45 is found in solution in a monomeric form contrary to p75DD (Vilar et al., 2014). In addition, the negative charge complements a positive patch in p75DD to facilitate the formation of the heterocomplex. The structural studies suggested that p45 forms a complex with p75^{NTR} through its helix α3 of the DD.

A comparison of the p45DD-binding site on p75DD with the p75DD homodimer-binding site (Fig. 7) show that the two sites overlap. The presence of a partially overlapping binding site on p75DD for homodimerization with p75DD and heterodimerization with p45DD suggests that p45DD may break a p75DD dimer by competition.

Another interesting feature of the sequence of p45DD is the presence of the RxDΦ motif (x, any residues; Φ, a hydrophobic residue) at the beginning of helix α6, which is typically observed in DEDs, such as the DED from PEA-15, FADD, Caspase-8, and others (Carrington et al., 2006; Hill, Vaidyanathan, Ramos, Ginsberg, & Werner, 2002). Although p75DD has a similar RxDΦ sequence, RADI, the analog residue of E160 involved in the salt bridge in p45 is absent in $p75^{NTR}$ and is replaced by a His residue.

The TM domain of p45 is needed to facilitate the binding to $p75^{NTR}$ and their inhibition (Vilar et al., 2014). This was explained by the finding that a cysteine–cysteine disulfide bond interaction within the TM domain of p45 (Cys58 in mouse sequence) and $p75^{NTR}$ (Cys257 in rat sequence) was required for stabilization of their heterodimer formation (Vilar et al., 2014). This constituted the first example of an inhibition of $p75^{NTR}$ by a heterodimer formation through the TM domain, supporting the dimer nature of the active receptor and the important role played by the Cys257 in $p75^{NTR}$ activation (Vilar et al., 2014). The results revealed a new mechanism of modulating p75-mediated inhibitory signaling via heterodimer formation with a member of the TNFR superfamily such as p45. p45 may constitute a general inhibitor of other TNFRSF members due to its ability to bind to the DED of FADD and promote the inhibition of cell death mediated by Fas activation in spinal cord injury (Sung et al., 2013).

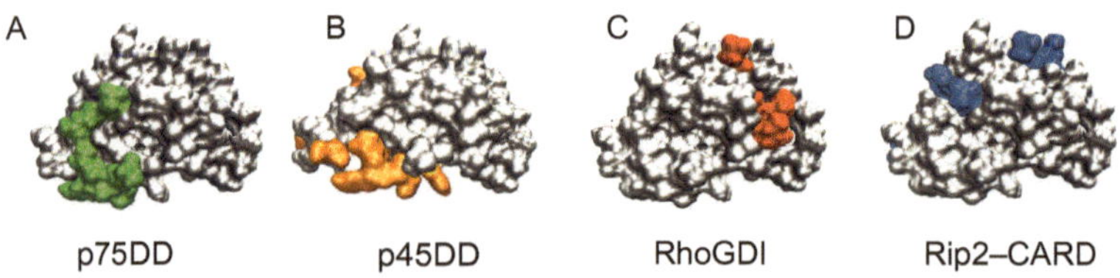

Fig. 7 p75DD as a protein–protein interaction hub. Interaction sites on the p75DD with the protein interactors discussed in the text. (A) p75DD homodimer sites. (B) p45DD. (C) RhoGDI. (D) RIP2-CARD. *Adapted from Lin, Z., Tann, J. Y., Goh, E. T., Kelly, C., Lim, K. B., Gao, J. F., & Ibanez, C. F. (2015). Structural basis of death domain signaling in the p75 neurotrophin receptor.* Elife, 4. *doi: 10.7554/eLife.11692.*

4.2.3 p75DD–RhoGDI Complex

The structural characterization of the p75DD/RhoGDI complex by the Ibañez group (Fig. 6D) provides the molecular mechanism of the RhoA activation by $p75^{NTR}$ (Lin et al., 2015). Although the laboratory of Dr. Yamashita initially described that RhoA binds directly to $p75^{NTR}$ (Yamashita, Higuchi, & Tohyama, 2002; Yamashita et al., 1999), soon later the same group proposed that $p75^{NTR}$ act as a displacement factor and RhoGDI, an activator of RhoA, competes with RhoA to bind to p75NTR (Yamashita & Tohyama, 2003). However, this model is incompatible with the fact that a ternary complex p75/RhoGDI/RhoA has been observed by coimmunoprecipitation (Lin et al., 2015). The solution structure of p75DD/RhoGDI provides an explanation for this discrepancy. The dimer interface is mediated by hydrophobic and electrostatic interactions formed by helices α1 and α6 of the $p75^{NTR}$ DD and β-strands S2, S3, S9, and α2-helix of RhoGDI (Fig. 6D). Based on the structure of the RhoGDI/RhoA complex (Longenecker et al., 1999), the authors proposed that p75-DD and RhoA bind on opposite sides of the RhoGDI molecule (see a model of the ternary complex in Fig. 6E), allowing the formation of a tripartite p75DD/RhoGDI/RhoA complex.

4.2.4 p75-DD/RIP2–CARD Complex

In 2001, the laboratory of Dr. M. Chao reported the identification of RIP2 as a new protein interactor of $p75^{NTR}$ (Khursigara et al., 2001). Receptor interacting protein 2 (RIP2) also called RICK is a serine/threonine kinase with a CARD domain in the C-terminus and was initially identified as a modulator of the NF-κB pathway (McCarthy, Ni, & Dixit, 1998) and later in innate immunity (Kobayashi et al., 2002) through its interaction with the CARD-containing proteins NOD1 and NOD2 (Abbott, Wilkins, Asara, & Cantley, 2004).

RIP2 interacts with $p75^{NTR}$ and influences the ability of NGF to regulate cell survival decisions (Khursigara et al., 2001). The interaction between RIP2 and $p75^{NTR}$ was NGF-dependent and the complex formation promoted NF-κB activity in both HEK 293 and Schwann cells (Khursigara et al., 2001). Mapping of the interactions using deletion constructs showed that the DD of $p75^{NTR}$ and the CARD domain of RIP2 were forming a direct complex, suggesting a heterotypic interaction. Interestingly the closely related protein RIP1, that contains a DD, instead of a CARD, does not interact with $p75^{NTR}$ (Khursigara et al., 2001). The recently reported structural characterization of the complex between p75DD/RIP2-CARD by NMR demonstrates the structural basis of their interaction (Fig. 6F) (Lin et al., 2015). In addition, the p75-DD/RIP2-CARD complex represents the first structural characterization of a heterotypic interaction in the DD superfamily. The main interface in the core structure of the p75DD/RIP2-CARD complex involved helices $\alpha 2$, $\alpha 3$, $\alpha 5$, $\alpha 6$, and $\alpha 5$–$\alpha 6$ loop of p75-DD and helices $\alpha 1$, $\alpha 3$–$\alpha 4$, and $\alpha 5$–$\alpha 6$ loops of RIP2 CARD (Fig. 6F). The interaction is mainly electrostatic, although some hydrophobic interactions also contribute to the binding interface.

The interaction between $p75^{NTR}$ and RhoGDI and RIP2 places the p75-DD in a key position in the bidirectional path leading to NF-κB or RhoA activation. It is known that neurotrophins inactivate RhoA and the myelin-derived ligands activate RhoA via the p75/NgR (or Troy/NgR) complex. The authors proposed that the activation of these pathways is governed by competitive protein–protein interactions (Lin et al., 2015). Because RhoGDI and RIP2 bind to an overlapping region in p75DD (Fig. 7), the binding of neurotrophins to p75NTR would induce the release of RhoGDI and the recruitment of RIP2. Free RhoGDI would remain in the cytosol complexing and inactivating RhoA (Lin et al., 2015). Recruitment of RhoGDI/RhoA to $p75^{NTR}$ would lessen the interaction between RhoGDI and RhoA, releasing RhoA, a lipid-modified protein, to the membrane and its activation by membrane-associated guanine nucleotide exchange factors (GEFs) (Lin et al., 2015).

Once RIP2 is recruited to p75DD, how is the NF-κB pathway activated? No molecular details are known regarding the connection between the RIP2 bound to p75 and the downstream mediators of the NF-κB pathway. However we can look into the NOD1/RIP2 pathway which has been recently elucidated. The CARD domain of RIP2 is connected to the kinase domain by a flexible region. Initial models suggested that the interaction of the CARD to the dimer of NOD1 would bring closer the kinase domains of RIP2 for its trans-activation. However recent data suggest a different mechanism and revealed that, although the oligomer formation of RIP2 is still necessary, ubiquitinylation of RIP2, and not its kinase activity, is necessary for NF-κB activation (Hasegawa et al., 2008). K63-ubiquitinylation of RIP2 mediated by an E3-ubiquitin ligase induced the recruitment of the kinase TAK1 to phosphorylate NEMO and then the activation of the NF-κB is initiated (Krieg et al., 2009; Tao et al., 2009). In the case of p75/NGF, the activation of TAK1 has not been yet reported. However, TRAF6 is recruited to the JTM of $p75^{NTR}$ upon NGF stimulation, and $p75^{NTR}$ is polyubiquitinated by the E3 ubiquitin ligase activity of TRAF6 in a NGF-dependent manner (Geetha et al., 2012). It is not established yet if

TRAF6 and RIP2 from part of the same complex with p75^{NTR} and if TRAF6 is able to ubiquitinate RIP2 to trigger the activation of the NF-κB pathway in a ligand-dependent manner. Future research is needed to elucidate the molecular details of this pathway.

5. CONCLUSIONS AND FUTURE DIRECTIONS

p75^{NTR} was the first NGFR to be cloned and the founding member of the TNFR superfamily. However its structural characterization has been unreasonable delayed with respect to the other TNFRSF members. This tendency is changing recently with the crystal and NMR structures of several p75^{NTR} complexes that we reviewed here. Still the current structural data do not fully explain how p75^{NTR} is able to signal upon neurotrophin binding and trigger all the repertoire of p75^{NTR} activities. Several questions still remain unanswered; How does a covalent p75 dimer respond to NT binding? How do p75^{NTR} and NgR interact? How p75 interact with TrkA to modulate Trk activities? How does TRAF6 bind to the p75-JTM and how does the p75/TRAF6/RIP2 trigger cell survival? Hopefully new structural studies will answer these and others questions in the coming future.

ACKNOWLEDGMENTS

The author acknowledges the support of the Ministerio de Economia y Competitividad (MINECO, grant: BFU2013/42746-P) from the Spanish Government.

REFERENCES

Abbott, D. W., Wilkins, A., Asara, J. M., & Cantley, L. C. (2004). The Crohn's disease protein, NOD2, requires RIP2 in order to induce ubiquitinylation of a novel site on NEMO. *Current Biology*, *14*(24), 2217–2227. http://dx.doi.org/10.1016/j.cub.2004.12.032.

Anastasia, A., Barker, P. A., Chao, M. V., & Hempstead, B. L. (2015). Detection of p75NTR trimers: Implications for receptor stoichiometry and activation. *Journal of Neuroscience*, *35*(34), 11911–11920. http://dx.doi.org/10.1523/JNEUROSCI.0591-15.2015.

Andera, L. (2009). Signaling activated by the death receptors of the TNFR family. *Biomedical Papers of the Medical Faculty of the University Palacky, Olomouc, Czechoslovakia*, *153*(3), 173–180.

Arch, R. H., Gedrich, R. W., & Thompson, C. B. (1998). Tumor necrosis factor receptor-associated factors (TRAFs)—A family of adapter proteins that regulates life and death. *Genes & Development*, *12*(18), 2821–2830.

Aurikko, J. P., Ruotolo, B. T., Grossmann, J. G., Moncrieffe, M. C., Stephens, E., Leppanen, V. M., … Blundell, T. L. (2005). Characterization of symmetric complexes of nerve growth factor and the ectodomain of the pan-neurotrophin receptor, p75NTR. *Journal of Biological Chemistry*, *280*(39), 33453–33460. http://dx.doi.org/10.1074/jbc.M503189200.

Baldwin, A. N., Bitler, C. M., Welcher, A. A., & Shooter, E. M. (1992). Studies on the structure and binding properties of the cysteine-rich domain of rat low affinity nerve growth factor receptor (p75NGFR). *Journal of Biological Chemistry*, *267*(12), 8352–8359.

Barker, P. A., Barbee, G., Misko, T. P., & Shooter, E. M. (1994). The low affinity neurotrophin receptor, p75LNTR, is palmitoylated by thioester formation through cysteine 279. *Journal of Biological Chemistry*, *269*(48), 30645–30650.

Bibel, M., Hoppe, E., & Barde, Y. A. (1999). Biochemical and functional interactions between the neurotrophin receptors trk and p75NTR. *EMBO Journal*, *18*(3), 616–622. http://dx.doi.org/10.1093/emboj/18.3.616.

Bodmer, J. L., Schneider, P., & Tschopp, J. (2002). The molecular architecture of the TNF superfamily. *Trends in Biochemical Sciences*, *27*(1), 19–26.

Bothwell, M. (1995). Functional interactions of neurotrophins and neurotrophin receptors. *Annual Review of Neuroscience*, *18*, 223–253. http://dx.doi.org/10.1146/annurev.ne.18.030195.001255.

Branschadel, M., Aird, A., Zappe, A., Tietz, C., Krippner-Heidenreich, A., & Scheurich, P. (2010). Dual function of cysteine rich domain (CRD) 1 of TNF receptor type 1: Conformational stabilization of CRD2 and control of receptor responsiveness. *Cellular Signalling*, *22*(3), 404–414. http://dx.doi.org/10.1016/j.cellsig.2009.10.011.

Bronfman, F. C. (2007). Metalloproteases and gamma-secretase: New membrane partners regulating p75 neurotrophin receptor signaling? *Journal of Neurochemistry*, *103*(Suppl. 1), 91–100. http://dx.doi.org/10.1111/j.1471-4159.2007.04781.x.

Cabal-Hierro, L., & Lazo, P. S. (2012). Signal transduction by tumor necrosis factor receptors. *Cellular Signalling*, *24*(6), 1297–1305. http://dx.doi.org/10.1016/j.cellsig.2012.02.006.

Carrington, P. E., Sandu, C., Wei, Y., Hill, J. M., Morisawa, G., Huang, T., … Werner, M. H. (2006). The structure of FADD and its mode of interaction with procaspase-8. *Molecular Cell*, *22*(5), 599–610. http://dx.doi.org/10.1016/j.molcel.2006.04.018.

Casademunt, E., Carter, B. D., Benzel, I., Frade, J. M., Dechant, G., & Barde, Y. A. (1999). The zinc finger protein NRIF interacts with the neurotrophin receptor p75(NTR) and participates in programmed cell death. *EMBO Journal*, *18*(21), 6050–6061. http://dx.doi.org/10.1093/emboj/18.21.6050.

Chan, F. K. (2007). Three is better than one: Pre-ligand receptor assembly in the regulation of TNF receptor signaling. *Cytokine*, *37*(2), 101–107. http://dx.doi.org/10.1016/j.cyto.2007.03.005.

Chan, F. K., Chun, H. J., Zheng, L., Siegel, R. M., Bui, K. L., & Lenardo, M. J. (2000). A domain in TNF receptors that mediates ligand-independent receptor assembly and signaling. *Science*, *288*(5475), 2351–2354.

Chao, M. V. (2003). Neurotrophins and their receptors: A convergence point for many signalling pathways. *Nature Reviews. Neuroscience*, *4*(4), 299–309. http://dx.doi.org/10.1038/nrn1078.

Chao, M. V., & Hempstead, B. L. (1995). p75 and Trk: A two-receptor system. *Trends in Neurosciences*, *18*(7), 321–326.

Chinnaiyan, A. M., O'Rourke, K., Tewari, M., & Dixit, V. M. (1995). FADD, a novel death domain-containing protein, interacts with the death domain of Fas and initiates apoptosis. *Cell*, *81*(4), 505–512.

Chittka, A., & Chao, M. V. (1999). Identification of a zinc finger protein whose subcellular distribution is regulated by serum and nerve growth factor. *Proceedings of the National Academy of Sciences of the United States of America*, *96*(19), 10705–10710.

Compaan, D. M., & Hymowitz, S. G. (2006). The crystal structure of the costimulatory OX40-OX40L complex. *Structure*, *14*(8), 1321–1330. http://dx.doi.org/10.1016/j.str.2006.06.015.

Coulson, E. J., Reid, K., Baca, M., Shipham, K. A., Hulett, S. M., Kilpatrick, T. J., & Bartlett, P. F. (2000). Chopper, a new death domain of the p75 neurotrophin receptor that mediates rapid neuronal cell death. *Journal of Biological Chemistry*, *275*(39), 30537–30545. http://dx.doi.org/10.1074/jbc.M005214200.

Croft, M., Benedict, C. A., & Ware, C. F. (2013). Clinical targeting of the TNF and TNFR superfamilies. *Nature Reviews. Drug Discovery*, *12*(2), 147–168. http://dx.doi.org/10.1038/nrd3930.

Croft, M., Duan, W., Choi, H., Eun, S. Y., Madireddi, S., & Mehta, A. (2012). TNF superfamily in inflammatory disease: Translating basic insights. *Trends in Immunology*, *33*(3), 144–152. http://dx.doi.org/10.1016/j.it.2011.10.004.

Dechant, G., Tsoulfas, P., Parada, L. F., & Barde, Y. A. (1997). The neurotrophin receptor p75 binds neurotrophin-3 on sympathetic neurons with high affinity and specificity. *Journal of Neuroscience*, *17*(14), 5281–5287.

Deinhardt, K., & Chao, M. V. (2014). Trk receptors. *Handbook of Experimental Pharmacology*, *220*, 103–119. http://dx.doi.org/10.1007/978-3-642-45106-5_5.

Esposito, D., Patel, P., Stephens, R. M., Perez, P., Chao, M. V., Kaplan, D. R., & Hempstead, B. L. (2001). The cytoplasmic and transmembrane domains of the p75 and Trk A receptors regulate high affinity binding to nerve growth factor. *Journal of Biological Chemistry*, *276*(35), 32687–32695. http://dx.doi.org/10.1074/jbc.M011674200.

Feng, D., Kim, T., Ozkan, E., Light, M., Torkin, R., Teng, K. K., … Garcia, K. C. (2010). Molecular and structural insight into proNGF engagement of p75NTR and sortilin. *Journal of Molecular Biology*, *396*(4), 967–984. http://dx.doi.org/10.1016/j.jmb.2009.12.030.

Ferrao, R., & Wu, H. (2012). Helical assembly in the death domain (DD) superfamily. *Current Opinion in Structural Biology*, *22*(2), 241–247. http://dx.doi.org/10.1016/j.sbi.2012.02.006.

Forsyth, P. A., Krishna, N., Lawn, S., Valadez, J. G., Qu, X., Fenstermacher, D. A., … Kenchappa, R. S. (2014). p75 neurotrophin receptor cleavage by alpha- and gamma-secretases is required for neurotrophin-mediated proliferation of brain tumor-initiating cells. *Journal of Biological Chemistry*, *289*(12), 8067–8085. http://dx.doi.org/10.1074/jbc.M113.513762.

Frade, J. M. (2005). Nuclear translocation of the p75 neurotrophin receptor cytoplasmic domain in response to neurotrophin binding. *Journal of Neuroscience*, *25*(6), 1407–1411. http://dx.doi.org/10.1523/JNEUROSCI.3798-04.2005.

Frankowski, H., Castro-Obregon, S., del Rio, G., Rao, R. V., & Bredesen, D. E. (2002). PLAIDD, a type II death domain protein that interacts with p75 neurotrophin receptor. *Neuromolecular Medicine*, *1*(3), 153–170. http://dx.doi.org/10.1385/NMM:1:3:153.

Fu, Q., Fu, T. M., Cruz, A. C., Sengupta, P., Thomas, S. K., Wang, S., … Chou, J. J. (2016). Structural basis and functional role of intramembrane trimerization of the Fas/CD95 death receptor. *Molecular Cell*, *61*(4), 602–613. http://dx.doi.org/10.1016/j.molcel.2016.01.009.

Geetha, T., Zheng, C., Unroe, B., Sycheva, M., Kluess, H., & Babu, J. R. (2012). Polyubiquitination of the neurotrophin receptor p75 directs neuronal cell survival. *Biochemical and Biophysical Research Communications*, *421*(2), 286–290. http://dx.doi.org/10.1016/j.bbrc.2012.04.001.

Gong, Y., Cao, P., Yu, H. J., & Jiang, T. (2008). Crystal structure of the neurotrophin-3 and p75NTR symmetrical complex. *Nature*, *454*(7205), 789–793. http://dx.doi.org/10.1038/nature07089.

Gowrishankar, K., Zeidler, M. G., & Vincenz, C. (2004). Release of a membrane-bound death domain by gamma-secretase processing of the p75NTR homolog NRADD. *Journal of Cell Science*, *117*(Pt. 18), 4099–4111. http://dx.doi.org/10.1242/jcs.01263.

Gray, P. W., Barrett, K., Chantry, D., Turner, M., & Feldmann, M. (1990). Cloning of human tumor necrosis factor (TNF) receptor cDNA and expression of recombinant soluble TNF-binding protein. *Proceedings of the National Academy of Sciences of the United States of America*, *87*(19), 7380–7384.

Grob, P. M., Berlot, C. H., & Bothwell, M. A. (1983). Affinity labeling and partial purification of nerve growth factor receptors from rat pheochromocytoma and human melanoma cells. *Proceedings of the National Academy of Sciences of the United States of America*, *80*(22), 6819–6823.

Grob, P. M., Ross, A. H., Koprowski, H., & Bothwell, M. (1985). Characterization of the human melanoma nerve growth factor receptor. *Journal of Biological Chemistry*, *260*(13), 8044–8049.

Harrington, A. W., Leiner, B., Blechschmitt, C., Arevalo, J. C., Lee, R., Morl, K., … Giehl, K. M. (2004). Secreted proNGF is a pathophysiological death-inducing ligand after adult CNS injury. *Proceedings of the National Academy of Sciences of the United States of America*, *101*(16), 6226–6230. http://dx.doi.org/10.1073/pnas.0305755101.

Hasegawa, M., Fujimoto, Y., Lucas, P. C., Nakano, H., Fukase, K., Nunez, G., & Inohara, N. (2008). A critical role of RICK/RIP2 polyubiquitination in Nod-induced NF-kappaB activation. *EMBO Journal*, *27*(2), 373–383. http://dx.doi.org/10.1038/sj.emboj.7601962.

He, X. L., & Garcia, K. C. (2004). Structure of nerve growth factor complexed with the shared neurotrophin receptor p75. *Science*, *304*(5672), 870–875. http://dx.doi.org/10.1126/science.1095190.

Hempstead, B. L. (2014). Deciphering proneurotrophin actions. *Handbook of Experimental Pharmacology*, *220*, 17–32. http://dx.doi.org/10.1007/978-3-642-45106-5_2.

Hempstead, B. L., Martin-Zanca, D., Kaplan, D. R., Parada, L. F., & Chao, M. V. (1991). High-affinity NGF binding requires coexpression of the trk proto-oncogene and the low-affinity NGF receptor. *Nature*, *350*(6320), 678–683. http://dx.doi.org/10.1038/350678a0.

Hill, J. M., Vaidyanathan, H., Ramos, J. W., Ginsberg, M. H., & Werner, M. H. (2002). Recognition of ERK MAP kinase by PEA-15 reveals a common docking site within the death domain and death effector domain. *EMBO Journal*, *21*(23), 6494–6504.

Hsu, H., Shu, H. B., Pan, M. G., & Goeddel, D. V. (1996). TRADD-TRAF2 and TRADD-FADD interactions define two distinct TNF receptor 1 signal transduction pathways. *Cell*, *84*(2), 299–308.

Huang, B., Eberstadt, M., Olejniczak, E. T., Meadows, R. P., & Fesik, S. W. (1996). NMR structure and mutagenesis of the Fas (APO-1/CD95) death domain. *Nature*, *384*(6610), 638–641. http://dx.doi.org/10.1038/384638a0.

Huang, E. J., & Reichardt, L. F. (2003). Trk receptors: Roles in neuronal signal transduction. *Annual Review of Biochemistry*, *72*, 609–642. http://dx.doi.org/10.1146/annurev.biochem.72.121801.161629.

Huber, L. J., & Chao, M. V. (1995). A potential interaction of p75 and trkA NGF receptors revealed by affinity crosslinking and immunoprecipitation. *Journal of Neuroscience Research*, *40*(4), 557–563. http://dx.doi.org/10.1002/jnr.490400415.

Iacaruso, M. F., Galli, S., Marti, M., Villalta, J. I., Estrin, D. A., Jares Erijman, E. A., & Pietrasanta, L. I. (2011). Structural model for p75(NTR)-TrkA intracellular domain interaction: A combined FRET and bioinformatics study. *Journal of Molecular Biology*, *414*(5), 681–698. http://dx.doi.org/10.1016/j.jmb.2011.09.022.

Jing, S., Tapley, P., & Barbacid, M. (1992). Nerve growth factor mediates signal transduction through trk homodimer receptors. *Neuron*, *9*(6), 1067–1079.

Johnson, D., Lanahan, A., Buck, C. R., Sehgal, A., Morgan, C., Mercer, E., … Chao, M. (1986). Expression and structure of the human NGF receptor. *Cell*, *47*(4), 545–554.

Jung, K. M., Tan, S., Landman, N., Petrova, K., Murray, S., Lewis, R., … Kim, T. W. (2003). Regulated intramembrane proteolysis of the p75 neurotrophin receptor modulates its association with the TrkA receptor. *Journal of Biological Chemistry*, *278*(43), 42161–42169. http://dx.doi.org/10.1074/jbc.M306028200.

Kanning, K. C., Hudson, M., Amieux, P. S., Wiley, J. C., Bothwell, M., & Schecterson, L. C. (2003). Proteolytic processing of the p75 neurotrophin receptor and two homologs generates C-terminal fragments with signaling capability. *Journal of Neuroscience*, *23*(13), 5425–5436.

Kaplan, D. R., Hempstead, B. L., Martin-Zanca, D., Chao, M. V., & Parada, L. F. (1991). The trk proto-oncogene product: A signal transducing receptor for nerve growth factor. *Science*, *252*(5005), 554–558.

Kersse, K., Verspurten, J., Vanden Berghe, T., & Vandenabeele, P. (2011). The death-fold superfamily of homotypic interaction motifs. *Trends in Biochemical Sciences*, *36*(10), 541–552. http://dx.doi.org/10.1016/j.tibs.2011.06.006.

Khursigara, G., Bertin, J., Yano, H., Moffett, H., DiStefano, P. S., & Chao, M. V. (2001). A prosurvival function for the p75 receptor death domain mediated via the caspase recruitment domain receptor-interacting protein 2. *Journal of Neuroscience*, *21*(16), 5854–5863.

Khursigara, G., Orlinick, J. R., & Chao, M. V. (1999). Association of the p75 neurotrophin receptor with TRAF6. *Journal of Biological Chemistry*, *274*(5), 2597–2600.

Kim, T., & Hempstead, B. L. (2009). NRH2 is a trafficking switch to regulate sortilin localization and permit proneurotrophin-induced cell death. *EMBO Journal*, *28*(11), 1612–1623. http://dx.doi.org/10.1038/emboj.2009.118.

Klein, R., Jing, S. Q., Nanduri, V., O'Rourke, E., & Barbacid, M. (1991). The trk proto-oncogene encodes a receptor for nerve growth factor. *Cell*, *65*(1), 189–197.

Kobayashi, K., Inohara, N., Hernandez, L. D., Galan, J. E., Nunez, G., Janeway, C. A., … Flavell, R. A. (2002). RICK/Rip2/CARDIAK mediates signalling for receptors of the innate and adaptive immune systems. *Nature*, *416*(6877), 194–199. http://dx.doi.org/10.1038/416194a.

Kraemer, B. R., Yoon, S. O., & Carter, B. D. (2014). The biological functions and signaling mechanisms of the p75 neurotrophin receptor. *Handbook of Experimental Pharmacology*, *220*, 121–164. http://dx.doi.org/10.1007/978-3-642-45106-5_6.

Kramer, J. M., Hanel, W., Shen, F., Isik, N., Malone, J. P., Maitra, A., … Gaffen, S. L. (2007). Cutting edge: Identification of a pre-ligand assembly domain (PLAD) and ligand binding site in the IL-17 receptor. *Journal of Immunology*, *179*(10), 6379–6383.

Krieg, A., Correa, R. G., Garrison, J. B., Le Negrate, G., Welsh, K., Huang, Z., … Reed, J. C. (2009). XIAP mediates NOD signaling via interaction with RIP2. *Proceedings of the National Academy of Sciences of the United States of America*, *106*(34), 14524–14529. http://dx.doi.org/10.1073/pnas.0907131106.

Lamballe, F., Klein, R., & Barbacid, M. (1991). The trk family of oncogenes and neurotrophin receptors. *Princess Takamatsu Symposia*, *22*, 153–170.

Lauri, A., Bertucci, P., & Arendt, D. (2016). Neurotrophin, p75, and Trk signaling module in the developing nervous system of the marine annelid Platynereis dumerilii. *BioMed Research International*, *2016*, 2456062. http://dx.doi.org/10.1155/2016/2456062.

Lee, K. F., Li, E., Huber, L. J., Landis, S. C., Sharpe, A. H., Chao, M. V., & Jaenisch, R. (1992). Targeted mutation of the gene encoding the low affinity NGF receptor p75 leads to deficits in the peripheral sensory nervous system. *Cell*, *69*(5), 737–749.

Li, J., Yin, Q., & Wu, H. (2013). Structural basis of signal transduction in the TNF receptor superfamily. *Advances in Immunology*, *119*, 135–153. http://dx.doi.org/10.1016/B978-0-12-407707-2.00005-9.

Liepinsh, E., Ilag, L. L., Otting, G., & Ibanez, C. F. (1997). NMR structure of the death domain of the p75 neurotrophin receptor. *EMBO Journal*, *16*(16), 4999–5005. http://dx.doi.org/10.1093/emboj/16.16.4999.

Lin, Z., Tann, J. Y., Goh, E. T., Kelly, C., Lim, K. B., Gao, J. F., & Ibanez, C. F. (2015). Structural basis of death domain signaling in the p75 neurotrophin receptor. *eLife*, *4*, e11692. http://dx.doi.org/10.7554/eLife.11692.

Locksley, R. M., Killeen, N., & Lenardo, M. J. (2001). The TNF and TNF receptor superfamilies: Integrating mammalian biology. *Cell, 104*(4), 487–501.

Longenecker, K., Read, P., Derewenda, U., Dauter, Z., Liu, X., Garrard, S., … Derewenda, Z. S. (1999). How RhoGDI binds Rho. *Acta Crystallographica. Section D: Biological Crystallography, 55*(Pt. 9), 1503–1515.

Mahadeo, D., Kaplan, L., Chao, M. V., & Hempstead, B. L. (1994). High affinity nerve growth factor binding displays a faster rate of association than p140trk binding. Implications for multi-subunit polypeptide receptors. *Journal of Biological Chemistry, 269*(9), 6884–6891.

Majdan, M., Lachance, C., Gloster, A., Aloyz, R., Zeindler, C., Bamji, S., … Barker, P. A. (1997). Transgenic mice expressing the intracellular domain of the p75 neurotrophin receptor undergo neuronal apoptosis. *Journal of Neuroscience, 17*(18), 6988–6998.

Matusica, D., Skeldal, S., Sykes, A. M., Palstra, N., Sharma, A., & Coulson, E. J. (2013). An intracellular domain fragment of the p75 neurotrophin receptor (p75(NTR)) enhances tropomyosin receptor kinase A (TrkA) receptor function. *Journal of Biological Chemistry, 288*(16), 11144–11154. http://dx.doi.org/10.1074/jbc.M112.436469.

McCarthy, J. V., Ni, J., & Dixit, V. M. (1998). RIP2 is a novel NF-kappaB-activating and cell death-inducing kinase. *Journal of Biological Chemistry, 273*(27), 16968–16975.

Mi, S., Lee, X., Shao, Z., Thill, G., Ji, B., Relton, J., … Pepinsky, R. B. (2004). LINGO-1 is a component of the Nogo-66 receptor/p75 signaling complex. *Nature Neuroscience*, 7(3), 221–228. http://dx.doi.org/10.1038/nn1188.

Mineev, K. S., Goncharuk, S. A., Kuzmichev, P. K., Vilar, M., & Arseniev, A. S. (2015). NMR dynamics of transmembrane and intracellular domains of p75NTR in lipid-protein nanodiscs. *Biophysical Journal, 109*(4), 772–782. http://dx.doi.org/10.1016/j.bpj.2015.07.009.

Mukai, Y., Nakamura, T., Yoshikawa, M., Yoshioka, Y., Tsunoda, S., Nakagawa, S., … Tsutsumi, Y. (2010). Solution of the structure of the TNF-TNFR2 complex. *Science Signaling, 3*(148), ra83. http://dx.doi.org/10.1126/scisignal.2000954.

Murray, S. S., Perez, P., Lee, R., Hempstead, B. L., & Chao, M. V. (2004). A novel p75 neurotrophin receptor-related protein, NRH2, regulates nerve growth factor binding to the TrkA receptor. *Journal of Neuroscience, 24*(11), 2742–2749. http://dx.doi.org/10.1523/JNEUROSCI.3960-03.2004.

Nadezhdin, K. D., Garcia-Carpio, I., Goncharuk, S. A., Mineev, K. S., Arseniev, A. S., & Vilar, M. (2016). Structural basis of p75 transmembrane domain dimerization. *Journal of Biological Chemistry, 291*(23), 12346–12357. http://dx.doi.org/10.1074/jbc.M116.723585.

Naismith, J. H., Devine, T. Q., Brandhuber, B. J., & Sprang, S. R. (1995). Crystallographic evidence for dimerization of unliganded tumor necrosis factor receptor. *Journal of Biological Chemistry, 270*(22), 13303–13307.

Naismith, J. H., Devine, T. Q., Kohno, T., & Sprang, S. R. (1996). Structures of the extracellular domain of the type I tumor necrosis factor receptor. *Structure, 4*(11), 1251–1262.

Naismith, J. H., & Sprang, S. R. (1998). Modularity in the TNF-receptor family. *Trends in Biochemical Sciences, 23*(2), 74–79.

Nykjaer, A., Lee, R., Teng, K. K., Jansen, P., Madsen, P., Nielsen, M. S., … Petersen, C. M. (2004). Sortilin is essential for proNGF-induced neuronal cell death. *Nature, 427*(6977), 843–848. http://dx.doi.org/10.1038/nature02319.

Park, H. H., Lo, Y. C., Lin, S. C., Wang, L., Yang, J. K., & Wu, H. (2007). The death domain superfamily in intracellular signaling of apoptosis and inflammation. *Annual Review of Immunology, 25*, 561–586. http://dx.doi.org/10.1146/annurev.immunol.25.022106.141656.

Parkhurst, C. N., Zampieri, N., & Chao, M. V. (2010). Nuclear localization of the p75 neurotrophin receptor intracellular domain. *Journal of Biological Chemistry*, *285*(8), 5361–5368. http://dx.doi.org/10.1074/jbc.M109.045054.

Qu, Q., Chen, J., Wang, Y., Gui, W., Wang, L., Fan, Z., & Jiang, T. (2013). Structural characterization of the self-association of the death domain of p75(NTR). *PLoS One*, *8*(3), e57839. http://dx.doi.org/10.1371/journal.pone.0057839.

Quistgaard, E. M., Groftehauge, M. K., Madsen, P., Pallesen, L. T., Christensen, B., Sorensen, E. S., … Thirup, S. S. (2014). Revisiting the structure of the Vps10 domain of human sortilin and its interaction with neurotensin. *Protein Science*, *23*(9), 1291–1300. http://dx.doi.org/10.1002/pro.2512.

Quistgaard, E. M., Madsen, P., Groftehauge, M. K., Nissen, P., Petersen, C. M., & Thirup, S. S. (2009). Ligands bind to Sortilin in the tunnel of a ten-bladed beta-propeller domain. *Nature Structural & Molecular Biology*, *16*(1), 96–98. http://dx.doi.org/10.1038/nsmb.1543.

Radeke, M. J., Misko, T. P., Hsu, C., Herzenberg, L. A., & Shooter, E. M. (1987). Gene transfer and molecular cloning of the rat nerve growth factor receptor. *Nature*, *325*(6105), 593–597. http://dx.doi.org/10.1038/325593a0.

Reichardt, L. F. (2006). Neurotrophin-regulated signalling pathways. *Philosophical Transactions of the Royal Society of London. Series B: Biological Sciences*, *361*(1473), 1545–1564. http://dx.doi.org/10.1098/rstb.2006.1894.

Riedl, S. J., & Shi, Y. (2004). Molecular mechanisms of caspase regulation during apoptosis. *Nature Reviews. Molecular Cell Biology*, *5*(11), 897–907. http://dx.doi.org/10.1038/nrm1496.

Rifkin, J. T., Todd, V. J., Anderson, L. W., & Lefcort, F. (2000). Dynamic expression of neurotrophin receptors during sensory neuron genesis and differentiation. *Developmental Biology*, *227*(2), 465–480. http://dx.doi.org/10.1006/dbio.2000.9841.

Rodriguez-Tebar, A., Dechant, G., & Barde, Y. A. (1990). Binding of brain-derived neurotrophic factor to the nerve growth factor receptor. *Neuron*, *4*(4), 487–492.

Rodriguez-Tebar, A., Dechant, G., Gotz, R., & Barde, Y. A. (1992). Binding of neurotrophin-3 to its neuronal receptors and interactions with nerve growth factor and brain-derived neurotrophic factor. *EMBO Journal*, *11*(3), 917–922.

Ross, A. H., Daou, M. C., McKinnon, C. A., Condon, P. J., Lachyankar, M. B., Stephens, R. M., … Wolf, D. E. (1996). The neurotrophin receptor, gp75, forms a complex with the receptor tyrosine kinase TrkA. *Journal of Cell Biology*, *132*(5), 945–953.

Ryden, M., Murray-Rust, J., Glass, D., Ilag, L. L., Trupp, M., Yancopoulos, G. D., … Ibanez, C. F. (1995). Functional analysis of mutant neurotrophins deficient in low-affinity binding reveals a role for p75LNGFR in NT-4 signalling. *EMBO Journal*, *14*(9), 1979–1990.

Schachtrup, C., Ryu, J. K., Mammadzada, K., Khan, A. S., Carlton, P. M., Perez, A., … Akassoglou, K. (2015). Nuclear pore complex remodeling by p75(NTR) cleavage controls TGF-beta signaling and astrocyte functions. *Nature Neuroscience*, *18*(8), 1077–1080. http://dx.doi.org/10.1038/nn.4054.

Skeldal, S., Matusica, D., Nykjaer, A., & Coulson, E. J. (2011). Proteolytic processing of the p75 neurotrophin receptor: A prerequisite for signalling? Neuronal life, growth and death signalling are crucially regulated by intra-membrane proteolysis and trafficking of p75(NTR). *Bioessays*, *33*(8), 614–625. http://dx.doi.org/10.1002/bies.201100036.

Skeldal, S., Sykes, A. M., Glerup, S., Matusica, D., Palstra, N., Autio, H., … Coulson, E. J. (2012). Mapping of the interaction site between sortilin and the p75 neurotrophin receptor reveals a regulatory role for the sortilin intracellular domain in p75 neurotrophin receptor shedding and apoptosis. *Journal of Biological Chemistry*, *287*(52), 43798–43809. http://dx.doi.org/10.1074/jbc.M112.374710.

Smith, C. A., Farrah, T., & Goodwin, R. G. (1994). The TNF receptor superfamily of cellular and viral proteins: Activation, costimulation, and death. *Cell*, *76*(6), 959–962.

Song, Y., & Buchwald, P. (2015). TNF superfamily protein–protein interactions: Feasibility of small-molecule modulation. *Current Drug Targets*, *16*(4), 393–408.

Sung, T. C., Chen, Z., Thuret, S., Vilar, M., Gage, F. H., Riek, R., & Lee, K. F. (2013). P45 forms a complex with FADD and promotes neuronal cell survival following spinal cord injury. *PLoS One*, *8*(7), e69286. http://dx.doi.org/10.1371/journal.pone.0069286.

Sykes, A. M., Palstra, N., Abankwa, D., Hill, J. M., Skeldal, S., Matusica, D., … Coulson, E. J. (2012). The effects of transmembrane sequence and dimerization on cleavage of the p75 neurotrophin receptor by gamma-secretase. *Journal of Biological Chemistry*, *287*(52), 43810–43824. http://dx.doi.org/10.1074/jbc.M112.382903.

Tanaka, K., Kelly, C. E., Goh, K. Y., Lim, K. B., & Ibáñez, C. F. (2016). Death domain signaling by disulfide-linked dimers of the p75 neurotrophin receptor mediates neuronal death in the CNS. *Journal of Neuroscience*, *36*(20), 5587–5595. http://dx.doi.org/10.1523/JNEUROSCI.4536-15.2016.

Tao, M., Scacheri, P. C., Marinis, J. M., Harhaj, E. W., Matesic, L. E., & Abbott, D. W. (2009). ITCH K63-ubiquitinates the NOD2 binding protein, RIP2, to influence inflammatory signaling pathways. *Current Biology*, *19*(15), 1255–1263. http://dx.doi.org/10.1016/j.cub.2009.06.038.

Underwood, C. K., Reid, K., May, L. M., Bartlett, P. F., & Coulson, E. J. (2008). Palmitoylation of the C-terminal fragment of p75(NTR) regulates death signaling and is required for subsequent cleavage by gamma-secretase. *Molecular and Cellular Neuroscience*, *37*(2), 346–358. http://dx.doi.org/10.1016/j.mcn.2007.10.005.

Vicario, A., Kisiswa, L., Tann, J. Y., Kelly, C. E., & Ibanez, C. F. (2015). Neuron-type-specific signaling by the p75NTR death receptor regulated by differential proteolytic cleavage. *Journal of Cell Science*, *128*, 1507–1517. http://dx.doi.org/10.1242/jcs.161745.

Vilar, M., Charalampopoulos, I., Kenchappa, R. S., Reversi, A., Klos-Applequist, J. M., Karaca, E., … Ibanez, C. F. (2009a). Ligand-independent signaling by disulfide-crosslinked dimers of the p75 neurotrophin receptor. *Journal of Cell Science*, *122*(Pt. 18), 3351–3357. http://dx.doi.org/10.1242/jcs.055061.

Vilar, M., Charalampopoulos, I., Kenchappa, R. S., Simi, A., Karaca, E., Reversi, A., … Ibanez, C. F. (2009b). Activation of the p75 neurotrophin receptor through conformational rearrangement of disulphide-linked receptor dimers. *Neuron*, *62*(1), 72–83. http://dx.doi.org/10.1016/j.neuron.2009.02.020.

Vilar, M., Sung, T. C., Chen, Z., Garcia-Carpio, I., Fernandez, E. M., Xu, J., … Lee, K. F. (2014). Heterodimerization of p45-p75 modulates p75 signaling: Structural basis and mechanism of action. *PLoS Biology*, *12*(8), e1001918. http://dx.doi.org/10.1371/journal.pbio.1001918.

Wajant, H., Grell, M., & Scheurich, P. (1999). TNF receptor associated factors in cytokine signaling. *Cytokine & Growth Factor Reviews*, *10*(1), 15–26.

Wang, K. C., Kim, J. A., Sivasankaran, R., Segal, R., & He, Z. (2002). P75 interacts with the Nogo receptor as a co-receptor for Nogo, MAG and OMgp. *Nature*, *420*(6911), 74–78. http://dx.doi.org/10.1038/nature01176.

Wang, L., Yang, J. K., Kabaleeswaran, V., Rice, A. J., Cruz, A. C., Park, A. Y., … Wu, H. (2010). The Fas-FADD death domain complex structure reveals the basis of DISC assembly and disease mutations. *Nature Structural & Molecular Biology*, *17*(11), 1324–1329. http://dx.doi.org/10.1038/nsmb.1920.

Weber, C. H., & Vincenz, C. (2001). The death domain superfamily: A tale of two interfaces? *Trends in Biochemical Sciences*, *26*(8), 475–481.

Wehrman, T., He, X., Raab, B., Dukipatti, A., Blau, H., & Garcia, K. C. (2007). Structural and mechanistic insights into nerve growth factor interactions with the TrkA and p75 receptors. *Neuron*, *53*(1), 25–38. http://dx.doi.org/10.1016/j.neuron.2006.09.034.

White, F. A., Silos-Santiago, I., Molliver, D. C., Nishimura, M., Phillips, H., Barbacid, M., & Snider, W. D. (1996). Synchronous onset of NGF and TrkA survival dependence in developing dorsal root ganglia. *Journal of Neuroscience*, *16*(15), 4662–4672.

Wiesmann, C., & de Vos, A. M. (2001). Nerve growth factor: Structure and function. *Cellular and Molecular Life Sciences*, *58*(5–6), 748–759.

Wiesmann, C., Ultsch, M. H., Bass, S. H., & de Vos, A. M. (1999). Crystal structure of nerve growth factor in complex with the ligand-binding domain of the TrkA receptor. *Nature*, *401*(6749), 184–188. http://dx.doi.org/10.1038/43705.

Wilson, K. H. (2009). The genome sequence of the protostome Daphnia pulex encodes respective orthologues of a neurotrophin, a Trk and a p75NTR: Evolution of neurotrophin signaling components and related proteins in the bilateria. *BMC Evolutionary Biology*, *9*, 243. http://dx.doi.org/10.1186/1471-2148-9-243.

Wolf, D. E., McKinnon, C. A., Daou, M. C., Stephens, R. M., Kaplan, D. R., & Ross, A. H. (1995). Interaction with TrkA immobilizes gp75 in the high affinity nerve growth factor receptor complex. *Journal of Biological Chemistry*, *270*(5), 2133–2138.

Wong, S. T., Henley, J. R., Kanning, K. C., Huang, K. H., Bothwell, M., & Poo, M. M. (2002). A p75(NTR) and Nogo receptor complex mediates repulsive signaling by myelin-associated glycoprotein. *Nature Neuroscience*, *5*(12), 1302–1308. http://dx.doi.org/10.1038/nn975.

Wu, H. (2004). Assembly of post-receptor signaling complexes for the tumor necrosis factor receptor superfamily. *Advances in Protein Chemistry*, *68*, 225–279. http://dx.doi.org/10.1016/S0065-3233(04)68007-7.

Wu, H., & Arron, J. R. (2003). TRAF6, a molecular bridge spanning adaptive immunity, innate immunity and osteoimmunology. *Bioessays*, *25*(11), 1096–1105. http://dx.doi.org/10.1002/bies.10352.

Wu, H., & Hymowitz, S. G. (2009). Structure and function of tumor necrosis factor (TNF) at the cell surface. In A. Ralph, A. Bradshaw, & E. A. Dennis (Eds.), *Vol. 1 Handbook of cell signaling* (pp. 265–275). Oxford: Academic Press.

Yamashita, T., Higuchi, H., & Tohyama, M. (2002). The p75 receptor transduces the signal from myelin-associated glycoprotein to Rho. *Journal of Cell Biology*, *157*(4), 565–570. http://dx.doi.org/10.1083/jcb.200202010.

Yamashita, T., & Tohyama, M. (2003). The p75 receptor acts as a displacement factor that releases Rho from Rho-GDI. *Nature Neuroscience*, *6*(5), 461–467. http://dx.doi.org/10.1038/nn1045.

Yamashita, T., Tucker, K. L., & Barde, Y. A. (1999). Neurotrophin binding to the p75 receptor modulates Rho activity and axonal outgrowth. *Neuron*, *24*(3), 585–593.

Yan, H., & Chao, M. V. (1991). Disruption of cysteine-rich repeats of the p75 nerve growth factor receptor leads to loss of ligand binding. *Journal of Biological Chemistry*, *266*(18), 12099–12104.

Yano, H., & Chao, M. V. (2000). Neurotrophin receptor structure and interactions. *Pharmaceutica Acta Helvetiae*, *74*(2–3), 253–260.

Ye, H., Arron, J. R., Lamothe, B., Cirilli, M., Kobayashi, T., Shevde, N. K., ... Wu, H. (2002). Distinct molecular mechanism for initiating TRAF6 signalling. *Nature*, *418*(6896), 443–447. http://dx.doi.org/10.1038/nature00888.

Ye, X., Mehlen, P., Rabizadeh, S., VanArsdale, T., Zhang, H., Shin, H., ... Bredesen, D. E. (1999). TRAF family proteins interact with the common neurotrophin receptor and modulate apoptosis induction. *Journal of Biological Chemistry*, *274*(42), 30202–30208.

CHAPTER FIVE

Releasing Mechanism of Neurotrophic Factors via Polysialic Acid

C. Sato[1]
Bioscience and Biotechnology Center, Nagoya University, Nagoya, Japan
[1]Corresponding author: e-mail address: chi@agr.nagoya-u.ac.jp

Contents

1. Introduction 90
2. Biochemical Features of BDNF 91
3. Regions and Cellular Distribution of BDNF 95
4. Extracellular Secretion of BDNF 97
   4.1 BDNF Release via Exocytosis 98
   4.2 BDNF Release via Polysialic Acid 100
5. Conclusion 105
Acknowledgments 105
References 106

Abstract

Neurotrophins are well-characterized neurologically active molecules in the central nervous system. The regulation of these signaling molecules, which are involved in cell growth, differentiation, and survival, is critical for normal brain function. Among the different types of neurotrophins, brain-derived neurotrophic factor (BDNF) is involved in various brain functions, including memory consolidation, synaptic plasticity, and adult neurogenesis, and is therefore a key molecule for understanding comprehensive brain function and neurodevelopmental and psychiatric diseases. The concentration of BDNF in body fluid is highly related to several neurodevelopmental and psychiatric diseases, including Alzheimer's diseases, depression, schizophrenia, and bipolar disorder. In the present review, the mechanisms by which BDNF is released from secretory vesicles are reviewed, with a particular focus on the recently described glycan-mediated release. In addition, the impact of glycan-mediated BDNF release on psychiatric disorders is also discussed.

Vitamins and Hormones, Volume 104
ISSN 0083-6729
http://dx.doi.org/10.1016/bs.vh.2016.11.004

1. INTRODUCTION

Neurotrophic factors are soluble molecules that bind to specific receptors to regulate cell growth, survival, differentiation, and maintenance in the central nervous system (CNS) and peripheral nervous systems in an autocrine and/or paracrine manner. In 1951, nerve growth factor (NGF) was first identified in sarcoma tissues and was shown to cause the outgrowth of fibers from sensory nerves (Levi-Montalcini & Hamburger, 1951). Several other neurotrophic factors have since been characterized and classified into three groups based on their structural and functional properties: the neurotrophin family, transforming growth factor (TGF) family (glial cell line-derived neurotrophic factor (GDNF) family), and ciliary neurotrophic factor (CNTF) family (Fig. 1).

The neurotrophin family consists of NGF, brain-derived neurotrophic factor (BDNF), neurotrophin (NT)-3, and NT-4, which are involved in various brain functions, including axon targeting, synapse maturation, and plasticity (Chao, 2003; Poo, 2001). The GDNF family is comprised of GDNF, neurturin (NRTN), artemin (ARTN), and persephin (PSPN). GDNF was identified in the supernatant of a rat glioma cell line as a growth factor promoting the survival of embryonic dopaminergic neurons of the midbrain (Lin, Doherty, Lile, Bektesh, & Collins, 1993). Members of this group have structural similarity with respect to the position of seven cysteine residues that form a total of three intramolecular disulfide bonds. Members of the GDNF family share ~50% sequence similarity, whereas that among TGF-β family is ~20%, although GDNF shares profound conformational similarity with TGF-β2 and bone morphogenetic protein (BMP)-7 (Eigenbrot & Gerber, 1997). GDNF binds to GDNF family receptor-α (GFRα) or GFRα–RET to form a signaling complex, which then forms a dimer to initiate signal transduction (Chen et al., 2001; Naveilhan et al., 1998). The CNTF family consists of leukemia inhibitory factor (LIF), interleukin (IL)-6, and CNTF, which was shown to be involved in the survival of ciliary neurons from chick embryonic extracts (Adler, Landa, Manthorpe, & Varon, 1979) and several types of neurons and neural stem cells (Müller, Chakrapani, Schwegler, Hofmann, & Kirsch, 2009). All members of this family adopt the all α structural conformation and form homodimers to bind CNTFR, LIFR, or gp130, which then dimerize to mediate signal transduction (Kass, 2011).

	Neurotrophin family	TGF-β (GDNF) family	CNTF family
Member	NGF, proNGF BDNF, proBDNF NT-3, NT-4	GDNF, NRTN, ARTN, PSPN	CNTF, LIF, IL-6
Structure (PDB number)	 (1BND)	 (1AG9)	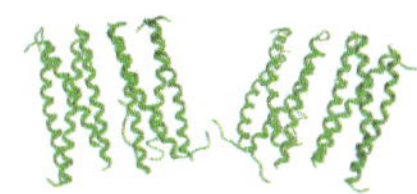 (1CNT)
Receptor	TrkA, B, C, $p75^{NTR}$ Sortilin	GFRα1–4 GFRα/RET	CNTFR, LIFR, gp130
Function	Survival Proliferation Differentiation Cell migration Plasticity, etc.	Survival Proliferation Differentiation Cell migration	Survival

Fig. 1 Classification of neurotrophic factors. Neurotrophic factors are classified into three groups depending on the structure and function. The members of neurotrophins are NGF, its proform, proNGF, BDNF, its proform, proBDNF, NT-3, and NT-4. In fish, NT-6/7 is also found. 1BND is the crystal structure of BDNF/NT-4 heterodimer. All these molecules are highly conserved, and all these structures are all β conformations. Neurotrophins function via binding to TrkB or $p75^{NTR}$ receptor in a variety of brain functions. Sortilin is a newly identified proneurotrophin receptor mainly expressed in TGN and is shown to be involved in the processing for mature neurotrophins. The TGF-β family, also called as GDNF family in brain, consists of GDNF, neurturin (NRTN), artemin (ARTN), and persephin (PSPN). 1AGQ is the crystal structure of GDNF dimer. The molecules bind to GFRα-1 or GFRα-1–RET to function and function as a dimer. CNTF family consists of CNTF, LIF, and IL-6. 1CNT is the crystal structure of CNTF. All these structures are all α conformations. Homodimers of these molecules bind to CNTFR, LIFR, or gp130, and receptor dimer formation is important for signal transduction.

The functions of these neurotrophic molecules have been well studied and described in the literatures. In the present review, we focus on the mechanism of BDNF secretion, particularly on a newly identified secretion mechanism that mediated by a unique glycan, polysialic acid that modifies neural cell adhesion molecule, NCAM.

2. BIOCHEMICAL FEATURES OF BDNF

BDNF is enriched in adult brains and was first isolated from pig brains. It was biochemically shown to have survival-promoting activity on dorsal root ganglion cells (Barde, Edgar, & Thoenen, 1982). BDNF is biosynthesized as a precursor protein, pre-proBDNF, which consists of 247 amino acid residues and is proteolytically cleaved to proBDNF (229 aa, 32 kDa). ProBDNF is further processed in secretory vesicles by proteases such as Furin in many tissue types and seven-membered proprotein convertase (PC7) in the hippocampus or amygdala in the brain (Khatib, Siegfried, Chrétien, Metrakos, & Seidah, 2002; Steiner, Smeekens, Ohagi, & Chan, 1992; Wetsel et al., 2013). These enzymes have similar specificity toward (R/K)-2Xn-R↓, where $n=0$–2, which when cleaved, generates mature BDNF (mBDNF) (119 aa, 13.5 kDa) and prodomain (proD) (110 aa, 18 kDa) (Fig. 2A). The amino acid sequence of mBDNF is conserved among mammalian animals (Tettamanti et al., 2010). Like other secreted proteins, pre-proBDNF has a highly conserved N-terminal signal peptide that is cleaved in the endoplasmic reticulum (ER). The resulting proBDNF forms noncovalent dimers that are transferred to the Golgi apparatus via intermediate nonclathrin-coated transport vesicles. In the Golgi, proBDNF is further cleaved by enzymes localized in the *trans*-Golgi network (TGN), as described earlier, generating mBDNF, which typically exists as a stable dimer via strong noncovalent interactions. While proBDNF is processed inside the cells, it can also be secreted in an activity-dependent manner (Hartmann, Drummond, Handberg, Ewell, & Pozzo-Miller, 2012). ProBDNF functions as a ligand of pan neurotrophin receptor of the tumor necrosis factor receptor family, $p75^{NTR}$, or it is processed extracellularly by the tissue plasminogen activator/plasmin system (Pang et al., 2004) or metalloproteinases (Cao et al., 2014; Yamamori et al., 2013) to yield mBDNF.

BDNF consists of three pairs of antiparallel β-sheets connected to four loops (Géral, Angelova, & Lesieur, 2013; Pattarawarapan & Burgess, 2003). This conformation is highly conserved among members of the neurotrophin family, with the N- and C-terminals and loops 1–4 playing an important role in receptor activation. BDNF binds to two different types of receptors with different affinities and resulting functions (Carter et al., 1996; Soppet et al., 1991). Tropomyosin-related kinase (Trk) receptor type B (TrkB) belongs to the tyrosine kinase receptor family and binds BDNF with high affinity ($K_D = \sim 10^{-11}$ M) and is expressed throughout the brain in the cerebral cortex, hippocampus, thalamus, choroid plexus, and granular layer of the cerebellum, brain stem, retina, and spinal cord. The second receptor is the pan neurotrophin receptor of $p75^{NTR}$, which is expressed in cholinergic cortical neurons, and binds BDNF with low affinity

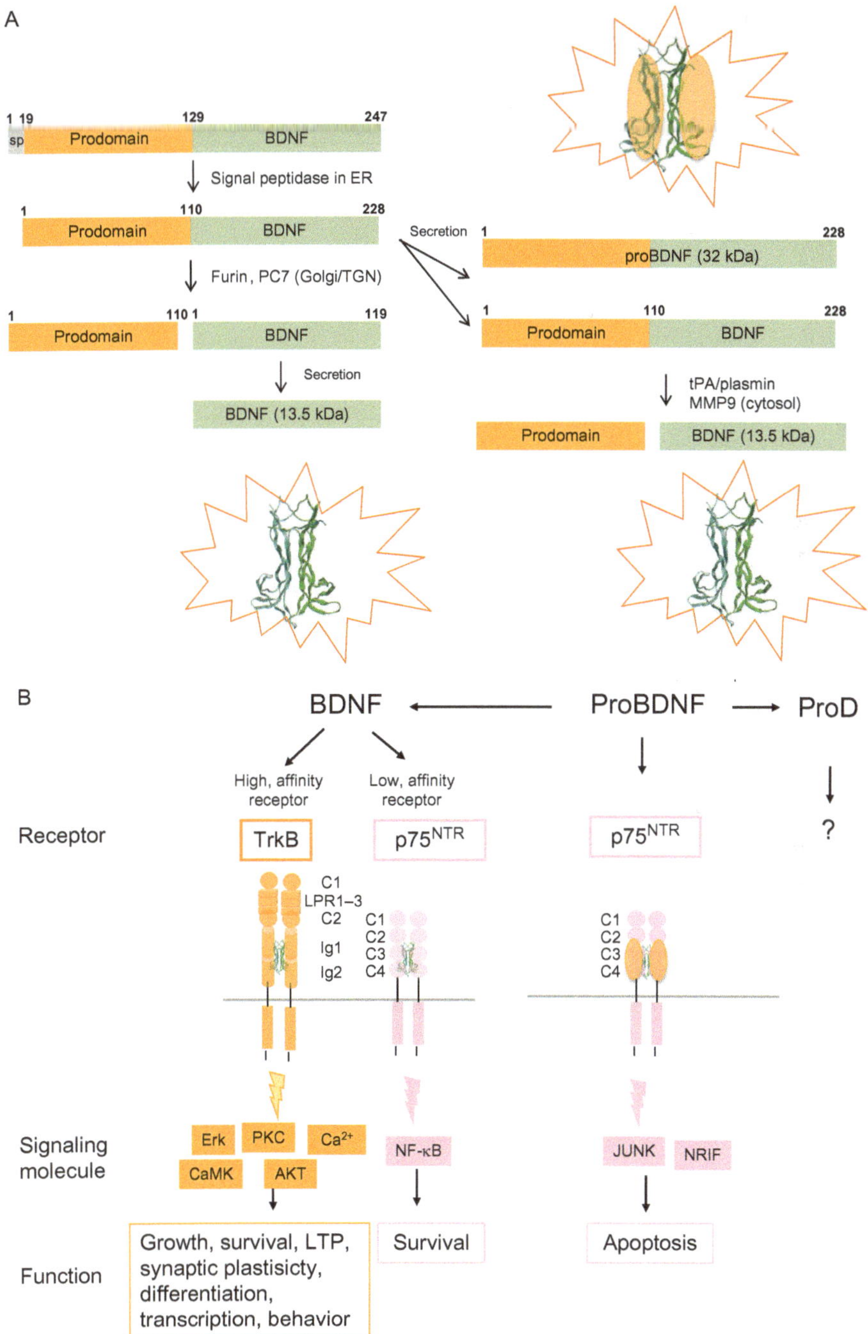

Fig. 2 See legend on next page.

($K_D = \sim 10^{-9}$ M) (Fig. 2B; Banfield et al., 2001; Meakin & Shooter, 1992). The BDNF–TrkB signaling pathway has important implications for adult brain functions, including neural cell growth, survival, differentiation, long-term potentiation (LTP), and synaptic plasticity (Andero, Choi, & Ressler, 2014).

Trk receptors have five homologous domains, three tandem leucine-rich motifs flanked by two cysteine cluster regions (domains 1–3; LPR1–3), and two immunoglobulin superfamily (Ig) extracellular regions (Ig1 and Ig2). The specificity of Trk receptors toward target ligands is dependent on the specific binding site within Ig2 to the loop structure of neurotrophins (Allen, Watson, Shoemark, Barua, & Patel, 2013). Three types of Trk receptors expressed in the brain have specificity for neurotrophins; TrkA is specific for NGF, TrkB is specific for BDNF and NT-4, and TrkC is specific for NT-3. In contrast, $p75^{NTR}$ lacks such specificity and is able to bind all neurotrophins with low affinity. However, proBDNF preferentially binds to $p75^{NTR}$ and activates different downstream pathways than those triggered by the BDNF–TrkB signaling cascade (Fig. 2B; Teng et al., 2005).

BDNF has diverse and critical functions, as demonstrated by the lethality of BDNF deficiency in postnatal mice and the profound defects observed in BDNF heterozygote mice (Chao, 2003). Based on accumulated evidence, secreted BDNF appears to be involved in various neural and brain functions, such as cell growth, survival, maintenance, differentiation and migration, synaptic plasticity, cognitive and mood-related behavior, and major psychiatric disorders, including schizophrenia, bipolar disorder, depression,

Fig. 2 Biochemical features of BDNF. (A) Processing of active forms of BDNF. BDNF is biosynthesized as a pre-proBDNF (247 aa), and pre-proBDNF is cleaved to proBDNF (229 aa, 32 kDa) by a signal peptidase. ProBDNF in secretory vesicles is further cleaved by proteases such as Furin and proprotein convertase localized in Golgi apparatus in brain cells, and mature BDNF (mBDNF) (119 aa, 13.5 kDa) and prodomain (proD) (110 aa, 18 kDa) are synthesized. mBDNF will be active after releasing from the vesicles. Some portion of proBDNF in vesicles present as dimer is also released in an activity-dependent manner. Secreted proBDNF functions by binding to its receptor or proBDNF is further processed by tPA/Plasmin or MMP9 to form mBDNF as an active form. (B) Active BDNF dimer can bind to two types of the BDNF receptor, a high-affinity receptor, TrkB and a low-affinity receptor, $p75^{NTR}$. Binding of BDNF to TrkB initiates dimerization of two receptors, and signal transduction starts via several signaling molecule, Erk, PKC, Ca^{2+}, CAMK2, and Akt for TrkB and NF-κB for $p75^{NTR}$. These signaling cascades lead to the variety of neural cell functions or brain functions such as survival, growth, synaptic plasticity, and behaviors. ProBDNF also binds to $p75^{NTR}$ and leads to the apoptosis via JUNK or NRIF.

addiction, Rett syndrome, eating disorders, and Alzheimer's diseases (Allen et al., 2013; Andero et al., 2014; Autry & Monteggia, 2012; Nagahara & Tuszynski, 2011).

3. REGIONS AND CELLULAR DISTRIBUTION OF BDNF

The detection of BDNF mRNA is frequently used as an indicator of the cellular localization of BDNF protein (Conner, Lauterborn, Yan, Gall, & Varon, 1997). BDNF protein is released from the biosynthesizing cells as a primary site. The mRNA for BDNF has been detected all over the tissues in throughout the CNS, including the olfactory bulb, cerebral cortex, hippocampus, amygdala, thalamus, hypothalamus, and spinal cord, and other nonneuronal tissues, such as the salivary gland, skeletal muscle, spleen, vascular endothelial cells, aorta, kidney, ovary, heart, lung, retina, and immune cells (Conner et al., 1997; Ernfors, Wetmore, Olson, & Persson, 1990; Hofer, Pagliusi, Hohn, Leibrock, & Barde, 1990; Maisonpierre et al., 1991; Naveilhan et al., 1998; Radka, Holst, Fritsche, & Altar, 1996; Schmidt-Kastner, Wetmore, & Olson, 1996). The BDNF proteins have been detected throughout the brain using immunohistochemical approaches. Within the CNS, the hippocampus and cerebral cortex are particularly enriched in BDNF. The hippocampus is a critical region of the brain for learning and memory, for which BDNF plays a critical role. In the hippocampus, CA1 pyramidal cells have been shown to contain BDNF mRNA (Conner et al., 1997; Schmidt-Kastner et al., 1996); however, there are conflicting reports on the detection of BDNF in the hippocampus using specific anti-BDNF antibodies (Lessmann, Gottmann, & Malcangio, 2003). These differences may have arisen from the conditions used for the immunohistochemical analysis, particularly during the fixation and permeabilization steps (Yan et al., 1997), and the state of the cells, as BDNF expression is dependent on several factors, including cellular electrical activity, environmental conditions, stress, and circadian rhythm (Chourbaji, Brandwein, & Gass, 2011). In addition, because BDNF binds to extracellular glycans such as polysialic acid (polySia or PSA) (Kanato, Kitajima, & Sato, 2008; Sato & Kitajima, 2013a, 2013b) and other glycosaminoglycan chains (Kanato, Ono, Kitajima, & Sato, 2009), which are frequently remodeled in response to the cellular state (Sumida et al., 2015), the staining patterns with anti-BDNF antibodies may be potentially altered based on glycan binding.

It has been reported that some amounts of BDNF are able to cross the blood–brain barrier (Pan, Banks, Fasold, Bluth, & Kastin, 1998), and serum BDNF is considered to be a marker for depression and stress (Autry & Monteggia, 2012; Karege et al., 2002; Shimizu et al., 2003). Although the serum levels of BDNF reflect those of brain BDNF, most BDNF in brain tissues do not cross the blood–brain or other brain barriers. Therefore, serum BDNF is also considered to originate from tissues other than brain. Notably, even though ~90% of serum BDNF is stored in platelets, platelets contain very low levels of BDNF mRNA, whereas peripheral tissues, such as skeletal muscle, vascular endothelial cells, aorta, and salivary glands, express high levels of BDNF mRNA, indicating that serum BDNF may originate from these tissues (Nakahashi et al., 2000; Radka et al., 1996). In 2012, MEG-01, a megakaryocytic cell line, was demonstrated to produce BDNF protein in the presence of thrombopoietin, suggesting that BDNF is a physiological autocrine regulator of megakaryocyte progenitors (Tamura et al., 2012). Recently, using the primary cultures of megakaryocytes that are the progenitor of platelets, gene expression of BDNF was detected. The expression features of mRNA transcripts in the cells were the same as that of neurons. In addition, in rat and human megakaryocytes, BDNF is shown to be stored in α-granules, although not in mouse. Therefore, the cellular origin of platelet BDNF is critically identified as megakaryocytes (Chacón-Fernández et al., 2016).

The distribution of mRNA within neuronal cells has been well characterized. BDNF mRNA is localized in somatic structure and dendritic compartments but is absent in axons (Tongiorgi, 2008; Tongiorgi & Baj, 2008; Tongiorgi, Righi, & Cattaneo, 1997). BDNF mRNA of dendritic compartment was detected in hippocampal CA1, CA2, and CA3 pyramidal cells, granule cells, cortical, and hypothalamic neurons (Tongiorgi, 2008). The level of BDNF mRNA increased in response to electrical activity (Tongiorgi, 2008) and dendritic localization of BDNF mRNA is regulated by the polyadenylation of mRNA transcripts in cortical and hippocampal neurons (An et al., 2008). In somata and dendrites in hippocampal, cortical, and hypothalamic neurons, BDNF protein is observed detectable at the same place as in regions containing BDNF mRNA (Aoki et al., 2000; Jakawich et al., 2010; Matsuda et al., 2009; Swanwick, Harrison, & Kapur, 2004), although several inconsistent observations of the distribution of BDNF protein in regions distinct from those containing BDNF mRNA have been reported. BDNF protein has also been detected in the axons of cultured and native neurons (Aoki et al., 2000; Jakawich et al., 2010; Matsuda et al., 2009;

Swanwick et al., 2004). The presence of BDNF in axon terminals of the lateral geniculate nucleus, dorsal root ganglia, amygdala, granule cells of dentate gyrus and CA3 pyramidal cells, and postsynaptic density and dendritic shafts of cerebral cortex has also been confirmed by electron microscopy (Aoki et al., 2000; Avwenagha, Bird, Lieberman, Yan, & Campbell, 2006; Dieni et al., 2012; Salio, Averill, Priestley, & Merighi, 2007). BDNF is often found within the ER and Golgi apparatus-like organelles, which are localized at extrasynaptic sites, as well as presynaptic and postsynaptic sites, in axons and dendrites (Brigadski, Hartmann, & Lessmann, 2005; Dean et al., 2012; Kolarow, Brigadski, & Lessmann, 2007; Wu et al., 2004).

4. EXTRACELLULAR SECRETION OF BDNF

Two types of secretion mechanisms for BDNF and proBDNF from secretory vesicles have been characterized: constitutive release mechanism (activity-independent manner) and regulatory release (activity-dependent manner) mechanism, which is activated in response to electrical activity (Edelmann, Lessmann, & Brigadski, 2014; Lessmann & Brigadski, 2009). Unlike other neurotrophins, such as NGF and NT-3, which are mainly secreted via the constitutive pathway, the majority of BDNF appears to be secreted by the regulatory release pathway. The mechanism of BDNF secretion has been studied using highly specific antibodies against BDNF by ELISA, and using eGFP-tagged BDNF and pH-sensitive eGFP-tagged BDNF combined with neuroimaging systems, which have allowed determination of the secretion sites with high sensitivity and temporal resolution.

In the rat hippocampus, BDNF secretion from somatic and dendritic sites following kainate stimulation was first detected using specific antibodies (Wetmore, Olson, & Bean, 1994) and was later shown to occur during LTP by using a soluble form of TrkB-Fc that scavenges BDNF function (Figurov, Pozzo-Miller, Olafsson, Wang, & Lu, 1996). In addition to the hippocampus, the cortex, amygdala, cerebellum, striatum, and spinal cord were also shown to release BDNF (Akaneya, Tsumoto, Kinoshita, & Hatanaka, 1997; Canudas, Pezzi, Canals, Pallàs, & Alberch, 2005; Kang, Welcher, Shelton, & Schuman, 1997; Kerr et al., 1999; Meis, Endres, & Lessmann, 2012); however, the time course of BDNF release remained unclear due to the low temporal resolution of the analysis. ELISA measurements have been used to quantify and improve the time resolution of BDNF release to ~30 s (Canossa et al., 1997), although the specific cell type or subcellular secretion sites cannot be accurately determined using this approach. However, the use of eGFP- or pHluorin-tagged BDNF allows for the cellular secretion sites of BDNF to be monitored by time-lapse imaging (Matsuda et al., 2009; Stenovec et al., 2015). Using this approach, BDNF secretion from extrasynaptic and postsynaptic regions of dendrites on hippocampal cultured neurons was observed (Hartmann, Heumann, & Lessmann, 2001). In addition, axonal secretion of BDNF by hippocampal and cortical neurons was also shown (Adachi, Kohara, & Tsumoto, 2005; Kojima et al., 2001). In addition, activity-dependent BDNF secretion has been reported from soma, dendrites, and axons of primary hippocampal cells treated with potassium (Brigadski et al., 2005), PCP (NMDA blocker) (Adachi et al., 2013), and HSF (Hartmann et al., 2012) as triggers. In astrocytes, exposure to glutamate (Jean, Lercher, & Dreyfus, 2008) and prostaglandin E2 (Hutchinson, Chou, Israel, Xu, & Regan, 2009) leads to changes in the cellular BDNF concentration, whereas the treatment of microglial cells with ATP (Coull et al., 2005) or morphine (Ferrini et al., 2013) stimulates BDNF release. Interestingly, hypoxia also triggers BDNF release from microvesicular endothelial cells (Wang, Ward, Boswell, & Katz, 2006).

Not only determining the molecules that trigger BDNF secretion, but also the time course of stimulated BDNF release is important for understanding the neurological functions of BDNF, such as LTP in the hippocampus during memory consolidation. The analyses of hippocampal, cortical, and dorsal horn slices clearly showed that the activity-dependent release of BDNF occurred 5–20 min after stimulation (Aicardi et al., 2004). BDNF release is regulated by intravesicular pH, with the neutralization of the normally acidic conditions in vesicles increasing the rate of release (Brigadski et al., 2005; Lessmann & Brigadski, 2009). In the case of microglia, peaks in the amount of BDNF release were observed 5 and 60 min after stimulation (Trang, Beggs, Wan, & Salter, 2009). The second release of BDNF is associated with increased BDNF protein expression. This second release was also regulated by the concentration of Ca^{2+} outside the cell and is considered to be important for the consolidation of BDNF-induced phenomena, such as fear learning in amygdala (Ou, Yeh, & Gean, 2010).

4.1 BDNF Release via Exocytosis

Among vesicle membrane stacks of TGN, two distinct types of secretory vesicles contain BDNF: constitutively secreted and stimulation- and activity-dependent secretory vesicles. Constitutively secreted vesicle is relatively small (50–100 nm) and fuses with the plasma membrane without any specific triggering factors in an Ca^{2+}-independent manner (Brigadski et al., 2005). A portion of cellular BDNF is constitutively released via these vesicles (Harter & Reinhard, 2000). In contrast, activity-dependent secretory vesicles, which are known as large dense-core vesicles, are relatively large (100–300 nm), and are involved in the stimulated release of BDNF. The majority of BDNF is secreted by this regulated system (Brigadski et al., 2005). The fusion of these vesicles with the plasma membrane occurs at lipid raft regions and is highly restricted in a Ca^{2+}-dependent manner (Lang et al., 2001). After pre-proBDNF is translated in the ER, it is transported into the Golgi apparatus and TGN where BDNF is packaged into dense-core vesicles (DCVs). The prodomain of proBDNF is involved in sorting of BDNF into secretory vesicles, particularly those of the regulated secretory pathway. A single-nucleotide polymorphism (SNP) at of nucleotide 196 in the proregion of the BDNF leads to one single amino acid change (Val66Met), and the resulting Val66Met-proBDNF protein is demonstrated to be not targeted effectively into the secretory vesicles (Chen et al., 2004; Egan et al., 2003). Domain-swapping experiments have also revealed that the prodomain is also important for the correct targeting of proNTs toward the constitutive or regulated secretory pathways (Brigadski et al., 2005). The prodomain of BDNF binds to a sorting receptor, carboxypeptidase E (CPE), which is localized at lipid rafts in the TGN (Lou et al., 2005). In addition, Golgi-resident sortilin is also a receptor for proBDNF and plays an important role in the sorting of secretory vesicles toward the regulated pathway via binding to prodomain (Chen et al., 2005).

Activity-dependent secretion is triggered by increased intracellular Ca^{2+} concentrations, whereas constitutive release does not require a trigger, although this process has not been well studied in the case of BDNF (Brigadski et al., 2005). BDNF is typically secreted from the cell soma, dendrites, and axons in an activity-dependent manner that requires the influx of extracellular Ca^{2+}. Ca^{2+} is considered to be involved in the following Ca^{2+} required fusion/docking molecules. BDNF-containing vesicles are thought to be docked on the inside the cell membrane via the SNARE system. The SNARE proteins (syntaxin-SNAP25-VAMP/synaptobrevin) are an essential molecule for the secretion of BDNF, as demonstrated by the fact that TTX, which cleaves synaptobrevin 2, completely inhibits BDNF secretion (Shimojo et al., 2015). In 2009, it was demonstrated that synaptotagmin 4 (Syt4), which is involved in the fusion of secretory vesicles to the cell membrane, inhibits BDNF secretion from axons and dendrites via

Ca^{2+}-independent binding to SNARE (Dean et al., 2009). The activity-dependent increase of cellular Ca^{2+} levels promotes the binding of Syt4 to SNARE. As Syt4-KO mice exhibit increased release of BDNF and Syt4-TG mice show decreased BDNF secretion, Syt4 is considered to be involved in the inhibition of BDNF release. CAPS2 protein is also involved in the activity-dependent secretion of BDNF (Shinoda et al., 2011). CAPS2 is located in chromogranin-containing DCVs and consists of a C2-like domain, membrane-bound PH-like domain, dense-core vesicle-binding domain, and Munc13-1 homologous domain, and has high homology to CAPS1, which is involved in the exocytosis of DCVs. Anti-CAPS2 beads were shown to capture BDNF and NT-3, and CAPS2-KO mice exhibit decreased BDNF secretion in an activity-dependent manner. Interestingly, CAPS2-KO mice also display impaired social behavior and SNPs within CAP2 are correlated with autism spectrum disorder (Sadakata et al., 2012).

4.2 BDNF Release via Polysialic Acid

BDNF binds directly to an unique acidic glycan, polysialic acid (polySia, PSA), which is a linear polymer of sialic acid with a degree of polymerization (DP) of 8–400 (Sato & Kitajima, 2013a; Fig. 3A). PolySia is linked to two N-linked glycans on the Ig5 domain of NCAM in a spatiotemporal manner (Fig. 3A). PolySia-NCAM is mainly expressed in embryonic brains. In adult brain, almost polySia chains on NCAM disappear; however, polySia expression is continuing detectable in restricted regions, including such as the hippocampus, olfactory bulb, amygdala, prefrontal cortex, and hypothalamus, where neurogenesis and neural modeling continue throughout life (Rutishauser, 2008). Due to the unique spatiotemporal expression of polySia, this unique polymer has been well studied with respect to function. PolySia is considered to function as an antiadhesive molecule due to its large exclusion volume and hydration properties (Yang, Major, & Rutishauser, 1994), which are predicted to act as a repulsive field to exclude most molecules nearby from polySia-modified NCAM (Rutishauser, 2008; Fig. 3B). PolySia negatively regulates interactions among not only CAMs but also among cells. PolySia-NCAM is also a well-known marker of adult neurogenesis, although the role of this molecule in neurogenesis is not well understood.

In 2000, BDNF was shown to rescue the impaired LTP in the CA1 region of the hippocampus of NCAM-KO mice (Muller et al., 2000), although the mechanism remains unknown. In 2008, BDNF was confirmed to bind directly to polySia chains of greater than 12 DP by gel-filtration and

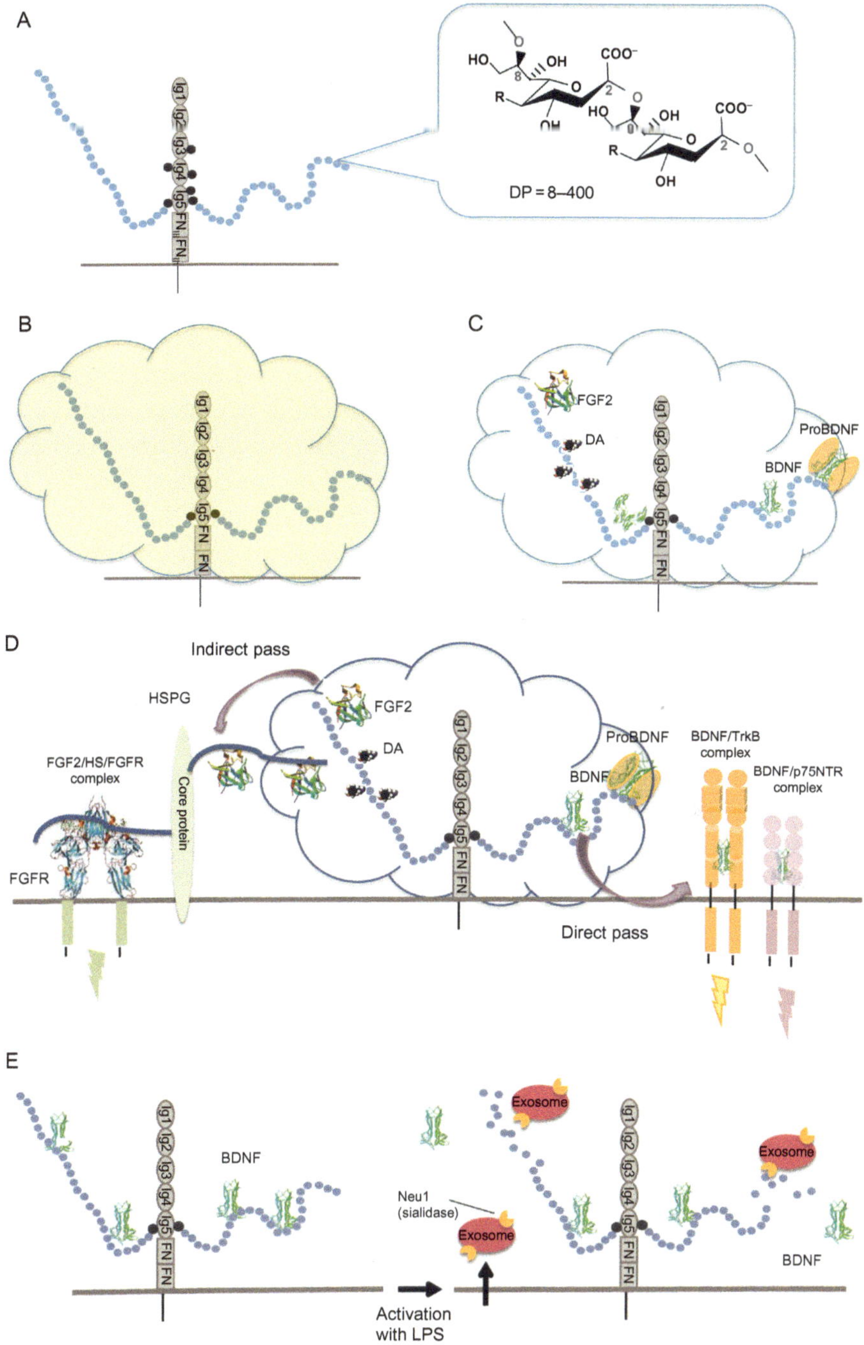

Fig. 3 See legend on next page.

native-PAGE analyses (Kanato et al., 2008). Several NTs other than BDNF and a number of neurologically active molecules, including FGF2 (Ono, Hane, Kitajima, & Sato, 2012) and dopamine (Isomura, Kitajima, & Sato, 2011), also bind to polySia directly (Sato & Kitajima, 2013a). Therefore, polySia functions as a multimolecule carrier with an attractive field for these neurologically active molecules (Fig. 3C), which are released via different molecular mechanisms (Fig. 3D).

The K_D values of polySia alone and polySia on NCAM for neurologically active molecules were comprehensively measured based on surface plasmon resonance (SPR) and were found to range from $\sim 10^{-9}$ to 10^{-10} M (Hane et al., 2015; Hane, Sumida, Kitajima, & Sato, 2012; Ono et al., 2012). Using gel-filtration analysis, BDNF in complex with polySia was shown to readily migrate toward the BDNF receptors TrkB and $p75^{NTR}$ (Kanato et al., 2008; Fig. 3D). Based on the reports that the K_D of BDNF toward TrkB and $p75^{NTR}$ was $\sim 10^{-11}$ and 10^{-9} M, respectively, the migration of the BDNF in the polySia chains toward receptors can be considered to be due to the affinity difference. Interestingly, FGF2 in complex with polySia cannot migrate directly toward FGFR unlike the migration of BDNF in complex with polySia toward BDNF-Rs, although the K_D value of FGF2 toward FGFR is sufficiently low (60 nM) (Ono et al., 2012). FGF2 requires heparan sulfate (HS) to migrate away from polySia chains,

Fig. 3 Releasing mechanism of BDNF via polysialic acid. (A) Structures of polySia and polySia-NCAM. α2,8-linked polySia structure modifies NCAM. NCAM has five immunoglobulin domains (Ig1–Ig5) and two fibronectin type-III (FN_{III}) domains in its extracellular domain. On the Ig5 domains, two of the three N-glycosylation sites are polysialylated. PolySia is indicated by *blue circle* (sialic acid), and N-glycosylation sites are indicated by *black circles*. (B) Repulsive field of polySia. PolySia-NCAM gives a repulsive field on the cell surface to regulate cell–cell interaction negatively by a large exclusion volume of polySia (shown in *yellow*). (C) Attractive field of polySia. PolySia functions as a reservoir of biologically active molecules. As an attractive field, polySia on NCAM directly binds to the biologically active molecules that are deeply involved in neural functions, such as neurotrophins, neurotransmitters, and growth factors. PolySia regulates their concentrations outside the cells, protection from protease and regulation of the processing from proBDNF to BDNF. (D) Releasing mechanism of polySia-1. BDNF in polySia chain can migrate toward TrkB or $p75^{NTR}$ via direct migration. On the other hands, FGF2 in polySia chain can migrate toward to heparan sulfate (HS) and HS–FGF2 complex binds to FGFR. Therefore, polySia release FGF2 toward FGFR indirectly. (E) Releasing mechanism of polysialic acid-2. BDNF in polySia (or other sialoglycan) can be released by the sialidase localized on the exosomes that are secreted from microglia cells after inflammatory stimulation.

and the resulting HS–FGF2 complexes bind to FGFR to form ternary complexes that enhance the signals. Therefore, FGF2 requires a coreceptor such as HS proteoglycan (HSPG) and polySia can release FGF2 to HSPG and pass the molecule to the FGFR indirectly (Fig. 3D). As polySia expression completely inhibits FGF2- and HS-driven cell growth using NIH-3T3 cells, polySia can be considered to be the primary reservoir of FGF2. Taken together, these findings indicate that polySia binds to multiple neurologically active molecules; however, BDNF has binding specificity to shorter chain (DP greater than 12), whereas FGF2 has binding specificity to longer chain polySia (DP greater than 17), although these molecules are released from polySia via different mechanisms. Another interesting biological feature of the polySia is the regulation of processing mechanism of proBDNF to BDNF and proD (Hane et al., 2015). It is clearly shown that polySia can bind to BDNF, but not proD at all, and that proBDNF has the ability to bind to oligo/polySia at restricted concentrations, 40–60 nM. In addition, plasmin, a key processing enzyme of proBDNF outside the cells, was shown to be significantly inhibited in the presence of polySia, but not HS due to the binding properties of proBDNF to polySia, indicating that polySia is also involved in the processing mechanism of proBDNF to BDNF in the extracellular space that is deeply involved in release of the surface BDNF.

PolySia was proved to present on the microglia cell surface recently, and BDNF in polySia chains was also shown to present by the ELISA measurement after treatment with endo-*N*-acylneuraminidase that cleaves polySia endogenously (Sumida et al., 2015). A similar amount of BDNF was released from polySia following exosialidase treatment, indicating that almost all BDNF detected by ELISA are present on polySia structure. Microglial cells secrete exosomes containing the lysosomal-derived enzyme neuramidase (Neu1) after stimulation with lipopolysaccharide (LPS) (Sumida et al., 2015). Notably, secreted exosomes from microglial cells display sialidase activity toward polySia on the surface of microglial cells and other polySia-expressing cells. The clearance of polySia from the cell surface is temporally related to BDNF secretion and is maximal at 5–10 min after LPS stimulation. PolySia clearance and BDNF secretion are inhibited by the treatment of cells with the sialidase inhibitor Neu5Ac2en. These findings were the first to demonstrate that cell surface-resident BDNF in polySia is released by exosomes secreted from microglial cells after activation (Sumida et al., 2015; Fig. 3E).

Sialic acids are a group of 2-keto-3-deoxy nononic acids that consists of *N*-acetylneuraminic acid (Neu5Ac), *N*-glycolneuraminic acid (Neu5Gc), and deaminoneuraminic acid (KDN). In nature, Neu5Ac and Neu5Gc are the major sialic acids and are found in almost all animals, with the exception of humans (Varki, Strobert, Dick, Benirschke, & Varki, 2011). Humans are unable to synthesize Neu5Gc because the activity of CMP-Neu5Ac hydroxylase (CMAH) that catalyzes the hydroxylation of CMP-Neu5Ac to CMP-Neu5Gc is lost due to the insertion of an Alu sequence in the CMAH gene (Chou et al., 2002). Since CMP-Sia is the donor substrate of sialyltransferases, almost all deuterostome animals can use both CMP-Neu5Ac and CMP-Neu5Gc to present both Neu5Ac and Neu5Gc in the tissues; however, especially in brains, only Neu5Ac is observed, indicating that brains exclude Neu5Gc. However, Neu5Gc can be obtained from foods, such as meat, and small amounts of Neu5Gc are incorporated into human tissues and probably into the brain tissues (Tangvoranuntakul et al., 2003). To examine if Neu5Gc incorporation in the brain leads to the impact on the BDNF release via polySia cleavage by Neu1 sialidase, microglial cells were cultured in Neu5Gc-containing medium and Neu5Gc was confirmed to be incorporated into polySia chains. At the same time, using chemically synthesized polySia-containing Neu5Gc, Neu1 was proved to cleave polySia-containing Neu5Gc with lower speed compared with polySia-containing Neu5Ac. As the same results of Neu1 specificity toward sialic acid species observed in vitro, the release of BDNF from Neu5Gc-containing cells after LPS treatment was delayed by 10–30 min compared to that of Neu5Ac-containing cells. Glycosidase-mediated BDNF secretion appears to be highly temporally regulated and may be affect brain functions (Sumida et al., 2015).

PolySia is biosynthesized by two polysialyltransferases, ST8SIA2 and ST8SIA4. Interestingly, ST8SIA2- and ST8SIA4-deficient mice have different phenotypes. LTP of CA1 in the hippocampus is impaired in ST8SIA4-KO mice, but proceeds normally in ST8SIA2-KO mice (Schnaar, Gerardy-Schahn, & Hildebrandt, 2014). However, adult neurogenesis in DG is impaired in ST8SIA2-KO mice, but not in ST8SIA4-KO mice (Angata et al., 2007). In humans, SNPs or genetic variations in the ST8SIA2 gene are related to schizophrenia and bipolar disorder (Sato, Hane, & Kitajima, 2016); however, the effects of these SNPs on polySia expression and function have not been well studied. In 2011, a single amino acid substitution in ST8SIA2 of a schizophrenia patient was shown to result in a shorter polySia chain compared with that of normal ST8SIA2. In addition, SPR analyses showed that the binding properties of polySia biosynthesized by the disease type of ST8SIA2 toward BDNF, proBDNF, and FGF2 were impaired (Hane et al., 2015, 2012; Isomura et al., 2011). Taken together, these data indicate that the binding and release of BDNF and other neurologically active molecules via cell surface glycans are involved in the various neural functions and that the impairments of the glycans genetically and/or environmentally influence the normal brain functions and may lead to the psychiatric disorders. It is important to consider that expression of the glycan on cell surface is highly and dynamically regulated.

5. CONCLUSION

BDNF is involved in a variety of brain functions that modify the behavior of animals. Once released from cells, BDNF must bind to its target receptors to activate the downstream signaling pathways. Therefore, to better understand the mechanism by which BDNF and other neurotrophic factors are released from cells, further studies of cell type, timing and affected pathways are critical. Recently, glycan-mediated storage of BDNF on the cell surface and glycosidase-mediated BDNF release has both been demonstrated. The composition and distribution of glycans on the cell surface change depending on the physiological conditions. Such changes to the composition of the cell surface atmosphere might be attributable to the contradicting findings related to detection of BDNF that have been reported in the literature. The modification of glycans on the neural cell surface appears to be related to the development of several psychiatric disease and neurodegenerative diseases. By studying cell surface glycans, we can obtain further insights into the functions of BDNF, NTs, and other soluble molecules that are intimately involved in neural function.

ACKNOWLEDGMENTS

This research was supported in part by Grants-in-Aid for Scientific Research (C) (15K06995) (to C.S.), Scientific Research on Innovative Areas (No. 23110002, Deciphering sugar chain-based signals regulating integrative neural functions) (26110708) (to C.S.) from the Ministry of Education, Science, Sports and Culture, and DAIKO foundation (to C.S.).

REFERENCES

Adachi, N., Kohara, K., & Tsumoto, T. (2005). Difference in trafficking of brain-derived neurotrophic factor between axons and dendrites of cortical neurons, revealed by live-cell imaging. *BMC Neuroscience*, *6*, 42.

Adachi, N., Numakawa, T., Kumamaru, E., Itami, C., Chiba, S., Iijima, Y., et al. (2013). Phencyclidine-induced decrease of synaptic connectivity via inhibition of BDNF secretion in cultured cortical neurons. *Cerebral Cortex*, *23*, 847–858.

Adler, R., Landa, K. B., Manthorpe, M., & Varon, S. (1979). Cholinergic neuronotrophic factors: Intraocular distribution of trophic activity for ciliary neurons. *Science*, *204*, 1434–1436.

Aicardi, G., Argilli, E., Cappello, S., Santi, S., Riccio, M., Thoenen, H., et al. (2004). Induction of long-term potentiation and depression is reflected by corresponding changes in secretion of endogenous brain-derived neurotrophic factor. *Proceedings of the National Academy of Sciences of the United States of America*, *101*, 15788–15792.

Akaneya, Y., Tsumoto, T., Kinoshita, S., & Hatanaka, H. (1997). Brain-derived neurotrophic factor enhances long-term potentiation in rat visual cortex. *The Journal of Neuroscience*, *17*, 6707–6716.

Allen, S. J., Watson, J. J., Shoemark, D. K., Barua, N. U., & Patel, N. K. (2013). GDNF, NGF and BDNF as therapeutic options for neurodegeneration. *Pharmacology & Therapeutics*, *138*, 155–175.

An, J. J., Gharami, K., Liao, G. Y., Woo, N. H., Lau, A. G., Vanevski, F., et al. (2008). Distinct role of long 3′ UTR BDNF mRNA in spine morphology and synaptic plasticity in hippocampal neurons. *Cell*, *134*, 175–187.

Andero, R., Choi, D. C., & Ressler, K. J. (2014). BDNF-TrkB receptor regulation of distributed adult neural plasticity, memory formation, and psychiatric disorders. *Progress in Molecular Biology and Translational Science*, *122*, 169–192.

Angata, K., Huckaby, V., Ranscht, B., Terskikh, A., Marth, J., & Fukuda, M. (2007). Polysialic acid-directed migration and differentiation of neural precursors are essential for mouse brain development. *Molecular and Cellular Biology*, *27*, 6659–6668.

Aoki, C., Wu, K., Elste, A., Len, G., Lin, S., McAuliffe, G., et al. (2000). Localization of brain-derived neurotrophic factor and TrkB receptors to postsynaptic densities of adult rat cerebral cortex. *Journal of Neuroscience Research*, *59*, 454–463.

Autry, A. E., & Monteggia, L. M. (2012). Brain-derived neurotrophic factor and neuropsychiatric disorders. *Pharmacological Reviews*, *64*, 238–258.

Avwenagha, O., Bird, M. M., Lieberman, A. R., Yan, Q., & Campbell, G. (2006). Patterns of expression of brain-derived neurotrophic factor and tyrosine kinase B mRNAs and distribution and ultrastructural localization of their proteins in the visual pathway of the adult rat. *Neuroscience*, *140*, 913–928.

Banfield, M. J., Naylor, R. L., Robertson, A. G., Allen, S. J., Dawbarn, D., & Brady, R. L. (2001). Specificity in Trk receptor:neurotrophin interactions: The crystal structure of TrkB-d5 in complex with neurotrophin-4/5. *Structure*, *9*, 1191–1199.

Barde, Y. A., Edgar, D., & Thoenen, H. (1982). Purification of a new neurotrophic factor from mammalian brain. *The EMBO Journal*, *1*, 549–553.

Brigadski, T., Hartmann, M., & Lessmann, V. (2005). Differential vesicular targeting and time course of synaptic secretion of the mammalian neurotrophins. *The Journal of Neuroscience*, *25*, 7601–7614.

Canossa, M., Griesbeck, O., Berninger, B., Campana, G., Kolbeck, R., & Thoenen, H. (1997). Neurotrophin release by neurotrophins: Implications for activity-dependent neuronal plasticity. *Proceedings of the National Academy of Sciences of the United States of America*, *94*, 13279–13286.

Canudas, A. M., Pezzi, S., Canals, J. M., Pallàs, M., & Alberch, J. (2005). Endogenous brain-derived neurotrophic factor protects dopaminergic nigral neurons against transneuronal degeneration induced by striatal excitotoxic injury. *Brain Research. Molecular Brain Research*, *134*, 147–154.

Cao, W., Duan, J., Wang, X., Zhong, X., Hu, Z., Huang, F., et al. (2014). Early enriched environment induces an increased conversion of proBDNF to BDNF in the adult rat's hippocampus. *Behavioural Brain Research*, *265*, 76–83.

Carter, B. D., Kaltschmidt, C., Kaltschmidt, B., Offenhäuser, N., Böhm-Matthaei, R., Baeuerle, P. A., et al. (1996). Selective activation of NF-kappa B by nerve growth factor through the neurotrophin receptor p75. *Science*, *272*, 542–545.

Chacón-Fernández, P., Säuberli, K., Colzani, M., Moreau, T., Ghevaert, C., & Barde, Y. A. (2016). Brain-derived neurotrophic factor in megakaryocytes. *The Journal of Biological Chemistry*, *291*, 9872–9881.

Chao, M. V. (2003). Neurotrophins and their receptors: A convergence point for many signalling pathways. *Nature Reviews. Neuroscience*, *4*, 299–309.

Chen, Z., Chai, Y., Cao, L., Huang, A., Cui, R., Lu, C., et al. (2001). Glial cell line-derived neurotrophic factor promotes survival and induces differentiation through the phosphatidylinositol 3-kinase and mitogen-activated protein kinase pathway respectively in PC12 cells. *Neuroscience*, *104*, 593–598.

Chen, Z. Y., Ieraci, A., Teng, H., Dall, H., Meng, C. X., Herrera, D. G., et al. (2005). Sortilin controls intracellular sorting of brain-derived neurotrophic factor to the regulated secretory pathway. *The Journal of Neuroscience*, *25*, 6156–6166.

Chen, Z. Y., Patel, P. D., Sant, G., Meng, C. X., Teng, K. K., Hempstead, B. L., et al. (2004). Variant brain-derived neurotrophic factor (BDNF) (Met66) alters the intracellular trafficking and activity-dependent secretion of wild-type BDNF in neurosecretory cells and cortical neurons. *The Journal of Neuroscience*, *24*, 4401–4411.

Chou, H. H., Hayakawa, T., Diaz, S., Krings, M., Indriati, E., Leakey, M., et al. (2002). Inactivation of CMP-N-acetylneuraminic acid hydroxylase occurred prior to brain expansion during human evolution. *Proceedings of the National Academy of Sciences of the United States of America*, *99*, 11736–11741.

Chourbaji, S., Brandwein, C., & Gass, P. (2011). Altering BDNF expression by genetics and/or environment: Impact for emotional and depression-like behaviour in laboratory mice. *Neuroscience and Biobehavioral Reviews*, *35*, 599–611.

Conner, J. M., Lauterborn, J. C., Yan, Q., Gall, C. M., & Varon, S. (1997). Distribution of brain-derived neurotrophic factor (BDNF) protein and mRNA in the normal adult rat CNS: Evidence for anterograde axonal transport. *The Journal of Neuroscience*, *17*, 2295–2313.

Coull, J. A., Beggs, S., Boudreau, D., Boivin, D., Tsuda, M., Inoue, K., et al. (2005). BDNF from microglia causes the shift in neuronal anion gradient underlying neuropathic pain. *Nature*, *438*, 1017–1021.

Dean, C., Liu, H., Dunning, F. M., Chang, P. Y., Jackson, M. B., & Chapman, E. R. (2009). Synaptotagmin-IV modulates synaptic function and long-term potentiation by regulating BDNF release. *Nature Neuroscience*, *12*, 767–776.

Dean, C., Liu, H., Staudt, T., Stahlberg, M. A., Vingill, S., Bückers, J., et al. (2012). Distinct subsets of Syt-IV/BDNF vesicles are sorted to axons versus dendrites and recruited to synapses by activity. *The Journal of Neuroscience*, *32*, 5398–5413.

Dieni, S., Matsumoto, T., Dekkers, M., Rauskolb, S., Ionescu, M. S., Deogracias, R., et al. (2012). BDNF and its pro-peptide are stored in presynaptic dense core vesicles in brain neurons. *The Journal of Cell Biology*, *196*, 775–788.

Edelmann, E., Lessmann, V., & Brigadski, T. (2014). Pre- and postsynaptic twists in BDNF secretion and action in synaptic plasticity. *Neuropharmacology*, *76*(Pt. C), 610–627.

Egan, M. F., Kojima, M., Callicott, J. H., Goldberg, T. E., Kolachana, B. S., Bertolino, A., et al. (2003). The BDNF val66met polymorphism affects activity-dependent secretion of BDNF and human memory and hippocampal function. *Cell, 112*, 257–269.

Eigenbrot, C., & Gerber, N. (1997). X-ray structure of glial cell-derived neurotrophic factor at 1.9 a resolution and implications for receptor binding. *Nature Structural Biology, 4*, 435–438.

Ernfors, P., Wetmore, C., Olson, L., & Persson, H. (1990). Identification of cells in rat brain and peripheral tissues expressing mRNA for members of the nerve growth factor family. *Neuron, 5*, 511–526.

Ferrini, F., Trang, T., Mattioli, T. A., Laffray, S., Del'Guidice, T., Lorenzo, L. E., et al. (2013). Morphine hyperalgesia gated through microglia-mediated disruption of neuronal Cl^- homeostasis. *Nature Neuroscience, 16*, 183–192.

Figurov, A., Pozzo-Miller, L. D., Olafsson, P., Wang, T., & Lu, B. (1996). Regulation of synaptic responses to high-frequency stimulation and LTP by neurotrophins in the hippocampus. *Nature, 381*, 706–709.

Géral, C., Angelova, A., & Lesieur, S. (2013). From molecular to nanotechnology strategies for delivery of neurotrophins: Emphasis on brain-derived neurotrophic factor (BDNF). *Pharmaceutics, 5*, 127–167.

Hane, M., Matsuoka, S., Ono, S., Miyata, S., Kitajima, K., & Sato, C. (2015). Protective effects of polysialic acid on proteolytic cleavage of FGF2 and proBDNF/BDNF. *Glycobiology, 25*, 1112–1124.

Hane, M., Sumida, M., Kitajima, K., & Sato, C. (2012). Structural and functional impairments of polysialic acid (polySia)-neural cell adhesion molecule (NCAM) synthesized by a mutated polysialyltransferase of a schizophrenic patient. *Pure and Applied Chemistry, 84*, 1895–1906.

Harter, C., & Reinhard, C. (2000). The secretory pathway from history to the state of the art. *Sub-Cellular Biochemistry, 34*, 1–38.

Hartmann, D., Drummond, J., Handberg, E., Ewell, S., & Pozzo-Miller, L. (2012). Multiple approaches to investigate the transport and activity-dependent release of BDNF and their application in neurogenetic disorders. *Neural Plasticity, 2012*. 203734. http://dx.doi.org/10.1155/2012/203734.

Hartmann, M., Heumann, R., & Lessmann, V. (2001). Synaptic secretion of BDNF after high-frequency stimulation of glutamatergic synapses. *The EMBO Journal, 20*, 5887–5897.

Hofer, M., Pagliusi, S. R., Hohn, A., Leibrock, J., & Barde, Y. A. (1990). Regional distribution of brain-derived neurotrophic factor mRNA in the adult mouse brain. *The EMBO Journal, 9*, 2459–2464.

Hutchinson, A. J., Chou, C. L., Israel, D. D., Xu, W., & Regan, J. W. (2009). Activation of EP2 prostanoid receptors in human glial cell lines stimulates the secretion of BDNF. *Neurochemistry International, 54*, 439–446.

Isomura, R., Kitajima, K., & Sato, C. (2011). Structural and functional impairments of polysialic acid by a mutated polysialyltransferase found in schizophrenia. *Journal of Biological Chemistry, 286*, 21535–21545.

Jakawich, S. K., Nasser, H. B., Strong, M. J., McCartney, A. J., Perez, A. S., Rakesh, N., et al. (2010). Local presynaptic activity gates homeostatic changes in presynaptic function driven by dendritic BDNF synthesis. *Neuron, 68*, 1143–1158.

Jean, Y. Y., Lercher, L. D., & Dreyfus, C. F. (2008). Glutamate elicits release of BDNF from basal forebrain astrocytes in a process dependent on metabotropic receptors and the PLC pathway. *Neuron Glia Biology, 4*, 35–42.

Kanato, Y., Kitajima, K., & Sato, C. (2008). Direct binding of polysialic acid to a brain-derived neurotrophic factor depends on the degree of polymerization. *Glycobiology, 18*, 1044–1053.

Kanato, Y., Ono, S., Kitajima, K., & Sato, C. (2009). Complex formation of a brain-derived neurotrophic factor and glycosaminoglycans. *Bioscience, Biotechnology, and Biochemistry, 73*, 2735–2741.

Kang, H., Welcher, A. A., Shelton, D., & Schuman, E. M. (1997). Neurotrophins and time: Different roles for TrkB signaling in hippocampal long-term potentiation. *Neuron, 19*, 653–664.

Karege, F., Perret, G., Bondolfi, G., Schwald, M., Bertschy, G., & Aubry, J. M. (2002). Decreased serum brain-derived neurotrophic factor levels in major depressed patients. *Psychiatry Research, 109*, 143–148.

Kass, D. J. (2011). Cytokine-like factor 1 (CLF1): Life after development? *Cytokine, 55*, 325–329.

Kerr, B. J., Bradbury, E. J., Bennett, D. L., Trivedi, P. M., Dassan, P., French, J., et al. (1999). Brain-derived neurotrophic factor modulates nociceptive sensory inputs and NMDA-evoked responses in the rat spinal cord. *The Journal of Neuroscience, 19*, 5138–5148.

Khatib, A. M., Siegfried, G., Chrétien, M., Metrakos, P., & Seidah, N. G. (2002). Proprotein convertases in tumor progression and malignancy: Novel targets in cancer therapy. *The American Journal of Pathology, 160*, 1921–1935.

Kojima, M., Takei, N., Numakawa, T., Ishikawa, Y., Suzuki, S., Matsumoto, T., et al. (2001). Biological characterization and optical imaging of brain-derived neurotrophic factor-green fluorescent protein suggest an activity-dependent local release of brain-derived neurotrophic factor in neurites of cultured hippocampal neurons. *Journal of Neuroscience Research, 64*, 1–10.

Kolarow, R., Brigadski, T., & Lessmann, V. (2007). Postsynaptic secretion of BDNF and NT-3 from hippocampal neurons depends on calcium calmodulin kinase II signaling and proceeds via delayed fusion pore opening. *The Journal of Neuroscience, 27*, 10350–10364.

Lang, T., Bruns, D., Wenzel, D., Riedel, D., Holroyd, P., Thiele, C., et al. (2001). SNAREs are concentrated in cholesterol-dependent clusters that define docking and fusion sites for exocytosis. *The EMBO Journal, 20*, 2202–2213.

Lessmann, V., & Brigadski, T. (2009). Mechanisms, locations, and kinetics of synaptic BDNF secretion: An update. *Neuroscience Research, 65*, 11–22.

Lessmann, V., Gottmann, K., & Malcangio, M. (2003). Neurotrophin secretion: Current facts and future prospects. *Progress in Neurobiology, 69*, 341–374.

Levi-Montalcini, R., & Hamburger, V. (1951). Selective growth stimulating effects of mouse sarcoma on the sensory and sympathetic nervous system of the chick embryo. *The Journal of Experimental Zoology, 116*, 321–361.

Lin, L. F., Doherty, D. H., Lile, J. D., Bektesh, S., & Collins, F. (1993). GDNF: A glial cell line-derived neurotrophic factor for midbrain dopaminergic neurons. *Science, 260*, 1130–1132.

Lou, H., Kim, S. K., Zaitsev, E., Snell, C. R., Lu, B., & Loh, Y. P. (2005). Sorting and activity-dependent secretion of BDNF require interaction of a specific motif with the sorting receptor carboxypeptidase e. *Neuron, 45*, 245–255.

Maisonpierre, P. C., Le Beau, M. M., Espinosa, R., Ip, N. Y., Belluscio, L., de la Monte, S. M., et al. (1991). Human and rat brain-derived neurotrophic factor and neurotrophin-3: Gene structures, distributions, and chromosomal localizations. *Genomics, 10*, 558–568.

Matsuda, N., Lu, H., Fukata, Y., Noritake, J., Gao, H., Mukherjee, S., et al. (2009). Differential activity-dependent secretion of brain-derived neurotrophic factor from axon and dendrite. *The Journal of Neuroscience, 29*, 14185–14198.

Meakin, S. O., & Shooter, E. M. (1992). The nerve growth factor family of receptors. *Trends in Neurosciences, 15*, 323–331.

Meis, S., Endres, T., & Lessmann, V. (2012). Postsynaptic BDNF signalling regulates long-term potentiation at thalamo-amygdala afferents. *The Journal of Physiology*, *590*, 193–208.

Müller, S., Chakrapani, B. P., Schwegler, H., Hofmann, H. D., & Kirsch, M. (2009). Neurogenesis in the dentate gyrus depends on ciliary neurotrophic factor and signal transducer and activator of transcription 3 signaling. *Stem Cells*, *27*, 431–441.

Muller, D., Djebbara-Hannas, Z., Jourdain, P., Vutskits, L., Durbec, P., Rougon, G., et al. (2000). Brain-derived neurotrophic factor restores long-term potentiation in polysialic acid-neural cell adhesion molecule-deficient hippocampus. *Proceedings of the National Academy of Sciences of the United States of America*, *97*, 4315–4320.

Nagahara, A. H., & Tuszynski, M. H. (2011). Potential therapeutic uses of BDNF in neurological and psychiatric disorders. *Nature Reviews. Drug Discovery*, *10*, 209–219.

Nakahashi, T., Fujimura, H., Altar, C. A., Li, J., Kambayashi, J., Tandon, N. N., et al. (2000). Vascular endothelial cells synthesize and secrete brain-derived neurotrophic factor. *FEBS Letters*, *470*, 113–117.

Naveilhan, P., Baudet, C., Mikaels, A., Shen, L., Westphal, H., & Ernfors, P. (1998). Expression and regulation of GFRalpha3, a glial cell line-derived neurotrophic factor family receptor. *Proceedings of the National Academy of Sciences of the United States of America*, *95*, 1295–1300.

Ono, S., Hane, M., Kitajima, K., & Sato, C. (2012). Novel regulation of fibroblast growth factor 2 (FGF2)-mediated cell growth by polysialic acid. *The Journal of Biological Chemistry*, *287*, 3710–3722.

Ou, L. C., Yeh, S. H., & Gean, P. W. (2010). Late expression of brain-derived neurotrophic factor in the amygdala is required for persistence of fear memory. *Neurobiology of Learning and Memory*, *93*, 372–382.

Pan, W., Banks, W. A., Fasold, M. B., Bluth, J., & Kastin, A. J. (1998). Transport of brain-derived neurotrophic factor across the blood-brain barrier. *Neuropharmacology*, *37*, 1553–1561.

Pang, P. T., Teng, H. K., Zaitsev, E., Woo, N. T., Sakata, K., Zhen, S., et al. (2004). Cleavage of proBDNF by tPA/plasmin is essential for long-term hippocampal plasticity. *Science*, *306*, 487–491.

Pattarawarapan, M., & Burgess, K. (2003). Molecular basis of neurotrophin-receptor interactions. *Journal of Medicinal Chemistry*, *46*, 5277–5291.

Poo, M. M. (2001). Neurotrophins as synaptic modulators. *Nature Reviews. Neuroscience*, *2*, 24–32.

Radka, S. F., Holst, P. A., Fritsche, M., & Altar, C. A. (1996). Presence of brain-derived neurotrophic factor in brain and human and rat but not mouse serum detected by a sensitive and specific immunoassay. *Brain Research*, *709*, 122–301.

Rutishauser, U. (2008). Polysialic acid in the plasticity of the developing and adult vertebrate nervous system. *Nature Reviews. Neuroscience*, *9*, 26–35.

Sadakata, T., Shinoda, Y., Oka, M., Sekine, Y., Sato, Y., Saruta, C., et al. (2012). Reduced axonal localization of a Caps2 splice variant impairs axonal release of BDNF and causes autistic-like behavior in mice. *Proceedings of the National Academy of Sciences of the United States of America*, *109*, 21104–21109.

Salio, C., Averill, S., Priestley, J. V., & Merighi, A. (2007). Costorage of BDNF and neuropeptides within individual dense-core vesicles in central and peripheral neurons. *Developmental Neurobiology*, *67*, 326–338.

Sato, C., Hane, M., & Kitajima, K. (2016). Relationship between ST8SIA2, polysialic acid and its binding molecules, and psychiatric disorders. *Biochimica et Biophysica Acta*, *1860*, 1739–1752.

Sato, C., & Kitajima, K. (2013a). Disialic, oligosialic and polysialic acids: Distribution, functions and related disease. *Journal of Biochemistry*, *154*, 115–136.

Sato, C., & Kitajima, K. (2013b). Impact of structural aberrancy of polysialic acid and its synthetic enzyme ST8SIA2 in schizophrenia. *Frontiers in Cellular Neuroscience*, *7*, 61. http://dx.doi.org/10.3389/fncel.2013.00061.

Schmidt-Kastner, R., Wetmore, C., & Olson, L. (1996). Comparative study of brain-derived neurotrophic factor messenger RNA and protein at the cellular level suggests multiple roles in hippocampus, striatum and cortex. *Neuroscience*, *74*, 161–183.

Schnaar, R. L., Gerardy-Schahn, R., & Hildebrandt, H. (2014). Sialic acids in the brain: Gangliosides and polysialic acid in nervous system development, stability, disease, and regeneration. *Physiological Reviews*, *94*, 461–518.

Shimizu, E., Hashimoto, K., Okamura, N., Koike, K., Komatsu, N., Kumakiri, C., et al. (2003). Alterations of serum levels of brain-derived neurotrophic factor (BDNF) in depressed patients with or without antidepressants. *Biological Psychiatry*, *54*, 70–75.

Shimojo, M., Courchet, J., Pieraut, S., Torabi-Rander, N., Sando, R., Polleux, F., et al. (2015). SNAREs controlling vesicular release of BDNF and development of callosal axons. *Cell Reports*, *11*, 1054–1066.

Shinoda, Y., Sadakata, T., Nakao, K., Katoh-Semba, R., Kinameri, E., Furuya, A., et al. (2011). Calcium-dependent activator protein for secretion 2 (CAPS2) promotes BDNF secretion and is critical for the development of GABAergic interneuron network. *Proceedings of the National Academy of Sciences of the United States of America*, *108*, 373–378.

Soppet, D., Escandon, E., Maragos, J., Middlemas, D. S., Reid, S. W., Blair, J., et al. (1991). The neurotrophic factors brain-derived neurotrophic factor and neurotrophin-3 are ligands for the trkB tyrosine kinase receptor. *Cell*, *65*, 895–903.

Steiner, D. F., Smeekens, S. P., Ohagi, S., & Chan, S. J. (1992). The new enzymology of precursor processing endoproteases. *The Journal of Biological Chemistry*, *267*, 23435–23438.

Stenovec, M., Lasič, E., Božić, M., Bobnar, S. T., Stout, R. F., Grubišić, V., et al. (2015). Ketamine inhibits ATP-evoked exocytotic release of brain-derived neurotrophic factor from vesicles in cultured Rat astrocytes. *Molecular Neurobiology*, *53*, 6882–6896.

Sumida, M., Hane, M., Yabe, U., Shimoda, Y., Pearce, O. M. T., Kiso, M., et al. (2015). Rapid trimming of cell surface polysialic acid (PolySia) by exovesicular sialidase triggers release of preexisting surface neurotrophin. *Journal of Biological Chemistry*, *290*, 13202–13214.

Swanwick, C. C., Harrison, M. B., & Kapur, J. (2004). Synaptic and extrasynaptic localization of brain-derived neurotrophic factor and the tyrosine kinase B receptor in cultured hippocampal neurons. *The Journal of Comparative Neurology*, *478*, 405–417.

Tamura, S., Nagasawa, A., Masuda, Y., Tsunematsu, T., Hayasaka, K., Matsuno, K., et al. (2012). BDNF, produced by a TPO-stimulated megakaryocytic cell line, regulates autocrine proliferation. *Biochemical and Biophysical Research Communications*, *427*, 542–546.

Tangvoranuntakul, P., Gagneux, P., Diaz, S., Bardor, M., Varki, N., Varki, A., et al. (2003). Human uptake and incorporation of an immunogenic nonhuman dietary sialic acid. *Proceedings of the National Academy of Sciences of the United States of America*, *100*, 12045–12050.

Teng, H. K., Teng, K. K., Lee, R., Wright, S., Tevar, S., Almeida, R. D., et al. (2005). ProBDNF induces neuronal apoptosis via activation of a receptor complex of p75NTR and sortilin. *The Journal of Neuroscience*, *25*, 5455–5463.

Tettamanti, G., Cattaneo, A. G., Gornati, R., de Eguileor, M., Bernardini, G., & Binelli, G. (2010). Phylogenesis of brain-derived neurotrophic factor (BDNF) in vertebrates. *Gene*, *450*, 85–93.

Tongiorgi, E. (2008). Activity-dependent expression of brain-derived neurotrophic factor in dendrites: Facts and open questions. *Neuroscience Research*, *61*, 335–346.

Tongiorgi, E., & Baj, G. (2008). Functions and mechanisms of BDNF mRNA trafficking. *Novartis Foundation Symposium*, *289*, 136–147. discussion 147–151, 193–135.

Tongiorgi, E., Righi, M., & Cattaneo, A. (1997). Activity-dependent dendritic targeting of BDNF and TrkB mRNAs in hippocampal neurons. *The Journal of Neuroscience, 17*, 9492–9505.

Trang, T., Beggs, S., Wan, X., & Salter, M. W. (2009). P2X4-receptor-mediated synthesis and release of brain-derived neurotrophic factor in microglia is dependent on calcium and p38-mitogen-activated protein kinase activation. *The Journal of Neuroscience, 29*, 3518–3528.

Varki, N. M., Strobert, E., Dick, E. J., Benirschke, K., & Varki, A. (2011). Biomedical differences between human and nonhuman hominids: Potential roles for uniquely human aspects of sialic acid biology. *Annual Review of Pathology, 6*, 365–393.

Wang, H., Ward, N., Boswell, M., & Katz, D. M. (2006). Secretion of brain-derived neurotrophic factor from brain microvascular endothelial cells. *The European Journal of Neuroscience, 23*, 1665–1670.

Wetmore, C., Olson, L., & Bean, A. J. (1994). Regulation of brain-derived neurotrophic factor (BDNF) expression and release from hippocampal neurons is mediated by non-NMDA type glutamate receptors. *The Journal of Neuroscience, 14*, 1688–1700.

Wetsel, W. C., Rodriguiz, R. M., Guillemot, J., Rousselet, E., Essalmani, R., Kim, I. H., et al. (2013). Disruption of the expression of the proprotein convertase PC7 reduces BDNF production and affects learning and memory in mice. *Proceedings of the National Academy of Sciences of the United States of America, 110*, 17362–17367.

Wu, Y. J., Krüttgen, A., Möller, J. C., Shine, D., Chan, J. R., Shooter, E. M., et al. (2004). Nerve growth factor, brain-derived neurotrophic factor, and neurotrophin-3 are sorted to dense-core vesicles and released via the regulated pathway in primary rat cortical neurons. *Journal of Neuroscience Research, 75*, 825–834.

Yamamori, H., Hashimoto, R., Ishima, T., Kishi, F., Yasuda, Y., Ohi, K., et al. (2013). Plasma levels of mature brain-derived neurotrophic factor (BDNF) and matrix metalloproteinase-9 (MMP-9) in treatment-resistant schizophrenia treated with clozapine. *Neuroscience Letters, 556*, 37–41.

Yan, Q., Rosenfeld, R. D., Matheson, C. R., Hawkins, N., Lopez, O. T., Bennett, L., et al. (1997). Expression of brain-derived neurotrophic factor protein in the adult rat central nervous system. *Neuroscience, 78*, 431–448.

Yang, P., Major, D., & Rutishauser, U. (1994). Role of charge and hydration in effects of polysialic acid on molecular interactions on and between cell membranes. *The Journal of Biological Chemistry, 269*, 23039–23044.

CHAPTER SIX

Emergent Role of Coronin-1a in Neuronal Signaling

M. Martorella, K. Barford, B. Winkler, C.D. Deppmann[1]
University of Virginia, Charlottesville, VA, United States
[1]Corresponding author: e-mail address: deppmann@virginia.edu

Contents

1. Introduction 114
2. The Coronin Family of Proteins 114
   2.1 Type I Coronin Family 115
   2.2 Type II Coronins 117
   2.3 Type III Coronin 118
3. Type I Coronins as Actin Regulators 118
4. Coronin-1a as a Signaling Effector Protein 121
   4.1 Coronin-1a in Pathogen–Host Signaling 121
   4.2 Coronin-1a as a Regulator of Internalization Mechanisms 122
   4.3 Coronin-1a Is Required for GPCR Activation in Excitatory Neurons 123
   4.4 Coronin-1a in NGF–TrkA Signaling 123
   4.5 Coronin-1a in Signaling Endosome Maturation 125
   4.6 Coronin-1a in Signaling Endosome Trafficking 126
   4.7 Coronin-1a in Axon Guidance and Target Innervation 127
5. Conclusions and Future Directions 128
Acknowledgments 129
References 129

Abstract

The Coronin family of proteins were first noted for their role in pathogen–host interactions and for modulating actin dynamics. Recently, however, Coronins have been found in a greater variety of cell types, and novel roles for the Coronins within the nervous system have been discovered. In the immune system, Coronin-1a enables *Mycobacterium tuberculosis* to evade lysosomal destruction. This activity appears to be analogous to protection of the NGF–TrkA signaling endosome during sympathetic nervous system development that is required for survival signaling. Similarly, others have implicated Coronin-1a in GPCR signaling during the formation of excitatory connections in the central nervous system. Its role in multiple signaling pathways suggests that it may influence cross talk between key pathways (TrkA, GPCRs) during neurodevelopment. Here, we review the role of Coronin-1a in neural development and function.

Vitamins and Hormones, Volume 104
ISSN 0083-6729
http://dx.doi.org/10.1016/bs.vh.2016.10.002

1. INTRODUCTION

Coronin family members have been principally described to play critical roles in regulating actin filament dynamics and cargo internalization (BoseDasgupta & Pieters, 2014a, 2014b; Chan, Creed, & Bear, 2011; Gandhi, Achard, Blanchoin, & Goode, 2009; Yan, Di Ciano-Oliveira, Grinstein, & Trimble, 2007). Beyond regulating cytoskeletal rearrangements, there is an emergent role for Coronin-1a in mediating a range of signaling events. A role for Coronin-1a in signal transduction was first described in the context of pathogen–host interaction by Pieters and colleagues who found Coronin-1a to be a phagosomal coat protein for *Mycobacterium tuberculosis* engulfed by macrophages (Ferrari, Langen, Naito, & Pieters, 1999; Jayachandran et al., 2007). By recruiting Coronin-1a, these pathogenic phagosomes induce calcium–calcineurin signaling to evade fusion with lysosomes, thereby allowing for propagation of the disease. Coronin family members are also highly expressed in the nervous system, yet their roles in neuronal signaling are only beginning to be described. We and others have found that in neurons Coronin-1a facilitates second messenger pathways such as IP3-calcium and cAMP-PKA. These findings appear to be unraveling new functions of Coronin within the nervous system. This chapter will review: (1) the Coronin family of proteins, (2) Coronin-1a as a cytoskeletal regulator, and (3) Coronin-1a as an effector protein of several different signaling pathways. The emphasis will be on signaling within the nervous system; however, we will draw on findings from nonneuronal cells such as fibroblasts and immune cells to illustrate conserved functions of the Coronin family and to elucidate future research directions.

2. THE CORONIN FAMILY OF PROTEINS

The highly conserved Coronin family of proteins has roles in cell motility, phagocytosis, vesicular trafficking, and cytokinesis (Chan et al., 2011; de Hostos et al., 1993; Yan et al., 2007). Humans express seven paralogs of Coronin, grouped into types I, II, and III. Structurally, all seven Coronin family members contain an N-terminal seven-bladed beta-propeller region made up of seven WD40 repeats followed by a variable linker region. Types I and II Coronins contain a coiled-coil C-terminus necessary for homodimer and trimer formation, while type III has an additional WD40 repeat domain at the C-terminus (Fig. 1A and B) (Appleton, Wu, &

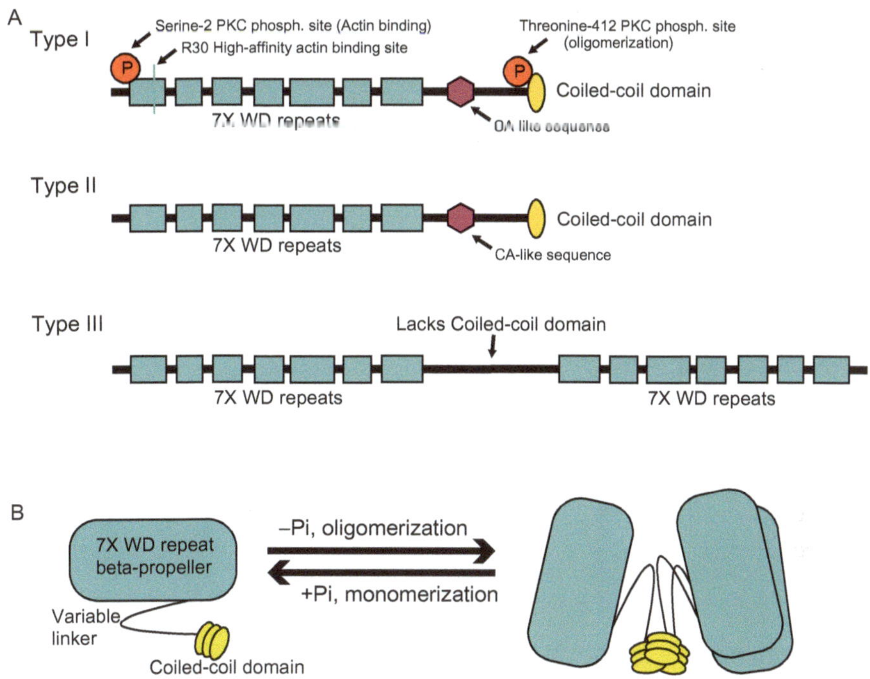

Fig. 1 (A) Structural characteristics of each Coronin family subtype. *Highlighted* are the conserved phosphorylation sites and high-affinity F-actin binding site (R30) on type I Coronins. (B) Coronin oligomerization occurs via coiled-coil domain interactions. Phosphorylation disrupts coiled-coil interactions and induces monomerization.

Wiesmann, 2006; BoseDasgupta & Pieters, 2014a, 2014b; Yee et al., 2014). de Hostos (1999) named Coronins 1–5, but the discovery of new Coronin family members and the expanding knowledge of their structure and function has led to the replacement of this nomenclature with the naming system established by The Human Genome Organization Committee. Coronin family members are now named based on their phylogenetic relationships. The official names and common alternatives are shown in Table 1. The structural differences, expression profile, subcellular localization, and general function of each Coronin type will be briefly elaborated upon here. However, the majority of this chapter will focus on Coronin-1a as a critical signaling effector protein in nervous system development and function.

2.1 Type I Coronin Family

Type I Coronin family members include Coronin-1a, -1b, -1c, and -6. Type I family members display broad overlapping and nonoverlapping

Table 1 Coronin Family Nomenclature

Species	HGNC Name	NCBI Gene ID	Gene Symbol	Type	Aliases	de Hostos Naming
Homo sapiens	Coronin-1a	11151	CORO1A	Type I	p57, IMD8, TACO, CLABP, ClipinA	Coronin1
	Coronin-1b	57175	CORO1B	Type I	Coronin-2	Coronin2
	Coronin-1c	23603	CORO1C	Type I	HCRNN4	Coronin3
	Coronin-2a	7464	CORO2A	Type II	IR10, WDR2, ClipinB	Coronin4
	Coronin-2b	10391	CORO2B	Type II	ClipinC	Coronin5
	Coronin-6	84940	CORO6	Type I	ClipinE, Coro1D	N/A
	Coronin-7	79585	CORO7	Type III	CRN7, POD1	N/A
Drosophila	POD-1	31620	pod1	Type III	Coronin-7, CG4532	N/A
	Coronin	35596	Coro	Type I-like	Coronin-8, CG9446	N/A

expression profiles. While Coronin-1b and -1c are ubiquitously expressed, Coronin-1a seems to have a more restricted expression profile. Of particular note is the expression of Coronin-1a in T cells, thymocytes, macrophages, neutrophils, neurons, and endothelial cells (Jayachandran & Pieters, 2015; Moshous & de Villartay, 2014; Suo et al., 2014). Interestingly, many of the cells expressing Coronin-1a are polarized and require active cytoskeleton rearrangement and protein transport.

The Coronin family members are structurally similar and contain a WD40-repeat containing beta-propeller surface capable of binding F-actin. This surface has been most highly characterized in type I Coronins. High-affinity F-actin binding in type I Coronins is conferred by conserved residues along the beta-propeller surface (Gandhi, Jangi, & Goode, 2010). Type I family members are typically localized to protrusions of the cell membrane where they have been reported to modulate actin dynamics in nonpolarized cells (Chan et al., 2011; Terzi, Kocaefe, & Ayter, 2014). The localization and activity of type I Coronins are typically regulated by protein kinase C (PKC) phosphorylation. Phosphorylation at an N-terminal serine residue disrupts interactions with actin (Ser2 in Coronin-1a, -1b, and -6), while phosphorylation at a threonine in the coiled-coil trimerization domain (Thr412 in Coronin-1a, variations in Coronin-1c and -6) prevents Coronin oligomerization (BoseDasgupta & Pieters, 2014a, 2014b; Cai, Holowecky, Schaller, & Bear, 2005; Oku et al., 2012; Yee et al., 2014) (Fig. 1B). Importantly, Ser2 phosphorylation has not been observed in Coronin-1c. This suggests that Coronin family members may have overlapping but distinct roles in part due to differences in the regulation of their activity. Other regulatory sites include acetylation at the N-terminus of Coronin-1b and Coronin-6 and at the C-terminus of Coronin-1c, but a specific regulating acetyltransferase has not been identified (Chan et al., 2011). As a monomer, Coronin is incapable of binding F-actin and is redistributed from the subcortical network to the cytosol. This cytosolic localization has been shown to impact its normal function in the immune response as well as to inactivate its ability to regulate actin dynamics (BoseDasgupta, Suzette, Jenoe, & Pieters, 2015; Chan et al., 2011). Implications of this redistribution will be further discussed in Section 4.

2.2 Type II Coronins

Type II Coronin family members include Coronin-2a and -2b. Coronin-2a is expressed ubiquitously, while Coronin-2b is enriched in the nervous system (Chan et al., 2011; Marshall, Aloor, & Bear, 2009). Though lacking the same high-affinity actin binding sites in the propeller region seen in type I Coronins, both Coronin-2a and -2b bind F-actin via uncharacterized beta-propeller surfaces and induce ADF/cofilin activity along ADP-actin-containing filaments while promoting Arp2/3 nucleation activity along ADP + Pi-actin filaments (Cai, Makhov, & Bear, 2007). This process is further elaborated upon in Section 3. Because of this dual role in actin dynamics, Coronin-2a is known to mediate focal adhesion turnover during cell migration and assist in chemokine-mediated T-cell migration (Huang et al., 2011; Shiow et al., 2008). Coronin-2b is not well characterized; however, in brain tissue samples, Coronin-2b coimmunoprecipitates with a focal adhesion protein, vinculin. Additionally, immunocytochemistry shows that Coronin-2b localizes to neurite tips (Nakamura, Takeuchi, Muraoka, Takezoe, & Mori, 1999). These data suggest that Coronin-2b may perform a similar role as Coronin-2a in actin reorganization, but acts within the nervous system to affect neurite growth and growth cone advance.

2.3 Type III Coronin

The Type III classification includes the long and structurally distinct isoform of Coronin, Coronin-7. Unlike type I and type II Coronins, Coronin-7 lacks a coiled-coil domain and instead contains two beta-propeller WD-repeat regions connected by a flexible linker (Rybakin et al., 2004). This Coronin family member is the most highly conserved across species, but interestingly has different functions in mammalian vs nonmammalian organisms. In Drosophila, Coronin-7 localizes to growth cones and cross-links actin filaments and microtubules (Rothenberg, Rogers, Vale, Jan, & Jan, 2003). Mammalian Coronin-7 does not bind actin, but instead has been localized to the *cis*-Golgi network and is suspected to play a role in secretory vesicle trafficking. This seems likely due to its WD-repeat sequence homologies with other secretory proteins, namely rabconnectin3 and α- and β′-COP (Rybakin et al., 2004). Coronin-7 also seems to be ubiquitously expressed, with the highest expression in the thymus and the brain. Specifically, Coronin-7 is most abundant in the hypothalamus, striatum, and *locus coeruleus*. Interestingly, its upregulation has been associated with increased food intake and obesity in mouse models (Eriksson et al., 2015).

3. TYPE I CORONINS AS ACTIN REGULATORS

Coronin-1a was first characterized by its role in modulating actin filament organization and dynamics (see Box 1 for general description of actin assembly) (de Hostos, Bradike, Lottspeich, Guggenheim, & Gerisch, 1991; de Hostos et al., 1993). Coronin promotes the formation of protrusions at the leading edge of migrating cells, aids in remodeling actin during the formation of cellular junctions, and is involved in endocytosis and endosomal tubulation (Chan et al., 2011). In order to enable these processes to occur, Coronin functions both to stabilize and to destabilize actin filaments (Gandhi et al., 2009; Liu, Needham, May, & Nolen, 2011; Mikati, Breitsprecher, Jansen, Reisler, & Goode, 2015). Coronin accomplishes these dual roles by influencing the binding of actin filament modifiers to filaments (e.g., cofilin and Arp2/3) in a manner that depends on the adenine nucleotide associated with individual actin subunits of filaments, and by recruiting the regulators of these actin modulators (e.g., Slingshot) (Cai et al., 2007; Gandhi et al., 2009). Coronin binds to ADP + Pi- and ADP-actin subunits via two different surfaces on its beta-propeller (Gandhi et al., 2009). When bound to ADP-actin, Coronin promotes actin destabilization through two

BOX 1 Actin Assembly

Actin assembly is a continuous, highly dynamic process that can be divided into three general phases: nucleation, elongation, and steady state (Cooper, 2000). Actin itself comes in two forms, a monomeric unit, G-actin, and a polymeric filament, F-actin. In both forms, actin subunits can be bound to ATP, ADP + Pi, or ADP, and these different ATP-binding states determine which actin modifiers are bound, and thus whether assembly or disassembly is favored (Rottner & Stradal, 2016). Actin monomers bound to ATP exhibit higher affinity for one another and typically promote assembly (Cooper, 2000). Because actin is an ATPase, it slowly hydrolyzes ATP to ADP, thus favoring its disassembly over time. Nucleation involves the polymerization of three or more ATP-bound G-actin monomers into an F-actin filament. This assembly is spontaneous, but it is facilitated in vivo by factors like the Arp2/3 complex, which, together with nucleation promoting factors (NPFs) such as WASP and Scar/WAVE, promotes filament initiation along the sides of existing actin filaments (Rottner & Stradal, 2016). Filament elongation is promoted largely by the formin family of proteins or Ena/VASP proteins. Some formins are involved in the reallocation of monomers to the plus end to establish an actin monomer concentration that favors actin assembly. New actin branches formed during elongation are stabilized by cortactin, an NPF subtype (Cooper, 2000). The third actin phase is steady state, also referred to as "treadmilling," and it occurs when monomers are removed from the minus end, at the same rate monomers are assembled at the plus end (Cooper, 2000). This "treadmilling" effect is considered an in vitro concept, however, because actin modulators will continuously alter this steady state. Actin disassembly is mediated by ADF/cofilin proteins, with accessory factors, that sever actin filaments, thereby increasing the number of minus ends and promoting (Mikati et al., 2015).

mechanisms. First, Coronin enhances filament severing by providing additional and more favorable binding surfaces for cofilin (Gandhi et al., 2009). Second, Coronin recruits the phosphatase activator of cofilin, Slingshot (Gandhi et al., 2009). The opposite effect has been observed in yeast when Coronin is bound to ADP + Pi-actin. The conformation Coronin adopts sterically hinders cofilin binding, and the C-terminal coiled-coil provides an additional handle for Arp2/3 binding (Liu et al., 2011; Mikati et al., 2015). Several other studies in yeast have demonstrated that removing the coiled-coil domain enhances cofilin activity while impairing Arp2/3 binding, confirming that the coiled-coil domain interacts differently with these actin modifiers (Gandhi et al., 2009; Mikati et al., 2015).

It should be noted that the majority of our understanding of the role of Coronin in actin modification has been established in yeast. Mammalian systems suggest that it may play a broader and more dynamic role. For one, Coronin stabilization of Arp2/3 binding at ADP + Pi-actin filaments has not been confirmed to occur in mammals; however, it has been shown that Coronin interacts with and replaces Arp2/3 at ADP + Pi filaments (Cai, Makhov, Schafer, & Bear, 2008). This would provide several advantages. For one, Coronin would act as a temporary stabilizer of newly formed branches that is also capable of recruiting Slingshot, thus enabling the dynamic nature of actin by allowing for the gradual disassembly of branches and turnover of actin filaments (Cai et al., 2008). Another key aspect of Coronin replacing Arp2/3 is its flexibility. Normally new actin branches form at a 70-degree angle when bound to Arp2/3, but stabilization of branches by Coronin allows for a greater variety of angles to be established (Cai et al., 2008). This is advantageous in that it allows the cell to rapidly adjust the actin network in order to accommodate newly formed cellular adhesions, trafficking of vesicles, or unique morphological events (such as those that occur during apoptosis) (Cai et al., 2008).

Whether the cellular functions Coronin promotes during actin dynamics bears a more significant and interrelated effect on cell signaling requires first that our understanding of its interactions with different actin modulators is more thoroughly understood. Evidence, thus far, heavily implicates Coronin, cofilin, and other actin modifiers in recycling and transcytosis events in neurons, but how these functions relate to neurotrophin signaling remains an open question (Box 2).

BOX 2 Neurotrophin Signaling in the Development of the PNS

In vertebrates, formation of a functional peripheral nervous system is achieved by an overproduction of neurons followed by a 50% die off. Which neurons are destined to live or die? Upon final target organ innervation, a competition ensues whereby neurons compete for target-derived trophic factors. The neurons that receive sufficient trophic factor survive and those that do not undergo apoptosis (Levi-Montalcini & Booker, 1960). Nerve growth factor is one such trophic factor that has been broadly studied in the context of sympathetic and sensory neuron development. NGF transduces its signal by binding to the TrkA receptor tyrosine kinase at the distal axon. For survival signaling to occur, the NGF–TrkA complex must internalize into what is commonly referred to as the signaling endosome,

BOX 2 Neurotrophin Signaling in the Development of the PNS—cont'd

and then traffic back to the cell body via dynein to induce survival pathways like the phosphorylation of CREB (Lonze & Ginty, 2002).

Given that axons often arrive to targets nearly simultaneously and have comparable access and responsiveness to trophic factor, what determines which neurons win and which neurons lose? We demonstrated that a series of positive feedback loops are critical to create a bistability-destined to live or die (Deppmann et al., 2008). In the first feedback loop, NGF signaling regulates the expression of its own receptor, TrkA. This affects the robustness of neurotrophin signaling. The second feedback loop found to be critical for competition for survival was one whereby NGF regulated its own signal duration. This was attributed to an increase in signaling endosome stability and Coronin-1a upregulation (Suo et al., 2014).

4. CORONIN-1A AS A SIGNALING EFFECTOR PROTEIN

A growing body of work suggests that Coronin-1a is required for a range of receptor-mediated signaling pathways. This section will focus primarily on how Coronin-1a mediates signaling from GPCRs (b2-adrenergic receptor) and RTKs (TrkA). The extent to which Coronin paralog functions overlap remains unclear; however, the phenotypes from knockout mice described later suggest that, at least in some cell types, Coronin-1a is not functionally redundant with other family members.

4.1 Coronin-1a in Pathogen–Host Signaling

The role of Coronin-1a as a signaling effector protein was first described in the context of pathogen–host interaction. Using *M. tuberculosis* infection of macrophages as a model, Pieters and colleagues have elucidated several facets of Coronin-1a signaling that are required for Coronin-1a function and which are also conserved in neural signaling.

Upon engulfment of *M. tuberculosis* by macrophages, pathogenic phagosomes recruit Coronin-1a to their membrane, allowing them to evade lysosomal fusion. Coronin-1a then mediates calcium release and calcineurin activation, and this pathway is required for protection from the lysosome. Loss of Coronin-1a, calcium, or calcineurin activity is sufficient to drive pathogenic phagosomes toward lysosomal fusion (Jayachandran & Pieters, 2015;

Jayachandran et al., 2007). Interestingly, Coronin-1a must be oligomerized into a trimer for this protection to occur. Disruption of trimerization using point mutants that either mimic constitutive phosphorylation or coiled-coil deletions render Coronin-1a unable to activate Ca^{2+}/calcineurin and protect the pathogenic phagosome from lysosomal fusion (BoseDasgupta & Pieters, 2014a, 2014b). It is unclear whether these mutants are capable of binding F-actin or membranes more generally. Because many Coronin functions require membrane association, this suggests that trimerization may be required for functions beyond cytoskeletal rearrangement.

4.2 Coronin-1a as a Regulator of Internalization Mechanisms

In the context of pathogen–host interaction, Coronin-1a not only prevents lysosomal fusion but is also an important regulator in the switch from phagocytosis to macropinocytosis (BoseDasgupta & Pieters, 2014a, 2014b). During the initial stages of an infection by *M. tuberculosis*, macrophages primarily internalize cargo, including mycobacterium, via phagocytosis. However, phagocytosis is an ineffective mechanism for clearing bacterial infections because of the ability of mycobacteria to evade phagosome/lysosomal fusion and because of its reliance on receptor-mediated internalization. Macropinocytosis, however, presents an efficient mechanism to clear infectious material as it is nonsaturatable and independent of receptor binding. Indeed, after prolonged exposure to infection the immune system mobilizes cytokine signaling which subsequently reprograms the major internalization mechanism from phagocytosis to macropinocytosis by stimulating PKC activity (BoseDasgupta & Pieters, 2014a, 2014b). PKC activity increases PI3K activity, and PI3K promotes macropinocytosis by generating PIP3 at the membrane. This helps create the membrane ruffles necessary for macropinocytosis and shifts internalization away from phagocytosis. Macropinosomes containing mycobacteria are able to fuse with the lysosome such that macrophages can effectively clear the infection.

Coronin-1a has emerged as a potentially key component in the switch to macropinocytosis. A phosphomimetic form of Coronin-1a, in the absence of cytokine signaling, is sufficient to initiate PI3K signaling and switch the internalization mechanism to macropinocytosis (BoseDasgupta & Pieters, 2014a, 2014b). As mentioned previously, PKC phosphorylates Coronin-1a, thereby disrupting its trimerization and preventing Ca^{2+}/calcineurin signaling (BoseDasgupta & Pieters, 2014a, 2014b). Additionally, PI3K, which is necessary for this switch to macropinocytosis, is known to be inhibited by Ca^{2+}/calcineurin signaling. At this time, it is unclear whether decreased Ca^{2+}/calcineurin signaling following Coronin-1a inactivation lifts inhibition of PI3K activity, or whether Coronin-1a interacts directly with PI3K to activate it after monomerizing and redistributing to the cytosol (BoseDasgupta et al., 2015). This relationship between PI3K and Coronin-1a has been observed in other systems, i.e., axon growth, which will be elaborated upon in Section 4; however, the mechanism by which this interaction occurs is equally elusive (Suo, Park, Young, Makita, & Deppmann, 2015).

4.3 Coronin-1a Is Required for GPCR Activation in Excitatory Neurons

Coronin-1a has recently been shown to impact the development of excitatory synaptic connections in the central nervous system. Coronin-1a knockout mice display a dramatic loss of excitatory connections resulting in decreased anxiety and social interactions, increased aggression, and learning deficits (Jayachandran et al., 2014). This effect is attributed to a requirement of Coronin-1a for activation of cAMP/PKA signaling after activation of G protein-coupled receptors (GPCRs). Without Coronin-1a, both cAMP/PKA signaling and downstream signaling pathways, such as CREB phosphorylation, were lost (Jayachandran et al., 2014). A deficit in CREB phosphorylation has also been observed in peripheral neurons isolated from Coronin-1a knockout mice, but a link to GPCR signaling was not investigated in those studies (Suo et al., 2014). Because Coronin-1a shares homology with the Gβ subunit, Pieters and colleagues suggest that Coronin-1a may activate Gα-s and adenylate cyclase by displacing Gβ after GPCR activation (Jayachandran et al., 2014) (Fig. 2). This claim is supported by coimmunoprecipitation of Coronin-1a with Gα-s but not Gα-i (Jayachandran et al., 2014). Notably, the behavioral phenotypes observed in the Coronin-1a knockout mouse could be rescued by infusing these animals with the cAMP analog, 8-Br-cAMP (Jayachandran et al., 2014).

4.4 Coronin-1a in NGF–TrkA Signaling

Coronin-1a has also been shown to be important for RTK signaling. In particular, we recently demonstrated that Coronin-1a is a novel effector for NGF–TrkA signaling (Fig. 2). This type of signaling is well established as being critical for several aspects of peripheral nervous system (PNS)

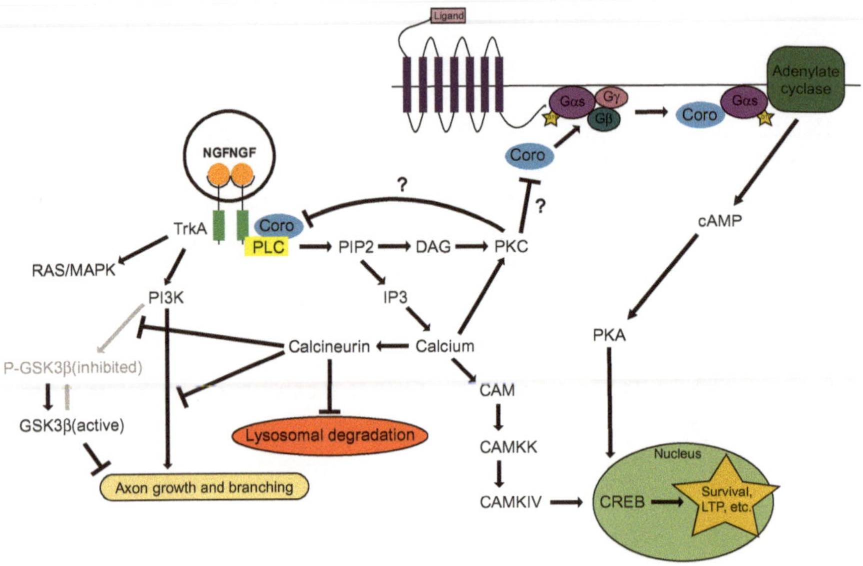

Fig. 2 Schematic of the role of Coronin-1a in NGF–TrkA and beta-adrenergic GPCR signaling pathways in neurons. It is not known whether activation of PKC by Coronin-1a results in a negative feedback loop in which PKC downregulates Coronin-1a activity.

development including survival, synapse formation, and axon growth (Ascaño, Richmond, Borden, & Kuruvilla, 2009; Levi-Montalcini & Booker, 1960; Sharma et al., 2010). Like other RTKs, NGF binding induces TrkA dimerization and autophosphorylation, which in turn elicits classic downstream pathways, including Ras-MAPK, PI3K, and PLC-γ/calcium signaling (Harrington & Ginty, 2013). Importantly, these pathways are persistently activated on the signaling endosome where they direct trafficking, maturation, and stability. We found that Coronin-1a binds to the signaling endosome once it arrives at the cell body from the distal axon (Suo et al., 2014). This recruitment was necessary for the signaling endosome to evade lysosomal fusion, which would extinguish TrkA signaling (Fig. 3). This is reminiscent of the role for Coronin-1a in protecting pathogenic phagosomes from lysosomal fusion (BoseDasgupta & Pieters, 2014a, 2014b). Indeed, the signaling mechanisms allowing *M. tuberculosis* to persist in macrophages are similarly involved in protecting the NGF–TrkA signaling endosome. For instance, Coronin-1a is required for NGF–TrkA-dependent calcium release and calcineurin activation (Suo et al., 2014). There is also a severe deficit in several NGF–TrkA specific pathways that rely on calcium signaling including phosphorylation of CREB, which is a

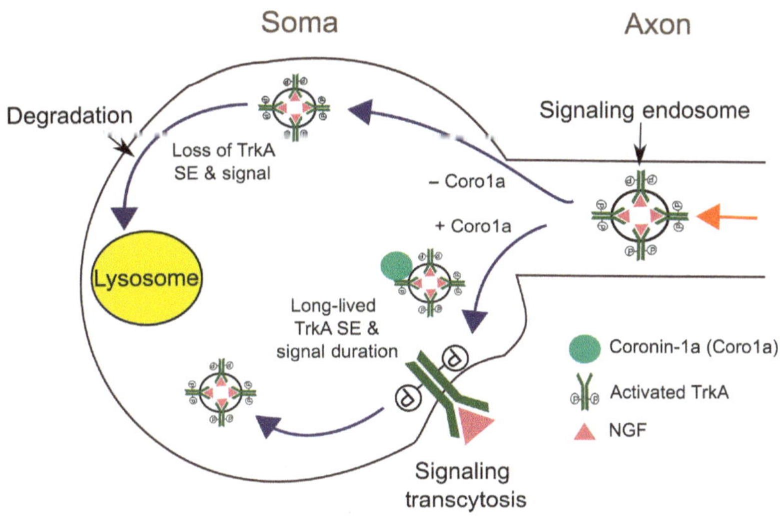

Fig. 3 Coronin-1a endows the NGF–TrkA signaling endosome with enhanced signaling and trafficking capabilities. By associating with Coronin-1a, the NGF–TrkA signaling endosome can undergo transcytosis and prolong NGF–TrkA signaling.

critical mediator of peripheral neuron survival (Lonze & Ginty, 2002; Riccio, Pierchala, Ciarallo, & Ginty, 1997). The implications of this are discussed later.

4.5 Coronin-1a in Signaling Endosome Maturation

The Coronin-1a-dependent signaling events endow the signaling endosome with unique maturation capabilities. For example, in Coronin-1a knockout mice the NGF–TrkA signaling endosome degrades via lysosomal fusion 6–10 times faster than in wild-type neurons, in which signaling endosomes successfully evade lysosomal fusion for up to 24 h (Suo et al., 2014). This loss of protection can be phenocopied by manipulating downstream signals of Coronin-1a, including the addition of antagonists of PLC-γ, calcium, or calcineurin. Likewise, the Coronin-1a knockout phenotype can be partially rescued by calcium ionophores. Based on the previous findings that PKC activity can cause Coronin-1a to become monomeric and dissociate from actin and phagosomes, it will be interesting to determine whether these treatments also change the association of Coronin-1a with TrkA and the signaling endosome (Fig. 2). Importantly, Coronin-1a does not influence the degradation rate of another RTK, EGF receptor. This suggests that Coronin-1a may be selectively recruited to signaling endosomes that require long-lived signaling, such as those that undergo long distance transport.

4.6 Coronin-1a in Signaling Endosome Trafficking

Coronin-1a provides the NGF–TrkA signaling endosome with unique trafficking capabilities (Fig. 3). Upon reaching the soma, Coronin-1a is recruited to the signaling endosome. This is required not only for prolonging the signal but also for enabling a newly described trafficking event that we refer to as signaling transcytosis. During signaling transcytosis, retrograde TrkA receptors are exocytosed onto somal or dendritic membranes and later reinternalized. This phenomenon was observed using a modified version of the TrkA receptor from mice in which a Flag epitope was fused to the extracellular domain and knocked into the endogenous locus (Sharma et al., 2010). Applying a mouse anti-Flag antibody to live cells allows for tracking of the TrkA receptor following internalization. Using a modified version of this assay in which a second Cy3-coupled antimouse secondary antibody was applied to the cell body compartment after the initial anti-Flag antibody feeding at the distal axon, we were able to observe that retrogradely transported TrkA (decorated with anti-FLAG antibody) recycles to the cell body plasma membrane and reinternalizes (decorated with anti-FLAG antibody and Cy3-antimouse antibody). Coronin-1a KO neurons are unable to undergo this recycling event (Suo et al., 2014). These experiments confirmed that Coronin-1a is a necessary component for TrkA signaling transcytosis at the soma.

The necessity of Coronin-1a in TrkA signaling and recycling suggests that it may be a critical component of the signaling endosome with an array of functions allowing for proper neural development. Shuttling of retrograde NGF–TrkA signaling endosomes into dendrites is an essential event during the formation of postsynaptic densities in the PNS that results in the clustering of postsynaptic density components (Sharma et al., 2010). So far, it can be said that Coronin-1a assists in this process by prolonging the signaling capabilities of the NGF–TrkA signaling endosome and enabling transcytosis to occur, but whether it may play a larger role in mediating actin dynamics or protein clustering, or whether other Coronin family members expressed in the nervous system are involved, is yet to be known. It is also possible that the NGF–TrkA signaling endosome may not directly undergo recycling but will instead form a multivesicular body (MVB). MVBs have the potential to recycle TrkA to dendritic or somal surfaces or release vesicles into the extracellular environment as exosomes. Currently, it remains an open question as to how presynaptic neurons are guided to postsynaptic neurons, and how postsynaptic neurons receiving comparable levels of NGF undergo different levels of survival signaling. Whether the stability engendered by Coronin-1a to NGF–TrkA endosomes allows for the trafficking necessary to produce paracrine messengers warrants future study.

It is worth noting that the divergent fates of the NGF–TrkA signaling endosome likely give rise to distinct signaling endosomes with unique molecular profiles and functions. For instance, a TrkA receptor that has undergone signaling transcytosis might very well be decorated with a different set of actin modifiers, second messengers, and Rabs than an NGF–TrkA signaling endosome recently trafficked from the distal axon. It also stands to reason that the NGF-dependent processes elicited by these different pools of signaling endosomes likely bear different and significant consequences on survival, growth, and morphology of developing neurons.

4.7 Coronin-1a in Axon Guidance and Target Innervation

Coronin-1a is also involved in NGF–TrkA-dependent axon growth cone dynamics of neurons approaching their target. In Coronin-1a knockout mice, growth cones show a fourfold decrease in area (Suo et al., 2015). This is significant because smaller growth cones are associated with accelerated axon growth, whereas larger growth cones result in decelerated growth but increased capacity for turning and branching (Carmeliet, 2003). Consistent with this, loss of Coronin-1a resulted in reduced NGF-dependent branching and overshooting of final targets (Suo et al., 2015). These phenotypes agree with the previously stated role of Coronin-1a in mediating actin reorganization and promoting lamellipodial protrusions. Because Coronin-1a expression is induced by final target innervation and exposure to NGF, this represents a critical molecular switch in axon patterning of targets like the heart (Suo et al., 2014). When sympathetic axons initially reach their target prior to Coronin-1a expression, NGF-dependent PI3K signaling promotes rapid axon extension (Fig. 4A). Upon Coronin-1a upregulation, exuberant axon growth and branching is suppressed downstream of NGF-PI3K signaling at the level of GSK3beta activation (Figs. 2 and 4B). We speculate that this throttling of axon growth and branching allows for appropriate arborization of the final target tissue by innervating axons. We have previously demonstrated that Coronin-1a does not associate with the signaling endosome until it reaches the cell body; however, the mechanism for this spatial selectivity is unknown. Therefore, we suggest that NGF–TrkA-Coronin-1a-mediated axon growth and branching phenotypes are likely to be driven by axonal plasma membrane signaling and perhaps directly effecting cytoskeletal dynamics.

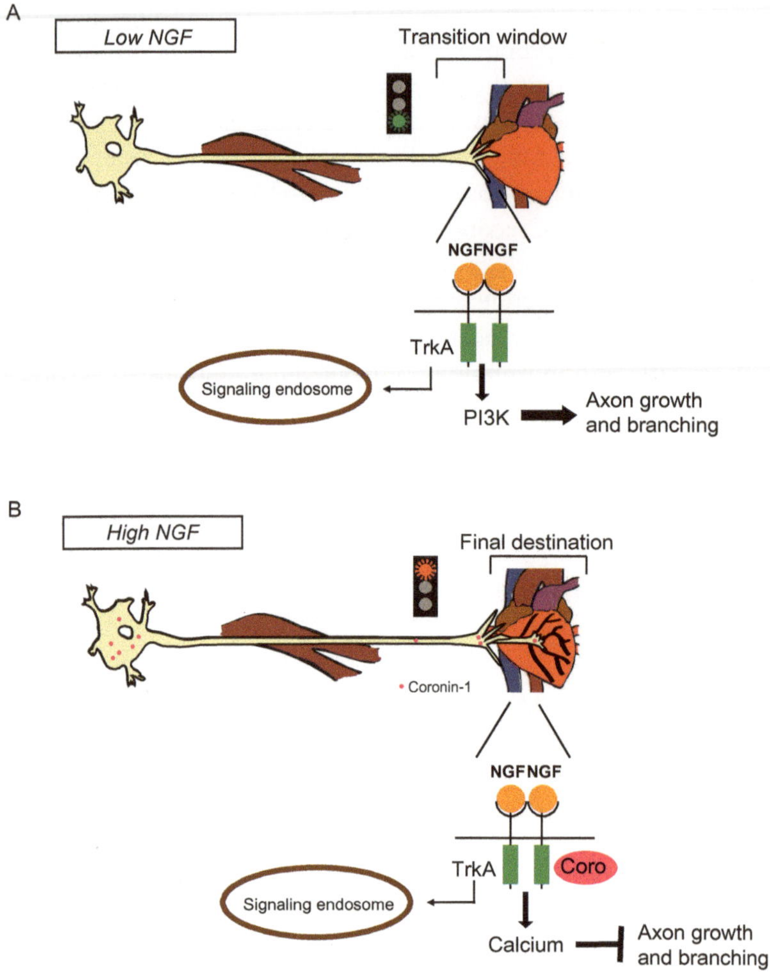

Fig. 4 (A) Upon initially encountering a target organ, NGF concentrations are low and NGF-induced genes are not yet expressed. This marks a "transition window," in which low NGF availability enhances axon growth and branching via PI3K signaling. (B) High NGF induces Coronin-1a expression, and PI3K signaling is dampened. Axon growth and branching slows.

5. CONCLUSIONS AND FUTURE DIRECTIONS

Despite the unveiling of Coronins as highly conserved actin modifiers and signaling mediators nearly 10 years ago, our understanding of their involvement in these processes across cell lineages remains limited. Structurally, the Coronin family is interesting, as the structural variations in each subtype suggest distinct roles, while the conserved WD repeats allow for potentially intersecting functions across a diverse array of signaling effectors. The details of Coronin subtype interplay as well as differences in its activity depending on oligomeric assembly are not fully understood. Indeed, Coronin appears to be a mediator of several receptor pathways involved in immune and nervous system development. The multifaceted roles of Coronin-1a in actin assembly and signal modulation suggest that it may act as a key mediator of neuron development and function. Already there are examples of its function in axon growth, branching, and growth cone morphology, but our understanding of the role for Coronin family members within nervous system development remains limited and many more functions are likely to be discovered. Understanding when and where Coronin participates in this diversity of functions provides an important substrate for future investigation.

ACKNOWLEDGMENTS

C.D.D. was supported by NIH-NINDS (1R01NS072388) and NSF-IOS-1453242. B.W. was supported by NIH-NINDS (R01NS083378). We are grateful to Dorothy Schafer for helpful discussion. The authors declare no competing financial interests.

REFERENCES

Appleton, B. A., Wu, P., & Wiesmann, C. (2006). The crystal structure of murine coronin-1: A regulator of actin cytoskeletal dynamics in lymphocytes. *Structure*, *14*(1), 87–96.

Ascaño, M., Richmond, A., Borden, P., & Kuruvilla, R. J. (2009). Axonal targeting of Trk receptors via transcytosis regulates sensitivity to neurotrophin responses. *The Journal of Neuroscience*, *29*(37), 11674–11685.

BoseDasgupta, S., & Pieters, J. (2014a). Coronin 1 trimerization is essential to protect pathogenic mycobacteria within macrophages from lysosomal delivery. *FEBS Letters*, *588*(21), 3898–3905.

BoseDasgupta, S., & Pieters, J. (2014b). Inflammatory stimuli reprogram macrophage phagocytosis to macropinocytosis for the rapid elimination of pathogens. *PLoS Pathogens*, *10*(1), e1003879.

BoseDasgupta, S., Suzette, M., Jenoe, P., & Pieters, J. (2015). Cytokine-induced macropinocytosis in macrophages is regulated by 14-3-3f through its interaction with serine-phosphorylated coronin 1. *FEBS Journal*, *282*(7), 1167–1181.

Cai, L., Holowecky, N., Schaller, M. D., & Bear, J. E. (2005). Phosphorylation of coronin 1B by protein kinase C regulates interaction with Arp2/3 and cell motility. *The Journal of Biological Chemistry*, *280*(36), 31913–31923.

Cai, L., Makhov, A. M., & Bear, J. E. (2007). F-actin binding is essential for coronin 1B function in vivo. *Journal of Cell Science*, *120*(Pt. 10), 1779–1790.

Cai, L., Makhov, A. M., Schafer, D. A., & Bear, J. E. (2008). Coronin 1B antagonizes cortactin and remodels Arp2/3-containing actin branches in lamellipodia. *Cell*, *134*, 828–842.

Carmeliet, P. (2003). Blood vessels and nerves: Common signals, pathways and diseases. *Nature Reviews. Genetics*, *4*, 710–720. CrossRef Medline.

Chan, K. T., Creed, S. J., & Bear, J. E. (2011). Unraveling the enigma: Progress towards understanding the Coronin family of actin regulators. *Trends in Cell Biology, 21*(8), 481–488.

Cooper, G. M. (2000). *The cell: A molecular approach* (2nd ed.). . Sunderland (MA): Sinauer Associates.

de Hostos, E. L. (1999). The coronin family of actin-associated proteins. *Trends in Cell Biology, 9*(9), 345–350.

de Hostos, E. L., Bradike, B., Lottspeich, F., Guggenheim, R., & Gerisch, G. (1991). Coronin, an actin-binding protein of Dictyostelium discoideum localized to cell surface projections, has sequence similarities to G protein/3 subunits. *The EMBO Journal, 10,* 4097–4104.

de Hostos, E. L., Rehfuess, C., Bradtke, B., Waddell, D. R., Albrecht, R., Murphy, J., et al. (1993). *Dictyostelium* mutants lacking the cytoskeletal protein coronin are defective in cytokinesis and cell motility. *The Journal of Cell Biology, 120*(1), 163–173.

Deppmann, C. D., Mihalas, S., Sharma, N., Lonze, B. E., Niebur, E., & Ginty, D. D. (2008). A model for neuronal competition during development. *Science, 320*(5874), 369–373.

Eriksson, A., Williams, M. J., Voisin, S., Hansson, I., Krishnan, A., Philippot, G., et al. (2015). Implication of coronin 7 in body weight regulation in humans, mice and flies. *BMC Neuroscience, 16,* 13.

Ferrari, G., Langen, H., Naito, M., & Pieters, J. (1999). A coat protein on phagosomes involved in the intracellular survival of mycobacteria. *Cell, 97*(4), 435–447.

Gandhi, M., Achard, V., Blanchoin, L., & Goode, B. L. (2009). Coronin switches roles in actin disassembly depending on the nucleotide state of actin. *Molecular Cell, 34*(3), 364–374.

Gandhi, M., Jangi, M., & Goode, B. L. (2010). Functional surfaces on the actin-binding protein coronin revealed by systematic mutagenesis. *The Journal of Biological Chemistry, 285*(45), 34899–34908.

Harrington, A. W., & Ginty, D. D. (2013). Long-distance retrograde neurotrophic factor signalling in neurons. *Nature Reviews. Neuroscience, 14,* 177–187.

Huang, W., Ghisletti, S., Saijo, K., Gandhi, M., Aouadi, M., Tesz, G. J., et al. (2011). Coronin 2A mediates actin-dependent de-repression of inflammatory response genes. *Nature, 470*(7334), 414–418.

Jayachandran, R., Liu, X., BoseDasgupta, S., Müller, P., Zhang, C.-L., Moshous, D., et al. (2014). Coronin 1 regulates cognition and behavior through modulation of cAMP/protein kinase A signaling. *PLoS Biology, 12*(3), e1001820.

Jayachandran, R., & Pieters, J. (2015). Regulation of immune cell homeostasis and function by coronin 1. *International Immunopharmacology, 28*(2), 825–828.

Jayachandran, R., Sundaramurthy, V., Combaluzier, B., Mueller, P., Korf, H., Huygen, K., et al. (2007). Survival of mycobacteria in macrophages is mediated by coronin 1-dependent activation of calcineurin. *Cell, 130*(1), 37–50.

Levi-Montalcini, R., & Booker, B. (1960). Destruction of the sympathetic ganglia in mammals by an antiserum to a nerve-growth protein. *Proceedings of the National Academy of Sciences of the United States of America, 46*(3), 384–391.

Liu, S. L., Needham, K. M., May, J. R., & Nolen, B. J. (2011). Mechanism of a concentration-dependent switch between activation and inhibition of Arp2/3 complex by coronin. *The Journal of Biological Chemistry, 286,* 17039–17046.

Lonze, B. E., & Ginty, D. D. (2002). Function and regulation of CREB family transcription factors in the nervous system. *Neuron, 35,* 605–623.

Marshall, T. W., Aloor, H. L., & Bear, J. E. (2009). Coronin 2A regulates a subset of focal-adhesion-turnover events through the cofilin pathway. *Journal of Cell Science, 122,* 3061–3069.

Mikati, M. A., Breitsprecher, D., Jansen, S., Reisler, E., & Goode, B. L. (2015). Coronin enhances actin filament severing by recruiting cofilin to filament sides and altering F-actin conformation. *Journal of Molecular Biology*, *427*(19), 3137–3147.

Moshous, D., & de Villartay, J. P. (2014). The expanding spectrum of human coronin 1A deficiency. *Current Allergy and Asthma Reports*, *14*(12), 481.

Nakamura, T., Takeuchi, K., Muraoka, S., Takezoe, N., & Mori, N. (1999). A neurally enriched coronin-like protein, ClipinC, is a novel candidate for an actin cytoskeleton-cortical membrane-linking protein. *Journal of Biological Chemistry*, *274*(19), 13322–13327.

Oku, T., Nakano, M., Kaneko, Y., Ando, Y., Kenmotsu, H., Itoh, S., et al. (2012). Constitutive turnover of phosphorylation at Thr-412 of human p57/coronin-1 regulates the interaction with actin. *The Journal of Biological Chemistry*, *287*(51), 42910–42920.

Riccio, A., Pierchala, B. A., Ciarallo, C. L., & Ginty, D. D. (1997). An NGF-TrkA-mediated retrograde signal to transcription factor CREB in sympathetic neurons. *Science*, *277*(5329), 1097–1100.

Rothenberg, M. E., Rogers, S. L., Vale, R. D., Jan, L. Y., & Jan, Y. (2003). Drosophila Pod-1 crosslinks both actin and microtubules and controls the targeting of axons. *Neuron*, *39*(5), 779–791. http://dx.doi.org/10.1016/s0896-6273(03)00508-7.

Rottner, K., & Stradal, T. E. B. (2016). How distinct Arp2/3 complex variants regulate actin filament assembly. *Nature Cell Biology*, *18*, 1–3.

Rybakin, V., Stumpf, M., Schulze, A., Majoul, I., Noegel, A., & Hasse, A. (2004). Coronin 7, the mammalian POD-1 homologue, localizes to the Golgi apparatus. *FEBS Letters*, *573*(1–3), 161–167.

Sharma, N., Deppmann, C. D., Harrington, A. W., St Hillaire, C., Chen, Z. Y., Lee, F. S., et al. (2010). Long-distance control of synapse assembly by target-derived NGF. *Neuron*, *67*(3), 422–434.

Shiow, L. R., Roadcap, D. W., Paris, K., Watson, S. R., Grigorova, I. L., Lebet, T., et al. (2008). The actin regulator coronin-1A is mutated in a thymic egress deficient mouse strain and in a T−B+NK+ SCID patient. *Nature Immunology*, *9*(11), 1307–1315.

Suo, D., Park, J., Harrington, A. W., Zweifel, L. S., Mihalas, S., & Deppmann, C. D. (2014). Coronin-1 is a neurotrophin endosomal effector required for developmental competition for survival. *Nature Neuroscience*, *17*(1), 36–45.

Suo, D., Park, J., Young, S., Makita, T., & Deppmann, C. D. (2015). Coronin-1 and calcium signaling governs sympathetic final target innervation. *Journal of Neuroscience*, *35*(9), 3893–3902.

Terzi, Y. K., Kocaefe, Y. C., & Ayter, S. (2014). Coronin 1A inhibits neurite outgrowth in PC12 cells. *Neuroscience Letters*, *582*, 38–42.

Yan, M., Di Ciano-Oliveira, C., Grinstein, S., & Trimble, W. S. (2007). Coronin function is required for chemotaxis and phagocytosis in human neutrophils. *The Journal of Immunology*, *178*, 5769–5778.

Yee, C. S. K., Ozden, S., Chou, J. S., Geha, R. S., Ayvaz, D., Aytekin, C., et al. (2014). Coronin-1A oligomerization is critical for host defense against viral pathogens. *Journal of Allergy and Clinical Immunology*, *133*(2), AB94.

CHAPTER SEVEN

Neuronal Roles of the Bicaudal D Family of Motor Adaptors

M. Budzinska*, K.B. Wicher†, M. Terenzio‡,1

*Molecular NeuroPathobiology Laboratory, UCL Institute of Neurology, University College London, London, United Kingdom
†Ossianix, Stevenage Bioscience Catalyst, Stevenage, United Kingdom
‡Weizmann Institute of Science, Rehovot, Israel
[1]Corresponding author: e-mail address: marco.terenzio@weizmann.ac.il

Contents

1. Introduction 134
2. The BICD Family of Cargo Adaptors 136
3. Role of BicD in *Drosophila* Development 138
4. Mammalian BICDs and Their Role in Mitotic Cells 139
5. Role of BICD in Neurons 141
6. BICDs and NT Signaling 143
7. Conclusions 146
Acknowledgments 147
References 147

Abstract

All cell types rely on active intracellular cargo transport to shuttle essential cellular components such as proteins, lipids, RNA, and even organelles from the center to the periphery and vice versa. Additionally, several signaling pathways take advantage of intracellular transport to propagate their signals by moving activated receptors and protein effectors to specific locations inside the cell. Neurons particularly, being a very polarized cell type, are highly dependent on molecular motors for the anterograde and retrograde delivery of essential cellular components and signaling molecules. For these reasons, motor adaptor proteins have been extensively investigated in regard to their role in physiology and pathology of the nervous system.

In this chapter, we will concentrate on a family of motor adaptor proteins, Bicaudal D (BICD), and their function in the context of the nervous system. BicD was originally described as essential for the correct localization of maternal mRNAs in *Drosophila*'s oocyte and a regulator of the Golgi to ER retrograde transport in mammalian cells. Both mammalian BICD1 and BICD2 are highly expressed in the nervous system during development, and their importance in neuronal homeostasis has been recently under scrutiny. Several mutations in BICD2 have been linked to the development of

Vitamins and Hormones, Volume 104
ISSN 0083-6729
http://dx.doi.org/10.1016/bs.vh.2016.11.005

neuromuscular diseases, and BICD2 knockout (KO) mice display migration defects of the radial cerebellar granule cells. More in line with the overall topic of this book, BICD1 was identified as a novel regulator of neurotrophin (NT) signaling as its deletion leads to defective sorting of ligand-activated NT receptors with dramatic consequences on the NT-mediated signaling pathway.

1. INTRODUCTION

One of the ways cells communicate with the outside world is through the activation of an array of versatile and unique membrane receptors that in turn initiate downstream signaling cascades. This communication route is common for many cell types, is highly conserved throughout evolution, and is often dependent on endocytosis of the receptors themselves, in a process of membrane engulfment and internalization, which can be clathrin-dependent or independent (Le Roy & Wrana, 2005). The process of endosome formation is well characterized and it is particularly important for neurons (Cosker & Segal, 2014), where it enables fast and effective renewal of the synaptic vesicle pool and efficient neurotransmitter release (Granseth, Odermatt, Royle, & Lagnado, 2006). Additionally, it is essential for the establishment of the developing nervous system and for its support throughout life by activation of prosurvival signaling cascades like neurotrophin (NT) signaling (Schmieg, Menendez, Schiavo, & Terenzio, 2014). Following endocytosis, signaling-competent receptor complexes are transported through the endosomal pathway, and this active intracellular transport is pivotal to their signaling.

Active intracellular cargo transport is indeed an essential cellular function and cells rely on it to localize and/or shuttle proteins, lipids, RNA, and even organelles to and from particular subcellular locations. As previously mentioned, a striking example of this dependence are neurons, a very polarized cell type, often exhibiting extreme size, and unique morphology, which utilize molecular motors for the delivery of essential components from the soma to growth cones and synapses (Gibbs, Greensmith, & Schiavo, 2015). This is particularly evident in the case of peripheral motor and sensory neurons, which are the biggest cells in the human body and can reach up to 1 m in length and for which perturbation of the intracellular cargo transport often leads to neuropathologies (Schreij, Fon, & McPherson, 2016).

The machinery regulating intracellular transport is relatively well characterized and generally conserved in all cell types. Depending on the cargo and its destination, two major groups of motor proteins, kinesins and dyneins, which are arranged in large macromolecular complexes, mediate trafficking along microtubules. Generally, kinesins move anterogradely toward the microtubule-positive end, while dyneins walk in the opposite direction toward the center of the cell. Interestingly, multiple different microtubule motors can be present at once on a single cargo to enable its delivery to difficult to reach and often distant parts of the cell (Maday, Twelvetrees, Moughamian, & Holzbaur, 2014) and facilitate direction reversals (Splinter et al., 2010). In fact, although cargos often move unidirectionally, some, like mitochondria, travel in both directions (Morris & Hollenbeck, 1993).

Our understanding of the nature of the dynein macromolecular complex has progressed considerably, especially in regard to its motor activity, including the modalities by which dynein hydrolyses ATP and interacts with multiple cofactors to progress along microtubules (Carter, Diamant, & Urnavicius, 2016). However, the mechanism by which this protein complex operates cargo recognition and sorting, though heavily investigated, it is still far from being exhaustively unraveled. Dynein cargos greatly vary among different cell types, ranging from RNA to organelles, and highly complex multiprotein assemblies such as mitochondria or signaling endosomes (Debaisieux, Encheva, Chakravarty, Snijders, & Schiavo, 2016). Another layer of complexity is represented by the notion that dynein cargos are transported through a plethora of distinct cellular compartments, e.g., Golgi network and endoplasmic reticulum (ER) or plasma membrane and lysosomes (Tan, Scherer, & Vallee, 2011), a feat that is often achieved by varying the composition of the relevant dynein complexes (Trokter, Mucke, & Surrey, 2012).

One of the ways cargoes are loaded onto the dynein complex is through their association to several families of molecular motor adaptors (reviewed in Akhmanova & Hammer, 2010; Fu & Holzbaur, 2014; Jolly & Gelfand, 2011; Schlager & Hoogenraad, 2009). Motor adaptors are often core components of much larger protein complexes, which comprise, among others, membrane-associated cargo receptors, scaffolding proteins, and signaling molecules (reviewed in Akhmanova & Hammer, 2010; Hinckelmann, Zala, & Saudou, 2013; Schlager & Hoogenraad, 2009). Adaptor proteins can modulate organelle-specific responses to environmental stimuli (Hoogenraad & Akhmanova, 2016), and this is crucial in neurons, where such processes are tightly regulated. In this chapter, we will attempt to summarize the state of the field in regard to what is known about a family of adaptor proteins, Bicaudal D (BICD), with a particular focus on its functions in neurons.

2. THE BICD FAMILY OF CARGO ADAPTORS

Even though BICD proteins display a widespread cytoplasmic distribution (Hoogenraad et al., 2001, 2003), they are enriched at the minus-end of microtubules (Fumoto, Hoogenraad, & Kikuchi, 2006; Hoogenraad et al., 2001; Mach & Lehmann, 1997), suggesting a role in regulating cytoskeleton function and/or intracellular transport. Indeed, BICDs have been shown in recent years to be pivotal regulators of dynein (Carter et al., 2016), as well as being important players in development, cellular homeostasis, and signaling (Cosker & Segal, 2014; Dienstbier & Li, 2009).

Recent publications suggest that BICDs are important to activate dynein motility (Hoogenraad et al., 2003). BICDs take part in processes like positioning of the centrosome and nucleus (Fig. 1B) (Splinter et al., 2010), which might explain their high expression throughout the organism in early development. In later life stages of many higher organisms like mammals, expression of BICD proteins is predominantly observed in organs like heart, muscle, spinal cord motor, and sensory neurons as well as brain (Baens & Marynen, 1997). In mammals, BICDs were shown to take part in cargo shuttle between Golgi and ER as well as regulation of exosomes secretion (Schlager et al., 2010) (Fig. 1A).

Important roles for BICDs in neuronal biology are also emerging, as several mutations in BICD2 are linked to the development of neuromuscular diseases (Neveling et al., 2013). In the following paragraphs, we will first describe the role of BICD proteins in *Drosophila* and mammalian nonneuronal cells and then concentrate on what is known regarding the functions of these proteins in the context of the nervous system.

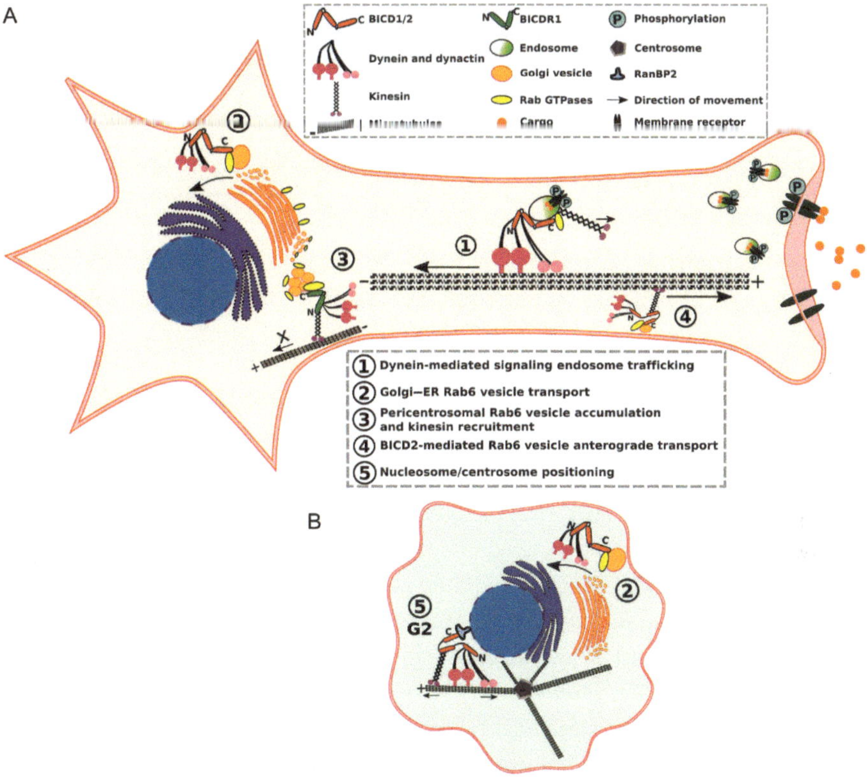

Fig. 1 BICDs main functions in mammalian neurons (A) and mitotic cells (B). In neurons (A), BICDs mediate trafficking of endosomes (1) and vesicles, enabling the integration of signaling outputs initiated by ligand-activation and internalization of various membrane receptors. BICD1/2 C-termini also bind cargos and other accessory proteins, including Rab GTPases, such as Rab6. BICDs and Rab6 association is instrumental in shuttling cargos between the Golgi apparatus and the endoplasmic reticulum (2), in COPI-independent manner. Additionally, during the phase of early neurogenesis, BICDR1 is highly expressed and promotes the accumulation of Rab6 exocytic vesicles in the pericentrosomal region, where they are stalled, leading to neurite outgrowth inhibition (3). Although both BICD2 and BICDR1 bind the dynein complex, the transport kinetics of the latter is much faster than the one mediated by BICD2. In later developmental stages, when BICDR1 expression ceases, BICD2 takes over and facilitates anterograde transport of Rab6 exocytic vesicles by binding to kinesin motors (4), thus restoring neurite outgrowth. Specific to mitotic cells (B), BICD2 helps the positioning of nucleus and centrosome before mitosis by binding to the large nuclear pore complex RanBP2 (5) in the G2 stage of the cell cycle. During that time, BICD2 abolishes its binding to Rab6 vesicles (2).

3. ROLE OF BicD IN *DROSOPHILA* DEVELOPMENT

BICD gene was first identified in *Drosophila melanogaster* as a major maternal-effect regulator of the anterior–posterior polarity in embryos (Mohler & Wieschaus, 1986; Steward & Nusslein-Volhard, 1986). It was found to play crucial role in the specification and differentiation of one of 16 interconnected cystocyte sister cells into an oocyte (Suter, Romberg, & Steward, 1989; Suter & Steward, 1991) and in restricting meiosis to this cell (Huynh & St Johnston, 2000) among others by regulating localization of cell cycle regulatory kinase Polo (Mirouse, Formstecher, & Couderc, 2006). BicD activity is required for establishment and maintenance of polarized microtubule network throughout fly oogenesis (Oh & Steward, 2001) and for proper transporting and localization of important morphogen mRNAs such as *oskar* and *nanos* (Wharton & Struhl, 1989), and *fs(1)K10* and *orb* (Swan & Suter, 1996) in developing oocyte. BicD mediates also apical localization of pair-rule transcripts such as *ftz* and *h* in blastoderm embryos (Bullock & Ish-Horowicz, 2001) and of *insc* mRNA in neuroblasts (Hughes, Bullock, & Ish-Horowicz, 2004). BicD exerts its RNA localization role by forming a complex with Egalitarian (Egl) protein (Mach & Lehmann, 1997), which binds to a microtubule motor protein dynein (Navarro, Puthalakath, Adams, Strasser, & Lehmann, 2004) and to various mRNAs (Dienstbier & Li, 2009).

In *Drosophila*, BicD plays role in the localization of various organelles. During oogenesis, its activity is required for proper positioning of nucleus to a prospective dorsoanterior corner of the growing oocyte in a process which also depends on the functional microtubule cytoskeleton and Lis1 (Swan, Nguyen, & Suter, 1999). Similarly, BicD together with Ste-20 like kinase and dynein mediate nucleus migration in precursors of R-cells in developing fly eye (Houalla, Hien Vuong, Ruan, Suter, & Rao, 2005). BicD interacts also with *Drosophila* homolog of Rab6 present on the Golgi membranes and thus regulates polarized delivery of secretory vesicles containing protein morphogens such as Gurken to the plasma membrane of oocyte (Januschke et al., 2007). In fly larvae neurons, BicD facilitates neurotransmitter release and recycling of synaptic membranes during high-frequency stimulation by interacting with clathrin heavy chain and thus possibly with the clathrin-mediated endocytic pathway (Li et al., 2010). Likewise, BicD regulates the localization of clathrin mRNA and protein in the oocyte, and both proteins are required to maintain enhanced endocytosis at the posterior of the oocyte (Vazquez-Pianzola et al., 2014).

4. MAMMALIAN BICDs AND THEIR ROLE IN MITOTIC CELLS

Although there is only one BicD protein in *Drosophila*, an entire BICD family exists in mammals (Baens & Marynen, 1997). The most characterized mammalian homologues are BICD1 and BICD2; in addition, two related and less conserved proteins, BICDR1 and BICDR2, were recently described (Schlager et al., 2010, 2014). Mammalian BICD family members are present in various parts of the cell and, though slightly differing in their roles, share many common structural features, such as highly conserved coiled-coil (CC) regions, and functions. While this chapter will address the major different roles of mammalian BICDs, an in-depth discussion about their structure is outside the scope; this topic has been, however, extensively covered in a recent review by Hoogenraad and Akhmanova (2016).

The first set of observations regarding BICD1 and BICD2 in mammalian mitotic cells described them as important players in COPI-independent retrograde transport between Golgi apparatus and ER (Fig. 1A), where they localize (Matanis et al., 2002). This transport route functions in parallel to the more known COPI-dependent route and might be involved in the intracellular shuttling of various proteins that lack a distinctive dilysine and KDEL retrieval signal, or cargos that cannot be captured within small COPI-positive vesicles (Girod et al., 1999; White et al., 1999). COPI-independent transport is slower than the COPI-mediated one, and it is still not fully understood, though it has been implicated in the delivery of long-lived Golgi-resident enzymes back to ER for quality control or recycling (Sengupta et al., 2015).

BICD1/2 interacts via its C-terminal coiled-coil domain with the Golgi-specific Rab6 GTPase, which can itself directly bind dynactin, and with dynein via its N-terminal domain (Matanis et al., 2002); this interaction is instrumental in attracting and localizing dynein complexes to Rab6 active zones on the Golgi (Matanis et al., 2002). BicD association through its C-terminus with Rab6 is also observed in *Drosophila*, and it is conserved in several species (review in Dienstbier & Li, 2009). A study by Hoogenraad et al. shows that BICD2 is specifically enriched in Golgi and ER regions as well as in intermediate compartments and recycling endosomes (Hoogenraad et al., 2001). In addition, overexpression of the C-terminal domain of BICD2 abolishes detection of endogenous proteins within the Golgi, while the organelle itself remains intact, suggesting that this domain may compete with endogenous BICD2 and that BICD2 itself is crucial for Golgi targeting. Overexpressed GFP-BICD2 N-terminal portion, on the other hand, was shown to have a diffuse cytoplasmic distribution, while overexpressing BICD2 without its N-terminal domain resulted in Golgi fragmentation (Hoogenraad et al., 2001). This phenomenon is also observed upon inhibition of dynein function, suggesting that BICD2 N-terminus indeed associates with the dynein motor complex and is important for its proper functioning (Hoogenraad et al., 2001).

The formation of a complex between BICD1 and dynein is facilitated by phosphorylation of BICD1 itself induced by GSK-3β (Fumoto et al., 2006). BICD1 and GSK-3β together form a stable nucleation center for the anchorage of microtubules. Consequently, their knockdown or alanine mutation of a specific GSK-3β phosphorylation site on BICD1/2 reduces centrosomally focused microtubules and induces the mislocalization of centrosomal proteins (Fumoto et al., 2006).

BICD2 has also been shown to be important for mitosis, where it ensures accurate positioning of the nucleus by interacting with dynein and kinesin motors as well as microtubules (Splinter et al., 2010) (Fig. 1B). In the G2 phase of the cell cycle, before the onset of mitosis, its C-terminus is recruited by the large scaffold nucleoporin protein RanBP2, whose main function is to shuttle proteins between nucleus and cytoplasm and to facilitate protein sumoylation (Sakin, Richter, Hsiao, Urlaub, & Melchior, 2015). This binding promotes the activation of the dynein complex, which in turn results in bringing the centrosome and the nucleus closely together (Splinter et al., 2010). BICD2 association with kinesin on the other hand, which happens through its middle coiled domain, results in the opposite effect in that it pushes the centrosome away from the nucleus (Splinter et al., 2010). In addition, BICD2 binds to individual nuclear pore complexes on the nuclear envelope prior to mitosis; this interaction, however, is abolished upon RanBP2 knockdown (Splinter et al., 2010). Intriguingly, the study by Splinter et al. also shows that BICD2 colocalization with RanBP2 is specific to the G2 stage, at which point BICD2 does not associate with Rab6 on the Golgi (Splinter et al., 2010). At the same time, if BICD2 is present on the Golgi and ER structures (Fig. 1B), it does not bind to RanBP2 on the nuclear pore (Splinter et al., 2010). This suggests that BICD2 has mutually exclusive functions depending on the stage of cell cycle and is important for cell reorganization prior to mitosis (Splinter et al., 2010).

5. ROLE OF BICD IN NEURONS

One of the key features of the nervous system organization is the establishment of the cell dendritic arborization pattern, which is crucial for the correct assembly of neuronal circuitry and the integration and processing of synaptic and sensory inputs. BICD has been shown to be central in this process by interacting with a few pivotal cargoes. One such cargo is the Fragile X mental retardation protein (FMRP), an RNA-binding protein that controls neuronal morphogenesis and plays a critical role in the Fragile X syndrome, the most-frequently inherited form of intellectual disability in human (Richter, Bassell, & Klann, 2015). It does so by mediating the translational regulation and localization of a significant number of mRNAs (Richter et al., 2015) in a way that is dependent on microtubule-based transport of FMRP complexes within neurites (Richter et al., 2015). A study of 2010 has shown that FMRP is complexed with BicD in *Drosophila* neurons and that both BicD and FMRP are required for efficient dendritic branching (Bianco, Dienstbier, Salter, Gatto, & Bullock, 2010). The binding of FMRP to BicD links it to dynein with the surprising result that both the motility and FMRP protein levels were dramatically reduced in BicD mutant *Drosophila* neurons (Bianco et al., 2010).

Bicd-1 was also identified as a key regulator of dendritic branch formation in an RNAi screen performed in *C. elegans* PVD neurons (Aguirre-Chen, Bulow, & Kaprielian, 2011), which are nociceptors that respond to harsh mechanical stimuli and cold temperatures (Chatzigeorgiou et al., 2010). In this context, bicd-1 operates in conjunction with dhc-1 and unc-116, components of the dynein and kinesin 1 complexes, respectively, while also interacting with the repulsive guidance receptor unc-5 (Aguirre-Chen et al., 2011). Specifically, loss of bicd-1 resulted in the appearance of ectopic dendrite branches posterior and proximal to the neuronal cell body, while reduction in the number of terminal branches was observed (Aguirre-Chen et al., 2011).

Another study conducted in *Drosophila* described clathrin heavy chain (Chc), a major player in endocytosis and vesicle coating, as the most abundant protein coprecipitated with BicD from head extracts (Li et al., 2010), suggesting that BicD interacts with important components of the Chc-mediated membrane pathway. Indeed, this study suggests that dynein-mediated transport is required for the proper homeostasis of the synaptic vesicle cycle and its perturbation in BicD mutants results in a mislocalization of

Chc at the level of the presynaptic boutons, which in turn affects the levels of neurotransmitter release during high-frequency electrical activity (Li et al., 2010).

The role of BICD in the mammalian nervous system is only now starting to emerge. The pattern of expression of BICD1 in the developing mouse embryo was recently described as confined to the ventral horn of the developing spinal cord, dorsal root ganglia, and brain at embryonic day 12 (Terenzio, Golding, Russell, et al., 2014). However, at embryonic day 14.5 BICD1 expression dramatically changes, being lost from the brain and spinal cord, yet still strongly retained in DRG and skin, skeletal muscle, and heart ventricle (Terenzio, Golding, Russell, et al., 2014). Interestingly, an older publication described the association between BICD1 and Rab6B, a brain-specific GTPase isoform of Rab6, which takes place at membranous structures on the Golgi and other vesicular organelles and determines their retrograde movement in neurites (Wanschers et al., 2007). This last result seems in line with the aforementioned observation of BicD being implicated in retrograde microtubule-based transport of cargoes, such as FMRP complexes, within *Drosophila* neurons (Richter et al., 2015).

While a BICD1 KO mouse has not yet been described in the literature, a BICD2 KO mouse has been generated and reported in 2014 (Jaarsma et al., 2014). These mice have a striking phenotype, and they develop defects in the laminar organization of the cerebral cortex, hippocampus, and cerebellar cortex, which are indicative of radial neuronal migration defects (Jaarsma et al., 2014). Surprisingly, depletion of BICD2 caused a decreased deposition of the extracellular matrix protein Tenascin C by the Bergmann glia cells, which was speculated to be the reason for the observed migration defect of the radial cerebellar granule cells (Jaarsma et al., 2014).

To further cement the important role exerted by BICD proteins in the homeostasis of the nervous system, BICDs have been implicated in the etiology of neuropathologies such as spinal muscular atrophy (SMA), a heterogeneous group of neuromuscular disorders caused by degeneration of lower motor neurons (Neveling et al., 2013). Indeed, in recent years, several heterozygous mutations present in many different regions of the BICD2 protein were described to be causative of autosomal dominant spinal muscular atrophy with lower extremity predominance-2 (SMALED2) and hereditary spastic paraplegia (Martinez-Carrera & Wirth, 2015). The authors speculate that these mutations may increase or decrease the affinity of BICD2 for either the molecular motor or its cargoes with possibly deleterious effects in cells like spinal motor neurons, which are highly dependent on retrograde transport (Martinez-Carrera & Wirth, 2015).

A recent addition to the mammalian BICD family, Bicaudal D-related protein 1 (BICDR1), was also shown to be important for the fine tuning of the process of neurogenesis (Schlager et al., 2010) (Fig. 1A). This protein was shown to associate with Rab6-positive exocytotic vesicles and recruit dynein and kinesin-3 motor complexes to the pericentrosomal region of developing neurons (Schlager et al., 2010). Its high expression in very early development suppresses neurite outgrowth, which is restored upon lowering expression of this homologue in later life stages and allowing efficient anterograde cargo transport and neurite growth (Schlager et al., 2010). Furthermore, BICDR1 enhances the velocity of dynein-mediated transport of Rab6-positive vesicles, while BICD2 does not, although both of them interact with dynein complex through the same conserved domain (Schlager et al., 2014).

6. BICDs AND NT SIGNALING

The NT signaling pathway is one of the most studied signaling pathways in neurons (Cosker & Segal, 2014). As the name implies, its main purpose is trophic or prosurvival support, which is crucial to the development of the nervous system and modulates events like dendritic branching, axonal growth, guidance, and differentiation (Ascano, Bodmer, & Kuruvilla, 2012; Emdal et al., 2015; Garcia et al., 2010). NTs have also nontrophic roles, as they have been described to activate the apoptotic pathway in a manner that is highly dependent on the stage of maturation of the nervous system and of its individual components (Anastasia, Barker, Chao, & Hempstead, 2015; Frade, Rodriguez-Tebar, & Barde, 1996). NT-dependent apoptotic pathway can also be triggered in neuronal tumor cells (Matsushima & Bogenmann, 1993; Muragaki, Chou, Kaplan, Trojanowski, & Lee, 1997).

The main effectors of NT signaling are growth factors, which include nerve growth factor (NGF), brain-derived neurotrophic factor (BDNF), neurotrophin-3 (NT-3), and NT4/5. As dimers, they bind to and activate two distinct classes of receptors. The first class, the tropomyosin receptor kinase (Trk), is related to receptor tyrosine kinases and includes TrkA, TrkB, and TrkC (Kaplan & Miller, 1997). Growth factor binding promotes receptor dimerization and autophosphorylation, which in turn leads to the activation of downstream signaling cascades by the endocytosed receptors complexes (Senger & Campenot, 1997). The low-affinity NGF receptor or $p75^{NTR}$ represent a second distinct class of NT receptors that can be activated by all four NTs (Roux & Barker, 2002). Unlike the other group, activated $p75^{NTR}$ mediates mainly activation of the cell death cascade by apoptosis (Frade et al., 1996). In addition, all Trk receptors have the ability to heterodimerize with $p75^{NTR}$, which modulates the activity of Trk receptors by increasing their specificity, affinity, and level of activation by mature growth factors (Nykjaer, Willnow, & Petersen, 2005).

The output created by growth factor binding to receptors is dependent on the class, composition, and availability of the receptors at the plasma membrane as well as the maturation stage of the ligand itself (Nykjaer et al., 2005). NTs, in fact, are secreted by many organs, including brain as well as target-innervated tissues, e.g., muscle, and are additionally released by neurons themselves as larger precursors, in which case are called pro-NTs and in their mature form after proteolytic cleavage, which can be achieved both intra- and extracellularly (Gray & Ellis, 2008). Both pro-NTs and mature NTs are important and play different roles in early development as well as in adulthood (Nykjaer et al., 2005).

Given the complexity of the NT signaling network, proper regulation in time and space is of great essence. There are several steps and "check points" that determine the establishment of a correct signaling output, many of which are well characterized and include endocytosis, axonal retrograde transport from the distal part of the axon toward the soma and sorting of activated receptors and NT ligands. NT receptors can either act locally at the cell membrane or in the soma where they elicit a transcriptional response through the activation of signaling intermediates like AKT and ERK1/2, which promote survival, differentiation and growth (Zhang, Moheban, Conway, Bhattacharyya, & Segal, 2000), or JNK, which induces apoptosis (Harrington, Kim, & Yoon, 2002).

Although many components of NT signaling have been studied and described extensively over the years, an in-depth understanding of processes like receptor transport and sorting is still missing. Recently, BICD1 was identified as an important player in NT receptor sorting and signaling in mouse embryonic stem cell-derived motor neurons by a high-throughput siRNA screen for molecular motors and protein adaptors (Terenzio, Golding, Russell, et al., 2014; Terenzio, Golding, & Schiavo, 2014). In essence, BICD1 was shown to control the trafficking of activated NT receptors to their appropriate degradation routes (Fig. 1). This controls were postulated to be in place in order to ensure an optimal response to NT stimulation, by modulating both the amplitude and the duration of the signal elicited by ligand-bound NT receptors (Terenzio, Golding, Russell, et al., 2014) (Fig. 1A(2)). This is achieved by modulating the balance between NT receptor degradation over recycling, where the receptor pool averted from degradation is targeted back to the plasma membrane by a retromer-mediated sorting process (Terenzio, Golding, Russell, et al., 2014).

In order to monitor the endocytic pathway, the authors made use of the nonpathogenic-binding fragment (H_CT) of tetanus neurotoxin (TeNT) and an antibody raised against $p75^{NTR}$. H_CT was chosen for being a reliable probe to monitor NT receptor uptake and intracellular trafficking since it was shown to share the same retrograde trafficking route of NTs and their receptors (Deinhardt et al., 2006; Lalli & Schiavo, 2002. This has been further validated by the finding that TeNT hijacks the NT internalization machinery and binds to the neuromuscular junction basement membrane protein Nidogen (Bercsenyi et al., 2014).

In spite of its role as dynein adaptor and its postulated involvement in retrograde transport in neuronal cell lines (Wanschers et al., 2007), BICD1 was shown not to be essential for the transport of signaling endosomes in motor neurons as their speed and frequency were not perturbed by BICD1 depletion (Terenzio, Golding, Russell, et al., 2014). Similarly, BICD1 depletion did not affect the morphology or branching complexity of motor neuron neurite network (Terenzio, Golding, Russell, et al., 2014). Transmission electron microscopy performed on motor neurons depleted of BICD1 revealed the presence of enlarged sorting endosome compartments in which H_CT accumulated in a spatially segregated manner (Terenzio, Golding, Russell, et al., 2014). These organelles were reminiscent of those shown to trap NGF in sympathetic neurons treated with lysosomal inhibitors (Claude, Hawrot, Dunis, & Campenot, 1982). The morphology of these organelles together with the spatial segregation of H_CT was highly suggestive of an impairment of the NT receptor targeting to lysosomes due to BICD1 depletion.

As previously mentioned, the rewiring of the endosomal pathway caused by BICD1 depletion ultimately affected NT signaling output, as shown by a reduced Akt1 and ERK1/2 activation downstream of TrkB signaling stimulated by BDNF (Terenzio, Golding, Russell, et al., 2014).

7. CONCLUSIONS

In this chapter, we have briefly summarized the essential functions described in the literature regarding the BICD family of protein adaptors with a focus on their prominent roles in neuronal homeostasis (Fig. 1). Importantly, correct development and maintenance of the nervous system are bound to the flux of signaling information, RNA and cell constituents that is orchestrated through a complex network of intracellular cargo transport. For this reason, motor adaptor proteins like BICDs are positioned to be master regulators of neuronal physiology.

Indeed, important roles for BICDs in the etiology of neuropathologies such as SMA are starting to emerge (Martinez-Carrera & Wirth, 2015; Neveling et al., 2013), which comprises a heterogeneous group of neuromuscular disorders caused by degeneration of lower motor neurons, a cell type that is highly dependent on NTs for its specification and maintenance in adulthood (Schmieg et al., 2014).

Unfortunately not much is known about the role of BICDs with regard to NT signaling and much is left for further investigation. Only one member of the BICD family, BICD1, has been clearly implicated in the regulation of this vital signaling pathway (Terenzio, Golding, Russell, et al., 2014), though it stands to logic that other members of the family might be involved in this process too. Sadly, a knock out mouse model for BICD1, which might lead to more insights into the role of this motor adaptor has not yet being described and should certainly be the focus of future efforts.

Interestingly, a very recent study taking advantage of stable isotope labeling with amino acids in cell culture in embryonic stem cell-derived motor neurons in conjunction with an affinity purification strategy using H_CT conjugated to monocrystalline iron oxide nanoparticles has allowed the magnetic isolation and subsequent mass-spectrometry analysis of retrograde NT signaling endosomes (Debaisieux et al., 2016). These carriers were found to be specifically enriched in proteins known to be involved in neurodegenerative diseases and neuroinfection (Debaisieux et al., 2016), and BICD1 was shown to associate to the same compartment through a similar magnetic isolation procedure (Terenzio, Golding, Russell, et al., 2014), reinforcing its potential role as a master regulator of these processes.

H_CT and TrkB were found to colocalize with sorting nexin 1 (SNX1) and the vacuolar protein sorting-associated protein 26 (VPS26) in motor neurons (Terenzio, Golding, Russell, et al., 2014), which are members of the retromer complex responsible for the recycling of membrane receptors back to the plasma membrane or Golgi (Haft et al., 2000). Interestingly, although SNX1 is involved in recycling pathway, it was also shown to be crucial for EGF receptor and the protease-activated receptor-1 (PAR1) targeting for lysosomal degradation (Gullapalli, Wolfe, Griffin, Magnuson, & Trejo, 2006). Intriguingly, PAR1, a G protein-coupled receptor with critical roles in cancer, angiogenesis, inflammation, and thrombosis, was shown in a yeast two-hybrid screen to be a direct interactor of BICD1 through its C-terminal cytoplasmic domain (Swift et al., 2010). In this context, BICD1 was found to inhibit PAR1 signaling and be a potent suppressor of PAR1-driven proliferation of breast carcinoma cells (Swift et al., 2010). Moreover, silencing of BICD1 expression impacted PAR1endocytosis suggesting that BICD1 might act as an adaptor protein in the transport of PAR1 from the plasma membrane to endosomal vesicles (Swift et al., 2010). This study clearly shows that BICD1 is able to regulate the signaling output of membrane receptors (Fig. 1) other than the NT ones and in a variety of cell type, which include cycling cells and neurons.

Overall more and more roles for BICDs are likely to emerge in regard to neuronal signaling in the near future, and they might become central in unraveling how intracellular cargo transport defect is at the hearth of several neuropathologies.

ACKNOWLEDGMENTS

We thank Dr. Sandip Koley for constructive comments and for critical reading of the document. Research on these topics is generously supported by the BBSRC Industrial CASE PhD Fellowship (515092 and 533334). The authors have no conflicting financial interests.

REFERENCES

Aguirre-Chen, C., Bulow, H. E., & Kaprielian, Z. (2011). C. elegans bicd-1, homolog of the Drosophila dynein accessory factor Bicaudal D, regulates the branching of PVD sensory neuron dendrites. *Development*, *138*(3), 507–518. http://dx.doi.org/10.1242/dev.060939.

Akhmanova, A., & Hammer, J. A., 3rd. (2010). Linking molecular motors to membrane cargo. *Current Opinion in Cell Biology*, *22*(4), 479–487. http://dx.doi.org/10.1016/j.ceb.2010.04.008.

Anastasia, A., Barker, P. A., Chao, M. V., & Hempstead, B. L. (2015). Detection of p75NTR Trimers: Implications for receptor stoichiometry and activation. *The Journal of Neuroscience*, *35*(34), 11911–11920. http://dx.doi.org/10.1523/JNEUROSCI.0591-15.2015.

Ascano, M., Bodmer, D., & Kuruvilla, R. (2012). Endocytic trafficking of neurotrophins in neural development. *Trends in Cell Biology*, *22*(5), 266–273. http://dx.doi.org/10.1016/j.tcb.2012.02.005.

Baens, M., & Marynen, P. (1997). A human homologue (BICD1) of the Drosophila bicaudal-D gene. *Genomics*, *45*(3), 601–606. http://dx.doi.org/10.1006/geno.1997.4971.

Bercsenyi, K., Schmieg, N., Bryson, J. B., Wallace, M., Caccin, P., Golding, M., … Schiavo, G. (2014). Tetanus toxin entry. Nidogens are therapeutic targets for the prevention of tetanus. *Science*, *346*(6213), 1118–1123. http://dx.doi.org/10.1126/science.1258138.

Bianco, A., Dienstbier, M., Salter, H. K., Gatto, G., & Bullock, S. L. (2010). Bicaudal-D regulates fragile X mental retardation protein levels, motility, and function during neuronal morphogenesis. *Current Biology*, *20*(16), 1487–1492. http://dx.doi.org/10.1016/j.cub.2010.07.016.

Bullock, S. L., & Ish-Horowicz, D. (2001). Conserved signals and machinery for RNA transport in Drosophila oogenesis and embryogenesis. *Nature*, *414*(6864), 611–616. http://dx.doi.org/10.1038/414611a.

Carter, A. P., Diamant, A. G., & Urnavicius, L. (2016). How dynein and dynactin transport cargos: A structural perspective. *Current Opinion in Structural Biology*, *37*, 62–70. http://dx.doi.org/10.1016/j.sbi.2015.12.003.

Chatzigeorgiou, M., Yoo, S., Watson, J. D., Lee, W. H., Spencer, W. C., Kindt, K. S., … Schafer, W. R. (2010). Specific roles for DEG/ENaC and TRP channels in touch and thermosensation in C. elegans nociceptors. *Nature Neuroscience*, *13*(7), 861–868. http://dx.doi.org/10.1038/nn.2581.

Claude, P., Hawrot, E., Dunis, D. A., & Campenot, R. B. (1982). Binding, internalization, and retrograde transport of 125I-nerve growth factor in cultured rat sympathetic neurons. *The Journal of Neuroscience*, *2*(4), 431–442.

Cosker, K. E., & Segal, R. A. (2014). Neuronal signaling through endocytosis. *Cold Spring Harbor Perspectives in Biology*, *6*(2), 11482–11488, http://dx.doi.org/10.1101/cshperspect.a020669.

Debaisieux, S., Encheva, V., Chakravarty, P., Snijders, A. P., & Schiavo, G. (2016). Analysis of signaling endosome composition and dynamics using SILAC in embryonic stem cell-derived neurons. *Molecular and Cellular Proteomics*, *15*(2), 542–557. http://dx.doi.org/10.1074/mcp.M115.051649.

Deinhardt, K., Salinas, S., Verastegui, C., Watson, R., Worth, D., Hanrahan, S., … Schiavo, G. (2006). Rab5 and Rab7 control endocytic sorting along the axonal retrograde transport pathway. *Neuron*, *52*(2), 293–305. http://dx.doi.org/10.1016/j.neuron.2006.08.018.

Dienstbier, M., & Li, X. (2009). Bicaudal-D and its role in cargo sorting by microtubule-based motors. *Biochemical Society Transactions*, *37*(Pt. 5), 1066–1071. http://dx.doi.org/10.1042/BST0371066.

Emdal, K. B., Pedersen, A. K., Bekker-Jensen, D. B., Tsafou, K. P., Horn, H., Lindner, S., … Olsen, J. V. (2015). Temporal proteomics of NGF-TrkA signaling identifies an inhibitory role for the E3 ligase Cbl-b in neuroblastoma cell differentiation. *Science Signaling*, *8*(374), ra40. http://dx.doi.org/10.1126/scisignal.2005769.

Frade, J. M., Rodriguez-Tebar, A., & Barde, Y. A. (1996). Induction of cell death by endogenous nerve growth factor through its p75 receptor. *Nature*, *383*(6596), 166–168. http://dx.doi.org/10.1038/383166a0.

Fu, M. M., & Holzbaur, E. L. (2014). Integrated regulation of motor-driven organelle transport by scaffolding proteins. *Trends in Cell Biology*, *24*(10), 564–574. http://dx.doi.org/10.1016/j.tcb.2014.05.002.

Fumoto, K., Hoogenraad, C. C., & Kikuchi, A. (2006). GSK-3beta-regulated interaction of BICD with dynein is involved in microtubule anchorage at centrosome. *The EMBO Journal*, *25*(24), 5670–5682. http://dx.doi.org/10.1038/sj.emboj.7601459.

Garcia, N., Santafe, M. M., Tomas, M., Lanuza, M. A., Besalduch, N., & Tomas, J. (2010). Involvement of brain-derived neurotrophic factor (BDNF) in the functional elimination of synaptic contacts at polyinnervated neuromuscular synapses during development. *Journal of Neuroscience Research*, *88*(7), 1406–1419. http://dx.doi.org/10.1002/jnr.22320.

Gibbs, K. L., Greensmith, L., & Schiavo, G. (2015). Regulation of axonal transport by protein kinases. *Trends in Biochemical Sciences*, *40*(10), 597–610. http://dx.doi.org/10.1016/j.tibs.2015.08.003.

Girod, A., Storrie, B., Simpson, J. C., Johannes, L., Goud, B., Roberts, L. M., … Pepperkok, R. (1999). Evidence for a COP-I-independent transport route from the Golgi complex to the endoplasmic reticulum. *Nature Cell Biology*, *1*(7), 423–430. http://dx.doi.org/10.1038/15658.

Granseth, B., Odermatt, B., Royle, S. J., & Lagnado, L. (2006). Clathrin-mediated endocytosis is the dominant mechanism of vesicle retrieval at hippocampal synapses. *Neuron*, *51*(6), 773–786. http://dx.doi.org/10.1016/j.neuron.2006.08.029.

Gray, K., & Ellis, V. (2008). Activation of pro-BDNF by the pericellular serine protease plasmin. *FEBS Letters*, *582*(6), 907–910. http://dx.doi.org/10.1016/j.febslet.2008.02.026.

Gullapalli, A., Wolfe, B. L., Griffin, C. T., Magnuson, T., & Trejo, J. (2006). An essential role for SNX1 in lysosomal sorting of protease-activated receptor-1: Evidence for retromer-, Hrs-, and Tsg101-independent functions of sorting nexins. *Molecular Biology of the Cell*, *17*(3), 1228–1238. http://dx.doi.org/10.1091/mbc.E05-09-0899.

Haft, C. R., de la Luz Sierra, M., Bafford, R., Lesniak, M. A., Barr, V. A., & Taylor, S. I. (2000). Human orthologs of yeast vacuolar protein sorting proteins Vps26, 29, and 35: Assembly into multimeric complexes. *Molecular Biology of the Cell*, *11*(12), 4105–4116.

Harrington, A. W., Kim, J. Y., & Yoon, S. O. (2002). Activation of Rac GTPase by p75 is necessary for c-jun N-terminal kinase-mediated apoptosis. *The Journal of Neuroscience*, *22*(1), 156–166.

Hinckelmann, M. V., Zala, D., & Saudou, F. (2013). Releasing the brake: Restoring fast axonal transport in neurodegenerative disorders. *Trends in Cell Biology*, *23*(12), 634–643. http://dx.doi.org/10.1016/j.tcb.2013.08.007.

Hoogenraad, C. C., & Akhmanova, A. (2016). Bicaudal D family of motor adaptors: Linking dynein motility to cargo binding. *Trends in Cell Biology*, *26*(5), 327–340. http://dx.doi.org/10.1016/j.tcb.2016.01.001.

Hoogenraad, C. C., Akhmanova, A., Howell, S. A., Dortland, B. R., De Zeeuw, C. I., Willemsen, R., … Galjart, N. (2001). Mammalian Golgi-associated Bicaudal-D2 functions in the dynein-dynactin pathway by interacting with these complexes. *The EMBO Journal*, *20*(15), 4041–4054. http://dx.doi.org/10.1093/emboj/20.15.4041.

Hoogenraad, C. C., Wulf, P., Schiefermeier, N., Stepanova, T., Galjart, N., Small, J. V., … Akhmanova, A. (2003). Bicaudal D induces selective dynein-mediated microtubule minus end-directed transport. *The EMBO Journal*, *22*(22), 6004–6015. http://dx.doi.org/10.1093/emboj/cdg592.

Houalla, T., Hien Vuong, D., Ruan, W., Suter, B., & Rao, Y. (2005). The Ste20-like kinase misshapen functions together with Bicaudal-D and dynein in driving nuclear migration in the developing drosophila eye. *Mechanisms of Development*, *122*(1), 97–108. http://dx.doi.org/10.1016/j.mod.2004.08.005.

Hughes, J. R., Bullock, S. L., & Ish-Horowicz, D. (2004). Inscuteable mRNA localization is dynein-dependent and regulates apicobasal polarity and spindle length in Drosophila neuroblasts. *Current Biology*, *14*(21), 1950–1956. http://dx.doi.org/10.1016/j.cub.2004.10.022.

Huynh, J. R., & St Johnston, D. (2000). The role of BicD, Egl, Orb and the microtubules in the restriction of meiosis to the Drosophila oocyte. *Development*, *127*(13), 2785–2794.

Jaarsma, D., van den Berg, R., Wulf, P. S., van Erp, S., Keijzer, N., Schlager, M. A., … Hoogenraad, C. C. (2014). A role for Bicaudal-D2 in radial cerebellar granule cell migration. *Nature Communications*, *5*, 3411. http://dx.doi.org/10.1038/ncomms4411.

Januschke, J., Nicolas, E., Compagnon, J., Formstecher, E., Goud, B., & Guichet, A. (2007). Rab6 and the secretory pathway affect oocyte polarity in Drosophila. *Development*, *134*(19), 3419–3425. http://dx.doi.org/10.1242/dev.008078.

Jolly, A. L., & Gelfand, V. I. (2011). Bidirectional intracellular transport: Utility and mechanism. *Biochemical Society Transactions*, *39*(5), 1126–1130. http://dx.doi.org/10.1042/BST0391126.

Kaplan, D. R., & Miller, F. D. (1997). Signal transduction by the neurotrophin receptors. *Current Opinion in Cell Biology*, *9*(2), 213–221.

Lalli, G., & Schiavo, G. (2002). Analysis of retrograde transport in motor neurons reveals common endocytic carriers for tetanus toxin and neurotrophin receptor p75NTR. *The Journal of Cell Biology*, *156*(2), 233–239. http://dx.doi.org/10.1083/jcb.200106142.

Le Roy, C., & Wrana, J. L. (2005). Clathrin- and non-clathrin-mediated endocytic regulation of cell signalling. *Nature Reviews. Molecular Cell Biology*, *6*(2), 112–126. http://dx.doi.org/10.1038/nrm1571.

Li, X., Kuromi, H., Briggs, L., Green, D. B., Rocha, J. J., Sweeney, S. T., & Bullock, S. L. (2010). Bicaudal-D binds clathrin heavy chain to promote its transport and augments synaptic vesicle recycling. *The EMBO Journal*, *29*(5), 992–1006. http://dx.doi.org/10.1038/emboj.2009.410.

Mach, J. M., & Lehmann, R. (1997). An Egalitarian-BicaudalD complex is essential for oocyte specification and axis determination in Drosophila. *Genes and Development*, *11*(4), 423–435.

Maday, S., Twelvetrees, A. E., Moughamian, A. J., & Holzbaur, E. L. (2014). Axonal transport: Cargo-specific mechanisms of motility and regulation. *Neuron*, *84*(2), 292–309. http://dx.doi.org/10.1016/j.neuron.2014.10.019.

Martinez-Carrera, L. A., & Wirth, B. (2015). Dominant spinal muscular atrophy is caused by mutations in BICD2, an important golgin protein. *Frontiers in Neuroscience*, *9*, 401. http://dx.doi.org/10.3389/fnins.2015.00401.

Matanis, T., Akhmanova, A., Wulf, P., Del Nery, E., Weide, T., Stepanova, T., … Hoogenraad, C. C. (2002). Bicaudal-D regulates COPI-independent Golgi-ER transport by recruiting the dynein-dynactin motor complex. *Nature Cell Biology*, *4*(12), 986–992. http://dx.doi.org/10.1038/ncb891.

Matsushima, H., & Bogenmann, E. (1993). Expression of trkA cDNA in neuroblastomas mediates differentiation in vitro and in vivo. *Molecular and Cellular Biology*, *13*(12), 7447–7456.

Mirouse, V., Formstecher, E., & Couderc, J. L. (2006). Interaction between Polo and BicD proteins links oocyte determination and meiosis control in Drosophila. *Development*, *133*(20), 4005–4013. http://dx.doi.org/10.1242/dev.02565.

Mohler, J., & Wieschaus, E. F. (1986). Dominant maternal-effect mutations of Drosophila melanogaster causing the production of double-abdomen embryos. *Genetics*, *112*(4), 803–822.

Morris, R. L., & Hollenbeck, P. J. (1993). The regulation of bidirectional mitochondrial transport is coordinated with axonal outgrowth. *Journal of Cell Science*, *104*(Pt. 3), 917–927.

Muragaki, Y., Chou, T. T., Kaplan, D. R., Trojanowski, J. Q., & Lee, V. M. (1997). Nerve growth factor induces apoptosis in human medulloblastoma cell lines that express TrkA receptors. *The Journal of Neuroscience*, *17*(2), 530–542.

Navarro, C., Puthalakath, H., Adams, J. M., Strasser, A., & Lehmann, R. (2004). Egalitarian binds dynein light chain to establish oocyte polarity and maintain oocyte fate. *Nature Cell Biology*, *6*(5), 427–435. http://dx.doi.org/10.1038/ncb1122.

Neveling, K., Martinez-Carrera, L. A., Holker, I., Heister, A., Verrips, A., Hosseini-Barkooie, S. M., … Wirth, B. (2013). Mutations in BICD2, which encodes a golgin and important motor adaptor, cause congenital autosomal-dominant spinal muscular atrophy. *The American Journal of Human Genetics*, *92*(6), 946–954. http://dx.doi.org/10.1016/j.ajhg.2013.04.011

Nykjaer, A., Willnow, T. E., & Petersen, C. M. (2005). p75NTR—Live or let die. *Current Opinion in Neurobiology*, *15*(1), 49–57. http://dx.doi.org/10.1016/j.conb.2005.01.004.

Oh, J., & Steward, R. (2001). Bicaudal-D is essential for egg chamber formation and cytoskeletal organization in drosophila oogenesis. *Developmental Biology*, *232*(1), 91–104. http://dx.doi.org/10.1006/dbio.2001.0170.

Richter, J. D., Bassell, G. J., & Klann, E. (2015). Dysregulation and restoration of translational homeostasis in fragile X syndrome. *Nature Reviews. Neuroscience*, *16*(10), 595–605. http://dx.doi.org/10.1038/nrn4001.

Roux, P. P., & Barker, P. A. (2002). Neurotrophin signaling through the p75 neurotrophin receptor. *Progress in Neurobiology*, *67*(3), 203–233.

Sakin, V., Richter, S. M., Hsiao, H. H., Urlaub, H., & Melchior, F. (2015). Sumoylation of the GTPase Ran by the RanBP2 SUMO E3 ligase complex. *The Journal of Biological Chemistry*, *290*(39), 23589–23602. http://dx.doi.org/10.1074/jbc.M115.660118.

Schlager, M. A., & Hoogenraad, C. C. (2009). Basic mechanisms for recognition and transport of synaptic cargos. *Molecular Brain*, *2*, 25. http://dx.doi.org/10.1186/1756-6606-2-25.

Schlager, M. A., Kapitein, L. C., Grigoriev, I., Burzynski, G. M., Wulf, P. S., Keijzer, N., … Hoogenraad, C. C. (2010). Pericentrosomal targeting of Rab6 secretory vesicles by Bicaudal-D-related protein 1 (BICDR-1) regulates neuritogenesis. *The EMBO Journal*, *29*(10), 1637–1651. http://dx.doi.org/10.1038/emboj.2010.51.

Schlager, M. A., Serra-Marques, A., Grigoriev, I., Gumy, L. F., Esteves da Silva, M., Wulf, P. S., … Hoogenraad, C. C. (2014). Bicaudal d family adaptor proteins control the velocity of Dynein-based movements. *Cell Reports*, *8*(5), 1248–1256. http://dx.doi.org/10.1016/j.celrep.2014.07.052.

Schmieg, N., Menendez, G., Schiavo, G., & Terenzio, M. (2014). Signalling endosomes in axonal transport: Travel updates on the molecular highway. *Seminars in Cell and Developmental Biology*, *27*, 32–43. http://dx.doi.org/10.1016/j.semcdb.2013.10.004.

Schreij, A. M., Fon, E. A., & McPherson, P. S. (2016). Endocytic membrane trafficking and neurodegenerative disease. *Cellular and Molecular Life Sciences*, *73*(8), 1529–1545. http://dx.doi.org/10.1007/s00018-015-2105-x.

Senger, D. L., & Campenot, R. B. (1997). Rapid retrograde tyrosine phosphorylation of trkA and other proteins in rat sympathetic neurons in compartmented cultures. *The Journal of Cell Biology*, *138*(2), 411–421.

Sengupta, P., Satpute-Krishnan, P., Seo, A. Y., Burnette, D. T., Patterson, G. H., & Lippincott-Schwartz, J. (2015). ER trapping reveals Golgi enzymes continually revisit the ER through a recycling pathway that controls Golgi organization. *Proceedings of the National Academy of Sciences of the United States of America*, *112*(49), E6752–E6761. http://dx.doi.org/10.1073/pnas.1520957112.

Splinter, D., Tanenbaum, M. E., Lindqvist, A., Jaarsma, D., Flotho, A., Yu, K. L., … Akhmanova, A. (2010). Bicaudal D2, dynein, and kinesin-1 associate with nuclear pore complexes and regulate centrosome and nuclear positioning during mitotic entry. *PLoS Biology*, *8*(4), e1000350. http://dx.doi.org/10.1371/journal.pbio.1000350.

Steward, R., & Nusslein-Volhard, C. (1986). The genetics of the dorsal-Bicaudal-D region of Drosophila melanogaster. *Genetics*, *113*(3), 665–678.

Suter, B., Romberg, L. M., & Steward, R. (1989). Bicaudal-D, a Drosophila gene involved in developmental asymmetry: Localized transcript accumulation in ovaries and sequence similarity to myosin heavy chain tail domains. *Genes and Development*, *3*(12A), 1957–1968.

Suter, B., & Steward, R. (1991). Requirement for phosphorylation and localization of the Bicaudal-D protein in Drosophila oocyte differentiation. *Cell*, *67*(5), 917–926.

Swan, A., Nguyen, T., & Suter, B. (1999). Drosophila Lissencephaly-1 functions with Bic-D and dynein in oocyte determination and nuclear positioning. *Nature Cell Biology*, *1*(7), 444–449. http://dx.doi.org/10.1038/15680.

Swan, A., & Suter, B. (1996). Role of Bicaudal-D in patterning the Drosophila egg chamber in mid-oogenesis. *Development*, *122*(11), 3577–3586.

Swift, S., Xu, J., Trivedi, V., Austin, K. M., Tressel, S. L., Zhang, L., … Kuliopulos, A. (2010). A novel protease-activated receptor-1 interactor, Bicaudal D1, regulates G protein signaling and internalization. *The Journal of Biological Chemistry*, *285*(15), 11402–11410. http://dx.doi.org/10.1074/jbc.M110.105403.

Tan, S. C., Scherer, J., & Vallee, R. B. (2011). Recruitment of dynein to late endosomes and lysosomes through light intermediate chains. *Molecular Biology of the Cell*, *22*(4), 467–477. http://dx.doi.org/10.1091/mbc.E10-02-0129.

Terenzio, M., Golding, M., Russell, M. R., Wicher, K. B., Rosewell, I., Spencer-Dene, B., … Schiavo, G. (2014a). Bicaudal-D1 regulates the intracellular sorting and signalling of neurotrophin receptors. *The EMBO Journal*, *33*(14), 1582–1598. http://dx.doi.org/10.15252/embj.201387579.

Terenzio, M., Golding, M., & Schiavo, G. (2014b). siRNA screen of ES cell-derived motor neurons identifies novel regulators of tetanus toxin and neurotrophin receptor trafficking. *Frontiers in Cellular Neuroscience*, *8*, 140. http://dx.doi.org/10.3389/fncel.2014.00140.

Trokter, M., Mucke, N., & Surrey, T. (2012). Reconstitution of the human cytoplasmic dynein complex. *Proceedings of the National Academy of Sciences of the United States of America*, *109*(51), 20895–20900. http://dx.doi.org/10.1073/pnas.1210573110.

Vazquez-Pianzola, P., Adam, J., Haldemann, D., Hain, D., Urlaub, H., & Suter, B. (2014). Clathrin heavy chain plays multiple roles in polarizing the Drosophila oocyte downstream of Bic-D. *Development*, *141*(9), 1915–1926. http://dx.doi.org/10.1242/dev.099432.

Wanschers, B. F., van de Vorstenbosch, R., Schlager, M. A., Splinter, D., Akhmanova, A., Hoogenraad, C. C., … Fransen, J. A. (2007). A role for the Rab6B Bicaudal-D1 interaction in retrograde transport in neuronal cells. *Experimental Cell Research*, *313*(16), 3408–3420. http://dx.doi.org/10.1016/j.yexcr.2007.05.032.

Wharton, R. P., & Struhl, G. (1989). Structure of the Drosophila BicaudalD protein and its role in localizing the the posterior determinant nanos. *Cell*, *59*(5), 881–892.

White, J., Johannes, L., Mallard, F., Girod, A., Grill, S., Reinsch, S., … Stelzer, E. H. (1999). Rab6 coordinates a novel Golgi to ER retrograde transport pathway in live cells. *The Journal of Cell Biology*, *147*(4), 743–760.

Zhang, Y., Moheban, D. B., Conway, B. R., Bhattacharyya, A., & Segal, R. A. (2000). Cell surface Trk receptors mediate NGF-induced survival while internalized receptors regulate NGF-induced differentiation. *The Journal of Neuroscience*, *20*(15), 5671–5678.

CHAPTER EIGHT

BDNF and Hippocampal Synaptic Plasticity

G. Leal*, C.R. Bramham†, C.B. Duarte*,‡,1
*CNC—Center for Neuroscience and Cell Biology, University of Coimbra, Coimbra, Portugal
†K.G. Jebsen Center for Neuropsychiatric Disorders, University of Bergen, Bergen, Norway
‡University of Coimbra, Coimbra, Portugal
[1]Corresponding author: e-mail address: cbduarte@ci.uc.pt

Contents

1. Introduction 154
2. Basic Mechanisms of LTP 157
3. The Neurotrophin Family and Its Receptors 159
4. TrkB Localization and Signaling 160
   4.1 Ras–MAPK Pathway 161
   4.2 PI3K Pathway 162
   4.3 PLCγ Pathway 162
5. Synthesis and Secretion of BDNF 162
6. BDNF and LTP in the Hippocampus 164
   6.1 Pre- and Postsynaptic Secretion and Actions of BDNF During LTP 166
   6.2 Mechanisms Underlying Pre- and Postsynaptic Effects of BDNF 169
   6.3 Modulatory Effects of BDNF on Synapse Structure 169
   6.4 BDNF-Mediated Dendritic Protein Synthesis and LTP 171
   6.5 BDNF–TrkB Signaling in Synaptic Tagging and Capture 175
   6.6 BDNF Effects at Nonassociative MF LTP 176
   6.7 BDNF as a Regulator of Neurogenesis 177
7. $p75^{NTR}$ and LTD 177
8. Interplay Between Glucocorticoids and BDNF to Determine Cellular Response to Stress 180
9. Concluding Remarks 182
References 183

Abstract

Brain-derived neurotrophic factor (BDNF) belongs to a family of small secreted proteins that also include nerve growth factor, neurotrophin 3, and neurotrophin 4. BDNF stands out among all neurotrophins by its high expression levels in the brain and its potent effects at synapses. Several aspects of BDNF biology such as transcription, processing, and secretion are regulated by synaptic activity. Such observations prompted the

Vitamins and Hormones, Volume 104
ISSN 0083-6729
http://dx.doi.org/10.1016/bs.vh.2016.10.004

suggestion that BDNF may regulate activity-dependent forms of synaptic plasticity such as long-term potentiation (LTP), a sustained enhancement of excitatory synaptic efficacy thought to underlie learning and memory. Here, we will review the evidence pointing to a fundamental role of this neurotrophin in LTP, especially within the hippocampus. Prominent questions in the field, including the release and action sites of BDNF during LTP, as well as the signaling and molecular mechanisms involved, will also be addressed. The diverse effects of BDNF at excitatory synapses are determined by the activation of TrkB receptors and downstream signaling pathways, and the functions, typically opposing in nature, of its immature form (proBDNF). The activation of $p75^{NTR}$ receptors by proBDNF and the implications for long-term depression will also be addressed. Finally, we discuss the synergy between TrkB and glucocorticoid receptor signaling to determine cellular responses to stress.

1. INTRODUCTION

With more than 100 billion neurons, the adult human brain represents a massive network for information processing, storage, and recall. Neurons are different from most other cells in the body in that they are highly polarized anatomically, with morphological regions subserving specialized functions. Neurons connect to each other at synapses to form neuronal networks. Most of the actions mediating information storage are thought to take place at the synapse, and synaptic alterations are critically involved in several fundamental physiological processes. On the other hand, several disease-related phenomena also involve synaptic dysfunction.

There are two distinct types of synaptic connections, chemical and electric synapses. The typical chemical synapse in the nervous system requires the following steps: (i) synthesis of a neurotransmitter molecule in the presynaptic cell and its storage in secretory vesicles; (ii) activity-dependent fusion of the neurotransmitter-containing vesicles with the plasma membrane with the concomitant release of the neurotransmitter from the presynaptic terminal; (iii) diffusion of these molecules across the synaptic cleft (space between the pre- and postsynaptic neurons); (iv) the activation of specific receptors localized in the postsynaptic cell eliciting a biochemical/ electric response; (v) and finally, a means of terminating the action of the released neurotransmitter. In the vertebrate central nervous system (CNS), the predominant mode of excitatory transmission is mediated by the neurotransmitter glutamate and the ionotropic glutamate receptors (Traynelis et al., 2010). Upon activation, ionotropic glutamate receptors open a channel which is more or less selective for certain ion species, depending on their molecular composition. While α-amino-3-hydroxy-5-methyl-4-isoxazole propionic acid (AMPA) receptors (AMPAR) support fast excitatory transmission, *N*-methyl-D-aspartate (NMDA) receptors (NMDAR) have been classically viewed as coincidence detectors for the induction of long-term synaptic plasticity, partly expressed as changes in AMPAR-mediated transmission.

Synaptic plasticity refers to a change in synaptic strength in response to use or disuse. Long-term changes in synaptic efficacy are often accompanied by structural as well as functional changes at the synapse. The vast majority of CNS pathways studied exhibit synaptic plasticity and, although this phenomenon has been primarily studied at excitatory and inhibitory synapses, multiple forms of plasticity exist across pathways with different underlying mechanisms. Plasticity at synapses can be mediated at the presynaptic level by changing the release of neurotransmitter molecules or postsynaptically by changing the number, type, or properties of neurotransmitter receptors and their coupling to the intracellular signaling machinery. Long-term potentiation (LTP) is the main form of synaptic plasticity reflecting the activity of synaptic information storage processes, and has been identified as the prime candidate to be the cellular correlate of learning and memory (Bliss, Collingridge, & Morris, 2007; Lynch, 2004; Malenka, 2003a; Padamsey & Emptage, 2014). Since its discovery, LTP has been studied in detail in several brain regions and much is known about the molecular mechanisms underlying this form of plasticity, especially in the hippocampus. The regular organization of hippocampal principal cell layers coupled with the laminar distribution of many of its inputs has facilitated the use of the hippocampus as a mammalian model system to study synaptic transmission. The hippocampal formation is comprised by three cytoarchitechtonically distinct regions: the dentate gyrus (DG), the subiculum, and the hippocampus proper (*Cornu Ammonis*, CA). The CA is subdivided into the CA1, CA2, and CA3 regions. Other regions of the hippocampal formation include the presubiculum, parasubiculum, and entorhinal cortex (Strange, Witter, Lein, & Moser, 2014; Witter, 2011). The subregions of the hippocampus are connected by sequential excitatory projection constituting the entorhinal-hippocampal loop. However, the circuitry is now recognized as more complex with extensive divergence of outputs from the entorhinal cortex and CA3, backprojections from CA3 to the DG, and anatomical specialization along the longitudinal axis of the hippocampal formation (Strange et al., 2014; Witter, 2011). Perforant path fibers provide most of the input from the entorhinal cortex to the DG (Fig. 1). Dentate granule cells, the

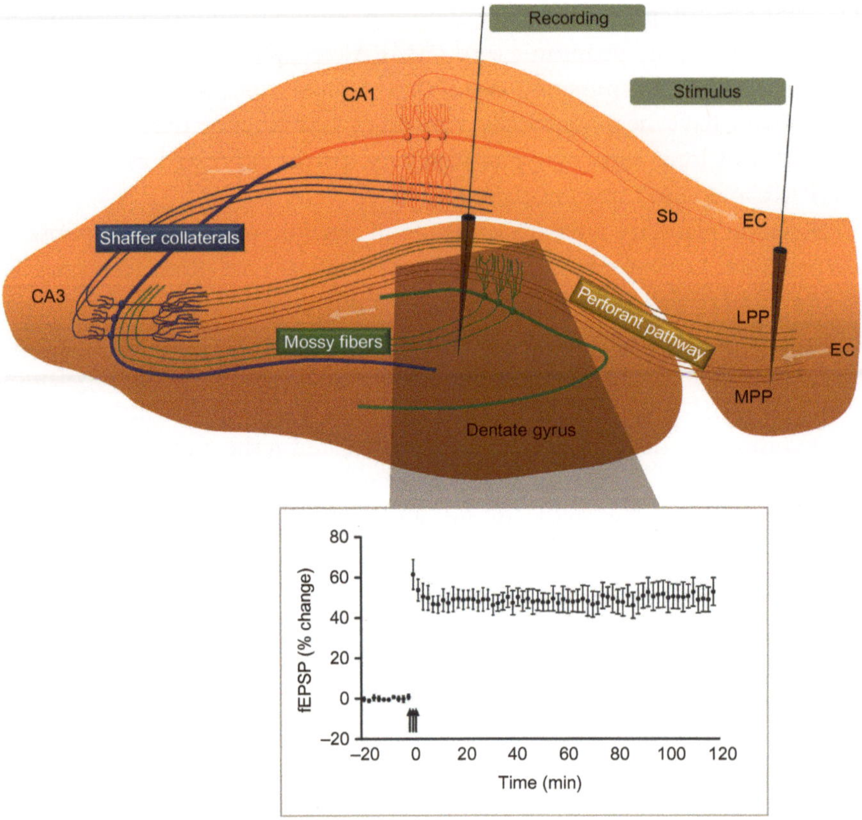

Fig. 1 Schematic representation of electrode placement in the hippocampus for the induction of LTP at medial perforant path-dentate gyrus synapses in live anesthetized rats. *Top*: Diagram showing the location of major circuits within the rodent hippocampus. *CA*, *Cornu Ammonis*; *CA3/CA1*, CA3 and CA1 hippocampal regions; *EC*, entorhinal cortex; LPP, lateral perforant path; *MPP*, medial perforant path; *Sb*, subiculum. Electrodes were positioned to stimulate unilaterally medial perforant fibers and to record evoked currents in the hilar region of the dentate gyrus. *Bottom*: High-frequency stimulation (HFS)-induced LTP. Population response size is measured by the evoked field excitatory postsynaptic potential (fEPSP) slope and is expressed in percentage of baseline. Time-course plots of medial perforant path-dentate gyrus evoked fEPSP recorded before and after HFS (indicated by *arrows*).

major cell type in the DG, do not project back and their axons (mossy fibers (MFs)) project into the CA3 region in the hippocampus. CA3 pyramidal neurons in turn constitute the major input into the CA1 region (Schaffer collateral; Fig. 1), but they also synapse back to the DG (Scharfman, 2007). CA1 projections are to some extent more elaborated; CA1 neurons constitute the major excitatory input into the subiculum region, but they also project into the entorhinal cortex (Fig. 1; Witter, 2011).

BDNF plays a fundamental role in hippocampal LTP and learning. This neurotrophin was shown to (i) regulate the induction and maintenance of a stable LTP, (ii) induce alterations in neurotransmitter release, (iii) modulate postsynaptic glutamate receptors, (iv) regulate protein synthesis, including in specific subcellular compartments such as synapses, (v) activate transcription, and (vi) modulate structural plasticity at dendritic spines (reviewed in Bramham & Messaoudi, 2005; Leal, Afonso, Salazar, & Duarte, 2015; Lu, Christian, & Lu, 2008; Minichiello, 2009; Panja & Bramham, 2014; Park & Poo, 2013; Waterhouse & Xu, 2009). In this chapter, we discuss the role of BDNF in synaptic plasticity and highlight the most important questions in the field.

2. BASIC MECHANISMS OF LTP

The concept of synaptic plasticity was postulated by Hebb in 1949 when he suggested that the experience can modify synapses favoring some neuronal pathways within a circuit and weakening others (Hebb, 1949). Hebb's theory postulated that "cells that fire together, wire together." Since then, an enormous effort was made to understand the mechanisms that contribute to synaptic strengthening.

The first full description of LTP was published in 1973 by Bliss and Lomo (1973). They reported that trains of high-frequency stimulation (HFS) to the rabbit perforant path induced a sustained increase in efficiency of synaptic transmission onto granule cells of the DG. Subsequent reports confirming the Hebbian nature of LTP induction gave weight to LTP's candidacy as a cellular mechanism for memory formation. LTP is characterized by three basic properties: (i) the cooperativity, which describes the existence of an intensity threshold for induction; (ii) LTP is associative in the sense that a weak input can be potentiated by other separated but convergent input; (iii) finally, LTP is input specific, meaning that potentiation is not shared to other pathways distinct from the potentiated synapse (Bliss & Collingridge, 1993).

LTP in the hippocampus is the most studied form of synaptic plasticity and is typically divided into at least three distinct and sequential phases: short-term potentiation, early LTP (E-LTP), and late LTP (L-LTP). Short-term potentiation and E-LTP are transient and involve the modification of preexisting proteins, whereas L-LTP requires changes in gene expression and de novo protein synthesis and lasts for hours or even days (Kandel, 2001; Sweatt, 1999).

Postsynaptic influx of Ca^{2+} mediated by the activation of NMDAR is necessary and sufficient for E-LTP induction (Bliss & Collingridge, 1993). The opening of NMDAR requires glutamate and glycine (or D-serine) binding, together with membrane depolarization which relieves the Mg^{2+} block. Activation of NMDAR allows the entry of Ca^{2+} into the cell and this rise in the intracellular Ca^{2+} concentration activates several enzymes that mediate E-LTP induction, including Ca^{2+}/calmodulin-dependent protein kinase II (CaMKII) and protein kinase C (PKC; Fig. 2; Sweatt, 1999). The sustained activation of these enzymes is also crucial for the maintenance of E-LTP (Bliss & Collingridge, 1993; Lynch, 2004). The generation of persistently active forms of PKC and CaMKII that become autonomously active and phosphorylate downstream targets, such as AMPAR, results in increased trafficking and insertion of these receptors

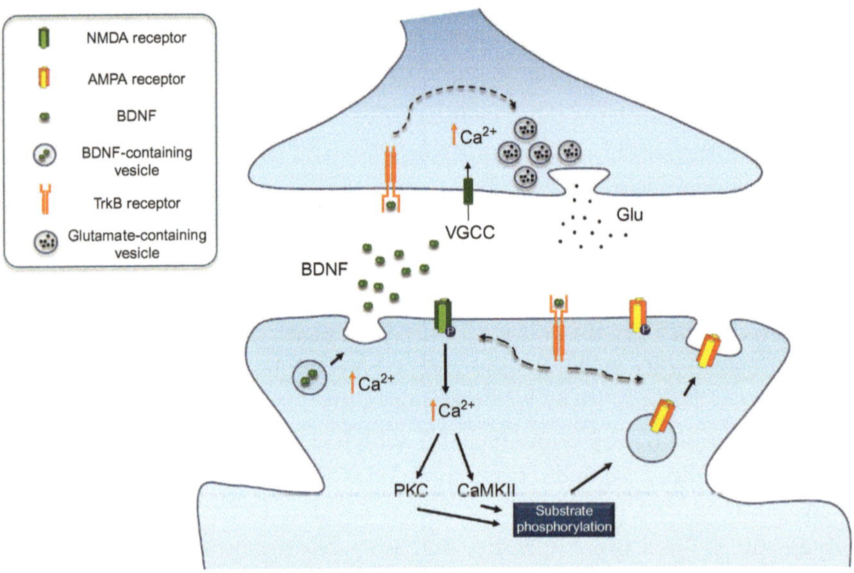

Fig. 2 BDNF-mediated regulation of excitatory synapses during early LTP. Synaptic BDNF is secreted in response to activity and can activate pre- and/or postsynaptic TrkB receptors. At presynaptic sites, BDNF increases the exocytosis of glutamate-containing synaptic vesicles. Postsynaptically, BDNF–TrkB signaling induces NMDA receptor (NMDAR) phosphorylation, which increases the opening probability of the ion channel. BDNF also increases the phosphorylation and membrane insertion of AMPA receptors (AMPAR). During E-LTP, calcium influx via NMDAR activates PKC and CaMKII that phosphorylate several substrates, including AMPAR, to maintain increased synaptic efficacy.

into the postsynaptic membrane, with a consequent upregulation in their activity (Fig. 2; Derkach, Oh, Guire, & Soderling, 2007; Lin et al., 2009; Malenka, 2003a, 2003b; Shepherd & Huganir, 2007). Increases in NMDAR surface expression were also observed upon E-LTP induction (Grosshans, Clayton, Coultrap, & Browning, 2002).

LTP may also be mediated by enhancement of synaptic vesicle exocytosis at presynaptic sites. Indeed, the apparent simple question of whether LTP is expressed pre- or postsynaptically (or both) resulted in an intense debate in the field (Lisman, 2009). Considerable amount of evidence supports both pre- and postsynaptic expression of LTP. Ultimately, one may say that LTP expression is determined by the induction protocol and the state and nature of the studied synapse (MacDougall & Fine, 2014).

E-LTP is transient and is followed by L-LTP which is believed to depend on protein synthesis in dendritic compartments (Bramham & Wells, 2007; Sutton & Schuman, 2006) and transcription activity (Adams & Dudek, 2005; Lynch, 2004). Several signaling molecules such as PKA, CaMKIV, and extracellular signal-regulated protein kinase (ERK) are involved in the activation of key transcription factors, including cyclic AMP-responsive element-binding protein (CREB) and the ternary complex factor (TCF) Elk-1 (Adams & Dudek, 2005; Lynch, 2004; Platenik, Kuramoto, & Yoneda, 2000). Such factors will ultimately promote the synthesis of proteins that mediate structural and/or functional changes at synapses, which are thought to be required for L-LTP maintenance. The augmented excitatory synaptic strength is presumed to be required for structural changes in dendritic spines. The actin-mediated enlargement of spine heads and an increase in spine number are believed to further support long-term changes in synaptic efficacy (Bailey, Kandel, & Harris, 2015; Bramham, 2008; Fukazawa et al., 2003; Honkura, Matsuzaki, Noguchi, Ellis-Davies, & Kasai, 2008; Krucker, Siggins, & Halpain, 2000).

It is important to note that multiple forms of LTP have been described in different brain regions and, as it will be clear in the following sections, even within the hippocampus several forms of LTP that differ in nature as well as in neurotrophin dependence can exist.

3. THE NEUROTROPHIN FAMILY AND ITS RECEPTORS

Nerve growth factor (NGF) was the first neurotrophin to be discovered in the early 1950s as a target-derived protein that promotes the growth and survival of sympathetic and sensory neurons during development (Cohen, Levi-Montalcini, & Hamburger, 1954). The establishment of the neurotrophin family came in the early 1980s when BDNF was purified from pig brain and showed similar neurotrophic actions in sensory neurons (Barde, Edgar, & Thoenen, 1982). Two more neurotrophins were discovered since then, neurotrophin 3 (NT3) and neurotrophin 4 (NT4) (Lewin & Barde, 1996).

A major advance in our understanding of neurotrophins came when, in a seminal review, several outstanding questions were raised about the biology of these secreted proteins (Thoenen, 1991). Interestingly, many of these issues are still current and under investigation at this time. It was also suggested that neurotrophins can be released upon activity (Thoenen, 1991). This idea inspired further investigations on the roles of neurotrophins in synaptic plasticity. A large amount of evidence now clearly supports the pleiotropic functions of neurotrophins in the nervous system, including in the control of neuronal survival and differentiation, in synaptogenesis, and in the regulation of activity-dependent forms of plasticity (Park & Poo, 2013).

Physiological responses to neurotrophins are mediated by the activation of two distinct classes of membrane-bound receptors. Neurotrophins bind to $p75^{NTR}$ receptors and to one of three tropomyosin-related kinase (Trk) receptors. NGF binds to TrkA, BDNF and NT4 bind to TrkB, and NT3 binds to TrkC (Chao, 2003; Reichardt, 2006). In contrast with Trk receptors, activation of $p75^{NTR}$ does not display the same specificity. $p75^{NTR}$ binds both the mature and the uncleaved forms (proneurotrophins) of neurotrophins (see later) (Lee, Kermani, Teng, & Hempstead, 2001; Teng et al., 2005).

4. TrkB LOCALIZATION AND SIGNALING

TrkB receptors have been detected in nerve terminals, axons, and dendritic spines of glutamatergic pyramidal and granule cells in the hippocampus using a combination of light and electron microscopy (Drake, Milner, & Patterson, 1999). Using a similar approach and subcellular fractionation, TrkB receptors were also found in dendritic spines of cerebrocortical neurons (Aoki et al., 2000). In a different study, the authors used subcellular fractionation and biotinylation approaches to show that TrkB receptors are present at glutamatergic nerve terminals in the hippocampus and are evenly distributed between the presynaptic active zone and the postsynaptic density (Pereira et al., 2006).

The binding of neurotrophins to TrkB receptors induces ligand–receptor dimerization and autophosphorylation of tyrosine residues in the intracellular kinase domain of the receptor. This leads in turn to the phosphorylation of tyrosine residues (Tyr 515 and Tyr 816) located in the juxtamembrane region and in the carboxyl terminus of the receptor. Phosphorylation of these two tyrosine residues mediates the interaction of the receptor with Shc (Src homology 2-containing protein) and phospholipase Cγ (PLCγ), respectively (Reichardt, 2006). Fibroblast growth factor receptor substrate 2 (FRS2) is another molecule that may compete with Shc adaptors for the binding to Trk receptors (Kouhara et al., 1997; Meakin, MacDonald, Gryz, Kubu, & Verdi, 1999). Additional adaptor proteins containing pleckstrin homology and SH2 domains can also activate Trk signaling through the binding to phosphotyrosine residues on the catalytic domain of Trk receptors (Qian, Riccio, Zhang, & Ginty, 1998).

Trk receptors can also be activated independent of ligand binding through transactivation. Such process has been reported upon G-protein-coupled receptor ligands such as adenosine (Lee & Chao, 2001; Rajagopal & Chao, 2006) and pituitary adenylate cyclase-activating polypeptide (Lee, Rajagopal, Kim, Chang, & Chao, 2002), and it has been proposed to be mediated by activation of Src family members (Lee & Chao, 2001; Lee, Rajagopal, et al., 2002; Rajagopal & Chao, 2006). Evidence of Zinc-mediated and Src-dependent transactivation of TrkB receptors has also been reported (Huang, Pan, Xiong, & McNamara, 2008).

Due to the high homology of the intracellular domains, the signaling cascades are highly conserved among Trk receptors (Atwal, Massie, Miller, & Kaplan, 2000). The three main intracellular signaling cascades activated by Trk receptors are the Ras–mitogen-activated protein kinase (MAPK) pathway, the phosphatidylinositol 3-kinase (PI3K)–Akt pathway, and the PLCγ–Ca^{2+} pathway (Kaplan & Miller, 2000).

4.1 Ras–MAPK Pathway

TrkB phosphorylation at Tyr 515 recruits and phosphorylates Shc. Phosphorylated Shc recruits the growth factor receptor-bound protein 2 (GRB2) and the guanine nucleotide exchange factor son of sevenless (SOS) resulting in the activation of the Ras–MAPK pathway (Reichardt, 2006). FRS2 is also recruited by phosphorylated Tyr 515 and may bind GRB2 to activate Ras through GRB2–SOS complex and independently of Shc (Minichiello, 2009). The Ras–MAPK pathway is coupled to the activation CREB, which enhances the transcription of essential prosurvival genes, and other downstream targets (Reichardt, 2006). For example, activation ERK coupled to CREB is required for BDNF-induced LTP at DG synapses in vivo (Ying et al., 2002).

4.2 PI3K Pathway

Recruitment of Shc to the active Trk receptors is also coupled to the stimulation of the PI3K pathway via GRB2 and other adaptor proteins (Holgado-Madruga, Moscatello, Emlet, Dieterich, & Wong, 1997; Yamada et al., 1997). Activation of PI3K generates 3-phosphoinositides that activate 3-phosphoinositide-dependent protein kinase 1 (PDK1) which in turn, together with the 3-phosphoinositides, activate protein kinase Akt (Crowder & Freeman, 1998). Activation of PI3K–Akt pathway is involved, for instance, in the BDNF-mediated activation of protein synthesis, via a signaling cascade acting on the mammalian target of rapamycin (mTOR) (Takei et al., 2004).

4.3 PLCγ Pathway

TrkB phosphorylation on Tyr 816 recruits PLCγ1 which is phosphorylated by the active receptor (Kaplan & Miller, 2000). Activated PLCγ1 hydrolyzes phosphatidylinositol-4,5-bisphosphate (PtdIns(4,5)P_2) to generate inositol-1,4,5-trisphosphate (Ins(1,4,5)P_3) and diacylglycerol. Ins(1,4,5)P_3 promotes the release of Ca^{2+} from internal stores, which results in the activation of Ca^{2+}-dependent enzymes such as Ca^{2+}/calmodulin-dependent protein kinases. Signaling through this pathway controls the expression and/or activity of many proteins, including ion channels and transcription factors, and plays an important role in TrkB-mediated LTP at CA3–CA1 synapses (Minichiello et al., 2002).

5. SYNTHESIS AND SECRETION OF BDNF

In accordance with the activity dependence of synaptic modulation by BDNF, the expression of the *bdnf* gene and the release of the neurotrophin are both modulated by neuronal activity. BDNF and the other neurotrophins are synthesized in the endoplasmic reticulum, as pre-pro-neurotrophins. Pre-proBDNF is then converted into proBDNF by removal of the signal peptide and further cleaved to generate the mature form of the neurotrophin. Whether BDNF is released in the precursor and/or in the mature form remains controversial (Leal et al., 2015; Leal, Comprido, & Duarte, 2014). Electron microscopy studies with immunogold staining and using antibodies against the pro- and mature forms of BDNF showed that the precursor and mature forms of the neurotrophin are colocalized in large dense core vesicles in excitatory presynaptic terminals of the adult mouse hippocampus (Dieni et al., 2012). This suggests that proBDNF is at least partly cleaved within the cell by furin, in the *trans*-Golgi, or by pro-protein convertase 1/3, in secretory granules (Mowla et al., 2001; Seidah, Benjannet, Pareek, Chretien, & Murphy, 1996), as previously suggested based on biochemical studies (Matsumoto et al., 2008). In contrast, other reports showed that a significant fraction of BDNF may be released in the precursor form (Mizoguchi et al., 2011; Nagappan et al., 2009; Pang et al., 2004; Woo et al., 2005; Yang et al., 2014, 2009); in this case, proBDNF is further processed in the extracellular compartment by metalloproteinases and by plasmin, giving rise to the mature neurotrophin. Accordingly, studies performed in cultured hippocampal neurons using epitope-tagged BDNF showed the extracellular accumulation of proBDNF following low-frequency electrical stimulation. HFS also induces the release of tissue plasminogen activator (tPA) which cleaves proBDNF to generate the mature form of BDNF in the extracellular compartment (Nagappan et al., 2009). This is particularly relevant considering the role played by tPA/plasminogen-mediated cleavage of proBDNF in the late phase of LTP in the hippocampal CA1 region (Pang et al., 2004). Matrix metalloproteinase-9 also plays a role in the BDNF-mediated seizure activity induced by pentylenetetrazole in mice, further suggesting that conditions of high neuronal activity favor the release of proBDNF which is then cleaved by extracellular proteases, giving rise to BDNF (Mizoguchi et al., 2011). A clear identification of the conditions that lead to the release and extracellular accumulation of proBDNF and the mature neurotrophin is important since the two forms of BDNF interact with distinct receptors and have different and sometimes opposing functions in the nervous system (see later). An additional question requiring further elucidation is the subcellular localization of endogenous BDNF in the adult brain.

BDNF produced in the cell body is targeted to vesicles of the regulated secretory pathway, which in cultured hippocampal neurons are delivered preferentially to the dendritic compartment (Hartmann, Heumann, & Lessmann, 2001; Kolarow, Brigadski, & Lessmann, 2007; Matsuda et al., 2009). Furthermore, BDNF is synthesized locally at the synapse after transport of the corresponding mRNAs along dendrites. Large dense core vesicles containing BDNF are also transported anterogradely along the axons before reaching the presynaptic region (Matsuda et al., 2009). The vesicles containing BDNF/proBDNF are exocytosed by an activity-dependent mechanism from the pre- and postsynaptic regions. The release of BDNF from the dendritic compartment following activation of glutamatergic synapses depends on the entry of Ca^{2+} through NMDA receptor channels and/or activation of postsynaptic voltage-gated Ca^{2+} channels (Hartmann et al., 2001; Kolarow et al., 2007; Matsuda et al., 2009). A recent study showed that extracellular BDNF may be endocytosed at the postsynaptic level, and this pool of the neurotrophin may also be released following repetitive neuronal spiking by a mechanism dependent on the activation of glutamate receptors (Wong, Lee, Xie, Cui, & Poo, 2015). Accordingly, TrkB-mediated endocytosed BDNF can be recycled and readily available for activity-dependent secretion in hippocampal neurons to maintain LTP (Santi et al., 2006). In a similar context, internalized TrkB receptors can also be targeted for degradation or recycled back to the plasma membrane by a BDNF-dependent mechanism (Chen, Ieraci, Tanowitz, & Lee, 2005; Huang et al., 2013).

Much higher levels of neuronal spiking are required to release BDNF from the axonal compartment in cultured hippocampal neurons (Matsuda et al., 2009), suggesting that under physiological conditions, neuronal activity is more likely to induce the exocytosis of BDNF stored in the postsynaptic compartment. However, at this point, we still have a limited understanding of the molecular mechanisms governing the maturation and secretion of BDNF in the adult brain.

6. BDNF AND LTP IN THE HIPPOCAMPUS

A large amount of evidence supports the fundamental role of BDNF in hippocampal synaptic plasticity. Seminal studies in the early 1990s showed that a stimulation paradigm that induces LTP in the CA1 region of the hippocampus also upregulates the BDNF mRNA (Patterson, Grover, Schwartzkroin, & Bothwell, 1992). Additional studies showed that in BDNF knockout mouse lines, LTP is impaired (Korte et al., 1995; Patterson et al., 1996) and reexpressing BDNF (Korte et al., 1996) or exogenous application of the neurotrophin (Patterson et al., 1996) can restore these defects. Studies using extracellular scavengers and antibodies have also been very helpful in understanding the role of endogenous neurotrophins, in particular of BDNF, in LTP. Preventing TrkB activation with TrkB–IgG fusion proteins reduces the magnitude of theta-burst (TBS)-induced LTP in hipppocampal CA1 synapses (Figurov, Pozzo-Miller, Olafsson, Wang, & Lu, 1996). This type of approach was also used to show that BDNF–TrkB signaling is required just around the period of synaptic stimulation to induce LTP and to demonstrate that the pattern of stimulation critically determines the neurotrophin dependence of LTP (Chen, Kolbeck, Barde, Bonhoeffer, & Kossel, 1999; Kang, Welcher, Shelton, & Schuman, 1997; Kossel, Cambridge, Wagner, & Bonhoeffer, 2001). Similar approaches were used to show that BDNF and not other ligands of TrkB are involved in LTP induction (Chen et al., 1999), and that BDNF is required for L-LTP in hippocampal CA1 synapses (Korte, Kang, Bonhoeffer, & Schuman, 1998). However, BDNF–TrkB signaling critical period in LTP varies across synapses and according to the stimulation paradigm (see Panja & Bramham, 2014). For example, TrkB ligands are required for up to 60 min following induction of HFS-LTP to maintain synaptic potentiation at CA1 synapses (Kang et al., 1997). On the other hand, LTP consolidation at DG synapses in live anesthetized rats requires sustained (up to hours) BDNF–TrkB signaling (Panja et al., 2014).

Exogenous application of BDNF has also been used to shed light on the mechanisms underlying the effects of this neurotrophin at excitatory synapses. Bath application of BDNF induces a long-lasting enhancement of synaptic transmission at CA3–CA1 synapses that requires protein synthesis in dendrites (Kang & Schuman, 1995, 1996). These results were hard to reproduce, but a recent study demonstrated that the way BDNF is delivered to cultured hippocampal neurons and hippocampal slices results in distinct modes of TrkB activation with the recruitment of distinct downstream signaling mechanisms (Ji et al., 2010). Fast delivery of BDNF affects basal synaptic transmission, whereas slow delivery facilitates LTP induction (Ji et al., 2010).

Exogenous BDNF was also shown to facilitate LTP when paired with synaptic stimulation that would normally result in short-term potentiation (Figurov et al., 1996). Furthermore, dendritic application of BDNF was found to facilitate LTP induced by weak burst synaptic stimulation in dentate granule cells through a postsynaptic mechanism (Kovalchuk, Hanse, Kafitz, & Konnerth, 2002). Intrahippocampal infusion of BDNF also induces a robust and sustained increase in synaptic transmission at the medial perforant path-dentate granule cells synapses in live anesthetized rats (Messaoudi, Bardsen, Srebro, & Bramham, 1998; Messaoudi, Ying,

Kanhema, Croll, & Bramham, 2002; Ying et al., 2002) through the MAPK–ERK pathway. In contrast with region CA1, BDNF-LTP in the DG is transcription dependent (Messaoudi et al., 2002) and associated with induction of a panel of immediate early genes (including *Arc*, *Narp*, *neuritin*, *Klf10/TIEG1*, and *Arl4l*) in granule cells (Wibrand et al., 2006). Arc mRNA is rapidly delivered to dendritic processes, and synthesis of Arc protein is required for persistent LTP induced by BDNF infusion or patterned HFS (Messaoudi et al., 2007; Ying et al., 2002).

Corroborating the crucial role of BDNF in synaptic plasticity, *trkB* forebrain-specific knockout mice display impaired LTP at CA3–CA1 hippocampal synapses (Minichiello et al., 1999). In the follow-up study, the authors used mouse lines with targeted mutations in the Shc or PLCγ docking sites at TrkB receptors to show that TrkB mediates LTP in hippocampal CA1 synapses via recruitment of the PLCγ signaling pathway (Minichiello, 2009; Minichiello et al., 2002). These findings contrast with BDNF effects on synaptic efficacy at rodent DG synapses in vivo which require MAPK–ERK activation (Ying et al., 2002). Additional studies are required to determine whether distinct pathways mediate the effects of TrkB in LTP of CA1 and DG synapses; the differential results reported in the literature may also be due to the use of distinct approaches to dissect the signaling pathways involved. Furthermore, despite the well-established role of BDNF in LTP, our knowledge concerning the synaptic release and action of BDNF during relevant patterns of synaptic activation is sparse and deserves further investigation.

6.1 Pre- and Postsynaptic Secretion and Actions of BDNF During LTP

It is becoming evident that BDNF release and actions are inherent to the LTP form and clearly determined by the nature and individual circumstances at specific synapses (Edelmann, Lessmann, & Brigadski, 2014). BDNF may have permissive or instructive roles in LTP (Bramham & Messaoudi, 2005; Poo, 2001). Permissive modulatory actions of BDNF involve mechanisms that make synapses capable to undergo LTP but which are not directly involved in generating LTP. For example, secretion of BDNF is important to maintain a readily available pool of neurotransmitter-containing vesicles that then can be recruited upon induction of LTP (Figurov et al., 1996). On the other hand, BDNF may also have instructive roles which refers to processes directly involved in generating LTP (see Bramham & Messaoudi, 2005; Park & Poo, 2013).

Instead of presenting a complete review of the literature, which has been recently provided (e.g., Edelmann et al., 2014), we will highlight the key findings regarding the synaptic secretion and actions of BDNF.

6.1.1 Secretion

It is clear that neurons have the ability to secret BDNF from both axons and dendrites in response to neuronal activity (Lessmann & Brigadski, 2009), but whether BDNF is secreted pre- or postsynaptically during LTP is still a matter of debate. A postsynaptic secretion of BDNF was suggested by several studies. The fact that impaired LTP in BDNF knockout mice can be restored by viral-mediated expression of BDNF in CA1 neurons (postsynapse) implies a postsynaptic source of BDNF during LTP (Korte et al., 1996). More recently, the use of transgenic animals carrying a truncated form of BDNF mRNA allowed to demonstrate that the lack of long-3′UTR transcripts impairs the dendritic targeting of BDNF mRNA and results in reduced LTP (An et al., 2008), supporting the role of postsynaptic secretion of BDNF in LTP. Another study demonstrated that synapse stimulation paired with postsynaptic spike induces the enlargement of spine heads in CA1 hippocampal neurons by a mechanism that was mediated by BDNF and required protein synthesis (Tanaka et al., 2008). The requirement for postsynaptic spiking in this paradigm suggests that BDNF is released postsynaptically.

Recently, an elegant study investigated the role of BDNF in different spike timing-dependent paradigms of LTP in CA3–CA1 synapses. The authors demonstrated that two distinct forms of LTP may be elicited depending on the paradigms of synaptic stimulation used. One is expressed presynaptically and does not require BDNF, whereas the second is BDNF dependent and expressed at the postsynapse (Edelmann et al., 2015). In the latter case, endogenous postsynaptic release of BDNF elicits LTP via AMPAR insertion (Edelmann et al., 2015). In contrast, using a genetic approach to disrupt BDNF production in all forebrain (CA3 and CA1) or in the postsynaptic neurons (CA1), Zakharenko and colleagues showed that the presynaptic release of BDNF recruits a presynaptic module of synaptic plasticity (Zakharenko et al., 2003). These findings correlate with the preferential presynaptic detection of BDNF at these synapses (Dieni et al., 2012).

In corticostriatal synapses, for instance, axonal NMDA receptors were found to play an essential role in mediating activity-induced presynaptic release of BDNF and LTP induction (Park, Popescu, & Poo, 2014). Together, these data highlight the need of a better understanding of the molecular mechanisms governing pre- and postsynaptic secretion of BDNF during LTP.

6.1.2 Actions

Since the majority of BDNF actions on LTP have been attributed to the activation of TrkB receptors, manipulating the levels or function of such receptors provides a good strategy to investigate where BDNF acts during LTP. Using two distinct lines of *trkB* mutant animals, one in which TrkB receptor expression is reduced throughout the brain and a second line with reduced levels of the receptor in the hippocampal CA1 region (postsynapse), Xu and colleagues concluded that BDNF may act presynaptically to modulate LTP (Xu et al., 2000). A presynaptic action of the neurotrophin had also been previously proposed in the seminal studies investigating the effects of BDNF in basal synaptic transmission and facilitation of LTP (see Gottmann, Mittmann, & Lessmann, 2009). In a separate study, viral-mediated expression of PLCγ pleckstrin homology domains was used to prevent PLCγ signaling in presynaptic CA3 or postsynaptic CA1 neurons. The authors showed that concurrent inhibition of pre- and postsynaptic TrkB signaling is required to reduce LTP to levels similar to those observed in BDNF knockout animals (Gartner & Staiger, 2002).

Several lines of evidence also point to a postsynaptic action of BDNF in hippocampal LTP. For example, the effects of BDNF in facilitating LTP in dentate granule cells were shown to require activation of postsynaptic Ca^{2+} channels and NMDAR (Kovalchuk et al., 2002). Spike timing-induced structural plasticity at single spines was also shown to require BDNF–TrkB signaling and protein synthesis (Tanaka et al., 2008). More recently, as previously mentioned, postsynaptic secretion of BDNF following spike timing-dependent stimulation was shown to act in an autocrine fashion to elicit postsynaptic LTP at CA3–CA1 synapses (Edelmann et al., 2015).

Together, the available evidence suggest that BDNF may act pre- and postsynaptically to support LTP. It is possible that under certain circumstances, determined by the stimulation protocol and by the individual state of potentiated synapses, BDNF may exert concomitant pre- and postsynaptic changes at the same synapse. Future research should focus on developing technology to allow the visualization and the precise spatiotemporal control of endogenous BDNF during physiological relevant patterns of synaptic activity.

6.2 Mechanisms Underlying Pre- and Postsynaptic Effects of BDNF

Historically, presynaptic effects of exogenous BDNF in enhancing neurotransmitter release were first described in *Xenopus* neuromuscular junction synapses (Lohof, Ip, & Poo, 1993). A convincing amount of evidence now clearly demonstrates that BDNF modulates the efficiency of vesicular glutamate release in mammalian synapses, including within the hippocampus (Tyler, Perrett, & Pozzo-Miller, 2002). Similar enhancements of presynaptic function by endogenous BDNF have been demonstrated (Fig. 2; see Gottmann et al., 2009; Tyler et al., 2006). The molecular mechanisms underlying such effects of the neurotrophin are not fully understood; however, they likely involve the phosphorylation of proteins that regulate the availability of vesicles to be released by the presynaptic terminal (Jovanovic, Czernik, Fienberg, Greengard, & Sihra, 2000) and the recruitment of motor complexes that move synaptic vesicles along actin cytoskeleton (Yano et al., 2006).

Postsynaptic functions of BDNF in synaptic plasticity have been greatly attributed to the modulatory actions exerted on glutamate receptors. BDNF increases the trafficking and synaptic delivery of AMPAR in hippocampal slices and hippocampal neuronal cultures (Fig. 2; Caldeira, Melo, Pereira, Carvalho, Correia, et al., 2007; Fortin et al., 2012). BDNF was also shown to modulate NMDA receptor trafficking, phosphorylation state, and open probability, to regulate synaptic transmission in hippocampal neurons (Fig. 2; Caldeira, Melo, Pereira, Carvalho, Carvalho, et al., 2007; Levine, Crozier, Black, & Plummer, 1998; Lin et al., 1998). A recent study showed that the expression of a BDNF-dependent postsynaptic form of LTP was marked by increased insertion of AMPAR (Edelmann et al., 2015). These results, although correlative, suggest that BDNF is likely to modulate glutamate receptors during synaptic plasticity.

6.3 Modulatory Effects of BDNF on Synapse Structure

LTP is also accompanied by activity-dependent structural changes at synapses which typically involve increased number and volume of dendritic spines (Kasai, Fukuda, Watanabe, Hayashi-Takagi, & Noguchi, 2010). These structural modifications are believed to support alterations in the levels of glutamate receptors and signaling molecules, and ultimately to sustain synaptic strength. The role of BDNF in structural plasticity is well documented and may occur at different levels. Several in vitro studies support the notion that in the hippocampus, BDNF increases neurite branching and growth in developing neurons, while it regulates spine density and morphology in mature neurons (Ji et al., 2010; Zagrebelsky & Korte, 2014). Exogenous application of BDNF was shown to increase spine density in CA1 pyramidal neurons via MAPK/ERK pathway and Ca^{2+} entry via transient receptor potential canonical (TRPC) subfamily channel 3 (Alonso, Medina, & Pozzo-Miller, 2004; Amaral & Pozzo-Miller, 2007). The role of endogenous BDNF has also been assessed by several different studies. The use of TrkB-Fc chimera allowed to demonstrate that TBS-induced LTP requires actin polymerization via BDNF–TrkB signaling-mediated recruitment of p21-activated kinase (PAK)-cofilin pathways (Rex et al., 2007). Activation of the Rho-GTPase Rac1 via a direct interaction of phosphorylated TrkB and the guanine exchange factor Tiam1 was shown to be required for BDNF- and glutamate-induced spine head enlargements (Lai et al., 2012). Furthermore, the authors showed that TrkB-coupled Tiam1 is required for the NMDAR-dependent activation of pathways regulating actin dynamics (PAK) and protein synthesis (S6 kinase) (Lai et al., 2012). In a different study, spike timing-induced spine enlargements at CA1 neurons were shown to require endogenous BDNF and protein synthesis (Tanaka et al., 2008).

In accordance with the evidence described earlier, exogenous BDNF was shown to facilitate LTP-induced spine enlargements (Bosch et al., 2014) and to boost activity-induced increases in Homer2b levels at synapses (previously demonstrated to depend on new protein synthesis) (Bosch et al., 2014). BDNF actions on structural plasticity may also rely on the ability of this neurotrophin to induce local synthesis of cytoskeleton-related proteins such as RhoA and LIMK1 (Briz et al., 2015; Schratt et al., 2006). Interestingly, synaptic activity was shown to modulate synaptic structure through the modulation of dendritic synthesis of BDNF, which may contribute to further enhance the local effects of the neurotrophin (Verpelli et al., 2010). To add a new layer of complexity to BDNF actions at the structural level, Ji and colleagues found that the mode of BDNF application and the development stage of neurons directly influence BDNF effects (Ji et al., 2010). To read a full description of BDNF-mediated structural changes at synapses, please see Zagrebelsky and Korte (2014).

From the evidence aforementioned, one may suggest that BDNF acts both at the cytoskeleton and at the translational level to mediate synaptic structural changes believed to support long-term changes in synaptic efficacy.

6.4 BDNF-Mediated Dendritic Protein Synthesis and LTP

Increasing evidence supports the involvement of local protein synthesis in synaptic plasticity and memory processes (Sutton & Schuman, 2006). Indeed, the role of BDNF in L-LTP and memory has been greatly attributed to the ability of this neurotrophin to modulate protein synthesis, including in specific subcellular compartments such as in dendrites (Leal et al., 2015; Leal, Comprido, et al., 2014; Lu et al., 2008; Panja & Bramham, 2014). Seminal studies showed that BDNF-induced LTP at CA1 hippocampal neurons requires dendritic protein synthesis (Kang & Schuman, 1996). Other studies that followed also demonstrate the protein synthesis dependence of BDNF-mediated LTP (e.g., Leal, Comprido, et al., 2014; Panja et al., 2014; Tanaka et al., 2008). Interestingly, one of the most convincing evidence provided so far relating dendritic protein synthesis and hippocampal synaptic plasticity implicated the dendritic translation of BDNF mRNA (An et al., 2008). In this study, mice carrying a truncated form of BDNF mRNA lacking the long 3′UTR have impaired dendritic localization of BDNF mRNA and impaired LTP at CA1 neurons (An et al., 2008), suggesting that dendritic synthesis and secretion of BDNF are fundamental for synaptic plasticity.

Stabilization of LTP in DG of live rodents requires a period of persistent BDNF–TrkB signaling lasting more than 8 h after LTP induction (Panja et al., 2014). TrkB signaling mediates sustained activation of ERK/MAPK and its substrate MAP kinase-activating kinase 1 (MNK1). In turn, MNK stimulates translation initiation complex formation and enhances protein synthesis. This enhancement occurs in mechanistically distinct early and late stages, involving sequential MNK-dependent regulation of distinct translational repressors. In agreement with these findings, BDNF-induced protein synthesis in cultured cortical neurons was shown to require MNK signaling (Genheden et al., 2015).

The repertoire of mRNAs present in dendritic compartments is much more diverse than initially thought (Cajigas et al., 2012), thus revealing the enormous potential of local translation to regulate the synaptic proteome and synapse function. Localization of mRNAs in dendrites results from a complex and orchestrated series of events. Transcripts must contain one or more signals that are recognized by specific *trans*-acting factors to be then packaged in large ribonucleoprotein complexes, usually termed RNA granules (Fig. 3; Swanger & Bassell, 2013). These granules are heterogeneous structures composed by RNA-binding proteins, mRNAs, ribosomal machinery, and proteins involved in mRNA localization, stabilization, and translational control (Elvira et al., 2006; Fritzsche et al., 2013; Kanai, Dohmae, & Hirokawa, 2004). RNA granules are then engaged with motor proteins for the active transport along the microtubule cytoskeleton in dendrites (Fig. 3; Bramham & Wells, 2007; Dictenberg, Swanger, Antar, Singer, & Bassell, 2008; Kanai et al., 2004). During this process, mRNAs are generally kept in a translational-silent state and then, by several means, translated upon appropriate stimulation at or nearby activated synapses (Fig. 3; Bramham & Wells, 2007; Doyle & Kiebler, 2011).

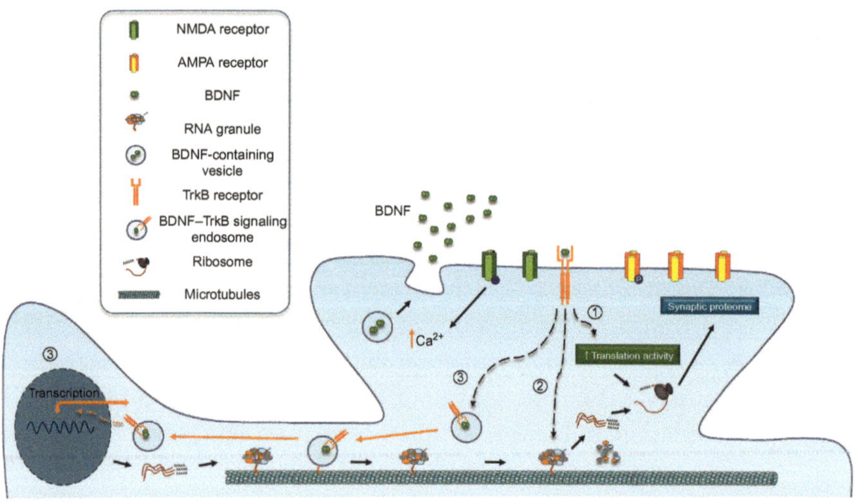

Fig. 3 BDNF as a regulator of L-LTP: neurotrophin-induced alterations in dendritic protein synthesis and gene expression. Secreted BDNF activates postsynaptic TrkB receptors. BDNF–TrkB signaling alters synaptic protein composition by multiple mechanisms. (1) BDNF acts at the translational machinery to increase protein synthesis locally in or in the vicinity of active synapses. (2) BDNF can regulate dendritic mRNA trafficking and induces the disassembly of RNA granules constituents with a concomitant increase in translation of their mRNA content. (3) Postsynaptic BDNF–TrkB complex is endocytosed and the signaling endosome translocates to the cell body to induce the transcription of specific targets thought to be required for L-LTP maintenance.

Pioneer studies using a construct carrying the 3′- and 5′UTR sequences of CAMKII-α mRNA allowed the visualization of protein synthesis in dendrites of hippocampal neurons following application of BDNF (Aakalu, Smith, Nguyen, Jiang, & Schuman, 2001). More recently, the use of a metabolic labeling approach that consisted in the incorporation of noncanonical amino acids into proteins followed by their chemical-mediated fluorescence tagging allowed the in situ visualization of BDNF-induced translation in dendrites of hippocampal neurons (Dieterich et al., 2010). BDNF may act at several different levels to modulate translation of target transcripts in dendrites:

(i) BDNF–TrkB signaling can induce the dendritic targeting of mRNAs. For example, exogenous application of BDNF increases the delivery of RNA-binding proteins to dendrites (Leal, Afonso, & Duarte, 2014) and the dendritic accumulation of BDNF and TrkB mRNAs (Righi, Tongiorgi, & Cattaneo, 2000). Furthermore, BDNF-LTP at the medial perforant path synapse is also accompanied by rapid and selective upregulation of Arc mRNA species in granule cell dendrites (Ying et al., 2002).

(ii) BDNF can alter RNA granule composition. This neurotrophin was shown to promote the dissociation of dendritic RNA granules coincident with increases in the translation of their mRNA content (Fig. 3; Shiina, Shinkura, & Tokunaga, 2005).

(iii) BDNF can also act at the translational machinery to modulate dendritic protein synthesis (Fig. 3). For example, BDNF was shown to recruit mTOR pathways to induce translation initiation in hippocampal neuron dendrites (Takei et al., 2004) and stimulate BDNF-LTP in the CA1 hippocampal regions (Cammalleri et al., 2003). In the DG, endogenous BDNF drives protein synthesis-dependent LTP through the activation of the ERK–MNK1 pathway (Panja et al., 2014).

(iv) Some actions of BDNF on dendritic protein synthesis may also result from the regulation of microRNA (miRNA)-mediated translation repression. Perhaps the most prominent example was provided by the report showing that BDNF-induced local translation of LIMK1 results from the relief of miR-134-mediated silencing (Schratt et al., 2006). Recent findings indeed demonstrate the high importance of miRNAs in conferring specificity to BDNF-induced translation (Huang, Ruiz, Eyler, Lin, & Meffert, 2012).

(v) Finally, BDNF may also regulate transcriptional processes which affect protein synthesis at the whole-cell level and also has the potential to impact the availability of mRNAs to be transported and translated in dendrites, as in the case of Arc (Bramham et al., 2010; Zheng, Luo, & Wang, 2009). Accordingly, activation of BDNF receptors in the dendritic compartment was shown to induce an anterograde signal that travels to the cell body, where it regulates gene expression (Fig. 3; Cohen, Bas Orth, Kim, Jeon, & Jaffrey, 2011).

Additional information about the mechanisms by which BDNF affects protein synthesis can be found elsewhere (Leal, Comprido, et al., 2014).

Among the dendritic targets synthesized following stimulation of hippocampal neurons with BDNF are proteins with well-established roles in synaptic plasticity such as GluA1 AMPAR subunits (Fortin et al., 2012; Schratt, Nigh, Chen, Hu, & Greenberg, 2004) and scaffold and cytoskeleton-related proteins such as Homer2, CAMKII-α, PSD-95, Arc, RhoA, and LIMK1 (Briz et al., 2015; Butko et al., 2012; Leal, Comprido, et al., 2014; Schratt et al., 2004, 2006; Takei et al., 2004; Yin, Edelman, & Vanderklish, 2002).

Confirming the role of BDNF as a master regulator of protein synthesis, a high-throughput proteomic study performed in a synaptoneurosome preparation derived from cultured cortical neurons that were treated with BDNF found widespread changes in the synaptic proteome (Liao et al., 2007). Of particular note was the fact that proteins modulating synaptic structure and function and components of the translation machinery were among the proteins upregulated following stimulation with BDNF (Liao et al., 2007). These observations suggest that augmented synaptic translation capacity may further contribute to BDNF-induced alterations in the synaptic proteome. Similar functional sets of proteins were found to be altered by BDNF in cultured hippocampal neurons (Manadas et al., 2009). Adapting and combining orthogonal noncanonical amino acid tagging (BONCAT) and stable isotope labeling by amino acid in cell culture (SILAC) methodologies, several proteins were found to be altered following BDNF treatment in hippocampal slices (Bowling et al., 2016). These proteins are involved in synaptic transmission, trafficking, and cation binding and differ significantly from proteins identified in a similar screening performed in cultured neurons. These findings are likely related with the heterogeneity of nonneuronal cells containing TrkB receptors present in hippocampal slices and the fact that cultured neurons and hippocampal slices are developmentally distinct.

One may speculate that dendritic protein synthesis in response to BDNF allows: first, a fast source of new proteins specifically where needed and on demand, and second, taking in consideration the nature and roles of the newly synthesized proteins, a way to support functionally and structurally long-term enhancements on synaptic strength.

A recent study showed that BDNF reduces the activity of the proteasome in hippocampal synapses, and this effect was shown to contribute to the BDNF-mediated facilitation of the early phase of LTP in hippocampal CA1 synapses (Santos et al., 2015). These results suggest that the accumulation of proteins due to proteasome inhibition may act together with the increased translation activity to regulate the synaptic proteome in the early phase of LTP. However, the identity of the proteins that are upregulated due to proteasome inhibition and contribute to synaptic potentiation remains to be investigated.

6.5 BDNF–TrkB Signaling in Synaptic Tagging and Capture

The synaptic tagging and capture hypothesis was postulated by Frey and Morris in 1997 as a working platform to explain how synapse specificity can be achieved during L-LTP persistence (Frey & Morris, 1997). This hypothesis implies that a weak event (E-LTP) can set a tag that will capture plasticity-related products (PRPs) synthesized by an independent and strong event (L-LTP). Later, revised forms of the tagging and capture theory were developed to consider compartmentalization in neurons and concepts like dendritic protein synthesis and structural alterations at synapses. One of these theories (clustered plasticity model) proposed that synapses located close to each other compete for PRPs and postulated dendritic branches as integrative units in long-term memory engram formation (Govindarajan, Kelleher, & Tonegawa, 2006).

Recent studies suggest that TrkB induced by E-LTP may work as a tag to capture PRPs (BDNF) induced by L-LTP in an independent pathway (Lu et al., 2011). A similar argument was used to explain long-term memory consolidation (behavior tagging) (Lu et al., 2011). In a different study, a dual role for pre- and postsynaptic released BDNF in tagging and capture (postsynapse), respectively, was also proposed (Barco et al., 2005). In a different approach, BDNF and PKMζ were also suggested as PRPs, and it was shown that prior activity of synapses within a compartment governs the threshold for the induction of long-term functional plasticity (Sajikumar & Korte, 2011). From the abovementioned evidence, it is becoming clear that BDNF is likely to play a prominent role in the formation of complex memory engrams involving multiple compartments and circuits.

6.6 BDNF Effects at Nonassociative MF LTP

MF-CA3 LTP is an unusual form of plasticity since it is nonassociative and generally accepted to be NMDAR independent and expressed presynaptically (Nicoll & Schmitz, 2005), although some NMDAR-dependent forms of plasticity at these synapses have also been reported (Rebola, Carta, Lanore, Blanchet, & Mulle, 2011). Despite the findings indicating that BDNF protein is highly expressed, and its receptor TrkB is also present at MF-CA3 synapses (Conner, Lauterborn, Yan, Gall, & Varon, 1997; Dieni et al., 2012; Drake et al., 1999), the role of BDNF–TrkB signaling in synaptic plasticity at this pathway is largely unknown. Early studies showed the modulation of GABAergic transmission by BDNF at immature MF synapses (Sivakumaran, Mohajerani, & Cherubini, 2009) and impairments in HFS-induced LTP at these synapses in hippocampal slices from conditional BDNF-null mice (Huang et al., 2008). In a different approach, intrahippocampal infusion of BDNF was shown to induce a long-lasting increase in MF-CA3 synaptic efficacy (Gomez-Palacio-Schjetnan & Escobar, 2008) and to regulate the ability of MF-CA3 synapses to undergo LTP following HFS (Schjetnan & Escobar, 2012). Furthermore, recent evidence indicates that LTP at MF-CA3 synapses induced by a 50 Hz stimulation protocol is reduced to a similar extent in heterozygous BDNF knockout mice and when scavenging endogenous BDNF with TrkB-Fc (Schildt, Endres, Lessmann, & Edelmann, 2013), suggesting for the first time a role for endogenous released BDNF in MF-LTP.

Mechanistically, little is known about the putative processes mediating BDNF actions at this nonassociative form of synaptic plasticity. A recent study provided the first hint, by showing that transcription of neuronal activity-regulated pentraxin (Narp) is enhanced by BDNF, and Narp was suggested to mediate BDNF-induced glutamate release in hippocampal cultures and at MF synapses in hippocampal slices (Mariga et al., 2015). Accord ingly, a significant impairment of MF-LTP was observed in slices from Narp knockout mice which, in contrast with what was observed in wild-type slices, did not undergo any further increase following BDNF treatment (Mariga et al., 2015). These results suggest that Narp expression contributes to MF-LTP and BDNF-induced enhancement of MF-LTP. Narp mRNA is also induced in dentate granule cells following BDNF-LTP induction (Wibrand et al., 2006), but it is not known whether Narp synthesis contributes to LTP formation in the DG.

6.7 BDNF as a Regulator of Neurogenesis

BDNF-mediated actions in the adult brain are not limited to modification of mature synapses. In the DG, birth of new functional neurons from adult neural progenitor cells persists throughout life, and this capacity is essential for learning and memory formation (Zhao, Deng, & Gage, 2008). BDNF has been shown to stimulate proliferation of NPCs and promote long-term survival of their progeny (Katoh-Semba et al., 2002; Lee, Duan, & Mattson, 2002; Sairanen, Lucas, Ernfors, Castren, & Castren, 2005; Scharfman et al., 2005). Both intrahippocampal infusion of BDNF and its peripheral injection have been associated with potent stimulation of hippocampal neurogenesis (Scharfman et al., 2005; Schmidt & Duman, 2010; Shirayama, Chen, Nakagawa, Russell, & Duman, 2002). Recent work also showed that BDNF-LTP in the DG is associated with enhanced neurogenesis (Kuipers et al., 2016). Inhibition of Arc translation blocks BDNF-LTP induction and the associated proneurogenic effects. Interestingly, basal rates of proliferation and newborn cell survival are unaltered in Arc knockout mice. While the mechanisms remain to be explored, these findings link the proneurogenic effects of acute BDNF infusion to induction of Arc-dependent LTP in the adult rodent DG.

7. p75^{NTR} AND LTD

In the precursor form, neurotrophins bind to a complex formed by p75^{NTR}, which belong to the tumor necrosis factor receptor superfamily, and different coreceptors. In this case, the neurotrophin mature domain binds to p75^{NTR}, while their prodomain binds sortilin, which acts as a coreceptor (Nykjaer et al., 2004; Nykjaer, Willnow, & Petersen, 2005). However, it remains to be determined whether each prodomain of the dimeric molecules binds to the same sortilin protein (Feng et al., 2010). More recently, the prodomain of NGF was also found to interact with the sortilin family member SorCS2, a transmembrane protein highly expressed in the developing and adult nervous system (Deinhardt et al., 2011; Siao et al., 2012; Willnow, Petersen, & Nykjaer, 2008).

The extracellular domain of p75^{NTR} contains four cysteine-rich domains, each with six cysteine residues, a single transmembrane domain, and an intracellular region. The latter sequence contains two domains responsible for the receptor signaling activity, the Chopper domain and the death domain (reviewed in Almeida & Duarte, 2014). The Chopper domain is located in the juxtamembrane region and was found to be necessary and sufficient to initiate neuronal cell death (Coulson et al., 2000; Underwood, Reid, May, Bartlett, & Coulson, 2008). The lack of catalytic activity in the intracellular domain of $p75^{NTR}$ indicates that these receptors are coupled to the activation of intracellular signaling mechanisms distinct from those induced by Trk receptors. The $p75^{NTR}$ signaling is mediated by binding partners that are constitutively bound to the receptors and/or recruited upon receptor activation. Accordingly, more than 20 different intracellular $p75^{NTR}$-binding proteins were identified up to now (Barker, 2004; Bronfman & Fainzilber, 2004; Dechant & Barde, 2002; Roux & Barker, 2002; Schecterson & Bothwell, 2010). The differences in the signaling mechanisms activated by Trk receptors and $p75^{NTR}$ are thought to account for the opposing functional roles that have been described for neurotrophins.

The first evidence pointing to opposing roles for neurotrophins in synaptic activity was obtained in studies using cultured superior cervical ganglion neurons cocultured with cardiac myocytes to reproduce in vitro heart sympathetic innervation. Acute application of norepinephrine increases the myocyte frequency of contraction, while opposite effects are induced by acetylcholine (Furshpan, MacLeish, O'Lague, & Potter, 1976). Neurotrophins were shown to regulate differentially the activity-dependent release of norepinephrine and acetylcholine by activating Trk and $p75^{NTR}$ (Lockhart, Turrigiano, & Birren, 1997; Yang, Slonimsky, & Birren, 2002). NGF-mediated activation of TrkA receptors potentiates excitatory noradrenergic synaptic transmission (Lockhart et al., 1997), while BDNF induces a shift from excitatory to inhibitory (cholinergic) neurotransmission by a mechanism dependent on the activation of $p75^{NTR}$ (Yang et al., 2002). The latter response was found to be mediated by CaMKII since presynaptic inhibition of the kinase or expression of a constitutively active form of CaMKII blocked or mimicked the BDNF-induced inhibitory neurotransmission, respectively (Slonimsky, Mattaliano, Moon, Griffith, & Birren, 2006).

Differential effects of Trk and $p75^{NTR}$ on synaptic plasticity have also been reported in hippocampal neurons. In contrast with the effect of BDNF–TrkB signaling, which induces the maturation of dendritic spines in cultured neurons (see earlier) (e.g., Ji et al., 2010), overexpression of $p75^{NTR}$ was found to decrease dendritic complexity and the number of spines in hippocampal pyramidal neurons (Zagrebelsky et al., 2005).

Accordingly, $p75^{NTR-/-}$ mice are characterized by an increased spine density and greater dendritic complexity when compared with wild-type mice (Zagrebelsky et al., 2005). Furthermore, $p75^{NTR}$-null mice showed impaired hippocampal LTD evoked by low frequency stimulation (LFS) of CA1 synapses, in addition to alterations in the total expression of GluA2 and GluA3 subunits of AMPAR, while LTP was not affected (Rosch, Schweigreiter, Bonhoeffer, Barde, & Korte, 2005). Similarly, LTD of CA1 synapses induced by repeated incubations with the group I metabotropic glutamate receptor agonist DHPG ((*S*)-3,5-dihydroxyphenylglycine) was shown to be mediated by $p75^{NTR}$ (Egashira et al., 2010; Sakuragi, Tominaga-Yoshino, & Ogura, 2013). In this paradigm, inhibition of $p75^{NTR}$ during the period of incubation with DHPG enhanced Trk receptor phosphorylation (activity) and induced LTP, suggesting that $p75^{NTR}$ activation dampens the activity of mature neurotrophins mediated by Trk receptors (Sakuragi et al., 2013).

In agreement with the observations described earlier, stimulation with proBDNF induces NMDAR-dependent LTD evoked by LFS of hippocampal CA1 synapses by a mechanism mediated by $p75^{NTR}$ and the GluN2B NMDAR subunit (Woo et al., 2005). This effect of proBDNF may be mediated by changes in trafficking and/or endocytosis of GluN2B subunits since GluN2B-mediated synaptic currents and expression are reduced in the hippocampus of $p75^{NTR-/-}$ mice (Woo et al., 2005). However, this study did not allow concluding whether proBDNF expressed under the control of its endogenous promoter can elicit similar effects. This question was addressed using knock-in mice in which the proconvertase/furin cleavage site of proBDNF was mutated and expressed under the control of endogenous BDNF promoter elements (Yang et al., 2014). The results obtained showed that proBDNF–$p75^{NTR}$-mediated signaling reduces dendritic complexity and spine formation in DG granule neurons, and a similar effect on spine density was observed in CA1 pyramidal neurons (Yang et al., 2014). In accordance with these observations, hippocampal slices from cleavage-resistant *probdnf* knock-in mice showed deficits in basal synaptic transmission and enhanced LTD induced by LFS of CA1 synapses (Yang et al., 2014). Together, the evidence suggests that proBDNF plays a role in the regulation of hippocampal structure, synaptic transmission, and plasticity.

The prodomain of BDNF is thought to play an important role in the folding of the protein (Kolbeck, Jungbluth, & Barde, 1994). However, the recent findings showing a higher expression of the BDNF propeptide in hippocampal lysates when compared with total proBDNF protein levels suggest that the free prodomain of the neurotrophin, resulting from the intracellular or extracellular cleavage of the proneurotrophin, may also be biologically active (Dieni et al., 2012). Accordingly, a recent study showed that the BDNF propeptide directly facilitates LFS-evoked LTD in CA1 synapses, by a mechanism dependent on the activity of GluN2B-containing NMDAR and $p75^{NTR}$. Furthermore, incubation with the BDNF propeptide enhanced the internalization of GluA2-containing AMPA receptors and promoted the NMDA-induced internalization of the receptors in cultured hippocampal neurons, by a mechanism dependent on the activation of $p75^{NTR}$ (Mizui et al., 2015). These effects of the BDNF prodomain on AMPAR may contribute to LTD induced by LFS. Similarly, recombinant proBDNF negatively regulates the activity of neuromuscular synapses by activation of presynaptic $p75^{NTR}$ (Yang et al., 2009).

In contrast with the compelling evidence pointing to a role of $p75^{NTR}$ and proBDNF in LTD, their impact in behavior has been difficult to establish. Some studies reported deficits in memory acquisition in $p75^{NTR}$-null mice when compared with control animals (Peterson, Dickinson-Anson, Leppert, Lee, & Gage, 1999; Wright, Alt, Turner, & Krueger, 2004), while others have shown an improved spatial learning performance in $p75^{NTR}$ knockout mice (Barrett et al., 2010; Greferath et al., 2000). Differences in the genetic background of the animals used in these studies may account for the conflicting results reported. Furthermore, $BDNF^{+/-}$ mice fail to induce LTD after LFS of CA1 synapses, and these results correlate with an impairment of object-place memory and object-place exploration (Aarse, Herlitze, & Manahan-Vaughan, 2016).

8. INTERPLAY BETWEEN GLUCOCORTICOIDS AND BDNF TO DETERMINE CELLULAR RESPONSE TO STRESS

Stressful situations elicit the release of steroid hormones that evoke widespread responses across the body including in the CNS. Among steroid hormones, glucocorticoids that are secreted by the adrenal cortex are prominently involved in coping with stress. However, prolonged exposure to such hormones as observed in chronic stress has deleterious effects in the brain, in particular within the hippocampus. Such effects include spine loss and dendritic atrophy (barrel cortex; Liston & Gan, 2011), loss and death of hippocampal neurons (Sapolsky, Krey, & McEwen, 1985; Sapolsky, Uno,

Rebert, & Finch, 1990), and impaired hippocampal LTP (Pavlides, Watanabe, & McEwen, 1993).

Chronic stress is typically associated with decreased levels of BDNF, although this regulation appears to be complex as it is determined by the nature and severity of the stimulus, differs across brain regions, and changes over time during and after the stressor (see Gray, Milner, & McEwen, 2013). Glucocorticoids and BDNF have strong effects on hippocampal structural and cellular plasticity that tend to be of opposing nature (reviewed in Gray et al., 2013; Numakawa, Adachi, Richards, Chiba, & Kunugi, 2013; Suri & Vaidya, 2013), leading to the hypothesis that some of the chronic effects of glucocorticoids may result from a decline in BDNF levels and signaling. Several lines of evidence support this view. For instance, exposure to dexamethasone (a synthetic glucocorticoid) suppresses BDNF-induced glutamate release in cultured cortical neurons (Numakawa et al., 2009). Additionally, exogenous corticosterone induces paired pulse facilitation deficits and attenuates tetanus-induced LTP at hippocampal CA1 synapses, with a concomitant decrease in BDNF mRNA levels (Zhou, Zhang, & Zhang, 2000). Importantly, coapplication of BDNF partly restored corticosterone-induced alterations in PPF (Zhou et al., 2000). In a separated study, chronic infusion of BDNF reverted chronic stress-induced impairments in LTP and memory (Radecki, Brown, Martinez, & Teyler, 2005), supporting the notion that lack of trophic support may underlie some of the aversive effects of chronic stress in the brain.

Recent findings demonstrate that the interplay between glucocorticoids and BDNF may also play a role in acute, fast, and adaptive responses to stress. For example, glucocorticoid receptor activation was shown to recruit BDNF signaling pathways to consolidate contextual fear memories (Chen, Bambah-Mukku, Pollonini, & Alberini, 2012; Revest et al., 2014). Such processes are likely mediated by the increased production, release and processing of BDNF (Revest et al., 2014), and/or transactivation of TrkB receptors (Jeanneteau, Garabedian, & Chao, 2008; Numakawa et al., 2009). The cross talk between glucocorticoids and BDNF was further explored in a recent study which showed that glucocorticoid receptors are also a prominent target of BDNF-mediated signaling (Arango-Lievano et al., 2015). The authors showed that BDNF induces the phosphorylation of glucocorticoid receptors and suggested that the coordinated/paired actions of BDNF and glucocorticoids are essential for neuroplasticity responses to stress (Arango-Lievano et al., 2015).

In summary, the interplay between glucocorticoids and BDNF, with the influence of other molecules (Gray et al., 2013), may work as a facilitator of plasticity in adaptive stress response. A fine-switch in this cross talk may also be deeply involved in maladaptive stress responses leading to cognitive impairment and disease states.

9. CONCLUDING REMARKS

It is unequivocal that BDNF has potent effects at synapses. Among those are structural and functional alterations that ultimately dictate how plastic synapses can be. Compelling amount of evidence implicates the BDNF–TrkB pathway in hippocampal LTP. However, our knowledge about the downstream signaling mechanisms involved is still limited. Future studies should address such issues, especially in in vivo systems during relevant patterns of synaptic activity. Similarly, whether BDNF is secreted at pre- or postsynaptic sites and if it acts in an autocrine or paracrine fashion are questions that require further research.

Synaptic plasticity is a ubiquitous property of synapses making the abovementioned premise valid for each synapse expressing the BDNF–TrkB signaling machinery and different activity-dependent forms of plasticity. It will be important to characterize the spatiotemporal regulation of the mechanisms governing activity-dependent synaptic secretion of endogenous BDNF and to investigate how different pools of BDNF (e.g., secreted vs recycled/endocytosed BDNF) cooperate together to modulate or mediate processes that support sustained increments in synaptic efficacy. BDNF regulates the synaptic composition in multiple ways. Dendritic protein synthesis and local control of protein degradation may provide a fast and selective way to control the synaptic proteome. Future experiments should focus on molecular mechanisms governing BDNF-induced local translation (and control of protein degradation) and ultimately investigate how these alterations in the synaptic proteome are important for LTP. Among prominent challenges in the future are the need of (i) a better understanding of signaling pathways involved in BDNF-mediated dendritic translation (e.g., BDNF effects on dendritic mRNA trafficking, RNA granules composition, and translation machinery), (ii) a clear picture of targets locally translated (or degraded) in response to endogenous released BDNF, and (iii) a better temporal resolution of such processes and how those correlate with alterations in synapse structure and function.

The role of p75^{NTR} in synaptic plasticity is still emerging, and additional studies are required to fully understand the mechanisms whereby proBDNF regulates synaptic activity. In particular, future studies should address the molecular mechanisms acting downstream of p75^{NTR} to cause synaptic depression. This will also depend on the improved understanding of the mechanisms of action of p75^{NTR}, which are not as well characterized as the TrkB-induced signaling activity.

Abnormal neurotrophin signaling and function may underlie several neurological and psychiatric disorders (Balaratnasingam & Janca, 2012), some of which characterized by defective neuronal plasticity. A better knowledge of fundamental processes of BDNF biology is required taking also in consideration that this neurotrophin has become an attractive target to better understand the progression of several CNS diseases, and with the aim of developing potential treatments for those disorders of the nervous system.

REFERENCES

Aakalu, G., Smith, W. B., Nguyen, N., Jiang, C., & Schuman, E. M. (2001). Dynamic visualization of local protein synthesis in hippocampal neurons. *Neuron*, *30*, 489–502.

Aarse, J., Herlitze, S., & Manahan-Vaughan, D. (2016). The requirement of BDNF for hippocampal synaptic plasticity is experience-dependent. *Hippocampus*, *26*, 739–751.

Adams, J. P., & Dudek, S. M. (2005). Late-phase long-term potentiation: Getting to the nucleus. *Nature Reviews. Neuroscience*, *6*, 737–743.

Almeida, R. D., & Duarte, C. B. (2014). p75NTR processing and signaling: Functional role. In R. M. Kostrzewa (Ed.), *Handbook of neurotoxicity* (pp. 1899–1923). New York: Springer Science.

Alonso, M., Medina, J. H., & Pozzo-Miller, L. (2004). ERK1/2 activation is necessary for BDNF to increase dendritic spine density in hippocampal CA1 pyramidal neurons. *Learning & Memory*, *11*, 172–178.

Amaral, M. D., & Pozzo-Miller, L. (2007). TRPC3 channels are necessary for brain-derived neurotrophic factor to activate a nonselective cationic current and to induce dendritic spine formation. *The Journal of Neuroscience*, *27*, 5179–5189.

An, J. J., Gharami, K., Liao, G. Y., Woo, N. H., Lau, A. G., Vanevski, F., et al. (2008). Distinct role of long 3′ UTR BDNF mRNA in spine morphology and synaptic plasticity in hippocampal neurons. *Cell*, *134*, 175–187.

Aoki, C., Wu, K., Elste, A., Len, G., Lin, S., McAuliffe, G., et al. (2000). Localization of brain-derived neurotrophic factor and TrkB receptors to postsynaptic densities of adult rat cerebral cortex. *Journal of Neuroscience Research*, *59*, 454–463.

Arango-Lievano, M., Lambert, W. M., Bath, K. G., Garabedian, M. J., Chao, M. V., & Jeanneteau, F. (2015). Neurotrophic-priming of glucocorticoid receptor signaling is essential for neuronal plasticity to stress and antidepressant treatment. *Proceedings of the National Academy of Sciences of the United States of America*, *112*, 15737–15742.

Atwal, J. K., Massie, B., Miller, F. D., & Kaplan, D. R. (2000). The TrkB-Shc site signals neuronal survival and local axon growth via MEK and P13-kinase. *Neuron*, *27*, 265–277.

Bailey, C. H., Kandel, E. R., & Harris, K. M. (2015). Structural components of synaptic plasticity and memory consolidation. *Cold Spring Harbor Perspectives in Biology*, 7, a021758.

Balaratnasingam, S., & Janca, A. (2012). Brain derived neurotrophic factor: A novel neurotrophin involved in psychiatric and neurological disorders. *Pharmacology & Therapeutics, 134*, 116–124.

Barco, A., Patterson, S. L., Alarcon, J. M., Gromova, P., Mata-Roig, M., Morozov, A., et al. (2005). Gene expression profiling of facilitated L-LTP in VP16-CREB mice reveals that BDNF is critical for the maintenance of LTP and its synaptic capture. *Neuron, 48*, 123–137.

Barde, Y. A., Edgar, D., & Thoenen, H. (1982). Purification of a new neurotrophic factor from mammalian brain. *The EMBO Journal, 1*, 549–553.

Barker, P. A. (2004). p75NTR is positively promiscuous: Novel partners and new insights. *Neuron, 42*, 529–533.

Barrett, G. L., Reid, C. A., Tsafoulis, C., Zhu, W., Williams, D. A., Paolini, A. G., et al. (2010). Enhanced spatial memory and hippocampal long-term potentiation in p75 neurotrophin receptor knockout mice. *Hippocampus, 20*, 145–152.

Bliss, T. V., & Collingridge, G. L. (1993). A synaptic model of memory: Long-term potentiation in the hippocampus. *Nature, 361*, 31–39.

Bliss, T., Collingridge, G., & Morris, R. (2007). Synaptic plasticity in the hippocampus. In R. Morris, P. Andersen, D. Amaral, T. Bliss, & J. O'Keefe (Eds.), *The hippocampus book* (pp. 343–474). New York: Oxford University Press.

Bliss, T. V., & Lomo, T. (1973). Long-lasting potentiation of synaptic transmission in the dentate area of the anaesthetized rabbit following stimulation of the perforant path. *The Journal of Physiology, 232*, 331–356.

Bosch, M., Castro, J., Saneyoshi, T., Matsuno, H., Sur, M., & Hayashi, Y. (2014). Structural and molecular remodeling of dendritic spine substructures during long-term potentiation. *Neuron, 82*, 444–459.

Bowling, H., Bhattacharya, A., Zhang, G., Lebowitz, J. Z., Alam, D., Smith, P. T., et al. (2016). BONLAC: A combinatorial proteomic technique to measure stimulus-induced translational profiles in brain slices. *Neuropharmacology, 100*, 76–89.

Bramham, C. R. (2008). Local protein synthesis, actin dynamics, and LTP consolidation. *Current Opinion in Neurobiology, 18*, 524–531.

Bramham, C. R., Alme, M. N., Bittins, M., Kuipers, S. D., Nair, R. R., Pai, B., et al. (2010). The Arc of synaptic memory. *Experimental Brain Research, 200*, 125–140.

Bramham, C. R., & Messaoudi, E. (2005). BDNF function in adult synaptic plasticity: The synaptic consolidation hypothesis. *Progress in Neurobiology, 76*, 99–125.

Bramham, C. R., & Wells, D. G. (2007). Dendritic mRNA: Transport, translation and function. *Nature Reviews. Neuroscience, 8*, 776–789.

Briz, V., Zhu, G., Wang, Y., Liu, Y., Avetisyan, M., Bi, X., et al. (2015). Activity-dependent rapid local RhoA synthesis is required for hippocampal synaptic plasticity. *The Journal of Neuroscience, 35*, 2269–2282.

Bronfman, F. C., & Fainzilber, M. (2004). Multi-tasking by the p75 neurotrophin receptor: Sortilin things out? *EMBO Reports, 5*, 867–871.

Butko, M. T., Yang, J., Geng, Y., Kim, H. J., Jeon, N. L., Shu, X., et al. (2012). Fluorescent and photo-oxidizing TimeSTAMP tags track protein fates in light and electron microscopy. *Nature Neuroscience, 15*, 1742–1751.

Cajigas, I. J., Tushev, G., Will, T. J., tom Dieck, S., Fuerst, N., & Schuman, E. M. (2012). The local transcriptome in the synaptic neuropil revealed by deep sequencing and high-resolution imaging. *Neuron, 74*, 453–466.

Caldeira, M. V., Melo, C. V., Pereira, D. B., Carvalho, R. F., Carvalho, A. L., & Duarte, C. B. (2007). BDNF regulates the expression and traffic of NMDA receptors in cultured hippocampal neurons. *Molecular and Cellular Neurosciences, 35*, 208–219.

Caldeira, M. V., Melo, C. V., Pereira, D. B., Carvalho, R., Correia, S. S., Backos, D. S., et al. (2007). Brain-derived neurotrophic factor regulates the expression and synaptic delivery of alpha-amino-3-hydroxy-5-methyl-4-isoxazole propionic acid receptor subunits in hippocampal neurons. *The Journal of Biological Chemistry, 282*, 12619–12628.

Cammalleri, M., Lutjens, R., Berton, F., King, A. R., Simpson, C., Francesconi, W., et al. (2003). Time-restricted role for dendritic activation of the mTOR-p70S6K pathway in the induction of late-phase long-term potentiation in the CA1. *Proceedings of the National Academy of Sciences of the United States of America*, *100*, 14368–14373.

Chao, M. V. (2003). Neurotrophins and their receptors: A convergence point for many signalling pathways. *Nature Reviews. Neuroscience*, *4*, 299–309.

Chen, D. Y., Bambah-Mukku, D., Pollonini, G., & Alberini, C. M. (2012). Glucocorticoid receptors recruit the CaMKIIalpha-BDNF-CREB pathways to mediate memory consolidation. *Nature Neuroscience*, *15*, 1707–1714.

Chen, Z. Y., Ieraci, A., Tanowitz, M., & Lee, F. S. (2005). A novel endocytic recycling signal distinguishes biological responses of Trk neurotrophin receptors. *Molecular Biology of the Cell*, *16*, 5761–5772.

Chen, G., Kolbeck, R., Barde, Y. A., Bonhoeffer, T., & Kossel, A. (1999). Relative contribution of endogenous neurotrophins in hippocampal long-term potentiation. *The Journal of Neuroscience*, *19*, 7983–7990.

Cohen, M. S., Bas Orth, C., Kim, H. J., Jeon, N. L., & Jaffrey, S. R. (2011). Neurotrophin-mediated dendrite-to-nucleus signaling revealed by microfluidic compartmentalization of dendrites. *Proceedings of the National Academy of Sciences of the United States of America*, *108*, 11246–11251.

Cohen, S., Levi-Montalcini, R., & Hamburger, V. (1954). A nerve growth-stimulating factor isolated from sarcom as 37 and 180. *Proceedings of the National Academy of Sciences of the United States of America*, *40*, 1014–1018.

Conner, J. M., Lauterborn, J. C., Yan, Q., Gall, C. M., & Varon, S. (1997). Distribution of brain-derived neurotrophic factor (BDNF) protein and mRNA in the normal adult rat CNS: Evidence for anterograde axonal transport. *The Journal of Neuroscience*, *17*, 2295–2313.

Coulson, E. J., Reid, K., Baca, M., Shipham, K. A., Hulett, S. M., Kilpatrick, T. J., et al. (2000). Chopper, a new death domain of the p75 neurotrophin receptor that mediates rapid neuronal cell death. *The Journal of Biological Chemistry*, *275*, 30537–30545.

Crowder, R. J., & Freeman, R. S. (1998). Phosphatidylinositol 3-kinase and Akt protein kinase are necessary and sufficient for the survival of nerve growth factor-dependent sympathetic neurons. *The Journal of Neuroscience*, *18*, 2933–2943.

Dechant, G., & Barde, Y. A. (2002). The neurotrophin receptor p75(NTR): Novel functions and implications for diseases of the nervous system. *Nature Neuroscience*, *5*, 1131–1136.

Deinhardt, K., Kim, T., Spellman, D. S., Mains, R. E., Eipper, B. A., Neubert, T. A., et al. (2011). Neuronal growth cone retraction relies on proneurotrophin receptor signaling through Rac. *Science Signaling*, *4*, ra82.

Derkach, V. A., Oh, M. C., Guire, E. S., & Soderling, T. R. (2007). Regulatory mechanisms of AMPA receptors in synaptic plasticity. *Nature Reviews. Neuroscience*, *8*, 101–113.

Dictenberg, J. B., Swanger, S. A., Antar, L. N., Singer, R. H., & Bassell, G. J. (2008). A direct role for FMRP in activity-dependent dendritic mRNA transport links filopodial-spine morphogenesis to fragile X syndrome. *Developmental Cell*, *14*, 926–939.

Dieni, S., Matsumoto, T., Dekkers, M., Rauskolb, S., Ionescu, M. S., Deogracias, R., et al. (2012). BDNF and its pro-peptide are stored in presynaptic dense core vesicles in brain neurons. *The Journal of Cell Biology*, *196*, 775–788.

Dieterich, D. C., Hodas, J. J., Gouzer, G., Shadrin, I. Y., Ngo, J. T., Triller, A., et al. (2010). In situ visualization and dynamics of newly synthesized proteins in rat hippocampal neurons. *Nature Neuroscience*, *13*, 897–905.

Doyle, M., & Kiebler, M. A. (2011). Mechanisms of dendritic mRNA transport and its role in synaptic tagging. *The EMBO Journal*, *30*, 3540–3552.

Drake, C. T., Milner, T. A., & Patterson, S. L. (1999). Ultrastructural localization of full-length trkB immunoreactivity in rat hippocampus suggests multiple roles in modulating activity-dependent synaptic plasticity. *The Journal of Neuroscience*, *19*, 8009–8026.

Edelmann, E., Cepeda-Prado, E., Franck, M., Lichtenecker, P., Brigadski, T., & Lessmann, V. (2015). Theta burst firing recruits BDNF release and signaling in postsynaptic CA1 neurons in spike-timing-dependent LTP. *Neuron, 86*, 1041–1054.

Edelmann, E., Lessmann, V., & Brigadski, T. (2014). Pre- and postsynaptic twists in BDNF secretion and action in synaptic plasticity. *Neuropharmacology, 76*(Pt. C), 610–627.

Egashira, Y., Tanaka, T., Soni, P., Sakuragi, S., Tominaga-Yoshino, K., & Ogura, A. (2010). Involvement of the p75(NTR) signaling pathway in persistent synaptic suppression coupled with synapse elimination following repeated long-term depression induction. *Journal of Neuroscience Research, 88*, 3433–3446.

Elvira, G., Wasiak, S., Blandford, V., Tong, X. K., Serrano, A., Fan, X., et al. (2006). Characterization of an RNA granule from developing brain. *Molecular & Cellular Proteomics, 5*, 635–651.

Feng, D., Kim, T., Ozkan, E., Light, M., Torkin, R., Teng, K. K., et al. (2010). Molecular and structural insight into proNGF engagement of p75NTR and sortilin. *Journal of Molecular Biology, 396*, 967–984.

Figurov, A., Pozzo-Miller, L. D., Olafsson, P., Wang, T., & Lu, B. (1996). Regulation of synaptic responses to high-frequency stimulation and LTP by neurotrophins in the hippocampus. *Nature, 381*, 706–709.

Fortin, D. A., Srivastava, T., Dwarakanath, D., Pierre, P., Nygaard, S., Derkach, V. A., et al. (2012). Brain-derived neurotrophic factor activation of CaM-kinase kinase via transient receptor potential canonical channels induces the translation and synaptic incorporation of GluA1-containing calcium-permeable AMPA receptors. *The Journal of Neuroscience, 32*, 8127–8137.

Frey, U., & Morris, R. G. (1997). Synaptic tagging and long-term potentiation. *Nature, 385*, 533–536.

Fritzsche, R., Karra, D., Bennett, K. L., Ang, F. Y., Heraud-Farlow, J. E., Tolino, M., et al. (2013). Interactome of two diverse RNA granules links mRNA localization to translational repression in neurons. *Cell Reports, 5*, 1749–1762.

Fukazawa, Y., Saitoh, Y., Ozawa, F., Ohta, Y., Mizuno, K., & Inokuchi, K. (2003). Hippocampal LTP is accompanied by enhanced F-actin content within the dendritic spine that is essential for late LTP maintenance in vivo. *Neuron, 38*, 447–460.

Furshpan, E. J., MacLeish, P. R., O'Lague, P. H., & Potter, D. D. (1976). Chemical transmission between rat sympathetic neurons and cardiac myocytes developing in microcultures: Evidence for cholinergic, adrenergic, and dual-function neurons. *Proceedings of the National Academy of Sciences of the United States of America, 73*, 4225–4229.

Gartner, A., & Staiger, V. (2002). Neurotrophin secretion from hippocampal neurons evoked by long-term-potentiation-inducing electrical stimulation patterns. *Proceedings of the National Academy of Sciences of the United States of America, 99*, 6386–6391.

Genheden, M., Kenney, J. W., Johnston, H. E., Manousopoulou, A., Garbis, S. D., & Proud, C. G. (2015). BDNF stimulation of protein synthesis in cortical neurons requires the MAP kinase-interacting kinase MNK1. *The Journal of Neuroscience, 35*, 972–984.

Gomez-Palacio-Schjetnan, A., & Escobar, M. L. (2008). In vivo BDNF modulation of adult functional and morphological synaptic plasticity at hippocampal mossy fibers. *Neuroscience Letters, 445*, 62–67.

Gottmann, K., Mittmann, T., & Lessmann, V. (2009). BDNF signaling in the formation, maturation and plasticity of glutamatergic and GABAergic synapses. *Experimental Brain Research, 199*, 203–234.

Govindarajan, A., Kelleher, R. J., & Tonegawa, S. (2006). A clustered plasticity model of long-term memory engrams. *Nature Reviews. Neuroscience*, 7, 575–583.

Gray, J. D., Milner, T. A., & McEwen, B. S. (2013). Dynamic plasticity: The role of glucocorticoids, brain-derived neurotrophic factor and other trophic factors. *Neuroscience, 239*, 214–227.

Greferath, U., Bennie, A., Kourakis, A., Bartlett, P. F., Murphy, M., & Barrett, G. L. (2000). Enlarged cholinergic forebrain neurons and improved spatial learning in p75 knockout mice. *The European Journal of Neuroscience*, *12*, 885–893.

Grosshans, D. R., Clayton, D. A., Coultrap, S. J., & Browning, M. D. (2002). LTP leads to rapid surface expression of NMDA but not AMPA receptors in adult rat CA1. *Nature Neuroscience*, *5*, 27–33.

Hartmann, M., Heumann, R., & Lessmann, V. (2001). Synaptic secretion of BDNF after high-frequency stimulation of glutamatergic synapses. *The EMBO Journal*, *20*, 5887–5897.

Hebb, D. O. (1949). *The organization of behavior*. New York: Wiley.

Holgado-Madruga, M., Moscatello, D. K., Emlet, D. R., Dieterich, R., & Wong, A. J. (1997). Grb2-associated binder-1 mediates phosphatidylinositol 3-kinase activation and the promotion of cell survival by nerve growth factor. *Proceedings of the National Academy of Sciences of the United States of America*, *94*, 12419–12424.

Honkura, N., Matsuzaki, M., Noguchi, J., Ellis-Davies, G. C., & Kasai, H. (2008). The subspine organization of actin fibers regulates the structure and plasticity of dendritic spines. *Neuron*, *57*, 719–729.

Huang, Y. Z., Pan, E., Xiong, Z. Q., & McNamara, J. O. (2008). Zinc-mediated transactivation of TrkB potentiates the hippocampal mossy fiber-CA3 pyramid synapse. *Neuron*, *57*, 546–558.

Huang, Y. W., Ruiz, C. R., Eyler, E. C., Lin, K., & Meffert, M. K. (2012). Dual regulation of miRNA biogenesis generates target specificity in neurotrophin-induced protein synthesis. *Cell*, *148*, 933–946.

Huang, S. H., Wang, J., Sui, W. H., Chen, B., Zhang, X. Y., Yan, J., et al. (2013). BDNF-dependent recycling facilitates TrkB translocation to postsynaptic density during LTP via a Rab11-dependent pathway. *The Journal of Neuroscience*, *33*, 9214–9230.

Jeanneteau, F., Garabedian, M. J., & Chao, M. V. (2008). Activation of Trk neurotrophin receptors by glucocorticoids provides a neuroprotective effect. *Proceedings of the National Academy of Sciences of the United States of America*, *105*, 4862–4867.

Ji, Y., Lu, Y., Yang, F., Shen, W., Tang, T. T., Feng, L., et al. (2010). Acute and gradual increases in BDNF concentration elicit distinct signaling and functions in neurons. *Nature Neuroscience*, *13*, 302–309.

Jovanovic, J. N., Czernik, A. J., Fienberg, A. A., Greengard, P., & Sihra, T. S. (2000). Synapsins as mediators of BDNF-enhanced neurotransmitter release. *Nature Neuroscience*, *3*, 323–329.

Kanai, Y., Dohmae, N., & Hirokawa, N. (2004). Kinesin transports RNA: Isolation and characterization of an RNA-transporting granule. *Neuron*, *43*, 513–525.

Kandel, E. R. (2001). The molecular biology of memory storage: A dialogue between genes and synapses. *Science*, *294*, 1030–1038.

Kang, H., & Schuman, E. M. (1995). Long-lasting neurotrophin-induced enhancement of synaptic transmission in the adult hippocampus. *Science*, *267*, 1658–1662.

Kang, H., & Schuman, E. M. (1996). A requirement for local protein synthesis in neurotrophin-induced hippocampal synaptic plasticity. *Science*, *273*, 1402–1406.

Kang, H., Welcher, A. A., Shelton, D., & Schuman, E. M. (1997). Neurotrophins and time: Different roles for TrkB signaling in hippocampal long-term potentiation. *Neuron*, *19*, 653–664.

Kaplan, D. R., & Miller, F. D. (2000). Neurotrophin signal transduction in the nervous system. *Current Opinion in Neurobiology*, *10*, 381–391.

Kasai, H., Fukuda, M., Watanabe, S., Hayashi-Takagi, A., & Noguchi, J. (2010). Structural dynamics of dendritic spines in memory and cognition. *Trends in Neurosciences*, *33*, 121–129.

Katoh-Semba, R., Asano, T., Ueda, H., Morishita, R., Takeuchi, I. K., Inaguma, Y., et al. (2002). Riluzole enhances expression of brain-derived neurotrophic factor with

consequent proliferation of granule precursor cells in the rat hippocampus. *The FASEB Journal, 16*, 1328–1330.
Kolarow, R., Brigadski, T., & Lessmann, V. (2007). Postsynaptic secretion of BDNF and NT-3 from hippocampal neurons depends on calcium calmodulin kinase II signaling and proceeds via delayed fusion pore opening. *The Journal of Neuroscience, 27*, 10350–10364.
Kolbeck, R., Jungbluth, S., & Barde, Y. A. (1994). Characterisation of neurotrophin dimers and monomers. *European Journal of Biochemistry, 225*, 995–1003.
Korte, M., Carroll, P., Wolf, E., Brem, G., Thoenen, H., & Bonhoeffer, T. (1995). Hippocampal long-term potentiation is impaired in mice lacking brain-derived neurotrophic factor. *Proceedings of the National Academy of Sciences of the United States of America, 92*, 8856–8860.
Korte, M., Griesbeck, O., Gravel, C., Carroll, P., Staiger, V., Thoenen, H., et al. (1996). Virus-mediated gene transfer into hippocampal CA1 region restores long-term potentiation in brain-derived neurotrophic factor mutant mice. *Proceedings of the National Academy of Sciences of the United States of America, 93*, 12547–12552.
Korte, M., Kang, H., Bonhoeffer, T., & Schuman, E. (1998). A role for BDNF in the late-phase of hippocampal long-term potentiation. *Neuropharmacology, 37*, 553–559.
Kossel, A. H., Cambridge, S. B., Wagner, U., & Bonhoeffer, T. (2001). A caged Ab reveals an immediate/instructive effect of BDNF during hippocampal synaptic potentiation. *Proceedings of the National Academy of Sciences of the United States of America, 98*, 14702–14707.
Kouhara, H., Hadari, Y. R., Spivak-Kroizman, T., Schilling, J., Bar-Sagi, D., Lax, I., et al. (1997). A lipid-anchored Grb2-binding protein that links FGF-receptor activation to the Ras/MAPK signaling pathway. *Cell, 89*, 693–702.
Kovalchuk, Y., Hanse, E., Kafitz, K. W., & Konnerth, A. (2002). Postsynaptic induction of BDNF-mediated long-term potentiation. *Science, 295*, 1729–1734.
Krucker, T., Siggins, G. R., & Halpain, S. (2000). Dynamic actin filaments are required for stable long-term potentiation (LTP) in area CA1 of the hippocampus. *Proceedings of the National Academy of Sciences of the United States of America, 97*, 6856–6861.
Kuipers, S. D., Trentani, A., Tiron, A., Mao, X., Kuhl, D., & Bramham, C. (2016). BDNF-induced LTP is associated with rapid Arc-dependent enhancement in adult hippocampal neurogenesis. *Scientific Reports, 6*, 21222.
Lai, K. O., Wong, A. S., Cheung, M. C., Xu, P., Liang, Z., Lok, K. C., et al. (2012). TrkB phosphorylation by Cdk5 is required for activity-dependent structural plasticity and spatial memory. *Nature Neuroscience, 15*, 1506–1515.
Leal, G., Afonso, P. M., & Duarte, C. B. (2014). Neuronal activity induces synaptic delivery of hnRNP A2/B1 by a BDNF-dependent mechanism in cultured hippocampal neurons. *PLoS One, 9*. e108175.
Leal, G., Afonso, P. M., Salazar, I. L., & Duarte, C. B. (2015). Regulation of hippocampal synaptic plasticity by BDNF. *Brain Research, 1621*, 82–101.
Leal, G., Comprido, D., & Duarte, C. B. (2014). BDNF-induced local protein synthesis and synaptic plasticity. *Neuropharmacology, 76*(Pt. C), 639–656.
Lee, F. S., & Chao, M. V. (2001). Activation of Trk neurotrophin receptors in the absence of neurotrophins. *Proceedings of the National Academy of Sciences of the United States of America, 98*, 3555–3560.
Lee, J., Duan, W., & Mattson, M. P. (2002). Evidence that brain-derived neurotrophic factor is required for basal neurogenesis and mediates, in part, the enhancement of neurogenesis by dietary restriction in the hippocampus of adult mice. *Journal of Neurochemistry, 82*, 1367–1375.
Lee, R., Kermani, P., Teng, K. K., & Hempstead, B. L. (2001). Regulation of cell survival by secreted proneurotrophins. *Science, 294*, 1945–1948.

Lee, F. S., Rajagopal, R., Kim, A. H., Chang, P. C., & Chao, M. V. (2002). Activation of Trk neurotrophin receptor signaling by pituitary adenylate cyclase-activating polypeptides. *The Journal of Biological Chemistry, 277*, 9096–9102.
Lessmann, V., & Brigadski, T. (2009). Mechanisms, locations, and kinetics of synaptic BDNF secretion: An update. *Neuroscience Research, 65*, 11–22.
Levine, E. S., Crozier, R. A., Black, I. B., & Plummer, M. R. (1998). Brain-derived neurotrophic factor modulates hippocampal synaptic transmission by increasing N-methyl-D-aspartic acid receptor activity. *Proceedings of the National Academy of Sciences of the United States of America, 95*, 10235–10239.
Lewin, G. R., & Barde, Y. A. (1996). Physiology of the neurotrophins. *Annual Review of Neuroscience, 19*, 289–317.
Liao, L., Pilotte, J., Xu, T., Wong, C. C., Edelman, G. M., Vanderklish, P., et al. (2007). BDNF induces widespread changes in synaptic protein content and up-regulates components of the translation machinery: An analysis using high-throughput proteomics. *Journal of Proteome Research, 6*, 1059–1071.
Lin, D. T., Makino, Y., Sharma, K., Hayashi, T., Neve, R., Takamiya, K., et al. (2009). Regulation of AMPA receptor extrasynaptic insertion by 4.1 N, phosphorylation and palmitoylation. *Nature Neuroscience, 12*, 879–887.
Lin, S. Y., Wu, K., Levine, E. S., Mount, H. T., Suen, P. C., & Black, I. B. (1998). BDNF acutely increases tyrosine phosphorylation of the NMDA receptor subunit 2B in cortical and hippocampal postsynaptic densities. *Brain Research Molecular Brain Research, 55*, 20–27.
Lisman, J. E. (2009). The pre/post LTP debate. *Neuron, 63*, 281–284.
Liston, C., & Gan, W. B. (2011). Glucocorticoids are critical regulators of dendritic spine development and plasticity in vivo. *Proceedings of the National Academy of Sciences of the United States of America, 108*, 16074–16079.
Lockhart, S. T., Turrigiano, G. G., & Birren, S. J. (1997). Nerve growth factor modulates synaptic transmission between sympathetic neurons and cardiac myocytes. *The Journal of Neuroscience, 17*, 9573–9582.
Lohof, A. M., Ip, N. Y., & Poo, M. M. (1993). Potentiation of developing neuromuscular synapses by the neurotrophins NT-3 and BDNF. *Nature, 363*, 350–353.
Lu, Y., Christian, K., & Lu, B. (2008). BDNF: A key regulator for protein synthesis-dependent LTP and long-term memory? *Neurobiology of Learning and Memory, 89*, 312–323.
Lu, Y., Ji, Y., Ganesan, S., Schloesser, R., Martinowich, K., Sun, M., et al. (2011). TrkB as a potential synaptic and behavioral tag. *The Journal of Neuroscience, 31*, 11762–11771.
Lynch, M. A. (2004). Long-term potentiation and memory. *Physiological Reviews, 84*, 87–136.
MacDougall, M. J., & Fine, A. (2014). The expression of long-term potentiation: Reconciling the priests and the postivists. *Philosophical Transactions of the Royal Society of London. Series B, Biological Sciences, 369*, 20130135.
Malenka, R. C. (2003a). The long-term potential of LTP. *Nature Reviews. Neuroscience, 4*, 923–926.
Malenka, R. C. (2003b). Synaptic plasticity and AMPA receptor trafficking. *Annals of the New York Academy of Sciences, 1003*, 1–11.
Manadas, B., Santos, A. R., Szabadfi, K., Gomes, J. R., Garbis, S. D., Fountoulakis, M., et al. (2009). BDNF-induced changes in the expression of the translation machinery in hippocampal neurons: Protein levels and dendritic mRNA. *Journal of Proteome Research, 8*, 4536–4552.
Mariga, A., Glaser, J., Mathias, L., Xu, D., Xiao, M., Worley, P., et al. (2015). Definition of a bidirectional activity-dependent pathway involving BDNF and narp. *Cell Reports, 13*, 1747–1756.

Matsuda, N., Lu, H., Fukata, Y., Noritake, J., Gao, H., Mukherjee, S., et al. (2009). Differential activity-dependent secretion of brain-derived neurotrophic factor from axon and dendrite. *The Journal of Neuroscience*, *29*, 14185–14198.

Matsumoto, T., Rauskolb, S., Polack, M., Klose, J., Kolbeck, R., Korte, M., et al. (2008). Biosynthesis and processing of endogenous BDNF: CNS neurons store and secrete BDNF, not pro-BDNF. *Nature Neuroscience*, *11*, 131–133.

Meakin, S. O., MacDonald, J. I., Gryz, E. A., Kubu, C. J., & Verdi, J. M. (1999). The signaling adapter FRS-2 competes with Shc for binding to the nerve growth factor receptor TrkA. A model for discriminating proliferation and differentiation. *The Journal of Biological Chemistry*, *274*, 9861–9870.

Messaoudi, E., Bardsen, K., Srebro, B., & Bramham, C. R. (1998). Acute intrahippocampal infusion of BDNF induces lasting potentiation of synaptic transmission in the rat dentate gyrus. *Journal of Neurophysiology*, *79*, 496–499.

Messaoudi, E., Kanhema, T., Soule, J., Tiron, A., Dagyte, G., da Silva, B., et al. (2007). Sustained Arc/Arg3.1 synthesis controls long-term potentiation consolidation through regulation of local actin polymerization in the dentate gyrus in vivo. *The Journal of Neuroscience*, *27*, 10445–10455.

Messaoudi, E., Ying, S. W., Kanhema, T., Croll, S. D., & Bramham, C. R. (2002). Brain-derived neurotrophic factor triggers transcription-dependent, late phase long-term potentiation in vivo. *The Journal of Neuroscience*, *22*, 7453–7461.

Minichiello, L. (2009). TrkB signalling pathways in LTP and learning. *Nature Reviews. Neuroscience*, *10*, 850–860.

Minichiello, L., Calella, A. M., Medina, D. L., Bonhoeffer, T., Klein, R., & Korte, M. (2002). Mechanism of TrkB-mediated hippocampal long-term potentiation. *Neuron*, *36*, 121–137.

Minichiello, L., Korte, M., Wolfer, D., Kuhn, R., Unsicker, K., Cestari, V., et al. (1999). Essential role for TrkB receptors in hippocampus-mediated learning. *Neuron*, *24*, 401–414.

Mizoguchi, H., Nakade, J., Tachibana, M., Ibi, D., Someya, E., Koike, H., et al. (2011). Matrix metalloproteinase-9 contributes to kindled seizure development in pentylenetetrazole-treated mice by converting pro-BDNF to mature BDNF in the hippocampus. *The Journal of Neuroscience*, *31*, 12963–12971.

Mizui, T., Ishikawa, Y., Kumanogoh, H., Lume, M., Matsumoto, T., Hara, T., et al. (2015). BDNF pro-peptide actions facilitate hippocampal LTD and are altered by the common BDNF polymorphism Val66Met. *Proceedings of the National Academy of Sciences of the United States of America*, *112*, E3067–E3074.

Mowla, S. J., Farhadi, H. F., Pareek, S., Atwal, J. K., Morris, S. J., Seidah, N. G., et al. (2001). Biosynthesis and post-translational processing of the precursor to brain-derived neurotrophic factor. *The Journal of Biological Chemistry*, *276*, 12660–12666.

Nagappan, G., Zaitsev, E., Senatorov, V. V., Jr., Yang, J., Hempstead, B. L., & Lu, B. (2009). Control of extracellular cleavage of ProBDNF by high frequency neuronal activity. *Proceedings of the National Academy of Sciences of the United States of America*, *106*, 1267–1272.

Nicoll, R. A., & Schmitz, D. (2005). Synaptic plasticity at hippocampal mossy fibre synapses. *Nature Reviews. Neuroscience*, *6*, 863–876.

Numakawa, T., Adachi, N., Richards, M., Chiba, S., & Kunugi, H. (2013). Brain-derived neurotrophic factor and glucocorticoids: Reciprocal influence on the central nervous system. *Neuroscience*, *239*, 157–172.

Numakawa, T., Kumamaru, E., Adachi, N., Yagasaki, Y., Izumi, A., & Kunugi, H. (2009). Glucocorticoid receptor interaction with TrkB promotes BDNF-triggered PLC-gamma signaling for glutamate release via a glutamate transporter. *Proceedings of the National Academy of Sciences of the United States of America*, *106*, 647–652.

Nykjaer, A., Lee, R., Teng, K. K., Jansen, P., Madsen, P., Nielsen, M. S., et al. (2004). Sortilin is essential for proNGF-induced neuronal cell death. *Nature, 427*, 843–848.

Nykjaer, A., Willnow, T. E., & Petersen, C. M. (2005). p75NTR—Live or let die. *Current Opinion in Neurobiology, 15*, 49–57.

Padamsey, Z., & Emptage, N. (2014). Two sides to long-term potentiation: A view towards reconciliation. *Philosophical Transactions of the Royal Society of London. Series B, Biological Sciences, 369*, 20130154.

Pang, P. T., Teng, H. K., Zaitsev, E., Woo, N. T., Sakata, K., Zhen, S., et al. (2004). Cleavage of proBDNF by tPA/plasmin is essential for long-term hippocampal plasticity. *Science, 306*, 487–491.

Panja, D., & Bramham, C. R. (2014). BDNF mechanisms in late LTP formation: A synthesis and breakdown. *Neuropharmacology, 76*(Pt. C), 664–676.

Panja, D., Kenney, J. W., D'Andrea, L., Zalfa, F., Vedeler, A., Wibrand, K., et al. (2014). Two-stage translational control of dentate gyrus LTP consolidation is mediated by sustained BDNF-TrkB signaling to MNK. *Cell Reports, 9*, 1430–1445.

Park, H., & Poo, M. M. (2013). Neurotrophin regulation of neural circuit development and function. *Nature Reviews. Neuroscience, 14*, 7–23.

Park, H., Popescu, A., & Poo, M. M. (2014). Essential role of presynaptic NMDA receptors in activity-dependent BDNF secretion and corticostriatal LTP. *Neuron, 84*, 1009–1022.

Patterson, S. L., Abel, T., Deuel, T. A., Martin, K. C., Rose, J. C., & Kandel, E. R. (1996). Recombinant BDNF rescues deficits in basal synaptic transmission and hippocampal LTP in BDNF knockout mice. *Neuron, 16*, 1137–1145.

Patterson, S. L., Grover, L. M., Schwartzkroin, P. A., & Bothwell, M. (1992). Neurotrophin expression in rat hippocampal slices: A stimulus paradigm inducing LTP in CA1 evokes increases in BDNF and NT-3 mRNAs. *Neuron, 9*, 1081–1088.

Pavlides, C., Watanabe, Y., & McEwen, B. S. (1993). Effects of glucocorticoids on hippocampal long-term potentiation. *Hippocampus, 3*, 183–192.

Pereira, D. B., Rebola, N., Rodrigues, R. J., Cunha, R. A., Carvalho, A. P., & Duarte, C. B. (2006). Trkb receptors modulation of glutamate release is limited to a subset of nerve terminals in the adult rat hippocampus. *Journal of Neuroscience Research, 83*, 832–844.

Peterson, D. A., Dickinson-Anson, H. A., Leppert, J. T., Lee, K. F., & Gage, F. H. (1999). Central neuronal loss and behavioral impairment in mice lacking neurotrophin receptor p75. *The Journal of Comparative Neurology, 404*, 1–20.

Platenik, J., Kuramoto, N., & Yoneda, Y. (2000). Molecular mechanisms associated with long-term consolidation of the NMDA signals. *Life Sciences, 67*, 335–364.

Poo, M. M. (2001). Neurotrophins as synaptic modulators. *Nature Reviews. Neuroscience, 2*, 24–32.

Qian, X., Riccio, A., Zhang, Y., & Ginty, D. D. (1998). Identification and characterization of novel substrates of Trk receptors in developing neurons. *Neuron, 21*, 1017–1029.

Radecki, D. T., Brown, L. M., Martinez, J., & Teyler, T. J. (2005). BDNF protects against stress-induced impairments in spatial learning and memory and LTP. *Hippocampus, 15*, 246–253.

Rajagopal, R., & Chao, M. V. (2006). A role for Fyn in Trk receptor transactivation by G-protein-coupled receptor signaling. *Molecular and Cellular Neurosciences, 33*, 36–46.

Rebola, N., Carta, M., Lanore, F., Blanchet, C., & Mulle, C. (2011). NMDA receptor-dependent metaplasticity at hippocampal mossy fiber synapses. *Nature Neuroscience, 14*, 691–693.

Reichardt, L. F. (2006). Neurotrophin-regulated signalling pathways. *Philosophical Transactions of the Royal Society of London. Series B, Biological Sciences, 361*, 1545–1564.

Revest, J. M., Le Roux, A., Roullot-Lacarriere, V., Kaouane, N., Vallee, M., Kasanetz, F., et al. (2014). BDNF-TrkB signaling through Erk1/2 MAPK phosphorylation mediates the enhancement of fear memory induced by glucocorticoids. *Molecular Psychiatry*, *19*, 1001–1009.

Rex, C. S., Lin, C. Y., Kramar, E. A., Chen, L. Y., Gall, C. M., & Lynch, G. (2007). Brain-derived neurotrophic factor promotes long-term potentiation-related cytoskeletal changes in adult hippocampus. *The Journal of Neuroscience*, *27*, 3017–3029.

Righi, M., Tongiorgi, E., & Cattaneo, A. (2000). Brain-derived neurotrophic factor (BDNF) induces dendritic targeting of BDNF and tyrosine kinase B mRNAs in hippocampal neurons through a phosphatidylinositol-3 kinase-dependent pathway. *The Journal of Neuroscience*, *20*, 3165–3174.

Rosch, H., Schweigreiter, R., Bonhoeffer, T., Barde, Y. A., & Korte, M. (2005). The neurotrophin receptor p75NTR modulates long-term depression and regulates the expression of AMPA receptor subunits in the hippocampus. *Proceedings of the National Academy of Sciences of the United States of America*, *102*, 7362–7367.

Roux, P. P., & Barker, P. A. (2002). Neurotrophin signaling through the p75 neurotrophin receptor. *Progress in Neurobiology*, *67*, 203–233.

Sairanen, M., Lucas, G., Ernfors, P., Castren, M., & Castren, E. (2005). Brain-derived neurotrophic factor and antidepressant drugs have different but coordinated effects on neuronal turnover, proliferation, and survival in the adult dentate gyrus. *The Journal of Neuroscience*, *25*, 1089–1094.

Sajikumar, S., & Korte, M. (2011). Metaplasticity governs compartmentalization of synaptic tagging and capture through brain-derived neurotrophic factor (BDNF) and protein kinase Mzeta (PKMzeta). *Proceedings of the National Academy of Sciences of the United States of America*, *108*, 2551–2556.

Sakuragi, S., Tominaga-Yoshino, K., & Ogura, A. (2013). Involvement of TrkB- and p75 (NTR)-signaling pathways in two contrasting forms of long-lasting synaptic plasticity. *Scientific Reports*, *3*, 3185.

Santi, S., Cappello, S., Riccio, M., Bergami, M., Aicardi, G., Schenk, U., et al. (2006). Hippocampal neurons recycle BDNF for activity-dependent secretion and LTP maintenance. *The EMBO Journal*, *25*, 4372–4380.

Santos, A. R., Mele, M., Vaz, S. H., Kellermayer, B., Grimaldi, M., Colino-Oliveira, M., et al. (2015). Differential role of the proteasome in the early and late phases of BDNF-induced facilitation of LTP. *The Journal of Neuroscience*, *35*, 3319–3329.

Sapolsky, R. M., Krey, L. C., & McEwen, B. S. (1985). Prolonged glucocorticoid exposure reduces hippocampal neuron number: Implications for aging. *The Journal of Neuroscience*, *5*, 1222–1227.

Sapolsky, R. M., Uno, H., Rebert, C. S., & Finch, C. E. (1990). Hippocampal damage associated with prolonged glucocorticoid exposure in primates. *The Journal of Neuroscience*, *10*, 2897–2902.

Scharfman, H. E. (2007). The CA3 "backprojection" to the dentate gyrus. *Progress in Brain Research*, *163*, 627–637.

Scharfman, H., Goodman, J., Macleod, A., Phani, S., Antonelli, C., & Croll, S. (2005). Increased neurogenesis and the ectopic granule cells after intrahippocampal BDNF infusion in adult rats. *Experimental Neurology*, *192*, 348–356.

Schecterson, L. C., & Bothwell, M. (2010). Neurotrophin receptors: Old friends with new partners. *Developmental Neurobiology*, *70*, 332–338.

Schildt, S., Endres, T., Lessmann, V., & Edelmann, E. (2013). Acute and chronic interference with BDNF/TrkB-signaling impair LTP selectively at mossy fiber synapses in the CA3 region of mouse hippocampus. *Neuropharmacology*, *71*, 247–254.

Schjetnan, A. G., & Escobar, M. L. (2012). In vivo BDNF modulation of hippocampal mossy fiber plasticity induced by high frequency stimulation. *Hippocampus*, *22*, 1–8.

Schmidt, H. D., & Duman, R. S. (2010). Peripheral BDNF produces antidepressant-like effects in cellular and behavioral models. *Neuropsychopharmacology*, *35*, 2378–2391.

Schratt, G. M., Nigh, E. A., Chen, W. G., Hu, L., & Greenberg, M. E. (2004). BDNF regulates the translation of a select group of mRNAs by a mammalian target of rapamycin-phosphatidylinositol 3-kinase-dependent pathway during neuronal development. *The Journal of Neuroscience*, *24*, 7366–7377.

Schratt, G. M., Tuebing, F., Nigh, E. A., Kane, C. G., Sabatini, M. E., Kiebler, M., et al. (2006). A brain-specific microRNA regulates dendritic spine development. *Nature*, *439*, 283–289.

Seidah, N. G., Benjannet, S., Pareek, S., Chretien, M., & Murphy, R. A. (1996). Cellular processing of the neurotrophin precursors of NT3 and BDNF by the mammalian proprotein convertases. *FEBS Letters*, *379*, 247–250.

Shepherd, J. D., & Huganir, R. L. (2007). The cell biology of synaptic plasticity: AMPA receptor trafficking. *Annual Review of Cell and Developmental Biology*, *23*, 613–643.

Shiina, N., Shinkura, K., & Tokunaga, M. (2005). A novel RNA-binding protein in neuronal RNA granules: Regulatory machinery for local translation. *The Journal of Neuroscience*, *25*, 4420–4434.

Shirayama, Y., Chen, A. C., Nakagawa, S., Russell, D. S., & Duman, R. S. (2002). Brain-derived neurotrophic factor produces antidepressant effects in behavioral models of depression. *The Journal of Neuroscience*, *22*, 3251–3261.

Siao, C. J., Lorentz, C. U., Kermani, P., Marinic, T., Carter, J., McGrath, K., et al. (2012). ProNGF, a cytokine induced after myocardial infarction in humans, targets pericytes to promote microvascular damage and activation. *The Journal of Experimental Medicine*, *209*, 2291–2305.

Sivakumaran, S., Mohajerani, M. H., & Cherubini, E. (2009). At immature mossy-fiber-CA3 synapses, correlated presynaptic and postsynaptic activity persistently enhances GABA release and network excitability via BDNF and cAMP-dependent PKA. *The Journal of Neuroscience*, *29*, 2637–2647.

Slonimsky, J. D., Mattaliano, M. D., Moon, J. I., Griffith, L. C., & Birren, S. J. (2006). Role for calcium/calmodulin-dependent protein kinase II in the p75-mediated regulation of sympathetic cholinergic transmission. *Proceedings of the National Academy of Sciences of the United States of America*, *103*, 2915–2919.

Strange, B. A., Witter, M. P., Lein, E. S., & Moser, E. I. (2014). Functional organization of the hippocampal longitudinal axis. *Nature Reviews. Neuroscience*, *15*, 655–669.

Suri, D., & Vaidya, V. A. (2013). Glucocorticoid regulation of brain-derived neurotrophic factor: Relevance to hippocampal structural and functional plasticity. *Neuroscience*, *239*, 196–213.

Sutton, M. A., & Schuman, E. M. (2006). Dendritic protein synthesis, synaptic plasticity, and memory. *Cell*, *127*, 49–58.

Swanger, S. A., & Bassell, G. J. (2013). Dendritic protein synthesis in the normal and diseased brain. *Neuroscience*, *232*, 106–127.

Sweatt, J. D. (1999). Toward a molecular explanation for long-term potentiation. *Learning & Memory*, *6*, 399–416.

Takei, N., Inamura, N., Kawamura, M., Namba, H., Hara, K., Yonezawa, K., et al. (2004). Brain-derived neurotrophic factor induces mammalian target of rapamycin-dependent local activation of translation machinery and protein synthesis in neuronal dendrites. *The Journal of Neuroscience*, *24*, 9760–9769.

Tanaka, J., Horiike, Y., Matsuzaki, M., Miyazaki, T., Ellis-Davies, G. C., & Kasai, H. (2008). Protein synthesis and neurotrophin-dependent structural plasticity of single dendritic spines. *Science*, *319*, 1683–1687.

Teng, H. K., Teng, K. K., Lee, R., Wright, S., Tevar, S., Almeida, R. D., et al. (2005). ProBDNF induces neuronal apoptosis via activation of a receptor complex of p75NTR and sortilin. *The Journal of Neuroscience*, *25*, 5455–5463.

Thoenen, H. (1991). The changing scene of neurotrophic factors. *Trends in Neurosciences, 14*, 165–170.

Traynelis, S. F., Wollmuth, L. P., McBain, C. J., Menniti, F. S., Vance, K. M., Ogden, K. K., et al. (2010). Glutamate receptor ion channels: Structure, regulation, and function. *Pharmacological Reviews, 62*, 405–496.

Tyler, W. J., Perrett, S. P., & Pozzo-Miller, L. D. (2002). The role of neurotrophins in neurotransmitter release. *The Neuroscientist, 8*, 524–531.

Tyler, W. J., Zhang, X. L., Hartman, K., Winterer, J., Muller, W., Stanton, P. K., et al. (2006). BDNF increases release probability and the size of a rapidly recycling vesicle pool within rat hippocampal excitatory synapses. *The Journal of Physiology, 574*, 787–803.

Underwood, C. K., Reid, K., May, L. M., Bartlett, P. F., & Coulson, E. J. (2008). Palmitoylation of the C-terminal fragment of p75(NTR) regulates death signaling and is required for subsequent cleavage by gamma-secretase. *Molecular and Cellular Neurosciences, 37*, 346–358.

Verpelli, C., Piccoli, G., Zibetti, C., Zanchi, A., Gardoni, F., Huang, K., et al. (2010). Synaptic activity controls dendritic spine morphology by modulating eEF2-dependent BDNF synthesis. *The Journal of Neuroscience, 30*, 5830–5842.

Waterhouse, E. G., & Xu, B. (2009). New insights into the role of brain-derived neurotrophic factor in synaptic plasticity. *Molecular and Cellular Neurosciences, 42*, 81–89.

Wibrand, K., Messaoudi, E., Havik, B., Steenslid, V., Lovlie, R., Steen, V. M., et al. (2006). Identification of genes co-upregulated with Arc during BDNF-induced long-term potentiation in adult rat dentate gyrus in vivo. *The European Journal of Neuroscience, 23*, 1501–1511.

Willnow, T. E., Petersen, C. M., & Nykjaer, A. (2008). VPS10P-domain receptors—Regulators of neuronal viability and function. *Nature Reviews. Neuroscience, 9*, 899–909.

Witter, M. P. (2011). Connectivity of the hippocampus. In V. Cutsuridis, B. P. Graham, S. Cobb, & I. Vida (Eds.), *Hippocampal microcircuits: Vol. 5.* (pp. 5–26). New York: Springer.

Wong, Y. H., Lee, C. M., Xie, W., Cui, B., & Poo, M. M. (2015). Activity-dependent BDNF release via endocytic pathways is regulated by synaptotagmin-6 and complexin. *Proceedings of the National Academy of Sciences of the United States of America, 112*, E4475–E4484.

Woo, N. H., Teng, H. K., Siao, C. J., Chiaruttini, C., Pang, P. T., Milner, T. A., et al. (2005). Activation of p75NTR by proBDNF facilitates hippocampal long-term depression. *Nature Neuroscience, 8*, 1069–1077.

Wright, J. W., Alt, J. A., Turner, G. D., & Krueger, J. M. (2004). Differences in spatial learning comparing transgenic p75 knockout, New Zealand Black, C57BL/6, and Swiss Webster mice. *Behavioural Brain Research, 153*, 453–458.

Xu, B., Gottschalk, W., Chow, A., Wilson, R. I., Schnell, E., Zang, K., et al. (2000). The role of brain-derived neurotrophic factor receptors in the mature hippocampus: Modulation of long-term potentiation through a presynaptic mechanism involving TrkB. *The Journal of Neuroscience, 20*, 6888–6897.

Yamada, M., Ohnishi, H., Sano, S., Nakatani, A., Ikeuchi, T., & Hatanaka, H. (1997). Insulin receptor substrate (IRS)-1 and IRS-2 are tyrosine-phosphorylated and associated with phosphatidylinositol 3-kinase in response to brain-derived neurotrophic factor in cultured cerebral cortical neurons. *The Journal of Biological Chemistry, 272*, 30334–30339.

Yang, J., Harte-Hargrove, L. C., Siao, C. J., Marinic, T., Clarke, R., Ma, Q., et al. (2014). proBDNF negatively regulates neuronal remodeling, synaptic transmission, and synaptic plasticity in hippocampus. *Cell Reports, 7*, 796–806.

Yang, F., Je, H. S., Ji, Y., Nagappan, G., Hempstead, B., & Lu, B. (2009). Pro-BDNF-induced synaptic depression and retraction at developing neuromuscular synapses. *The Journal of Cell Biology, 185*, 727–741.

Yang, B., Slonimsky, J. D., & Birren, S. J. (2002). A rapid switch in sympathetic neurotransmitter release properties mediated by the p75 receptor. *Nature Neuroscience*, *5*, 539–545.
Yano, H., Ninan, I., Zhang, H., Milner, T. A., Arancio, O., & Chao, M. V. (2006). BDNF-mediated neurotransmission relies upon a myosin VI motor complex. *Nature Neuroscience*, *9*, 1009–1018.
Yin, Y., Edelman, G. M., & Vanderklish, P. W. (2002). The brain-derived neurotrophic factor enhances synthesis of Arc in synaptoneurosomes. *Proceedings of the National Academy of Sciences of the United States of America*, *99*, 2368–2373.
Ying, S. W., Futter, M., Rosenblum, K., Webber, M. J., Hunt, S. P., Bliss, T. V., et al. (2002). Brain-derived neurotrophic factor induces long-term potentiation in intact adult hippocampus: Requirement for ERK activation coupled to CREB and upregulation of Arc synthesis. *The Journal of Neuroscience*, *22*, 1532–1540.
Zagrebelsky, M., Holz, A., Dechant, G., Barde, Y. A., Bonhoeffer, T., & Korte, M. (2005). The p75 neurotrophin receptor negatively modulates dendrite complexity and spine density in hippocampal neurons. *The Journal of Neuroscience*, *25*, 9989–9999.
Zagrebelsky, M., & Korte, M. (2014). Form follows function: BDNF and its involvement in sculpting the function and structure of synapses. *Neuropharmacology*, *76*(Pt. C), 628–638.
Zakharenko, S. S., Patterson, S. L., Dragatsis, I., Zeitlin, S. O., Siegelbaum, S. A., Kandel, E. R., et al. (2003). Presynaptic BDNF required for a presynaptic but not postsynaptic component of LTP at hippocampal CA1-CA3 synapses. *Neuron*, *39*, 975–990.
Zhao, C., Deng, W., & Gage, F. H. (2008). Mechanisms and functional implications of adult neurogenesis. *Cell*, *132*, 645–660.
Zheng, F., Luo, Y., & Wang, H. (2009). Regulation of brain-derived neurotrophic factor-mediated transcription of the immediate early gene Arc by intracellular calcium and calmodulin. *Journal of Neuroscience Research*, *87*, 380–392.
Zhou, J., Zhang, F., & Zhang, Y. (2000). Corticosterone inhibits generation of long-term potentiation in rat hippocampal slice: Involvement of brain-derived neurotrophic factor. *Brain Research*, *885*, 182–191.

CHAPTER NINE

Neurotrophins in the Brain: Interaction With Alcohol Exposure During Development

K.E. Boschen, A.Y. Klintsova[1]
University of Delaware, Newark, DE, United States
[1]Corresponding author: e-mail address: aklintsova@psych.udel.edu

Contents

1. Introduction 198
2. Neurotrophins and the Developing Brain 198
3. Alcohol as a Teratogenic Agent 199
 3.1 Fetal Alcohol Spectrum Disorders 200
 3.2 Modeling Developmental Alcohol Effects in Experimental Animals 202
4. Neurotrophins as Targets of Alcohol Exposure 204
 4.1 BDNF 205
 4.2 Other Neurotrophic Factors: NGF, NT-3, VEGF 207
5. Interactions Between Developmental Alcohol Exposure and Neurotrophins in the Brain: Molecular and Intracellular Effects 208
 5.1 Glutamatergic NMDA Receptors and GABA Receptors 208
 5.2 Oxidative Stress 211
 5.3 Neuroinflammation 213
6. Interactions Between Developmental Alcohol Exposure and Neurotrophins: Neuroanatomical and Behavioral Effects 215
 6.1 Neuroanatomical Effects 215
 6.2 Behavioral Effects 222
7. Neurotrophins as Compensatory/Neuroprotective Molecules Following Developmental Alcohol Exposure 225
8. Conclusions 228
Acknowledgments 228
References 228

Abstract

Fetal alcohol spectrum disorders (FASDs) are a result of the teratogenic effects of alcohol on the developing fetus. Decades of research examining both individuals with FASDs and animal models of developmental alcohol exposure have revealed the devastating effects of alcohol on brain structure, function, behavior, and cognition. Neurotrophic factors have an important role in guiding normal brain development and cellular

Vitamins and Hormones, Volume 104
ISSN 0083-6729
http://dx.doi.org/10.1016/bs.vh.2016.10.008

plasticity in the adult brain. This chapter reviews the current literature showing that alcohol exposure during the developmental period impacts neurotrophin production and proposes avenues through which alcohol exposure and neurotrophin action might interact. These areas of overlap include formation of long-term potentiation, oxidative stress processes, neuroinflammation, apoptosis and cell loss, hippocampal adult neurogenesis, dendritic morphology and spine density, vasculogenesis and angiogenesis, and behaviors related to spatial memory, anxiety, and depression. Finally, we discuss how neurotrophins have the potential to act in a compensatory manner as neuroprotective molecules that can combat the deleterious effects of in utero alcohol exposure.

1. INTRODUCTION

Neurotrophins, members of one of the three families of neurotrophic factors, play a key role in proper brain development and are important mediators of synaptic plasticity involved in memory formation during adulthood. Disruption of neurotrophins or receptor function during development could result in long-term changes to cognition, memory formation, and mood. Prenatal alcohol exposure has long been known to cause severe cognitive and behavioral deficits due to functional and anatomical changes within the brain. The deficits caused by alcohol exposure in utero are collectively known as fetal alcohol spectrum disorders (FASDs), which manifest with a variety of physical, intellectual, social, and executive functioning impairments. Recent evidence suggests that alcohol–neurotrophin interactions during development could contribute to the damaging effects of prenatal alcohol exposure on neuroplasticity, learning, and memory measures. Specifically, disruption of neurotrophin signaling and prenatal alcohol exposure not only results in overlapping detrimental effects, but there is also emerging evidence that alcohol may act directly on neurotrophin pathways. In this chapter, we review the role of neurotrophins during brain development, how prenatal alcohol exposure affects brain function and structure through the life span, and potential ways in which alcohol and neurotrophins interact to result in alcohol-related deficits.

2. NEUROTROPHINS AND THE DEVELOPING BRAIN

The proper timing of the birth, survival, and death of neurons is carefully orchestrated with the help of neurotrophin signaling. Each neurotrophin exhibits a unique expression profile throughout development in the brain and periphery, with alterations to expression causing significant and long-term consequences on cell structure and function. Studies in chicken embryos show neurotrophin-3 (NT-3) as the earliest neurotrophin found in the developing embryo, with expression of its receptor, tropomyosin receptor kinase C (TrkC), beginning in the neural plate (Bernd, 2008). Brain-derived neurotrophic factor (BDNF) and its receptor, TrkB, emerge later, appearing first in the neural tube. Nerve growth factor (NGF) and its receptor, TrkA, is first expressed in peripheral sensory neurons later in embryonic development, suggesting a role for this neurotrophin in later developmental stages rather than initial cell proliferation and migration processes. Another set of neurotrophins, vascular endothelial growth factors or VEGF, is critical for embryonic vasculogenesis. VEGF-A, VEGF-B, VEGF-C, and VEGF-B have three receptors, VEGFR1–3, with VEGF-A and its receptor VEGFR2 being the most plentiful in the brain. Exposure to toxins or stress during gastrulation and neurulation could influence expression of these neurotrophins in a timing-dependent manner.

Neurotrophins also show overlapping but distinctive patterns of expression during later brain development. In rat embryos, levels of NT-3, NGF, and BDNF increase to coincide with mass neurogenesis in the developing central and peripheral nervous system around embryonic days (E) 11–12 (Maisonpierre et al., 1990). During these first waves of neurogenesis, NT-3 is the most abundantly produced neurotrophin, with BDNF being far less plentiful, suggesting a prominent role of NT-3 during initial neurogenesis. In addition to being developmentally regulated, neurotrophic expression is also brain region specific, with expression of NT-3, NGF, and BDNF decreasing in concentration in the adult vs prenatal spinal cord; conversely, expression of NGF and BDNF increases into adulthood in the hippocampus (Karege, Schwald, & Cisse, 2002; Katoh-Semba, Semba, Takeuchi, & Kato, 1998; Timmusk et al., 1993), while expression of NT-3 peaks in the early postnatal period (equivalent to third trimester of human pregnancy). Based on these expression profiles, NT-3 might play a more important role in early neuronal proliferation and migration, with NGF and BDNF contributing to later developmental processes, such as neuronal survival, neurite outgrowth, and synaptic plasticity.

3. ALCOHOL AS A TERATOGENIC AGENT

During prenatal development, the fetus could be exposed to a variety of environmental or infectious agents, called teratogens, which impact normal growth and survival, ultimately resulting in malformations to the fetus or "birth defects." Common teratogens include pesticides, infections, drugs of abuse, and prescription medication. Teratogens exert their deleterious effects through a variety of pathways and the resulting defects can range from severe (limb loss following maternal use of thalidomide) to subtle (delayed language skills following prenatal cocaine exposure). One of the most widely used teratogenic agents is alcohol, which will be the focus of this chapter.

Alcohol (ethyl alcohol or ethanol, specifically) is a commonly used psychoactive drug with sedative properties. Primarily used recreationally, over 70% of adults over age 18 report using alcohol in the past year (NIAAA, 2015). Binge drinking (usually defined as >4 drinks in one sitting) is also prevalent, with over 24% of adults reporting at least one episode of binge drinking in the past month. Alcohol exerts its sedative effects through inhibition of NMDA-subtype glutamate receptor activity and enhancement of chloride ion flow through $GABA_A$ receptor ion channels. In adults, even low doses of alcohol can result in impaired motor control, judgment, and reaction time due to the widespread and potent effects of alcohol on multiple brain regions.

Pregnant women also report relatively high rates of alcohol consumption, with 20–30% of women reporting some degree of alcohol use during their pregnancy, despite well-publicized campaigns to warn women of potential danger to the developing fetus. When a pregnant woman ingests alcohol, the alcohol easily passes through the placenta, meaning that blood alcohol concentrations (BACs) in the fetus are comparable to or even higher than those recorded from the mother (Nava-Ocampo, Velazquez-Armenta, Brien, & Koren, 2004; van Faassen & Niemelä, 2011). While alcohol is easily metabolized in adults through the action of alcohol dehydrogenase in the stomach and liver; however, fetuses lack the enzymes necessary for the breakdown of alcohol. Thus, the alcohol is only metabolized after being circulated back into the maternal blood supply, meaning that concentrations of alcohol within the placenta could be higher than the mother's BAC and remain elevated for a prolonged period.

3.1 Fetal Alcohol Spectrum Disorders

Fetal exposure to alcohol can cause a wide range of long-lasting physiological and behavioral effects, collectively referred to as FASDs. The most severe form of fetal alcohol effects is fetal alcohol syndrome (FAS), which

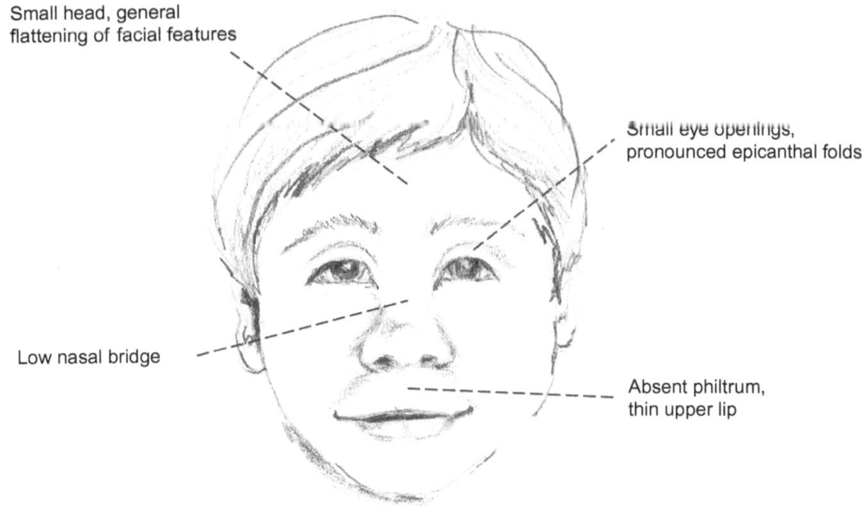

Fig. 1 Classic craniofacial dysmorphologies associated with fetal alcohol syndrome (FAS). All illustrations drawn by P.T. Boschen.

manifests with craniofacial malformations (Fig. 1), including short palpebral fissure length, smooth philtrum, and a thin upper lip, reduced body and brain weight, intellectual disabilities, and cognitive and behavioral impairments. Prevalence statistics for the United States have been historically difficult to obtain due to the stigma associated with maternal drinking, making pregnant women less likely to accurately report rates of alcohol consumption. The current statistics estimate that up to 5% of live births each year in the Unites States are affected by FASD, while FAS occurs in approximately 0.1% of live births (Centers for Disease Control, 2015; May et al., 2009). Children with FASD, including FAS, can require lifelong care, both in terms of extra medical expenditures in childhood (Amendah, Grosse, & Bertrand, 2011) and access to specialized services to help cope with learning disabilities and other behavioral needs. Estimates from 2002 indicated that care for children with FAS costs on average $2 million across the life span, and that services for these individuals cost the United States approximately $4 billion a year (1998 estimate; Lupton, Burd, & Harwood, 2004). More recent estimates or costs including the full range of FASD are not readily available, but it can be inferred that care for these individuals is a significant use of resources, both for families and society as a whole.

Cost estimates for individuals with FAS and the full spectrum of FASDs are difficult to calculate due to the range of primary and secondary disabilities caused by prenatal alcohol exposure. Primary disabilities in children with FASDs refer to changes to the brain that directly result in impaired mental function, such as decreased volume of both white and gray matter structures which cause deficient performance on reading, writing, and arithmetic tasks (Mattson, Riley, Delis, Stern, & Jones, 1996; Mattson, Riley, Sowell, et al., 1996). These children often perform poorly on working memory and IQ tests (Mattson, Goodman, Caine, Delis, & Riley, 1999; Mattson, Riley, Gramling, Delis, & Jones, 1997; Rasmussen, Soleimani, & Pei, 2011) and are more likely to require special education services. These disabilities often extend beyond scholastic performance and can affect social interactions, executive functioning, impulse control, and emotion regulation, often resulting in higher rates of incarceration and mental illness in adulthood (Franklin, Deitz, Jirikowic, & Astley, 2008; Irner, Teasdale, & Olofsson, 2012; Kodituwakku, Handmaker, Cutler, Weathersby, & Handmaker, 1995; Pei, Job, Kully-Martens, & Rasmussen, 2011; Stevens et al., 2012; Streissguth et al., 2004).

Public safety campaigns, including announcements from the American Academy of Pediatrics and the CDC in 2015 and 2016 officially recommending no alcohol use during pregnancy, have ultimately not eradicated the existence of a completely preventable cause of mental retardation, leading researchers to continue to investigate (1) the molecular, cellular, and system-wide damage caused by prenatal alcohol exposure; and (2) potential behavioral and pharmacological interventions to reverse the detrimental cognitive and behavioral effects. This pursuit has led to a vast number of animal models of FASDs which have been critical in understanding how the dose, timing, and pattern of alcohol exposure can impact behavioral and neuroanatomical outcomes.

3.2 Modeling Developmental Alcohol Effects in Experimental Animals

Rats and mice are the most commonly used for FASD models, though sheep and primates are sometimes used as well. Rat and mouse pups are born at a developmentally earlier stage than humans, meaning that brain development occurring during the third trimester of human pregnancy takes place over the first 2 postnatal weeks in rodents (Dobbing & Sands, 1979). Alcohol exposure during specific points in development results in damage to whatever developmental process is occurring during this time. For example, administration of alcohol during the first trimester equivalent (gastrulation or neurulation) would cause craniofacial dysmorphologies analogous to those seen in children with FAS, but administration during the third trimester equivalent would produce neuroanatomical damage to late-developing brain structures such as the hippocampus, prefrontal cortex, or cerebellum. As mentioned earlier, neurotrophins also display developmentally regulated expression patterns, suggesting that interactions between alcohol and neurotrophin function would likely depend on developmental stage.

FASD models can be split into two broad categories: prenatal and postnatal exposure. Prenatal models mimic alcohol exposure during approximately the first two trimesters of human pregnancy, while postnatal models mimic exposure during the third trimester. In prenatal models, alcohol is administered to the dam either through maternal voluntary drinking (Fig. 2A), intraperitoneal or subcutaneous injection or intragastric gavage. Voluntary drinking paradigms are relatively low stress for the dam; however, they usually model low-to-moderate levels of drinking due to the amount of

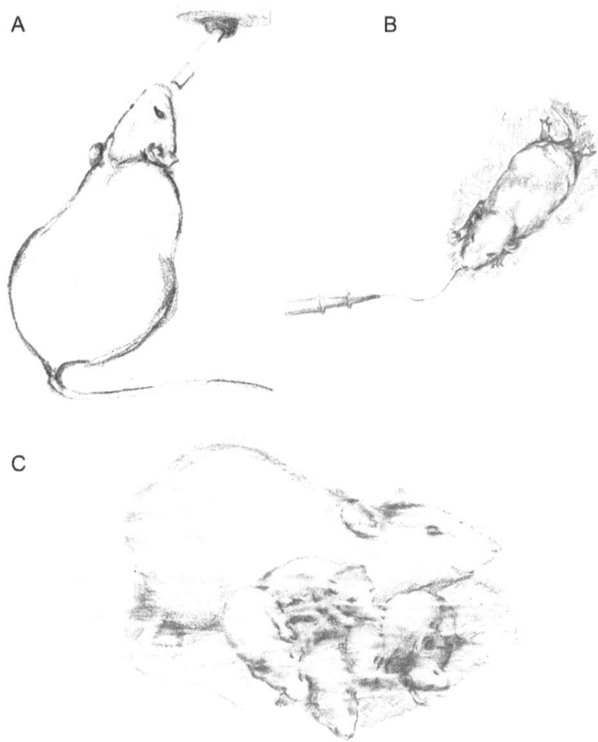

Fig. 2 Representative illustration of alcohol exposure paradigms. (A) Maternal drinking during pregnancy, (B) postnatal gavage/intragastric intubation, and (C) postnatal exposure using a vapor chamber (dam with pups).

alcohol being ingested are under the control of the dam rather than the experimenter directly. In contrast, injections and gavage are able to achieve higher BACs and the relative amount of alcohol administered is constant across animals; however, injections and gavage are inherently stressful procedures. Postnatal models deliver the alcohol directly to the pups, enabling control and alcohol-exposed pups to be assigned to the same litter, which leaves maternal care of the pups unaffected. Intragastric intubation (Fig. 2B) and injections are commonly used methods of alcohol delivery for neonatal pups, achieving high BACs in a controlled dose, with the negative consequence of increased stress and morbidity rate. Vapor inhalation paradigms (Fig. 2C) are also commonly used, with pups being placed in a chamber filled with ethanol vapor for 1–4 h at a time. Long periods of maternal separation can be detrimental to pup development, so dams are often placed in the vapor chamber with the pups. As a result, this method usually models low-to-moderate levels of alcohol exposure in order to leave the dam unimpaired and able to sufficiently care for the pups. While ethanol vapor can cause irritation to mucous membranes, the stress induced by this method is minimal.

Two important differences distinguish the models described earlier: developmental time point (trimester equivalent) and level of BAC achieved. These two variables can determine the type and severity of alcohol-induced damage depending on what developmental processes are occurring. For instance, alcohol exposure during gastrulation can affect neural tube closure, while alcohol exposure during the third trimester equivalent could impact synaptogenesis and cell differentiation in the prefrontal cortex. These developmental windows also have unique expression patterns of neurotrophins, meaning that alcohol would have differing interactions with neurotrophins based on exposure time point and level of alcohol administered. In the remainder of the chapter, we will discuss how rodent models of FASD target neurotrophin expression during different developmental periods, and other avenues through which alcohol and neurotrophins may interact, from molecules to behavior. Finally, we will discuss cases in which neurotrophins may act in a neuroprotective capacity following developmental alcohol exposure.

4. NEUROTROPHINS AS TARGETS OF ALCOHOL EXPOSURE

Alcohol during development can have direct consequences on neurotrophin signaling, even long after cessation of the exposure. The effects of alcohol exposure on neurotrophin and receptor expression are highly dependent on a number of variables, including the alcohol dose and developmental window targeted, timing of the tissue collection following the exposure, route of administration, and the brain region in question. Inter-study variability on these factors can make the literature difficult to parse. The most well-studied neurotrophin–alcohol relationship is the interaction with BDNF and its receptor TrkB, though numerous papers have assessed NGF as well. NT-3 and VEGF have been less explored, though it is likely that more resources will be put toward investigating VEGF in more detail in the future due to an emerging understanding of its role in neuroplasticity.

4.1 BDNF

In recent years, numerous studies have been published trying to determine how developmental alcohol exposure affects BDNF levels in the brain, and often these studies have been at odds with one another due to the range of alcohol exposure paradigms, rodent species, and timing of the tissue analysis utilized, as well as differences in which brain regions were examined. While there is little doubt rodent models of FASDs do impact BDNF, questions remain regarding the directionality and stability of the changes. It is not known if alcohol interacts with the BDNF molecule or TrkB receptor directly; it seems more likely that BDNF levels are affected through action of alcohol on NMDA or $GABA_A$ receptors. For example, in cerebellar granule cell cultures, BDNF was increased following NMDA treatment; however, this enhancement was blocked following pretreatment with ethanol (Bhave, Ghoda, & Hoffman, 1999).

Models of prenatal alcohol exposure have consistently demonstrated changes to BDNF signaling in various brain regions. Following exposure from gestational days 5–20, BDNF protein and mRNA were reduced in the rat cortex and hippocampus when assessed on PD7-8 (Feng, Yan, & Yan, 2005). Mice exposed to alcohol prenatally show decreased levels of BDNF protein, total and exon III-, IV-, and VI-driven *Bdnf* mRNA transcripts in adulthood in the medial prefrontal cortex (mPFC; Caldwell et al., 2008); at 18 months of age, levels of BDNF were depleted in the liver and elevated in the hippocampus of prenatally exposed mice, suggesting an interaction with developmental alcohol exposure and natural aging processes (Ceccanti et al., 2012). Prenatal exposure also has been shown to alter levels of the TrkB receptor, inhibiting the phosphorylation of TrkB on PD7-8 while leaving the total amount of TrkB unchanged (Feng et al., 2005). In another study, male rat pups exposed to alcohol in utero also showed decreased levels of TrkB receptors on postnatal day 1 in the hippocampus, no changes to TrkB in the septum or cerebellum, and increased levels in cortex (Moore, Madorsky, Paiva, & Barrow Heaton, 2004). Interestingly, female pups showed a different pattern of changes, with levels of TrkB decreased in septum, no changes in hippocampus, and consistent increases in TrkB in cortex. Most alterations to receptor number had returned to baseline by postnatal day 10. These studies highlight the variety of changes to BDNF signaling which can vary by brain region, sex of animal, and time point. In general, the consensus of these prenatal studies seems to point to decreased BDNF production in hippocampus following prenatal alcohol exposure, with more variability reported in other brain regions.

BDNF and TrkB receptor expression have also been shown to be altered in postnatal alcohol exposure models. Studies assessing BDNF 24 h or less following neonatal exposure have found increased protein levels in hippocampus and cortex (PD2–10 or PD4; Heaton, Mitchell, Paiva, & Walker, 2000; Heaton et al., 2003), increased BDNF and TrkB protein, *Bdnf* total and exon I- and IV-driven gene expression in hippocampus (PD4–9; Boschen, Criss, Palamarchouk, Roth, & Klintsova, 2015), and decreased *Bdnf* and *Trkb* gene expression in cerebellum (PD2–3 and PD4 exposures; Heaton et al., 2003; Light, Ge, & Belcher, 2001). Another report assessing *Bdnf* mRNA and downstream signaling pathways on PD8 following PD5–8 alcohol exposure found decreased levels of *Bdnf* and downregulation of the MAPK and Akt pathways in the cerebral cortex (Fattori, Abe, Kobayashi, Costa, & Tsuji, 2008). Postnatal exposure (PD10–15) via vapor inhalation disrupted normal age-related fluctuations in *Bdnf* gene expression, with alcohol-exposed rats having higher expression in the hippocampus on PD16 and 20 and decreased expression on PD60 compared to control animals, again supporting an interaction between alcohol exposure and age on alterations in neurotrophin signaling (Miki et al., 2008). Aside from age-related fluctuations, BDNF production is also linked to the circadian cycle. Allen, West, Chen, and Earnest (2004) reported that PD4–9 alcohol exposure significantly decreased overall levels of BDNF protein in the suprachiasmatic nucleus (SCN) and blunted the circadian rhythmicity of expression when the animals were 5–6 months of age. BDNF is thought to contribute to circadian regulation through action in the SCN (Liang, Allen, & Earnest, 2000) and disruptions to circadian rhythms have been reported in both children with FASDs and in rodent neonatal exposure models (Chen, Olson, Picciano, Starr, & Owens, 2012; Sakata-Haga et al., 2006).

4.2 Other Neurotrophic Factors: NGF, NT-3, VEGF

The interaction between alcohol and NGF has been well studied with alcohol addiction models in adult rodents, with levels of NGF often increasing during or immediately following alcohol intake and decreasing during alcohol withdrawal periods (Aloe, Bracci-Laudiero, & Tirassa, 1993; Gericke, Schulte-Herbruggen, Arendt, & Hellweg, 2006; Heberlein et al., 2008). In developmental models, prenatal exposure significantly increased levels of NGF in a cortex/striatum homogenate on PD1, though levels returned to baseline by PD10 (Heaton et al., 2000). Levels of the NGF receptor TrkA were decreased in the male hippocampus and cerebellum and increased in the cortex of male and female pups on PD1 after prenatal exposure (Moore et al., 2004). TrkA levels were also increased in the thymus and spleen of prenatally exposed neonatal mice, possibly contributing to alcohol-related immunodeficiency (Gauthier, 2015; Gottesfeld, Morgan, & Perez-Polo, 1990). Following PD4-10 exposure, NGF protein was upregulated in cortex/striatum on PD10 and remained elevated at PD21, returning to baseline levels by PD60 (Heaton et al., 2000). However, *NGF* gene expression was unchanged in cerebral cortex on PD8 following PD5-8 alcohol exposure in another study (Fattori et al., 2008), possibly due to differences in route of administration (vapor vs intubation) or specific brain region included in the analyses (cortex/striatum vs cerebral cortex only). The interaction between NGF and withdrawal during the developmental period remains to be elucidated, but work with chronic alcohol administration models in adulthood suggests that there may be a strong relationship that could further propagate tissue damage, making this avenue of research worthy of further investigation.

Limited work has examined how models of FASD impact production on NT-3, and work that has been done has found few alcohol-induced changes. Heaton et al. (2000) found no significant alterations to NT-3 in any brain region following prenatal or postnatal alcohol exposure, and similar findings were reported following PD5-8 exposure (Fattori et al., 2008). Prenatal alcohol exposure did affect levels of the TrkC receptor protein (Moore et al., 2004). TrkC was decreased in the male and female hippocampus and male cerebellum and increased in the male cortex on PD1. On PD10, there were no significant differences for either sex. These results are consistent with the idea that NT-3 plays a large role in early embryonic development and is less critical during the later stages, such as the time points assessed in these studies.

Recent work investigating VEGF has found some evidence that developmental alcohol exposure alters VEGF regulation, which could in turn impact angiogenesis, microvasculature structure, and cell proliferation. Prenatal alcohol exposure decreased VEGF levels in the neonatal mouse cortex on PD2 and induced retraction and reorganization of cortical microvasculature in humans with FAS or partial FAS (Jegou et al., 2012). Age might interact with alcohol's effects on VEGF, as neonatal alcohol exposure significantly upregulated VEGF levels in the adult mouse hippocampus and cortex (Ceccanti et al., 2012). Since VEGF has a role in both angiogenesis and cytogenesis, enhanced levels of this protein could be a double-edged sword with potential beneficial effects on adult neurogenesis and maintenance of the neurogenic niche, but also possible negative health-related consequences such as increased tumor risk.

5. INTERACTIONS BETWEEN DEVELOPMENTAL ALCOHOL EXPOSURE AND NEUROTROPHINS IN THE BRAIN: MOLECULAR AND INTRACELLULAR EFFECTS

Beyond the direct interaction of developmental alcohol exposure and neurotrophins, there are numerous molecular processes within and between cells affected by both in utero alcohol exposure and neurotrophic factors. This section will discuss how alcohol and neurotrophin signaling either influence or are influenced by NMDA and $GABA_A$ receptors, including long-term potentiation (LTP). Oxidative stress and neuroinflammation are two other areas of possible overlap, with both of these processes representing potential sources of secondary tissue damage following developmental alcohol exposure.

5.1 Glutamatergic NMDA Receptors and GABA Receptors

Alcohol acts as a glutamatergic NMDA receptor antagonist and $GABA_A$ receptor agonist. These receptor types are also affected by neurotrophins and present a potential area of crossover between neurotrophic factor and alcohol action. Both NMDA and GABA receptors have widespread distribution throughout the brain and are found on most, but not all, neurons. Since glutamate and GABA are the predominant excitatory and inhibitory neurotransmitters, respectively, alterations to their signaling pathways can impact not only immediate cell and system function but also long-term structural and functional plasticity. Importantly, NMDA receptors are necessary for induction and maintenance of LTP, a critical process for learning and memory formation.

While neurotrophins do not act directly on glutamate receptors, NMDA- and non-NMDA-mediated glutamatergic activity can enhance synthesis of BDNF and NGF in the rodent hippocampus (Lindholm, Castren, Berzaghi, Blochl, & Thoenen, 1994; Zafra, Hengerer, Leibrock, Thoenen, & Lindholm, 1990). NT-3 seems to be unaffected by NMDA receptor activity. The interaction between NMDA receptors and neurotrophins is important as increased activity in a neural system could induce neurotrophic factor synthesis and release to supplement other plasticity-related processes. Activity-dependent BDNF release caused through NMDA receptors can regulate antiapoptotic signaling cascades through the TrkB receptor; blockade of this receptor disrupts the antiapoptotic effect (Bhave et al., 1999). Neurotrophic factors, namely BDNF, have also been shown to be necessary for protein synthesis-dependent, late-phase LTP in the neonatal and adult rodent brain. Even in the presence of protein synthesis inhibitors, administration of BDNF can maintain LTP in vitro (Lu & Chow, 1999) and inhibition of BDNF through antagonists or BDNF scavengers can eliminate previously stable LTP. In the neonatal rat, high-frequency stimulation often only produces short-term potentiation due to synaptic fatigue; however, administration of BDNF allows for induction of LTP in the neonatal brain (Gottschalk, Pozzo-Miller, Figurov, & Lu, 1998). Some authors have posited that it is no coincidence that levels of BDNF increase in the neonatal brain alongside the ability to maintain LTP (Figurov, Pozzo-Miller, Olafsson, Wang, & Lu, 1996). One way in which BDNF is thought to enhance maintenance of LTP is by increasing vesicle docking and neurotransmitter release from the presynaptic terminal (as opposed to direction interaction with AMPA or NMDA receptors). Mutant BDNF knockout mice display fewer docked vesicles and decreased levels of vesicle-docking proteins synaptophysin and synaptobrevin (Pozzo-Miller et al., 1999) and treatment with BDNF attenuates these effects. Increases in region-specific due to BDNF have functional significance, resulting in increased frequency of AMPA-mediated spikes which are connected to vesicle, and thus neurotransmitter, release (Tyler & Pozzo-Miller, 2001). Aside from BDNF, there is evidence that other neurotrophic factors could affect LTP. Recent work has shown that VEGF may play a role in LTP and memory formation, as overexpression of VEGF in vivo enhanced

LTP but blockade of VEGF reduced LTP to normal levels (Licht et al., 2011). These data suggest importance of VEGF for enhancement of LTP but not necessarily induction and maintenance of "normal" LTP. The mechanisms through which VEGF affects LTP remain to be determined.

Alcohol is known to act as an NMDA receptor antagonist by reducing ion channel opening probability and open time. Though the precise nature of its antagonistic properties is still unclear, a direct interaction of alcohol with the NMDA receptor is likely due to the rapid reduction in channel activity (<100 ms) (Ron & Wang, 2009). NMDA receptors help regulate neuronal migration and differentiation in the developing brain, making disruption of NMDA receptor activity due to alcohol exposure all the more devastating during this time period (Komuro & Rakic, 1993). A short-term blockade of NMDA receptors in the late fetal or early neonatal period results in widespread apoptosis with a pattern of damage mimicking developmental alcohol exposure (Ikonomidou et al., 1999; Thomas, Fleming, & Riley, 2001). NMDA receptors are also hypothesized to be particularly sensitive to damage through excitotoxic stimulation during the neonatal period (Ikonomidou, Mosinger, Salles, Labruyere, & Olney, 1989), though it is not known if this hypersensitivity could extend to antagonism via alcohol exposure. Rodent models of FASD have also demonstrated that alcohol exposure affects LTP formation in the developing rat hippocampus, though this effect seems to be dose and timing dependent. Administration of alcohol during the third trimester equivalent disrupts LTP in neonatal CA1 in vivo, and in vitro models show that this effect is likely due to action on both the NMDA and AMPA receptor ion channels (Puglia & Valenzuela, 2010a, 2010b). Age of the animal during LTP recording also likely makes a difference, as measurements of LTP in the adult rat brain following prenatal alcohol exposure tell a relatively different story, one that is highly dependent on sex of the animal. While male rats show reduced LTP in adulthood, female animals either exhibit no effect of alcohol exposure or an enhancement of LTP (Sickmann et al., 2014; Titterness & Christie, 2012). This observation in females seems to be independent of circulating estradiol (estrogen), as ovariectomized female rats present with the same pattern; some of the authors suggest increased synthesis of glutamate in females as a potential explanation.

Neurotrophins, such as BDNF, can alter $GABA_A$ receptor distribution and reduce inhibitory postsynaptic currents. Administration of BDNF in culture rapidly downregulates postsynaptic inhibitory current which is $GABA_A$ receptor mediated; this reduction is due to the presence of fewer postsynaptic $GABA_A$ receptors (Brunig, Penschuck, Berninger, Benson, & Fritschy, 2001). BDNF also attracts more glutamatergic terminals to the postsynaptic cell, while at the same time reducing the number of GABAergic terminals, shifting the balance from inhibitory to excitatory (Singh et al., 2006). On the other side of the coin, GABAergic activity can downregulate BDNF and NGF synthesis (Berninger et al., 1995; Lindholm et al., 1994). Interestingly, it has been suggested that GABA signaling during embryonic development, a window during which GABA acts as an excitatory rather than an inhibitory neurotransmitter, might actually enhance BDNF synthesis; the switch to downregulation of BDNF by GABA would arise after GABA changes to an inhibitory role (Zucca & Valenzuela, 2010).

Alcohol binds directly to the $GABA_A$ receptor and mimics the effects of GABA by opening the ion channel which allows hyperpolarizing chloride to enter the cell. $GABA_A$ agonists, including alcohol, benzodiazepines, and barbiturates, induce apoptosis throughout the developing brain, compounding the cell death related to NMDA receptor antagonism following alcohol exposure. $GABA_A$ hyperactivation seems to have the opposite effect during development as it does in the adult brain, where activation of this receptor acts as a protective mechanism against NMDA-related excitotoxic apoptosis (Ikonomidou et al., 2000; Olney et al., 1991, Olney, Wozniak, Jevtovic-Todorovic, & Ikonomidou, 2001). Thus, the fetus is far more sensitive to exposure to GABAmimetic agents compared to adults and alcohol presents a source of cell death on two fronts that work in concert. Normal neuronal maturation relies on a synchrony of signaling pathways, including GABA, glutamate, and neurotrophic factors, to achieve proper cell differentiation, neurite outgrowth, and synaptic integration. GABA is particularly important during development prior to formation of glutamatergic synapses and has an excitatory effect in maturing cells; administration of alcohol to cultured neonatal hippocampal pyramidal cells disrupts GABA-mediated LTP in these cells (Zucca & Valenzuela, 2010).

5.2 Oxidative Stress

Oxidative stress damages cells and DNA through the production of peroxidases and other free radical molecules, leading to increased cell death. Neurotrophins have been shown to combat oxidative damage in various disease models through upregulation of antioxidant molecules. Alternatively, reactive oxygen species (ROS) inhibit neurotrophins (Gardiner, Barton, Overall, & Marc, 2009). Diets high in saturated fat increase oxidative stress-related damage and reduce BDNF. Vitamin E supplementation decreased tissue damage and upregulated BDNF, possibly through phosphorylation of synapsin I and activation of CREB signaling pathways (Wu, Ying, & Gomez-Pinilla, 2004). NGF has also been shown to reduce oxidative damage to synaptic transporters by β-amyloid and iron molecules (Guo & Mattson, 2000). Postmortem analysis of brain tissue from autistic donors found increased levels of the ROS 3-nitrotyrosine in the frontal cortex and cerebellum; levels of NT-3 were also found to be increased in some of the same regions; however, more work is needed to determine any type of causation in this association. Cultured hybrid cells made from mitochondrial DNA of Parkinson's disease patients had enhanced vulnerability to oxidative damage and reduce glutathione presence; however, administration of either BDNF or glial cell line-derived neurotrophic factor protected the cells from oxidative stress-induced death (Onyango, Tuttle, & Bennett, 2005). Finally, most relevant to the current topic, BDNF, NGF, and NT-5 all showed neuroprotective effects on oxidative damage in the neonatal brain (Kirschner et al., 1996). Administration of 1-methyl-4-phenylpyridinium (MPP+) into the PD8 striatum caused significant hypoxic lesions to the tissue which were reduced in volume by systemic administration of neurotrophic factors prior to MPP+ injection. While the exact mechanisms of the neurotrophin–oxidative stress interaction are still unclear, it has been proposed that a positive feedback loop between neurotrophin and antioxidant production exists in the healthy brain (Gardiner et al., 2009).

Developmental alcohol exposure has been repeatedly demonstrated to increase markers of oxidative stress in the neonatal and adult rodent brain. Prenatal exposure increased levels of ROS and apoptosis in fetal rat cortical neurons, while antioxidant pretreatment of the cells prevented the apoptosis, meaning ROS production directly contributed to increased alcohol-induced cell death (Ramachandran et al., 2003). Increased levels of the oxidative stress marker lipid peroxidase were found in cultured PC12 (dopaminergic) neurons following acute alcohol administration (Sun, Chen, James-Kracke, Wixom, & Cheng, 1997). Lipid peroxidase was also upregulated throughout the brain for up to 12 weeks following postnatal alcohol exposure (Petkov, Stoianovski, Petkov, & Vyglenova, 1992) and until at least PD60 following perinatal exposure (Brocardo et al., 2012). Perinatal alcohol also increased protein oxidation in the hippocampus and cerebellum and increased anxiety- and depressive-like behaviors in the mice. Animals exposed to aerobic exercise during adolescence had increased levels of the antioxidant glutathione and ameliorated aberrant behavioral performance; however, it is not known if these two results were through overlapping or unique exercise-induced mechanisms.

5.3 Neuroinflammation

Microglial colonization of the CNS occurs during the embryonic stages in rodents, with phenotypically mature microglia comprising the majority of these immune cells by PD14 (Bilbo & Schwarz, 2009). Dysregulation of the developing immune system has the capacity to shape long-term microglial morphological and chemical activation patterns to immune challenges, possibly contributing to the emergence of neurological and psychological disorders in adulthood, including schizophrenia and depression. Immune activation is often modeled through systemic injection of lipopolysaccharide (LPS), molecules which comprise part of the cell wall of gram-negative bacteria, or the bacteria *Escherichia coli*. While LPS does not pass the blood–brain barrier, it can still induce a microglial response and upregulation of cytokine production. Immune activation through either of these methods has been shown to reduce levels of neurotrophins in various brain regions, possibly through a negative feedback loop with the hypothalamic–pituitary–adrenal (HPA) axis (Calabrese et al., 2014). Injection of LPS significantly decreased BDNF and NGF in the rat hippocampus and throughout the cortex, though the reductions were region specific (Guan & Fang, 2006). NT-3 was only decreased in frontal cortex, suggesting a differential vulnerability of neurotrophin response to infection. Further study found decreased levels of both pro- and mature BDNF isoforms at synapses in the mouse brain, which corresponded with reductions in *Bdnf* gene expression (Schnydrig et al., 2007). Chapman et al. (2012) demonstrated that *E. coli* affects CA1 and DG *Bdnf* gene expression in an exon-specific pattern, with exon I-, II-, and IV-driven transcripts showing the largest reduction. A direct interaction between the proinflammatory cytokine IL-1β and BDNF was reported over two decades ago; systemic injection of IL-1β decreased hippocampal *Bdnf* expression, though the exact mechanism of this interaction is still unclear (Lapchak, Araujo, & Hefti, 1993). Further studies are needed to clarify exactly how infection or cytokines impede neurotrophin production, the exact role of the HPA axis in this process, and whether the findings discussed here also hold true in the developing brain.

Children with in utero exposure to alcohol have a higher risk of developing secondary health complications, such as respiratory illnesses, in part due to increased likelihood of premature birth and nutritional deficiencies in alcohol-exposed infants (Gauthier, 2015). Recent work in animal models has begun to delve into the complex neuroinflammatory response induced by developmental alcohol exposure and investigation of this response as a secondary source of damage during alcohol exposure and withdrawal. To date, most in vivo work has focused on postnatal FASD models and the short-term impact of alcohol on markers of neuroinflammation. PD4-9 exposure in rat and mouse intragastric intubation models increases microglial activation and proinflammatory cytokine gene expression (IL-1β, TNF-α, CCL2, CCL4, CD11b) in the hippocampus (Boschen, Ruggiero, & Klintsova, 2016), cerebellum, and cortex on PD10 (Drew, Johnson, Douglas, Phelan, & Kane, 2015). PD3-5 alcohol vapor inhalation also induced proinflammatory cytokine gene expression in the hippocampus and cerebellum of the neonatal pups, with particular upregulation of cytokine expression during withdrawal periods (Topper, Baculis, & Valenzuela, 2015). Ethanol administration to cultured cerebellar microglia has potentially neurotoxic effects, reducing the number of microglia while simultaneously increasing activation and production of proinflammatory cytokines in the remaining cells (Kane et al., 2011). Tiwari and Chopra (2011) showed that aberrant activation of the developing immune system can have long-term consequences on cytokine expression. PD7-9 alcohol exposure via intubation resulted in enhanced levels of TNF-α and IL-1β in the cerebral cortex and hippocampus in adulthood. This finding is important as aberrant activation of the immune system early in development could result in overactivation of microglia and overproduction of inflammatory molecules following further immune challenges later in life and contribute to pathological conditions (Bilbo & Schwarz, 2009); overexpression of cytokines at baseline conditions in the alcohol-exposed brain could indicate disruption of normal immune function in these animals. Interestingly, there is limited evidence that postnatal alcohol exposure might also enhance levels of cytokines with antiinflammatory properties. Levels of TGF-β mRNA were increased in the hippocampus on PD10 following PD4-9 alcohol exposure (Boschen et al., 2016) and in the adult cortex and hippocampus following PD7-9 exposure (Tiwari & Chopra, 2011). However, Topper et al. (2015) reported no changes to TGF-β or another antiinflammatory cytokine, IL-10, in the neonatal brain following PD3-5 vapor inhalation; these differences could be due to the model or timing of exposure used. Further investigation of how antiinflammatory cytokines might react to alcohol-induced damage in the developing brain is needed and considered as a potential therapeutic intervention to mitigate apoptosis.

6. INTERACTIONS BETWEEN DEVELOPMENTAL ALCOHOL EXPOSURE AND NEUROTROPHINS: NEUROANATOMICAL AND BEHAVIORAL EFFECTS

6.1 Neuroanatomical Effects

Through action on the molecular mechanisms discussed earlier, alcohol exposure during development exerts deleterious effects on neuroanatomical measures including cell number, dendritic morphology and spine density, the process of adult neurogenesis in the hippocampus, and structure of microvasculature. All of these measures can speak to the general health of the neural circuit and give information about potential dysfunction leading to impaired behavioral outcomes. Section 6.1.1 primarily focuses on alcohol's detrimental effects on brain region volume and cell number. The importance of neurotrophins for cell proliferation and migration cannot be overstated and will be addressed in more detail in the section discussing adult neurogenesis (Section 6.1.2), in addition to evidence that developmental alcohol exposure disrupts these processes. The health of neurons can also be assessed through other means, including the complexity and morphology of the dendritic trees and spine density on the dendrites (Section 6.1.3). Section 6.1.4 reviews how neurotrophins and alcohol impact angiogenesis and microvasculature.

6.1.1 Brain Region Volume and Alcohol-Induced Apoptosis

With the emergence of noninvasive technologies to study brain structure and function in humans, brain region volume, and cortical thickness became a few neuroanatomical measures well characterized in children with prenatal alcohol exposure. Significant reductions in overall brain volumes and in specific brain regions in children with FASDs were reported in several studies (Archibald et al., 2001; Autti-Ramo et al., 2002; Gautam et al., 2015; Johnson, Swayze, Sato, & Andreasen, 1996; Sowell et al., 2002; Willoughby, Sheard, Nash, & Rovet, 2008). Cortical thickness, white matter, and subcortical gray matter in children with prenatal alcohol exposure are also subjects of change; however, the changes in these parameters are not unidirectional: greater cortical thickness was observed in parietal and posterior temporal regions (Fernandez-Jaen et al., 2011; Sowell et al., 2008, 2001, 2002; Yang et al., 2012). In contrast, reduced cortical thickness was reported by other groups (Treit et al., 2014; Zhou et al., 2011), especially at ages when cortical thickness is expected to be at its maximum (Robertson et al., 2016). Other brain areas such as the hippocampus (Willoughby et al., 2008), caudate nucleus (Archibald et al., 2001; Cortese et al., 2006; Fryer et al., 2012), and cerebellar vermis (Autti-Ramo et al., 2002; Fryer et al., 2012) had decreased volume, suggesting significant regional loss of cells and neuropil. In addition, results from studies using diffusion tensor imaging (DTI) demonstrate that individuals exposed to alcohol in utero have less organized white matter fiber tracts (Wozniak & Muetzel, 2011) and significantly decreased fractional anisotropy in white matter regions (Fryer et al., 2012; Sowell et al., 2008) compared to control subjects. Decreases in overall and region-specific brain volume are well documented in animal models of FASDs (Bonthius & West, 1988, 1990; Maier, Chen, & West, 1996; Maier, Miller, Blackwell, & West, 1999; Napper & West, 1995). More recently, both ex vivo MRI and DTI studies confirmed that animal models of FASDs cause decreased overall brain and cortical volume (Leigland, Budde, Cornea, & Kroenke, 2013; Leigland, Ford, Lerch, & Kroenke, 2013; O'Leary-Moore et al., 2010; Parnell, Holloway, Baker, Styner, & Sulik, 2014; Parnell et al., 2013), as well as abnormally high cerebral cortical functional anisotropy that indicates reduced morphological complexity (Leigland, Budde, et al., 2013; Leigland, Ford, et al., 2013).

One mechanism through which alcohol alters neurodevelopment is by inducing apoptosis, or programmed cell death (Creeley & Olney, 2013; Kerr, Wyllie, & Currie, 1972). A critical process in normal development, apoptosis is a naturally occurring phenomenon in the developing mammalian brain both pre- and postnatally (Johnston et al., 2009). Apoptosis exhibits region specificity, showing variations in both quantity and latency (Nijhawan, Honarpour, & Wang, 2000; Rabinowicz, de Courten-Myers, Petetot, Xi, & de los Reyes, 1996; Rakic & Zecevic, 2000). As many as half of the cells generated during early development die by apoptosis, making the interplay between the successful establishment of appropriate connections with synaptic targets and the competition for antiapoptotic factors extremely important. Neurotrophic factors represent one group of factors capable of protecting developing neurons against apoptotic cascades (Nijhawan et al., 2000; Stiles & Jernigan, 2010).

Alcohol is a teratogen that is known to induce widespread apoptosis in CNS tissue at various developmental stages (Climent, Pascual, Renau-Piqueras, & Guerri, 2002; Dunty, Chen, Zucker, Dehart, & Sulik, 2001; Ikonomidou et al., 2000; Olney et al., 2002). Exposure to alcohol during the gestational period in primates or the early postnatal period in rodents produces massive stimulation of apoptosis in the immature brain (Farber,

Creeley, & Olney, 2010; Ikonomidou et al., 2000; Olney et al., 2002). The mechanism of apoptosis includes the alcohol-driven decrease in neuronal activity in the developing brain through a suppression of glutamatergic NMDA receptors and potentiation of GABAergic receptors (discussed in Section 5.1), resulting in suppressed spontaneous network activity and the silencing of neurons in a dose-dependent manner (Lebedeva et al., 2015). Since synaptic activity is essential for neuronal survival during the gestational/early postnatal period, such silencing can result in activity-dependent apoptosis. Indeed, Ikonomidou et al. (2000) demonstrated a wide range of regional susceptibilities to alcohol-induced neurodegeneration following a single-day binge-like developmental exposure to ethanol. Furthermore, increased rates of apoptosis were found in hippocampi of rats exposed to alcohol in either a single-day or 2-day binge during third trimester equivalent (Smith, Guevremont, Williams, & Napper, 2015). Hippocampal cells displayed increased sensitivity to alcohol-induced neurodegeneration during the neonatal period (Heaton et al., 2003; Ikonomidou et al., 2000; Livy, Miller, Maier, & West, 2003; Wozniak et al., 2004). Our lab's unpublished data show that the rate of hippocampal apoptosis on PD5 (following a single day of binge alcohol exposure on PD4) is over 250% of the rate present in the normally developing brain at that age. Remarkably, the hippocampus has been shown to compensate for damage induced by ethanol through subsequent decreases in cell death and increases in adult neurogenesis. Smith and colleagues (2015) showed that animals exposed to ethanol on both PD6 and 8 displayed a reduced magnitude of apoptosis on PD8 relative to animals exposed to ethanol on PD8 only. Our data support these findings, indicating that apoptosis is decreased in CA1 of alcohol-exposed animals on PD8 (4 days after exposure). It is important to keep in mind that these compensatory events are not nearly as massive as the increase in cell death seen in the days immediately following the alcohol exposure. For the discussion of the role of neurotrophins in neuroprotection against apoptosis, see Section 7.

6.1.2 Adult Neurogenesis: Proliferation, Differentiation, Migration, and Cell Survival

Adult neurogenesis is the process through which the brain continues to generate new neurons throughout the life span in two specific regions: the subventricular zone (SVZ) of the lateral ventricle and the subgranular zone (SGZ) of the hippocampal dentate gyrus. Cells from the SVZ migrate via the rostral migratory stream to the olfactory bulb where they mature into interneurons, while most proliferating cells in the SGZ differentiate into dentate gyrus granule cells and migrate into the granule cell layer and integrate fully into the hippocampal trisynaptic pathway. This section will primarily focus on how neurotrophins and developmental alcohol exposure affect hippocampal adult neurogenesis, as this process is thought to be important for behaviors and cognition associated with the hippocampus, including spatial learning, context memory, pattern separation and completion, and emotion regulation. Performance on tasks that involve these dimensions of learning are often impaired in animal models of developmental alcohol exposure, and humans with FASDs often have significant learning disabilities (discussed in Section 6.2).

Adult neurogenesis is separated into stages of cell maturation: (1) proliferation, (2) differentiation, (3) migration, (4) maturation, and (5) long-term survival (dependent on successful synaptic integration). The entire process from cell birth to full integration takes up to 3 months. Neurotrophins are highly expressed in the adult rodent hippocampus and play an important role in all of these stages and work in synchrony for proper development, though certain neurotrophins seem to be more critical during certain maturation processes (Maisonpierre et al., 1990). BDNF is likely more important for later cell maturation, not initial proliferation, based on work showing that expression of the TrkB receptor is low during the first week following cell birth and increased as the cell ages (Donovan, Yamaguchi, & Eisch, 2008). Sairanen, Lucas, Ernfors, Castren, and Castren (2005) also reported that BDNF mutant mice displayed disrupted cell survival, even following administration of a selective serotonin reuptake inhibitor (SSRI) which increased new cell survival in wild-type mice. BDNF mutant mice also demonstrate stalled neuronal maturation and migration and reduced dendritic arborization in adult-born dentate gyrus cells (Chan, Cordeira, Calderon, Iyer, & Rios, 2008). VEGF has also been shown to play a role in adult neurogenesis, both indirectly through establishment of a neurogenic niche and by directly enhancing cell maturation processes (Licht et al., 2011).

Alcohol exposure during early development can impact both the health of neural progenitors in the adult rodent brain and the ability of these newly born granule cells to successfully mature and integrate into the hippocampal trisynaptic circuit, though the most consistent effects are reported on neuronal survival. Prenatal alcohol exposure decreased the size of the progenitor pool and number of proliferating neurons in one report (Redila et al., 2006) and a single binge on PD7 was shown to disrupt cell proliferation (Ieraci & Herrera, 2007). However, several other studies did not demonstrate an effect of alcohol on cell proliferation (Choi, Allan, & Cunningham, 2005; Gil-Mohapel et al., 2011; Hamilton et al., 2011; Klintsova et al., 2007). On the other hand, deficits in neuronal maturation processes and survival have been shown repeatedly (Gil-Mohapel et al., 2011; Hamilton et al., 2011; Ieraci & Herrera, 2007; Klintsova et al., 2007; Singh, Gupta, Jiang, Younus, & Ramzan, 2009), suggesting that alcohol exposure, particularly during the brain growth spurt period, permanently alters signaling pathways or gene expression necessary for successful maturation and synaptic integration. Ultimately, inability of the hippocampus to produce and integrate the number of new granule cells necessary for new memory formation could contribute to the cognitive deficits in children with FASDs and the behavioral impairments observed in animals models.

6.1.3 Dendritic Morphology and Spine Density

One avenue through which alcohol exposure could disrupt the ability of new cells in the hippocampus to successfully integrate is through alterations to dendritic spine density and complexity of the dendritic trees. Extension of neurites (dendrites toward the molecular layer and entorhinal cortex and axons toward CA3) is an essential step of granule cell maturation which begins when the cells are classified as immature neurons and can be immunolabeled with doublecortin (DCX). Formation of excitatory synapses with neurons located in the medial entorhinal cortex is a competitive process and a more complex dendritic tree increases the chances of functional connections being made. Even once a neuron is successfully integrated, decreased dendritic complexity and synapse number can impact efficiency of signal transduction within a brain region or circuit.

Neurotrophins play an important and well-orchestrated role in dendritic morphology and spine density throughout the brain. In the cortex, BDNF and NT-3 have been shown to oppose one another's impact on dendritic structure depending on the cortical layer (McAllister, Katz, & Lo, 1997). In Layer IV, BDNF enhanced dendritic growth while NT-3 administration blocked this effect. In Layer VI, the roles were reversed, with BDNF inhibiting and NT-3 administration stimulating dendritic growth. BDNF has been repeatedly shown to enhance dendritic length and complexity in the hippocampus and in cell culture (Alonso et al., 2002; Singh et al., 2006; Tolwani et al., 2002) and conditional knockdown of BDNF in the mouse hippocampus decreases granule cell dendritic arborization (Chan et al., 2008). BDNF also increases spine density on CA1 dendrites (Alonso et al., 2002; Tyler & Pozzo-Miller, 2001). Aside from increasing spinogenesis, BDNF administration can cause spines to take on a more unstable, immature phenotype (Horch, Krüttgen, Portbury, & Katz, 1999; Shimada, Mason, & Morrison, 1998), possibly priming the dendrite for upcoming changes to intracellular signaling. The enhancing effects of BDNF are likely through action at the TrkB receptor, as activation of the p75 receptor seems to have the opposite effect of BDNF administration by decreasing dendritic complexity and spine density (Zagrebelsky et al., 2005). VEGF has recently been reported to be necessary for dendritic growth in new olfactory bulb neurons arising from progenitors in the SVZ, suggesting another new role for this neurotrophin (Licht et al., 2010).

The exact mechanism through which developmental alcohol exposure compromises dendritic complexity in the adult brain is not well understood, but it is thought that these alterations contribute significantly to cognitive deficits in children with FAS. In fact, in cases of mental retardation due to either genetic or environmental factors, "dendritic abnormalities are the most consistent anatomical correlates" reported (Kaufmann & Moser, 2000). Third trimester-equivalent alcohol exposure has been shown to decrease basilar dendritic complexity in Layer II/III pyramidal neurons of the mPFC (Hamilton, Criss, & Klintsova, 2015; Hamilton, Whitcher, & Klintsova, 2010). Spine morphology on the basilar dendrites also shifted to a more mature, less plastic phenotype, though spine density was unchanged. On the other hand, the apical dendrites are affected in an almost opposite pattern by this model of alcohol exposure: dendritic tree morphology was stable but spine density was significantly decreased in alcohol-exposed animals (Whitcher & Klintsova, 2008). Perinatal alcohol exposure (combining pre- and postnatal exposure) had limited effect on dendritic structure in the nucleus accumbens, but reduced spine density on Layer II/III pyramidal cell dendrites in the mPFC (Lawrence, Otero, & Kelly, 2012). More recently, our lab has examined dendritic complexity of immature DCX+ dentate gyrus granule cells in the hippocampus of postnatally alcohol-exposed male Long–Evans rats using the Sholl analysis (Fig. 3). We have found that alcohol exposure (via intragastric intubation on PD4-9) significantly decreased the number of dendritic intersections and bifurcations per radius and decreased the amount of dendritic material per radius compared to control conditions (sham intubated and suckle control/undisturbed). Deficits in dendritic outgrowth at this stage of cell maturation could directly affect the neuron's ability to successfully integrate into the hippocampal circuit and survive, in part contributing to the decreases in neuronal survival observed following third trimester-equivalent alcohol

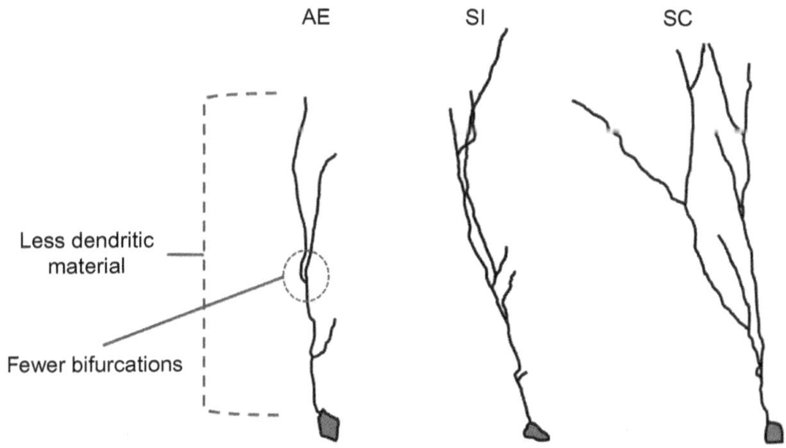

Fig. 3 Third trimester-equivalent (postnatal days 4–9) alcohol exposure decreases dendritic complexity in immature dentate gyrus granule cells in the adult rat hippocampus. *AE*, alcohol-exposed (5.25 g/kg/day, binge-like exposure, intragastric intubation); *SC*, suckle control (undisturbed); *SI*, sham intubated (no liquid).

exposure (Klintsova et al., 2007). Prenatal alcohol exposure has not been found to affect spine density in hippocampal CA1 neurons in rats housed in isolation; however, alcohol-exposed rats housed in a complex "enriched" environment (which increases spine density in normal animals) display no change in spine density (Berman, Hannigan, Sperry, & Zajac, 1996). These findings suggest that timing of the alcohol exposure dictates whether baseline changes to dendritic complexity and spine density are present and that deficits might not be apparent until the system is challenged in some way. Overall, the abnormalities in dendritic morphology and spine density in alcohol-exposed animals likely contribute significantly to the behavioral deficits observed in rodent models of FASDs.

6.1.4 Microvasculature

Establishment and maintenance of the blood–brain barrier is essential to brain development, function, and plasticity. Neurotrophins enhance microvasculature and promote blood–brain barrier permeability. Treatment with BDNF prior to a spinal cord injury can even offer limited protection against injury through opening of the blood–spinal cord barrier (Sharma, 2003). BDNF is thought to have dual roles in enhancement of angiogenesis: promotion of endothelial cell survival through action at the TrkB receptor and through recruitment of immune cells to the site of an injury (Kermani & Hempstead, 2007). BDNF is an important part of ischemic injury repair in the skin, arterial, and cardiac tissues. NGF has been show to support angiogenesis in certain preparations as well (Cantarella et al., 2002). Perhaps the most important neurotrophin for the microcirculatory system and angiogenesis is VEGF, particularly for vasculogenesis during development. In addition, VEGF is critical in the formation and restructuring of blood vessels following injury, establishment of the neurogenic niche in the hippocampus, and tumorigenesis (Alvarez-Buylla & Lim, 2004; Ferrara, Gerber, & LeCouter, 2003; Rosenstein & Krum, 2004). VEGF has been tested as a protective agent in hypoxic or ischemic injury and VEGF antagonists have been put through trials to inhibit tumor growth (Ellis & Hicklin, 2008), though the success of these trials has been limited, not at least in part due to the complex timing of events during hypoxic injuries and mechanistic underpinnings of VEGF signaling that remain to be addressed.

Hypoperfusion of blood to target organs during embryonic development has been hypothesized to contribute to low birth weight and the cognitive impairments observed in children with FAS. This hypothesis has been supported by studies showing that children with FAS have lower cerebral blood flow, with considerable hypoperfusion of the temporal lobe specifically (Bhatara et al., 2002; Riikonen, Salonen, Partanen, & Verho, 1999). Similar findings have been borne out using animal models. Alcohol administration in adult rats decreased vasodilatory response to a variety of drugs that increase microvessel diameter in the healthy brain (Mayhan, 1992). In utero, alcohol-exposed fetuses experience reduced placental and arterial blood flow to the brain (Bake, Tingling, & Miranda, 2012; Jones, Leichter, & Lee, 1981). Inspection of brain sections taken from prenatally exposed mice revealed a disorganized and dying microvasculature (Jegou et al., 2012); death of the microvessels was prevented with administration of VEGF, supporting a neuroprotective role for this neurotrophin. Interestingly, cerebellar hyperperfusion was observed in fetal sheep following third trimester exposure (Parnell et al., 2007), which correlated with a reduction in Purkinje cell number. While most studies have found reduced cranial blood flow following prenatal alcohol exposure, the relationship between alcohol and the microcirculatory system is not always clear cut, and even areas that receive transiently increased blood flow can still show signs of alcohol-induced damage.

6.2 Behavioral Effects

Through actions on the molecular and anatomical systems of the brain, neurotrophin levels are correlated with a wide range of behavioral phenotypes.

BDNF release in the hippocampus has been shown in multiple animal models to be required for short- and long-term contextual and spatial memory formation (Alonso et al., 2002; Griffin, Bechara, Birch, & Kelly, 2009; Lu & Chow, 1999). BDNF is rapidly upregulated during contextual memory formation (Hall, Thomas, & Everitt, 2000) and blockade of hippocampal BDNF release impairs the ability of mice to form hippocampus-dependent memories (Heldt, Stanek, Chhatwal, & Ressler, 2007). Exercise, which robustly enhances neurotrophin production in the hippocampus and perirhinal cortex, also improves performance on both spatial and nonspatial memory tasks (Griffin et al., 2009). Supporting the role of BDNF in the behavioral enhancements, injection of BDNF directly into the cerebroventricular system resulted in the same augmentation to memory. Mechanistically, BDNF increases levels of docked synaptic vesicles and neurotransmitter release, leading to enhanced LTP induction and maintenance, making BDNF a critical variable in the synaptic and cellular changes underlying memory formation (Fig. 4).

Neurotrophin dysregulation has also been implicated in anxiety- and depressive-like behaviors in rodent models of mood disorders (Martinowich, Manji, & Lu, 2007; Fig. 4). Some studies have found that decreases in BDNF have been correlated with behavioral phenotypes of depression and anxiety, but often these effects are small and inconsistent between models and differ between male and female animals. More convincing is the upregulation of BDNF following antidepressant treatment with SSRIs, and evidence suggests that regulation of BDNF expression and signaling is a downstream target of these drugs. *Bdnf* mutant mice that express increased anxiety-like behaviors did not show improvements following treatment with SSRIs, suggesting BDNF upregulation as a key component of antidepressant action (Chen et al., 2006). In humans, the Val66Met polymorphism has been linked to reduced hippocampal volume and the development of psychopathologies including depression, anxiety, and schizophrenia (Martinowich et al., 2007; Sanchez et al., 2011). This single-nucleotide polymorphism refers to a methionine (Met) substitution for valine (Val) in the prodomain (codon 66) of the BDNF protein and disrupts activity-dependent release of BDNF (Chen et al., 2005). While the Val66Met polymorphism increases an individual's genetic risk for developing psychopathology, experiencing early life stress or trauma can enhance this risk (Gatt et al., 2009). Mouse models of this polymorphism ($BDNF^{Met/Met}$) also show increased anxiety-related behaviors, including reductions in time spent in the center of an open field and in the open arms of an elevated plus maze (Chen et al., 2006). These behavioral changes were not reversed with administration of an SSRI and were accompanied by decreased hippocampal volume and dendritic complexity.

Children with FASDs exhibit both scholastic and intellectual disabilities (Mattson et al., 1999; Mattson, Riley, Delis, et al., 1996; Mattson et al., 1997; Mattson, Riley, Sowell, et al., 1996; Rasmussen et al., 2011), as well as abnormal social and emotional interactions. As adults, individuals with FASDs are more likely to have trouble with the law, in part due to executive functioning and impulse control deficits (Franklin et al., 2008; Irner et al., 2012; Kodituwakku et al., 1995; Pei et al., 2011; Stevens et al., 2012; Streissguth et al., 2004). To some degree, the learning and memory dysfunctions observed in humans with FASDs (Uecker & Nadel, 1996) have been modeled in animals as well. In animal models, alcohol-induced alterations to behavioral endpoints are largely dependent on the timing and dose of the exposure; however, both prenatal and postnatal models can significantly disrupt hippocampal-associated behavior (Berman & Hannigan, 2000; Clements, Girard, Ellard, & Wainwright, 2005; Hamilton, Boschen, Goodlett, Greenough, & Klintsova, 2012; Hunt, Jacobson, & Torok, 2009; O'Leary-Moore, McMechan, Mathison, Berman, & Hannigan, 2006; Thomas, Weinert, Sharif, & Riley, 1997). These deficits include impairment on spontaneous alternation tasks, the Morris Water Maze, the radial arm maze, discrimination and reversal learning, and variants of fear conditioning. Third trimester-equivalent alcohol exposure impairs the ability of rats to learn a novel context and reduces novelty-associated c-Fos expression in hippocampal CA1 (Murawski, Klintsova, & Stanton, 2012).

Beyond performance on spatial memory tasks, alcohol-exposed animals also show alterations to social and anxiety behaviors (Fig. 4). Third trimester-equivalent alcohol exposure reduced exploratory and play behaviors in adolescent rats housed in a complex environment with enhanced social and novelty-related stimulation (Boschen, Hamilton, Delorme, & Klintsova, 2014), and alcohol-exposed rats also show altered play fighting and decreased social recognition behaviors in other social interaction paradigms (Kelly & Dillingham, 1994; Kelly, Leggett, & Cronise, 2009; Middleton, Varlinskaya, & Mooney, 2012; Mooney & Varlinskaya, 2011). Children with FAS likewise show social behavior deficits which present independent of changes to IQ (Thomas, Kelly, Mattson, & Riley, 1998). There is some evidence that alcohol-exposed animals also show anxiety- and depressive-like behaviors, such as increased learned helplessness and immobility during the forced swim test, which were correlated with downregulation of *Bdnf* expression in the mPFC and hippocampus

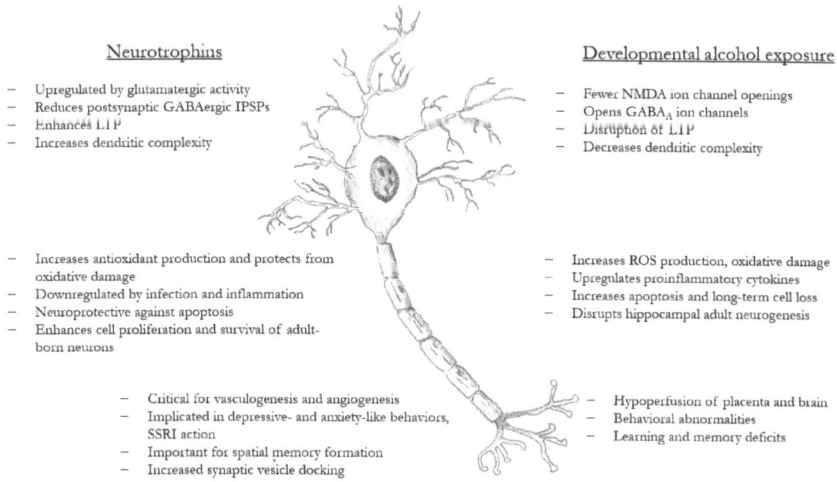

Fig. 4 Schematic representation of the overlapping effects of alcohol exposure during development and neurotrophin action on brain structure and function. *Numbers* indicate the section of the chapter's text which include references for each point.

(Caldwell et al., 2008; Hellemans, Verma, Yoon, Yu, & Weinberg, 2008). Depression is also more commonly diagnosed in children with documented prenatal alcohol exposure compared to age-matched controls (Roebuck, Mattson, & Riley, 1999). Increased presentation of psychopathology is apparent from early childhood and the risk continues through adolescence, at the least (O'Connor & Kasari, 2000; O'Connor & Paley, 2006; Steinhausen & Spohr, 1998). Like social functioning deficits, anxiety and depression are associated with prenatal alcohol exposure even if the children had a normal IQ and performance other measures of intelligence. Whether the increased propensity of children with prenatal alcohol exposure to develop depression and anxiety is related to altered neurotrophin functioning within the prefrontal cortex or limbic areas remains to be investigated.

7. NEUROTROPHINS AS COMPENSATORY/ NEUROPROTECTIVE MOLECULES FOLLOWING DEVELOPMENTAL ALCOHOL EXPOSURE

While BDNF and other neurotrophins have been long investigated for their role in neuroprotection, future work must continue to explore neurotrophins as compensatory molecules during or following developmental alcohol exposure (Davis, 2008; Idrus & Thomas, 2011; Pezet & Malcangio, 2004). As discussed earlier, alcohol exposure induces widespread waves of apoptosis throughout the developing brain shortly following administration. BDNF prevents apoptosis in a variety of experimental paradigms, including excitotoxicity and ischemic hypoxia (Almeida et al., 2005; Han & Holtzman, 2000). There is also evidence that pharmacological modulation of NGF and BDNF can protect against cytotoxicity and apoptosis following alcohol exposure, though this effect has only been directly demonstrated in vitro. Both BDNF and NGF reduced cytotoxicity as measured by a reduction in 3-(4,5-dimethylthiazol-2-yl)-2,5-diphenyl tetrazolium bromide in a model of ethanol exposure combined with hypoxia and hypoglycemia (Mitchell, Paiva, Walker, & Heaton, 1999). Administration of alcohol to cultured human neuroblastomas decreased cell viability (mitochondrial activity) and BDNF and CREB levels; exogenously applied BDNF improved mitochondrial activity in alcohol-exposed cells (Sakai et al., 2005). Alcohol induced apoptosis in cerebellar granule cells, but the upregulation of cell death was blocked by BDNF (Bhave et al., 1999). Finally, Climent et al. (2002) showed that BDNF activated anti-apoptotic and procell survival pathways in the postnatal cerebral cortex; however, this activation is disrupted by prenatal alcohol exposure. Further research needs to be done to determine if these results hold true in vivo. Numerous studies have found enhanced levels of neurotrophins in the hippocampus and cortex within 24 h of alcohol administration, suggesting a possible endogenous mechanism of neuroprotection (Boschen et al., 2015; Heaton et al., 2000, 2003). This enhancement may be timing and region specific and temporally restricted, with possible links to alcohol withdrawal. It is not known if blocking this upregulation of neurotrophins would further alcohol-related cell loss or which isoform of BDNF protein is increased (pro vs mature BDNF).

While administration of exogenous neurotrophic factors to cultured cells has shown some therapeutic promise, little work has been done exploring administration of neurotrophins to alcohol-exposed rodents, possibly due to logistical or technical issues (Davis, 2008). The neuroprotective effect of neurotrophic factors would depend largely on timing of administration following the alcohol exposure. To achieve true protection from the negative effects of alcohol, the neurotrophin would be administered either shortly before or during the alcohol exposure, or possibly during the withdrawal period. BDNF infused intracerebroventricularly to mice exposed prenatally to alcohol and stress has positive benefits on anxiety-like behavior (marble burying) and aberrant sexual and social interactions (Popova, Morozova, & Naumenko, 2011); however, these effects were only apparent following 7–10 days of daily infusions. Incerti, Vink, Roberson, Benassou, et al. (2010) and Incerti, Vink, Roberson, Wood, et al. (2010) reported that embryos exposed to alcohol on GD8 had decreased levels of BDNF after 6 h and enhanced levels after 24 h; BDNF levels were normalized with neuropeptides related to vasoactive intestinal peptide (VIP). Administration of these neuropeptides in adulthood reversed alcohol-induced learning deficits (Incerti, Vink, Roberson, Benassou, et al., 2010; Incerti, Vink, Roberson, Wood, et al., 2010). VIP and BDNF have been shown to reciprocally enhance expression of one another in certain cell types (Cellerino, Arango-Gonzalez, Pinzon-Duarte, & Kohler, 2003; Pellegri, Magistretti, & Martin, 1998). Recently, other pharmacological agents including peroxisome proliferator-activated receptor-γ (PPAR-γ), which suppresses production of proinflammatory cytokines (Drew et al., 2015) and lithium (Luo, 2010) have shown promise in protecting against alcohol-induced cytotoxic damage. Lithium, for one, is known to enhance neurotrophin levels, and activation of the neuroimmune response can downregulate BDNF expression, calling for future research to determine if these drugs exert action, at least in part, through neurotrophin signaling.

Neuropharmacological approaches to increase neurotrophin availability in order to promote neuroprotection against alcohol toxicity provide a quick, but invasive possible solution. Other therapeutic interventions known to naturally enhance neurotrophin expression in the brain present a very valuable avenue to have the brain potentially heal itself. Such therapies include the "usual suspects": exercise, learning experience, environmental stimulation, and dietary restriction (Bekinschtein, Oomen, Saksida, & Bussey, 2011; Cotman & Berchtold, 2007; Lee, Seroogy, & Mattson, 2002; Russo-Neustadt, Beard, & Cotman, 1999). Indeed, the first three interventions have been extensively in animal models of developmental alcohol exposure to produce improvements in behavioral outcomes and neuroanatomical parameters (Hamilton et al., 2012, 2014; Hannigan, Berman, & Zajac, 1993; Helfer, Goodlett, Greenough, & Klintsova, 2009; Klintsova et al., 2002; Redila et al., 2006; Thomas, Sather, & Whinery, 2008). It has been demonstrated that ethanol-mediated damage can be powerfully modulated by neurotrophins (Heaton, Madorsky, Paiva, & Mayer, 2002), which are upregulated by the interventions listed. Incorporating exercise, increased novelty and social stimulation, and careful diet regulation as part of a therapeutic program for children with FASDs could help the brain produce more neurotrophins, promote neuroplasticity, and mitigate at least some of the alcohol-induced deficits to cognition and behavior.

8. CONCLUSIONS

FASDs are devastating for children, families, and communities, and completely preventable through the avoidance of alcohol during pregnancy. Despite increased public awareness and safety campaigns from government and medical agencies, the rate of FASDs has not decreased significantly in recent years. Development of pharmacological and behavioral interventions for children with FASDs is an important step in helping these individuals gain functional independence and alleviate the emotional and monetary burden of care from the families and society. Exposure to alcohol during development can impact the production and function of neurotrophins, which represents both a mechanism of alcohol-induced brain damage and a potential avenue of therapeutic intervention which remains to be fully explored. There are also numerous processes which seem to be affected by both alcohol and neurotrophins or which could represent mechanisms through which alcohol-induced changes could then alter neurotrophin levels, such as oxidative stress and neuroinflammation. Both of these areas of research have many open questions remaining. Further investigation of neurotrophins as neuroprotective molecules in the alcohol-damaged brain could lead to the development of new therapeutic alternatives for children with FASDs.

ACKNOWLEDGMENTS

The authors would like to thank P.T. Boschen for the illustrations in this chapter and K.J. Criss and T.F. Gamage for assistance in editing the manuscript. The dendritic morphology and apoptosis work was supported by NIH/NIGMS COBRE: The Delaware Center for Neuroscience Research Grant 1P20GM103653 to A.Y.K.

REFERENCES

Allen, G. C., West, J. R., Chen, W. J., & Earnest, D. J. (2004). Developmental alcohol exposure disrupts circadian regulation of BDNF in the rat suprachiasmatic nucleus. *Neurotoxicology and Teratology*, *26*(3), 353–358. http://dx.doi.org/10.1016/j.ntt.2004.02.003.

Almeida, R. D., Manadas, B. J., Melo, C. V., Gomes, J. R., Mendes, C. S., Graos, M. M., … Duarte, C. B. (2005). Neuroprotection by BDNF against glutamate-induced apoptotic cell death is mediated by ERK and PI3-kinase pathways. *Cell Death and Differentiation*, *12*(10), 1329–1343. http://dx.doi.org/10.1038/sj.cdd.4401662.

Aloe, L., Bracci-Laudiero, L., & Tirassa, P. (1993). The effect of chronic ethanol intake on brain NGF level and on NGF-target tissues of adult mice. *Drug and Alcohol Dependence*, *31*(2), 159–167.

Alonso, M., Vianna, M. R., Depino, A. M., Mello e Souza, T., Pereira, P., Szapiro, G., … Medina, J. H. (2002). BDNF-triggered events in the rat hippocampus are required for both short- and long-term memory formation. *Hippocampus*, *12*(4), 551–560. http://dx.doi.org/10.1002/hipo.10035.

Alvarez-Buylla, A., & Lim, D. A. (2004). For the long run: Maintaining germinal niches in the adult brain. *Neuron, 41*(5), 683–686.

Amendah, D. D., Grosse, S. D., & Bertrand, J. (2011). Medical expenditures of children in the United States with fetal alcohol syndrome. *Neurotoxicology and Teratology, 33*(2), 322–324. http://dx.doi.org/10.1016/j.ntt.2010.10.008.

Archibald, S. L., Fennema-Notestine, C., Gamst, A., Riley, E. P., Mattson, S. N., & Jernigan, T. L. (2001). Brain dysmorphology in individuals with severe prenatal alcohol exposure. *Developmental Medicine and Child Neurology, 43*(3), 148–154.

Autti-Ramo, I., Autti, T., Korkman, M., Kettunen, S., Salonen, O., & Valanne, L. (2002). MRI findings in children with school problems who had been exposed prenatally to alcohol. *Developmental Medicine and Child Neurology, 44*(2), 98–106.

Bake, S., Tingling, J. D., & Miranda, R. C. (2012). Ethanol exposure during pregnancy persistently attenuates cranially directed blood flow in the developing fetus: Evidence from ultrasound imaging in a murine second trimester equivalent model. *Alcoholism, Clinical and Experimental Research, 36*(5), 748–758. http://dx.doi.org/10.1111/j.1530-0277.2011.01676.x.

Bekinschtein, P., Oomen, C. A., Saksida, L. M., & Bussey, T. J. (2011). Effects of environmental enrichment and voluntary exercise on neurogenesis, learning and memory, and pattern separation: BDNF as a critical variable? *Seminars in Cell & Developmental Biology, 22*(5), 536–542. http://dx.doi.org/10.1016/j.semcdb.2011.07.002.

Berman, R. F., & Hannigan, J. H. (2000). Effects of prenatal alcohol exposure on the hippocampus: Spatial behavior, electrophysiology, and neuroanatomy. *Hippocampus, 10*(1), 94–110. http://dx.doi.org/10.1002/(sici)1098-1063(2000)10:1<94::aid-hipo11>3.0.co;2-t.

Berman, R. F., Hannigan, J. H., Sperry, M. A., & Zajac, C. S. (1996). Prenatal alcohol exposure and the effects of environmental enrichment on hippocampal dendritic spine density. *Alcohol, 13*(2), 209–216. http://dx.doi.org/10.1016/0741-8329(95)02049-7.

Bernd, P. (2008). The role of neurotrophins during early development. *Gene Expression, 14*(4), 241–250. https://doi.org/10.3727/105221608786883799.

Berninger, B., Marty, S., Zafra, F., da Penha Berzaghi, M., Thoenen, H., & Lindholm, D. (1995). GABAergic stimulation switches from enhancing to repressing BDNF expression in rat hippocampal neurons during maturation in vitro. *Development, 121*(8), 2327–2335.

Bhatara, V. S., Lovrein, F., Kirkeby, J., Swayze, V., 2nd, Unruh, E., & Johnson, V. (2002). Brain function in fetal alcohol syndrome assessed by single photon emission computed tomography. *South Dakota Journal of Medicine, 55*(2), 59–62.

Bhave, S. V., Ghoda, L., & Hoffman, P. L. (1999). Brain-derived neurotrophic factor mediates the anti-apoptotic effect of NMDA in cerebellar granule neurons: Signal transduction cascades and site of ethanol action. *The Journal of Neuroscience, 19*(9), 3277–3286.

Bilbo, S. D., & Schwarz, J. M. (2009). Early-life programming of later-life brain and behavior: A critical role for the immune system. *Frontiers in Behavioral Neuroscience, 3*, 14. http://dx.doi.org/10.3389/neuro.08.014.2009.

Bonthius, D. J., & West, J. R. (1988). Blood alcohol concentration and microencephaly: A dose-response study in the neonatal rat. *Teratology, 37*(3), 223–231. http://dx.doi.org/10.1002/tera.1420370307.

Bonthius, D. J., & West, J. R. (1990). Alcohol-induced neuronal loss in developing rats: Increased brain damage with binge exposure. *Alcoholism, Clinical and Experimental Research, 14*(1), 107–118.

Boschen, K. E., Criss, K. J., Palamarchouk, V., Roth, T. L., & Klintsova, A. Y. (2015). Effects of developmental alcohol exposure vs. intubation stress on BDNF and TrkB expression in the hippocampus and frontal cortex of neonatal rats. *International Journal of Developmental Neuroscience, 43*, 16–24. http://dx.doi.org/10.1016/j.ijdevneu.2015.03.008.

Boschen, K. E., Hamilton, G. F., Delorme, J. E., & Klintsova, A. Y. (2014). Activity and social behavior in a complex environment in rats neonatally exposed to alcohol. *Alcohol*, *48*(6), 533–541. http://dx.doi.org/10.1016/j.alcohol.2014.07.005.

Boschen, K. E., Ruggiero, M. J., & Klintsova, A. Y. (2016). Neonatal binge alcohol exposure increases microglial activation in the developing rat hippocampus. *Neuroscience*, *324*, 355–366.

Brocardo, P. S., Boehme, F., Patten, A., Cox, A., Gil-Mohapel, J., & Christie, B. R. (2012). Anxiety- and depression-like behaviors are accompanied by an increase in oxidative stress in a rat model of fetal alcohol spectrum disorders: Protective effects of voluntary physical exercise. *Neuropharmacology*, *62*(4), 1607–1618. http://dx.doi.org/10.1016/j.neuropharm.2011.10.006.

Brunig, I., Penschuck, S., Berninger, B., Benson, J., & Fritschy, J. M. (2001). BDNF reduces miniature inhibitory postsynaptic currents by rapid downregulation of GABA(A) receptor surface expression. *The European Journal of Neuroscience*, *13*(7), 1320–1328.

Calabrese, F., Rossetti, A. C., Racagni, G., Gass, P., Riva, M. A., & Molteni, R. (2014). Brain-derived neurotrophic factor: A bridge between inflammation and neuroplasticity. *Frontiers in Cellular Neuroscience*, *8*, 430. http://dx.doi.org/10.3389/fncel.2014.00430.

Caldwell, K. K., Sheema, S., Paz, R. D., Samudio-Ruiz, S. L., Laughlin, M. H., Spence, N. E., … Allan, A. M. (2008). Fetal alcohol spectrum disorder-associated depression: Evidence for reductions in the levels of brain-derived neurotrophic factor in a mouse model. *Pharmacology, Biochemistry, and Behavior*, *90*(4), 614–624. http://dx.doi.org/10.1016/j.pbb.2008.05.004.

Cantarella, G., Lempereur, L., Presta, M., Ribatti, D., Lombardo, G., Lazarovici, P., … Bernardini, R. (2002). Nerve growth factor-endothelial cell interaction leads to angiogenesis in vitro and in vivo. *FASEB Journal*, *16*(10), 1307–1309. http://dx.doi.org/10.1096/fj.01-1000fje.

Ceccanti, M., Mancinelli, R., Tirassa, P., Laviola, G., Rossi, S., Romeo, M., & Fiore, M. (2012). Early exposure to ethanol or red wine and long-lasting effects in aged mice. A study on nerve growth factor, brain-derived neurotrophic factor, hepatocyte growth factor, and vascular endothelial growth factor. *Neurobiology of Aging*, *33*(2), 359–367. http://dx.doi.org/10.1016/j.neurobiolaging.2010.03.005.

Cellerino, A., Arango-Gonzalez, B., Pinzon-Duarte, G., & Kohler, K. (2003). Brain-derived neurotrophic factor regulates expression of vasoactive intestinal polypeptide in retinal amacrine cells. *The Journal of Comparative Neurology*, *467*(1), 97–104. http://dx.doi.org/10.1002/cne.10908.

Centers for Disease Control. (2015). *Fetal alcohol spectrum disorders (FASDs)*. Retrieved from http://www.cdc.gov/ncbddd/fasd/index.html.

Chan, J. P., Cordeira, J., Calderon, G. A., Iyer, L. K., & Rios, M. (2008). Depletion of central BDNF in mice impedes terminal differentiation of new granule neurons in the adult hippocampus. *Molecular and Cellular Neurosciences*, *39*(3), 372–383. http://dx.doi.org/10.1016/j.mcn.2008.07.017.

Chapman, T. R., Barrientos, R. M., Ahrendsen, J. T., Hoover, J. M., Maier, S. F., & Patterson, S. L. (2012). Aging and infection reduce expression of specific brain-derived neurotrophic factor mRNAs in hippocampus. *Neurobiology of Aging*. *33*(4). http://dx.doi.org/10.1016/j.neurobiolaging.2011.07.015. 832.e1–14.

Chen, Z. Y., Ieraci, A., Teng, H., Dall, H., Meng, C. X., Herrera, D. G., … Lee, F. S. (2005). Sortilin controls intracellular sorting of brain-derived neurotrophic factor to the regulated secretory pathway. *The Journal of Neuroscience*, *25*(26), 6156–6166. http://dx.doi.org/10.1523/jneurosci.1017-05.2005.

Chen, Z. Y., Jing, D., Bath, K. G., Ieraci, A., Khan, T., Siao, C. J., … Lee, F. S. (2006). Genetic variant BDNF (Val66Met) polymorphism alters anxiety-related behavior. *Science*, *314*(5796), 140–143. http://dx.doi.org/10.1126/science.1129663.

Chen, M. L., Olson, H. C., Picciano, J. F., Starr, J. R., & Owens, J. (2012). Sleep problems in children with fetal alcohol spectrum disorders. *Journal of Clinical Sleep Medicine, 8*(4), 421–429. http://dx.doi.org/10.5664/jcsm.2038.

Choi, I. Y., Allan, A. M., & Cunningham, L. A. (2005). Moderate fetal alcohol exposure impairs the neurogenic response to an enriched environment in adult mice. *Alcoholism, Clinical and Experimental Research, 29*(11), 2053–2062.

Clements, K. M., Girard, T. A., Ellard, C. G., & Wainwright, P. E. (2005). Short-term memory impairment and reduced hippocampal c-Fos expression in an animal model of fetal alcohol syndrome. *Alcoholism, Clinical and Experimental Research, 29*(6), 1049–1059.

Climent, E., Pascual, M., Renau-Piqueras, J., & Guerri, C. (2002). Ethanol exposure enhances cell death in the developing cerebral cortex: Role of brain-derived neurotrophic factor and its signaling pathways. *Journal of Neuroscience Research, 68*(2), 213–225.

Cortese, B. M., Moore, G. J., Bailey, B. A., Jacobson, S. W., Delaney-Black, V., & Hannigan, J. H. (2006). Magnetic resonance and spectroscopic imaging in prenatal alcohol-exposed children: Preliminary findings in the caudate nucleus. *Neurotoxicology and Teratology, 28*(5), 597–606. http://dx.doi.org/10.1016/j.ntt.2006.08.002.

Cotman, C. W., & Berchtold, N. C. (2007). Physical activity and the maintenance of cognition: Learning from animal models. *Alzheimer's & Dementia: The Journal of the Alzheimer's Association, 3*(2 Suppl.), S30–S37. http://dx.doi.org/10.1016/j.jalz.2007.01.013.

Creeley, C. E., & Olney, J. W. (2013). Drug-induced apoptosis: Mechanism by which alcohol and many other drugs can disrupt brain development. *Brain Sciences, 3*(3), 1153–1181. http://dx.doi.org/10.3390/brainsci3031153.

Davis, M. I. (2008). Ethanol-BDNF interactions: Still more questions than answers. *Pharmacology & Therapeutics, 118*(1), 36–57. http://dx.doi.org/10.1016/j.pharmthera.2008.01.003.

Dobbing, J., & Sands, J. (1979). Comparative aspects of the brain growth spurt. *Early Human Development, 3*(1), 79–83.

Donovan, M. H., Yamaguchi, M., & Eisch, A. J. (2008). Dynamic expression of TrkB receptor protein on proliferating and maturing cells in the adult mouse dentate gyrus. *Hippocampus, 18*(5), 435–439. http://dx.doi.org/10.1002/hipo.20410.

Drew, P. D., Johnson, J. W., Douglas, J. C., Phelan, K. D., & Kane, C. J. M. (2015). Pioglitazone blocks ethanol induction of microglial activation and immune responses in the hippocampus, cerebellum, and cerebral cortex in a mouse model of fetal alcohol spectrum disorders. *Alcoholism, Clinical and Experimental Research, 39*, 445–454. http://dx.doi.org/10.1111/acer.12639.

Dunty, W. C., Jr., Chen, S. Y., Zucker, R. M., Dehart, D. B., & Sulik, K. K. (2001). Selective vulnerability of embryonic cell populations to ethanol-induced apoptosis: Implications for alcohol-related birth defects and neurodevelopmental disorder. *Alcoholism, Clinical and Experimental Research, 25*(10), 1523–1535.

Ellis, L. M., & Hicklin, D. J. (2008). VEGF-targeted therapy: Mechanisms of anti-tumour activity. *Nature Reviews. Cancer, 8*(8), 579–591. http://dx.doi.org/10.1038/nrc2403.

Farber, N. B., Creeley, C. E., & Olney, J. W. (2010). Alcohol-induced neuroapoptosis in the fetal macaque brain. *Neurobiology of Disease, 40*(1), 200–206. http://dx.doi.org/10.1016/j.nbd.2010.05.025.

Fattori, V., Abe, S., Kobayashi, K., Costa, L. G., & Tsuji, R. (2008). Effects of postnatal ethanol exposure on neurotrophic factors and signal transduction pathways in rat brain. *Journal of Applied Toxicology, 28*(3), 370–376. http://dx.doi.org/10.1002/jat.1288.

Feng, M. J., Yan, S. E., & Yan, Q. S. (2005). Effects of prenatal alcohol exposure on brain-derived neurotrophic factor and its receptor tyrosine kinase B in offspring. *Brain Research, 1042*(2), 125–132. http://dx.doi.org/10.1016/j.brainres.2005.02.017.

Fernandez-Jaen, A., Fernandez-Mayoralas, D. M., Quinones Tapia, D., Calleja-Perez, B., Garcia-Segura, J. M., Arribas, S. L., & Munoz Jareno, N. (2011). Cortical thickness in fetal alcohol syndrome and attention deficit disorder. *Pediatric Neurology*, *45*(6), 387–391. http://dx.doi.org/10.1016/j.pediatrneurol.2011.09.004.

Ferrara, N., Gerber, H. P., & LeCouter, J. (2003). The biology of VEGF and its receptors. *Nature Medicine*, *9*(6), 669–676. http://dx.doi.org/10.1038/nm0603-669.

Figurov, A., Pozzo-Miller, L. D., Olafsson, P., Wang, T., & Lu, B. (1996). Regulation of synaptic responses to high-frequency stimulation and LTP by neurotrophins in the hippocampus. *Nature*, *381*(6584), 706–709. http://dx.doi.org/10.1038/381706a0.

Franklin, L., Deitz, J., Jirikowic, T., & Astley, S. (2008). Children with fetal alcohol spectrum disorders: Problem behaviors and sensory processing. *The American Journal of Occupational Therapy*, *62*(3), 265–273.

Fryer, S. L., Mattson, S. N., Jernigan, T. L., Archibald, S. L., Jones, K. L., & Riley, E. P. (2012). Caudate volume predicts neurocognitive performance in youth with heavy prenatal alcohol exposure. *Alcoholism, Clinical and Experimental Research*, *36*(11), 1932–1941. http://dx.doi.org/10.1111/j.1530-0277.2012.01811.x.

Gardiner, J., Barton, D., Overall, R., & Marc, J. (2009). Neurotrophic support and oxidative stress: Converging effects in the normal and diseased nervous system. *The Neuroscientist*, *15*(1), 47–61. http://dx.doi.org/10.1177/1073858408325269.

Gatt, J. M., Nemeroff, C. B., Dobson-Stone, C., Paul, R. H., Bryant, R. A., Schofield, P. R., … Williams, L. M. (2009). Interactions between BDNF Val66Met polymorphism and early life stress predict brain and arousal pathways to syndromal depression and anxiety. *Molecular Psychiatry*, *14*(7), 681–695. http://dx.doi.org/10.1038/mp.2008.143.

Gautam, P., Lebel, C., Narr, K. L., Mattson, S. N., May, P. A., Adnams, C. M., … Sowell, E. R. (2015). Volume changes and brain-behavior relationships in white matter and subcortical gray matter in children with prenatal alcohol exposure. *Human Brain Mapping*, *36*(6), 2318–2329. http://dx.doi.org/10.1002/hbm.22772.

Gauthier, T. W. (2015). Prenatal alcohol exposure and the developing immune system. *Alcohol Research: Current Reviews*, *37*(2), 279–285.

Gericke, C. A., Schulte-Herbruggen, O., Arendt, T., & Hellweg, R. (2006). Chronic alcohol intoxication in rats leads to a strong but transient increase in NGF levels in distinct brain regions. *Journal of Neural Transmission (Vienna, Austria: 1996)*, *113*(7), 813–820. http://dx.doi.org/10.1007/s00702-005-0361-x.

Gil-Mohapel, J., Boehme, F., Patten, A., Cox, A., Kainer, L., Giles, E., … Christie, B. R. (2011). Altered adult hippocampal neuronal maturation in a rat model of fetal alcohol syndrome. *Brain Research*, *1384*, 29–41. http://dx.doi.org/10.1016/j.brainres.2011.01.116.

Gottesfeld, Z., Morgan, B., & Perez-Polo, J. R. (1990). Prenatal alcohol exposure alters the development of sympathetic synaptic components and of nerve growth factor receptor expression selectivity in lymphoid organs. *Journal of Neuroscience Research*, *26*(3), 308–316. http://dx.doi.org/10.1002/jnr.490260307.

Gottschalk, W., Pozzo-Miller, L. D., Figurov, A., & Lu, B. (1998). Presynaptic modulation of synaptic transmission and plasticity by brain-derived neurotrophic factor in the developing hippocampus. *The Journal of Neuroscience*, *18*(17), 6830–6839.

Griffin, E. W., Bechara, R. G., Birch, A. M., & Kelly, A. M. (2009). Exercise enhances hippocampal-dependent learning in the rat: Evidence for a BDNF-related mechanism. *Hippocampus*, *19*(10), 973–980. http://dx.doi.org/10.1002/hipo.20631.

Guan, Z., & Fang, J. (2006). Peripheral immune activation by lipopolysaccharide decreases neurotrophins in the cortex and hippocampus in rats. *Brain, Behavior, and Immunity*, *20*(1), 64–71. http://dx.doi.org/10.1016/j.bbi.2005.04.005.

Guo, Z. H., & Mattson, M. P. (2000). Neurotrophic factors protect cortical synaptic terminals against amyloid and oxidative stress-induced impairment of glucose transport, glutamate transport and mitochondrial function. *Cerebral Cortex*, *10*(1), 50–57.

Hall, J., Thomas, K. L., & Everitt, B. J. (2000). Rapid and selective induction of BDNF expression in the hippocampus during contextual learning. *Nature Neuroscience*, *3*(6), 533–535. http://dx.doi.org/10.1038/75698.

Hamilton, G. F., Boschen, K. E., Goodlett, C. R., Greenough, W. T., & Klintsova, A. Y. (2012). Housing in environmental complexity following wheel running augments survival of newly generated hippocampal neurons in a rat model of binge alcohol exposure during the third trimester equivalent. *Alcoholism, Clinical and Experimental Research*, *36*(7), 1196–1204. http://dx.doi.org/10.1111/j.1530-0277.2011.01726.x.

Hamilton, G. F., Criss, K. J., & Klintsova, A. Y. (2015). Voluntary exercise partially reverses neonatal alcohol-induced deficits in mPFC layer II/III dendritic morphology of male adolescent rats. *Synapse*, *69*(8), 405–415. http://dx.doi.org/10.1002/syn.21827.

Hamilton, G. F., Jablonski, S. A., Schiffino, F. L., St Cyr, S. A., Stanton, M. E., & Klintsova, A. Y. (2014). Exercise and environment as an intervention for neonatal alcohol effects on hippocampal adult neurogenesis and learning. *Neuroscience*, *265*, 274–290. http://dx.doi.org/10.1016/j.neuroscience.2014.01.061.

Hamilton, G. F., Murawski, N. J., St Cyr, S. A., Jablonski, S. A., Schiffino, F. L., Stanton, M. E., & Klintsova, A. Y. (2011). Neonatal alcohol exposure disrupts hippocampal neurogenesis and contextual fear conditioning in adult rats. *Brain Research*, *1412*, 88–101. http://dx.doi.org/10.1016/j.brainres.2011.07.027.

Hamilton, G. F., Whitcher, L. T., & Klintsova, A. Y. (2010). Postnatal binge-like alcohol exposure decreases dendritic complexity while increasing the density of mature spines in mPFC Layer II/III pyramidal neurons. *Synapse*, *64*(2), 127–135. http://dx.doi.org/10.1002/syn.20711.

Han, B. H., & Holtzman, D. M. (2000). BDNF protects the neonatal brain from hypoxic-ischemic injury in vivo via the ERK pathway. *The Journal of Neuroscience*, *20*(15), 5775–5781.

Hannigan, J. H., Berman, R. F., & Zajac, C. S. (1993). Environmental enrichment and the behavioral effects of prenatal exposure to alcohol in rats. *Neurotoxicology and Teratology*, *15*(4), 261–266.

Heaton, M. B., Madorsky, I., Paiva, M., & Mayer, J. (2002). Influence of ethanol on neonatal cerebellum of BDNF gene-deleted animals: Analyses of effects on Purkinje cells, apoptosis-related proteins, and endogenous antioxidants. *Journal of Neurobiology*, *51*(2), 160–176.

Heaton, M. B., Mitchell, J. J., Paiva, M., & Walker, D. W. (2000). Ethanol-induced alterations in the expression of neurotrophic factors in the developing rat central nervous system. *Brain Research. Developmental Brain Research*, *121*(1), 97–107.

Heaton, M. B., Moore, D. B., Paiva, M., Madorsky, I., Mayer, J., & Shaw, G. (2003). The role of neurotrophic factors, apoptosis-related proteins, and endogenous antioxidants in the differential temporal vulnerability of neonatal cerebellum to ethanol. *Alcoholism, Clinical and Experimental Research*, *27*(4), 657–669. http://dx.doi.org/10.1097/01.alc.0000060527.55252.71.

Heberlein, A., Bleich, S., Bayerlein, K., Frieling, H., Groschl, M., Kornhuber, J., & Hillemacher, T. (2008). NGF plasma levels increase due to alcohol intoxication and decrease during withdrawal. *Psychoneuroendocrinology*, *33*(7), 999–1003. http://dx.doi.org/10.1016/j.psyneuen.2008.05.011.

Heldt, S. A., Stanek, L., Chhatwal, J. P., & Ressler, K. J. (2007). Hippocampus-specific deletion of BDNF in adult mice impairs spatial memory and extinction of aversive memories. *Molecular Psychiatry*, *12*(7), 656–670. http://dx.doi.org/10.1038/sj.mp.4001957.

Helfer, J. L., Goodlett, C. R., Greenough, W. T., & Klintsova, A. Y. (2009). The effects of exercise on adolescent hippocampal neurogenesis in a rat model of binge alcohol exposure during the brain growth spurt. *Brain Research*, *1294*, 1–11. http://dx.doi.org/10.1016/j.brainres.2009.07.090.

Hellemans, K. G., Verma, P., Yoon, E., Yu, W., & Weinberg, J. (2008). Prenatal alcohol exposure increases vulnerability to stress and anxiety-like disorders in adulthood. *Annals of the New York Academy of Sciences*, *1144*, 154–175. http://dx.doi.org/10.1196/annals.1418.016.

Horch, H. W., Krüttgen, A., Portbury, S. D., & Katz, L. C. (1999). Destabilization of cortical dendrites and spines by BDNF. *Neuron*, *23*(2), 353–364. http://dx.doi.org/10.1016/S0896-6273(00)80785-0.

Hunt, P. S., Jacobson, S. E., & Torok, E. J. (2009). Deficits in trace fear conditioning in a rat model of fetal alcohol exposure: Dose-response and timing effects. *Alcohol*, *43*(6), 465–474. http://dx.doi.org/10.1016/j.alcohol.2009.08.004.

Idrus, N. M., & Thomas, J. D. (2011). Fetal alcohol spectrum disorders: Experimental treatments and strategies for intervention. *Alcohol Research & Health*, *34*(1), 76–85.

Ieraci, A., & Herrera, D. G. (2007). Single alcohol exposure in early life damages hippocampal stem/progenitor cells and reduces adult neurogenesis. *Neurobiology of Disease*, *26*(3), 597–605. http://dx.doi.org/10.1016/j.nbd.2007.02.011.

Ikonomidou, C., Bittigau, P., Ishimaru, M. J., Wozniak, D. F., Koch, C., Genz, K., … Olney, J. W. (2000). Ethanol-induced apoptotic neurodegeneration and fetal alcohol syndrome. *Science*, *287*(5455), 1056–1060.

Ikonomidou, C., Bosch, F., Miksa, M., Bittigau, P., Vockler, J., Dikranian, K., … Olney, J. W. (1999). Blockade of NMDA receptors and apoptotic neurodegeneration in the developing brain. *Science*, *283*(5398), 70–74.

Ikonomidou, C., Mosinger, J. L., Salles, K. S., Labruyere, J., & Olney, J. W. (1989). Sensitivity of the developing rat brain to hypobaric/ischemic damage parallels sensitivity to N-methyl-aspartate neurotoxicity. *The Journal of Neuroscience*, *9*(8), 2809–2818.

Incerti, M., Vink, J., Roberson, R., Benassou, I., Abebe, D., & Spong, C. Y. (2010a). Prevention of the alcohol-induced changes in brain-derived neurotrophic factor expression using neuroprotective peptides in a model of fetal alcohol syndrome. *American Journal of Obstetrics and Gynecology*, *202*(5). http://dx.doi.org/10.1016/j.ajog.2010.03.050. 457.e451-454.

Incerti, M., Vink, J., Roberson, R., Wood, L., Abebe, D., & Spong, C. Y. (2010b). Reversal of alcohol-induced learning deficits in the young adult in a model of fetal alcohol syndrome. *Obstetrics and Gynecology*, *115*(2 Pt. 1), 350–356. http://dx.doi.org/10.1097/AOG.0b013e3181cb59da.

Irner, T. B., Teasdale, T. W., & Olofsson, M. (2012). Cognitive and social development in preschool children born to women using substances. *Journal of Addictive Diseases*, *31*(1), 29–44. http://dx.doi.org/10.1080/10550887.2011.642766.

Jegou, S., El Ghazi, F., de Lendeu, P. K., Marret, S., Laudenbach, V., Uguen, A., … Gonzalez, B. J. (2012). Prenatal alcohol exposure affects vasculature development in the neonatal brain. *Annals of Neurology*, *72*(6), 952–960. http://dx.doi.org/10.1002/ana.23699.

Johnson, V. P., Swayze, V. W., II, Sato, Y., & Andreasen, N. C. (1996). Fetal alcohol syndrome: Craniofacial and central nervous system manifestations. *American Journal of Medical Genetics*, *61*(4), 329–339. http://dx.doi.org/10.1002/(sici)1096-8628(19960202)61:4<329::aid-ajmg6>3.0.co;2-p.

Johnston, M. V., Ishida, A., Ishida, W. N., Matsushita, H. B., Nishimura, A., & Tsuji, M. (2009). Plasticity and injury in the developing brain. *Brain Dev*, *31*(1), 1–10. http://dx.doi.org/10.1016/j.braindev.2008.03.014.

Jones, P. J., Leichter, J., & Lee, M. (1981). Placental blood flow in rats fed alcohol before and during gestation. *Life Sciences*, *29*(11), 1153–1159.

Kane, C. J., Phelan, K. D., Han, L., Smith, R. R., Xie, J., Douglas, J. C., & Drew, P. D. (2011). Protection of neurons and microglia against ethanol in a mouse model of fetal alcohol spectrum disorders by peroxisome proliferator-activated receptor-gamma agonists. *Brain, Behavior, and Immunity*, *25*(Suppl. 1), S137–S145. http://dx.doi.org/10.1016/j.bbi.2011.02.016.

Karege, F., Schwald, M., & Cisse, M. (2002). Postnatal developmental profile of brain-derived neurotrophic factor in rat brain and platelets. *Neuroscience Letters*, *328*(3), 261–264.

Katoh-Semba, R., Semba, R., Takeuchi, I. K., & Kato, K. (1998). Age-related changes in levels of brain-derived neurotrophic factor in selected brain regions of rats, normal mice and senescence-accelerated mice: A comparison to those of nerve growth factor and neurotrophin-3. *Neuroscience Research*, *31*(3), 227–234.

Kaufmann, W. E., & Moser, H. W. (2000). Dendritic anomalies in disorders associated with mental retardation. *Cerebral Cortex*, *10*(10), 981–991.

Kelly, S. J., & Dillingham, R. R. (1994). Sexually dimorphic effects of perinatal alcohol exposure on social interactions and amygdala DNA and DOPAC concentrations. *Neurotoxicology and Teratology*, *16*(4), 377–384.

Kelly, S. J., Leggett, D. C., & Cronise, K. (2009). Sexually dimorphic effects of alcohol exposure during development on the processing of social cues. *Alcohol and Alcoholism*, *44*(6), 555–560. http://dx.doi.org/10.1093/alcalc/agp061.

Kermani, P., & Hempstead, B. (2007). Brain-derived neurotrophic factor: a newly described mediator of angiogenesis. *Trends in Cardiovascular Medicine*, *17*(4), 140–143. http://dx.doi.org/10.1016/j.tcm.2007.03.002.

Kerr, J. F., Wyllie, A. H., & Currie, A. R. (1972). Apoptosis: A basic biological phenomenon with wide-ranging implications in tissue kinetics. *British Journal of Cancer*, *26*(4), 239–257.

Kirschner, P. B., Jenkins, B. G., Schulz, J. B., Finkelstein, S. P., Matthews, R. T., Rosen, B. R., & Beal, M. F. (1996). NGF, BDNF and NT-5, but not NT-3 protect against MPP+ toxicity and oxidative stress in neonatal animals. *Brain Research*, *713*(1–2), 178–185.

Klintsova, A. Y., Helfer, J. L., Calizo, L. H., Dong, W. K., Goodlett, C. R., & Greenough, W. T. (2007). Persistent impairment of hippocampal neurogenesis in young adult rats following early postnatal alcohol exposure. *Alcoholism, Clinical and Experimental Research*, *31*(12), 2073–2082. http://dx.doi.org/10.1111/j.1530-0277.2007.00528.x.

Klintsova, A. Y., Scamra, C., Hoffman, M., Napper, R. M., Goodlett, C. R., & Greenough, W. T. (2002). Therapeutic effects of complex motor training on motor performance deficits induced by neonatal binge-like alcohol exposure in rats: II. A quantitative stereological study of synaptic plasticity in female rat cerebellum. *Brain Research*, *937*(1), 83–93. http://dx.doi.org/10.1016/S0006-8993(02)02492-7.

Kodituwakku, P. W., Handmaker, N. S., Cutler, S. K., Weathersby, E. K., & Handmaker, S. D. (1995). Specific impairments in self-regulation in children exposed to alcohol prenatally. *Alcoholism, Clinical and Experimental Research*, *19*(6), 1558–1564.

Komuro, H., & Rakic, P. (1993). Modulation of neuronal migration by NMDA receptors. *Science*, *260*(5104), 95–97.

Lapchak, P. A., Araujo, D. M., & Hefti, F. (1993). Systemic interleukin-1 beta decreases brain-derived neurotrophic factor messenger RNA expression in the rat hippocampal formation. *Neuroscience*, *53*(2), 297–301.

Lawrence, R. C., Otero, N. K., & Kelly, S. J. (2012). Selective effects of perinatal ethanol exposure in medial prefrontal cortex and nucleus accumbens. *Neurotoxicology and Teratology*, *34*(1), 128–135. http://dx.doi.org/10.1016/j.ntt.2011.08.002.

Lebedeva, J., Zakharov, A., Ogievetsky, E., Minlebaeva, A., Kurbanov, R., Gerasimova, E., ... Khazipov, R. (2015). Inhibition of cortical activity and apoptosis caused by ethanol in neonatal rats in vivo. *Cerebral Cortex*, http://dx.doi.org/10.1093/cercor/bhv293.

Lee, J., Seroogy, K. B., & Mattson, M. P. (2002). Dietary restriction enhances neurotrophin expression and neurogenesis in the hippocampus of adult mice. *Journal of Neurochemistry*, *80*(3), 539–547.

Leigland, L. A., Budde, M. D., Cornea, A., & Kroenke, C. D. (2013). Diffusion MRI of the developing cerebral cortical gray matter can be used to detect abnormalities in tissue microstructure associated with fetal ethanol exposure. *NeuroImage*, *83*, 1081–1087. http://dx.doi.org/10.1016/j.neuroimage.2013.07.068.

Leigland, L. A., Ford, M. M., Lerch, J. P., & Kroenke, C. D. (2013). The influence of fetal ethanol exposure on subsequent development of the cerebral cortex as revealed by magnetic resonance imaging. *Alcoholism, Clinical and Experimental Research*, *37*(6), 924–932. http://dx.doi.org/10.1111/acer.12051.

Liang, F. Q., Allen, G., & Earnest, D. (2000). Role of brain-derived neurotrophic factor in the circadian regulation of the suprachiasmatic pacemaker by light. *The Journal of Neuroscience*, *20*(8), 2978–2987.

Licht, T., Eavri, R., Goshen, I., Shlomai, Y., Mizrahi, A., & Keshet, E. (2010). VEGF is required for dendritogenesis of newly born olfactory bulb interneurons. *Development*, *137*(2), 261–271. http://dx.doi.org/10.1242/dev.039636.

Licht, T., Goshen, I., Avital, A., Kreisel, T., Zubedat, S., Eavri, R., ... Keshet, E. (2011). Reversible modulations of neuronal plasticity by VEGF. *Proceedings of the National Academy of Sciences of the United States of America*, *108*(12), 5081–5086. http://dx.doi.org/10.1073/pnas.1007640108.

Light, K. E., Ge, Y., & Belcher, S. M. (2001). Early postnatal ethanol exposure selectively decreases BDNF and truncated TrkB-T2 receptor mRNA expression in the rat cerebellum. *Brain Research Molecular Brain Research*, *93*(1), 46–55.

Lindholm, D., Castren, E., Berzaghi, M., Blochl, A., & Thoenen, H. (1994). Activity-dependent and hormonal regulation of neurotrophin mRNA levels in the brain—Implications for neuronal plasticity. *Journal of Neurobiology*, *25*(11), 1362–1372. http://dx.doi.org/10.1002/neu.480251105.

Livy, D. J., Miller, E. K., Maier, S. E., & West, J. R. (2003). Fetal alcohol exposure and temporal vulnerability: Effects of binge-like alcohol exposure on the developing rat hippocampus. *Neurotoxicology and Teratology*, *25*(4), 447–458.

Lu, B., & Chow, A. (1999). Neurotrophins and hippocampal synaptic transmission and plasticity. *Journal of Neuroscience Research*, *58*(1), 76–87.

Luo, J. (2010). Lithium-mediated protection against ethanol neurotoxicity. *Frontiers in Neuroscience*, *4*, 41. http://dx.doi.org/10.3389/fnins.2010.00041.

Lupton, C., Burd, L., & Harwood, R. (2004). Cost of fetal alcohol spectrum disorders. *American Journal of Medical Genetics Part C, Seminars in Medical Genetics*, *127c*(1), 42–50. http://dx.doi.org/10.1002/ajmg.c.30015.

Maier, S., Chen, W., & West, J. (1996). The effects of timing and duration of alcohol exposure on development of the fetal brain. In E. L. Abel (Ed.), *Fetal alcohol syndrome: From mechanism to prevention*. pp. 27–50. Boca Raton: CRC Press.

Maier, S. E., Miller, J. A., Blackwell, J. M., & West, J. R. (1999). Fetal alcohol exposure and temporal vulnerability: Regional differences in cell loss as a function of the timing of binge-like alcohol exposure during brain development. *Alcoholism, Clinical and Experimental Research*, *23*(4), 726–734.

Maisonpierre, P. C., Belluscio, L., Friedman, B., Alderson, R. F., Wiegand, S. J., Furth, M. E., ... Yancopoulos, G. D. (1990). NT-3, BDNF, and NGF in the developing rat nervous system: Parallel as well as reciprocal patterns of expression. *Neuron*, *5*(4), 501–509.

Martinowich, K., Manji, H., & Lu, B. (2007). New insights into BDNF function in depression and anxiety. *Nature Neuroscience, 10*(9), 1089–1093. http://dx.doi.org/10.1038/nn1971.

Mattson, S. N., Goodman, A. M., Caine, C., Delis, D. C., & Riley, E. P. (1999). Executive functioning in children with heavy prenatal alcohol exposure. *Alcoholism, Clinical and Experimental Research, 23*(11), 1808–1815.

Mattson, S. N., Riley, E. P., Delis, D. C., Stern, C., & Jones, K. L. (1996a). Verbal learning and memory in children with fetal alcohol syndrome. *Alcoholism, Clinical and Experimental Research, 20*(5), 810–816.

Mattson, S. N., Riley, E. P., Gramling, L., Delis, D. C., & Jones, K. L. (1997). Heavy prenatal alcohol exposure with or without physical features of fetal alcohol syndrome leads to IQ deficits. *The Journal of Pediatrics, 131*(5), 718–721.

Mattson, S. N., Riley, E. P., Sowell, E. R., Jernigan, T. L., Sobel, D. F., & Jones, K. L. (1996b). A decrease in the size of the basal ganglia in children with fetal alcohol syndrome. *Alcoholism, Clinical and Experimental Research, 20*(6), 1088–1093.

May, P. A., Gossage, J. P., Kalberg, W. O., Robinson, L. K., Buckley, D., Manning, M., & Hoyme, H. E. (2009). Prevalence and epidemiologic characteristics of FASD from various research methods with an emphasis on recent in-school studies. *Developmental Disabilities Research Reviews, 15*(3), 176–192. http://dx.doi.org/10.1002/ddrr.68.

Mayhan, W. G. (1992). Responses of cerebral arterioles during chronic ethanol exposure. *The American Journal of Physiology, 262*(3 Pt. 2), H787–H791.

McAllister, A. K., Katz, L. C., & Lo, D. C. (1997). Opposing roles for endogenous BDNF and NT-3 in regulating cortical dendritic growth. *Neuron, 18*(5), 767–778.

Middleton, F. A., Varlinskaya, E. I., & Mooney, S. M. (2012). Molecular substrates of social avoidance seen following prenatal ethanol exposure and its reversal by social enrichment. *Developmental Neuroscience, 34*(2–3), 115–128. http://dx.doi.org/10.1159/000337858.

Miki, T., Kuma, H., Yokoyama, T., Sumitani, K., Matsumoto, Y., Kusaka, T., … Takeuchi, Y. (2008). Early postnatal ethanol exposure induces fluctuation in the expression of BDNF mRNA in the developing rat hippocampus. *Acta Neurobiologiae Experimentalis (Wars), 68*(4), 484–493.

Mitchell, J. J., Paiva, M., Walker, D. W., & Heaton, M. B. (1999). BDNF and NGF afford in vitro neuroprotection against ethanol combined with acute ischemia and chronic hypoglycemia. *Developmental Neuroscience, 21*(1), 68–75. http://dx.doi.org/10.1159/000017368.

Mooney, S. M., & Varlinskaya, E. I. (2011). Acute prenatal exposure to ethanol and social behavior: Effects of age, sex, and timing of exposure. *Behavioural Brain Research, 216*(1), 358–364. http://dx.doi.org/10.1016/j.bbr.2010.08.014.

Moore, D. B., Madorsky, I., Paiva, M., & Barrow Heaton, M. (2004). Ethanol exposure alters neurotrophin receptor expression in the rat central nervous system: Effects of neonatal exposure. *Journal of Neurobiology, 60*(1), 114–126. http://dx.doi.org/10.1002/neu.20010.

Murawski, N. J., Klintsova, A. Y., & Stanton, M. E. (2012). Neonatal alcohol exposure and the hippocampus in developing male rats: Effects on behaviorally induced CA1 c-Fos expression, CA1 pyramidal cell number, and contextual fear conditioning. *Neuroscience, 206*, 89–99. http://dx.doi.org/10.1016/j.neuroscience.2012.01.006.

Napper, R. M., & West, J. R. (1995). Permanent neuronal cell loss in the inferior olive of adult rats exposed to alcohol during the brain growth spurt: A stereological investigation. *Alcoholism, Clinical and Experimental Research, 19*(5), 1321–1326.

National Institute of Alcohol Abuse and Alcoholism. (2015). *Fetal alcohol exposure*. Retrieved from: http://pubs.niaaa.nih.gov/publications/FASDFactsheet/FASDfact.htm.

Nava-Ocampo, A. A., Velazquez-Armenta, Y., Brien, J. F., & Koren, G. (2004). Elimination kinetics of ethanol in pregnant women. *Reproductive Toxicology, 18*(4), 613–617. http://dx.doi.org/10.1016/j.reprotox.2004.02.012.

Nijhawan, D., Honarpour, N., & Wang, X. (2000). Apoptosis in neural development and disease. *Annual Review of Neuroscience, 23*, 73–87. http://dx.doi.org/10.1146/annurev.neuro.23.1.73.

O'Connor, M. J., & Kasari, C. (2000). Prenatal alcohol exposure and depressive features in children. *Alcoholism, Clinical and Experimental Research, 24*(7), 1084–1092.

O'Connor, M. J., & Paley, B. (2006). The relationship of prenatal alcohol exposure and the postnatal environment to child depressive symptoms. *Journal of Pediatric Psychology, 31*(1), 50–64. http://dx.doi.org/10.1093/jpepsy/jsj021.

O'Leary-Moore, S. K., McMechan, A. P., Mathison, S. N., Berman, R. F., & Hannigan, J. H. (2006). Reversal learning after prenatal or early postnatal alcohol exposure in juvenile and adult rats. *Alcohol, 38*(2), 99–110. http://dx.doi.org/10.1016/j.alcohol.2006.05.005.

O'Leary-Moore, S. K., Parnell, S. E., Godin, E. A., Dehart, D. B., Ament, J. J., Khan, A. A., ... Sulik, K. K. (2010). Magnetic resonance microscopy-based analyses of the brains of normal and ethanol-exposed fetal mice. *Birth Defects Research. Part A, Clinical and Molecular Teratology, 88*(11), 953–964. http://dx.doi.org/10.1002/bdra.20719.

Olney, J. W., Labruyere, J., Wang, G., Wozniak, D. F., Price, M. T., & Sesma, M. A. (1991). NMDA antagonist neurotoxicity: Mechanism and prevention. *Science, 254*(5037), 1515–1518.

Olney, J. W., Tenkova, T., Dikranian, K., Qin, Y. Q., Labruyere, J., & Ikonomidou, C. (2002). Ethanol-induced apoptotic neurodegeneration in the developing C57BL/6 mouse brain. *Brain Research. Developmental Brain Research, 133*(2), 115–126.

Olney, J. W., Wozniak, D. F., Jevtovic-Todorovic, V., & Ikonomidou, C. (2001). Glutamate signaling and the fetal alcohol syndrome. *Mental Retardation and Developmental Disabilities Research Reviews*, 7(4), 267–275. http://dx.doi.org/10.1002/mrdd.1037.

Onyango, I. G., Tuttle, J. B., & Bennett, J. P., Jr. (2005). Brain-derived growth factor and glial cell line-derived growth factor use distinct intracellular signaling pathways to protect PD cybrids from H_2O_2-induced neuronal death. *Neurobiology of Disease, 20*(1), 141–154. http://dx.doi.org/10.1016/j.nbd.2005.02.009.

Parnell, S. E., Holloway, H. E., Baker, L. K., Styner, M. A., & Sulik, K. K. (2014). Dysmorphogenic effects of first trimester-equivalent ethanol exposure in mice: A magnetic resonance microscopy-based study. *Alcoholism, Clinical and Experimental Research, 38*(7), 2008–2014. http://dx.doi.org/10.1111/acer.12464.

Parnell, S. E., Holloway, H. T., O'Leary-Moore, S. K., Dehart, D. B., Paniaqua, B., Oguz, I., ... Sulik, K. K. (2013). Magnetic resonance microscopy-based analyses of the neuroanatomical effects of gestational day 9 ethanol exposure in mice. *Neurotoxicology and Teratology, 39*, 77–83. http://dx.doi.org/10.1016/j.ntt.2013.07.009.

Parnell, S. E., Ramadoss, J., Delp, M. D., Ramsey, M. W., Chen, W. J., West, J. R., & Cudd, T. A. (2007). Chronic ethanol increases fetal cerebral blood flow specific to the ethanol-sensitive cerebellum under normoxaemic, hypercapnic and acidaemic conditions: Ovine model. *Experimental Physiology, 92*(5), 933–943. http://dx.doi.org/10.1113/expphysiol.2007.038091.

Pei, J., Job, J., Kully-Martens, K., & Rasmussen, C. (2011). Executive function and memory in children with fetal alcohol spectrum disorder. *Child Neuropsychology, 17*(3), 290–309. http://dx.doi.org/10.1080/09297049.2010.544650.

Pellegri, G., Magistretti, P. J., & Martin, J. L. (1998). VIP and PACAP potentiate the action of glutamate on BDNF expression in mouse cortical neurones. *The European Journal of Neuroscience, 10*(1), 272–280.

Petkov, V. V., Stoianovski, D., Petkov, V. D., & Vyglenova, I. (1992). Lipid peroxidation changes in the brain in fetal alcohol syndrome. *Biulleten' Eksperimental'noĭ Biologii i Meditsiny, 113*(5), 500–502.

Pezet, S., & Malcangio, M. (2004). Brain-derived neurotrophic factor as a drug target for CNS disorders. *Expert Opinion on Therapeutic Targets*, *8*(5), 391–399. http://dx.doi.org/10.1517/14728222.8.5.391.

Popova, N. K., Morozova, M. V., & Naumenko, V. S. (2011). Ameliorative effect of BDNF on prenatal ethanol and stress exposure-induced behavioral disorders. *Neuroscience Letters*, *505*(2), 82–86. http://dx.doi.org/10.1016/j.neulet.2011.09.066.

Pozzo-Miller, L. D., Gottschalk, W., Zhang, L., McDermott, K., Du, J., Gopalakrishnan, R., ... Lu, B. (1999). Impairments in high-frequency transmission, synaptic vesicle docking, and synaptic protein distribution in the hippocampus of BDNF knockout mice. *The Journal of Neuroscience*, *19*(12), 4972–4983.

Puglia, M. P., & Valenzuela, C. F. (2010a). Ethanol acutely inhibits ionotropic glutamate receptor-mediated responses and long-term potentiation in the developing CA1 hippocampus. *Alcoholism, Clinical and Experimental Research*, *34*(4), 594–606. http://dx.doi.org/10.1111/j.1530-0277.2009.01128.x.

Puglia, M. P., & Valenzuela, C. F. (2010b). Repeated third trimester-equivalent ethanol exposure inhibits long-term potentiation in the hippocampal CA1 region of neonatal rats. *Alcohol*, *44*(3), 283–290. http://dx.doi.org/10.1016/j.alcohol.2010.03.001.

Rabinowicz, T., de Courten-Myers, G. M., Petetot, J. M., Xi, G., & de los Reyes, E. (1996). Human cortex development: Estimates of neuronal numbers indicate major loss late during gestation. *Journal of Neuropathology and Experimental Neurology*, *55*(3), 320–328.

Rakic, S., & Zecevic, N. (2000). Programmed cell death in the developing human telencephalon. *The European Journal of Neuroscience*, *12*(8), 2721–2734.

Ramachandran, V., Watts, L. T., Maffi, S. K., Chen, J., Schenker, S., & Henderson, G. (2003). Ethanol-induced oxidative stress precedes mitochondrially mediated apoptotic death of cultured fetal cortical neurons. *Journal of Neuroscience Research*, *74*(4), 577–588. http://dx.doi.org/10.1002/jnr.10767.

Rasmussen, C., Soleimani, M., & Pei, J. (2011). Executive functioning and working memory deficits on the CANTAB among children with prenatal alcohol exposure. *Journal of Population Therapeutics and Clinical Pharmacology*, *18*(1), e44–e53.

Redila, V. A., Olson, A. K., Swann, S. E., Mohades, G., Webber, A. J., Weinberg, J., & Christie, B. R. (2006). Hippocampal cell proliferation is reduced following prenatal ethanol exposure but can be rescued with voluntary exercise. *Hippocampus*, *16*(3), 305–311. http://dx.doi.org/10.1002/hipo.20164.

Riikonen, R., Salonen, I., Partanen, K., & Verho, S. (1999). Brain perfusion SPECT and MRI in foetal alcohol syndrome. *Developmental Medicine and Child Neurology*, *41*(10), 652–659.

Robertson, F. C., Narr, K. L., Molteno, C. D., Jacobson, J. L., Jacobson, S. W., & Meintjes, E. M. (2016). Prenatal alcohol exposure is associated with regionally thinner cortex during the preadolescent period. *Cerebral Cortex*, *26*, 3083–3095. http://dx.doi.org/10.1093/cercor/bhv131.

Roebuck, T. M., Mattson, S. N., & Riley, E. P. (1999). Behavioral and psychosocial profiles of alcohol-exposed children. *Alcoholism, Clinical and Experimental Research*, *23*(6), 1070–1076.

Ron, D., & Wang, J. (2009). The NMDA receptor and alcohol addiction. In A. M. Van Dongen (Ed.), *Frontiers in neuroscience. Biology of the NMDA receptor* (pp. 60–70). Boca Raton, FL: CRC Press/Taylor & Francis Taylor & Francis Group, LLC.

Rosenstein, J. M., & Krum, J. M. (2004). New roles for VEGF in nervous tissue—Beyond blood vessels. *Experimental Neurology*, *187*(2), 246–253. http://dx.doi.org/10.1016/j.expneurol.2004.01.022.

Russo-Neustadt, A., Beard, R. C., & Cotman, C. W. (1999). Exercise, antidepressant medications, and enhanced brain derived neurotrophic factor expression. *Neuropsychopharmacology*, *21*(5), 679–682. http://dx.doi.org/10.1016/s0893-133x(99)00059-7.

Sairanen, M., Lucas, G., Ernfors, P., Castren, M., & Castren, E. (2005). Brain-derived neurotrophic factor and antidepressant drugs have different but coordinated effects on neuronal turnover, proliferation, and survival in the adult dentate gyrus. *The Journal of Neuroscience*, *25*(5), 1089–1094. http://dx.doi.org/10.1523/jneurosci.3741-04.2005.

Sakai, R., Ukai, W., Sohma, H., Hashimoto, E., Yamamoto, M., Ikeda, H., & Saito, T. (2005). Attenuation of brain derived neurotrophic factor (BDNF) by ethanol and cytoprotective effect of exogenous BDNF against ethanol damage in neuronal cells. *Journal of Neural Transmission (Vienna, Austria: 1996)*, *112*(8), 1005–1013. http://dx.doi.org/10.1007/s00702-004-0246-4.

Sakata-Haga, H., Dominguez, H. D., Sei, H., Fukui, Y., Riley, E. P., & Thomas, J. D. (2006). Alterations in circadian rhythm phase shifting ability in rats following ethanol exposure during the third trimester brain growth spurt. *Alcoholism, Clinical and Experimental Research*, *30*(5), 899–907. http://dx.doi.org/10.1111/j.1530-0277.2006.00105.x.

Sanchez, M. M., Das, D., Taylor, J. L., Noda, A., Yesavage, J. A., & Salehi, A. (2011). BDNF polymorphism predicts the rate of decline in skilled task performance and hippocampal volume in healthy individuals. *Translational Psychiatry*, *1*. e51. http://dx.doi.org/10.1038/tp.2011.47.

Schnydrig, S., Korner, L., Landweer, S., Ernst, B., Walker, G., Otten, U., & Kunz, D. (2007). Peripheral lipopolysaccharide administration transiently affects expression of brain-derived neurotrophic factor, corticotropin and proopiomelanocortin in mouse brain. *Neuroscience Letters*, *429*(1), 69–73. http://dx.doi.org/10.1016/j.neulet.2007.09.067.

Sharma, H. S. (2003). Neurotrophic factors attenuate microvascular permeability disturbances and axonal injury following trauma to the rat spinal cord. *Acta Neurochirurgica. Supplement*, *86*, 383–388.

Shimada, A., Mason, C. A., & Morrison, M. E. (1998). TrkB signaling modulates spine density and morphology independent of dendrite structure in cultured neonatal Purkinje cells. *The Journal of Neuroscience*, *18*(21), 8559–8570.

Sickmann, H. M., Patten, A. R., Morch, K., Sawchuk, S., Zhang, C., Parton, R., … Christie, B. R. (2014). Prenatal ethanol exposure has sex-specific effects on hippocampal long-term potentiation. *Hippocampus*, *24*(1), 54–64. http://dx.doi.org/10.1002/hipo.22203.

Singh, A. K., Gupta, S., Jiang, Y., Younus, M., & Ramzan, M. (2009). In vitro neurogenesis from neural progenitor cells isolated from the hippocampus region of the brain of adult rats exposed to ethanol during early development through their alcohol-drinking mothers. *Alcohol and Alcoholism*, *44*(2), 185–198. http://dx.doi.org/10.1093/alcalc/agn109.

Singh, B., Henneberger, C., Betances, D., Arevalo, M. A., Rodriguez-Tebar, A., Meier, J. C., & Grantyn, R. (2006). Altered balance of glutamatergic/GABAergic synaptic input and associated changes in dendrite morphology after BDNF expression in BDNF-deficient hippocampal neurons. *The Journal of Neuroscience*, *26*(27), 7189–7200. http://dx.doi.org/10.1523/jneurosci.5474-05.2006.

Smith, C. C., Guevremont, D., Williams, J. M., & Napper, R. M. (2015). Apoptotic cell death and temporal expression of apoptotic proteins Bcl-2 and Bax in the hippocampus, following binge ethanol in the neonatal rat model. *Alcoholism, Clinical and Experimental Research*, *39*(1), 36–44. http://dx.doi.org/10.1111/acer.12606.

Sowell, E. R., Mattson, S. N., Kan, E., Thompson, P. M., Riley, E. P., & Toga, A. W. (2008). Abnormal cortical thickness and brain-behavior correlation patterns in individuals with heavy prenatal alcohol exposure. *Cerebral Cortex*, *18*(1), 136–144. http://dx.doi.org/10.1093/cercor/bhm039.

Sowell, E. R., Thompson, P. M., Mattson, S. N., Tessner, K. D., Jernigan, T. L., Riley, E. P., & Toga, A. W. (2001). Voxel-based morphometric analyses of the brain in children and adolescents prenatally exposed to alcohol. *Neuroreport*, *12*(3), 515–523.

Sowell, E. R., Thompson, P. M., Mattson, S. N., Tessner, K. D., Jernigan, T. L., Riley, E. P., & Toga, A. W. (2002). Regional brain shape abnormalities persist into adolescence after heavy prenatal alcohol exposure. *Cerebral Cortex, 12*(8), 856–865.

Steinhausen, H. C., & Spohr, H. L. (1998). Long-term outcome of children with fetal alcohol syndrome: Psychopathology, behavior, and intelligence. *Alcoholism, Clinical and Experimental Research, 22*(2), 334–338.

Stevens, S. A., Major, D., Rovet, J., Koren, G., Fantus, E., Nulman, I., & Desrocher, M. (2012). Social problem solving in children with fetal alcohol spectrum disorders. *Journal of Population Therapeutics and Clinical Pharmacology, 19*(1), e99–e110.

Stiles, J., & Jernigan, T. L. (2010). The basics of brain development. *Neuropsychology Review, 20*(4), 327–348. http://dx.doi.org/10.1007/s11065-010-9148-4.

Streissguth, A. P., Bookstein, F. L., Barr, H. M., Sampson, P. D., O'Malley, K., & Young, J. K. (2004). Risk factors for adverse life outcomes in fetal alcohol syndrome and fetal alcohol effects. *Journal of Developmental and Behavioral Pediatrics, 25*(4), 228–238.

Sun, A. Y., Chen, Y. M., James-Kracke, M., Wixom, P., & Cheng, Y. (1997). Ethanol-induced cell death by lipid peroxidation in PC12 cells. *Neurochemical Research, 22*(10), 1187–1192.

Thomas, J. D., Fleming, Sl, & Riley, E. P. (2001). MK-801 can exacerbate or attenuate behavioral alterations associated with neonatal alcohol exposure in the rat, depending on the timing of administration. *Alcoholism, Clinical and Experimental Research, 25*(5), 764–773.

Thomas, S. E., Kelly, S. J., Mattson, S. N., & Riley, E. P. (1998). Comparison of social abilities of children with fetal alcohol syndrome to those of children with similar IQ scores and normal controls. *Alcoholism, Clinical and Experimental Research, 22*(2), 528–533.

Thomas, J. D., Sather, T. M., & Whinery, L. A. (2008). Voluntary exercise influences behavioral development in rats exposed to alcohol during the neonatal brain growth spurt. *Behavioral Neuroscience, 122*(6), 1264–1273. http://dx.doi.org/10.1037/a0013271.

Thomas, J. D., Weinert, S. P., Sharif, S., & Riley, E. P. (1997). MK-801 administration during ethanol withdrawal in neonatal rat pups attenuates ethanol-induced behavioral deficits. *Alcoholism, Clinical and Experimental Research, 21*(7), 1218–1225.

Timmusk, T., Palm, K., Metsis, M., Reintam, T., Paalme, V., Saarma, M., & Persson, H. (1993). Multiple promoters direct tissue-specific expression of the rat BDNF gene. *Neuron, 10*(3), 475–489.

Titterness, A. K., & Christie, B. R. (2012). Prenatal ethanol exposure enhances NMDAR-dependent long-term potentiation in the adolescent female dentate gyrus. *Hippocampus, 22*(1), 69–81. http://dx.doi.org/10.1002/hipo.20849.

Tiwari, V., & Chopra, K. (2011). Resveratrol prevents alcohol-induced cognitive deficits and brain damage by blocking inflammatory signaling and cell death cascade in neonatal rat brain. *Journal of Neurochemistry, 117*(4), 678–690. http://dx.doi.org/10.1111/j.1471-4159.2011.07236.x.

Tolwani, R. J., Buckmaster, P. S., Varma, S., Cosgaya, J. M., Wu, Y., Suri, C., & Shooter, E. M. (2002). BDNF overexpression increases dendrite complexity in hippocampal dentate gyrus. *Neuroscience, 114*(3), 795–805.

Topper, L. A., Baculis, B. C., & Valenzuela, C. F. (2015). Exposure of neonatal rats to alcohol has differential effects on neuroinflammation and neuronal survival in the cerebellum and hippocampus. *Journal of Neuroinflammation, 12*, 160. http://dx.doi.org/10.1186/s12974-015-0382-9.

Treit, S., Zhou, D., Lebel, C., Rasmussen, C., Andrew, G., & Beaulieu, C. (2014). Longitudinal MRI reveals impaired cortical thinning in children and adolescents prenatally exposed to alcohol. *Human Brain Mapping, 35*(9), 4892–4903. http://dx.doi.org/10.1002/hbm.22520.

Tyler, W. J., & Pozzo-Miller, L. D. (2001). BDNF enhances quantal neurotransmitter release and increases the number of docked vesicles at the active zones of hippocampal excitatory synapses. *The Journal of Neuroscience*, *21*(12), 4249–4258.

Uecker, A., & Nadel, L. (1996). Spatial locations gone awry: object and spatial memory deficits in children with fetal alcohol syndrome. *Neuropsychologia*, *34*(3), 209–223. http://dx.doi.org/10.1016/0028-3932(95)00096-8.

van Faassen, E., & Niemelä, O. (2011). *Biochemistry of prenatal alcohol exposure*. New York: NOVA Science Publishers.

Whitcher, L. T., & Klintsova, A. Y. (2008). Postnatal binge-like alcohol exposure reduces spine density without affecting dendritic morphology in rat mPFC. *Synapse*, *62*(8), 566–573. http://dx.doi.org/10.1002/syn.20532.

Willoughby, K. A., Sheard, E. D., Nash, K., & Rovet, J. (2008). Effects of prenatal alcohol exposure on hippocampal volume, verbal learning, and verbal and spatial recall in late childhood. *Journal of the International Neuropsychological Society*, *14*(6), 1022–1033. http://dx.doi.org/10.1017/s1355617708081368.

Wozniak, D. F., Hartman, R. E., Boyle, M. P., Vogt, S. K., Brooks, A. R., Tenkova, T., … Muglia, L. J. (2004). Apoptotic neurodegeneration induced by ethanol in neonatal mice is associated with profound learning/memory deficits in juveniles followed by progressive functional recovery in adults. *Neurobiology of Disease*, *17*(3), 403–414. http://dx.doi.org/10.1016/j.nbd.2004.08.006.

Wozniak, J. R., & Muetzel, R. L. (2011). What does diffusion tensor imaging reveal about the brain and cognition in fetal alcohol spectrum disorders? *Neuropsychology Review*, *21*(2), 133–147. http://dx.doi.org/10.1007/s11065-011-9162-1.

Wu, A., Ying, Z., & Gomez-Pinilla, F. (2004). Dietary omega-3 fatty acids normalize BDNF levels, reduce oxidative damage, and counteract learning disability after traumatic brain injury in rats. *Journal of Neurotrauma*, *21*(10), 1457–1467. http://dx.doi.org/10.1089/neu.2004.21.1457.

Yang, Y., Roussotte, F., Kan, E., Sulik, K. K., Mattson, S. N., Riley, E. P., … Sowell, E. R. (2012). Abnormal cortical thickness alterations in fetal alcohol spectrum disorders and their relationships with facial dysmorphology. *Cerebral Cortex*, *22*(5), 1170–1179. http://dx.doi.org/10.1093/cercor/bhr193.

Zafra, F., Hengerer, B., Leibrock, J., Thoenen, H., & Lindholm, D. (1990). Activity dependent regulation of BDNF and NGF mRNAs in the rat hippocampus is mediated by non-NMDA glutamate receptors. *EMBO J*, *9*(11), 3545–3550.

Zagrebelsky, M., Holz, A., Dechant, G., Barde, Y. A., Bonhoeffer, T., & Korte, M. (2005). The p75 neurotrophin receptor negatively modulates dendrite complexity and spine density in hippocampal neurons. *The Journal of Neuroscience*, *25*(43), 9989–9999. http://dx.doi.org/10.1523/jneurosci.2492-05.2005.

Zhou, D., Lebel, C., Lepage, C., Rasmussen, C., Evans, A., Wyper, K., … Beaulieu, C. (2011). Developmental cortical thinning in fetal alcohol spectrum disorders. *NeuroImage*, *58*(1), 16 25. http://dx.doi.org/10.1016/j.neuroimage.2011.06.026.

Zucca, S., & Valenzuela, C. F. (2010). Low concentrations of alcohol inhibit BDNF-dependent GABAergic plasticity via L-type Ca^{2+} channel inhibition in developing CA3 hippocampal pyramidal neurons. *The Journal of Neuroscience*, *30*(19), 6776–6781. http://dx.doi.org/10.1523/jneurosci.5405-09.2010.

CHAPTER TEN

Neurotrophins: Role in Placental Growth and Development

A.S. Sahay, D.P. Sundrani, S.R. Joshi[1,2]
Interactive Research School for Health Affairs, Bharati Vidyapeeth University, Pune, India
[1]Corresponding author: e-mail address: srjoshi62@gmail.com

Contents

1. Introduction 244
2. Role of Neurotrophins in Placental Growth and Development 246
3. Neurotrophins in Different Regions of the Placenta 248
4. Cross Talk of Neurotrophins With Other Growth Factors and Proteases Involved in Placental Development 250
5. Role of Neurotrophins in Pregnancy 252
   5.1 Neurotrophins and Preeclampsia 252
   5.2 Neurotrophins and Preterm Birth 253
   5.3 Neurotrophins and Other Pregnancy Complications 254
6. Conclusion and Future Directions 255
Acknowledgments 256
References 256

Abstract

Neurotrophins, a family of closely related proteins, were originally identified as growth factors for survival, development, and function of neurons in both the central and peripheral nervous systems. Subsequently, neurotrophins have been shown to have functions in immune and reproductive systems. Neurotrophins like nerve growth factor and brain-derived neurotrophic factor (BDNF) are known to play an important role during pregnancy in the process of placental angiogenesis and maturation. Several studies have demonstrated the presence of neurotrophins in the human placenta. The current chapter reviews studies demonstrating the role of neurotrophins during pregnancy particularly in placental development. This chapter also focuses on the regional changes in neurotrophins in the human placenta and its interactions with other growth factors. Future research is needed to understand the mechanisms through which neurotrophins influence the growth and development of the placenta and pregnancy outcome.

[2] Senior author.

Vitamins and Hormones, Volume 104
ISSN 0083-6729
http://dx.doi.org/10.1016/bs.vh.2016.11.002

1. INTRODUCTION

Neurotrophins are a family of proteins which regulate the generation, survival, proliferation, differentiation, and death of neurons in the peripheral and central nervous system (Huang & Reichardt, 2001). The neurotrophins are synthesized as proneurotrophins and are then cleaved to give the C-terminal mature neurotrophins (Lee, Kermani, Teng, & Hempstead, 2001). In recent years, neurotrophins have been shown to exert their effects in a wide range of tissues, which were originally thought to be limited to the nervous system (Tometten, Blois, & Arck, 2005). The effects of neurotrophins are exerted via two different receptors: the tropomyosin-related kinase (trk) receptors and the structurally unrelated p75 neurotrophin receptor (p75NTR) (Hempstead, 2002).

Nerve growth factor (NGF), brain-derived neurotrophic factor (BDNF), neurotrophin-3 (NT-3), and neurotrophin-4 (NT-4) are the four family members of neurotrophins (Lee, Kim, Khursigara, & Chao, 2001). NGF was the first neurotrophin family member to be identified in the 1950s (Cowan, 2001). It is produced by the cleavage of its precursor (pro-NGF) whose function is different from that of mature NGF (Wang, Wang, & Thrimawithana, 2014). NGF promotes growth, differentiation, and function of nerve cells and maintenance of neurons (Toti et al., 2006; Wang et al., 2014). The effects of NGF are mediated by binding to its high-affinity catalytically active receptor, trkA receptor which is located on the cell surface (Huang & Reichardt, 2003; Vera, Tapia, Vega, & Romero, 2014). Binding of NGF to trkA activates cytosolic/endosomal pathways such as Ras-mitogen-activated protein kinase, phospholipase C, extracellular-signal-regulated kinase (ERK), and phosphatidylinositol 3-kinase (PI3K) (Klesse & Parada, 1999; Reichardt, 2006). NGF has been associated with many brain disorders, such as depression (Wiener et al., 2015), schizophrenia (Ajami, Hosseini, Taghipour, & Khalilian, 2014; Martinotti et al., 2012), and autism (Li et al., 2015). However, in the last few years, NGF has also been shown to have pleiotropic roles. A recent study reports increased expression of NGF in experimental and human pulmonary hypertension, where it promotes vascular cell proliferation and migration, secretion of proinflammatory cytokines, and pulmonary arterial hyperactivity (Freund-Michel et al., 2015). NGF has been shown to be involved in various processes such as apoptosis (la Sala, Corinti, Federici, Saragovi, & Girolomoni, 2000), inflammation (McMahon, 1996;

Prencipe et al., 2014), and neuronal angiogenesis (Calza, Giardino, Giuliani, Aloe, & Levi-Montalcini, 2001).

Another member of the neurotrophin family, BDNF is the most studied in the central nervous system. It was purified from pig brain in 1982 and was shown to support the survival of a subpopulation of dorsal root ganglion neurons (Barde, Edgar, & Thoenen, 1982). In response to neuronal activity pro-BDNF (precursor) is transported and processed via the secretory pathway to form mature-BDNF (Benarroch, 2015). BDNF mediates its role by binding to its high-affinity receptor trkB (Barbacid, 1994) and helps in the development of visual cortex, survival and growth of neurons, and development of brain (Binder & Scharfman, 2004). The effect of physical activity and exercise on BDNF levels has been extensively discussed and it is suggested that exercise may be used to enhance BDNF levels (Szuhany, Bugatti, & Otto, 2015). A recent review describes the three isoforms of BDNF and their influence on neuronal activity which can impact human behavior, viz, depression, hyperphagia, memory, and anxiety (Hempstead, 2015).

Other members of neurotrophin family are NT-3 and NT-4/5. NT-3 was discovered after NGF and BDNF. The mRNA levels of this neurotrophin in the central nervous system have been shown to be higher during fetal development as compared to the adult brain. NT-3 has been shown to promote neurite outgrowth from both nodose ganglion and sympathetic ganglion suggesting its broader specificity (Maisonpierre et al., 1990). NT-3 binds with greatest affinity to trkC and it also binds with less affinity to trkB (Razavi et al., 2015), whereas NT-4, like BDNF, binds with trkB (Bothwell, 2014). NT-3 and NT-4/5 have been reported to be increased in acutely depressed bipolar disorder subjects in comparison to healthy controls (Loch et al., 2015).

It is suggested that neurotrophins especially NT-3 and BDNF play a critical role during early development, wherein trkC (receptor for NT-3) has been the earliest to be detected in neural plate and neural tube (Bernd, 2008). Several studies have highlighted the critical role of neurotrophins such as NGF and BDNF during pregnancy (Bienertova-Vasku et al., 2013; Toti et al., 2006). It is suggested that neurotrophins like NGF and BDNF are involved in the process of placental development and maturation. In view of this, the focus of the current chapter is to review the role of neurotrophins in placental growth and development in normal pregnancy as well as in various pregnancy complications such as preeclampsia. Further, studies indicate that the placenta is a heterogenous structure and has functional differences in different regions of the placenta. Hence, this chapter also discusses the regional differences in placental development and the possible mechanisms affecting the levels of neurotrophins in different regions of the placenta.

2. ROLE OF NEUROTROPHINS IN PLACENTAL GROWTH AND DEVELOPMENT

The placenta is the central temporary organ uniquely designed to carry out exchange of nutrients, gases, antibodies, hormones, growth factors, and waste products between the mother and the fetus. In the past few years several studies have shown that neurotrophins may play an important role in the growth and development of the placenta and the fetus which are discussed later. Neurotrophins have been suggested to influence the development and maturation of the fetoplacental unit and affect the fetal growth trajectories. It is postulated that during gestation, different functions like angiogenesis, energy homeostasis may be regulated by neurotrophins that could promote the fetal growth and development (Mayeur et al., 2011). The source of neurotrophins for the placental and fetal growth and development can be either from the mother or they may be expressed in the placenta. A study by Kodomari and colleagues in BDNF-gene deficient mice demonstrated absence of BDNF mRNA expression in BDNF (−/−) foetuses, whereas, exogenous BDNF administration in pregnant dams led to an increase in BDNF protein levels in the fetal brain in a dose-dependent manner. The authors therefore suggest that maternal BDNF is transported to the fetus across the uteroplacental barrier (Kodomari, Wada, Nakamura, & Wada, 2009). Another recent study reports BDNF and TrkB expression in both rat and human placenta. This study also demonstrates that in pregnant women serum BDNF levels are lower in the first trimester as compared to second and third trimester. The authors speculate that at the start of pregnancy transplacental transport for fetal neurodevelopment leads to a decrease in maternal BDNF serum levels and as pregnancy progresses the placenta and fetus assume a role in the production of BDNF (Garces et al., 2014).

BDNF has been reported to potentiate embryo implantation, development of the placenta, and growth of the fetus from mid gestation to late gestational stages by increasing trophoblast cell growth and survival in mice suggesting it may be critical for the development of the fetoplacental unit (Kawamura et al., 2009). During the preimplantation stage, BDNF has been demonstrated to promote blastocyst outgrowth and after implantation BDNF is reported to suppress apoptosis in the embryo and thus help in early embryonic development (Kawamura et al., 2007). BDNF expression has also been reported in human membranous chorion and villous tissue. The BDNF/TrkB signaling has been suggested to have an antiapoptotic role in response to the oxidative stress suggesting their protective role under unfavorable conditions (Fujita et al., 2011). Another study focusing on the BDNF/TrkB system in rat and human placentas from pregnancies with fetal growth restriction also suggest that this system may play a crucial role in the placental development and may be involved in placental pathologies and disturbances of fetal growth (Mayeur et al., 2010). The occurrence of NT-3 transcript and protein was reported for the first time in normal and pathological placenta in different trimesters of pregnancy suggesting their role in the development of the placenta and growth of the fetus (Casciaro et al., 2009).

NGF mRNA and peptide expression has also been demonstrated in human placenta in first and third trimester of pregnancy suggesting that NGF may regulate implantation, help in the regulation of maternal immunity or may modulate maternal vascular growth (Toti et al., 2006). A recent study in high-abortion rate mouse model reported that NGF administration during early pregnancy boosts spontaneous abortion rates by increasing the local inflammatory response via recruitment of NGF-expressing immune cells, increased decidual innervation with substance P+ nerve fibers, and a shift toward Th1 cytokine. Further in a normal-pregnancy mice model, treatment with a NGF-neutralizing antibody also induced abortions. This study also reported that in humans there is an enhanced synthesis and an irregular distribution of NGF in placental tissue in case of spontaneous abortion. The authors therefore suggest that a balanced level of NGF expression is necessary for healthy pregnancy which if exceeded may result in fetal rejection possibly because of the increased inflammation (Frank et al., 2014).

In view of the above-mentioned studies it is clear that neurotrophins have important functions during implantation, placental development, and in modulating fetal growth. The role of neurotrophins in the mother, placenta, and fetus during pregnancy has been described in Fig. 1. In Section 3, we describe the current status on neurotrophins in different regions of the placenta.

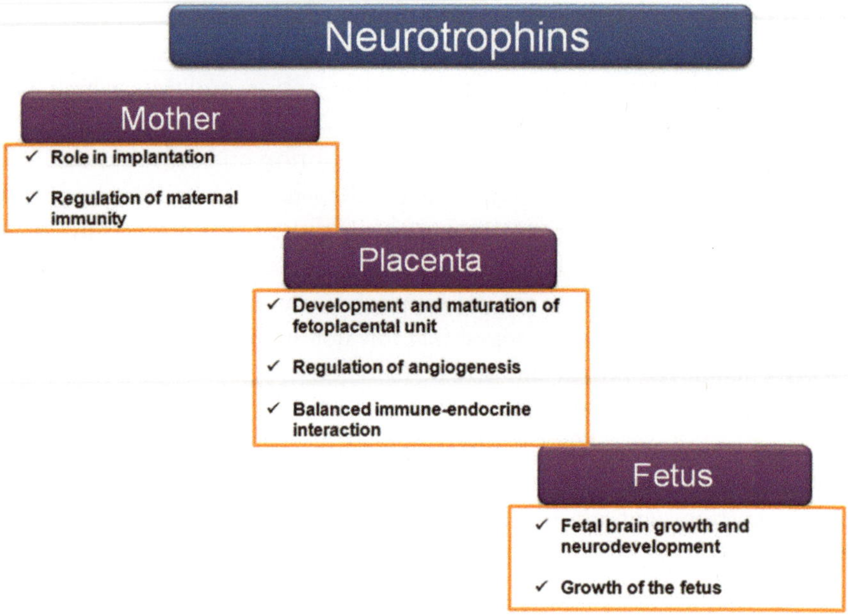

Fig. 1 Role of neurotrophins in mother, placenta, and fetus during pregnancy.

3. NEUROTROPHINS IN DIFFERENT REGIONS OF THE PLACENTA

The human placenta has two plates, the chorionic plate (representing fetal surface) and the basal plate (representing maternal surface). The pattern of blood flow and the placental architecture vary in different regions of the placenta and are nonuniform across the placental disk. Reports indicate that the differences in placental vascularization result in differences of oxygen availability (Hempstock et al., 2003; Matijevic, Meekins, Walkinshaw, Neilson, & McFadyen, 1995). In normal pregnancies the establishment of a continuous intervillous circulation is an incremental phenomenon which starts at the periphery and then gradually progresses to the rest of the placenta. In contrast, the early blood flow in case of complicated pregnancies was most common in central regions or throughout the placenta (Jauniaux, Hempstock, Greenwold, & Burton, 2003). These differences in the blood flow, maturity, and the architecture are likely to result in regional differences of oxidative stress which could further lead to regional differences in various growth factors in the placenta. Various studies have reported regional differences in the oxidative stress markers such as malondialdehyde and antioxidant enzymes such as superoxide dismutase, catalase, and glutathione peroxidase (Hempstock et al., 2003; Mistry et al., 2010; Roland et al., 2010; Sahay, Sundrani, Wagh, Mehendale, & Joshi, 2015b).

It is known that oxidative stress regulates the expression of the neurotrophins (Jain, Banerjee, Ahmed, Arora, & Mediratta, 2013; Wu, Ying, & Gomez-Pinilla, 2004). Therefore, it is very much likely that neurotrophins will vary in different regions of the placenta. However, there are only few studies which have examined neurotrophins in different regions of the placenta. A recent study reports higher NGF and BDNF levels in the central maternal region (area around the cord insertion site from the maternal/basal plate) of preterm preeclamptic placenta as compared to that of term preeclamptic placenta and normotensive control placenta, respectively. This indicated that the central maternal region of the placenta might be the most important regulator of placental functions (Sahay, Sundrani, Wagh, Mehendale, & Joshi, 2015a). In another study, immunostaining of human placenta showed NT-3 expression in the decidual stromal cells and epithelial cells of endometrial gland on the maternal side and the staining of fetal vessels in first trimester placenta (Casciaro et al., 2009). Further, a study by Toti et al. shows that NGF is mainly localized in trophoblast cells and only rarely in the villous stroma. The authors also report staining of maternal decidual vessel but not fetal vessel in the first trimester which suggests a potential role of NGF in initial events like implantation, vasculogenesis, and angiogenesis for a successful pregnancy (Toti et al., 2006). A study examined all the four neurotrophins, i.e., NGF, BDNF, NT-3, and NT-4, in separate fetal membranes (amnion, membranous chorion, and deciduas) and villous tissue samples in the third trimester. All neurotrophins except NT-4 were detected in amnion and membranous chorion. Higher BDNF mRNA levels were found in the membranous chorion and villous tissue as compared to amnion and deciduas (Fujita et al., 2011). Amnion and choriodecidual membranes from women delivering at term not-in-labor, term spontaneous labor, and preterm labor with and without chorioamnionitis have also shown differential the expression of neurotrophic factors. This implicates a role for neurotrophins in pregnancy processes (Marvin, Keelan, Eykholt, Sato, & Mitchell, 2002). The above-mentioned studies suggest that different regions of the placenta may have different specialized functions and this requires more extensive studies. Regional differences in the expression of neurotrophins may be influenced by various other growth factors and proteases. In Section 4, we discuss the interaction of neurotrophins with other growth factors like the angiogenic growth factors and proteases like matrix metalloproteinases (MMPs).

4. CROSS TALK OF NEUROTROPHINS WITH OTHER GROWTH FACTORS AND PROTEASES INVOLVED IN PLACENTAL DEVELOPMENT

Neurotrophins have been given the name "angioneurins" based on their angiogenic effects (Zacchigna, Lambrechts, & Carmeliet, 2008). Placental neurotrophins may help in modulating angiogenesis during pregnancy (Lewin & Barde, 1996). NGF has also been suggested to enhance angiogenesis by increasing vascular endothelial growth factor (VEGF) mRNA and protein levels in cultured neonatal rat ovaries (Julio-Pieper, Lara, Bravo, & Romero, 2006). NGF has been shown to promote ovarian angiogenesis by stimulating the production of VEGF via trkA- and ERK-2-dependent mechanism in vitro (Julio-Pieper et al., 2009). Similarly, treatment of NGF has been shown to induce neoangiogenesis by increasing VEGF expression in newborn rats (Calza et al., 2001). In neuroblastoma cells BDNF has been shown to induce VEGF via hypoxia-inducible factor-1 alpha (Nakamura et al., 2006) and thus BDNF may regulate angiogenesis through VEGF. The interaction of neurotrophins and angiogenesis under various pathological conditions, such as cancer and inflammation, has been elaborately discussed (Caporali & Emanueli, 2009). It has been suggested that NGF and VEGF act in concert to play a role in the regulation of angiogenesis. NGF can directly induce angiogenesis via its receptor or it can control angiogenesis indirectly through VEGF although it still remains to be elucidated (Nico, Mangieri, Benagiano, Crivellato, & Ribatti, 2008). A study in human umbilical vein endothelial cells has reported that these cells express trkA and p75 receptors. Treatment of these cells with NGF led to rapid phosphorylation of trkA and ERK. The authors also reported that NGF exerted a direct angiogenic effect in vivo when delivered to chorioallantoic membrane of the chicken embryo (Cantarella et al., 2002). It is suggested that neurotrophins like BDNF and NT-3 are important players for angiogenesis and vessel stabilization during embryogenesis (Kermani & Hempstead, 2007). Taken together all these studies indicate that neurotrophins may regulate angiogenesis in the placenta in normal pregnancy as well as during pregnancy complications. Differential regulation of neurotrophins may therefore affect different angiogenic factors in women with preeclampsia which is characterized by altered angiogenesis.

In vitro studies suggest that the expression of MMPs (a family of endopeptidases involved in extracellular matrix degradation) is stimulated by neurotrophins like NGF and BDNF through signal transduction pathways via activation of their specific receptors, i.e., TrkA and TrkB, respectively (Lin, Chang, Fong, Hsu, & Tang, 2013; Park et al., 2007). The role of NGF in epithelial healing process which is mediated by upregulation of MMP-9 has been recently reported (Blanco-Mezquita et al., 2013). Stimulation with BDNF has been shown to upregulate MMP-9 mRNA and protein levels in rat primary cortical neurons. This upregulation was determined by extracellular-signal-regulated kinase 1/2 (ERK1/2) pathway critical for activating neuronal plasticity (Kuzniewska et al., 2013). The role of neurotrophins and their receptors in regulating placental angiogenic factors and MMPs during pregnancy has been described in Fig. 2.

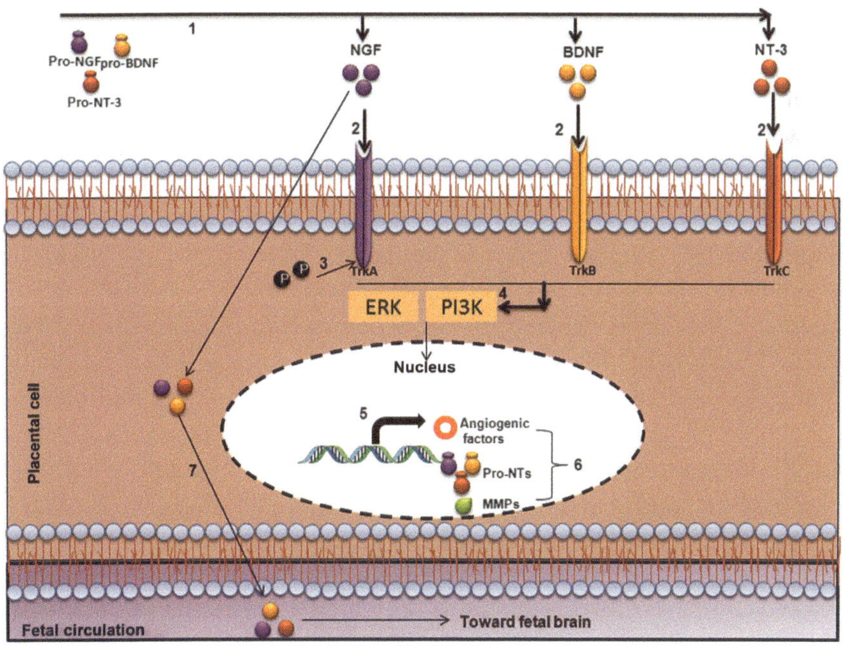

Fig. 2 Role of neurotrophins and their receptors in regulating placental angiogenic factors and matrix metalloproteinases during pregnancy. 1. Proneurotrophins are processed to give the respective mature neurotrophins; 2. Mature neurotrophins bind to their respective receptors; 3. Dimerization and autophosphorylation of the receptors; 4. Activation of various signaling pathways; 5. Transcription of various angiogenic factors like vascular endothelial growth factor, neurotrophins, and matrix metalloproteinases; 6. The growth factors and matrix metalloproteinases carry out normal placental development; 7. The maternal neurotrophins are also transferred to the fetus via the placenta; *BDNF*, brain-derived neurotrophic factor; *ERK*, extracellular signal-regulated kinases; *MMPs*, matrix metalloproteinases; *NGF*, nerve growth factor; *NT-3*, neurotrophin-3; *NT-4*, neurotrophin-4; *NTs*, neurotrophins; *PI3K*, phosphoinositide 3-kinase; *Trk*, the tropomyosin-related kinase family of receptors.

5. ROLE OF NEUROTROPHINS IN PREGNANCY

Several studies have attempted to examine the maternal levels of neurotrophins so as to understand their role during pregnancy and pregnancy complications which are discussed in Sections 5.1, 5.2, and 5.3. These sections focus on the role of neurotrophins in adverse pregnancy outcomes like preeclampsia, preterm birth, and intrauterine growth restriction (IUGR).

5.1 Neurotrophins and Preeclampsia

Preeclampsia is a pregnancy-specific hypertensive disorder manifesting after 20 weeks of gestation and is a major cause of maternal and fetal morbidity and mortality. Various studies indicate that shallow trophoblast invasion and impaired spiral artery remodeling are associated with the development of preeclampsia. Different factors contribute to the pathophysiology of preeclampsia such as oxidative stress (Hung & Burton, 2006), altered angiogenesis (Levine et al., 2004), and inflammation (Borzychowski, Sargent, & Redman, 2006). Despite extensive research, the pathophysiology of preeclampsia is not completely elucidated. Few studies have examined the maternal levels of neurotrophins in women with preeclampsia. Lower maternal plasma BDNF levels in women with preeclampsia as compared to normotensive control women have been reported (D'Souza, Kilari, et al., 2014). In contrast, a study reports higher maternal plasma BDNF level in normotensive small for gestational age and in women with preeclampsia as compared to control women in the third trimester (Fujita et al., 2011). Another study reports no difference in maternal plasma BDNF levels between preeclamptic women and controls. However, in the same study, when cord BDNF levels were compared, higher cord blood BDNF levels were found in preeclampsia cases as compared to normotensive control and the authors hypothesize that under unfavorable conditions like oxidative stress, the activation of BDNF/TrkB system may lead to the development of preeclampsia (Bienertova-Vasku et al., 2013). Lower maternal plasma NGF levels are also reported in the preeclamptic group as compared to the control group (Dangat, Kilari, Mehendale, Lalwani, & Joshi, 2013; Kilari et al., 2011). The above-mentioned studies have been carried out at delivery when preeclampsia had already progressed. Therefore, these changes at delivery may be the secondary effects when the pathology has progressed for many weeks and months. To better understand the role played by neurotrophins in pathophysiology of preeclampsia studies have also examined the levels of neurotrophins prospectively beginning from early gestation.

Prospective studies examining maternal plasma NGF and BDNF levels in the Indian population report no changes in NGF levels while lower BDNF levels at different time points across pregnancy in preeclampsia women as compared to the normotensive control women (D'Souza et al., 2015; D'Souza, Patil, et al., 2014). However, there is a need for additional longitudinal studies on larger numbers to understand the role of neurotrophins across gestation especially in pregnancy complications.

Studies have also reported neurotrophin expression in placenta of normal and pathological pregnancies. A study reports higher TrkB mRNA in preeclamptic placentas as compared to control placentas. This indicates that during pregnancy the BDNF/TrkB system may be activated at the fetomaternal interface under unfavorable conditions (Fujita et al., 2011). A study reports no difference in placental NT-3 expression in women with preeclampsia as compared to control women (Casciaro et al., 2009). The levels of neurotrophins in the umbilical cord may reflect the transfer of neurotrophins from the mother and placenta to the fetus. Therefore, examining the levels of neurotrophins in the umbilical cord is important. However, there are limited studies which report neurotrophin levels in the umbilical cord. A higher BDNF and NGF levels in umbilical cord in the preeclamptic group as compared to control group has been reported suggesting that these may have a compensatory role in the fetoplacental circulation in preeclampsia (D'Souza, Kilari, et al., 2014; D'Souza et al., 2015). It is necessary to follow-up children born to mothers with preeclampsia to understand the association of cord neurotrophins and cognitive performance at later ages.

5.2 Neurotrophins and Preterm Birth

Preterm birth is defined as birth before completing 37 weeks of gestation (Loftin et al., 2010) and is one of the largest direct causes of neonatal deaths (Blencowe et al., 2013). Recent evidence suggests that preterm birth may lead to neurodevelopmental impairment in children (Rogers & Velten, 2011). A study comparing maternal serum BDNF in term and preterm deliveries demonstrates no difference in the BDNF levels but reports a higher maternal serum BDNF level than their preterm infants (Malamitsi-Puchner,

Economou, Rigopoulou, & Boutsikou, 2004). In contrast, another study reports increased maternal plasma BDNF levels in women delivering preterm as compared to those delivering at term (Dhobale, Mehendale, Pisal, D'Souza, & Joshi, 2012).

Placental BDNF and NGF levels as well as cord plasma NGF levels were shown to be lower in the preterm group as compared to the term group. The authors discuss that this may be a result of altered global DNA methylation patterns and may affect fetal growth which can have consequences for neurodevelopmental disorders in later life (Dhobale, Mehendale, Pisal, Nimbargi, & Joshi, 2012; Dhobale, Pisal, Mehendale, & Joshi, 2013). In another study, levels of neurotrophins (BDNF and NT-3) in human umbilical cord blood from infants at different gestational ages were examined. The authors demonstrated an increase in BDNF but not NT-3 with progression of gestation and they also report lower cord blood BDNF levels in infants with severe intraventricular hemorrhage as compared to their normal counterparts. They suggest that cord blood BDNF levels may reflect the degree of neural maturity in the infants (Chouthai, Sampers, Desai, & Smith, 2003). Cord NT-3 levels have been reported to be lower in preterm infants with placental inflammation as compared to those without inflammation (Kumar et al., 2011). Similarly, the concentration of NT-3 has been shown to be lower in case of preterm birth and is suggested to reflect developmental immaturity (Matoba et al., 2009). Another study by Malamitsi-Puchner et al. demonstrated lower serum BDNF levels in umbilical cord and neonates on postpartum day 1 and postpartum day 4 in preterm cases as compared to term cases. Further, in preterm cases, the BDNF levels were higher in neonates at postpartum day 1 as compared to their umbilical cord levels. These results show immaturity of the preterm infant immune system and differential degree of central and peripheral nervous system maturity (Malamitsi-Puchner et al., 2004). The present literature suggests that changes in maternal and placental neurotrophin levels may have consequences for children born preterm in their later life. Longitudinal studies examining neurotrophins right from early pregnancy and follow-up of children born preterm for their neurodevelopmental functioning are needed.

5.3 Neurotrophins and Other Pregnancy Complications

Neurotrophins have also been studied in pregnancies with IUGR. BDNF and its TrkB receptor mRNA levels have been shown to be higher in IUGR pregnancies as compared to control. This implicates that the BDNF/TrkB system is involved in maintaining the growth and development under adverse conditions (Mayeur et al., 2010). A study reports maternal, fetal, neonatal day 1 and day 4 levels of NGF, BDNF, NT-3, and NT-4 in IUGR, and appropriate for gestation age (AGA) groups. The study demonstrates no significant differences in BDNF, NT-3, and NT-4 levels in any of the groups. However, higher NGF levels were found in AGA maternal, fetal, neonatal day-1 and neonatal day-4 samples as compared to their respective IUGR samples (Malamitsi-Puchner et al., 2007). A study including women with generalized anxiety disorder and women without any psychiatric disorder reported higher cord blood BDNF levels in healthy neonates as compared to those born to women with generalized anxiety disorder. The authors suggest that lower levels of circulating BDNF may negatively affect the neurodevelopment of the fetus born to mothers with generalized anxiety disorder (Uguz et al., 2013). A study in appropriate for gestational age neonates and fetuses examined NGF, BDNF, NT-3, and NT-4 from the umbilical cord and infant on day 1 and day 4 after birth. This study indicates that there is a gradual decrease in the levels of NT-3 and NT-4 from umbilical cord to day 4 in neonates. In contrast, NGF and BDNF demonstrated an increase from umbilical cord to day 4 in neonates. This suggests that NT-3 and NT-4 are involved in early stages of neurodevelopment (Nikolaou et al., 2006). Further extensive studies examining neurotrophins in different pregnancy complications and outcomes like small for gestational age and low birth weight are warranted.

6. CONCLUSION AND FUTURE DIRECTIONS

This chapter has described the role of neurotrophins in placental growth and development. It is well known that transport of the nutrients to the fetus is majorly determined by the placenta. The placenta is the primary interface between the mother and the fetus and any changes in placental development may influence the growth of the fetus. Further, any alterations in the expression of neurotrophins may induce critical alterations in the function of the placenta. It is therefore necessary to examine the role of neurotrophins and its interactions with angiogenic factors and MMPs in the placenta. In addition it is necessary to examine the changes of the above factors in various regions of the placenta. This will help to better understand the placental pathology of pregnancy complications like preeclampsia and adverse birth outcomes like preterm birth and IUGR. These studies will provide a plausible cellular and molecular mechanism through which neurotrophins regulate placental growth and development.

ACKNOWLEDGMENTS

The authors thank Department of Science and Technology, Government of India for providing "INSPIRE senior research fellowship" to A.S.S. and Indian Council of Medical Research (ICMR), India for providing research associate (RA) fellowship to D.P.S.

REFERENCES

Ajami, A., Hosseini, S. H., Taghipour, M., & Khalilian, A. (2014). Changes in serum levels of brain derived neurotrophic factor and nerve growth factor-beta in schizophrenic patients before and after treatment. *Scandinavian Journal of Immunology*, *80*(1), 36–42. http://dx.doi.org/10.1111/sji.12158.

Barbacid, M. (1994). The Trk family of neurotrophin receptors. *Journal of Neurobiology*, *25*(11), 1386–1403. http://dx.doi.org/10.1002/neu.480251107.

Barde, Y. A., Edgar, D., & Thoenen, H. (1982). Purification of a new neurotrophic factor from mammalian brain. *The EMBO Journal*, *1*(5), 549–553.

Benarroch, E. E. (2015). Brain-derived neurotrophic factor: Regulation, effects, and potential clinical relevance. *Neurology*, *84*(16), 1693–1704. http://dx.doi.org/10.1212/wnl.0000000000001507.

Bernd, P. (2008). The role of neurotrophins during early development. *Gene Expression*, *14*(4), 241–250.

Bienertova-Vasku, J., Bienert, P., Zlamal, F., Splichal, Z., Tomandl, J., Tomandlova, M., … Vasku, A. (2013). Brain-derived neurotrophic factor and ciliary neurotrophic factor in maternal plasma and umbilical cord blood from pre-eclamptic and physiological pregnancies. *Journal of Obstetrics and Gynaecology*, *33*(4), 359–363. http://dx.doi.org/10.3109/01443615.2013.776026.

Binder, D. K., & Scharfman, H. E. (2004). Brain-derived neurotrophic factor. *Growth Factors*, *22*(3), 123–131.

Blanco-Mezquita, T., Martinez-Garcia, C., Proenca, R., Zieske, J. D., Bonini, S., Lambiase, A., & Merayo-Lloves, J. (2013). Nerve growth factor promotes corneal epithelial migration by enhancing expression of matrix metalloprotease-9. *Investigative Ophthalmology & Visual Science*, *54*(6), 3880–3890. http://dx.doi.org/10.1167/iovs.12-10816.

Blencowe, H., Cousens, S., Chou, D., Oestergaard, M., Say, L., Moller, A. B., … Lawn, J. (2013). Born too soon: The global epidemiology of 15 million preterm births. *Reproductive Health*, *10*(Suppl. 1), S2. http://dx.doi.org/10.1186/1742-4755-10-s1-s2.

Borzychowski, A. M., Sargent, I. L., & Redman, C. W. (2006). Inflammation and pre-eclampsia. *Seminars in Fetal and Neonatal Medicine*, *11*(5), 309–316. http://dx.doi.org/10.1016/j.siny.2006.04.001.

Bothwell, M. (2014). NGF, BDNF, NT3, and NT4. *Handbook of Experimental Pharmacology*, *220*, 3–15. http://dx.doi.org/10.1007/978-3-642-45106-5_1.

Calza, L., Giardino, L., Giuliani, A., Aloe, L., & Levi-Montalcini, R. (2001). Nerve growth factor control of neuronal expression of angiogenetic and vasoactive factors. *Proceedings of the National Academy of Sciences of the United States of America*, *98*(7), 4160–4165. http://dx.doi.org/10.1073/pnas.051626998.

Cantarella, G., Lempereur, L., Presta, M., Ribatti, D., Lombardo, G., Lazarovici, P., … Bernardini, R. (2002). Nerve growth factor-endothelial cell interaction leads to angiogenesis in vitro and in vivo. *FASEB Journal*, *16*(10), 1307–1309. http://dx.doi.org/10.1096/fj.01-1000fje.

Caporali, A., & Emanueli, C. (2009). Cardiovascular actions of neurotrophins. *Physiological Reviews*, *89*(1), 279–308. http://dx.doi.org/10.1152/physrev.00007.2008.

Cacciaro, A., Arcuri, F., Occhini, R., Toti, M. S., De Felice, C., & Toti, P. (2009). Expression of placental neurotrophin-3 (NT-3) in physiological pregnancy, preeclampsia and chorioamnionitis. *Clinical Medicine. Pathology*, *2*, 9–15.

Chouthai, N. S., Sampers, J., Desai, N., & Smith, G. M. (2003). Changes in neurotrophin levels in umbilical cord blood from infants with different gestational ages and clinical conditions. *Pediatric Research*, *53*(6), 965–969. http://dx.doi.org/10.1203/01.pdr.0000061588.39652.26.

Cowan, W. M. (2001). Viktor Hamburger and Rita Levi-Montalcini: The path to the discovery of nerve growth factor. *Annual Review of Neuroscience*, *24*, 551–600. http://dx.doi.org/10.1146/annurev.neuro.24.1.551.

Dangat, K., Kilari, A., Mehendale, S., Lalwani, S., & Joshi, S. (2013). Higher levels of brain derived neurotrophic factor but similar nerve growth factor in human milk in women with preeclampsia. *International Journal of Developmental Neuroscience*, *31*(3), 209–213. http://dx.doi.org/10.1016/j.ijdevneu.2012.12.007.

Dhobale, M., Mehendale, S., Pisal, H., D'Souza, V., & Joshi, S. (2012a). Association of brain-derived neurotrophic factor and tyrosine kinase B receptor in pregnancy. *Neuroscience*, *216*, 31–37. http://dx.doi.org/10.1016/j.neuroscience.2012.04.016.

Dhobale, M., Mehendale, S., Pisal, H., Nimbargi, V., & Joshi, S. (2012b). Reduced maternal and cord nerve growth factor levels in preterm deliveries. *International Journal of Developmental Neuroscience*, *30*(2), 99–103. http://dx.doi.org/10.1016/j.ijdevneu.2011.12.007.

Dhobale, M. V., Pisal, H. R., Mehendale, S. S., & Joshi, S. R. (2013). Differential expression of human placental neurotrophic factors in preterm and term deliveries. *International Journal of Developmental Neuroscience*, *31*(8), 719–723. http://dx.doi.org/10.1016/j.ijdevneu.2013.09.006.

D'Souza, V. A., Kilari, A. S., Joshi, A. A., Mehendale, S. S., Pisal, H. M., & Joshi, S. R. (2014). Differential regulation of brain-derived neurotrophic factor in term and preterm preeclampsia. *Reproductive Sciences*, *21*(2), 230–235. http://dx.doi.org/10.1177/1933719113493512.

D'Souza, V., Kilari, A., Pisal, H., Patil, V., Mehendale, S., Wagh, G., … Joshi, S. (2015). Maternal nerve growth factor levels during pregnancy in women with preeclampsia: A longitudinal study. *International Journal of Developmental Neuroscience*, *47*(Pt. B), 340–346. http://dx.doi.org/10.1016/j.ijdevneu.2015.08.003.

D'Souza, V., Patil, V., Pisal, H., Randhir, K., Joshi, A., Mehendale, S., … Joshi, S. (2014). Levels of brain derived neurotrophic factors across gestation in women with preeclampsia. *International Journal of Developmental Neuroscience*, *37*, 36–40. http://dx.doi.org/10.1016/j.ijdevneu.2014.06.008.

Frank, P., Barrientos, G., Tirado-Gonzalez, I., Cohen, M., Moschansky, P., Peters, E. M., … Blois, S. M. (2014). Balanced levels of nerve growth factor are required for normal pregnancy progression. *Reproduction*, *148*(2), 179–189. http://dx.doi.org/10.1530/rep-14-0112.

Freund-Michel, V., Cardoso Dos Santos, M., Guignabert, C., Montani, D., Phan, C., Coste, F., … Muller, B. (2015). Role of nerve growth factor in development and persistence of experimental pulmonary hypertension. *American Journal of Respiratory and Critical Care Medicine*, *192*(3), 342–355. http://dx.doi.org/10.1164/rccm.201410-1851OC.

Fujita, K., Tatsumi, K., Kondoh, E., Chigusa, Y., Mogami, H., Fujii, T., … Konishi, I. (2011). Differential expression and the anti-apoptotic effect of human placental neurotrophins and their receptors. *Placenta*, *32*(10), 737–744. http://dx.doi.org/10.1016/j.placenta.2011.07.001.

Garces, M. F., Sanchez, E., Torres-Sierra, A. L., Ruiz-Parra, A. I., Angel-Muller, E., Alzate, J. P., … Caminos, J. E. (2014). Brain-derived neurotrophic factor is expressed in rat and human placenta and its serum levels are similarly regulated throughout pregnancy in both species. *Clinical Endocrinology*, *81*(1), 141–151. http://dx.doi.org/10.1111/cen.12391.

Hempstead, B. L. (2002). The many faces of p75NTR. *Current Opinion in Neurobiology*, *12*(3), 260–267.

Hempstead, B. L. (2015). Brain-derived neurotrophic factor: Three ligands, many actions. *Transactions of the American Clinical and Climatological Association*, *126*, 9–19.

Hempstock, J., Bao, Y. P., Bar-Issac, M., Segaren, N., Watson, A. L., Charnock-Jones, D. S., … Burton, G. J. (2003). Intralobular differences in antioxidant enzyme expression and activity reflect the pattern of maternal arterial bloodflow within the human placenta. *Placenta*, *24*(5), 517–523. http://dx.doi.org/10.1053/plac.2002.0955.

Huang, E. J., & Reichardt, L. F. (2001). Neurotrophins: Roles in neuronal development and function. *Annual Review of Neuroscience*, *24*, 677–736. http://dx.doi.org/10.1146/annurev.neuro.24.1.677.

Huang, E. J., & Reichardt, L. F. (2003). Trk receptors: Roles in neuronal signal transduction. *Annual Review of Biochemistry*, *72*, 609–642. http://dx.doi.org/10.1146/annurev.biochem.72.121801.161629.

Hung, T. H., & Burton, G. J. (2006). Hypoxia and reoxygenation: A possible mechanism for placental oxidative stress in preeclampsia. *Taiwanese Journal of Obstetrics & Gynecology*, *45*(3), 189–200. http://dx.doi.org/10.1016/s1028-4559(09)60224-2.

Jain, S., Banerjee, B. D., Ahmed, R. S., Arora, V. K., & Mediratta, P. K. (2013). Possible role of oxidative stress and brain derived neurotrophic factor in triazophos induced cognitive impairment in rats. *Neurochemical Research*, *38*(10), 2136–2147. http://dx.doi.org/10.1007/s11064-013-1122-0.

Jauniaux, E., Hempstock, J., Greenwold, N., & Burton, G. J. (2003). Trophoblastic oxidative stress in relation to temporal and regional differences in maternal placental blood flow in normal and abnormal early pregnancies. *The American Journal of Pathology*, *162*(1), 115–125. http://dx.doi.org/10.1016/s0002-9440(10)63803-5.

Julio-Pieper, M., Lara, H. E., Bravo, J. A., & Romero, C. (2006). Effects of nerve growth factor (NGF) on blood vessels area and expression of the angiogenic factors VEGF and TGFbeta1 in the rat ovary. *Reproductive Biology and Endocrinology*, *4*, 57. http://dx.doi.org/10.1186/1477-7827-4-57.

Julio-Pieper, M., Lozada, P., Tapia, V., Vega, M., Miranda, C., Vantman, D., … Romero, C. (2009). Nerve growth factor induces vascular endothelial growth factor expression in granulosa cells via a trkA receptor/mitogen-activated protein kinase-extracellularly regulated kinase 2-dependent pathway. *The Journal of Clinical Endocrinology and Metabolism*, *94*(8), 3065–3071. http://dx.doi.org/10.1210/jc.2009-0542.

Kawamura, K., Kawamura, N., Fukuda, J., Kumagai, J., Hsueh, A. J., & Tanaka, T. (2007). Regulation of preimplantation embryo development by brain-derived neurotrophic factor. *Developmental Biology*, *311*(1), 147–158. http://dx.doi.org/10.1016/j.ydbio.2007.08.026.

Kawamura, K., Kawamura, N., Sato, W., Fukuda, J., Kumagai, J., & Tanaka, T. (2009). Brain-derived neurotrophic factor promotes implantation and subsequent placental development by stimulating trophoblast cell growth and survival. *Endocrinology*, *150*(8), 3774–3782. http://dx.doi.org/10.1210/en.2009-0213.

Kermani, P., & Hempstead, B. (2007). Brain-derived neurotrophic factor: A newly described mediator of angiogenesis. *Trends in Cardiovascular Medicine*, *17*(4), 140–143. http://dx.doi.org/10.1016/j.tcm.2007.03.002.

Kilari, A., Mehendale, S., Pisal, H., Panchanadikar, T., Kale, A., & Joshi, S. (2011). Nerve growth factor, birth outcome and pre-eclampsia. *International Journal of Developmental Neuroscience*, *29*(1), 71–75. http://dx.doi.org/10.1016/j.ijdevneu.2010.09.001.

Klesse, L. J., & Parada, L. F. (1999). Trks: Signal transduction and intracellular pathways. *Microscopy Research and Technique*, *45*(4-5), 210–216. http://dx.doi.org/10.1002/(sici)1097-0029(19990515/01)45:4/5<210::aid-jemt4>3.0.co;2-f.

Kodomari, I., Wada, E., Nakamura, S., & Wada, K. (2009). Maternal supply of BDNF to mouse fetal brain through the placenta. *Neurochemistry International*, *54*(2), 95–98. http://dx.doi.org/10.1016/j.neuint.2008.11.005.

Kumar, P., Venners, S. A., Fu, L., Pearson, C., Ortiz, K., & Wang, X. (2011). Association of antenatal steroid use with cord blood immune biomarkers in preterm births. *Early Human Development*, *87*(8), 559–564. http://dx.doi.org/10.1016/j.earlhumdev.2011.04.013.

Kuzniewska, B., Rejmak, E., Malik, A. R., Jaworski, J., Kaczmarek, L., & Kalita, K. (2013). Brain-derived neurotrophic factor induces matrix metalloproteinase 9 expression in neurons via the serum response factor/c-Fos pathway. *Molecular and Cellular Biology*, *33*(11), 2149–2162. http://dx.doi.org/10.1128/mcb.00008-13.

la Sala, A., Corinti, S., Federici, M., Saragovi, H. U., & Girolomoni, G. (2000). Ligand activation of nerve growth factor receptor TrkA protects monocytes from apoptosis. *Journal of Leukocyte Biology*, *68*(1), 104–110.

Lee, R., Kermani, P., Teng, K. K., & Hempstead, B. L. (2001). Regulation of cell survival by secreted proneurotrophins. *Science*, *294*(5548), 1945–1948. http://dx.doi.org/10.1126/science.1065057.

Lee, F. S., Kim, A. H., Khursigara, G., & Chao, M. V. (2001). The uniqueness of being a neurotrophin receptor. *Current Opinion in Neurobiology*, *11*(3), 281–286.

Levine, R. J., Maynard, S. E., Qian, C., Lim, K. H., England, L. J., Yu, K. F., … Karumanchi, S. A. (2004). Circulating angiogenic factors and the risk of preeclampsia. *The New England Journal of Medicine*, *350*(7), 672–683. http://dx.doi.org/10.1056/NEJMoa031884.

Lewin, G. R., & Barde, Y. A. (1996). Physiology of the neurotrophins. *Annual Review of Neuroscience*, *19*, 289–317. http://dx.doi.org/10.1146/annurev.ne.19.030196.001445.

Li, Q., Chen, C. F., Wang, D. Y., Lu, Y. T., Huan, Y., Liu, M., … Hu, X. (2015). Transplantation of umbilical cord blood mononuclear cells increases levels of nerve growth factor in the cerebrospinal fluid of patients with autism. *Genetics and Molecular Research*, *14*(3), 8725–8732. http://dx.doi.org/10.4238/2015.July.31.21.

Lin, C. Y., Chang, S. L., Fong, Y. C., Hsu, C. J., & Tang, C. H. (2013). Apoptosis signal-regulating kinase 1 is involved in brain-derived neurotrophic factor (BDNF)-enhanced cell motility and matrix metalloproteinase 1 expression in human chondrosarcoma cells. *International Journal of Molecular Sciences*, *14*(8), 15459–15478. http://dx.doi.org/10.3390/ijms140815459.

Loch, A. A., Zanetti, M. V., de Sousa, R. T., Chaim, T. M., Serpa, M. H., Gattaz, W. F., … Machado-Vieira, R. (2015). Elevated neurotrophin-3 and neurotrophin 4/5 levels in unmedicated bipolar depression and the effects of lithium. *Progress in Neuropsychopharmacology and Biological Psychiatry*, *56*, 243–246. http://dx.doi.org/10.1016/j.pnpbp.2014.09.014.

Loftin, R. W., Habli, M., Snyder, C. C., Cormier, C. M., Lewis, D. F., & Defranco, E. A. (2010). Late preterm birth. *Reviews in Obstetrics & Gynecology*, *3*(1), 10–19.

Maisonpierre, P. C., Belluscio, L., Squinto, S., Ip, N. Y., Furth, M. E., Lindsay, R. M., & Yancopoulos, G. D. (1990). Neurotrophin-3: A neurotrophic factor related to NGF and BDNF. *Science*, *247*(4949 Pt. 1), 1446–1451.

Malamitsi-Puchner, A., Economou, E., Rigopoulou, O., & Boutsikou, T. (2004). Perinatal changes of brain-derived neurotrophic factor in pre- and fullterm neonates. *Early Human Development*, *76*(1), 17–22.

Malamitsi-Puchner, A., Nikolaou, K. E., Economou, E., Boutsikou, M., Boutsikou, T., Kyriakakou, M., … Hassiakos, D. (2007). Intrauterine growth restriction and circulating neurotrophin levels at term. *Early Human Development*, *83*(7), 465–469. http://dx.doi.org/10.1016/j.earlhumdev.2006.09.001.

Martinotti, G., Di Iorio, G., Marini, S., Ricci, V., De Berardis, D., & Di Giannantonio, M. (2012). Nerve growth factor and brain-derived neurotrophic factor concentrations in schizophrenia: A review. *Journal of Biological Regulators and Homeostatic Agents*, *26*(3), 347–356.

Marvin, K. W., Keelan, J. A., Eykholt, R. L., Sato, T. A., & Mitchell, M. D. (2002). Expression of angiogenic and neurotrophic factors in the human amnion and choriodecidua. *American Journal of Obstetrics and Gynecology*, *187*(3), 728–734. http://dx.doi.org/10.1067/mob.2002.126286.

Matijevic, R., Meekins, J. W., Walkinshaw, S. A., Neilson, J. P., & McFadyen, I. R. (1995). Spiral artery blood flow in the central and peripheral areas of the placental bed in the second trimester. *Obstetrics and Gynecology*, *86*(2), 289–292.

Matoba, N., Yu, Y., Mestan, K., Pearson, C., Ortiz, K., Porta, N., … Wang, X. (2009). Differential patterns of 27 cord blood immune biomarkers across gestational age. *Pediatrics*, *123*(5), 1320–1328. http://dx.doi.org/10.1542/peds.2008-1222.

Mayeur, S., Lukaszewski, M. A., Breton, C., Storme, L., Vieau, D., & Lesage, J. (2011). Do neurotrophins regulate the feto-placental development? *Medical Hypotheses*, *76*(5), 726–728. http://dx.doi.org/10.1016/j.mehy.2011.02.008.

Mayeur, S., Silhol, M., Moitrot, E., Barbaux, S., Breton, C., Gabory, A., … Lesage, J. (2010). Placental BDNF/TrkB signaling system is modulated by fetal growth disturbances in rat and human. *Placenta*, *31*(9), 785–791. http://dx.doi.org/10.1016/j.placenta.2010.06.008.

McMahon, S. B. (1996). NGF as a mediator of inflammatory pain. *Philosophical Transactions of the Royal Society of London. Series B, Biological Sciences*, *351*(1338), 431–440. http://dx.doi.org/10.1098/rstb.1996.0039.

Mistry, H. D., Kurlak, L. O., Williams, P. J., Ramsay, M. M., Symonds, M. E., & Broughton Pipkin, F. (2010). Differential expression and distribution of placental glutathione peroxidases 1, 3 and 4 in normal and preeclamptic pregnancy. *Placenta*, *31*(5), 401–408. http://dx.doi.org/10.1016/j.placenta.2010.02.011.

Nakamura, K., Martin, K. C., Jackson, J. K., Beppu, K., Woo, C. W., & Thiele, C. J. (2006). Brain-derived neurotrophic factor activation of TrkB induces vascular endothelial growth factor expression via hypoxia-inducible factor-1alpha in neuroblastoma cells. *Cancer Research*, *66*(8), 4249–4255. http://dx.doi.org/10.1158/0008-5472.can-05-2789.

Nico, B., Mangieri, D., Benagiano, V., Crivellato, E., & Ribatti, D. (2008). Nerve growth factor as an angiogenic factor. *Microvascular Research*, *75*(2), 135–141. http://dx.doi.org/10.1016/j.mvr.2007.07.004.

Nikolaou, K. E., Malamitsi-Puchner, A., Boutsikou, T., Economou, E., Boutsikou, M., Puchner, K. P., … Hassiakos, D. (2006). The varying patterns of neurotrophin changes in the perinatal period. *The Annals of the New York Academy of Sciences*, *1092*, 426–433. http://dx.doi.org/10.1196/annals.1365.041.

Park, M. J., Kwak, H. J., Lee, H. C., Yoo, D. H., Park, I. C., Kim, M. S., … Hong, S. I. (2007). Nerve growth factor induces endothelial cell invasion and cord formation by promoting matrix metalloproteinase-2 expression through the phosphatidylinositol 3-kinase/Akt signaling pathway and AP-2 transcription factor. *The Journal of Biological Chemistry*, *282*(42), 30485–30496. http://dx.doi.org/10.1074/jbc.M701081200.

Prencipe, G., Minnone, G., Strippoli, R., De Pasquale, L., Petrini, S., Caiello, I., … Bracci-Laudiero, L. (2014). Nerve growth factor downregulates inflammatory response in human monocytes through TrkA. *The Journal of Immunology*, *192*(7), 3345–3354. http://dx.doi.org/10.4049/jimmunol.1300825.

Razavi, S., Nazem, G., Mardani, M., Esfandiari, E., Salehi, H., & Esfahani, S. H. (2015). Neurotrophic factors and their effects in the treatment of multiple sclerosis. *Advanced Biomedical Research*, *4*, 53. http://dx.doi.org/10.4103/2277-9175.151570.

Reichardt, L. F. (2006). Neurotrophin-regulated signalling pathways. *Philosophical Transactions of the Royal Society of London. Series B, Biological Sciences*, *361*(1473), 1545–1564. http://dx.doi.org/10.1098/rstb.2006.1894.

Rogers, L. K., & Velten, M. (2011). Maternal inflammation, growth retardation, and preterm birth: Insights into adult cardiovascular disease. *Life Sciences*, *89*(13-14), 417–421. http://dx.doi.org/10.1016/j.lfs.2011.07.017.

Roland, L., Beauchemin, D., Acteau, G., Fradette, C., St-Pierre, I., & Bilodeau, J. F. (2010). Effects of labor on placental expression of superoxide dismutases in preeclampsia *Placenta*, *31*(5), 392–400. http://dx.doi.org/10.1016/j.placenta.2010.02.007.

Sahay, A. S., Sundrani, D. P., Wagh, G. N., Mehendale, S. S., & Joshi, S. R. (2015a). Neurotrophin levels in different regions of the placenta and their association with birth outcome and blood pressure. *Placenta*, *36*(8), 938–943. http://dx.doi.org/10.1016/j.placenta.2015.06.006.

Sahay, A. S., Sundrani, D. P., Wagh, G. N., Mehendale, S. S., & Joshi, S. R. (2015b). Regional differences in the placental levels of oxidative stress markers in pre-eclampsia. *International Journal of Gynaecology and Obstetrics*, *129*(3), 213–218. http://dx.doi.org/10.1016/j.ijgo.2015.03.001.

Szuhany, K. L., Bugatti, M., & Otto, M. W. (2015). A meta-analytic review of the effects of exercise on brain-derived neurotrophic factor. *Journal of Psychiatric Research*, *60*, 56–64. http://dx.doi.org/10.1016/j.jpsychires.2014.10.003.

Tometten, M., Blois, S., & Arck, P. C. (2005). Nerve growth factor in reproductive biology: Link between the immune, endocrine and nervous system? *Chemical Immunology and Allergy*, *89*, 135–148. http://dx.doi.org/10.1159/000087962.

Toti, P., Ciarmela, P., Florio, P., Volpi, N., Occhini, R., & Petraglia, F. (2006). Human placenta and fetal membranes express nerve growth factor mRNA and protein. *Journal of Endocrinological Investigation*, *29*(4), 337–341.

Uguz, F., Sonmez, E. O., Sahingoz, M., Gokmen, Z., Basaran, M., Gezginc, K., … Tasyurek, E. (2013). Maternal generalized anxiety disorder during pregnancy and fetal brain development: A comparative study on cord blood brain-derived neurotrophic factor levels. *Journal of Psychosomatic Research*, *75*(4), 346–350. http://dx.doi.org/10.1016/j.jpsychores.2013.04.010.

Vera, C., Tapia, V., Vega, M., & Romero, C. (2014). Role of nerve growth factor and its TRKA receptor in normal ovarian and epithelial ovarian cancer angiogenesis. *Journal of Ovarian Research*, 7, 82. http://dx.doi.org/10.1186/s13048-014-0082-6.

Wang, H., Wang, R., & Thrimawithana, T. (2014). The nerve growth factor signaling and its potential as therapeutic target for glaucoma. *BioMed Research International*, *2014*, 759473. http://dx.doi.org/10.1155/2014/759473.

Wiener, C. D., de Mello Ferreira, S., Pedrotti Moreira, F., Bittencourt, G., de Oliveira, J. F., Lopez Molina, M., … Oses, J. P. (2015). Serum levels of nerve growth factor (NGF) in patients with major depression disorder and suicide risk. *Journal of Affective Disorders*, *184*, 245–248. http://dx.doi.org/10.1016/j.jad.2015.05.067.

Wu, A., Ying, Z., & Gomez-Pinilla, F. (2004). The interplay between oxidative stress and brain-derived neurotrophic factor modulates the outcome of a saturated fat diet on synaptic plasticity and cognition. *The European Journal of Neuroscience*, *19*(7), 1699–1707. http://dx.doi.org/10.1111/j.1460-9568.2004.03246.x.

Zacchigna, S., Lambrechts, D., & Carmeliet, P. (2008). Neurovascular signalling defects in neurodegeneration. *Nature Reviews. Neuroscience*, *9*(3), 169–181. http://dx.doi.org/10.1038/nrn2336.

CHAPTER ELEVEN

Development of Precision Small-Molecule Proneurotrophic Therapies for Neurodegenerative Diseases

J. Janssens*[,1], D. Lu[†,1], B. Ni[†,1], W. Chadwick[†], S. Siddiqui[†], A. Azmi*, H. Etienne*, A. Jushaj*, J. van Gastel*, B. Martin[‡], S. Maudsley*[,†,2]

*Translational Neurobiology Group, University of Antwerp, Antwerpen, Belgium
[†]Receptor Pharmacology Unit, National Institute on Aging, National Institutes of Health, Baltimore MD United States
[‡]Metabolism Unit, National Institute on Aging, National Institutes of Health, Baltimore MD United States
[2]Corresponding author: e-mail address: stuart.maudsley@molgen.vib-ua.be

Contents

1. Introduction 264
2. Mammalian Neurotrophin Ligand–Receptor System 265
   2.1 Neurotrophin Ligands 266
   2.2 Neurotrophin Receptors 268
3. Neurotrophin Ligand–Receptor Biological Functionality 270
   3.1 Neurotrophin Processing and Cell Survival 271
   3.2 Regulation of Synaptic Plasticity by Neurotrophins 274
   3.3 Neurotrophins and Age-Related Neurodegeneration 276
4. Proneurotrophin-Based Therapy for Age-Related Neurodegeneration 280
   4.1 Effective Drug Design-Related Considerations for Refining Neurotrophic Activity 282
   4.2 Small-Molecule Neurotrophic Groups and Lead Compounds 287
   4.3 Clinical Activity of Small-Molecule Neurotrophic Therapeutics 292
5. Future Developments in Small-Molecule Neurotrophic Agents 294
References 295

Abstract

Age-related neurodegenerative diseases, such as Alzheimer's disease, will represent one of the largest future burdens on worldwide healthcare systems due to the increasing proportion of elderly in our society. As deficiencies in neurotrophins are implicated in the pathogenesis of many age-related neurodegenerative disorders, it is reasonable to consider that global neurotrophin resistance may also become a major healthcare

[1] Authors contributed equally to this manuscript.

Vitamins and Hormones, Volume 104
ISSN 0083-6729
http://dx.doi.org/10.1016/bs.vh.2016.10.006

threat. Central nervous system networks are effectively maintained through aging by neuroprotective and neuroplasticity signaling mechanisms which are predominantly controlled by neurotrophin receptor signaling. Neurotrophin receptors are single pass receptor tyrosine kinases that form dimeric structures upon ligand binding to initiate cellular signaling events that control many protective and plasticity-related pathways. Declining functionality of the neurotrophin ligand–receptor system is considered one of the hallmarks of neuropathological aging. Therefore, it is imperative to develop effective therapeutic strategies to contend with this significant issue. While the therapeutic applications of cognate ligands for neurotrophin receptors are limited, the development of nonpeptidergic, small-molecule ligands can overcome these limitations, and productively regulate this important receptor system with beneficial effects. Using our advanced knowledge of the high-dimensionality complexity of receptor systems, the future generation of precision medicines targeting these systems will be an attainable goal.

1. INTRODUCTION

Neurotrophins are large peptidergic growth factors that are essential for the differentiation, growth, and dynamic development of neuronal cells in the vertebrate nervous system. In the central nervous system (CNS), neurotrophins and their respective cognate tyrosine kinase receptors modulate multiple signaling pathways that promote neuronal survival, neurite outgrowth, and synaptic plasticity. Hence, it is not surprising that deficiencies in the functionality of the neurotrophin signaling system are implicated in the pathogenesis of many age-related neurodegenerative disorders, such as Alzheimer's disease (AD), frontotemporal dementia (FTD), Huntington's disease (HD), and Parkinson's disease (PD). With increasing age, most organs encounter difficulties in maintaining their regenerative and protective capacities to cope with injury or disease. Neuronal cells of the CNS are especially vulnerable for the age-related degradation due to their high energy demands and postmitotic nature. The maintenance of this sensitive postmitotic state is physiologically important as inadvertent cell cycle reentry of these cells is actually linked to the onset of neurodegenerative diseases such as AD (Maudsley & Mattson, 2006). Neurodegeneration can occur when neuronal cells fail to protectively respond to age-related increases in oxidative, metabolic, and/or ionic stress which results in the accumulation of damage to proteins, lipids, and DNA. The presence of either a surfeit or a deficit of neurotrophic signaling in these circumstances often determines whether neurons are able to resist or succumb to the disease pathogenesis (Ichim, Tauszig-Delamasure, & Mehlen, 2012; Mattson, 2004; Maudsley & Mattson, 2006).

In this regard, it is clear that the effective therapeutic targeting of the neurotrophin receptor system may represent one of the most important future pharmacological targets. Recent research has also demonstrated that not only do neurotrophin receptor ligands control neurophysiology but also they contribute significantly to whole somatic energy metabolism (Mou et al., 2015), cardiovascular health (Golden et al., 2010), and psychosocial behavior (Terracciano et al., 2010)—therefore, the generation of effective and tractable agents to control this system across lifespan will create a truly systems level therapeutic agent (Janssens et al., 2014).

Neurotrophic factors can be classified into three main subgroups including (i) the neurotrophin family, (ii) the glial cell line-derived neurotrophic factor (GDNF) protein family, and (iii) neuropoietic cytokines (Harandi, Lindquist, Kolan, Brannstrom, & Liu, 2014). In addition, growth factors such as insulin-like growth factor 1 (IGF-1), ciliary neurotrophic factor (CNTF), and vascular endothelial growth factor (VEGF) have also been described in literature to be involved in the pathogenesis of the aging brain (Shetty, Hattiangady, & Shetty, 2005). In this review, however, we specifically focus on the signaling functions mediated by members of the neurotrophin protein family and their potential exploitation for treatment of neurodegenerative disorders. Since neurotrophins are thought to play an important role in the etiology of different neurological disorders, these signaling molecules offer the promise to provide primary or adjunctive therapy for these devastating diseases. Therefore, we will elaborate on the current status of therapeutic interventions using neurotrophins as well as novel therapeutic approaches including peptide ligands or small-molecule mimetics.

2. MAMMALIAN NEUROTROPHIN LIGAND–RECEPTOR SYSTEM

The current understanding of the complex molecular interplay between neurotrophins and their receptors under physiological and pathological conditions is evolving rapidly. The era of neurotrophin research was triggered with the initial discovery of nerve growth factor (NGF) in the 1950s (Levi-Montalcini, 1987). This finding led to the formulation of the "*neurotrophic factor hypothesis*" which states that "*developing neurons compete with each other for a limited supply of neurotrophic factors provided by the target tissue.*" Essentially, survival of these neurons is strongly dependent on the amount of a neurotrophic factor that these cells can receive, via the magnitude or type of either the ligand stimulus or the generated protective signal. Therefore, neurons possessing an enhanced sensitivity/responsivity to neurotrophic support survive, while those that do not undergo cellular atrophy and death. Subsequent research over the past years, however, has demonstrated that neurotrophic actions are considerably more complex and diverse than initially predicted (Dawbarn & Allen, 2003). The neurotrophic hypothesis has been broadened by the demonstration that neurotrophins are central to many facets of CNS function and especially to neuronal signaling promoting survival and differentiation in both the developing and adult nervous system. In addition to these classical effects on neuronal survival, neurotrophins also modulate signaling pathways involved in the regulation of axonal/dendritic growth, neurotransmitter release, long-term potentiation (LTP), organization of synaptic connections, and synaptic plasticity. The surprising discovery that neurotrophins and their receptors do not exist in *Caenorhabditis elegans* and *Drosophila melanogaster*, implies that these proteins are primarily involved in higher-order organization of complex neuronal signaling rather than required for the development of neuronal circuits (Chao, 2003; Huang & Reichardt, 2003). It is well established that the overall levels of neurotrophins have to be correctly maintained as these levels will determine the balance between cell survival and apoptosis. Additionally, altered neurotrophin levels have profound effects on neuronal phenomena including myelination, regeneration, pain, depression, and neurodegeneration.

2.1 Neurotrophin Ligands

Secreted neurotrophins are essential for neuronal survival, differentiation, and axon localization toward target sites in the CNS and peripheral nervous system (PNS) (Ichim et al., 2012). Today, the mammalian neurotrophin family comprises four well-described neurotrophins including NGF, brain-derived neurotrophic factor (BDNF), neurotrophin-3 (NT3), and neurotrophin-4 (NT4) (also known as NT5) (Chao, 2003). It should be noted that NT4 was initially found in *Xenopus* ovary and may represent only a species variation of NT5, which has been characterized in higher vertebrates (Ip et al., 1992). This is the reason why the terminology NT4/5 is often applied. Neurotrophin-6 (NT-6) and neurotrophin-7 (NT-7) are also known but have only been described in fish and not in the mammalian brain (Gotz et al., 1994). The first member of the neurotrophins, NGF was discovered as a target-derived protein that promotes the survival and growth of sympathetic and sensory neurons during development (Levi-Montalcini, 1987). BDNF was initially found in pig brain and is the most widely distributed neurotrophin in the CNS, where it plays a crucial role in regulating axonal and dendritic growth/guidance, LTP, and participation in neurotransmitter release (Barde, Edgar, & Thoenen, 1982; Chao, 2003; Tapia-Arancibia, Aliaga, Silhol, & Arancibia, 2008). The other neurotrophins, NT3 and NT4 were identified several years later and their discovery completed the intricate network of prosurvival neurotrophins that orchestrate the generation and maintenance of diverse neuronal populations (Ibanez, 1995).

Neurotrophins are closely related molecules that share a sequence identity of approximately 50% and are highly conserved among human, mouse, and rat with 90% homology (Bradshaw et al., 1994; Wiesmann & de Vos, 2001). A common feature shared by all neurotrophins is that they are initially synthetized as long precursors (prepronurotrophins) in the endoplasmic reticulum. Prepronurotrophins have an N-terminal signal peptide, which is necessary for intracellular transport and subsequent secretion of a neurotrophic factor (Halban & Irminger, 1994). Cleavage of the signal peptide yields immature precursors or proneurotrophins (30–35 kDa) which can undergo further posttranslational modifications (i.e., glycosylation, amidation, etc.) in the Golgi apparatus before ending up in secretory vesicles (Lessmann & Brigadski, 2009). Complete maturation of neurotrophins is achieved by proteolytic cleavage of proneurotrophins to produce mature neurotrophins (12–13 kDa) by different proprotein convertases either at intra- or extracellular locations (Lessmann, Gottmann, & Malcangio, 2003; Nykjaer, Willnow, & Petersen, 2005). Proteolytic processing of proneurotrophins can be executed by a range of convertases expressed in different cell types including proprotein convertase subtilisin kexin (PCSK) enzymes and furin (also known as PCSK3). These convertases bind to specific recognition motifs (RxKR) in the amino acid sequence which are present in the propeptides of neurotrophins. The amount of expressed convertases depends mainly on the cell type. Furin is expressed in almost every tissue, whereas PCSK1 or PCSK2 are principally expressed in neuronal cells (Bruno & Cuello, 2006; Lessmann & Brigadski, 2009; Longo & Massa, 2013). As mentioned earlier, formation of mature neurotrophins, like NGF and BDNF, can also occur extracellularly. In this case, however, proneurotrophins are processed by the serine protease plasmin or by matrix-metalloproteinases (MMPs) like MMP7 and MMP9 (Bruno & Cuello, 2006; Lessmann & Brigadski, 2009). The most significant protease that cleaves proneurotrophins is plasmin, which is generally expressed as an inactive zymogen, plasminogen, and activated by proteolytic cleavage of the tissue plasminogen activator (tPA) (Lee & Chao, 2001). Although plasminogen is widely expressed in the nervous system, it is exclusively expressed by neurons in the brain (Lu et al., 2005; Plow, Herren, Redlitz, Miles, & Hoover-Plow, 1995; Tsirka, 1997).

Once the biological active form of neurotrophins, i.e., either the precursor or mature protein, is secreted in the extracellular space, a stable noncovalently bound homodimer is formed (Rattenholl et al., 2001). Interestingly, secreted forms of proneurotrophins also serve as signaling molecules that often exhibit opposing biological effects compared to the activities mediated by mature neurotrophins. This implies that proteolytic processing of proneurotrophins can be considered as a powerful cellular mechanism to control the direct downstream signaling functions of neurotrophins (Lu, Pang, & Woo, 2005).

2.2 Neurotrophin Receptors

In order to elicit a downstream signaling response, both pro- and mature neurotrophins bind as dimers to two distinct classes of transmembrane receptors: the p75 neurotrophin receptor ($p75^{NTR}$) and different members of the tyrosine receptor kinase (Trk) family, which comprise TrkA, TrkB, and TrkC (Dechant & Barde, 2002; Huang & Reichardt, 2003). Both receptor classes demonstrate independent signaling properties which can elicit individual physiological responses dependent on the downstream signal transduction pathways that are activated upon neurotrophin binding (Hempstead, 2002; Patapoutian & Reichardt, 2001). Each mature neurotrophin shows binding specificity for a particular Trk receptor: NGF preferentially binds to TrkA, BDNF and NT4/5 to TrkB, and NT3 to TrkC (Klein, Jing, Nanduri, O'Rourke, & Barbacid, 1991; Klein, Lamballe, Bryant, & Barbacid, 1992; Klein, Nanduri, et al., 1991; Lamballe, Klein, & Barbacid, 1991). TrkB and TrkC receptors are widely expressed in a large number of neuronal cell populations of the CNS or PNS, whereas the expression of the TrkA receptor is more restricted to specific neuronal subtypes including sympathetic neurons and basal forebrain cholinergic neurons (BFCNs) (Sobreviela et al., 1994). All neurotrophins show different expression patterns through the PNS and CNS with NGF being the most dominant neurotrophic factor in the PNS for sympathetic and sensory neurons, whereas BDNF is considered to be the most abundantly expressed factor in the CNS corresponding to TrkA and TrkB expression patterns, respectively. Despite overlapping expression patterns of neurotrophins and neurotrophin receptors in different brain regions, certain neuronal mechanisms are modulated by specific receptor–ligand combinations including BDNF–TrkB-mediated promotion of synaptic plasticity and NT3–TrkC-mediated survival of peripheral proprioceptive neurons (Ernfors, Lee, Kucera, & Jaenisch, 1994; McMahon, Armanini, Ling, & Phillips, 1994; Poo, 2001).

Upon ligand binding, dimerization of the Trk receptors is induced followed by their autophosphorylation at specific tyrosine residues in the cytoplasmic domain. The phosphotyrosine residues subsequently function as specific binding sites for recruiting cytoplasmic signaling proteins that further activate downstream signaling cascades (Longo & Massa, 2013). Trk signaling modulated by binding of mature neurotrophins elicits predominantly neuronal survival and differentiation through activation of three principal signaling pathways, being the mitogen-activated protein kinase (MAPK), the Ras-dependent extracellular signal-regulated kinase (ERK) pathway, the phosphatidylinositol 3-kinase (PI3K)/protein kinase B (Akt) pathway, and the phospholipase Cγ1 (PLC-γ1) pathway (Longo et al., 2007; Reichardt, 2006; Weissmiller & Wu, 2012).

In contrast to the Trk receptor class, the $p75^{NTR}$ receptor binds both pro and mature forms of neurotrophins unselectively. $p75^{NTR}$ is a member of the tumor necrosis factor (TNF) receptor family and is structurally different from Trk receptors as it contains a cytoplasmic "death domain" involved in apoptosis and lacks known intrinsic kinase activity (Dawbarn & Allen, 2003; Liepinsh, Ilag, Otting, & Ibanez, 1997; Schecterson & Bothwell, 2010). To initiate cellular responses, the $p75^{NTR}$ receptor can cooperate with different coreceptors to form multimeric receptor complexes. Coreceptors that have been identified so far include Trk receptors, sortilin, Leucine-rich repeat and immunoglobulin-like domain-containing protein 1 (LINGO-1), and the Nogo-66 receptor (Lu et al., 2013). Interestingly, expression of the $p75^{NTR}$ receptor enhances the interaction between TrkA and NGF which indicates that $p75^{NTR}$ can confer ligand selectivity for the Trk receptors. In addition to these extracellular interactions, the intracellular domain (ICD) of the $p75^{NTR}$ receptor can also recruit different adaptor and signaling molecules. These proteins include neurotrophin receptor-interacting MAGE (melanoma-associated antigen) homologue (NRAGE), neurotrophin-associated cell death executor (NADE), TNF receptor-associated factors 2 and 6 (TRAF2 and TRAF6), and neurotrophin receptor interacting factor (NRIF) (Shu, Lu, Wei, Xiao, & Wang, 2015). $p75^{NTR}$-mediated signaling involves activation of numerous pathways such as nuclear factor κB (NF-κB),

JUN N-terminal kinase (JNK), Ras homolog gene family, member A (RhoA), MAPK, and PI3K–Akt signaling pathways. Dependent of the composition of the p75^{NTR} receptor complex and hence the expression levels of sortilin, Trk receptors, and adaptor proteins, neurotrophin signaling via these pathways induces diverse effects ranging from myelination, cell survival, cell death, inhibition of neurite outgrowth to regulation of proliferation (Longo & Massa, 2013).

3. NEUROTROPHIN LIGAND–RECEPTOR BIOLOGICAL FUNCTIONALITY

Similar to other forms of receptor tyrosine kinases, such as the epidermal growth factor receptor (EGFR), neurotrophin binding to Trk receptors canonically induces dimerization that promotes transphosphorylation of opposing monomers to create specific phosphotyrosine docking sites for the associated signaling adaptors (Siddiqui et al., 2012). The coherent assembly of signaling factors to the activated Trk species then stimulates several downstream signaling pathways classically associated with neurotrophic activity, including ERK, PI3K, and PLC-γ pathways. While standard dimerization is considered the basal signaling unit, it is highly likely that multimeric complexes of Trk receptors also exist that are potentially employed for creating additional diversity to neurotrophin-induced signaling output. Not only can the multimeric Trk complexes generate signaling diversity, the respective transphosphorylation sites can engender a bias in the output signaling function of the activated Trk species. This is exemplified by the phosphorylation of Tyr^{490} in TrkA which is most strongly associated with stimulation of ERK and PI3K pathways in contrast to phosphorylation of the Tyr^{785} site which entrains a bias toward PLC-y-based signaling (Huang & Reichardt, 2003). Analogous to the near continual activity status demonstrated by transmembrane G protein-coupled receptor (GPCR) systems (Luttrell, Maudsley, & Bohn, 2015; Maudsley, Martin, & Luttrell, 2005), the spatiotemporal formation of ligand-independent Trk dimers is a well-characterized phenomenon and is suggested to induce a form of basal ligand-independent activity (Shen & Maruyama, 2012). While receptor signaling systems are studied primarily at a proximal level, involving the activation of classical signal transduction cascades, a more relevant appreciation of receptor system functionality is found through investigation of the more (patho)-physiological ramifications of neurotrophin signaling. This aspect of receptor biology will be addressed in the following sections.

3.1 Neurotrophin Processing and Cell Survival

The complexity of neurotrophin receptor–ligand interfaces allows Trk and $p75^{NTR}$ receptors to present different binding sites with varying affinities to particular neurotrophins. The wide range of receptor–adaptor protein combinations enables neurotrophin receptor complexes to initiate a specific downstream response. Regulation of neuronal survival and apoptosis are key responses promoted by different neurotrophins and their specific maturation state. For decades researchers have focused on the prosurvival aspects of mature neurotrophins on neuronal cells as proneurotrophins were initially thought to be functionally inactive precursors. Pioneering work of Hempstead and colleagues, however, broke with this traditional view and demonstrated that proneurotrophins were active inducers of apoptosis by activating the $p75^{NTR}$-mediated receptor signaling (Lee, Kermani, Teng, & Hempstead, 2001). Hence, it became apparent that both pro- and mature neurotrophins can exert diametrically opposing effects through binding on their own cognate receptor. The proforms typically induce neuronal apoptosis during development but also under pathological conditions or aging of the nervous system. By contrast, the mature forms rather act as trophic factors to support survival, growth, and differentiation of neurons. In addition, precursor and mature forms may differentially affect cognitive functions as shown for (pro)-BDNF (Nykjaer & Willnow, 2012).

Mature neurotrophins preferentially bind to the Trk receptors and their concomitant activation and downstream signaling are well known for its prosurvival effects. Upon activation by neurotrophin binding, Trk family members recruit and increase the phosphorylation of PLC-γ and Src homologous and collagen-like (Shc) adaptor proteins, which leads to activation of PI3K–Akt and MAPK–ERK cascades. These signaling pathways are the principal pathways responsible for neuronal survival. The PI3K–Akt pathway suppresses cell death by inhibiting the apoptotic activities of B-cell lymphoma-associated death protein (BAD) and activation of the survival-promoting serine–threonine protein kinase Akt (Chung, Grammer, Lemon, Kazlauskas, & Blenis, 1994; del Peso, Gonzalez-Garcia, Page, Herrera, & Nunez, 1997). The other signaling cascade, MAPK–ERK, stimulates the activity and expression of antiapoptotic proteins, including the B-cell lymphoma 2 and the transcription factor cyclic AMP responsive element-binding protein (CREB). CREB-induced transcriptional effects influence cell cycle, neurite outgrowth, and synaptic plasticity (Lonze, Riccio, Cohen, & Ginty, 2002).

In contrast to the Trk receptors, the interactions of mature neurotrophins with $p75^{NTR}$ receptors produce more variable signaling outcomes depending on the cellular context. The signaling variety of $p75^{NTR}$ is mediated by its interactions with a wide range of protein partners, including coreceptors and adaptor proteins, to form multimeric receptor complexes that produce a specific cellular response ranging from apoptosis to myelination (Lu, Pang, & Woo, 2005). When Trk receptors are not ligated or present, mature neurotrophins can promote neural apoptosis and activation of the NF-κB transcription factors and JNK by binding to the $p75^{NTR}$ receptor (Casaccia-Bonnefil, Carter, Dobrowsky, & Chao, 1996; Huang & Reichardt, 2003). NF-κB proteins have emerged as major regulators of programmed cell death whether via apoptosis or necrosis (Dutta, Fan, Gupta, Fan, & Gelinas, 2006). The mechanisms, by which $p75^{NTR}$ initiates these intracellular death signals, are thought to involve interactions with different intracellular adaptor proteins and in particular members of the TRAF family (Lu, Pang, & Woo, 2005; Ye et al., 1999). Although seven TRAF proteins have been described in mammals (Inoue et al., 2000), not every TRAF protein has the same effect on cell death induction. Coexpression of TRAF2 with $p75^{NTR}$ is able to enhance cell death, whereas coexpression of TRAF6 was protective for the cell. Moreover, TRAF4 is able to inhibit the NF-κB response, whereas TRAF2 and TRAF6 enhanced $p75^{NTR}$-induced NF-κB activation (Ye et al., 1999). In addition to the TRAF proteins, $p75^{NTR}$-mediated apoptosis can also occur via interactions of $p75^{NTR}$ with NRAGE or NADE when NGF is bound, but not when BDNF, NT-3, or NT-4/5 were bound. NRAGE blocks the physical association of $p75^{NTR}$ with the TrkA receptor which drives signaling toward neuronal apoptosis (Salehi, Verhaagen, Dijkhuizen, & Swaab, 1996). NADE-mediated apoptosis involves the induction of caspase-2 and caspase-3 activities and concomitant fragmentation of nuclear DNA. Interestingly, in the absence of $p75^{NTR}$ receptor, NADE failed to induce apoptosis, suggesting that NADE expression is necessary but insufficient for $p75^{NTR}$-mediated apoptosis (Mukai et al., 2000).

Compared to mature neurotrophins, the principal mode of action of proneurotrophins is to promote cell death by interacting with the $p75^{NTR}$ receptor at the neuronal cell surface. Surprisingly, selective activation of cell death-related pathways by proneurotrophins is relying on the interaction of $p75^{NTR}$ with sortilin (also known as vacuolar sorting protein 10; VPS10) as the primary coreceptor to transmit intracellular death signals. Sortilin specifically binds the prodomain of both proNGF and proBDNF which elicit cell death in conditions of acute or chronic stress of the nervous system, but also during normal development and senescence of the nervous system (Chen et al., 2010; Nielsen et al., 2001). Interestingly, the expression profile of sortilin during aging in the nervous system is altered and aged neurons are more prone to pro-NGF-induced apoptosis compared to young neurons (Al-Shawi et al., 2008; Nykjaer & Willnow, 2012). Intracellular signal transduction following activation of the sortilin/p75^{NTR} receptor complex requires the regulated intramembrane proteolysis of the p75^{NTR} receptor by metalloproteinases, such as α- and γ-secretases (Skeldal, Matusica, Nykjaer, & Coulson, 2011). Upon release of the ICD of p75^{NTR}, the signaling pathway that induces apoptosis consists of different steps involving c-JNK3 signaling and caspase activation, but also translocation of the cleaved ICD to the nucleus. Several adaptor proteins, such as TRAFs and NRIF, assist during the translocation to the nucleus. How the released ICD of p75^{NTR} potentiates apoptosis in the nucleus is currently not completely understood but it is assumed that the ICD functions as a transcriptional modifier in the nucleus where it activates or represses neurotrophin-related genes (Skeldal et al., 2011). Regulation of gene expression represents a major mechanism by which proneurotrophins-mediated cell death can be controlled. Notably, expression alterations have mainly been observed for proneurotrophins and p75^{NTR}, whereas sortilin levels were less affected. However, control of intracellular trafficking of sortilin might provide an alternative route to sensitize neurons for death as approximately 90% of sortilin populates intracellular compartments such as the trans-Golgi network and early endosomes (Chen et al., 2005; Morris et al., 1998; Vaegter et al., 2011).

Besides the proapoptotic actions of proneurotrophins, recent reports have indicated that these neurotrophin precursors can also promote, or at least not impair, cell survival through stimulation of Trk receptors. Although these interactions occur with lower affinity, the prosurvival signaling potential is primary dependent on the cellular context, and more specifically on the ratio of Trk/p75^{NTR} receptors expressed in the membrane (Masoudi et al., 2009). Cells with a high receptor ratio are more prone to survive while those with a low ratio are more likely to undergo apoptosis (Clewes et al., 2008; Volosin et al., 2006).

Hence, these observations suggest that sortilin can act as a molecular switch which determines whether a neuronal cell, expressing Trk or p75^{NTR} receptors, will respond to proneurotrophins by activating pro- or antiapoptotic signaling pathways. In the presence of the sortilin receptor, proNGF is able to bind the p75^{NTR} receptor to elicit cell death, whereas in the absence of sortilin this is not possible. In this condition, proNGF is further processed into mature NGF which stimulates survival signals via TrkA after recruitment of p75^{NTR} (Nykjaer et al., 2004; Willnow, Petersen, & Nykjaer, 2008).

Consequently, it remains a challenging task for a cell to establish the exact signaling conditions that are responsible for cell survival or apoptosis, as both neurotrophin receptor classes are expressed in the nervous system and often by the same cell type. Moreover, the spatial and temporal balance between the two receptor classes (i.e., p75^{NTR} and Trk receptors) present on the membrane, as well as the expression status of secreted neurotrophins (pro- or mature forms), receptors, and convertases in- or outside the cell function as critical cellular regulators that determine whether a neuronal cell will survive or undergo apoptosis.

3.2 Regulation of Synaptic Plasticity by Neurotrophins

Besides their well-established regulatory functions in neuronal survival and differentiation, emerging evidence is pointing also toward a key role for neurotrophins in synaptic plasticity and hence in learning and memory functions. Among all neurotrophins, NT3 and especially BDNF are by far the best characterized neurotrophins for their role in coordinating long-term synaptic modifications. BDNF and NT3 exert specific and partially opposing effects on the dendritic and axonal morphology of central neurons, with the effects depending on the specific brain region, neuron type, and cellular layer in which the neurons are located (Vicario-Abejon, Owens, McKay, & Segal, 2002). The molecular effects of BDNF and NT-3 on synaptic plasticity can be of a permissive or an instructive nature (Gomez-Palacio-Schjetnan & Escobar, 2013). In the resting or permissive mode, neurotrophins are secreted at low levels which provide trophic regulation of synaptic functions, including the ability to produce LTP. In the active or instructive mode, neurotrophins are secreted at a higher level in response to intense synaptic activity which directly modifies the communication and morphology of synaptic contacts (Poo, 2001). BDNF is suggested to be a permissive factor required for the induction and maintenance of LTP in the hippocampus and neocortex of mammals. The actions carried out by BDNF include its capacity to contribute to the stabilization and maturation of already-existing synapses, as well as to generate new synaptic contacts. This is supported by the finding that LTP induction results in increased levels of BDNF and its TrkB receptor (Figurov, Pozzo-Miller, Olafsson, Wang, & Lu, 1996; Gomez-Palacio-Schjetnan & Escobar, 2013; Vicario-Abejon et al., 2002). In line with the stringent regulation of neuronal survival by pro- and mature neurotrophins, the conversion of proBDNF to its mature form also influences synaptic plasticity and the development of LTP in the hippocampus and other brain regions (Cowansage, LeDoux, & Monfils, 2010; Leal, Comprido, & Duarte, 2014). Proteolytic cleavage of proBDNF is performed by the extracellular protease plasmin (Lee & Chao, 2001), which is activated by tPA, an enzyme known to be important in LTP. Notably, the increased proBDNF levels reported in tPA and plasminogen deficient mice as well as the rescue of LTP by application of mature BDNF, indicate that the conversion of proBDNF into mature BDNF by the tPA/plasmin cascade is a critical step in the induction of LTP (Pang et al., 2004). Conversely, can one expect opposing functional roles of proBDNF when it remains unprocessed in the synaptic cleft? Given the fact that pro- and mature neurotrophins bind different receptor complexes in the regulation of cell survival, one can hypothesize that proBDNF might facilitate long-term depression (LTD) through binding of the $p75^{NTR}$ receptor (Woo et al., 2005). Although the biological role of the $p75^{NTR}$ receptor in synaptic plasticity is not completely understood, it has been demonstrated that proBDNF facilitated LTD through the activation of the GluN2B-containing NMDA (N-methyl-D-aspartate)-type receptors and the pan-neurotrophin receptor $p75^{NTR}$ (Mizui et al., 2015). The role of $p75^{NTR}$ in synaptic depression involves enhanced expression of the NMDAR subunit NR2B, a key molecule that is implicated in LTD but not in LTP (Liu et al., 2004; Massey et al., 2004). ProBDNF also enhanced the trafficking of AMPA (α-amino-3-hydroxy-5-methyl-4-isoxazolepropionic acid) receptors (AMPARs), a synaptic mechanism crucial for LTD regulation (Mizui et al., 2015).

The induction of either LTP or LTD is often accompanied by structural changes of synapses, and particularly the morphology and density of dendritic spines are altered. Morphological alterations following the induction of hippocampal LTP are characterized by an increased spine density, formation of new spines or new spine synapses (Engert & Bonhoeffer, 1999; Maletic-Savatic, Malinow, & Svoboda, 1999; Toni, Buchs, Nikonenko, Bron, & Muller, 1999; Trommald, Hulleberg, & Andersen, 1996). In contrast to the LTP-associated changes, shrinkage or retraction of dendritic spines seems to be linked to LTD induction (Jovanovic, Czernik, Fienberg, Greengard, & Sihra, 2000; Stoop & Poo, 1995). Extensive research on hippocampal neurons has indicated that proBDNF and mature BDNF can modulate the dendritic morphology and spine density. Long-term exposure of these neuronal cultures with BDNF has promoted axonal branching, dendritic growth, and refinement of synapses (Boulanger & Poo, 1999; McAllister, Katz, & Lo, 1996; Meyer-Franke, Kaplan, Pfrieger, & Barres, 1995; Patterson et al., 1996), whereas a preliminary study indicated that $p75^{NTR}$ can negatively modulate morphology and spine number (Patterson et al., 1996). Although the role of neurotrophins in synaptic function and plasticity is well established, the signaling pathways that modulate the accompanying morphological changes are not well understood.

3.3 Neurotrophins and Age-Related Neurodegeneration

It is well known that maintaining the overall levels of pro- and mature neurotrophins is crucial to control the balance between cell survival and apoptosis, as well as modulating synaptic plasticity. Indeed, given the idea that neurotrophins are key regulators of these functions in neuronal cells, it might not be surprising that alterations in the expression level of neurotrophins or their respective receptors are associated with a significant number of neurodegenerative disorders, including AD, FTD, HD, and PD, but also with psychiatric disorders such as depression, bipolar disorder, and schizophrenia. Among all dementia subtypes, AD has received perhaps more international research interest than any other CNS neurodegenerative disorder. It is therefore not surprising that our appreciation of neurotrophin functionality in the brain is closely interwoven with the neurotrophic deficits found in AD patients.

AD is one of the most common forms of dementia as it accounts for up to 75% of dementia cases. AD patients clinically present with progressive loss of memory and disturbance of additional cognitive functions such as word finding, reasoning, and problem solving (McKhann et al., 2011). Accumulation of extracellular diffuse and neuritic amyloid plaques, composed of amyloid β-peptide (Aβ), and intraneuronal neurofibrillary tangles (NFTs) composed of hyperphosphorylated protein tau (p-tau) are considered pathological hallmarks of the disease. The Aβ peptides are generated through amyloid precursor protein (APP) proteolysis and both form the cornerstone of the "amyloid cascade hypothesis" (Barnham et al., 2003; Gerlai, 2001; Hardy & Selkoe, 2002). This hypothesis states that the accumulation of Aβ is the causative agent of AD by destroying the synapses, inducing the formation of NFTs, and ultimately inducing neuronal loss (Struble, Ala,

Patrylo, Brewer, & Yan, 2010). Processing of APP is performed by three different protease enzymes: α-, β-, and γ-secretases (Mattson, 2004; Shoji et al., 1992). Two of them (β- and γ-secretase) are constitutively active and they are involved in the amyloidogenic pathway. Altered amyloid processing causes amyloid burden in the brain regions involved in learning, memory, and emotional behavior such as the entorhinal cortex, hippocampus, basal forebrain and amygdala tissue, and it may cause dementia due to cell death (Struble et al., 2010). To date, genetic analysis of familial early onset AD patients (EOAD, <65 years) has demonstrated high-penetrant mutations in the three EOAD genes, coding for the APP and the presenilins 1 and 2 (PSEN1 and PSEN2) (Alzheimer's Disease and Frontotemporal Dementia Mutation Database, 2011. http://www.molgen.ua.ac.be/ADMutations). All mutations associated with APP and PSEN proteins can lead to an increase in the production of Aβ peptides, specifically the more amyloidogenic form, Aβ(42) (Campion et al., 1995; Chapman et al., 1995; Chartier-Harlin et al., 1991; Cruts et al., 1995; Goate et al., 1991; Mullan et al., 1992; Sherrington et al., 1995; Wasco et al., 1995). Despite affecting millions of people in westernized societies, no prophylactic or therapeutic cure has been developed for AD unfortunately. Considering this issue, it has been known for nearly 20 years that the homeostasis of multiple neurotrophins is indeed altered during AD pathogenesis (Conner, Lauterborn, Yan, Gall, & Varon, 1997). The potential use of the well-characterized neurotrophins as potential therapeutic targets for AD patients has been explored heavily.

To date, several mouse models that express reduced neurotrophin levels have been developed and characterized, which confirm the role of neurotrophins in mediating the onset of cognitive phenotypes associated with these neurodegenerative diseases. Mice deficient for neurotrophins such as BDNF and NGF develop severe effects on adult brain functions and behavior. Moreover, mice that completely lack neurotrophic support do not survive the first weeks after birth, whereas heterozygous mice are viable. Lowering levels of NGF in mice caused severe deficits in memory acquisition and retention (Chen et al., 1997). These memory impairments correlated well with decreased cholinergic activity and loss of choline acetyltransferase (ChAT)-positive neurons in the frontal and temporal lobes of the cortex, reminiscent of AD patients (Hellweg, Gericke, Jendroska, Hartung, & Cervos-Navarro, 1998). Notably, age-related memory impairment was shown to be accompanied by elevated levels of proNGF, $p75^{NTR}$, and sortilin in the frontal cortex and hippocampus (Terry, Kutiyanawalla, &

Pillai, 2011). These results are in line with the proapoptotic activities of proNGF (Al-Shawi et al., 2008). In line with loss of NGF, the absence of normal BDNF levels cause increased aggressiveness, hyperactivity, and hyperphagia as well as serotonin-related memory impairments. This is consistent with defects in LTP that are present in the hippocampus upon loss of BDNF. In addition to the existing links between NGF and AD, an extensive body of evidence implicates altered levels of BDNF in the pathogenesis of AD. The first evidence that BDNF deficiency was implicated in AD was provided by the detection of reduced expression levels of BDNF and TrkB in the hippocampus and the temporal cortex of AD patients (Conner et al., 1997; Zuccato & Cattaneo, 2009). In addition to AD, studies investigating BDNF levels in patients with FTD, Lewy body dementia (LBD), and vascular dementia indicated a comparable decrease (Budni, Bellettini-Santos, Mina, Garcez, & Zugno, 2015; Ventriglia et al., 2013). The identification of genetic variations in the BDNF gene further implicated BDNF in the complex pathophysiology of the neurodegenerative process, especially because BDNF is a highly conserved protein. Moreover, it has been shown that a valine to methionine change at position 66 (Val66Met) in the prodomain of BDNF is responsible for alterations in the secretion and sorting of BDNF (Egan et al., 2003). To date, several studies have implicated sortilin in the control of proBDNF secretion. The decreased secretion of sortilin mediated by the Val66Met polymorphism might originate from the impaired binding of BDNF to sortilin (Chen et al., 2005; Evans et al., 2011). Follow-up examination of this polymorphism in patients has demonstrated that the Met-residue at position 66 is associated with poor episodic memory and cognition, but also with increased insulin resistance and metabolic disruption in AD and bipolar disorder/schizophrenia, respectively (Bonaccorso et al., 2015; Cai et al., 2012).

3.3.1 NGF/TrkA Functionality in AD-Related CNS Neurodegeneration

In the CNS, the highest levels of NGF are present in the hippocampus and in the areas containing BFCNs (Korsching, Auburger, Heumann, Scott, & Thoenen, 1985), which led to the formulation of the "*NGF hypothesis of AD*" (Calissano, Matrone, & Amadoro, 2010; Yuen, Howe, Li, Holtzman, & Mobley, 1996). pro-NGF is the predominant form of NGF in the human and rodent brain, whereas mature NGF can be hardly detected. In mild cognitive impairment (MCI) and AD, pro-NGF is increased in frontal and occipital cortices (Crutcher, Scott, Liang, Everson, & Weingartner, 1993; Peng, Wuu, Mufson, & Fahnestock, 2004) and in the hippocampus (Hock, Heese, Hulette, Rosenberg, & Otten, 2000; Narisawa-Saito, Wakabayashi, Tsuji, Takahashi, & Nawa, 1996), while a loss is observed in the basal forebrain (Mufson, Conner, & Kordower, 1995; Scott, Mufson, Weingartner, Skau, & Crutcher, 1995). This is suggestive that defective cleavage of pro-NGF to mature NGF can be associated with AD (Cuello, Bruno, Allard, Leon, & Iulita, 2010). In addition to this pro-NGF interaction with $p75^{NTR}$ has been proven to contribute to AD pathology (Capsoni & Cattaneo, 2006) and intracerebral administration of pro-NGF has also been found to cause AD-like learning impairments in mice (Kichev et al., 2009). In the adult CNS, $p75^{NTR}$ is expressed primarily by basal forebrain cholinergic, but also by hippocampal pyramidal, entorhinal cortex, and cortical layer V neurons—all regions vulnerable to AD pathology (Dougherty & Milner, 1999; Miller & Pitts, 2000; Mufson & Kordower, 1992). The detrimental cellular effects of pro-NGF isolated from AD brain tissue can actually be prevented by antibody blocking of the $p75^{NTR}$ (Pedraza et al., 2005). Moreover, $p75^{NTR}$ expression is upregulated in vitro by Aβ (Zhang et al., 2003) and in AD cortical and hippocampal tissues (Hu et al., 2002; Mufson, Kroin, Sendera, & Sobreviela, 1999). With respect to the cognate receptor for NGF, TrkA expression has been reported to be downregulated in the AD parietal cortex and BFCNs (Boissiere, Faucheux, Ruberg, Agid, & Hirsch, 1997; Salehi et al., 1996). Concomitant increased $p75^{NTR}$ and decreased TrkA levels in AD suggest that an increased $p75^{NTR}$:TrkA ratio plays a relevant role in the etiology of AD (Capsoni, Tiveron, Vignone, Amato, & Cattaneo, 2010). Furthermore, deficits in TrkA receptor phosphorylation contribute to age-related deficits in LTP (Aboulkassim et al., 2011; Kelly, Maguire, & Lynch, 2000), one of the major cellular mechanisms involved in learning and memory in AD (Bliss & Collingridge, 1993).

3.3.2 BDNF/TrkB Functionality in AD-Related CNS Neurodegeneration

Significant reductions in BDNF levels in hippocampal, temporal cortex, and other cortical regions are evident in AD brains and specific BDNF polymorphisms are associated with sporadic AD (Murer, Yan, & Raisman-Vozari, 2001; Tsai, 2003). BDNF stimulation of TrkB attenuates cell death and promotes neurite status of hippocampal, cortical, and basal forebrain neurons and protects them from insults including energy deprivation, excitotoxins, and free oxygen radicals, all factors promoting AD pathology (Murer et al., 2001). Loss of synaptic function is an early and fundamental process in AD (Kim, Tang, & Sisodia, 2006) and BDNF is a well established, potent promoter of synaptic function and plasticity (Kang, Welcher, Shelton, & Schuman, 1997). Recent studies have revealed the possibility that the selective targeting of TrkB can achieve neurotrophic effects beyond those achieved by BDNF through concomitant $p75^{NTR}$/TrkB signaling. A dimeric synthetic peptide corresponding to selected domains of BDNF and targeted to TrkB was shown to promote neurite outgrowth in the presence of an inhibitory environment created by the presence of MAG (myelin-associated glycoprotein), while in the same conditions BDNF had a markedly less neurite promoting effect (Williams et al., 2005). These observations raise the possibility, that in certain paradigms small molecules that activate TrkB while avoiding BDNF–$p75^{NTR}$ interactions might provide advantageous therapeutic profiles. Activation of TrkB can reverse neuritic degeneration caused by exposure to Aβ (Kimura, Takahashi, Tashiro, & Terao, 2006). This observation is consistent with the recent finding that BDNF promotes tau dephosphorylation (Elliott, Atlas, Lange, & Ginzburg, 2005) and the likely role of excessive tau phosphorylation in neuritic dystrophy.

4. PRONEUROTROPHIN-BASED THERAPY FOR AGE-RELATED NEURODEGENERATION

The eminent complex nature of ligand–receptor systems that primarily control disease processes underlying different age-related or neurodegenerative disorders often precludes the facile development of pharmacotherapeutic agents. As happened with the age-related insulin pathology of Type II diabetes mellitus (T2DM), companies mainly focused on the often time-consuming and elaborate development of synthetic proteins or protein analogs to ameliorate the disease etiology (Maudsley, Martin, & Egan, 2011). The central role neurotrophins play in the etiology of AD and other age-related neurodegenerative disease, creates a similar issue ofligand or drug creation. If a neurotrophic deficit or resistance is primarily responsible for neurodegeneration, then supplementation of dementia patients with a bioavailable, synthesized exogenous form of the specific neurotrophin seems the simplest therapeutic solution. However, the large scale and reliable generation of peptides as large as NGF, BDNF, NT3, or NT4 are incredibly difficult and problematic, especially from a standpoint of protein synthesis, quality control, stability, half-life, and means of delivery.

The seemingly simple pharmacological issue of effective delivery is often the greatest hindrance for peptidergic agents targeting CNS sites as they have minimal blood–brain barrier penetration and restricted diffusion within the CNS parenchyma. In contrast, most "small molecule" agents (<500 Da) often demonstrate a strong ability to penetrate this barrier, thus enabling the generation of effective, steady state concentrations in the CNS. In addition to the simple pharmacological superiority of small-molecule neurotrophic agents from an ADME (*A*bsorption, *D*istribution, *M*etabolism, *E*xcretion) point of view, the cost of unit synthesis for these compounds is miniscule compared to that required for complex peptidergic entities. Hence, from multiple angles, the future of CNS-targeting anti-neurodegenerative therapeutics is likely to reside with small-molecule neurotrophic agents. Small-molecule compounds can be designed with nearly limitless structural variation (facilitating a greater capacity for derivation of "precision medicines"), in complete contrast to the highly constrained and conserved molecular structure of the endogenous neurotrophins. This ability to generate many more forms of small-molecule ligands therefore also allows an enormous scope for signaling selectivity and the rational generation of "precision medicines" that possess a specific high-dimensionality efficacy profile and a minimized contra-indication profile (Maudsley, 2012; Maudsley et al., 2005; Maudsley, Martin, & Luttrell, 2007; Maudsley, Patel, Park, Luttrell, & Martin, 2012; Stranahan, Zhou, Martin, & Maudsley, 2009; Wang, Martin, Brenneman, Luttrell, & Maudsley, 2009). The well-characterized functions of neurotrophins in controlling cell survival provide potential opportunities for scientist to investigate their translational benefit to treat neurodegenerative disorders that are associated with loss of trophic support. To date, several clinical trials have been initiated that involve exogenous neurotrophin delivery to patients of several neurodegenerative disorders, including—but not limited to—AD, PD, and HD. The results of these trials, however, have largely been disappointing with insignificant therapeutic effects and undesired side effects (Thoenen & Sendtner, 2002).

We shall address in following sections how this tremendous potential for selective receptor efficacy is one of the most important aspects of small-molecule activity at neurotrophic receptors in neurodegenerative disorders. We shall also elaborate on a number of challenges that have been explored to deliver effective neurotrophic support in order to restore the neurological functions in these disorders.

4.1 Effective Drug Design-Related Considerations for Refining Neurotrophic Activity

To facilitate the development of novel agents that possess some degree of effective neurotrophic activity it is first important to understand the actual signaling diversity that neurotrophins encompass. In recent years, especially for the transmembrane GPCRs our understanding of signaling diversity has become an important topic for future therapeutic research. For GPCRs, their spectrum of activity, both ligand dependent or ligand independent, appears to be a result of the differential organization of the GPCRs into discrete multiprotein complexes termed "receptorsomes" that demonstrate a preferential precoupling to a specific downstream signaling cascade (Dohlman, 2015; Jean-Alphonse et al., 2014; Maudsley et al., 2005). In a similar manner, receptor-based signaling specificity is also a key factor in receptor tyrosine kinase signaling (Chadwick et al., 2008; Lemmon & Schlessinger, 2010; Maudsley et al., 2000; Tsou et al., 2014). Such structure-based signaling diversity has engendered the functional concept that these distinct signaling isoforms of the receptors can be pharmacologically targeted in a differential manner to generate engineered, or "biased," drug efficacy profiles (Luttrell et al., 2015; Maudsley, 2012).

In addition to specificity being determined by receptorsome-like structures dynamic exchanges between intracellular signaling cascades is also an important concept to heed for therapeutic design. Especially with respect to receptor tyrosine kinase signaling pathways, many molecules are shared in parallel signaling pathways (Martin et al., 2009). The question is: How does the system know exactly which signaling molecules to recruit in order to trigger downstream signaling? The answer is twofold and it involves (i) spatial and temporal insulation (Whitmarsh & Davis, 1998) and (ii) mutual inhibition, which is important in negating spurious signals (McClean, Mody, Broach, & Ramanathan, 2007). Spatial insulation can be achieved by localizing pathways to different cellular compartments or by incorporating the shared component into distinct macromolecular complexes through scaffolding complexes. If insulation is temporal, signals generate noncoincidental activation of the shared component on signaling pathways (Albert & Oltvai, 2007). The study in 2007 by McClean et al. reported that pathways acquire specificity by being able to filter out irrelevant crosstalk via mutual inhibition (McClean et al., 2007). Furthermore, they suggest that cell heterogeneity facilitates the spectrum of the decisions made. They studied two models which achieve specificity to any one signal by insulation and mutual inhibition, respectively. In the context of the developing nervous system, Wang, Liu, Chen, and Aihara (2011) have reported the importance of *trans*-activation and *cis*-inhibition in neural fate decision involving Notch signaling (Wang et al., 2011). Notch in a cell can be activated by Deltalike-1 (*Dll1*), a proneural gene, in neighboring cells (*trans*-activation) and it can be inhibited by *Dll* within the same cell (*cis*-inhibition).

Within this "neurotrophic spectrum" it is important to identify which specific component of this multifactorial effect is the most important to target therapeutically for a desired physiological effect. In this section, we will therefore discuss the known canonical signaling modalities associated with neurotropism and then suggest drug targeting mechanisms that could help "fine-tune" this efficacy requirement for novel small-molecule therapeutics. In general terms, the major pathways activated by the Trk receptors are Ras, RhoA, PI3K, PLC-γ1, and their downstream effectors. It is important for future therapeutic design of potentially pluripotent small-molecule ligands that a comprehensive parallel appreciation of neurotrophin signaling activity is considered. Therefore, we shall detail the most important pathways indicative of neurotrophic activity that will form the center of future precision drug efficacy engineering.

4.1.1 Multiple Dimensions of Neurotrophic Signal Transduction Efficacy

Activation of Ras is required for normal neuronal cell differentiation and also promotes survival of neurons. Transient activation of Ras is mediated by the Shc adaptor protein, which is recruited to its recognition sites on activated Trk receptors via interactions with the Shc PTB domain (Nimnual, Yatsula, & Bar-Sagi, 1998). This sets in motion the Ras-MAPK cascade. Trk-mediated phosphorylation of Shc creates a phosphotyrosine site on Shc that recruits another SH2 domain-containing protein, Grb2 (growth factor receptor-bound protein 2), which is bound to the nucleotide exchange factor SOS (Son of Sevenless). A Grb2–SOS complex translocates to the plasma membrane, where SOS activates the small G protein p21*ras* (Segal & Greenberg, 1996). Active Ras stimulates signaling through the protein kinases Raf-1, β-Raf, MAP–ERK kinase (MEK), and p38 MAP kinase (Wood, Sarnecki, Roberts, & Blenis, 1992). Activation of one or more of these kinases leads to the phosphorylation and activation of MEK and the MAP kinase isoforms, ERK-1 and ERK-2 (Lange-Carter & Johnson, 1994; Wood et al., 1992). Ras also triggers a signaling cascade through sequential activation of Wnk1, MEKK2, and MEK5 which results in activation of ERK5 (Watson et al., 2001). Each of these MAP kinase-associated cascades has overlapping, as well as distinct targets within the cell (Pearson et al., 2001). For example, Rsks and MEKK2 each phosphorylate transcription factors such as CREB, which activate transcription of genes essential for differentiation and survival of neurons (Hill & Treisman, 1995), and ERK5 activates myocyte enhancer factor 2 (MEF2) (Marinissen, Chiariello, Pallante, & Gutkind, 1999), while ERK1/ERK2 activates ELK-1 (Gille, Sharrocks, & Shaw, 1992).

Generation of P3-phosphorylated phosphoinositides by PI3K promotes survival of many neurons, and there is evidence to suggest that Ras can signal through PI3K (Kauffmann-Zeh et al., 1997). In many neurons, Ras-mediated activation of PI3K initiates the major pathways through which survival signals are conveyed (Vaillant et al., 1999). Both couplings appear to occur via two common components, Shc and Grb-2, with Grb-2 docking further to pathway-specific proteins. Recruitment of Gab1 (Grb2-associated binder-1) by phosphorylated Grb2 permits subsequent binding and activation of PI3K (Holgado-Madruga, Moscatello, Emlet, Dieterich, & Wong, 1997). PI3K may mediate neurotrophin survival effects, for example, that of NGF on PC12 cell survival (Yao & Osada, 1997). The 3-phosphoinositides generated by PI3K trigger the activation of two protein kinases, the serine–threonine kinase Akt and the p70 ribosomal protein S6 kinase (p70S6K) (Chung et al., 1994). Akt controls the activities of several proteins important in promoting cell survival, including the enzyme glycogen synthase kinase-3 (GSK-3β) and even substrates that directly regulate the caspase cascade, such as the proapoptotic factor—BAD (Yuan, Lipinski, & Degterev, 2003). Phosphorylated BAD is sequestered by 14-3-3 proteins, preventing its proapoptotic actions. Experiments with pharmacological inhibitors, as well as expression of wild type and dominant-inhibitory forms of Akt, demonstrate that Akt but not p70S6K mediates PI3K-dependent survival (Dudek et al., 1997; Skaper, Floreani, Negro, Facci, & Giusti, 1998).

It has been well documented that activation of Trk kinase results in PLC-γ1 recruitment to a docking site similar on all Trk receptors. The docked PLC-γ1 is activated via Trk-mediated phosphorylation and then hydrolyses phosphatidylinositol 4,5-bisphosphate to inositol trisphosphate (IP3) and diacylglycerol (DAG). IP3 triggers mobilization of Ca^{2+} cytoplasmic stores while DAG stimulates DAG-regulated isoforms of protein kinase C (PKC). These two signaling factors activate numerous intracellular enzymes, including all PKC isoforms, Ca^{2+}-calmodulin-dependent protein kinases, and other Ca^{2+}-calmodulin-regulated targets. PLC-γ1 activation results in activation of, e.g., protein kinase C-δ, which mediates NGF-promoted activation of MEK and ERK1/ERK2 (Corbit, Foster, & Rosner, 1999). The activity of PLC-γ1 has also been demonstrated to be involved in the ability of TrkB receptors to modulate synaptic transmission and LTP (Matsumoto et al., 2001; Minichiello et al., 2002).

RhoA is a monomeric G protein whose activity is strongly associated with the regulation of cytoskeletal dynamics, transcription, cell cycle progression, and actin stress fiber formation. As with other members of the Ras family RhoA possesses a strong relationship with specific downstream effectors that allow its functionality to be biased toward specific cellular processes. The most widely studied effectors for RhoA are diaphanous homologue 1 (DIAPH1) and Rho-associated coiled-coil containing protein kinase 1 (ROCK1). A considerable body of research has already demonstrated that RhoA GTPase-associated signaling activity, often under the direct control of $p75^{NTR}$, is strongly linked to proneurodegenerative neurosynaptic effects in HD (Brito et al., 2014) as well as in AD (Petratos et al., 2008). Confirming the potentially deleterious effects of RhoA signaling, it has been demonstrated that pharmacological inhibition of one of its primary downstream effectors, ROCK2, is beneficial with respect to suppression of amyloid deposition in murine models of AD (Herskowitz et al., 2013).

While neurotrophin binding to $p75^{NTR}$ can entrain neurophysiologically deleterious effects, e.g., complex and diverse RhoA modulation, at the same time activation of opposing cell-protective signaling paradigms is also in evidence. For example, $p75^{NTR}$ activation has long been demonstrated to stimulate NF-κB pathways associated with signaling intermediates such as Rho guanine nucleotide exchange factor 28 (RIP2), TRAF6, interleukin 1 receptor-associated kinase 1 (IRAK), protein tyrosine phosphatase, and nonreceptor type 13 (FAP1) (Charalampopoulos et al., 2012). Neurotrophin-based stimulation of this neuroprotective pathway can therefore often result in seemingly diverse complex outputs (Yoon, Casaccia-Bonnefil, Carter, & Chao, 1998) when placed in the context of other forms of prodegenerative signaling that can be entrained by neurotrophin receptors stimulating factors such as JNK (c-JNK; Jung & Kim, 2010) and also the caspase proapoptotic pathway (Pham et al., 2016).

Given the complexity of downstream signaling activities of neurotrophin receptors, it is likely that there are different "preorganized" receptor-scaffolding systems that allow a more discrete propagation of selective signals, in a manner similar to robust and coherent signaling signatures from

GPCRs (Maudsley et al., 2015). In this regard, considerable research has been performed to understand the diversity of the intracellular signaling and scaffolding proteins that physically associate with neurotrophin receptors (Dechant & Barde, 2002; Gentry, Rutkoski, Burke, & Carter, 2004; Roux & Barker, 2002). Therefore, it will be important for the development of future neurotrophin-targeting therapeutics to aspire the specific range of neurotrophic activity which minimizes prodegenerative signaling and favors activation of prosurvival pathways.

4.1.2 GPCR-Based Manipulation of Neurotrophin Signaling Paradigms

In addition to the neurotrophin-mediated regulation of complex signal transduction pathways outlined in Section 4.1.1, it has been demonstrated that GPCR signaling systems can also stimulate neurotrophin receptors through a so-called "transactivation" process. Compared to direct modulation of the neurotrophin receptors, indirect activation of neurotrophin receptors offers a more selective approach for activating specific subsets of Trk receptors. Multiple forms of transmembrane receptor tyrosine kinases can be transactivated, i.e., in a cognate neurotrophic ligand-independent manner, by stimulated GPCRs in certain conditions (Eguchi et al., 1998; Maudsley et al., 2000; Rajagopal, Chen, Lee, & Chao, 2004). During transactivation, the Trk receptors are activated by phosphorylation in the absence of neurotrophin ligands via alternative signaling pathways, such as GPCR activation. Activation by this mechanism has been demonstrated for both TrkA and TrkB receptors in response to either adenosine or pituitary adenylate cyclase-activating polypeptide acting via adenosine-A2A (ADORA2A) and vasoactive intestinal peptide-1 (VPAC1) receptors, respectively (Lee & Chao, 2001; Lee, Rajagopal, Kim, Chang, & Chao, 2002; Ravni et al., 2006). Trk receptor transactivation does not take place at the cell surface but within the cytoplasm and Golgi apparatus (Rajagopal et al., 2004) resulting in increased cell survival in a neurotrophin-free environment. Antidepressant drugs have shown to transactivate TrkB neurotrophin receptors (Chadwick et al., 2011; Jang et al., 2009), independently of BDNF and monoamine reuptake blockade, and emphasizes the need of an intact tissue context for the ability of antidepressant drugs to induce TrkB activation in the brain. Previous studies have shown that two Gα-coupled GPCRs, the ADORA2A and the pituitary adenylate cyclase-activating peptide (PACAP) receptors, can transactivate endogenous immature intracellular TrkA/B receptors in PC12 cells and hippocampal or cortical neurons (Rajagopal et al., 2004; Swift et al., 2011). Lee et al. (2001)

demonstrating a mechanism for the neuroprotective actions of adenosine involving engagement of the ADORA2A receptor, transactivation of Trk tyrosine kinase receptors, and selective induction of the PI3K/Akt pathway as observed with immunoprecipitation (Lee & Chao, 2001). Furthermore, receptor tyrosine kinase transactivation by dopamine receptors is compatible with normal expression levels of these receptors and is not restricted to a single cell type. Spatial intensity distribution analysis (SpIDA) is an image analysis method that can directly measure oligomerization and trafficking of endogenous proteins in single cells. It has been shown to quantify the dimerization/activation state and surface transport of TrkB receptors at early stages of their transactivation by several GPCRs. The SpIDA method is based on the fitting of fluorescence intensity histograms calculated from conventional laser-scanning confocal images to monitor and to quantify protein oligomerization in cells and tissues in a time and space resolved manner. Furthermore, it provides a unique approach to perform single cell multivariate quantification of signaling processes involving changes in protein interaction, trafficking, and activity. Indeed, SpIDA has been shown to detect transactivation of TrkB by dopamine receptors in intact neurons (Swift et al., 2011).

4.2 Small-Molecule Neurotrophic Groups and Lead Compounds

Although the therapeutic potential of neurotrophins has generated high expectations over the past decade, poor pharmacokinetics, and the generation of unwanted side effects as well as disconcerting efficacy profiles have limited the clinical application of neurotrophic factors (Apfel et al., 2000; Thoenen & Sendtner, 2002; Wellmer, Misra, Sharief, Kopelman, & Anand, 2001). Therefore, development of small-molecule compounds that mimic neurotrophic signaling and overcome the pharmacokinetic and side-effect barriers may have great therapeutic potential, especially with the consideration of creating "precision" signaling behavior that potentiates prosurvival and attenuates prodegenerative signaling activities.

4.2.1 Small-Molecule Modulators of Trk Receptor Activity

In the last few years, there has been a considerable surge to create more therapeutically tractable compounds that are able to manipulate the Trk receptor signaling system. Low molecular mass lead agents have been generated through multiple rounds of screening for mimicry of Trk-dependent signaling. From these pipelines, multiple small-molecule modulators/activators of both TrkA and TrkB receptors have been developed. Small-molecule TrkA activators include 1H5, an asterriquinone, and a 5E5 (a mono-indolylquinone) that activate TrkA through binding to a presumed intracellular site (Lin et al., 2007). Also the colorizing factor for gambooge resin, gambogic amide (xanthonoid), has demonstrated TrkA-dependent prosurvival activity in cell models (Jang et al., 2007).

The classical tricyclic antidepressant Amitriptyline (AMI) appears to function in a broad spectrum manner as it positively regulates both TrkA and TrkB activity (Chadwick et al., 2011; Jang et al., 2009). It is interesting to note that indeed the ability to control both of these neurotrophic pathways is associated with its capacity to induce molecular heterodimerization between TrkA and TrkB receptors. With respect to its potential use in neurodegenerative conditions, the signaling functionality of AMI is highly reminiscent of BDNF. Indicating the importance of developing small molecules that can control these proneurotrophic pathways, the functional signaling output (measured at both high-dimensionality transcriptomic and classical signaling levels) of AMI was more resistant to stress/aging/neurodegenerative cell contexts than either NGF or BDNF activity (Martin et al., 2013). Therefore, as a potential prophylactic agent, AMI would likely retain a greater signaling efficacy compared to both endogenous neurotrophins. With respect to its therapeutic potential for AD, AMI was able to be highly effective at restoring genetically induced cognitive capacity in a murine model of AD (Chadwick et al., 2011). This observation coincided with a seemingly contradictory elevation in cortical and hippocampal amyloid beta levels. In this context, it appeared that the BDNF-mimicking and proneurotrophic activity of AMI was sufficient to offset any potentially negative effects on cognitive performance or neuronal survival of β-amyloid. This divergent functionality of AMI was shown to originate from its ability to disrupt the integrity of a highly novel molecular species: a γ-secretase-TrkB receptor complex. The γ-secretase complex is a multimeric transmembrane protease consisting of PSEN-1, presenilin enhancer-2 (PEN-2), nicastrin, and anterior pharynx-defective 1 (Aph-1) that acts as one of the primary controllers of β-amyloid synthesis in the brain. In this experimental scenario, it appeared that AMI-induced disruption of the γ-secretase-TrkB heterocomplex coincided with an increase in both γ-secretase activity (elevating β-amyloid) as well as an increase in TrkB activity (enhanced TrkB phosphotyrosine content) which resulted in a net positive cognitive recovery of the mice. Interestingly, it appeared that the effect of AMI upon the TrkB system was linked with its ability to functionally interact with and promote β-arrestin recruitment to the GPCR-like Aph-1 subunit of γ-secretase (Martin et al., 2013). This intriguing intersection between neurotrophic activity and amyloid processing has also recently been confirmed in additional AD models such as the 5XFAD mouse model (Devi & Ohno, 2015), where functional divergences in the regulation of amyloidosis and neurotrophic signaling were apparent. Therefore, within the neurotrophin system AMI appears to possess multiple levels of nuanced and beneficial activity.

With respect to agents that selectively target TrkB-dependent functionality, 7,8-dihydroxyflavone (7,8-DHF), has been shown to effect TrkB tyrosine phosphorylation, induce ERK/Akt-1 activation, and protect neuronal cells from death with a similar efficacy to BDNF (Jang, Liu, Yepes, et al., 2010). Demonstrating the potential for complex connectivity between receptor signaling activity and downstream signal transduction behavior, the small-molecule agent diosmetin (5,7,3′-trihydroxy-4′-methoxyflavone) exerts strong neuronal antiapoptotic effects (activating Akt-1 and ERK) in a similar manner to BDNF but with a much lower capacity to induce TrkB tyrosine phosphorylation compared to the endogenous ligand. Similar signal transduction disconnects, with respect to canonical signaling pathways entrained by BDNF at the TrkB receptor, have been observed for the multifactorial neuroprotective compounds pinocembrin and quercetin (Jang, Liu, Yepes, et al., 2010). Molecular screening approaches using ectopic expression models of the TrkB receptor resulted in the identification of the tetranortriterpenoid deoxygedunin (Jang, Liu, Chan, et al., 2010). This agent stimulates selective dimerization and autophosphorylation of TrkB, protects in vitro cell lines from oxidative/metabolic stressors, and promotes the survival of vestibular ganglion neurons in BDNF knockout mice (Jang, Liu, Chan, et al., 2010).

Several small bicyclic peptidomimetic compounds, capable of promoting receptor dimerization have been identified (Scarpi et al., 2012). Scarpi et al. (2012) demonstrated that the lead agent MT2 was able to in-part recapitulate NGF-like activity (i.e., induce robust TrkA-Tyr^{490} phosphorylation) but constituted a differing efficacy profile to NGF with respect to its lower ability to phosphorylate TrkA $Tyr^{674,675,785}$ phosphorylation sites (Scarpi et al., 2012). Creating small proneurotrophic molecules to mimic specific domains of neurotrophins such as the loop 2 domain of BDNF has met with some success. A selective TrkB-interacting molecule, LM22A-4 was developed in this manner. LM22A-4 has demonstrated a strong neuroprotective activity upon application to in vitro models of AD, HD, and PD, where it prevented neuronal death at efficacies similar to BDNF (Massa et al., 2010). In addition to these effects in classical neurodegenerative disease models, application of LM22A-4 to a murine model of the BDNF-associated Rett Syndrome resulted in an enhanced adult neurogenesis functionality (Schmid et al., 2012).

While mimicking the prosurvival activities of the major physiological neurotrophic factors NGF and BDNF with small molecules, represents potentially the simplest mechanism of engineering drug-like agents. Moreover, the effective inhibition of the cytotoxic effects of $p75^{NTR}$ may also be an important therapeutic strategy. To this end, small-molecule agents have been developed that inhibit the deleterious effects of $p75^{NTR}$ activity. The agents LM11A-31 and LM11A-24 are capable of inhibiting cultured neuronal cell death at picomolar concentrations in a $p75^{NTR}$-specific manner (Massa et al., 2006). These agents appeared to mimic many of the traditional neurotrophic functions as they were able to physically recruit IRAK to the $p75^{NTR}$ as well activate downstream Akt-1 and NF-κB signaling. As $p75^{NTR}$ functionality is strongly associated with AD pathophysiology it is important to note that in vitro testing of these agents demonstrated their ability to inhibit amyloid-β-induced cytotoxicity, i.e., the activation of calpain–CDK5 (cyclin-dependent kinase 5), glycogen synthase kinase 3β (GSK3β), and JNK signaling in cultured hippocampal neurons and tissue slices (Yang et al., 2008). At the in vivo level, daily administration of LM11A-31 during 3 months reversed the behavioral deficits (cognitive impairment) and inhibited neurodegenerative pathophysiology (amyloid plaques) in the $hAPP_{Lond/Swe}$ mouse model (Knowles et al., 2013).

4.2.2 Potentiation and Modulation of Endogenous Neurotrophin/Neurotrophic Functionality

In addition to small molecule-based modulation of Trk receptors, there are several additional mechanisms that are potentially important for therapeutic exploitation. For example, there are multiple tractable agents that are capable of enhancing the levels or activities of endogenous neurotrophins, e.g., leteprinim (Potkin et al., 2002; Ramirez et al., 2002), retinoid agents (Arrieta et al., 2005; Mey & McCaffery, 2004), and AMPA receptor positive modulators (so-called "Ampakines"), including aniracetam, Org24448, LY503430, LY404187, and CX-614/CX-516 (Dicou, Rangon, Guimiot, Spedding, & Gressens, 2003; O'Neill et al., 2005; Wu et al., 2004). Within this group of molecules, Org24448 and CX516 are currently tested in clinical trials for a number of CNS disorders, including depression, schizophrenia, and AD. This group of small-molecular agents that, in-part appear to support neurotrophin receptor signaling through endogenously based mechanisms, also includes multiple antidepressant agents such as selective serotonin reuptake inhibitors and tricyclic antidepressants (Anderson, 2000; Castren, 2004; Malberg & Blendy, 2005). In addition, the effective small molecule modulation of downstream Trk-related signaling can also affect significant proneurotrophic activity. For example, the RhoA-associated ROCK pathway mediates neuronal cell contraction and growth-cone collapse downstream of signaling mediated by ephrins, semaphorins, chondroitin sulfate proteoglycans, and Nogo family members (Mueller, Mack, & Teusch, 2005). The inhibition of ROCK in vivo promotes neurotrophic effects in the spinal cord, which may be of clinical relevance for the treatment of neuropathic pain and AD (Mueller et al., 2005). Recently, clinical trials investigating the putative neuroprotective and neuroregenerative properties ascribed to the Rho GTPase antagonist, Cethrin (Baptiste & Fehlings, 2006), showed good safety, tolerability, and dose-dependent neurological outcomes from a 6-week follow-up trial of 37 patients who suffered a complete thoracic or cervical injury (i.e., American Spinal Injury Association, Grade A = having no sensory or motor function). In addition to agents that can provide support for specific neurotrophin receptor-mediated pathways, there are many small-molecule agents that can engender neuroprotective and generalized neurotrophic activities through less well-defined mechanisms. For example, proneurotrophic functionalities have been demonstrated for multiple additional small molecules including, GPI-1046, JNJ460, V-1111, FK1706 (Bova, Shownkeen, Goldberg, Horowitz, & Azar-Kia, 2000; Heightman et al., 2005; Steiner et al., 1997; Voda, Yamaji, & Gold, 2005), CEP-1347 (Roux et al., 2002), and NB-506 (Pollack et al., 1999).

4.2.3 Parallel Modulation of Synergistic Growth Factor Pathways

An additional level of therapeutic potential involves the exploitation of synergistic mechanisms between the signaling cascades of neurotrophins and other growth factors. This is particularly true for CNS neurons, which typically depend on more than one growth factor for their survival. Combinations of neurotrophins and other growth factor families with neurotrophic activity, such as GDNF, fibroblast growth factors (FGFs), IGFs, or CNTF, may be more effective than a single growth factor. Such rational combinations of neurotrophins and growth factors have already been shown to act synergistically on motor neurons (Haase et al., 1997; Sagot, Rosse, Vejsada, Perrelet, & Kato, 1998). AIT-082, a derivative of the purine hypoxanthine, was shown to increase the expression of various neurotrophic factors including NGF and

TGF-β in the brain and spinal cord (Rathbone, Middlemiss, Crocker, et al., 1999; Rathbone, Middlemiss, Gysbers, et al., 1999). AIT-082 has been demonstrated to induce neurite outgrowth from primary neurons and nerve fiber regeneration with functional recovery in several rat nerve injury models (Rathbone, Middlemiss, Crocker, et al., 1999). The compound was also reported to show memory enhancement in several animal models and patients. In addition, AIT-082 exhibited neuroprotective effects against glutamate and NMDA toxicity in cultured neurons where the compound acted both acutely, protecting against excitotoxic effects (possibly via increases in ATP production), and chronically, through neuroregenerative mechanisms involving neurotrophic factor upregulation in astrocytes (Rathbone, Middlemiss, Crocker, et al., 1999).

This complex cross-receptor platform modulation has also been studied at the molecular level in neuronal cell lines where the simultaneous in vitro coapplication of either epidermal growth factor (EGF) or IGF-1 with NGF was employed (Martin et al., 2009). With respect to ERK-mediated signaling, the simultaneous addition of either EGF or IGF-1 with NGF resulted in a positive synergism, elevating cellular ERK activation above the level of either ligand by itself. In contrast, for the cellular activation of Akt-1, the EGF–NGF combination resulted in a synergistic potentiation of Akt-1 activation while the IGF-1–NGF combination reduced total Akt-1 activation levels below those engendered by each individual agent. Interestingly a similar biased profile, i.e., EGF–NGF being synergistic and IGF-1–NGF being antagonistic, was also observed for GSK-3β activation as well. Signaling pathways, entrained by both neurotrophin and growth factor receptors were once thought to be linear but are now known to be highly branched, and transduction modules formerly thought to operate independently participate in a substantial degree of preorganized signaling routes. Our present findings suggest that the complexity and preorganization of signal transduction scaffolds may be structured into coherent modules that allow the directed flux from multiple growth factor receptors in a discrete and selective manner that specifically allow the temporal and spatial encryption of the original signal input.

4.3 Clinical Activity of Small-Molecule Neurotrophic Therapeutics

A large body of evidence supports the feasibility of developing small molecules with advantageous pharmaceutical profiles that function either as specific antagonists, partial agonists, agonists, reverse agonists or “superagonists” at $p75^{NTR}$,

Trk, or sortilin receptors to modulate underlying disease mechanisms. Some small-molecule therapeutics increase neurotrophin levels, both in vitro and in vivo, as a downstream consequence of their main mechanisms of action. These molecular families include α-amino-3-hydroxy-5-methyl-4-isoxazolepropionic acid (AMPA) receptor agonists and modulators as well as antidepressants, such as selective serotonin reuptake inhibitors (SSRI), serotonin/norepinephrine reuptake inhibitors (SNRI), and tricyclic antidepressants (TCA), such as imipramine. Within this group of molecules, CX516 and Org24448 (another "Ampakine"), are currently in clinical development for a number of CNS disorders, including depression, schizophrenia, and AD.

Xaliproden (SR57746A, Sanofi-Aventis) is a combined NGF potentiator and serotonin 5-HT1A receptor agonist that potentiates neurite outgrowth in PC12 cells in the presence of NGF (Pradines et al., 1995). Xaliproden has completed a phase IIb study in AD patients, supporting the safety of the drug in elderly patients. A phase III, 18-month study (randomized, double-blind, and placebocontrolled) of the efficacy of xaliproden in patients with mild to moderate AD was initiated in the United States and Canada in September 2005 and was completed at the end of 2006 (Clinical Trials database). Xaliproden also protects motor neuron cultures via activation of the MAPK pathway, in a similar fashion to BDNF (Appert-Collin et al., 2005).

Clinical trials for xaliproden in AD and amyotrophic lateral sclerosis (ALS) were however discontinued in 2007 following cross-platform analysis of phase III data. While xaliproden did demonstrate some efficacy with respect to maintenance of hippocampal volume (suggesting potential retardation of disease progression or enhanced neurogenesis), there was insufficient evidence for efficacy in attenuation of cognitive decline in AD patients. As with the demonstration of partial efficacy in AD, xaliproden was able to generate some ALS-ameliorative activity but overall for both ALS and AD the risk to benefit ratio for this agent was not considered significant enough to warrant further trial use beyond phase III. Xaliproden is still however in clinical development for multiple sclerosis and peripheral sensory neuropathy (chemotherapy-induced neuropathy; SR57746A; www.clinicaltrials.gov). A phase III trial showed that xaliproden significantly reduced the incidence of severe oxaliplatin-related periphery sensory neuropathy. The developmental strategies for this compound support the broad clinical potential of a small-molecule neurotrophin mimetic. Indeed, the antidepressants imipramine (tricyclic) and fluoxetine (SSRI) enhance NGF-induced neuronal sprouting through the nonopioid, nonphencyclidine sigma-1 receptors (Takebayashi, Hayashi, & Su, 2002), indicating that these compounds have complex pharmacological activity.

Future directions in NGF and BDNF-based therapies may target the neurotrophin receptors. Modulation of p75^{NTR} by a small-molecule ligand has already been suggested as a way of preventing p75^{NTR}-mediated cell death. LM11A-24 and -31 are small molecules which block the effect of this receptor with low nanomolar potency and are reported to prevent Aβ-induced cell death and inhibit tau phosphorylation (Longo et al., 2007). Derivatives of these compounds have been described as candidates for clinical development. In addition, small-molecule mimetics of NGF or BDNF have also been suggested as Alzheimer therapeutics (Dawbarn et al., 2006; Massa et al., 2010). Using in silico screening with a BDNF loop–domain pharmacophore, Longo and colleagues have identified small molecules with nanomolar neurotrophic activity at TrkB, which prevented neuronal degeneration and improved motor learning after traumatic brain injury in rats. These are being taken forward as lead compounds for future treatments (Massa et al., 2010). In addition, since SSRI antidepressants are known to elevate BDNF levels, research and clinical trials are being carried out to investigate the effect of these in AD; one such antidepressant, Citalopram, is currently undergoing phase II clinical trials.

5. FUTURE DEVELOPMENTS IN SMALL-MOLECULE NEUROTROPHIC AGENTS

As discussed in this review, there is a clearly productive future in the development of small-molecule neurotrophic agents for the treatment of neurodegenerative disorders such as AD. The clear advantages of small-molecule ligands over peptidergic agents, in bioavailability, structural diversity, and signal-selective efficacy as well as simplicity of synthesis make this novel therapeutic class an important field of growth in the next decade. With the ability to create a diverse and broad small-molecule neurotrophic pharmacopeia will also likely allow the more effective management of not only AD but also other multiple related neurodegenerative disorders. As we have shown, the creation of this pharmacopeia can be effected by de novo drug design as well as the effective resourcing of previously FDA-approved agents such as amitriptyline. Given the enormous medicinal chemistry-based flexibility of neurotrophic small molecules, the promise of future "precision" medicines, in which selective signal transduction efficacy profiles (and eventual phenotypic cellular effects) can be strongly associated with molecular structure, seems more and more likely. In this respect, however, caution should be noted concerning the true nature of eventual so-called "precision" medicines. The generation of individually based agents is clearly a theoretical dream, absent some ground-breaking advances in medicinal chemistry. However, the rational "molecular stratification" of patients with neurodegenerative diseases into therapeutically tractable groups that then can be "matched" with the signaling efficacy profiles of compounds with rationally screened transduction "signatures" that can reverse these pathological profiles (Luttrell et al., 2015; Maudsley et al., 2015) is likely to be the drug design paradigm that will expedite future therapeutic breakthroughs.

REFERENCES

Aboulkassim, T., Tong, X. K., Tse, Y. C., Wong, T. P., Woo, S. B., Neet, K. E., … Saragovi, H. U. (2011). Ligand-dependent TrkA activity in brain differentially affects spatial learning and long-term memory. *Molecular Pharmacology*, *80*(3), 498–508. http://dx.doi.org/10.1124/mol.111.071332.

Albert, R., & Oltvai, Z. N. (2007). Shaping specificity in signaling networks. *Nature Genetics*, *39*(3), 286–287. http://dx.doi.org/10.1038/ng0307-286.

Al-Shawi, R., Hafner, A., Olsen, J., Chun, S., Raza, S., Thrasivoulou, C., … Cowen, T. (2008). Neurotoxic and neurotrophic roles of proNGF and the receptor sortilin in the adult and ageing nervous system. *The European Journal of Neuroscience*, *27*(8), 2103–2114. http://dx.doi.org/10.1111/j.1460-9568.2008.06152.x.

Anderson, I. M. (2000). Selective serotonin reuptake inhibitors versus tricyclic antidepressants: A meta-analysis of efficacy and tolerability. *Journal of Affective Disorders*, *58*(1), 19–36.

Apfel, S. C., Schwartz, S., Adornato, B. T., Freeman, R., Biton, V., Rendell, M., … Dyck, P. J. (2000). Efficacy and safety of recombinant human nerve growth factor in patients with diabetic polyneuropathy: A randomized controlled trial. rhNGF Clinical Investigator Group. *JAMA*, *284*(17), 2215–2221.

Appert-Collin, A., Duong, F. H., Passilly Degrace, P., Warter, J. M., Poindron, P., & Gies, J. P. (2005). MAPK activation via 5-hydroxytryptamine 1A receptor is involved in the neuroprotective effects of xaliproden. *International Journal of Immunopathology and Pharmacology*, *18*(1), 21–31.

Arrieta, O., Garcia-Navarrete, R., Zuniga, S., Ordonez, G., Ortiz, A., Palencia, G., … Sotelo, J. (2005). Retinoic acid increases tissue and plasma contents of nerve growth factor and prevents neuropathy in diabetic mice. *European Journal of Clinical Investigation*, *35*(3), 201–207. http://dx.doi.org/10.1111/j.1365-2362.2005.01467.x.

Baptiste, D. C., & Fehlings, M. G. (2006). Pharmacological approaches to repair the injured spinal cord. *Journal of Neurotrauma*, *23*(3–4), 318–334. http://dx.doi.org/10.1089/neu.2006.23.318.

Barde, Y. A., Edgar, D., & Thoenen, H. (1982). Purification of a new neurotrophic factor from mammalian brain. *The EMBO Journal*, *1*(5), 549–553.

Barnham, K. J., Ciccotosto, G. D., Tickler, A. K., Ali, F. E., Smith, D. G., Williamson, N. A., … Cappai, R. (2003). Neurotoxic, redox-competent Alzheimer's beta-amyloid is released from lipid membrane by methionine oxidation. *The Journal of Biological Chemistry*, *278*(44), 42959–42965. http://dx.doi.org/10.1074/jbc.M305494200.

Bliss, T. V., & Collingridge, G. L. (1993). A synaptic model of memory: Long-term potentiation in the hippocampus. *Nature, 361*(6407), 31–39. http://dx.doi.org/10.1038/361031a0.

Boissiere, F., Faucheux, B., Ruberg, M., Agid, Y., & Hirsch, E. C. (1997). Decreased TrkA gene expression in cholinergic neurons of the striatum and basal forebrain of patients with Alzheimer's disease. *Experimental Neurology, 145*(1), 245–252. doi: S0014-4886(97)96443-8 [pii].

Bonaccorso, S., Sodhi, M., Li, J., Bobo, W. V., Chen, Y., Tumuklu, M., … Meltzer, H. Y. (2015). The brain-derived neurotrophic factor (BDNF) Val66Met polymorphism is associated with increased body mass index and insulin resistance measures in bipolar disorder and schizophrenia. *Bipolar Disorders, 17*(5), 528–535. http://dx.doi.org/10.1111/bdi.12294.

Boulanger, L. M., & Poo, M. M. (1999). Presynaptic depolarization facilitates neurotrophin-induced synaptic potentiation. *Nature Neuroscience, 2*(4), 346–351. http://dx.doi.org/10.1038/7258.

Bova, D., Shownkeen, H., Goldberg, K., Horowitz, S., & Azar-Kia, B. (2000). Delayed transient neurologic toxicity due to tacrolimus: CT and MRI. *Neuroradiology, 42*(9), 666–668.

Bradshaw, R. A., Murray-Rust, J., Ibanez, C. F., McDonald, N. Q., Lapatto, R., & Blundell, T. L. (1994). Nerve growth factor: Structure/function relationships. *Protein Science, 3*(11), 1901–1913. http://dx.doi.org/10.1002/pro.5560031102.

Brito, V., Giralt, A., Enriquez-Barreto, L., Puigdellivol, M., Suelves, N., Zamora-Moratalla, A., … Gines, S. (2014). Neurotrophin receptor p75(NTR) mediates Huntington's disease-associated synaptic and memory dysfunction. *The Journal of Clinical Investigation, 124*(10), 4411–4428. http://dx.doi.org/10.1172/JCI74809.

Bruno, M. A., & Cuello, A. C. (2006). Activity-dependent release of precursor nerve growth factor, conversion to mature nerve growth factor, and its degradation by a protease cascade. *Proceedings of the National Academy of Sciences of the United States of America, 103*(17), 6735–6740. http://dx.doi.org/10.1073/pnas.0510645103.

Budni, J., Bellettini-Santos, T., Mina, F., Garcez, M. L., & Zugno, A. I. (2015). The involvement of BDNF, NGF and GDNF in aging and Alzheimer's disease. *Aging and Disease, 6*(5), 331–341. http://dx.doi.org/10.14336/AD.2015.0825.

Cai, H., Cong, W. N., Ji, S., Rothman, S., Maudsley, S., & Martin, B. (2012). Metabolic dysfunction in Alzheimer's disease and related neurodegenerative disorders. *Current Alzheimer Research, 9*(1), 5–17.

Calissano, P., Matrone, C., & Amadoro, G. (2010). Nerve growth factor as a paradigm of neurotrophins related to Alzheimer's disease. *Developmental Neurobiology, 70*(5), 372–383. http://dx.doi.org/10.1002/dneu.20759.

Campion, D., Flaman, J. M., Brice, A., Hannequin, D., Dubois, B., Martin, C., et al. (1995). Mutations of the presenilin I gene in families with early-onset Alzheimer's disease. *Human Molecular Genetics, 4*(12), 2373–2377.

Capsoni, S., & Cattaneo, A. (2006). On the molecular basis linking Nerve Growth Factor (NGF) to Alzheimer's disease. *Cellular and Molecular Neurobiology, 26*(4–6), 619–633. http://dx.doi.org/10.1007/s10571-006-9112-2.

Capsoni, S., Tiveron, C., Vignone, D., Amato, G., & Cattaneo, A. (2010). Dissecting the involvement of tropomyosin-related kinase A and p75 neurotrophin receptor signaling in NGF deficit-induced neurodegeneration. *Proceedings of the National Academy of Sciences of the United States of America, 107*(27), 12299–12304. http://dx.doi.org/10.1073/pnas.1007181107.

Casaccia-Bonnefil, P., Carter, B. D., Dobrowsky, R. T., & Chao, M. V. (1996). Death of oligodendrocytes mediated by the interaction of nerve growth factor with its receptor p75. *Nature, 383*(6602), 716–719. http://dx.doi.org/10.1038/383716a0.

Castren, E. (2004). Neurotrophic effects of antidepressant drugs. *Current Opinion in Pharmacology*, *4*(1), 58–64. http://dx.doi.org/10.1016/j.coph.2003.10.004.

Chadwick, W., Magnus, T., Martin, B., Keselman, A., Mattson, M. P., & Maudsley, S. (2008). Targeting TNF-alpha receptors for neurotherapeutics. *Trends in Neurosciences*, *31*(10), 504–511. http://dx.doi.org/10.1016/j.tins.2008.07.005.

Chadwick, W., Mitchell, N., Caroll, J., Zhou, Y., Park, S. S., Wang, L., ... Maudsley, S. (2011). Amitriptyline-mediated cognitive enhancement in aged 3xTg Alzheimer's disease mice is associated with neurogenesis and neurotrophic activity. *PLoS One*, *6*(6), e21660. http://dx.doi.org/10.1371/journal.pone.0021660.

Chao, M. V. (2003). Neurotrophins and their receptors: A convergence point for many signalling pathways. *Nature Reviews. Neuroscience*, *4*(4), 299–309. http://dx.doi.org/10.1038/nrn1078.

Chapman, J., Asherov, A., Wang, N., Treves, T. A., Korczyn, A. D., & Goldfarb, L. G. (1995). Familial Alzheimer's disease associated with S182 codon 286 mutation. *Lancet*, *346*(8981), 1040.

Charalampopoulos, I., Vicario, A., Pediaditakis, I., Gravanis, A., Simi, A., & Ibanez, C. F. (2012). Genetic dissection of neurotrophin signaling through the p75 neurotrophin receptor. *Cell Reports*, *2*(6), 1563–1570. http://dx.doi.org/10.1016/j.celrep.2012.11.009.

Chartier-Harlin, M. C., Crawford, F., Houlden, H., Warren, A., Hughes, D., Fidani, L., et al. (1991). Early-onset Alzheimer's disease caused by mutations at codon 717 of the beta-amyloid precursor protein gene. *Nature*, *353*(6347), 844–846. http://dx.doi.org/10.1038/353844a0.

Chen, K. S., Nishimura, M. C., Armanini, M. P., Crowley, C., Spencer, S. D., & Phillips, H. S. (1997). Disruption of a single allele of the nerve growth factor gene results in atrophy of basal forebrain cholinergic neurons and memory deficits. *The Journal of Neuroscience*, *17*(19), 7288–7296.

Chen, S. C., Lu, G., Chan, C. Y., Chen, Y., Wang, H., Yew, D. T., ... Kung, H. F. (2010). Microarray profile of brain aging-related genes in the frontal cortex of SAMP8. *Journal of Molecular Neuroscience*, *41*(1), 12–16. http://dx.doi.org/10.1007/s12031-009-9215-6.

Chen, Z. Y., Ieraci, A., Teng, H., Dall, H., Meng, C. X., Herrera, D. G., ... Lee, F. S. (2005). Sortilin controls intracellular sorting of brain-derived neurotrophic factor to the regulated secretory pathway. *The Journal of Neuroscience*, *25*(26), 6156–6166. http://dx.doi.org/10.1523/JNEUROSCI.1017-05.2005.

Chung, J., Grammer, T. C., Lemon, K. P., Kazlauskas, A., & Blenis, J. (1994). PDGF- and insulin-dependent pp70S6k activation mediated by phosphatidylinositol-3-OH kinase. *Nature*, *370*(6484), 71–75. http://dx.doi.org/10.1038/370071a0.

Clewes, O., Fahey, M. S., Tyler, S. J., Watson, J. J., Seok, H., Catania, C., ... Allen, S. J. (2008). Human ProNGF: Biological effects and binding profiles at TrkA, P75NTR and sortilin. *Journal of Neurochemistry*, *107*(4), 1124–1135. http://dx.doi.org/10.1111/j.1471-4159.2008.05698.x.

Conner, J. M., Lauterborn, J. C., Yan, Q., Gall, C. M., & Varon, S. (1997). Distribution of brain-derived neurotrophic factor (BDNF) protein and mRNA in the normal adult rat CNS: Evidence for anterograde axonal transport. *The Journal of Neuroscience*, *17*(7), 2295–2313.

Corbit, K. C., Foster, D. A., & Rosner, M. R. (1999). Protein kinase Cdelta mediates neurogenic but not mitogenic activation of mitogen-activated protein kinase in neuronal cells. *Molecular and Cellular Biology*, *19*(6), 4209–4218.

Cowansage, K. K., LeDoux, J. E., & Monfils, M. H. (2010). Brain-derived neurotrophic factor: A dynamic gatekeeper of neural plasticity. *Current Molecular Pharmacology*, *3*(1), 12–29.

Crutcher, K. A., Scott, S. A., Liang, S., Everson, W. V., & Weingartner, J. (1993). Detection of NGF-like activity in human brain tissue: Increased levels in Alzheimer's disease. *The Journal of Neuroscience*, *13*(6), 2540–2550.

Cruts, M., Backhovens, H., Wang, S. Y., Van Gassen, G., Theuns, J., De Jonghe, C. D., et al. (1995). Molecular genetic analysis of familial early-onset Alzheimer's disease linked to chromosome 14q24.3. *Human Molecular Genetics*, *4*(12), 2363–2371.

Cuello, A. C., Bruno, M. A., Allard, S., Leon, W., & Iulita, M. F. (2010). Cholinergic involvement in Alzheimer's disease. A link with NGF maturation and degradation. *Journal of Molecular Neuroscience*, *40*(1–2), 230–235. http://dx.doi.org/10.1007/s12031-009-9238-z.

Dawbarn, D., & Allen, S. J. (2003). Neurotrophins and neurodegeneration. *Neuropathology and Applied Neurobiology*, *29*(3), 211–230.

Dawbarn, D., Fahey, M., Watson, J., Tyler, S., Shoemark, D., Sessions, R., … Allen, S. J. (2006). NGF receptor TrkAd5: Therapeutic agent and drug design target. *Biochemical Society Transactions*, *34*(Pt. 4), 587–590. http://dx.doi.org/10.1042/BST0340587.

Dechant, G., & Barde, Y. A. (2002). The neurotrophin receptor p75(NTR): Novel functions and implications for diseases of the nervous system. *Nature Neuroscience*, *5*(11), 1131–1136. http://dx.doi.org/10.1038/nn1102-1131.

del Peso, L., Gonzalez-Garcia, M., Page, C., Herrera, R., & Nunez, G. (1997). Interleukin-3-induced phosphorylation of BAD through the protein kinase Akt. *Science*, *278*(5338), 687–689.

Devi, L., & Ohno, M. (2015). TrkB reduction exacerbates Alzheimer's disease-like signaling aberrations and memory deficits without affecting beta-amyloidosis in 5XFAD mice. *Translational Psychiatry*, *5*, e562. http://dx.doi.org/10.1038/tp.2015.55.

Dicou, E., Rangon, C. M., Guimiot, F., Spedding, M., & Gressens, P. (2003). Positive allosteric modulators of AMPA receptors are neuroprotective against lesions induced by an NMDA agonist in neonatal mouse brain. *Brain Research*, *970*(1–2), 221–225.

Dohlman, H. G. (2015). Thematic minireview series: New directions in G protein-coupled receptor pharmacology. *The Journal of Biological Chemistry*, *290*(32), 19469–19470. http://dx.doi.org/10.1074/jbc.R115.675728.

Dougherty, K. D., & Milner, T. A. (1999). p75NTR immunoreactivity in the rat dentate gyrus is mostly within presynaptic profiles but is also found in some astrocytic and postsynaptic profiles. *The Journal of Comparative Neurology*, *407*(1), 77–91. http://dx.doi.org/10.1002/(SICI)1096-9861(19990428)407:1<77::AID-CNE6>3.0.CO;2-S [pii].

Dudek, H., Datta, S. R., Franke, T. F., Birnbaum, M. J., Yao, R., Cooper, G. M., … Greenberg, M. E. (1997). Regulation of neuronal survival by the serine-threonine protein kinase Akt. *Science*, *275*(5300), 661–665.

Dutta, J., Fan, Y., Gupta, N., Fan, G., & Gelinas, C. (2006). Current insights into the regulation of programmed cell death by NF-kappaB. *Oncogene*, *25*(51), 6800–6816. http://dx.doi.org/10.1038/sj.onc.1209938.

Egan, M. F., Kojima, M., Callicott, J. H., Goldberg, T. E., Kolachana, B. S., Bertolino, A., … Weinberger, D. R. (2003). The BDNF val66met polymorphism affects activity-dependent secretion of BDNF and human memory and hippocampal function. *Cell*, *112*(2), 257–269.

Eguchi, S., Numaguchi, K., Iwasaki, H., Matsumoto, T., Yamakawa, T., Utsunomiya, H., … Inagami, T. (1998). Calcium-dependent epidermal growth factor receptor transactivation mediates the angiotensin II-induced mitogen-activated protein kinase activation in vascular smooth muscle cells. *Journal of Biological Chemistry*, *273*(15), 8890–8896.

Elliott, E., Atlas, R., Lange, A., & Ginzburg, I. (2005). Brain-derived neurotrophic factor induces a rapid dephosphorylation of tau protein through a PI-3 Kinase signalling mechanism. *The European Journal of Neuroscience*, *22*(5), 1081–1089. doi: EJN4290 [pii].

Engert, F., & Bonhoeffer, T. (1999). Dendritic spine changes associated with hippocampal long-term synaptic plasticity. *Nature*, *399*(6731), 66–70. http://dx.doi.org/10.1038/19978.

Ernfors, P., Lee, K. F., Kucera, J., & Jaenisch, R. (1994). Lack of neurotrophin-3 leads to deficiencies in the peripheral nervous system and loss of limb proprioceptive afferents. *Cell*, 77(4), 503–512.

Evans, S. F., Irmady, K., Ostrow, K., Kim, T., Nykjaer, A., Saftig, P., … Hempstead, B. L. (2011). Neuronal brain-derived neurotrophic factor is synthesized in excess, with levels regulated by sortilin-mediated trafficking and lysosomal degradation. *The Journal of Biological Chemistry*, *286*(34), 29556–29567. http://dx.doi.org/10.1074/jbc.M111.219675.

Figurov, A., Pozzo-Miller, L. D., Olafsson, P., Wang, T., & Lu, B. (1996). Regulation of synaptic responses to high-frequency stimulation and LTP by neurotrophins in the hippocampus. *Nature*, *381*(6584), 706–709. http://dx.doi.org/10.1038/381706a0.

Gentry, J. J., Rutkoski, N. J., Burke, T. L., & Carter, B. D. (2004). A functional interaction between the p75 neurotrophin receptor interacting factors, TRAF6 and NRIF. *The Journal of Biological Chemistry*, *279*(16), 16646–16656. http://dx.doi.org/10.1074/jbc.M309209200.

Gerlai, R. (2001). Alzheimer's disease: Beta-amyloid hypothesis strengthened!. *Trends in Neurosciences*, *24*(4), 199.

Gille, H., Sharrocks, A. D., & Shaw, P. E. (1992). Phosphorylation of transcription factor p62TCF by MAP kinase stimulates ternary complex formation at c-fos promoter. *Nature*, *358*(6385), 414–417. http://dx.doi.org/10.1038/358414a0.

Goate, A., Chartier-Harlin, M. C., Mullan, M., Brown, J., Crawford, F., Fidani, L., et al. (1991). Segregation of a missense mutation in the amyloid precursor protein gene with familial Alzheimer's disease. *Nature*, *349*(6311), 704–706. http://dx.doi.org/10.1038/349704a0.

Golden, E., Emiliano, A., Maudsley, S., Windham, B. G., Carlson, O. D., Egan, J. M., … Mattson, M. P. (2010). Circulating brain-derived neurotrophic factor and indices of metabolic and cardiovascular health: Data from the Baltimore Longitudinal Study of Aging. *PLoS One*, *5*(4), e10099. http://dx.doi.org/10.1371/journal.pone.0010099.

Gomez-Palacio-Schjetnan, A., & Escobar, M. L. (2013). Neurotrophins and synaptic plasticity. *Current Topics in Behavioral Neurosciences*, *15*, 117–136. http://dx.doi.org/10.1007/7854_2012_231.

Gotz, R., Koster, R., Winkler, C., Raulf, F., Lottspeich, F., Schartl, M., & Thoenen, H. (1994). Neurotrophin-6 is a new member of the nerve growth factor family. *Nature*, *372*(6503), 266–269. http://dx.doi.org/10.1038/372266a0.

Haase, G., Kennel, P., Pettmann, B., Vigne, E., Akli, S., Revah, F., … Kahn, A. (1997). Gene therapy of murine motor neuron disease using adenoviral vectors for neurotrophic factors. *Nature Medicine*, *3*(4), 429–436.

Halban, P. A., & Irminger, J. C. (1994). Sorting and processing of secretory proteins. *The Biochemical Journal*, *299*(Pt. 1), 1–18.

Harandi, V. M., Lindquist, S., Kolan, S. S., Brannstrom, T., & Liu, J. X. (2014). Analysis of neurotrophic factors in limb and extraocular muscles of mouse model of amyotrophic lateral sclerosis. *PLoS One*, *9*(10), e109833. http://dx.doi.org/10.1371/journal.pone.0109833.

Hardy, J., & Selkoe, D. J. (2002). The amyloid hypothesis of Alzheimer's disease: Progress and problems on the road to therapeutics. *Science*, *297*(5580), 353–356. http://dx.doi.org/10.1126/science.1072994.

Heightman, T. D., Gaster, L. M., Pardoe, S. L., Pilleux, J. P., Hadley, M. S., Middlemiss, D. N., … Cutler, L. (2005). 8-Piperazinyl-2,3-dihydropyrrolo[3,2-g]isoquinolines: Potent, selective, orally bioavailable 5-HT1 receptor ligands. *Bioorganic & Medicinal Chemistry Letters*, *15*(19), 4370–4374. http://dx.doi.org/10.1016/j.bmcl.2005.06.042.

Hellweg, R., Gericke, C. A., Jendroska, K., Hartung, H. D., & Cervos-Navarro, J. (1998). NGF content in the cerebral cortex of non-demented patients with amyloid-plaques and in symptomatic Alzheimer's disease. *International Journal of Developmental Neuroscience, 16*(7–8), 787–794.

Hempstead, B. L. (2002). The many faces of p75NTR. *Current Opinion in Neurobiology, 12*(3), 260–267.

Herskowitz, J. H., Feng, Y., Mattheyses, A. L., Hales, C. M., Higginbotham, L. A., Duong, D. M., … Lah, J. J. (2013). Pharmacologic inhibition of ROCK2 suppresses amyloid-beta production in an Alzheimer's disease mouse model. *The Journal of Neuroscience, 33*(49), 19086–19098. http://dx.doi.org/10.1523/JNEUROSCI.2508-13.2013.

Hill, C. S., & Treisman, R. (1995). Transcriptional regulation by extracellular signals: Mechanisms and specificity. *Cell, 80*(2), 199–211.

Hock, C., Heese, K., Hulette, C., Rosenberg, C., & Otten, U. (2000). Region-specific neurotrophin imbalances in Alzheimer disease: Decreased levels of brain-derived neurotrophic factor and increased levels of nerve growth factor in hippocampus and cortical areas. *Archives of Neurology, 57*(6), 846–851. doi: noc90083 [pii].

Holgado-Madruga, M., Moscatello, D. K., Emlet, D. R., Dieterich, R., & Wong, A. J. (1997). Grb2-associated binder-1 mediates phosphatidylinositol 3-kinase activation and the promotion of cell survival by nerve growth factor. *Proceedings of the National Academy of Sciences of the United States of America, 94*(23), 12419–12424.

Hu, X. Y., Zhang, H. Y., Qin, S., Xu, H., Swaab, D. F., & Zhou, J. N. (2002). Increased p75 (NTR) expression in hippocampal neurons containing hyperphosphorylated tau in Alzheimer patients. *Experimental Neurology, 178*(1), 104–111. doi: S0014488602980180 [pii].

Huang, E. J., & Reichardt, L. F. (2003). Trk receptors: Roles in neuronal signal transduction. *Annual Review of Biochemistry, 72*, 609–642. http://dx.doi.org/10.1146/annurev.biochem.72.121801.161629.

Ibanez, C. F. (1995). Neurotrophic factors: From structure–function studies to designing effective therapeutics. *Trends in Biotechnology, 13*(6), 217–227. doi: S0167-7799(00)88949-0 [pii].

Ichim, G., Tauszig-Delamasure, S., & Mehlen, P. (2012). Neurotrophins and cell death. *Experimental Cell Research, 318*(11), 1221–1228. http://dx.doi.org/10.1016/j.yexcr.2012.03.006.

Inoue, J., Ishida, T., Tsukamoto, N., Kobayashi, N., Naito, A., Azuma, S., & Yamamoto, T. (2000). Tumor necrosis factor receptor-associated factor (TRAF) family: Adapter proteins that mediate cytokine signaling. *Experimental Cell Research, 254*(1), 14–24. http://dx.doi.org/10.1006/excr.1999.4733.

Ip, N. Y., Ibanez, C. F., Nye, S. H., McClain, J., Jones, P. F., Gies, D. R., et al. (1992). Mammalian neurotrophin-4: Structure, chromosomal localization, tissue distribution, and receptor specificity. *Proceedings of the National Academy of Sciences of the United States of America, 89*(7), 3060–3064.

Jang, S. W., Liu, X., Chan, C. B., France, S. A., Sayeed, I., Tang, W., … Ye, K. (2010a). Deoxygedunin, a natural product with potent neurotrophic activity in mice. *PLoS One, 5*(7), e11528. http://dx.doi.org/10.1371/journal.pone.0011528.

Jang, S. W., Liu, X., Chan, C. B., Weinshenker, D., Hall, R. A., Xiao, G., & Ye, K. (2009). Amitriptyline is a TrkA and TrkB receptor agonist that promotes TrkA/TrkB heterodimerization and has potent neurotrophic activity. *Chemistry & Biology, 16*(6), 644–656. http://dx.doi.org/10.1016/j.chembiol.2009.05.010.

Jang, S. W., Liu, X., Yepes, M., Shepherd, K. R., Miller, G. W., Liu, Y., … Ye, K. (2010b). A selective TrkB agonist with potent neurotrophic activities by 7,8-dihydroxyflavone. *Proceedings of the National Academy of Sciences of the United States of America, 107*(6), 2687–2692. http://dx.doi.org/10.1073/pnas.0913572107.

Jang, S. W., Okada, M., Sayeed, I., Xiao, G., Stein, D., Jin, P., & Ye, K. (2007). Gambogic amide, a selective agonist for TrkA receptor that possesses robust neurotrophic activity, prevents neuronal cell death. *Proceedings of the National Academy of Sciences of the United States of America*, *104*(41), 16329–16334. http://dx.doi.org/10.1073/pnas.0706662104.

Janssens, J., Etienne, H., Idriss, S., Azmi, A., Martin, B., & Maudsley, S. (2014). Systems-level G protein-coupled receptor therapy across a neurodegenerative continuum by the GLP-1 receptor system. *Frontiers in Endocrinology (Lausanne)*, *5*, 142. http://dx.doi.org/10.3389/fendo.2014.00142.

Jean-Alphonse, F., Bowersox, S., Chen, S., Beard, G., Puthenveedu, M. A., & Hanyaloglu, A. C. (2014). Spatially restricted G protein-coupled receptor activity via divergent endocytic compartments. *The Journal of Biological Chemistry*, *289*(7), 3960–3977. http://dx.doi.org/10.1074/jbc.M113.526350.

Jovanovic, J. N., Czernik, A. J., Fienberg, A. A., Greengard, P., & Sihra, T. S. (2000). Synapsins as mediators of BDNF-enhanced neurotransmitter release. *Nature Neuroscience*, *3*(4), 323–329. http://dx.doi.org/10.1038/73888.

Jung, E. J., & Kim, D. R. (2010). Control of TrkA-induced cell death by JNK activation and differential expression of TrkA upon DNA damage. *Molecules and Cells*, *30*(2), 121–125. http://dx.doi.org/10.1007/s10059-010-0096-x.

Kang, H., Welcher, A. A., Shelton, D., & Schuman, E. M. (1997). Neurotrophins and time: Different roles for TrkB signaling in hippocampal long-term potentiation. *Neuron*, *19*(3), 653–664. doi: S0896-6273(00)80378-5 [pii].

Kauffmann-Zeh, A., Rodriguez-Viciana, P., Ulrich, E., Gilbert, C., Coffer, P., Downward, J., & Evan, G. (1997). Suppression of c-Myc-induced apoptosis by Ras signalling through PI(3)K and PKB. *Nature*, *385*(6616), 544–548. http://dx.doi.org/10.1038/385544a0.

Kelly, A., Maguire, C., & Lynch, M. A. (2000). Deficits in nerve growth factor release and tyrosine receptor kinase phosphorylation are associated with age-related impairment in long-term potentiation in the dentate gyrus. *Neuroscience*, *95*(2), 359–365. doi: S0306452299004601 [pii].

Kichev, A., Ilieva, E. V., Pinol-Ripoll, G., Podlesniy, P., Ferrer, I., Portero-Otin, M., … Espinet, C. (2009). Cell death and learning impairment in mice caused by in vitro modified pro-NGF can be related to its increased oxidative modifications in Alzheimer disease. *The American Journal of Pathology*, *175*(6), 2574–2585. http://dx.doi.org/10.2353/ajpath.2009.090018.

Kim, S. H., Tang, Y. P., & Sisodia, S. S. (2006). Abeta star: A light onto synaptic dysfunction? *Nature Medicine*, *12*(7), 760–761. discussion 761, doi: nm0706-760 [pii].

Kimura, N., Takahashi, M., Tashiro, T., & Terao, K. (2006). Amyloid beta up-regulates brain-derived neurotrophic factor production from astrocytes: Rescue from amyloid beta-related neuritic degeneration. *Journal of Neuroscience Research*, *84*(4), 782–789. http://dx.doi.org/10.1002/jnr.20984.

Klein, R., Jing, S. Q., Nanduri, V., O'Rourke, E., & Barbacid, M. (1991). The trk proto-oncogene encodes a receptor for nerve growth factor. *Cell*, *65*(1), 189–197.

Klein, R., Lamballe, F., Bryant, S., & Barbacid, M. (1992). The trkB tyrosine protein kinase is a receptor for neurotrophin-4. *Neuron*, *8*(5), 947–956.

Klein, R., Nanduri, V., Jing, S. A., Lamballe, F., Tapley, P., Bryant, S., … Barbacid, M. (1991). The trkB tyrosine protein kinase is a receptor for brain-derived neurotrophic factor and neurotrophin-3. *Cell*, *66*(2), 395–403.

Knowles, J. K., Simmons, D. A., Nguyen, T. V., Vander Griend, L., Xie, Y., Zhang, H., … Longo, F. M. (2013). Small molecule p75NTR ligand prevents cognitive deficits and neurite degeneration in an Alzheimer's mouse model. *Neurobiology of Aging*, *34*(8), 2052–2063. http://dx.doi.org/10.1016/j.neurobiolaging.2013.02.015.

Korsching, S., Auburger, G., Heumann, R., Scott, J., & Thoenen, H. (1985). Levels of nerve growth factor and its mRNA in the central nervous system of the rat correlate with cholinergic innervation. *The EMBO Journal*, *4*(6), 1389–1393.

Lamballe, F., Klein, R., & Barbacid, M. (1991). trkC, a new member of the trk family of tyrosine protein kinases, is a receptor for neurotrophin-3. *Cell*, *66*(5), 967–979.

Lange-Carter, C. A., & Johnson, G. L. (1994). Ras-dependent growth factor regulation of MEK kinase in PC12 cells. *Science*, *265*(5177), 1458–1461.

Leal, G., Comprido, D., & Duarte, C. B. (2014). BDNF-induced local protein synthesis and synaptic plasticity. *Neuropharmacology*, *76*(Pt. C), 639–656. http://dx.doi.org/10.1016/j.neuropharm.2013.04.005.

Lee, F. S., & Chao, M. V. (2001). Activation of Trk neurotrophin receptors in the absence of neurotrophins. *Proceedings of the National Academy of Sciences of the United States of America*, *98*(6), 3555–3560. http://dx.doi.org/10.1073/pnas.061020198.

Lee, F. S., Rajagopal, R., Kim, A. H., Chang, P. C., & Chao, M. V. (2002). Activation of Trk neurotrophin receptor signaling by pituitary adenylate cyclase-activating polypeptides. *Journal of Biological Chemistry*, *277*(11), 9096–9102. http://dx.doi.org/10.1074/jbc.M107421200.

Lee, R., Kermani, P., Teng, K. K., & Hempstead, B. L. (2001). Regulation of cell survival by secreted proneurotrophins. *Science*, *294*(5548), 1945–1948. http://dx.doi.org/10.1126/science.1065057.

Lemmon, M. A., & Schlessinger, J. (2010). Cell signaling by receptor tyrosine kinases. *Cell*, *141*(7), 1117–1134. http://dx.doi.org/10.1016/j.cell.2010.06.011.

Lessmann, V., & Brigadski, T. (2009). Mechanisms, locations, and kinetics of synaptic BDNF secretion: An update. *Neuroscience Research*, *65*(1), 11–22. http://dx.doi.org/10.1016/j.neures.2009.06.004.

Lessmann, V., Gottmann, K., & Malcangio, M. (2003). Neurotrophin secretion: Current facts and future prospects. *Progress in Neurobiology*, *69*(5), 341–374.

Levi-Montalcini, R. (1987). The nerve growth factor 35 years later. *Science*, *237*(4819), 1154–1162.

Liepinsh, E., Ilag, L. L., Otting, G., & Ibanez, C. F. (1997). NMR structure of the death domain of the p75 neurotrophin receptor. *The EMBO Journal*, *16*(16), 4999–5005. http://dx.doi.org/10.1093/emboj/16.16.4999.

Lin, B., Pirrung, M. C., Deng, L., Li, Z., Liu, Y., & Webster, N. J. (2007). Neuroprotection by small molecule activators of the nerve growth factor receptor. *The Journal of Pharmacology and Experimental Therapeutics*, *322*(1), 59–69. http://dx.doi.org/10.1124/jpet.106.118034.

Liu, L., Wong, T. P., Pozza, M. F., Lingenhoehl, K., Wang, Y., Sheng, M., … Wang, Y. T. (2004). Role of NMDA receptor subtypes in governing the direction of hippocampal synaptic plasticity. *Science*, *304*(5673), 1021–1024. http://dx.doi.org/10.1126/science.1096615.

Longo, F. M., & Massa, S. M. (2013). Small-molecule modulation of neurotrophin receptors: A strategy for the treatment of neurological disease. *Nature Reviews. Drug Discovery*, *12*(7), 507–525.

Longo, F. M., Yang, T., Knowles, J. K., Xie, Y., Moore, L. A., & Massa, S. M. (2007). Small molecule neurotrophin receptor ligands: Novel strategies for targeting Alzheimer's disease mechanisms. *Current Alzheimer Research*, *4*(5), 503–506.

Lonze, B. E., Riccio, A., Cohen, S., & Ginty, D. D. (2002). Apoptosis, axonal growth defects, and degeneration of peripheral neurons in mice lacking CREB. *Neuron*, *34*(3), 371–385.

Lu, B., Pang, P. T., & Woo, N. H. (2005). The yin and yang of neurotrophin action. *Nature Reviews. Neuroscience*, *6*(8), 603–614. http://dx.doi.org/10.1038/nrn1726.

Lu, C. S., Horizon, A. A., Hwang, K. K., FitzGerald, J., Lin, W. S., Hahn, B. H., … Chen, P. P. (2005). Identification of polyclonal and monoclonal antibodies against tissue plasminogen activator in the antiphospholipid syndrome. *Arthritis and Rheumatism*, *52*(12), 4018–4027. http://dx.doi.org/10.1002/art.21485.

Lu, Y., Liu, X., Zhou, J., Huang, A., Zhou, J., & He, C. (2013). TROY interacts with Rho guanine nucleotide dissociation inhibitor alpha (RhoGDIalpha) to mediate Nogo-induced inhibition of neurite outgrowth. *The Journal of Biological Chemistry*, *288*(47), 34276–34286. http://dx.doi.org/10.1074/jbc.M113.519744.

Luttrell, L. M., Maudsley, S., & Bohn, L. M. (2015). Fulfilling the promise of "biased" G protein-coupled receptor agonism. *Molecular Pharmacology*, *88*(3), 579–588. http://dx.doi.org/10.1124/mol.115.099630.

Malberg, J. E., & Blendy, J. A. (2005). Antidepressant action: To the nucleus and beyond. *Trends in Pharmacological Sciences*, *26*(12), 631–638. http://dx.doi.org/10.1016/j.tips.2005.10.005.

Maletic-Savatic, M., Malinow, R., & Svoboda, K. (1999). Rapid dendritic morphogenesis in CA1 hippocampal dendrites induced by synaptic activity. *Science*, *283*(5409), 1923–1927.

Marinissen, M. J., Chiariello, M., Pallante, M., & Gutkind, J. S. (1999). A network of mitogen-activated protein kinases links G protein-coupled receptors to the c-jun promoter: A role for c-Jun NH2-terminal kinase, p38s, and extracellular signal-regulated kinase 5. *Molecular and Cellular Biology*, *19*(6), 4289–4301.

Martin, B., Brenneman, R., Golden, E., Walent, T., Becker, K. G., Prabhu, V. V., … Maudsley, S. (2009). Growth factor signals in neural cells: Coherent patterns of interaction control multiple levels of molecular and phenotypic responses. *The Journal of Biological Chemistry*, *284*(4), 2493–2511. http://dx.doi.org/10.1074/jbc.M804545200.

Martin, B., Chen, H., Daimon, C. M., Chadwick, W., Siddiqui, S., & Maudsley, S. (2013). Plurigon: Three dimensional visualization and classification of high-dimensionality data. *Frontiers in Physiology*, *4*, 190. http://dx.doi.org/10.3389/fphys.2013.00190.

Masoudi, R., Ioannou, M. S., Coughlin, M. D., Pagadala, P., Neet, K. E., Clewes, O., … Fahnestock, M. (2009). Biological activity of nerve growth factor precursor is dependent upon relative levels of its receptors. *The Journal of Biological Chemistry*, *284*(27), 18424–18433. http://dx.doi.org/10.1074/jbc.M109.007104.

Massa, S. M., Xie, Y., Yang, T., Harrington, A. W., Kim, M. L., Yoon, S. O., … Longo, F. M. (2006). Small, nonpeptide p75NTR ligands induce survival signaling and inhibit proNGF-induced death. *The Journal of Neuroscience*, *26*(20), 5288–5300. http://dx.doi.org/10.1523/JNEUROSCI.3547-05.2006.

Massa, S. M., Yang, T., Xie, Y., Shi, J., Bilgen, M., Joyce, J. N., … Longo, F. M. (2010). Small molecule BDNF mimetics activate TrkB signaling and prevent neuronal degeneration in rodents. *The Journal of Clinical Investigation*, *120*(5), 1774–1785. http://dx.doi.org/10.1172/JCI41356.

Massey, P. V., Johnson, B. E., Moult, P. R., Auberson, Y. P., Brown, M. W., Molnar, E., … Bashir, Z. I. (2004). Differential roles of NR2A and NR2B-containing NMDA receptors in cortical long-term potentiation and long-term depression. *The Journal of Neuroscience*, *24*(36), 7821–7828. http://dx.doi.org/10.1523/JNEUROSCI.1697-04.2004.

Matsumoto, T., Numakawa, T., Adachi, N., Yokomaku, D., Yamagishi, S., Takei, N., & Hatanaka, H. (2001). Brain-derived neurotrophic factor enhances depolarization-evoked glutamate release in cultured cortical neurons. *Journal of Neurochemistry*, *79*(3), 522–530.

Mattson, M. P. (2004). Pathways towards and away from Alzheimer's disease. *Nature*, *430*(7000), 631–639. http://dx.doi.org/10.1038/nature02621.

Maudsley, S. (2012). G protein-coupled receptor biased agonism: Development towards future selective therapeutics. *Mini Reviews in Medicinal Chemistry*, *12*(9), 803.

Maudsley, S., Martin, B., & Egan, J. M. (2011). To be or not to be—Obese. *Endocrinology*, *152*(10), 3592–3596. 152/10/3592 [pii].

Maudsley, S., Martin, B., Gesty-Palmer, D., Cheung, H., Johnson, C., Patel, S., … Luttrell, L. M. (2015). Delineation of a conserved arrestin-biased signaling repertoire in vivo. *Molecular Pharmacology, 87*(4), 706–717. http://dx.doi.org/10.1124/mol.114.095224.

Maudsley, S., Martin, B., & Luttrell, L. M. (2005). The origins of diversity and specificity in g protein-coupled receptor signaling. *The Journal of Pharmacology and Experimental Therapeutics, 314*(2), 485–494. http://dx.doi.org/10.1124/jpet.105.083121.

Maudsley, S., Martin, B., & Luttrell, L. M. (2007). G protein-coupled receptor signaling complexity in neuronal tissue: Implications for novel therapeutics. *Current Alzheimer Research, 4*(1), 3–19.

Maudsley, S., & Mattson, M. P. (2006). Protein twists and turns in Alzheimer disease. *Nature Medicine, 12*(4), 392–393. http://dx.doi.org/10.1038/nm0406-392.

Maudsley, S., Patel, S. A., Park, S. S., Luttrell, L. M., & Martin, B. (2012). Functional signaling biases in G protein-coupled receptors: Game Theory and receptor dynamics. *Mini Reviews in Medicinal Chemistry, 12*(9), 831–840. doi: MRMC-EPUB-20120611-30 [pii].

Maudsley, S., Pierce, K. L., Zamah, A. M., Miller, W. E., Ahn, S., Daaka, Y., … Luttrell, L. M. (2000). The beta(2)-adrenergic receptor mediates extracellular signal-regulated kinase activation via assembly of a multi-receptor complex with the epidermal growth factor receptor. *Journal of Biological Chemistry, 275*(13), 9572–9580.

McAllister, A. K., Katz, L. C., & Lo, D. C. (1996). Neurotrophin regulation of cortical dendritic growth requires activity. *Neuron, 17*(6), 1057–1064.

McClean, M. N., Mody, A., Broach, J. R., & Ramanathan, S. (2007). Cross-talk and decision making in MAP kinase pathways. *Nature Genetics, 39*(3), 409–414. doi: ng1957 [pii].

McKhann, G. M., Knopman, D. S., Chertkow, H., Hyman, B. T., Jack, C. R., Jr., Kawas, C. H., … Phelps, C. H. (2011). The diagnosis of dementia due to Alzheimer's disease: Recommendations from the National Institute on Aging-Alzheimer's Association workgroups on diagnostic guidelines for Alzheimer's disease. *Alzheimers Dement,* 7(3), 263–269. http://dx.doi.org/10.1016/j.jalz.2011.03.005.

McMahon, S. B., Armanini, M. P., Ling, L. H., & Phillips, H. S. (1994). Expression and coexpression of Trk receptors in subpopulations of adult primary sensory neurons projecting to identified peripheral targets. *Neuron, 12*(5), 1161–1171.

Mey, J., & McCaffery, P. (2004). Retinoic acid signaling in the nervous system of adult vertebrates. *The Neuroscientist, 10*(5), 409–421. http://dx.doi.org/10.1177/1073858404263520.

Meyer-Franke, A., Kaplan, M. R., Pfrieger, F. W., & Barres, B. A. (1995). Characterization of the signaling interactions that promote the survival and growth of developing retinal ganglion cells in culture. *Neuron, 15*(4), 805–819.

Miller, M. W., & Pitts, F. A. (2000). Neurotrophin receptors in the somatosensory cortex of the mature rat: Co-localization of p75, trk, isoforms and c-neu. *Brain Research, 852*(2), 355–366. doi: S0006899399021769 [pii].

Minichiello, L., Calella, A. M., Medina, D. L., Bonhoeffer, T., Klein, R., & Korte, M. (2002). Mechanism of TrkB-mediated hippocampal long-term potentiation. *Neuron, 36*(1), 121–137.

Mizui, T., Ishikawa, Y., Kumanogoh, H., Lume, M., Matsumoto, T., Hara, T., … Kojima, M. (2015). BDNF pro-peptide actions facilitate hippocampal LTD and are altered by the common BDNF polymorphism Val66Met. *Proceedings of the National Academy of Sciences of the United States of America, 112*(23), E3067–E3074. http://dx.doi.org/10.1073/pnas.1422336112.

Morris, N. J., Ross, S. A., Lane, W. S., Moestrup, S. K., Petersen, C. M., Keller, S. R., & Lienhard, G. E. (1998). Sortilin is the major 110-kDa protein in GLUT4 vesicles from adipocytes. *The Journal of Biological Chemistry, 273*(6), 3582–3587.

Mou, Z., Hyde, T. M., Lipska, B. K., Martinowich, K., Wei, P., Ong, C. J., ... Han, J. C. (2015). Human obesity associated with an intronic SNP in the brain-derived neurotrophic factor locus. *Cell Reports*, *13*(6), 1073–1080. http://dx.doi.org/10.1016/j.celrep.2015.09.065.

Mueller, B. K., Mack, H., & Teusch, N. (2005). Rho kinase, a promising drug target for neurological disorders. *Nature Reviews. Drug Discovery*, *4*(5), 387–398. http://dx.doi.org/10.1038/nrd1719.

Mufson, E. J., Conner, J. M., & Kordower, J. H. (1995). Nerve growth factor in Alzheimer's disease: Defective retrograde transport to nucleus basalis. *Neuroreport*, *6*(7), 1063–1066.

Mufson, E. J., & Kordower, J. H. (1992). Cortical neurons express nerve growth factor receptors in advanced age and Alzheimer disease. *Proceedings of the National Academy of Sciences of the United States of America*, *89*(2), 569–573.

Mufson, E. J., Kroin, J. S., Sendera, T. J., & Sobreviela, T. (1999). Distribution and retrograde transport of trophic factors in the central nervous system: Functional implications for the treatment of neurodegenerative diseases. *Progress in Neurobiology*, *57*(4), 451–484. doi: S0301-0082(98)00059-8 [pii].

Mukai, J., Hachiya, T., Shoji-Hoshino, S., Kimura, M. T., Nadano, D., Suvanto, P., ... Sato, T. A. (2000). NADE, a p75NTR-associated cell death executor, is involved in signal transduction mediated by the common neurotrophin receptor p75NTR. *The Journal of Biological Chemistry*, *275*(23), 17566–17570. http://dx.doi.org/10.1074/jbc.C000140200.

Mullan, M., Houlden, H., Windelspecht, M., Fidani, L., Lombardi, C., Diaz, P., et al. (1992). A locus for familial early-onset Alzheimer's disease on the long arm of chromosome 14, proximal to the alpha 1-antichymotrypsin gene. *Nature Genetics*, *2*(4), 340–342. http://dx.doi.org/10.1038/ng1292-340.

Murer, M. G., Yan, Q., & Raisman-Vozari, R. (2001). Brain-derived neurotrophic factor in the control human brain, and in Alzheimer's disease and Parkinson's disease. *Progress in Neurobiology*, *63*(1), 71–124. doi: S0301-0082(00)00014-9 [pii].

Narisawa-Saito, M., Wakabayashi, K., Tsuji, S., Takahashi, H., & Nawa, H. (1996). Regional specificity of alterations in NGF, BDNF and NT-3 levels in Alzheimer's disease. *Neuroreport*, 7(18), 2925–2928.

Nielsen, M. S., Madsen, P., Christensen, E. I., Nykjaer, A., Gliemann, J., Kasper, D., ... Petersen, C. M. (2001). The sortilin cytoplasmic tail conveys Golgi-endosome transport and binds the VHS domain of the GGA2 sorting protein. *The EMBO Journal*, *20*(9), 2180–2190. http://dx.doi.org/10.1093/emboj/20.9.2180.

Nimnual, A. S., Yatsula, B. A., & Bar-Sagi, D. (1998). Coupling of Ras and Rac guanosine triphosphatases through the Ras exchanger SoS. *Science*, *279*(5350), 560–563.

Nykjaer, A., Lee, R., Teng, K. K., Jansen, P., Madsen, P., Nielsen, M. S., ... Petersen, C. M. (2004). Sortilin is essential for proNGF-induced neuronal cell death. *Nature*, *427*(6977), 843–848. http://dx.doi.org/10.1038/nature02319.

Nykjaer, A., & Willnow, T. E. (2012). Sortilin: A receptor to regulate neuronal viability and function. *Trends in Neurosciences*, *35*(4), 261–270. http://dx.doi.org/10.1016/j.tins.2012.01.003.

Nykjaer, A., Willnow, T. E., & Petersen, C. M. (2005). p75NTR—Live or let die. *Current Opinion in Neurobiology*, *15*(1), 49–57. http://dx.doi.org/10.1016/j.conb.2005.01.004.

O'Neill, M. J., Murray, T. K., Clay, M. P., Lindstrom, T., Yang, C. R., & Nisenbaum, E. S. (2005). LY503430: Pharmacology, pharmacokinetics, and effects in rodent models of Parkinson's disease. *CNS Drug Reviews*, *11*(1), 77–96.

Pang, P. T., Teng, H. K., Zaitsev, E., Woo, N. T., Sakata, K., Zhen, S., ... Lu, B. (2004). Cleavage of proBDNF by tPA/plasmin is essential for long-term hippocampal plasticity. *Science*, *306*(5695), 487–491. http://dx.doi.org/10.1126/science.1100135.

Patapoutian, A., & Reichardt, L. F. (2001). Trk receptors: Mediators of neurotrophin action. *Current Opinion in Neurobiology*, *11*(3), 272–280.

Patterson, S. L., Abel, T., Deuel, T. A., Martin, K. C., Rose, J. C., & Kandel, E. R. (1996). Recombinant BDNF rescues deficits in basal synaptic transmission and hippocampal LTP in BDNF knockout mice. *Neuron*, *16*(6), 1137–1145.

Pearson, G., Robinson, F., Beers Gibson, T., Xu, B. E., Karandikar, M., Berman, K., & Cobb, M. H. (2001). Mitogen-activated protein (MAP) kinase pathways: Regulation and physiological functions. *Endocrine Reviews*, *22*(2), 153–183.

Pedraza, C. E., Podlesniy, P., Vidal, N., Arevalo, J. C., Lee, R., Hempstead, B., … Espinet, C. (2005). Pro-NGF isolated from the human brain affected by Alzheimer's disease induces neuronal apoptosis mediated by p75NTR. *The American Journal of Pathology*, *166*(2), 533–543. doi: S0002-9440(10)62275-4 [pii].

Peng, S., Wuu, J., Mufson, E. J., & Fahnestock, M. (2004). Increased proNGF levels in subjects with mild cognitive impairment and mild Alzheimer disease. *Journal of Neuropathology and Experimental Neurology*, *63*(6), 641–649. doi: 04-013 [pii].

Petratos, S., Li, Q. X., George, A. J., Hou, X., Kerr, M. L., Unabia, S. E., … Small, D. H. (2008). The beta-amyloid protein of Alzheimer's disease increases neuronal CRMP-2 phosphorylation by a Rho-GTP mechanism. *Brain*, *131*(Pt. 1), 90–108. http://dx.doi.org/10.1093/brain/awm260.

Pham, D. D., Do, H. T., Bruelle, C., Kukkonen, J. P., Eriksson, O., Mogollon, I., … Lindholm, D. (2016). p75 neurotrophin receptor signaling activates sterol regulatory element-binding protein-2 in hepatocyte cells via p38 mitogen-activated protein kinase and caspase-3. *The Journal of Biological Chemistry*, *291*(20), 10747–10758. http://dx.doi.org/10.1074/jbc.M116.722272.

Plow, E. F., Herren, T., Redlitz, A., Miles, L. A., & Hoover-Plow, J. L. (1995). The cell biology of the plasminogen system. *The FASEB Journal*, *9*(10), 939–945.

Pollack, S., Young, L., Bilsland, J., Wilkie, N., Ellis, S., Hefti, F., … Harper, S. (1999). The staurosporine-like compound L-753,000 (NB-506) potentiates the neurotrophic effects of neurotrophin-3 by acting selectively at the TrkA receptor. *Molecular Pharmacology*, *56*(1), 185–195.

Poo, M. M. (2001). Neurotrophins as synaptic modulators. *Nature Reviews. Neuroscience*, *2*(1), 24–32. http://dx.doi.org/10.1038/35049004.

Potkin, S. G., Alva, G., Keator, D., Carreon, D., Fleming, K., & Fallon, J. H. (2002). Brain metabolic effects of Neotrofin in patients with Alzheimer's disease. *Brain Research*, *951*(1), 87–95.

Pradines, A., Magazin, M., Schiltz, P., Le Fur, G., Caput, D., & Ferrara, P. (1995). Evidence for nerve growth factor-potentiating activities of the nonpeptidic compound SR 57746A in PC12 cells. *Journal of Neurochemistry*, *64*(5), 1954–1964.

Rajagopal, R., Chen, Z. Y., Lee, F. S., & Chao, M. V. (2004). Transactivation of Trk neurotrophin receptors by G-protein-coupled receptor ligands occurs on intracellular membranes. *Journal of Neuroscience*, *24*(30), 6650–6658. http://dx.doi.org/10.1523/JNEUROSCI.0010-04.2004.

Ramirez, J. J., Parakh, T., George, M. N., Freeman, L., Thomas, A. A., White, C. C., & Becton, A. (2002). The effects of Neotrofin on septodentate sprouting after unilateral entorhinal cortex lesions in rats. *Restorative Neurology and Neuroscience*, *20*(1–2), 51–59.

Rathbone, M. P., Middlemiss, P. J., Crocker, C. E., Glasky, M. S., Juurlink, B. H., Ramirez, J. J., … Caciagli, F. (1999a). AIT-082 as a potential neuroprotective and regenerative agent in stroke and central nervous system injury. *Expert Opinion on Investigational Drugs*, *8*(8), 1255–1262. http://dx.doi.org/10.1517/13543784.8.8.1255.

Rathbone, M. P., Middlemiss, P. J., Gysbers, J. W., Andrew, C., Herman, M. A., Reed, J. K., … Caciagli, F. (1999b). Trophic effects of purines in neurons and glial cells. *Progress in Neurobiology*, *59*(6), 663–690.

Rattenholl, A., Ruoppolo, M., Flagiello, A., Monti, M., Vinci, F., Marino, G., … Rudolph, R. (2001). Pro-sequence assisted folding and disulfide bond formation of human nerve growth factor. *Journal of Molecular Biology*, *305*(3), 523–533. http://dx.doi.org/10.1006/jmbi.2000.4295.

Ravni, A., Bourgault, S., Lebon, A., Chan, P., Galas, L., Fournier, A., … Vaudry, D. (2006). The neurotrophic effects of PACAP in PC12 cells: Control by multiple transduction pathways. *Journal of Neurochemistry*, *98*(2), 321–329. doi: JNC3884 [pii].

Reichardt, L. F. (2006). Neurotrophin-regulated signalling pathways. *Philosophical Transactions of the Royal Society of London. Series B, Biological Sciences*, *361*(1473), 1545–1564. http://dx.doi.org/10.1098/rstb.2006.1894.

Roux, P. P., & Barker, P. A. (2002). Neurotrophin signaling through the p75 neurotrophin receptor. *Progress in Neurobiology*, *67*(3), 203–233.

Roux, P. P., Dorval, G., Boudreau, M., Angers-Loustau, A., Morris, S. J., Makkerh, J., & Barker, P. A. (2002). K252a and CEP1347 are neuroprotective compounds that inhibit mixed-lineage kinase-3 and induce activation of Akt and ERK. *The Journal of Biological Chemistry*, *277*(51), 49473–49480. http://dx.doi.org/10.1074/jbc.M203428200.

Sagot, Y., Rosse, T., Vejsada, R., Perrelet, D., & Kato, A. C. (1998). Differential effects of neurotrophic factors on motoneuron retrograde labeling in a murine model of motoneuron disease. *The Journal of Neuroscience*, *18*(3), 1132–1141.

Salehi, A., Verhaagen, J., Dijkhuizen, P. A., & Swaab, D. F. (1996). Co-localization of high-affinity neurotrophin receptors in nucleus basalis of Meynert neurons and their differential reduction in Alzheimer's disease. *Neuroscience*, *75*(2), 373–387.

Scarpi, D., Cirelli, D., Matrone, C., Castronovo, G., Rosini, P., Occhiato, E. G., … Cozzolino, F. (2012). Low molecular weight, non-peptidic agonists of TrkA receptor with NGF-mimetic activity. *Cell Death & Disease*, *3*, e339. http://dx.doi.org/10.1038/cddis.2012.80.

Schecterson, L. C., & Bothwell, M. (2010). Neurotrophin receptors: Old friends with new partners. *Developmental Neurobiology*, *70*(5), 332–338. http://dx.doi.org/10.1002/dneu.20767.

Schmid, D. A., Yang, T., Ogier, M., Adams, I., Mirakhur, Y., Wang, Q., … Katz, D. M. (2012). A TrkB small molecule partial agonist rescues TrkB phosphorylation deficits and improves respiratory function in a mouse model of Rett syndrome. *The Journal of Neuroscience*, *32*(5), 1803–1810. http://dx.doi.org/10.1523/JNEUROSCI.0865-11.2012.

Scott, S. A., Mufson, E. J., Weingartner, J. A., Skau, K. A., & Crutcher, K. A. (1995). Nerve growth factor in Alzheimer's disease: Increased levels throughout the brain coupled with declines in nucleus basalis. *The Journal of Neuroscience*, *15*(9), 6213–6221.

Segal, R. A., & Greenberg, M. E. (1996). Intracellular signaling pathways activated by neurotrophic factors. *Annual Review of Neuroscience*, *19*, 463–489. http://dx.doi.org/10.1146/annurev.ne.19.030196.002335.

Shen, J., & Maruyama, I. N. (2012). Brain-derived neurotrophic factor receptor TrkB exists as a preformed dimer in living cells. *Journal of Molecular Signaling*, 7(1), 2. http://dx.doi.org/10.1186/1750-2187-7-2.

Sherrington, R., Rogaev, E. I., Liang, Y., Rogaeva, E. A., Levesque, G., Ikeda, M., … St George-Hyslop, P. H. (1995). Cloning of a gene bearing missense mutations in early-onset familial Alzheimer's disease. *Nature*, *375*(6534), 754–760. http://dx.doi.org/10.1038/375754a0.

Shetty, A. K., Hattiangady, B., & Shetty, G. A. (2005). Stem/progenitor cell proliferation factors FGF-2, IGF-1, and VEGF exhibit early decline during the course of aging in the hippocampus: Role of astrocytes. *Glia*, *51*(3), 173–186. http://dx.doi.org/10.1002/glia.20187.

Shoji, M., Golde, T. E., Ghiso, J., Cheung, T. T., Estus, S., Shaffer, L. M., et al. (1992). Production of the Alzheimer amyloid beta protein by normal proteolytic processing. *Science, 258*(5079), 126–129.

Shu, Y. H., Lu, X. M., Wei, J. X., Xiao, L., & Wang, Y. T. (2015). Update on the role of p75NTR in neurological disorders: A novel therapeutic target. *Biomedicine & Pharmacotherapy, 76*, 17–23. http://dx.doi.org/10.1016/j.biopha.2015.10.010.

Siddiqui, S., Fang, M., Ni, B., Lu, D., Martin, B., & Maudsley, S. (2012). Central role of the EGF receptor in neurometabolic aging. *International Journal of Endocrinology, 2012*, 739428. http://dx.doi.org/10.1155/2012/739428.

Skaper, S. D., Floreani, M., Negro, A., Facci, L., & Giusti, P. (1998). Neurotrophins rescue cerebellar granule neurons from oxidative stress-mediated apoptotic death: Selective involvement of phosphatidylinositol 3-kinase and the mitogen-activated protein kinase pathway. *Journal of Neurochemistry, 70*(5), 1859–1868.

Skeldal, S., Matusica, D., Nykjaer, A., & Coulson, E. J. (2011). Proteolytic processing of the p75 neurotrophin receptor: A prerequisite for signalling? Neuronal life, growth and death signalling are crucially regulated by intra-membrane proteolysis and trafficking of p75(NTR). *Bioessays, 33*(8), 614–625. http://dx.doi.org/10.1002/bies.201100036.

Sobreviela, T., Clary, D. O., Reichardt, L. F., Brandabur, M. M., Kordower, J. H., & Mufson, E. J. (1994). TrkA-immunoreactive profiles in the central nervous system: Colocalization with neurons containing p75 nerve growth factor receptor, choline acetyltransferase, and serotonin. *The Journal of Comparative Neurology, 350*(4), 587–611. http://dx.doi.org/10.1002/cne.903500407.

Steiner, J. P., Hamilton, G. S., Ross, D. T., Valentine, H. L., Guo, H., Connolly, M. A., … Suzdak, P. D. (1997). Neurotrophic immunophilin ligands stimulate structural and functional recovery in neurodegenerative animal models. *Proceedings of the National Academy of Sciences of the United States of America, 94*(5), 2019–2024.

Stoop, R., & Poo, M. M. (1995). Potentiation of transmitter release by ciliary neurotrophic factor requires somatic signaling. *Science, 267*(5198), 695–699.

Stranahan, A. M., Zhou, Y., Martin, B., & Maudsley, S. (2009). Pharmacomimetics of exercise: Novel approaches for hippocampally-targeted neuroprotective agents. *Current Medicinal Chemistry, 16*(35), 4668–4678. CMC—AbsEpub—076 [pii].

Struble, R. G., Ala, T., Patrylo, P. R., Brewer, G. J., & Yan, X. X. (2010). Is brain amyloid production a cause or a result of dementia of the Alzheimer's type? *Journal of Alzheimer's Disease, 22*(2), 393–399. http://dx.doi.org/10.3233/JAD-2010-100846.

Swift, J. L., Godin, A. G., Dore, K., Freland, L., Bouchard, N., Nimmo, C., … Beaulieu, J. M. (2011). Quantification of receptor tyrosine kinase transactivation through direct dimerization and surface density measurements in single cells. *Proceedings of the National Academy of Sciences of the United States of America, 108*(17), 7016–7021. doi: 1018280108 [pii].

Takebayashi, M., Hayashi, T., & Su, T. P. (2002). Nerve growth factor-induced neurite sprouting in PC12 cells involves sigma-1 receptors: Implications for antidepressants. *The Journal of Pharmacology and Experimental Therapeutics, 303*(3), 1227–1237. http://dx.doi.org/10.1124/jpet.102.041970.

Tapia-Arancibia, L., Aliaga, E., Silhol, M., & Arancibia, S. (2008). New insights into brain BDNF function in normal aging and Alzheimer disease. *Brain Research Reviews, 59*(1), 201–220. http://dx.doi.org/10.1016/j.brainresrev.2008.07.007.

Terracciano, A., Martin, B., Ansari, D., Tanaka, T., Ferrucci, L., Maudsley, S., … Costa, P. T., Jr. (2010). Plasma BDNF concentration, Val66Met genetic variant and depression-related personality traits. *Genes, Brain, and Behavior, 9*(5), 512–518. http://dx.doi.org/10.1111/j.1601-183X.2010.00579.x.

Terry, A. V., Jr., Kutiyanawalla, A., & Pillai, A. (2011). Age-dependent alterations in nerve growth factor (NGF)-related proteins, sortilin, and learning and memory in rats. *Physiology & Behavior*, *102*(2), 149–157. http://dx.doi.org/10.1016/j.physbeh.2010.11.005.

Thoenen, H., & Sendtner, M. (2002). Neurotrophins: From enthusiastic expectations through sobering experiences to rational therapeutic approaches. *Nature Neuroscience*, *5*(Suppl.), 1046–1050. http://dx.doi.org/10.1038/nn938.

Toni, N., Buchs, P. A., Nikonenko, I., Bron, C. R., & Muller, D. (1999). LTP promotes formation of multiple spine synapses between a single axon terminal and a dendrite. *Nature*, *402*(6760), 421–425. http://dx.doi.org/10.1038/46574.

Trommald, M., Hulleberg, G., & Andersen, P. (1996). Long-term potentiation is associated with new excitatory spine synapses on rat dentate granule cells. *Learning & Memory*, *3*(2–3), 218–228.

Tsai, S. J. (2003). Brain-derived neurotrophic factor: A bridge between major depression and Alzheimer's disease? *Medical Hypotheses*, *61*(1), 110–113. doi: S0306987703001415 [pii].

Tsirka, S. E. (1997). Clinical implications of the involvement of tPA in neuronal cell death. *Journal of Molecular Medicine (Berlin, Germany)*, *75*(5), 341–347.

Tsou, W. I., Nguyen, K. Q., Calarese, D. A., Garforth, S. J., Antes, A. L., Smirnov, S. V., … Kotenko, S. V. (2014). Receptor tyrosine kinases, TYRO3, AXL, and MER, demonstrate distinct patterns and complex regulation of ligand-induced activation. *The Journal of Biological Chemistry*, *289*(37), 25750–25763. http://dx.doi.org/10.1074/jbc.M114.569020.

Vaegter, C. B., Jansen, P., Fjorback, A. W., Glerup, S., Skeldal, S., Kjolby, M., … Nykjaer, A. (2011). Sortilin associates with Trk receptors to enhance anterograde transport and neurotrophin signaling. *Nature Neuroscience*, *14*(1), 54–61. http://dx.doi.org/10.1038/nn.2689.

Vaillant, A. R., Mazzoni, I., Tudan, C., Boudreau, M., Kaplan, D. R., & Miller, F. D. (1999). Depolarization and neurotrophins converge on the phosphatidylinositol 3-kinase-Akt pathway to synergistically regulate neuronal survival. *The Journal of Cell Biology*, *146*(5), 955–966.

Ventriglia, M., Zanardini, R., Bonomini, C., Zanetti, O., Volpe, D., Pasqualetti, P., … Bocchio-Chiavetto, L. (2013). Serum brain-derived neurotrophic factor levels in different neurological diseases. *BioMed Research International*, *2013*, 901082. http://dx.doi.org/10.1155/2013/901082.

Vicario-Abejon, C., Owens, D., McKay, R., & Segal, M. (2002). Role of neurotrophins in central synapse formation and stabilization. *Nature Reviews. Neuroscience*, *3*(12), 965–974. http://dx.doi.org/10.1038/nrn988.

Voda, J., Yamaji, T., & Gold, B. G. (2005). Neuroimmunophilin ligands improve functional recovery and increase axonal growth after spinal cord hemisection in rats. *Journal of Neurotrauma*, *22*(10), 1150–1161. http://dx.doi.org/10.1089/neu.2005.22.1150.

Volosin, M., Song, W., Almeida, R. D., Kaplan, D. R., Hempstead, B. L., & Friedman, W. J. (2006). Interaction of survival and death signaling in basal forebrain neurons: Roles of neurotrophins and proneurotrophins. *The Journal of Neuroscience*, *26*(29), 7756–7766. http://dx.doi.org/10.1523/JNEUROSCI.1560-06.2006.

Wang, L., Martin, B., Brenneman, R., Luttrell, L. M., & Maudsley, S. (2009). Allosteric modulators of g protein-coupled receptors: Future therapeutics for complex physiological disorders. *The Journal of Pharmacology and Experimental Therapeutics*, *331*(2), 340–348. doi: jpet.109.156380 [pii].

Wang, R., Liu, K., Chen, L., & Aihara, K. (2011). Neural fate decisions mediated by trans-activation and cis-inhibition in Notch signaling. *Bioinformatics*, *27*(22), 3158–3165. http://dx.doi.org/10.1093/bioinformatics/btr551.

Wasco, W., Pettingell, W. P., Jondro, P. D., Schmidt, S. D., Gurubhagavatula, S., Rodes, L., et al. (1995). Familial Alzheimer's chromosome 14 mutations. *Nature Medicine*, *1*(9), 848.

Watson, F. L., Heerssen, H. M., Bhattacharyya, A., Klesse, L., Lin, M. Z., & Segal, R. A. (2001). Neurotrophins use the Erk5 pathway to mediate a retrograde survival response. *Nature Neuroscience*, *4*(10), 981–988. http://dx.doi.org/10.1038/nn720.

Weissmiller, A. M., & Wu, C. (2012). Current advances in using neurotrophic factors to treat neurodegenerative disorders. *Translational Neurodegeneration*, *1*(1), 14. http://dx.doi.org/10.1186/2047-9158-1-14.

Wellmer, A., Misra, V. P., Sharief, M. K., Kopelman, P. G., & Anand, P. (2001). A double-blind placebo-controlled clinical trial of recombinant human brain-derived neurotrophic factor (rhBDNF) in diabetic polyneuropathy. *Journal of the Peripheral Nervous System*, *6*(4), 204–210.

Whitmarsh, A. J., & Davis, R. J. (1998). Structural organization of MAP-kinase signaling modules by scaffold proteins in yeast and mammals. *Trends in Biochemical Sciences*, *23*(12), 481–485. doi: S0968-0004(98)01309-7 [pii].

Wiesmann, C., & de Vos, A. M. (2001). Nerve growth factor: Structure and function. *Cellular and Molecular Life Sciences*, *58*(5–6), 748–759.

Williams, G., Williams, E. J., Maison, P., Pangalos, M. N., Walsh, F. S., & Doherty, P. (2005). Overcoming the inhibitors of myelin with a novel neurotrophin strategy. *The Journal of Biological Chemistry*, *280*(7), 5862–5869. doi: M411121200 [pii].

Willnow, T. E., Petersen, C. M., & Nykjaer, A. (2008). VPS10P-domain receptors—Regulators of neuronal viability and function. *Nature Reviews. Neuroscience*, *9*(12), 899–909. http://dx.doi.org/10.1038/nrn2516.

Woo, N. H., Teng, H. K., Siao, C. J., Chiaruttini, C., Pang, P. T., Milner, T. A., ... Lu, B. (2005). Activation of p75NTR by proBDNF facilitates hippocampal long-term depression. *Nature Neuroscience*, *8*(8), 1069–1077. http://dx.doi.org/10.1038/nn1510.

Wood, K. W., Sarnecki, C., Roberts, T. M., & Blenis, J. (1992). ras mediates nerve growth factor receptor modulation of three signal-transducing protein kinases: MAP kinase, Raf-1, and RSK. *Cell*, *68*(6), 1041–1050.

Wu, X., Zhu, D., Jiang, X., Okagaki, P., Mearow, K., Zhu, G., ... Marini, A. M. (2004). AMPA protects cultured neurons against glutamate excitotoxicity through a phosphatidylinositol 3-kinase-dependent activation in extracellular signal-regulated kinase to upregulate BDNF gene expression. *Journal of Neurochemistry*, *90*(4), 807–818. http://dx.doi.org/10.1111/j.1471-4159.2004.02526.x.

Yang, T., Knowles, J. K., Lu, Q., Zhang, H., Arancio, O., Moore, L. A., ... Longo, F. M. (2008). Small molecule, non-peptide p75 ligands inhibit Abeta-induced neurodegeneration and synaptic impairment. *PLoS One*, *3*(11), e3604. http://dx.doi.org/10.1371/journal.pone.0003604.

Yao, R., & Osada, H. (1997). Induction of neurite outgrowth in PC12 cells by gamma-lactam-related compounds via Ras-MAP kinase signaling pathway independent mechanism. *Experimental Cell Research*, *234*(2), 233–239. http://dx.doi.org/10.1006/excr.1997.3615.

Ye, X., Mehlen, P., Rabizadeh, S., VanArsdale, T., Zhang, H., Shin, H., ... Bredesen, D. E. (1999). TRAF family proteins interact with the common neurotrophin receptor and modulate apoptosis induction. *The Journal of Biological Chemistry*, *274*(42), 30202–30208.

Yoon, S. O., Casaccia-Bonnefil, P., Carter, B., & Chao, M. V. (1998). Competitive signaling between TrkA and p75 nerve growth factor receptors determines cell survival. *The Journal of Neuroscience*, *18*(9), 3273–3281.

Yuan, J., Lipinski, M., & Degterev, A. (2003). Diversity in the mechanisms of neuronal cell death. *Neuron*, *40*(2), 401–413.

Yuen, E. C., Howe, C. L., Li, Y., Holtzman, D. M., & Mobley, W. C. (1996). Nerve growth factor and the neurotrophic factor hypothesis. *Brain Dev*, *18*(5), 362–368. doi: 0387760496000514 [pii].

Zhang, Y., Hong, Y., Bounhar, Y., Blacker, M., Roucou, X., Tounekti, O., … LeBlanc, A. (2003). p75 neurotrophin receptor protects primary cultures of human neurons against extracellular amyloid beta peptide cytotoxicity. *The Journal of Neuroscience*, *23*(19), 7385–7394. 23/19/7385 [pii].

Zuccato, C., & Cattaneo, E. (2009). Brain-derived neurotrophic factor in neurodegenerative diseases. *Nature Reviews. Neurology*, *5*(6), 311–322. http://dx.doi.org/10.1038/nrneurol.2009.54.

CHAPTER TWELVE

The Role of Neurotrophins in Inflammation and Allergy

S. Manti*, P. Brown*, M.K. Perez*,†, G. Piedimonte*,†,1
*Center for Pediatric Research, Cleveland Clinic Foundation, Cleveland, OH, United States
†Pediatric Institute and Children's Hospital, Cleveland Clinic Foundation, Cleveland, OH, United States
[1]Corresponding author: e-mail address: piedimg@ccf.org

Contents

1. Introduction 314
   1.1 The Neurotrophin Family 314
   1.2 The NT Receptors 317
2. NTs and Allergic Disease 320
3. NTs and Inflammatory Responses 323
   3.1 Eosinophils 323
   3.2 Macrophages 328
   3.3 Mast Cells 329
   3.4 Lymphocytes 331
   3.5 Nerves and Structural Cells 333
4. Conclusions 335
References 336

Abstract

Allergic inflammation is the result of a specific pattern of cellular and humoral responses leading to the activation of the innate and adaptive immune system, which, in turn, results in physiological and structural changes affecting target tissues such as the airways and the skin. Eosinophil activation and the production of soluble mediators such as IgE antibodies are pivotal features in the pathophysiology of allergic diseases. In the past few years, however, convincing evidence has shown that neurons and other neurosensory structures are not only a target of the inflammatory process but also participate in the regulation of immune responses by actively releasing soluble mediators. The main products of these activated sensory neurons are a family of protein growth factors called neurotrophins. They were first isolated in the central nervous system and identified as important factors for the survival and differentiation of neurons during fetal and postnatal development as well as neuronal maintenance later in life. Four members of this family have been identified and well defined: nerve growth factor, brain-derived neurotrophic factor, neurotrophin 3, and neurotrophin 4/5. Neurotrophins play a critical role in the bidirectional signaling mechanisms between immune cells and the neurosensory network structures in the airways and the skin. Pruritus and airway

Vitamins and Hormones, Volume 104
ISSN 0083-6729
http://dx.doi.org/10.1016/bs.vh.2016.10.010

hyperresponsiveness, two major features of atopic dermatitis and asthma, respectively, are associated with the disruption of the neurosensory network activities. In this chapter, we provide a comprehensive description of the neuroimmune interactions underlying the pathophysiological mechanisms of allergic and inflammatory diseases.

1. INTRODUCTION

1.1 The Neurotrophin Family

Neurotrophins (NTs) are a family of growth factors comprising nerve growth factor (NGF), brain-derived neurotrophic factor (BDNF), neurotrophin 3 (NT-3), and neurotrophin 4/5 (NT-4/5) and were originally described as critical trophic factors involved in the regulation of neuronal development, survival, and function including synapse formation and plasticity (Bibel & Barde, 2000; Dechant & Barde, 2002; Huang & Reichardt, 2001; Lewin & Barde, 1996; Lu, Pang, & Woo, 2005) (Fig. 1). However, their functions have been shown to extend beyond these roles to the regeneration of injured neurons and the support of aberrant innervation in many diseases associated with inflammation (Lu et al., 2005). NGF is the

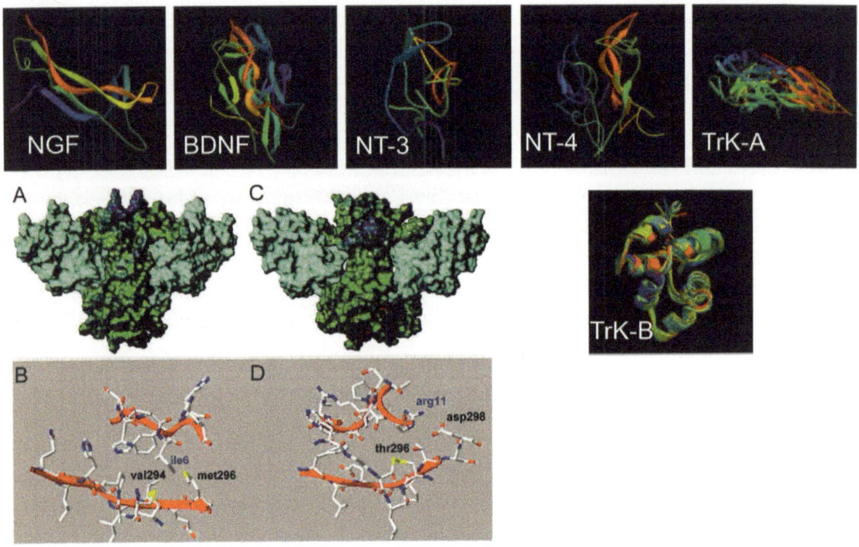

Fig. 1 The figure shows the 3D structure of each identified neurotrophins and the X-ray crystallographic structures of NGF-TrkAD5 (A) and NT-4-TrkBD5 (C), and interactions between the N-terminus of the neurotrophin with the D5 domain of the receptors (B and D).

prototypical NT and was first discovered during a search to identify the main mechanisms responsible for the survival of motor and sensory neurons following the deletions of their target tissues (Levi-Montalcini, 1987; Shooter, 2001), whereas the second characterized NT, BDNF, was extracted from pig brain and identified as a survival factor for neuronal populations not responsive to NGF (Barde, Edgar, & Thoenen, 1982). The analysis and characterization of their common protein structures led to the subsequent identification of the additional members of the NTs family, NT-3 and NT-4/5 (Table 1).

The NTs and their genes share a similar structure and function as well as the organization of the genomic sequences adjacent to these genes. Collectively, these observations have provided convincing evidence that the NT genes were the result of successive duplication of a genomic sequence derived from a common ancestral progenitor (Hallbook, 1999). Each protein product includes a signal sequence and a prodomain, followed by the mature NT sequence indicating that the formation of the mature product requires a posttranscriptional proteolytic step (Fig. 2) (Lu et al., 2005). This mechanism has been found to be of critical importance in regulating their function and conferring specificity to their biological actions (Lee et al., 2001).

Table 1 The Neurotrophins and Their Receptors

<table>
<tr><th>Neurotrophin</th><th>Receptors</th><th>Reported Functions</th></tr>
<tr><td>Nerve growth factor (NGF)</td><td>TrkA, p75NTR</td><td rowspan="2">– Growth and survival of neurons
– Growth, differentiation and neuronal survival
– Synaptic plasticity and neuronal morphogenesis</td></tr>
<tr><td>Brain-derived neurotrophic factor (BDNF)</td><td>TrkB, p75NTR</td></tr>
<tr><td>Neurotrophin 3 (NT-3)</td><td>TrkC, p75NTR</td><td>– Local growth of sympathetic axons
– Increased synaptic strength in nerve-muscle synapses
– Maturation of nociceptive and proprioceptive neurons
• Expression of specific ion channels and neuropeptides
– Increased excitatory postsynaptic currents in hippocampal regions</td></tr>
<tr><td>Neurotrophin 4/5 (NT-4/5)</td><td>TrkB, p75NTR</td><td>– Growth and survival of sensory neurons
– Survival of motoneuron, dopaminergic, and cholinergic neurons</td></tr>
</table>

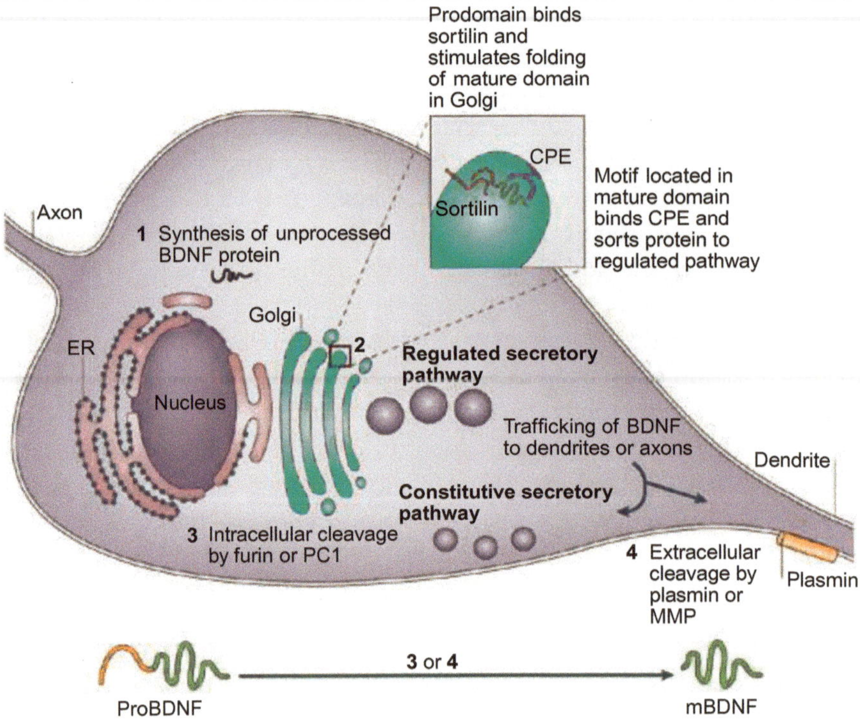

Fig. 2 A schematic drawing showing the synthesis and sorting of brain-derived neurotrophic factor (BDNF) in a typical neuron. First synthesized in the endoplasmic reticulum (ER) (1), proBDNF (precursor of BDNF) binds to intracellular sortilin in the Golgi to facilitate proper folding of the mature domain (2). A motif in the mature domain of BDNF binds to carboxypeptidase E (CPE), an interaction that sorts BDNF into large dense core vesicles, which are a component of the regulated secretory pathway. In the absence of this motif, BDNF is sorted into the constitutive pathway. After the binary decision of sorting, BDNF is transported to the appropriate site of release, either in dendrites or in axons. Because, in some cases, the prodomain is not cleaved intracellularly by furin or protein convertases (such as protein convertase 1, PC1) (3), proBDNF can be released by neurons. Extracellular proteases, such as metalloproteinases and plasmin, can subsequently cleave the proregion to yield mature BDNF (mBDNF) (4) (Lu et al., 2005).

1.2 The NT Receptors

The NT-mediated mechanisms are the result of their interactions with two distinct classes of receptors. The first of these NT receptors to be discovered, called the p75 neurotrophin receptor (p75NTR), was initially identified as a low-affinity receptor for NGF although it was later shown to also bind each of the NTs and their precursor forms. Classified as belonging to the tumor necrosis factor (TNF) receptor superfamily, p75NTR interacts with several proteins that regulate neuronal survival, differentiation, and synaptic plasticity (Frade & Barde, 1998; Rodriguez-Tebar, Dechant, & Barde, 1990). As a member of the TNF receptor superfamily, p75NTR is an indispensable receptor best known for its role as a low-affinity pan-neurotrophin receptor working in coordination with the tropomyosin-related kinase (Trk) family, neurite outgrowth inhibitor (Nogo), and sortilin receptors. In this manner, p75NTR is responsible for transducing the signals from BDNF, NGF, NT-3, and NT-4/5 to regulate a broad array of biological activities such as apoptosis, survival, neurite outgrowth, migration, and cell cycle arrest (Blochl & Blochl, 2007).

Structurally, p75NTR is constituted by the following domains: (1) the extracellular region, characterized by four cysteine-rich domains responsible for binding the NTs; (2) the transmembrane domain, which is responsible for linking the monomers; (3) the juxtamembrane region, which induces "death domain" clustering; and (4) the "death domains" themselves, which are two intracellular regions required for apoptosis and/or cell survival (Iacaruso et al., 2011). Although the ability of p75NTR to promote apoptosis is well known, the process of signaling through this receptor is complex and remains poorly understood. Research has shown that p75NTR modulates cell survival through several pathways: stimulation of the stress kinase c-Jun N-terminal kinase and the transcription factor NF-κB, advancement of the mitochondrial cascade, and interaction with a number of cytosolic proteins such as the TNF receptor-associated factors (TRAFs), the lipid signaling molecule ceramide, and the Trks receptors (Longo & Massa, 2013). The latter of these, the Trk receptors, are the other major class of NT receptors, represented by three members TrkA, TrkB, and TrkC (Chao, 2003; Huang & Reichardt, 2003) (Fig. 3).

The Trk receptors display specific interactions with the four members of the NTs family. NGF binds and activates TrkA, whereas BDNF and NT-4 interact with TrkB and NT-3 with TrkC. The specificity of Trk receptors-mediated biological actions is conferred by specific residues in the NTs and

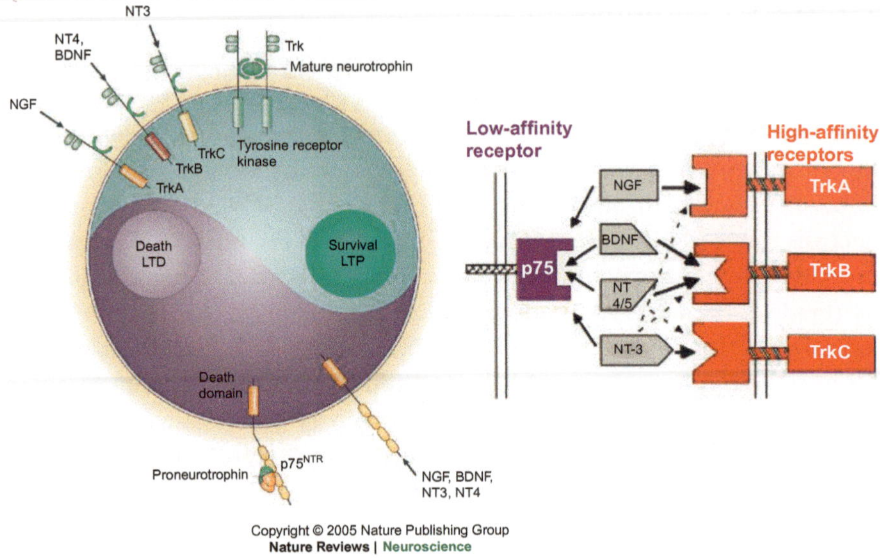

Fig. 3 The actions of neurotrophins are mediated by two principal transmembrane-receptor signaling systems. Each neurotrophin receptor—TrkA, TrkB, TrkC, and the p75 neurotrophin receptor ($p75^{NTR}$)—is characterized by specific affinities for the neurotrophins nerve growth factor (NGF), brain-derived neurotrophic factor (BDNF), neurotrophin 3 (NT-3), and NT-4. An emerging concept is that the two distinct receptor classes, Trk (top) and $p75^{NTR}$ (bottom), preferentially bind mature and proneurotrophins (neurotrophin precursors), respectively, to elicit opposing biological responses. *LTD*, long-term depression; *LTP*, long-term potentiation (Lu et al., 2005).

the Trk molecular structures that lie within the membrane-proximal immunoglobulin-like domain (Fig. 4) (Chao, 2003; Urfer et al., 1998; Wiesmann & de Vos, 2001). Following NTs binding, the Trk receptors dimerize and are autophosphorylated on intracellular tyrosine residues which results in rapid activation of multiple downstream signaling cascades including the extracellular-regulated kinase, phosphoinositide-3 kinase (PI3K)/Akt (protein kinase B a.k.a. PKB), and phospholipase Cγ (PLC) pathways. These vital intracellular signaling cascades can then activate cell-specific transcription factors involved in the regulation of a host of cellular functions known to be critical to neuronal development including cell survival, axon and dendrite growth, membrane trafficking, glial differentiation, and interactions with adjacent neurons (Reichardt, 2006).

In addition to performing these functions separately, p75NTR and Trks can interact with each other in regulation of cell survival as well. In particular, the affinity and selectivity of Trks for NTs as well as survival signaling

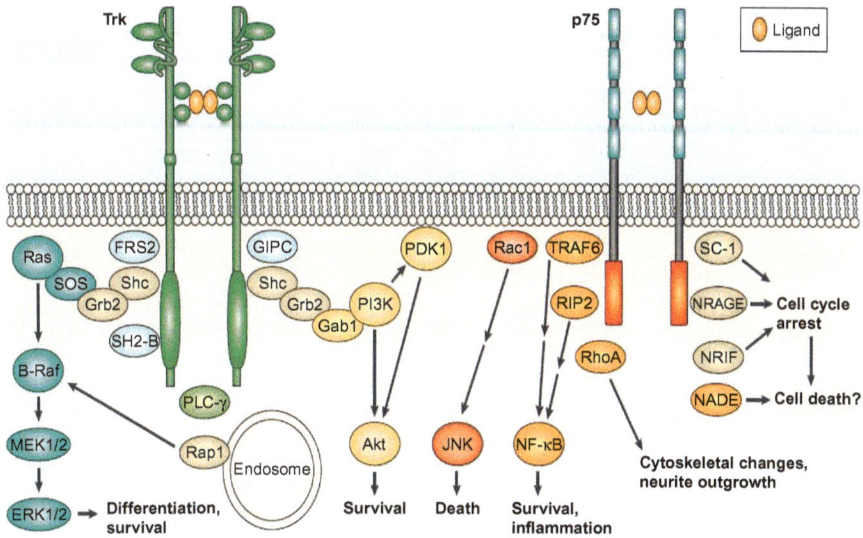

Fig. 4 Trk receptors mediate differentiation and survival signaling through extracellular signal-regulated kinase (ERK), phosphatidylinositol 3-kinase (PI3K), and phospholipase Cγ (PLC-γ) pathways 3 and 5. Trk family members recruit and increase the phosphorylation of PLC-γ and Src homologous and collagen-like adaptor protein (Shc), which leads to activation of PI3K and ERK. Rap1 exerts its actions from an endosomal location 136. The p75 receptor predominantly signals to activate NF-κB and Jun N-terminal kinase (JNK), and modulates RhoA activity. These responses are mediated through adaptor proteins that bind to the cytoplasmic domain of p75, including neurotrophin–receptor interacting factor (NRIF), neurotrophin-associated cell death executor (NADE), neurotrophin–receptor interacting MAGE homologue (NRAGE), Schwann cell 1 (SC1) and receptor-interacting protein 2 (RIP2) 35,36, which can exert effects on apoptosis, survival, neurite elongation, and growth arrest. *Akt*, protein kinase B; *FRS2*, fibroblast growth factor receptor substrate 2; *Gab1*, Grb2-associated binder-1; *Grb2*, growth factor receptor-bound protein 2; *GIPC*, GAIP-interacting protein, C-terminus; *MEK*, mitogen-activated protein kinase (MAPK)/ERK kinase; *PDK1*, phosphoinositide-dependent kinase 1; *SH2B*, Src homology 2-B; *SOS*, son of sevenless; *TRAF6*, tumor necrosis factor receptor-associated factor 6 (Chao, 2003).

are strongly potentiated by coactivation of the receptors (Longo & Massa, 2013). Lastly, although comprehensive description of NT-mediated signaling pathways responsible for the critical and complex survival and differentiation of neuronal cells can be found in several reviews (Chao, 2003; Chao & Lee, 2004; Huang & Reichardt, 2001; Lu et al., 2005), signaling via NT receptors is complex and remains poorly understood. This complexity is due to the pleiotropic role of NT receptors acting as critical sensors for several intracellular and/or extracellular pathways. This chapter will focus on the role of the NTs and their cognate receptors on the regulation of neuroinflammatory responses and the resulting pathophysiological conditions leading to the development of allergic and inflammatory diseases.

2. NTs AND ALLERGIC DISEASE

The pathophysiology of allergic diseases involves a complex integrated response in the target tissues such as the airways or the skin that is characterized by a unique activation sequence of the innate and adaptive immune system which, in turn, results in the local recruitment of inflammatory cells and the synthesis of inflammatory mediators and IgE antibodies. Allergic inflammation has long been known to be responsible for triggering neuronal responses and, in so doing, potentiating the resulting inflammatory mechanisms in the diseased tissues. However, over the years, clinical and experimental research has produced convincing and novel evidence that neurosensory mechanisms are not just activated in response to allergic stimuli but also play an important role in regulating and influencing the allergic response through the release of proinflammatory cytokines and soluble mediators. For example, elevated serum NT levels have been found in patients with various allergic diseases, including asthma, allergic rhinitis, and atopic dermatitis (AD) (Bonini et al., 1996; Raap et al., 2006; Sanico et al., 2000; Toyoda et al., 2002; Virchow et al., 1998) and have also been shown to mediate bronchial hyperresponsiveness (BHR) and airway inflammation in animal models (Braun et al., 1998; Path et al., 2002) (Table 2). Importantly, as evidenced within the clinical setting, patients affected by asthma exhibit neurologically mediated airway inflammation, remodeling, and hyperresponsiveness (Nassenstein et al., 2006) very closely resembling that seen in the aforementioned animal models.

By possessing the ability to modulate both physiological and pathophysiological neuroinflammatory responses, NTs strongly contribute to the link between allergic inflammation and neuronal dysfunction (Nassenstein et al., 2006). In addition to constitutive expression of NTs in different pulmonary compartments (e.g., vascular endothelium, bronchial and alveolar epithelium, smooth muscle, fibroblasts, immune cells, and nerves) (Watanabe et al., 2015), several animal models of allergic asthma have shown that NTs are upregulated in cases of allergic airway inflammation and induce airway hyperresponsiveness (AHR) and airway obstruction (Watanabe et al., 2015). It is believed that NTs exercise their pathophysiological effects by interfering with several regulatory signaling pathways. These include the

Table 2 The Neurotrophins in Allergic Disease

Allergic Disease	Pathophysiological Role of NTs
Allergic rhinitis	– Elevated NT levels in nasal secretions – Enhance nerve hyperresponsiveness in nasal cavities
Asthma	– Elevated NT levels in bronchoalveolar fluid – Blood NT levels correlate with airflow limitation – Induce AHR and airway obstruction trough: • Production of ROS- and MAPK-mediated allergen-induced airway inflammation • Modulation of neurite formation and cellular contractility • Proinflammatory cytokine release – Airway smooth muscle proliferation and matrix metalloproteinase induction
Atopic dermatitis	– Elevated serum NT levels correlate with disease severity – Contribute to cutaneous inflammation by promoting: • Endothelial-barrier permeability • Plasma extravasation • Cellular infiltration • Proinflammatory cytokine release • Nerve fiber sprouting • Recruitment keratinocytes and mast cells secreting NTs

cellular signaling pathways that (a) facilitate the production of reactive oxygen species (ROS)- and mitogen-activated protein kinase (MAPK)-mediated allergen-induced airway inflammation (Ye et al., 2011); (b) cause significant airway irritability by modulating both neurite formation and cellular contractility through affecting intracellular calcium levels; (c) stimulate the release of proinflammatory cytokines (Prakash et al., 2010); (d) facilitate tissue remodeling by driving the proliferation of airway smooth muscle cells and induction of matrix metalloproteinases (Yang et al., 2013); and (e) facilitate the induction of chronic pulmonary damage by promoting epithelial cell proliferation, migration, and subepithelial fibrosis (Kilic et al., 2011). The relevance of these multiple cell signaling pathways and their interplay with NTs in promoting airway dysfunction has been confirmed by research demonstrating that siRNAs targeting NTs led to decreased serum NGF and OVA-specific IgE levels as well as to improvements in AHR severity (Chen et al., 2014; Ogawa et al., 2012; Yang et al., 2013).

The pivotal role of NTs in exacerbating airway disease has also been recognized in several reported human studies (Muller et al., 2010; Virchow et al., 1998). Compared to healthy subjects, asthmatic patients exhibit significantly higher NT levels in serum and bronchoalveolar lavage fluid, and this correlated with the airway dysfunction observed and therefore reflected the clinical disease severity (Muller et al., 2010; Virchow et al., 1998). In support of these findings, inhaled corticosteroid treatment was also shown to decrease NT levels (Noga et al., 2001). The importance of inhaled corticosteroids showing the ability to modulate NT levels is significant as this is an established therapeutic approach for the treatment of asthma and thereby further underscores the role of NTs in airway dysfunction. Similarly, the more recent discovery of a genetic association between single-nucleotide polymorphisms in NT genes, elevated serum NT levels, and development of allergic asthma has drawn even greater scrutiny on this family of growth factors as they relate to the onset of atopic diseases (Jesenak et al., 2015; Wang et al., 2015).

To date, experimental and clinical studies have attributed elevated AHR severity as resulting from increased levels of NT-mediated neural activity. Although NGF, p75NTR, and TrkA are expressed constitutively in the epithelial layer of the nasal mucosa (Nockher & Renz, 2006) as well as within glands, nerves, and eosinophils, they are released in greater amounts during allergic inflammation. Increased NGF levels in nasal secretions have been reported in patients affected by allergic rhinitis, especially after allergen-specific nasal provocation testing (Sanico et al., 2000). The discovery of NT-associated genetic variants in patients affected by allergic rhinitis thus further support the findings of elevated nasopharyngeal NGF secretion in the study referenced prior (Andiappan et al., 2011; Jin et al., 2015; Noga et al., 2002). However, p75NTR expression within the mucosal nerves was unexpectedly decreased in allergic rhinitis patients, making the role of NT signaling via this mechanism unclear (Prakash et al., 2010). Conversely, TrkA was shown to be localized in the nasal epithelium, mucosal nerves, and the submucosa, which suggests that NGF may be able to act on all these airway tissue structures.

Further validation that NTs are involved in promoting the pathological sequelae of atopic diseases can be derived from studies of patients affected by AD, in which these individuals exhibited elevated plasma NT levels correlating with disease severity (Raap et al., 2005). Moreover, in the dermis of patients affected by AD, the relationship between the immune system and neuropeptidergic nerve fibers was shown to cause the increased mast cell- and keratinocyte-derived NGF levels observed (Tominaga et al., 2007). This complimentary neuroimmune relationship was also demonstrated to be capable of promoting endothelial barrier permeability, plasma extravasation, cellular infiltration, and the release of proinflammatory cytokines which all culminate in the NT-mediated induction of a vicious cycle resulting in heightened allergic skin inflammation (Liezmann, Klapp, & Peters, 2011). Collectively, these studies confirm the role of NTs as part of an integrated biological response connecting immune cells, neuronal cells, and structural cells of the airways together to act as a singular functional unit. In the following section, we will analyze the different components of this integrated response and the major mechanisms underlying their regulation and function. Examples of the neuroimmune interactions which occur in both the skin and the airways, and result in the typical manifestations of asthma and AD, are provided to further understanding of this important topic (Fig. 5) (Peters et al., 2006).

3. NTs AND INFLAMMATORY RESPONSES

One of the central paradigms of the complex interactions between NTs and the immune system is the active role played by immune cells in expressing both NTs and their receptors and, in doing so, their capability to act as cytokines affecting and regulating the development of immune responses (Bonini et al., 2003). As a result, this concept has led to a central change in our understanding of the pathophysiology of several immune diseases. The shift in dogma has fostered the notion that NTs are not only reacting to local inflammatory stimuli but are also capable of promoting the activation and release of cytokines and proinflammatory mediators in both an autocrine and/or paracrine fashion (Brown, Schneeberger, & Piedimonte, 2015). Therefore, this inherent functional dichotomy possessed by NTs positions them in a vital role as key components necessary for the generation of a competent host immune response. Within this section we will examine the interplay between the NTs and specific cells of the immune system as well as the physiological results of these interactions.

3.1 Eosinophils

Eosinophil granulocytes have long been considered one of the main effector cells in allergic airway diseases and their role is critical in both initiating and maintaining the allergic airway responses. The recruitment of eosinophils into inflammatory sites is associated with both acute and chronic responses following airway exposure to allergen which is a peculiar feature of allergic asthma (Kay, 1991; Lee, Gelfand, & Lee, 2001). Activated eosinophils

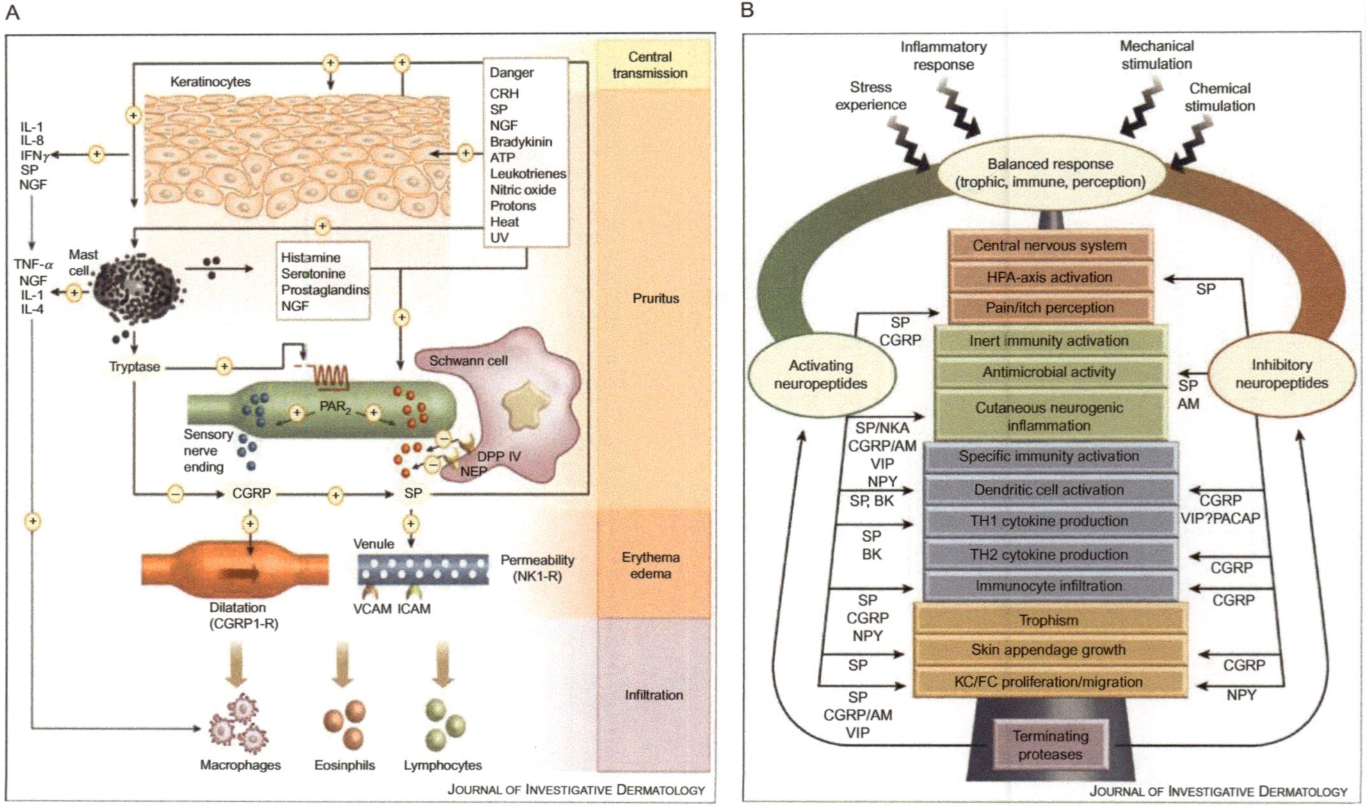

Fig. 5 See legend on opposite page.

release a large array of cytokines including IL-1α, IL-2, IL-6, IL-8, IL-10, IL-12, IL-16, IFN-γ, GM-CSF, TGF-α, and TGF-β, lipid mediators such as platelet activating factor, leukotriene C_4, and leukotriene D_4, prostaglandins (PGs), ROS, and several other mediators that are associated with the classical inflammatory responses to noxious stimuli (Raap & Wardlaw, 2008). In addition, they have been shown to be important modulators of acute phase and innate inflammatory processes as well as immunity-associated with both Th1 and Th2 responses (Jacobsen et al., 2007). However, their ability to interact with nerves in a bidirectional manner represents a paradigm shift in the pathophysiology of allergic diseases (Table 3). Eosinophils can be found in close proximity with sensory nerves and their ability to respond and release neuromediators is evidence of their critical role in modulating organ function in diseases such as asthma (Costello et al., 1997) and AD (Kubanov, Katunina, & Chikin, 2015).

In particular, during the acute inflammatory response, several mediators promote the maturation and recruitment of eosinophils from the bone marrow to their targeted tissues (Walsh, 2000). However, once the recruited eosinophils reach a designated site of inflammation, they must receive

Fig. 5 (A) This schematic shows neurogenic inflammatory signaling pathways and underscores the importance of neurotrophin-associated neuropeptide in the pathophysiological mechanism(s) regulating neurogenic inflammation. Neuropeptides are involved in cutaneous inflammation on several levels including central transmission, neurogenic inflammation, and cellular infiltration. SP and CGRP released from sensory nerve endings upon provocation mediate mast cell degranulation and regulate blood flow and plasma extravasation. Cytokines are recruited from keratinocytes and mast cells to initiate subsequent cutaneous infiltration. This process is maintained by subsequent feedback loops involving proteinase-activated receptors on sensory nerve endings and neurotrophins. Abbreviations: *ATP*, adenosine triphosphate; *CGRP*, calcitonin gene-related peptide; *CRH*, corticotropin releasing hormone; *DPP IV*, dipeptidyl peptidase IV; *ICAM*, intercellular adhesion molecule; *IFN*, interferon; *IL*, interleukin; *NEP*, neutral endopeptidase; *NGF*, nerve growth factor; *PAR*, proteinase-activated receptor; *SP*, substance P; *TNF*, tumor necrosis factor; *UV*, ultraviolet light; *VCAM*, vascular cell adhesion molecule (Peters et al., 2006). (B) This schematic drawing illustrates the functions of neuropeptide circuits in the control of tissue maintenance and inflammation. The bidirectional interactions between neurotrophin-associated neuropeptides and proinflammatory mediators contribute to keep the local neurogenic inflammatory response in a balance. Many neuropeptides hereby act in a dual fashion depending on the target and signaling context. SP, BK, CGRP, and VIP/PACAP, for example, all activate mast cells in the process of neurogenic skin inflammation. By contrast, SP inhibits HPA-axis activation and CGRP and VIP/PACAP inhibit DC activation (Peters et al., 2006). *(A) Modified from Steinhoff, M., et al. (2013). Modern aspects of cutaneous neurogenic inflammation.* Archives of Dermatology, 139*(11), 1479–1488.*

Table 3 The Role of Neurotrophins in Neuroimmune Interactions

Immune Cell	Effect of NTs
Eosinophils	– Prolong eosinophil survival – Accelerate recruitment of eosinophils expressing cell surface NT receptors – Stimulate additional NT release once activated
Lymphocytes	– Promote dendritic cell differentiation – Favor LPS-induced CD80 and/or CD86 cell surface expression – Induce proinflammatory cytokine release – Enhance ability of DCs to stimulate T-cell and Th2 immune responses
Macrophages	– Conflicting and reciprocal relationship between NTs and macrophages: • able to modify macrophage phenotype • Not able to modify macrophage antigen presentation capacity or proinflammatory cytokine production • Modulate TLR ligand-activated pathways on monocytes • Inhibit monocyte migration and antigen presentation
Mast cells	– Regulate the homing and survival of mast cells – Promote expression of IgE receptor type I, chymase, and mast cell-specific tryptase – Induce expression of proinflammatory mediators regulating mast cell degranulation and serotonin release – Promote mast cell tissue infiltration and activation
Nerves and structural cells	– Within vascular endothelial cells: • Promote endothelial cell proliferation and survival • Increase vasoconstrictor-induced $[Ca^{2+}]_i$ responses • Intensify endothelin-1-induced contractions – Within human bronchial smooth muscle cells: • Modulate cytokine- and $[Ca^{2+}]_i$-mediated gene and/or nongene expression – Within airway epithelium: • Cause hyperinnervation • Increase the susceptibility of the airway to both allergens and viruses

additional extracellular activation and survival signals in order to prevent their own apoptosis and unintentionally contribute to the triggering of an allergic response. These activation and survival signals are generated in the absence of any direct eosinophil–airway epithelial cell interactions which suggests the airway epithelium is not a key contributor to the longevity of these recruited eosinophils. With this point in mind, in addition to some of the well-known eosinophil-activating cytokines, such as IL-5 and GM-CSF, several experimental studies reported that NTs are also able to prolong the survival of eosinophils (Noga et al., 2002; Walsh, 2000). Thus, in accordance with their maturation and/or activation stages, eosinophils must also express cell surface receptors for the NTs for this signaling to occur (Noga et al., 2002).

The process of facilitating survival of eosinophils within sites of inflammation occurs as the eosinophils bind NTs with their extracellular cognate receptors. After binding with NTs occurs, the eosinophils become the target of various antiapoptotic cellular signals which are mediated by either autocrine or paracrine production of additional NTs. This heightened signaling and production of NTs leads to the induction of an increased halftime and accelerated recruitment of eosinophils which, in turn, initiate and maintain the allergic response (Noga et al., 2002). Upon activation, eosinophils represent an important cellular source of proinflammatory cytokines and NTs such as NGF and NT-3 (Kobayashi et al., 2002). The production and secretion of these proinflammatory cytokines and NTs by activated eosinophils has been shown to initiate a repetitive cycle of pathological effects (Michel et al., 2011). Confirming these findings, when eosinophils purified from bronchoalveolar lavage (BAL) samples were cultured in the presence of NTs, an antiapoptotic effect produced by the NTs was observed as well as increased eosinophilic tissue infiltration (Huang et al., 2015). Conversely, the levels of NGF, proinflammatory cytokines, and infiltrating eosinophils were decreased in the airway after the administration of small interfering RNAs against NGF (Chen et al., 2014). Therefore, these findings clearly demonstrate that allergic eosinophilia is influenced through NGF signaling within the inflamed lung.

In line with the previously discussed findings, elevated levels of NGF have been found in blood eosinophils from patients with AD compared to healthy subjects. Moreover, these blood eosinophils were shown to increase their release of IL-4 upon activation (Toyoda et al., 2003). Similarly, peripheral blood eosinophils from AD patients also release BDNF in an autocrine fashion (Raap et al., 2005), and importantly, this effect has not been observed in healthy subjects. It is well known that patients with AD experience strong itching sensations as a by-product of the disease.

Although the underlying mechanisms for enhanced itching sensations in AD patients are still unclear, the eosinophil–NT interaction was considered as a potential mechanism to explain the onset and persistence of itch. Within the papillary and upper dermis, researchers observed a greater density of nerves in the lesional form of AD compared to the nonlesional form (Peters et al., 2012). Through the neuronal expression of ICAM-1, VCAM-1, and Eotaxin-1, eosinophils are actively recruited to the nerves where they drive heightened sensory neuron branching after undergoing the process of degranulation (Foster et al., 2011). Likewise, serum concentrations of eosinophil-derived granule proteins have been shown to correlate with the number of nerves in the epidermis as well as the degree of AD severity (Taniuchi et al., 2001). In addition, treatment with anti-NGF downregulated p75NTR receptor expression on eosinophils, reduced NGF levels, and led to a decrease in the extent of cutaneous eosinophil granulocyte infiltration (Peters et al., 2011).

Perhaps the most important evidence for the existence of a direct interaction between eosinophils and neuroimmune mechanisms was demonstrated by Kobayashi et al. (2002), who revealed the presence of increased levels of NGF in the supernatant of cultured eosinophils stimulated with IgA immune complex and IL-5. When the eosinophil culture supernatant was added to PC-12 pheochromocytoma cells, it stimulated neurite extension and this effect was abolished by pretreatment with anti-NGF neutralizing antibody (Kobayashi et al., 2002). Collectively, these data underscore the importance of a bidirectional pattern influencing the interactions between NTs and eosinophils in the pathophysiology of allergic diseases. Since the release of sensory neuropeptides has long been shown to be involved in AHR, airway obstruction, and eosinophil-associated inflammatory responses—all of which are critical features in allergic asthma and AD (Dunzendorfer et al., 2002)—the importance of the relationship between eosinophils and NTs should not be underestimated.

3.2 Macrophages

Similar to eosinophils, both macrophages and monocytes have been shown to constitutively express NGF and BDNF in human studies and animal models (Nockher & Renz, 2006). The same is true of the TrkA and p75NTR receptors, which both respond to cell stimulation by specific inducers with a similar pattern of signaling and activation (Jin et al., 2015). In humans, alveolar macrophages recovered from the BAL of asthmatic patients were shown to produce cytokines associated with allergic immune responses involving IL-4 (Pouliot et al., 2005). Additionally, macrophages recovered from the BAL of allergen-sensitized animals produced large amounts of NGF and BDNF in a murine model of asthma (Braun et al., 1998, 1999). Furthermore, the release of NGF from the macrophages and monocytes of allergic patients is significantly higher compared to healthy controls (Rost et al., 2005). Taken together, these findings confirm the important and bidirectional mechanism of interaction between a specific effector cell of the innate immune system and sensory nerve-associated NTs.

Conversely, while some studies in the literature concluded that NTs were able to modify macrophage phenotype, others have shown that NTs are not able to modify the antigen-presenting capacity of macrophages or their production of proinflammatory cytokines (Samah, Porcheray, & Gras, 2008). In one particular report, the study authors identified a novel NGF/TrkA-mediated mechanism to regulate inflammatory responses (Prencipe et al., 2014). Generally, after TLRs on monocytes bind a ligand, activation signals are sent through the NF-κB/AP1-dependent pathway which leads to increased cytokine and chemokine expression and the initiation and maintenance of a chronic inflammatory response. However, in the aforementioned study, it was shown that NGF, through activating TrkA constitutively expressed on monocytes, modulated TLR ligand-activated pathways which led to a decrease in the release of proinflammatory cytokines while promoting endogenous antiinflammatory processes (Prencipe et al., 2014). More recently, findings yielded from other experimental studies demonstrate that NGF exerts an additional antiinflammatory functionality through the inhibition of monocyte migration and antigen presentation (Townley et al., 2002), an altered T-helper balance, downregulation of IFN-γ release, and upregulation of antiinflammatory cytokines such as IL-10 (Barada et al., 2007). In conclusion, the conflicting and reciprocal relationship between NTs and macrophages suggests the need to investigate this complex and bidirectional neuroimmune interaction further.

3.3 Mast Cells

Mast cells are known to be major contributors to the immediate hypersensitivity reactions following sensitization with IgE and specific antigens (Nockher & Renz, 2006). Their number is greatly increased within the inflamed tissues of allergic patients and their responses are mediated by the rapid release of a large variety of inflammatory mediators among which histamine, PGs, and leukotrienes all play a major role (Nockher & Renz,

2006). Not surprisingly, NGF has been shown to regulate the homing and survival of mast cells which, in addition to stem cell factor, acts as a cofactor in the prevention of mast cells apoptosis (Kanbe et al., 2000). Also important, NGF is known to be a promoter of the expression of typical mast cell proteins such as IgE receptor type I, chymase, and mast cell-specific tryptase (Kanbe et al., 2000).

In allergic diseases such as asthma, the number of mast cells recovered from the BAL of patients is dramatically increased compared to the airways of healthy subjects. This effect is associated with a correspondent elevation of NGF in the airways as well (Kassel et al., 2001). Moreover, NGF can act alone or as a cofactor with IL-3 (Kritas et al., 2014) to induce the expression of IL-6, a typical proinflammatory mediator associated with atopy and asthma that also regulates mast cell degranulation and serotonin release (Horigome et al., 1993). Conversely, other researchers hypothesized that in mast cells, observed increases in IL-1α, IL-1β, and IL-10 mRNA levels could instead be the result of cellular stress during the mast cell isolation procedure (Bullock & Johnson, 1996). In support of a mast cell–NT interaction, when NGF antibodies were used prior to sensitization of the nasal mucosa with specific allergens, the acute phase response, which is mediated by mast cell degranulation, was completely abolished (Braun et al., 1999). Likewise, in transgenic mice overexpressing lung NGF, the allergen-induced acute phase reaction was worsened compared to that observed in wild type animals (Path et al., 2002).

These observations provide compelling evidence in support of a bidirectional interaction between NTs and the key cell types which facilitate the allergic response. However, they are inherently flawed in that they all mainly address the role of the ligands (NGF, BDNF, NT-3, and -4/5). Interestingly, recent observations have focused on the role of the low-affinity p75NTR and its differential role played in acute and chronic inflammatory states. In AD patients, for example, Fischer et al. (2008) have shown that mast cell expression of both NGF and p75NTR is increased. The authors propose the existence of an interesting mechanism, in which allergen-induced upregulation of NGF may act on its original cellular source via autocrine signaling and activation of mast cell-expressed p75NTR.

In support of the previously discussed findings, NT-expressing nerves are in direct anatomic contact with mast cells located within the intraepithelial and intramuscular spaces (Siebenhaar et al., 2008). Not surprisingly then, it has been documented that mast cell-driven skin inflammation is attenuated in the absence of sensory nerves (Siebenhaar et al., 2008). Similarly, enhanced NGF synthesis by keratinocytes also promotes the dermal infiltration of mast cells as well as their activation within inflamed skin (Esche, de Benedetto, & Beck, 2004). Further supporting the existence of mast cell–NT interactions, allergic inflammation promotes the release of histamine from mast cells, which in turn induces substance P (SP) production by nerve fibers (de Vries et al., 2006) and contributes to what is known as the "itch-scratch cycle" in AD as well as to the extent of hyperinnervation observed in allergic asthma.

Fischer et al. (2008) also analyzed the role of NGF and NT-3, both of which are known ligands of the p75NTR, in chronic inflammation. They demonstrated that while NGF seems to be upregulated in chronic inflammatory conditions, NT-3 is actually downregulated. This unexpected finding suggested that the possibility of disease- and cell type-specific modulation of the expression of both NTs and their receptors exists and proved to be vital in the understanding of the pathophysiology of chronic obstructive pulmonary disease (Fischer et al., 2008). More recently, lung mast cells were identified as a key source of the aberrant release of NT-4, a NT whose levels are necessary for both airway smooth muscle hyperinnervation/hyperreactivity and allergen-induced neuroplasticity during the onset of allergic asthma (Patel et al., 2016).

3.4 Lymphocytes

The expression of NGF, BDNF, and NT-3 has been shown to be increased following stimulation with concavalin A in murine splenic T-cells (Barouch et al., 2000) as well as in T lymphocytes recovered from the airways in an experimental model of allergic asthma (Braun et al., 1998, 1999). Although the mechanism is well established, the amount of mediators released as well as the resulting response is critically dependent on the level of activation (Nockher & Renz, 2006). It is also well known that the inflammatory process resulting from an immunological disease is driven by the hypersensitivity response to a specific allergen, whereas the onset of major symptoms in several allergic disorders including asthma, AD, and rhinitis is associated with the activation of neurosensory pathways via an NT-mediated mechanism (Undem et al., 2000). A recent study by Veres et al. in a mouse model of allergic airway inflammation (Veres et al., 2009) has provided important insights into the mechanism(s) of interaction between the immune system and the nervous system, and, in particular, their specific roles during antigen presentation to the T-cells.

This process first requires the activation of dendritic cells (DCs) located in the airway mucosa whose role is to recognize, through the TLRs, the presence of any harmful biological stimuli and present them to the T-cells in order to initiate an appropriate immune response (Iwasaki, 2007). In particular, pathogen-associated molecular patterns present on the foreign organism are recognized by the DCs, thus leading to the upregulation of p75NTR expression which, in turn, leads to the induction of DC maturation through the activation of the TLR4/p38 MAPK/NF-κB signaling pathway axis (Jiang et al., 2008). Interestingly, NGF was confirmed to promote DC differentiation within the bone marrow as well as DC maturation in response to LPS. Importantly, these same studies demonstrated that NGF signaling within DCs heavily favors LPS-induced expression of CD80 and/or CD86, the release of proinflammatory cytokines (e.g., IL-1, IL-6, IL-12, TNF-α, and TSLP) (Jiang et al., 2008; Pan, Tan, & Li, 2016; Xia et al., 2015), and the T-cell- and Th2 immune response-stimulating abilities of DCs (Braun et al., 1998).

Only recently, Qin et al. (2014) reported that NGF reduced the surface expression of CD80, CD86, and major histocompatibility complex (MHC) class II molecules on DCs. These researchers also observed that the majority of $CD103^+$ DCs within the lung were undergoing maturation and polarization toward a Th2 phenotype and more readily contributed to a Th2 response in allergic asthma. The biological effects of NGF on DCs were abolished when functional inhibitors of p75NTR or NF-κB were administered and likewise when the LPS/TLR4 axis was inhibited (Pan et al., 2016). Notably, anti-NGF therapy was shown to significantly reduce the number of lung-resident DCs while at the same time leading to reduced IL-4 and increased IFN-γ levels in the serum and therefore could prove to be a potentially helpful treatment for asthmatic patients (Qin et al., 2014).

Currently however, the results from multiple in vitro studies investigating whether NGF promotes the maturation of DCs have drawn a number of conflicting conclusions (Jiang et al., 2008; Noga et al., 2007). For example, Noga et al. (2007) reported that following stimulation with NGF, no significant upregulation of CD80 and/or CD86 expression was observed in human DCs. Similarly, it was also shown that NGF-stimulated DCs were unable to induce T-cell proliferation (Jiang et al., 2008). One possible explanation of these seemingly contradictory findings is that a variety of nonoverlapping cell signaling pathways are involved in the onset of atopic diseases and therefore the ambiguity of these different signaling pathways can lead to very different results depending on which may be activated to a larger predominance.

Alternatively, the localized tissue microenvironment that the DCs being investigated may have been derived from, or currently reside in, could also lead to a high degree of variability in their functional activation responses which could also explain these contradictions. In their study, Veres et al. (2009) showed that DC–T-cell interactions occurred mainly beneath the smooth muscle layer and not in the epithelium. Interestingly, 10% of the DCs in this study were in contact with sensory neurons represented by C-fibers and clustered with up to approximately 4% of the T-cells observed. C-fibers, when activated, release proinflammatory neuropeptides including SP and calcitonin-related peptide (CGRP) whose expression and synthesis, in turn, are regulated by NTs such as NGF (Veres et al., 2009). However, and of significant importance, only the T-cells derived from allergic airways formed the observed clusters with nerve-associated DCs. Collectively, these observations support the existence of a novel bidirectional mechanism in which DC–T-cell interactions are modulated by neuropeptides.

3.5 Nerves and Structural Cells

Skin keratinocytes and lower airway epithelial cells have gained increasing importance as key players in the pathophysiology of allergic diseases. This vital importance stems from their bidirectional interaction with the neurosensory system which is of paramount importance in the pathophysiology of AD, allergic asthma, and rhinitis (Veres et al., 2009). Keratinocytes are the major source of NGF in the skin and their response to inflammatory stimuli results in the upregulation of NGF expression that has been observed in psoriasis and AD (Raychaudhuri, Jiang, & Farber, 1998). In addition, the onset of the pruritic lesions characteristic of allergic skin conditions is also associated with an increased expression of NT-4 (Grewe et al., 2000) and TrkA (Matsumura et al., 2015) by keratinocytes.

The airway epithelium has also now been shown to be a major source of NGF and BDNF as demonstrated by their elevated levels in the lung of mice following repeated allergen challenges (Iwasaki, 2007) and in bronchial biopsies of asthmatics (Iwasaki, 2007). As evidenced by in vitro experiments, NGF expression is dramatically increased following stimulation of cultured human airway epithelial cells with proinflammatory cytokines such as IL-1β and TNF-α (Pons et al., 2001). Moreover, vascular changes were shown to occur simultaneously with epithelial remodeling, especially so in chronically inflamed tissues. Not surprisingly, NTs have also been detected within vascular endothelial cells (Nakahashi et al., 2000), which can then exercise local autocrine or paracrine effects after binding to their cognate NT receptor expressed on the intima and adventitia (Ricci et al., 2000).

Multiple ongoing experimental and human studies have demonstrated that BDNF and NT-3 can induce nitric oxide (NO) production; promote endothelial cell proliferation, and survival (Caporali & Emanueli, 2009); increase vasoconstrictor-induced $[Ca^{2+}]_i$ responses; and intensify endothelin-1-induced contractions (Ricci et al., 2004). Evidence supporting the concept that airway smooth muscle is also a potential source of NTs has been reported as well. In support of this, the NTs NGF, BDNF, NT-4, and NT-3 as well as the cognate TrkB, TrkC, and p75NTR receptors have all been detected in human bronchial smooth muscle cells. Importantly, human bronchial smooth muscle cells were shown to not only produce NTs but were also shown to be capable of responding to the binding of the neurotransmitter SP. In this manner, the NTs are able to exert their influence over airway contractility by two distinct mechanisms: (1) cytokine- and $[Ca^{2+}]_i$-mediated nongene expression level modulation (Prakash, Thompson, & Pabelick, 2009) and (2) gene expression level effects such as those which mediate cellular growth and proliferation (Wang et al., 2000).

The presence and role of the NTs and their cognate receptors in the innervation of the lung has been documented as well (Freund-Michel & Frossard, 2008). In mouse airway epithelium, overexpression of NGF substantially increases the response to capsaicin and causes the development of hyperinnervation by SP-containing nerves. Similarly, administration of exogenous NGF results in the development of AHR, even in the absence of an asthma phenotype, in animals subjected to allergen sensitization and challenge (Freund-Michel & Frossard, 2008). On the contrary, in mice lacking p75NTR that were sensitized by allergen, AHR induced by capsaicin was severely blunted (Braun et al., 2004).

A chronic and proinflammatory status can induce neuropeptide synthesis to an even greater extent in vagal afferents, increasing the susceptibility of the airway to both allergens and viruses (Othumpangat et al., 2009). Recently, Othumpangat et al. studied the in vitro effects of a common airway pathogen, respiratory syncytial virus (RSV), on the expression of NGF in cultured human epithelial cells specifically derived from either the trachea, bronchi, or upper airways (nasal region) (Othumpangat et al., 2009). Their findings indicated that the RSV-induced upregulation of NGF expression is especially relevant in bronchial cells compared to either the nasal or tracheal epithelial cells. Importantly, that differential effects occur based on the specific tissue microenvironment locale is a critical and novel finding. Interestingly, this effect was also found to be associated with an antiapoptotic effect which, in turn, may represent an essential paracrine or autocrine survival factor allowing RSV to linger for an extended period of time within the infected airways of an individual and therefore has relevant clinical implications (Othumpangat et al., 2009).

4. CONCLUSIONS

While NTs have been studied extensively within the nervous system as critical survival factors with a wide range of actions, where they play a critical role in nerve development, axon growth, guidance, dendrite structure, synapse formation, and plasticity, their contribution to the pathophysiology of inflammatory and allergic diseases is a relatively novel concept. The more definitive characterization of the different NTs and their cognate receptors, in particular, has dramatically advanced the understanding of their biologic functions and signaling pathways. However, it took several years to start grasping the importance of these growth factors as key contributors within the more complex pathophysiological mechanisms underlying inflammatory and allergic diseases. In particular, an important step has been the understanding of a critical bidirectional interaction between the network of neurosensory structures, which is ubiquitous throughout most mammalian organs, and their impact on the activation and maturation of constantly evolving immune responses.

The initial role proposed for the NTs was that they served as passive spectators of inflammatory and allergic responses. This outdated concept is now being rapidly substituted with an overwhelming amount of evidence pointing to the nerves as actively engaged in a constant feedback mechanism cycle which leads to the development of multiple integrated functional responses. These numerous responses then culminate to result in the onset of inflammatory and allergic diseases. Specifically, increased levels of NGF and BDNF have consistently been found in the airways of asthmatic patients as well as within the nasal passages and the subepithelial structures of the skin.

Their increased expression following the receipt of harmful biological stimuli has been shown to be associated with the development of diseases such as allergic asthma, chronic rhinitis, and AD. In addition, recent observations suggest that different populations of neurons as well as different NTs may impact the development of inflammatory and allergic conditions in different manners depending on the specific structures they are associated with. Further studies focused on specific cell populations and/or tissue structures are certainly needed in order to further advance our understanding of the integrated functions of nerves and immunological responses.

REFERENCES

Andiappan, A. K., et al. (2011). Genetic variation in BDNF is associated with allergic asthma and allergic rhinitis in an ethnic Chinese population in Singapore. *Cytokine, 56*(2), 218–223.

Barada, K. A., et al. (2007). Up-regulation of nerve growth factor and interleukin-10 in inflamed and non-inflamed intestinal segments in rats with experimental colitis. *Cytokine, 37*(3), 236–245.

Barde, Y. A., Edgar, D., & Thoenen, H. (1982). Purification of a new neurotrophic factor from mammalian brain. *The EMBO Journal, 1*(5), 549–553.

Barouch, R., et al. (2000). Differential regulation of neurotrophin expression by mitogens and neurotransmitters in mouse lymphocytes. *Journal of Neuroimmunology, 103*(2), 112–121.

Bibel, M., & Barde, Y. A. (2000). Neurotrophins: Key regulators of cell fate and cell shape in the vertebrate nervous system. *Genes & Development, 14*(23), 2919–2937.

Blochl, A., & Blochl, R. (2007). A cell-biological model of p75NTR signaling. *Journal of Neurochemistry, 102*(2), 289–305.

Bonini, S., et al. (1996). Circulating nerve growth factor levels are increased in humans with allergic diseases and asthma. *Proceedings of the National Academy of Sciences of the United States of America, 93*(20), 10955–10960.

Bonini, S., et al. (2003). Nerve growth factor: Neurotrophin or cytokine? *International Archives of Allergy and Immunology, 131*(2), 80–84.

Braun, A., et al. (1998). Role of nerve growth factor in a mouse model of allergic airway inflammation and asthma. *European Journal of Immunology, 28*(10), 3240–3251.

Braun, A., et al. (1999). Cellular sources of enhanced brain-derived neurotrophic factor production in a mouse model of allergic inflammation. *American Journal of Respiratory Cell and Molecular Biology, 21*(4), 537–546.

Braun, A., et al. (2004). Brain-derived neurotrophic factor (BDNF) contributes to neuronal dysfunction in a model of allergic airway inflammation. *British Journal of Pharmacology, 141*(3), 431–440.

Brown, P. M., Schneeberger, D. L., & Piedimonte, G. (2015). Biomarkers of respiratory syncytial virus (RSV) infection: Specific neutrophil and cytokine levels provide increased accuracy in predicting disease severity. *Paediatric Respiratory Reviews, 16*(4), 232–240.

Bullock, E. D., & Johnson, E. M., Jr. (1996). Nerve growth factor induces the expression of certain cytokine genes and bcl-2 in mast cells. Potential role in survival promotion. *The Journal of Biological Chemistry, 271*(44), 27500–27508.

Caporali, A., & Emanueli, C. (2009). Cardiovascular actions of neurotrophins. *Physiological Reviews, 89*(1), 279–308.

Chao, M. V. (2003). Neurotrophins and their receptors: A convergence point for many signalling pathways. *Nature Reviews Neuroscience, 4*(4), 299–309.

Chao, M. V., & Lee, F. S. (2004). Neurotrophin survival signaling mechanisms. *Journal of Alzheimer's Disease, 6*(6 Suppl.), S7–S11.

Chen, Y. L., et al. (2014). Small interfering RNA targeting nerve growth factor alleviates allergic airway hyperresponsiveness. *Molecular Therapy Nucleic Acids, 3*. e158.

Costello, R. W., et al. (1997). Localization of eosinophils to airway nerves and effect on neuronal M2 muscarinic receptor function. *The American Journal of Physiology*, *273*(1 Pt. 1), L93–L103.

de Vries, A., et al. (2006). Airway hyper-responsiveness in allergic asthma in guinea-pigs is mediated by nerve growth factor via the induction of substance P: A potential role for trkA. *Clinical and Experimental Allergy*, *36*(9), 1192–1200.

Dechant, G., & Barde, Y. A. (2002). The neurotrophin receptor p75(NTR): Novel functions and implications for diseases of the nervous system. *Nature Neuroscience*, *5*(11), 1131–1136.

Dunzendorfer, S., et al. (2002). Neuropeptide-induced inhibition of IL-16 release from eosinophils. *Neuroimmunomodulation*, *10*(4), 217–223.

Esche, C., de Benedetto, A., & Beck, L. A. (2004). Keratinocytes in atopic dermatitis: Inflammatory signals. *Current Allergy and Asthma Reports*, *4*(4), 276–284.

Fischer, T. C., et al. (2008). Pan-neurotrophin receptor p75NTR expression is strongly induced in lesional atopic mast cells. *Clinical and Experimental Allergy*, *38*(7), 1168–1173.

Foster, E. L., et al. (2011). Eosinophils increase neuron branching in human and murine skin and in vitro. *PloS One*, *6*(7)e22029.

Frade, J. M., & Barde, Y. A. (1998). Nerve growth factor: Two receptors, multiple functions. *Bioessays*, *20*(2), 137–145.

Freund-Michel, V., & Frossard, N. (2008). The nerve growth factor and its receptors in airway inflammatory diseases. *Pharmacology & Therapeutics*, *117*(1), 52–76.

Grewe, M., et al. (2000). Neurotrophin-4 production by human epidermal keratinocytes: Increased expression in atopic dermatitis. *The Journal of Investigative Dermatology*, *114*(6), 1108–1112.

Hallbook, F. (1999). Evolution of the vertebrate neurotrophin and Trk receptor gene families. *Current Opinion in Neurobiology*, *9*(5), 616–621.

Horigome, K., et al. (1993). Mediator release from mast cells by nerve growth factor. Neurotrophin specificity and receptor mediation. *The Journal of Biological Chemistry*, *268*(20), 14881–14887.

Huang, E. J., & Reichardt, L. F. (2001). Neurotrophins: Roles in neuronal development and function. *Annual Review of Neuroscience*, *24*, 677–736.

Huang, E. J., & Reichardt, L. F. (2003). Trk receptors: Roles in neuronal signal transduction. *Annual Review of Biochemistry*, *72*, 609–642.

Huang, L. W., et al. (2015). Inhibition of nerve growth factor/tyrosine kinase receptor A signaling ameliorates airway remodeling in chronic allergic airway inflammation. *European Review for Medical and Pharmacological Sciences*, *19*(12), 2261–2268.

Iacaruso, M. F., et al. (2011). Structural model for p75(NTR)-TrkA intracellular domain interaction: A combined FRET and bioinformatics study. *Journal of Molecular Biology*, *414*(5), 681–698.

Iwasaki, A. (2007). Mucosal dendritic cells. *Annual Review of Immunology*, *25*, 381–418.

Jacobsen, E. A., et al. (2007). Eosinophils: Singularly destructive effector cells or purveyors of immunoregulation? *The Journal of Allergy and Clinical Immunology*, *119*(6), 1313–1320.

Jesenak, M., et al. (2015). Val66Met polymorphism in the BDNF gene in children with bronchial asthma. *Pediatric Pulmonology*, *50*(7), 631–637.

Jiang, Y., et al. (2008). TLR4 signaling induces functional nerve growth factor receptor p75NTR on mouse dendritic cells via p38MAPK and NF-kappa B pathways. *Molecular Immunology*, *45*(6), 1557–1566.

Jin, P., et al. (2015). A functional brain-derived neurotrophic factor (BDNF) gene variant increases the risk of moderate-to-severe allergic rhinitis. *The Journal of Allergy and Clinical Immunology*, *135*(6), 1486–1493. e8.

Kanbe, N., et al. (2000). Nerve growth factor prevents apoptosis of cord blood-derived human cultured mast cells synergistically with stem cell factor. *Clinical and Experimental Allergy*, *30*(8), 1113–1120.

Kassel, O., et al. (2001). Local increase in the number of mast cells and expression of nerve growth factor in the bronchus of asthmatic patients after repeated inhalation of allergen at low-dose. *Clinical and Experimental Allergy, 31*(9), 1432–1440.

Kay, A. B. (1991). Asthma and inflammation. *The Journal of Allergy and Clinical Immunology, 87*(5), 893–910.

Kilic, A., et al. (2011). Nerve growth factor induces type III collagen production in chronic allergic airway inflammation. *The Journal of Allergy and Clinical Immunology, 128*(5), 1058–1066. e1–e4.

Kobayashi, H., et al. (2002). Human eosinophils produce neurotrophins and secrete nerve growth factor on immunologic stimuli. *Blood, 99*(6), 2214–2220.

Kritas, S. K., et al. (2014). Nerve growth factor interactions with mast cells. *International Journal of Immunopathology and Pharmacology, 27*(1), 15–19.

Kubanov, A. A., Katunina, O. R., & Chikin, V. V. (2015). Expression of neuropeptides, neurotrophins, and neurotransmitters in the skin of patients with atopic dermatitis and psoriasis. *Bulletin of Experimental Biology and Medicine, 159*(3), 318–322.

Lee, N. A., Gelfand, E. W., & Lee, J. J. (2001). Pulmonary T cells and eosinophils: Coconspirators or independent triggers of allergic respiratory pathology? *The Journal of Allergy and Clinical Immunology, 107*(6), 945–957.

Lee, R., et al. (2001). Regulation of cell survival by secreted proneurotrophins. *Science, 294*(5548), 1945–1948.

Levi-Montalcini, R. (1987). The nerve growth factor 35 years later. *Science, 237*(4819), 1154–1162.

Lewin, G. R., & Barde, Y. A. (1996). Physiology of the neurotrophins. *Annual Review of Neuroscience, 19*, 289–317.

Liezmann, C., Klapp, B., & Peters, E. M. (2011). Stress, atopy and allergy: A re-evaluation from a psychoneuroimmunologic perspective. *Dermato-endocrinology, 3*(1), 37–40.

Longo, F. M., & Massa, S. M. (2013). Small-molecule modulation of neurotrophin receptors: A strategy for the treatment of neurological disease. *Nature Reviews. Drug Discovery, 12*(7), 507–525.

Lu, B., Pang, P. T., & Woo, N. H. (2005). The yin and yang of neurotrophin action. *Nature Reviews. Neuroscience, 6*(8), 603–614.

Matsumura, S., et al. (2015). Th2 cytokines enhance TrkA expression, upregulate proliferation, and downregulate differentiation of keratinocytes. *Journal of Dermatological Science, 78*(3), 215–223.

Michel, T., et al. (2011). Increased Th2 cytokine secretion, eosinophilic airway inflammation, and airway hyperresponsiveness in neurturin-deficient mice. *Journal of Immunology, 186*(11), 6497–6504.

Muller, G. C., et al. (2010). Plasma brain-derived neurotrophic factor levels are associated with clinical severity in school age children with asthma. *Clinical and Experimental Allergy, 40*(12), 1755–1759.

Nakahashi, T., et al. (2000). Vascular endothelial cells synthesize and secrete brain-derived neurotrophic factor. *FEBS Letters, 470*(2), 113–117.

Nassenstein, C., et al. (2006). Neuro-immune interaction in allergic asthma: Role of neurotrophins. *Biochemical Society Transactions, 34*(Pt. 4), 591–593.

Nockher, W. A., & Renz, H. (2006). Neurotrophins in allergic diseases: From neuronal growth factors to intercellular signaling molecules. *The Journal of Allergy and Clinical Immunology, 117*(3), 583–589.

Noga, O., et al. (2001). The influence of inhalative corticosteroids on circulating nerve growth factor, brain-derived neurotrophic factor and neurotrophin-3 in allergic asthmatics. *Clinical and Experimental Allergy, 31*(12), 1906–1912.

Noga, O., et al. (2002). Activation of the specific neurotrophin receptors TrkA, TrkB and TrkC influences the function of eosinophils. *Clinical and Experimental Allergy, 32*(9), 1348–1354.

Noga, O., et al. (2007). Differential activation of dendritic cells by nerve growth factor and brain-derived neurotrophic factor. *Clinical and Experimental Allergy*, *37*(11), 1701–1708.
Ogawa, H., et al. (2012). Nerve growth factor derived from bronchial epithelium after chronic mite antigen exposure contributes to airway hyperresponsiveness by inducing hyperinnervation, and is inhibited by in vivo siRNA. *Clinical and Experimental Allergy*, *42*(3), 460–470.
Othumpangat, S., et al. (2009). NGF is an essential survival factor for bronchial epithelial cells during respiratory syncytial virus infection. *PloS One*, *4*(7). e6444.
Pan, P., Tan, H., & Li, H. (2016). Nerve growth factor promotes the maturation and cytokine release of dendritic cells induced by ovalbumin with low-dose lipopolysaccharide via the NF-kB pathway. *American Journal of Respiratory and Critical Care Medicine*, *193*, A4928.
Patel, K. R., et al. (2016). Mast cell-derived neurotrophin 4 mediates allergen-induced airway hyperinnervation in early life. *Mucosal Immunology*, *9*(6), 1466–1476.
Path, G., et al. (2002). Augmentation of allergic early-phase reaction by nerve growth factor. *American Journal of Respiratory and Critical Care Medicine*, *166*(6), 818–826.
Peters, E. M., et al. (2006). Neuropeptide control mechanisms in cutaneous biology: Physiological and clinical significance. *The Journal of Investigative Dermatology*, *126*(9), 1937–1947.
Peters, E. M., et al. (2011). Nerve growth factor partially recovers inflamed skin from stress-induced worsening in allergic inflammation. *The Journal of Investigative Dermatology*, *131*(3), 735–743.
Peters, E. M., et al. (2012). The neuroimmune connection interferes with tissue regeneration and chronic inflammatory disease in the skin. *Annals of the New York Academy of Sciences*, *1262*, 118–126.
Pons, F., et al. (2001). Nerve growth factor secretion by human lung epithelial A549 cells in pro- and anti-inflammatory conditions. *European Journal of Pharmacology*, *428*(3), 365–369.
Pouliot, P., et al. (2005). Interleukin-4 production by human alveolar macrophages. *Clinical and Experimental Allergy*, *35*(6), 804–810.
Prakash, Y. S., Thompson, M. A., & Pabelick, C. M. (2009). Brain-derived neurotrophic factor in TNF-alpha modulation of Ca2+ in human airway smooth muscle. *American Journal of Respiratory Cell and Molecular Biology*, *41*(5), 603–611.
Prakash, Y., et al. (2010). Neurotrophins in lung health and disease. *Expert Review of Respiratory Medicine*, *4*(3), 395–411.
Prencipe, G., et al. (2014). Nerve growth factor downregulates inflammatory response in human monocytes through TrkA. *Journal of Immunology*, *192*(7), 3345–3354.
Qin, Q., et al. (2014). Lung dendritic cells undergo maturation and polarization towards a T helper type 2-stimulating phenotype in a mouse model of asthma: Role of nerve growth factor. *Experimental and Therapeutic Medicine*, *8*(5), 1402–1408.
Raap, U., & Wardlaw, A. J. (2008). A new paradigm of eosinophil granulocytes: Neuroimmune interactions. *Experimental Dermatology*, *17*(9), 731–738.
Raap, U., et al. (2005). Brain-derived neurotrophic factor is increased in atopic dermatitis and modulates eosinophil functions compared with that seen in nonatopic subjects. *Journal of Allergy and Clinical Immunology*, *115*(6), 1268–1275.
Raap, U., et al. (2006). Circulating levels of brain-derived neurotrophic factor correlate with disease severity in the intrinsic type of atopic dermatitis. *Allergy*, *61*(12), 1416–1418.
Raychaudhuri, S. P., Jiang, W. Y., & Farber, E. M. (1998). Psoriatic keratinocytes express high levels of nerve growth factor. *Acta Dermato-Venereologica*, *78*(2), 84–86.
Reichardt, L. F. (2006). Neurotrophin-regulated signalling pathways. *Philosophical Transactions of the Royal Society of London Series B, Biological Sciences*, *361*(1473), 1545–1564.

Ricci, A., et al. (2000). Neurotrophins and neurotrophin receptors in human pulmonary arteries. *Journal of Vascular Research*, *37*(5), 355–363.

Ricci, A., et al. (2004). Elevated neurotrophin and neurotrophin receptor expression in spontaneously hypertensive rat lungs. *Growth Factors*, *22*(3), 195–205.

Rodriguez-Tebar, A., Dechant, G., & Barde, Y. A. (1990). Binding of brain-derived neurotrophic factor to the nerve growth factor receptor. *Neuron*, *4*(4), 487–492.

Rost, B., et al. (2005). Monocytes of allergics and non-allergics produce, store and release the neurotrophins NGF, BDNF and NT-3. *Regulatory Peptides*, *124*(1-3), 19–25.

Samah, B., Porcheray, F., & Gras, G. (2008). Neurotrophins modulate monocyte chemotaxis without affecting macrophage function. *Clinical and Experimental Immunology*, *151*(3), 476–486.

Sanico, A. M., et al. (2000). Nerve growth factor expression and release in allergic inflammatory disease of the upper airways. *American Journal of Respiratory and Critical Care Medicine*, *161*(5), 1631–1635.

Shooter, E. M. (2001). Early days of the nerve growth factor proteins. *Annual Review of Neuroscience*, *24*, 601–629.

Siebenhaar, F., et al. (2008). Mast cell-driven skin inflammation is impaired in the absence of sensory nerves. *The Journal of Allergy and Clinical Immunology*, *121*(4), 955–961.

Taniuchi, S., et al. (2001). Serum eosinophil derived neurotoxin may reflect more strongly disease severity in childhood atopic dermatitis than eosinophil cationic protein. *Journal of Dermatological Science*, *26*(1), 79–82.

Tominaga, M., et al. (2007). A hypothetical mechanism of intraepidermal neurite formation in NC/Nga mice with atopic dermatitis. *Journal of Dermatological Science*, *46*(3), 199–210.

Townley, S. L., et al. (2002). Nerve growth factor, neuropeptides, and mast cells in ultraviolet-B-induced systemic suppression of contact hypersensitivity responses in mice. *The Journal of Investigative Dermatology*, *118*(3), 396–401.

Toyoda, M., et al. (2002). Nerve growth factor and substance P are useful plasma markers of disease activity in atopic dermatitis. *The British Journal of Dermatology*, *147*(1), 71–79.

Toyoda, M., et al. (2003). Localization and content of nerve growth factor in peripheral blood eosinophils of atopic dermatitis patients. *Clinical and Experimental Allergy*, *33*(7), 950–955.

Undem, B. J., et al. (2000). Neural integration and allergic disease. *The Journal of Allergy and Clinical Immunology*, *106*(5 Suppl.), S213–S220.

Urfer, R., et al. (1998). High resolution mapping of the binding site of TrkA for nerve growth factor and TrkC for neurotrophin-3 on the second immunoglobulin-like domain of the Trk receptors. *The Journal of Biological Chemistry*, *273*(10), 5829–5840.

Veres, T. Z., et al. (2009). Dendritic cell-nerve clusters are sites of T cell proliferation in allergic airway inflammation. *The American Journal of Pathology*, *174*(3), 808–817.

Virchow, J. C., et al. (1998). Neurotrophins are increased in bronchoalveolar lavage fluid after segmental allergen provocation. *American Journal of Respiratory and Critical Care Medicine*, *158*(6), 2002–2005.

Walsh, G. M. (2000). Eosinophil apoptosis: Mechanisms and clinical relevance in asthmatic and allergic inflammation. *British Journal of Haematology*, *111*(1), 61–67.

Wang, S., et al. (2000). p75(NTR) mediates neurotrophin-induced apoptosis of vascular smooth muscle cells. *The American Journal of Pathology*, *157*(4), 1247–1258.

Wang, J. Y., et al. (2015). Association between a functional single nucleotide polymorphism in the brain-derived neurotrophic factor gene and risk of child asthma. *Genetics and Molecular Research*, *14*(4), 16233–16240.

Watanabe, T., et al. (2015). Brain-derived neurotrophic factor expression in asthma. Association with severity and type 2 inflammatory processes. *American Journal of Respiratory Cell and Molecular Biology*, *53*(6), 844–852.

Wiesmann, C., & de Vos, A. M. (2001). Nerve growth factor: Structure and function. *Cellular and Molecular Life Sciences*, *58*(5-6), 748–759.

Xia, Q., et al. (2015). Role of nerve growth factor in differentiation of dendritic cells. *Zhong Nan Da Xue Xue Bao. Yi Xue Ban*, *40*(8), 829–836.

Yang, Y. G., et al. (2013). Nerve growth factor exacerbates allergic lung inflammation and airway remodeling in a rat model of chronic asthma. *Experimental and Therapeutic Medicine*, *6*(5), 1251–1258.

Ye, Y. L., et al. (2011). Dermatophagoides pteronyssinus 2 regulates nerve growth factor release to induce airway inflammation via a reactive oxygen species-dependent pathway. *American Journal of Physiology. Lung Cellular and Molecular Physiology*, *300*(2), L216–L224.

CHAPTER THIRTEEN

Neurotrophic Factors and Maternal Nutrition During Pregnancy

M. Dhobale[1]
BioTRaK Research and Diagnostics Centre, Navi Mumbai, India
[1]Corresponding author: e-mail address: madhavij14@gmail.com

Contents

1. Introduction 344
2. Neurotrophic Factors 345
   2.1 Nerve Growth Factor (NGF) 345
   2.2 Brain-Derived Neurotrophic Factor (BDNF) 346
   2.3 Neurotrophin-3 (NT-3) and Neurotrophin-4/5 (NT-4/5) 347
3. Maternal Nutrition 347
   3.1 Macronutrients 348
   3.2 Micronutrients 350
4. Association Between Maternal Nutrition and Neurotrophic Factors Through One Carbon Cycle 355
   4.1 One Carbon Cycle and Neurotrophic Factors 355
   4.2 One Carbon Cycle and Epigenetics 357
5. Consequences of Altered Maternal Nutrition in Pregnancy 358
References 359

Abstract

Maternal nutrition is one of the major determinants of pregnancy outcome. It has been suggested that reduced intakes or lack of specific nutrients during pregnancy influences the length of gestation, proper placental and fetal growth during pregnancy. Maternal nutrition, particularly micronutrients such as folate and vitamin B_{12}, and long-chain polyunsaturated fatty acids (LCPUFA) are the major determinants of the one carbon cycle and are suggested to be at the heart of intrauterine programming of diseases in adult life. LCPUFA play a key role in the normal feto-placental development, as well as in the development and functional maturation of the brain and central nervous system and also regulate the levels of neurotrophic factors. These neurotrophic factors are known to regulate the development of the placenta at the materno–fetal interface and act in a paracrine and endocrine manner. Neurotrophic factors like brain-derived neurotrophic factor and nerve growth factor are proteins involved in angiogenesis and potentiate the

Vitamins and Hormones, Volume 104
ISSN 0083-6729
http://dx.doi.org/10.1016/bs.vh.2016.10.011

placental development. This chapter mainly focuses on micronutrients since they play a main physiological role during pregnancy.

ABBREVIATIONS

AA arachidonic acid
ALA alpha linolenic acid
BDNF brain-derived neurotrophic factor
DHA docosahexaenoic acid
DNA deoxyribonucleic acid
EPA eicosapentaenoic acid
IUGR intrauterine growth restriction
LA linoleic acid
LCPUFAs long-chain PUFAs
MS methionine synthase
MTHFR methylene tetrahydrofolate reductase
NGF nerve growth factor
NT-3 neurotrophin-3
NT-4 neurotrophin-4
PPROM preterm premature rupture of membrane
PUFAs polyunsaturated fatty acids
SGA small for gestational age
TCbR transcobalamin receptor
THF tetrahydrofolate

1. INTRODUCTION

Influences of maternal nutrition on epigenetic programming are most important during prenatal and early postnatal development, when epigenetic mechanisms undergo establishment and maturation. Disruption of normal gene-specific methylation patterns by perturbations in maternal nutrition may affect the pregnancy outcome having long-term implications in the offspring. Thus, studies in pregnancy complications and in first episode psychotic patients suggest that altered metabolism of micronutrients like folic acid, vitamin B_{12}, and docosahexaenoic acid (DHA) through the one carbon cycle may epigenetically regulate neurotrophic factors. This may further contribute to risk for neurodevelopmental disorders in infants. Based on all these studies, both in humans and animals, we have recently published a hypothesis which proposes a role for altered maternal micronutrient metabolism contributing to preterm deliveries (Dhobale & Joshi, 2012).

Neurotrophic factors are growth factors and consist of four members: nerve growth factor (NGF), brain-derived neurotrophic factor (BDNF), neurotrophic factor 3 (NT-3), and neurotrophic factor-4 (NT-4) (Chao, Rajagopal, & Lee, 2006). They influence the proliferation, differentiation, survival, and death of neuronal and nonneuronal cells (Chao et al., 2006). Neurotrophic factors play a critical role in the development and maintenance of the mammalian nervous system. BDNF and NT-3 are highly expressed in cortical and hippocampal structures and have been linked to the survival and functioning of multiple neuronal populations. NGF and BDNF promote neuronal growth and differentiation in the central and peripheral nervous system during fetal development (Tapia-Arancibia, Rage, Givalois, & Arancibia, 2004; Tometten, Blois, & Arck, 2005). Alterations in neurotrophic factor levels have been implicated in neurodegenerative and psychiatric disorders. In addition, BDNF potentiates placental development and plays an important role in cytotrophoblast differentiation, proliferation, and survival (Kawamura et al., 2011).

2. NEUROTROPHIC FACTORS

There are four major types of neurotrophins: NGF, BDNF, NT-3, and NT-4. Among these, BDNF and NGF are suggested to play an important role in placental and fetal growth and development (Mayeur et al., 2010; Nico, Mangieri, Benagiano, Crivellato, & Ribatti, 2008; Toti et al., 2006; Zheng & Shao, 2012).

2.1 Nerve Growth Factor (NGF)

The NGF consists of a dimer of 13-kDa polypeptide chains, each of which has three intrachain disulfide bridges. The crystal structure of NGF has been resolved (McDonald et al., 1991). The NGF gene is located on human chromosome one and is expressed as two major splice variants (Edwards et al., 1988). TrkA and p75NTR are the two receptors for NGF that are involved in events of receptor tyrosine kinase (Trk) signaling (Gigante et al., 2003). These include receptor dimerization and transphosphorylation of activation tyrosines leading to activation of kinase activity, followed by autophosphorylation of tyrosines outside of the activation loop. Subsequent phosphorylation and activation of accessory proteins lead to the generation of a cascade of receptor-independent signaling pathways (Ras (rat sarcoma), phosphoinositide 3-kinase (PKC), phospholipase C (PI3) kinase pathway) (Sofroniew, Howe, & Mobley, 2001).

NGF and its receptors are suggested to play an important role during organogenesis (Miralles, Philippe, Czernichow, & Scharfmann, 1998) and act as an angiogenic factor in endothelial cells by autocrine and/or paracrine mechanisms (Nico et al., 2008). NGF mRNA is expressed in both fetal and maternal tissues (trophoblast, amnion/chorion, and maternal deciduas) in early gestation and at term (Toti et al., 2006). In mouse, NGF plays a vital role in placentation during the postimplantation period (Kanai-Azuma et al., 1997). In addition, NGF has been associated with functional activities of cells that include immune and endocrine systems and acts as an inflammatory mediator (Berry, Bindocci, & Alleva, 2012).

2.2 Brain-Derived Neurotrophic Factor (BDNF)

The human BDNF gene has seven noncoding exons that are associated with distinct promoters and one coding exon that encodes the mature BDNF proteins (Liu et al., 2005). BDNF protein contains a signal peptide following the initiation codon and a proregion containing an N-linked glycosylation site (Binder & Scharfman, 2004). BDNF protein shares about 50% amino acid identity with NGF, NT-3, and NT-4/5 and has a distinctive three-dimensional structure containing two pairs of antiparallel β-strands and cysteine residues in a cystine knot motif.

Human BDNF transcripts are expressed in the brain and in nonneural tissues like placenta and heart (Pruunsild, Kazantseva, Aid, Palm, & Timmusk, 2007). BDNF binds to its specific receptor, Trk B, results in kinase activation; subsequent receptor autophosphorylation creates specific binding sites for intracellular target proteins (PLC-γ1 (phospholipase C), p85 (the noncatalytic subunit of PI-3 kinase), and Shc (SH2-containing sequence)), which bind to the activated receptor via SH2 domains (Barbacid, 1995; Patapoutian & Reichardt, 2001; Thoenen, 1995). This activation can then lead to a variety of intracellular signaling cascades such as the Ras–MAP (mitogen-activated protein) kinase cascade and phosphorylation of cyclic AMP-response element-binding protein (CREB) (Patapoutian & Reichardt, 2001; Segal, 2003).

In the placenta, BDNF plays an important role in cytotrophoblast differentiation, proliferation, and survival (Kawamura et al., 2009, 2011; Mayeur et al., 2010). It has been reported that BDNF regulates angiogenesis and protects the endothelial progenitor cells by increasing the expression of superoxide dismutase (He & Katusic, 2012; Jiang, Liu, Zhang, & Chen, 2011). BDNF/TrkB-stimulated intracellular signaling is critical for neuronal survival, morphogenesis, and plasticity in the brain (Numakawa et al., 2010).

2.3 Neurotrophin-3 (NT-3) and Neurotrophin-4/5 (NT-4/5)

NT-3 has a high affinity for TrkC as a chemokine receptor (Chen et al., 2013) but it binds to TrkA and TrkB with low affinity (Skaper, 2008). NT-3 contains a tertiary fold and cysteine knot and its promoter contributes to the dimer to form heterodimers (Robinson, Radziejewski, Stuart, & Jones, 1995). The distribution of NT-3 messenger RNA and its biological activity on a variety of neuronal populations clearly distinguish NT-3 from NGF and BDNF.

NT-4 has been molecularly cloned from *Xenopus* (Ip et al., 1992). Structurally, mammalian NT-4 is quite unusual compared to the previously described neurotrophins because of a truncated proregion and a seven-amino acid insertion within its mature region. NT-4/5 bind to the TrkB receptor corresponding with the onset of neurogenesis in the neural tube during brain development (Bartkowska, Turlejski, & Djavadian, 2010).

It has been reported that NT-3 functions in the regulation of placental and fetal brain development for maternal inflammatory responses (Casciaro et al., 2009). NT-3 plays an essential role in the development of both the neural-crest-derived peripheral nervous system and the central nervous system (Chalazonitis, 2004). Kawamura et al. (2009) have reported the expression of neurotrophin 4/5 and their receptors TrkB in trophoblast cells and placentas during different stages (d10—delivery of pregnancy) of pregnancy in mice.

A proper placental development is essential during pregnancy since it forms the interface between the maternal-fetal circulations and is critical for fetal nutrition and oxygenation. Animal studies have demonstrated altered levels and expression of BDNF and NGF in the brain of the offspring at birth and in later life as a consequence of altered maternal micronutrients (folic acid, vitamin B_{12}) (Sable, Dangat, Joshi, & Joshi, 2012; Sable, Dangat, Kale, & Joshi, 2011). These studies highlight the importance of altered early life nutrition serving as a predisposition to noncommunicable diseases and neurodevelopmental disorders in adult life.

3. MATERNAL NUTRITION

Proper intrauterine environment is required for placental as well as fetal growth during pregnancy. Maternal nutrition is an important

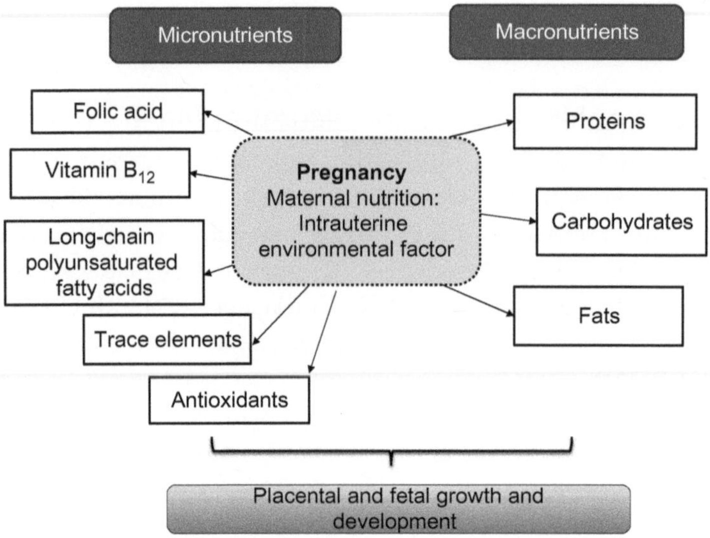

Fig 1 Maternal nutrition an essential intrauterine environmental factor.

intrauterine environmental factor affecting the development of the feto-placental unit. Adequate maternal nutrition plays an essential role in protecting the embryo from deficiencies, which may be inherent in the mother's circulation especially during the early stages of pregnancy when the placenta is not yet formed. However, the basis of biochemical mechanisms and pathophysiological events related to maternal-fetal nutrition and offspring health is not clearly understood (Berti et al., 2016). As per the requirement of maternal nutrients to the fetus is divided into two major categories macro and micronutrients (Fig. 1).

3.1 Macronutrients

The nutrients required in big or large amounts are called macronutrients. They are required for growth, metabolism, and to provide calories or energy. There are three broad classes of macronutrients: proteins, carbohydrates, and fats.

3.1.1 Proteins

Proteins are needed for growth, tissue repair, and immune function for the production of essential enzymes and hormones during feto-placental development. Meat, poultry, fish, meat substitutes, cheese, milk, nuts, and legumes are dietary rich sources of proteins. Starchy foods and vegetables contain smaller quantities of protein. During pregnancy there is an increased demand for energy and protein to enable the fetus and placenta to grow. Current recommendations suggest an allowance for pregnancy of 6–10 g protein per day (Duggleby & Jackson, 2002). A balanced protein energy supplementation to pregnant women is known to reduce the risk for preterm birth, small for gestational age (SGA) babies, and intrauterine growth restriction (IUGR) (Imdad & Bhutta, 2011; 2013; Ota, Tobe-Gai, Mori, & Farrar, 2012). Proteins are transported to the fetus as amino acids by specific amino acid transporter proteins. Amino acids are the building blocks for protein and also main regulators of metabolic pathways in feto-placental development (Rezaei, Wu, Hou, Bazer, & Wu, 2016). A report indicates that IUGR is associated with a reduction in placental amino acid transport capacity and metabolic pathways within the embryonic/fetal development (Lin et al., 2014).

3.1.2 Carbohydrates

Starchy foods (grain and potatoes), fruits, milk, and yogurt are dietary rich sources of carbohydrates. Vegetables, beans, nuts, and seeds also contain carbohydrates but in lesser amounts. Maternal dietary carbohydrates are suggested to be a valuable tool in the management of pregnancies (Clapp, 2002) since they are the main source of fuel to produce energy for the body. Carbohydrate is transported to the fetus as glucose which is taken up from the maternal plasma by the specific glucose transporter proteins (GLUTs) (Hay, 1994).

In normal early pregnancy, glucose tolerance is normal or slightly improved. Further, the peripheral sensitivity to insulin and hepatic basal glucose production is normal (Butte, 2000). The impairments in the carbohydrate metabolism, in particular gestational diabetes mellitus are the most frequent observed during pregnancy (Ianatchkova, Chaveeva, & Shterev, 2015). Also, it has been reported that children exposed to gestational diabetes *in utero* could have a higher future risk of glucose intolerance and diabetes mellitus (Wilk et al., 2015). Fetal exposure to excess or insufficient nutrition during vulnerable developmental periods appears to result in a lifelong predisposition to adult disease, such as type 2 diabetes (Carolan-Olah, Duarte-Gardea, & Lechuga, 2015).

3.1.3 Fats

Fats consist of a wide group of compounds including triglycerides, triesters of glycerol, and fatty acids (saturated, monounsaturated, and polyunsaturated). Dietary rich sources of fat are edible fats and oils, meat, fat spreads, milk, and milk products (Cashel & Greenfield, 1994). Among these, fatty acids are the main components of lipid classes (triacylglycerols, phospholipids, cholesteryl esters, nonesterified fatty acids) (Tvrzicka, Kremmyda, Stankova, & Zak, 2011). They are important energy substrates comprising of around 30% of the total energy intake in humans and can be stored in the adipose tissue (Jo et al., 2009). Fatty acids play an important role in fetal growth and development during pregnancy (Herrera, 2002). The polyunsaturated fatty acids (PUFAs), particularly long-chain PUFAs (LCPUFAs), play an important and vital physiologic role in the offspring who receive this fatty acid during critical periods of development (Mennitti et al., 2015).

Higher levels of triglycerides during pregnancy have been shown to be associated with the risk of preeclampsia and preterm birth (Ghio, Bertolotto, Resi, Volpe, & Di Cianni, 2011). It has been reported that an intake of trans fat during the second trimester of pregnancy is related to the risk for preeclampsia (Chavarro, Halldorsson, Leth, Bysted, & Olsen, 2011). A report suggests that imbalances in fatty acid intake during pregnancy and lactation may result in permanent changes in neuroendocrine function and energy metabolism in the fetus (Kabaran & Besler, 2015). Recently, it has been reported that the poor maternal diet quality during pregnancy increases neonatal adiposity (Shapiro et al., 2016).

3.2 Micronutrients

Micronutrients are nutrients required by humans throughout life in small quantities to coordinate a range of physiological functions. These micronutrients include vitamins (folic acid, vitamin B_{12}, etc.) and trace minerals (zinc, copper, cobalt, manganese, selenium, iodine, etc.).

3.2.1 Antioxidants

Antioxidants are substances that protect the cells against the effects of free radicals and are found in many fruits and vegetables. The common antioxidants include vitamin A, C, and E, selenium, beta-carotene, flavonoids, and polyphenols. An antioxidant is a molecule capable of slowing or preventing the oxidation of other molecules. Ascorbic acid, a naturally occurring organic compound, plays an important role in maternal health as well as fetal development (Dror & Allen, 2012). It has been reported that the plasma ascorbic acid concentrations are inversely related to SGA babies (Rajasingam, Seed, Briley, Shennan, & Poston, 2009). Human studies have shown high plasma ascorbic acid levels to be associated with preterm birth (Joshi et al., 2008). Recent studies also suggest that supplementation of vitamin E during pregnancy may reduce the risk for preterm birth (Bártfai et al., 2012).

Reports indicate that women who had total vitamin C intakes of <10th percentile of the requirement before conception had twice the risk of preterm delivery (Siega-Riz, Promislow, Savitz, Thorp, & McDonald, 2003). It has been suggested that low serum zinc levels may be associated with suboptimal outcome of pregnancy such as prolonged labor, pregnancy-induced hypertension, and preterm labor (Mahomed, Williams, King, Mudzamiri, & Woelk, 2007). Supplementation with vitamins C and E has been shown to synergistically protect the fetal membranes and decrease the risk of preterm premature rupture of membrane (PPROM) (Wall, Pressman, & Woods, 2002).

3.2.2 Folate

The word folate is derived from the Latin word folium, i.e., leaf. The term "folic acid" usually describes a group of compounds derived from 5,6,7,8-tetrahydropteroyl-g-glutamate more commonly referred as tetrahydrofolate (THF) (Stanger, 2002). Folate is a water soluble vitamin and is found in leafy vegetables such as spinach, asparagus, and lettuce. It is also present in dried or fresh beans, peas, lentils, and egg yolks. Dietary folic acid occurs mainly in polyglutamate forms (5-methyl-THF and 10-formyl-THF). These cannot cross the cell membrane and have to be converted to monoglutamate folate which is absorbed in the cell (Zhao, Diop-Bove, Visentin, & Goldman, 2011). The dietary folate intake requirement during pregnancy is 450 folate/d. During pregnancy, women are encouraged to consume foods which are high in folate in addition to folic acid fortified foods.

Folic acid has recently gained considerable attention because of its role in the pathogenesis of birth defects, cardiovascular disease, cancer, and neuropsychiatric disorders (Ebesunun & Obajobi, 2012; Molloy, Kirke, Brody, Scott, & Mills, 2008). Folate is essential for the synthesis of deoxyribonucleic acid (DNA), cellular division and proliferation, development of the placenta, and uterine enlargement (Antony, 2007).

Low folate concentrations in early pregnancy have been shown to adversely influence placentation, increase the risk for preterm birth, and lead to fetal growth retardation (Bergen et al., 2012). Evidence suggests that folic acid supplementation and fortification are effective in preventing neural tube defects (Blencowe, Cousens, Modell, & Lawn, 2010; Czeizel, Dudás, Paput, & Bánhidy, 2011). Folic acid plays a key role in the early stages of placental development, including extra villous trophoblast invasion, angiogenesis, and secretion of the matrix metalloproteinases (MMPs) (Williams, Bulmer, Innes, & Broughton Pipkin, 2011). Folate deficiency has been shown to be associated with an increased risk of placental abruption or abnormalities (Pickell et al., 2009; Ray & Laskin, 1999).

3.2.3 Vitamin B_{12}

Vitamin B_{12}, also called cobalamin, is a water soluble vitamin which plays a key role in the normal functioning of the brain and nervous system (Hanna, Lachover, & Rajarethinam, 2009). It is involved in the metabolism of every cell of the body, affecting DNA and fatty acid synthesis. Vitamin B_{12} is found in animal foods such as fish and shellfish, meat (especially liver), poultry, eggs, milk, and milk products (Watanabe, 2007). It has been suggested that women should have vitamin B_{12} levels >300 pg/mL before conception (Molloy et al., 2009).

Vitamin B_{12} binds to transcobalamin and then together they bind to the transcobalamin receptor (TCbR) (Fig. 2). After binding, this complex is taken up throughout the body by endocytosis (Quadros, Nakayama, & Sequeira, 2009). Cobalamin releases from transcobalamin in the lysosome and then free cobalamin is transferred to the cytoplasm. However, this process of release in the cytoplasm is unclear. Vitamin B_{12} is required in two reactions: first for the conversion of methylmalonyl CoA to succinyl CoA and second for the conversion of homocysteine to methionine in the one carbon cycle (Anderson, Sant, & Dolinoy, 2012; Li, Watkins, & Rosenblatt, 2009).

The majority of the Indian population consume a vegetarian diet. Hence, they are at a risk for lower levels of vitamin B_{12}, thereby necessitating supplementation during pregnancy (Hovdenak & Haram, 2012; Yajnik et al., 2006). Inadequate maternal vitamin B_{12} status has been shown to be associated with increased risk for neural tube defects during pregnancy (Molloy et al., 2009; Vanderjagt et al., 2011).

3.2.4 Trace Minerals

Chemical elements such as iron, manganese, zinc, copper, and iodine are required in very small amounts in humans and are essential for physiological and biochemical processes. Deficiencies of trace minerals like zinc, copper, and manganese have been implicated in various reproductive events like

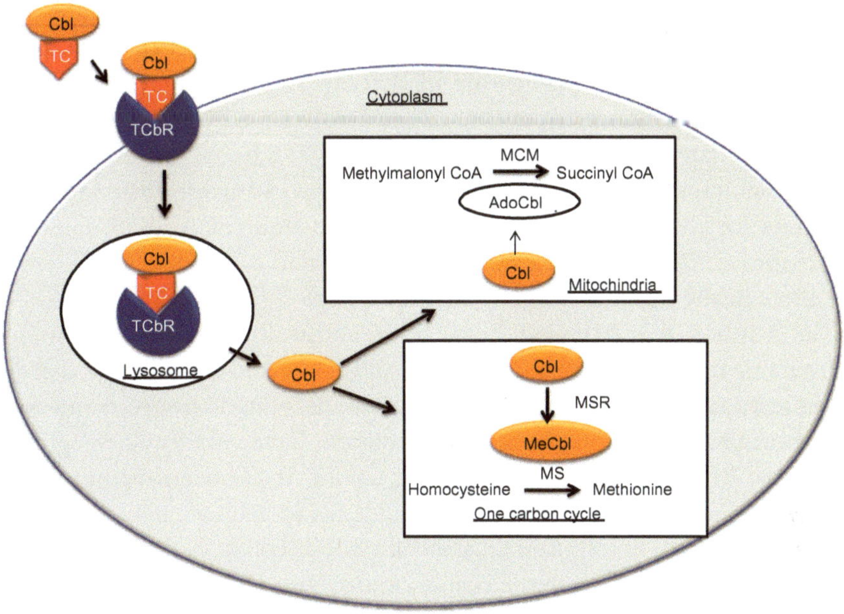

Fig. 2 Cellular vitamin B_{12} metabolism. *AdoCbl*, adenosylcobalamin; *Cbl*, cobalamin; *MCM*, methylmalonyl CoA mutase; *MeCbl*, methylcobalamin; *MS*, methionine synthase; *MSR*, methionine synthase reductase; *TC*, transcobalamin; *TCbR*, transcobalamin receptor. *Li, F., Watkins, D., & Rosenblatt, D.S. (2009). Vitamin B(12) and birth defects.* Molecular Genetics and Metabolism, *98, 166–172.*

infertility, pregnancy wastage, congenital anomalies, pregnancy-induced hypertension, placental abruption, premature rupture of membrane (PROM), still births, and low birth weight (LBW) (Pathak & Kapil, 2004). Zinc is an important element performing a range of functions in the body, as it is a cofactor for the synthesis of a number of enzymes and plays a key role in cell replication, tissue repair, and growth (Lowe, Lowe, Fraser, & Jackson, 2002). Copper is essential for accepting and donating electrons and is central to many energy-dependent chemical reactions (Gambling, Kennedy, & McArdle, 2011).

3.2.5 Long-Chain Polyunsaturated Fatty Acids (LCPUFA)

LCPUFA are important for fetal growth and development (Cetin et al., 2009). There are two types of LCPUFA namely, omega 3 and omega 6 fatty acids. Omega 3 fatty acids are the family of unsaturated fatty acids that have a double bond (C=C) at the ω-3 position and are commonly found in marine and plant oils (Carlson, Fallon, Kalish, Gura, & Puder, 2013). Adequate consumption of omega 3 fatty acids during pregnancy is critical since they are the building blocks of fetal brain and retina (Chang, Ke, & Chen, 2009) and are also known to play a role in determining the length of gestation (Coletta, Bell, & Roman, 2010). Alpha linolenic acid (ALA; 18:3, ω-3) an omega 3 fatty acid cannot be synthesized by the human body and hence is called an essential fatty acid (Greenberg, Bell, & Ausdal, 2008). ALA is the substrate for the synthesis of eicosapentaenoic acid (EPA; 20:5, ω-3) and docosahexaenoic acid (DHA; 22:6, ω-3) (Burdge & Calder, 2006) which are biologically the most active forms of omega 3 fatty acids.

ALA is found in flax seed, black gram, and walnuts. The most biologically active omega 3 fatty acids are EPA and DHA and are found in fish and other seafoods (Coletta et al., 2010). Vegetarians, who do not consume meat or fish and have a low intake of dairy products, have very low or negligible intakes of EPA and DHA (Rosell et al., 2005). It has been recommended that pregnant and lactating women should aim to achieve an average daily intake of at least 200 mg DHA (Koletzko et al., 2008).

Omega 6 fatty acids are the family of unsaturated fatty acids that have a double bond (C=C) at the ω-6 position. Linoleic acid (LA; 18:2, ω-6) an essential fatty acid is converted to arachidonic acid (AA; 20:4, ω-6). AA is physiologically an important omega 6 fatty acid released from membrane phospholipids by the enzyme phospholipase A2 and C. AA can also be cleaved from diacylglycerol and is a minor product of LA metabolism (Allaj, Guo, & Nie, 2013). It is essential for normal growth and development (Carlson, Cooke, Werkman, & Tolley, 1992) and is the precursor of eicosanoids (Ecker, 2012). AA and DHA are synthesized from their respective precursors, LA and ALA with the help of desaturase enzymes. Desaturases are key enzymes in the regulation of unsaturated fatty acid biosynthesis and are mainly present in the liver (Rodriguez et al., 1998).

LA is found in grains, meats, seeds, and various oils, including corn, safflower, cottonseed, and sunflower (Blasbalg, Hibbeln, Ramsden, Majchrzak, & Rawlings, 2011). Omega 6 fatty acids are substrates for eicosanoid and cytokine synthesis and play a role in modulating inflammation, immune function, and vascular system development (Weiss, Chambers, Gonzalez, Hagey, & Jones, 2012). Evidence from human studies indicates that both omega 3 and omega 6 series and their eicosanoid metabolites play a vital role in determining the duration of gestation and parturition (Allen & Harris, 2001; Greenberg et al., 2008).

In the body, the conversion of LA and ALA to gamma linolenic acid (GLA; 18:3, ω-6) and stearidonic acid (18:4, ω-3) is determined by the enzymatic activity of delta (Δ-6 desaturase (Rodriguez-Cruz, Tovar, Palacios-

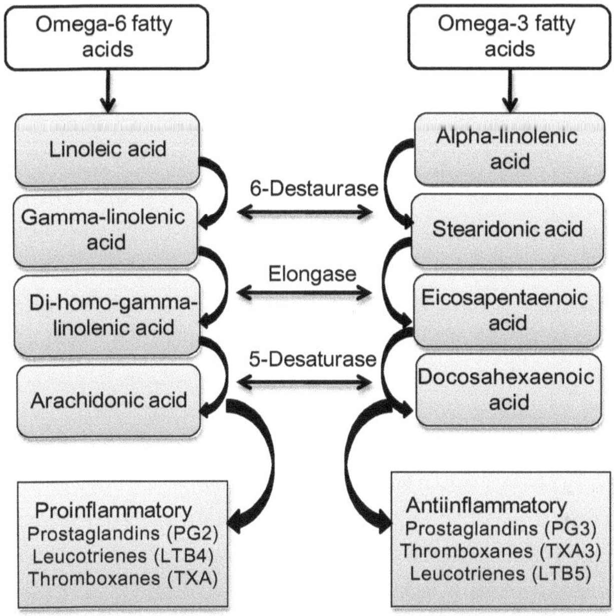

Fig. 3 LPUFA synthesis. *Dhobale, M., & Joshi, S. (2012). Altered maternal micronutrients (folic acid, vitamin B(12)) and omega 3 fatty acids through oxidative stress may reduce neurotrophic factors in preterm pregnancy.* Journal of Maternal-Fetal and Neonatal Medicine, *25(4), 317–323.*

González, Del Prado, & Torres, 2006). The Δ-5 desaturase introduces another double bond at the Δ-5 position of the 20 carbon fatty acids dihomo-gamma-linolenic acid (DGLA; 20:3, ω-6) and eicosatetraenoic acid (ETA; 20:4, ω-3) to synthesize AA and EPA, respectively. Further Δ-6 desaturase is also required for the synthesis of DHA from EPA (Fig. 3). In olden days, the ratio of omega 6 to omega 3 fatty acids was 1:1. Today, this ratio is about 25:1, indicating that modern day diets are deficient in omega 3 fatty acids as compared to the diet on which humans evolved (Simopoulos, 2011). It is recommended that pregnant and lactating women consume 300 mg/day of EPA + DHA (Huffman, Harika, Eilander, & Osendarp, 2011).

4. ASSOCIATION BETWEEN MATERNAL NUTRITION AND NEUROTROPHIC FACTORS THROUGH ONE CARBON CYCLE

4.1 One Carbon Cycle and Neurotrophic Factors

Folate and methionine are the key components of the one carbon cycle that provide the methyl groups for numerous methyl transferase reactions

(Kalhan & Marczewski, 2012). In the methionine cycle, dietary folate is converted to 5,10-methyltetrahydrofolate (5,10-MTHF). This 5,10-MTHF is converted to 5-MTHF with the help of the enzyme methylene tetrahydrofolate reductase (MTHFR) (Fox & Stover, 2008; Selhub, 2002). Further, this methyl group is transferred to homocysteine and converted to methionine with the help of the enzyme methionine synthase (MS) and cofactor vitamin B_{12}. Methionine is converted to *S*-adenosylmethionine (SAM) which is the major methyl donor for the methylation of phospholipids (e.g., phosphatidyl ethanolamine-DHA to phosphatidyl choline-DHA), chromatin (DNA and histones), and neurotransmitters. When folate levels are low, SAM levels decrease and *S*-adenosylhomocysteine (SAH) levels increase (Kim, Friso, & Choi, 2009; Kim, Hong, Lee, Lee, & Chang, 2009). Higher levels of SAH cause hyperhomocysteinemia due to the inhibition of DNA methyltransferases (Yi et al., 2000). Homocysteine is metabolized by two pathways: first by the remethylation pathway to methionine which requires folate and vitamin B_{12} and second by the transsulfuration pathway to cystathionine which requires vitamin B_6. The two pathways are coordinated by SAM, which acts as an allosteric inhibitor for the MTHFR reaction and as an activator of cystathionine beta-synthase (CBS) (Finkelstein, 1998; Selhub, 1999). SAM modulates epigenetic components leading to early programming and altered epigenetic marks, which persist into adulthood (McKay, Waltham, Williams, & Mathers, 2011). The impairments in the one carbon metabolism are known to be associated with altered feto-placental growth, cancers, vascular disease, cognitive decline, and neural tube defects (Beaudin & Stover, 2007; Smythies, Gottfries, & Regland, 1997; Stover, 2009).

Our recent human studies have demonstrated the association between altered maternal micronutrients like folate and vitamin B_{12}; reduced DHA levels; increased homocysteine and oxidative stress in preterm deliveries and preeclampsia (Dhobale et al., 2012; Kulkarni et al., 2011). Based on this, we have recently hypothesized that increased oxidative stress and decreased DHA may alter the levels and expression of neurotrophins in preterm deliveries, since balance between neurotrophins and oxidative stress is critical for normal growth and development (Dhobale & Joshi, 2012).

We have hypothesized and published "Altered metabolism of maternal micronutrients (folic acid and vitamin B_{12}) and omega 3 fatty acid lead to increased oxidative stress affecting protein and mRNA levels of neurotrophic factors (NGF and BDNF) in preterm pregnancy. This may further

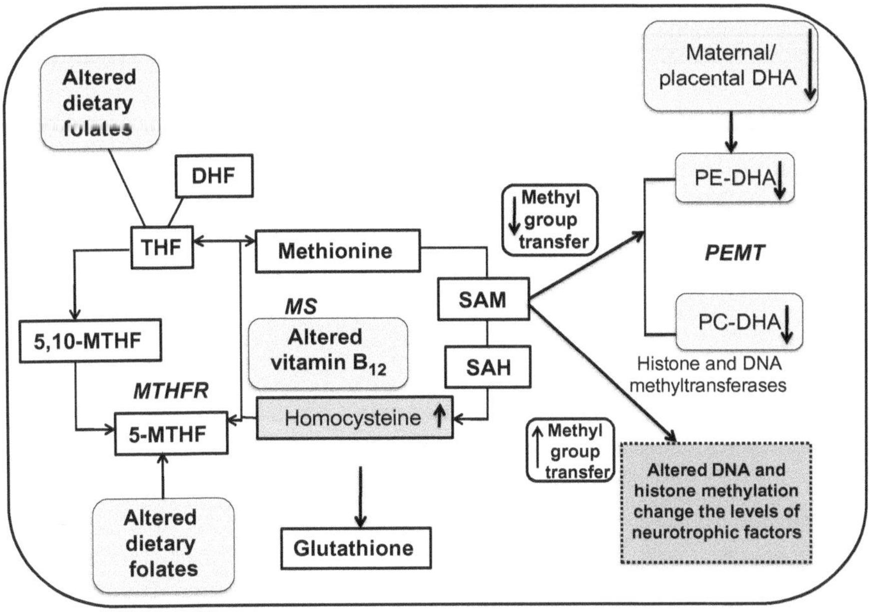

Fig. 4 Interaction between micronutrients in one carbon cycle with impact on oxidative stress and epigenesis in pregnancy. *DHF*, dihydrofolate; *MS*, methionine synthase; *MTHF*, methyl tetrahydrofolate; *MTHFR*, methylene tetrahydrofolate reductase; *PC-DHA*, phosphatidylcholine DHA; *PE-DHA*, phosphatidylethanolamine DHA; *PEMT*, phosphatidylethanolamine *N*-methyl transferase; *SAH*, *S*-adenosyl homocysteine; *SAM*, *S*-adenosyl methionine; *THF*, tetrahydrofolate. *Dhobale, M., & Joshi, S. (2012). Altered maternal micronutrients (folic acid, vitamin B(12)) and omega 3 fatty acids through oxidative stress may reduce neurotrophic factors in preterm pregnancy. Journal of Maternal-Fetal and Neonatal Medicine, 25(4), 317–323.*

contribute to risk for neurodevelopmental disorders in children born preterm (Fig. 4). (Dhobale & Joshi, 2012)"

4.2 One Carbon Cycle and Epigenetics

Maternal nutrition, particularly micronutrients such as folate and vitamin B_{12}, are the major determinants of the one carbon cycle and are suggested to be at the heart of intrauterine programming of diseases in adult life (Yajnik & Deshmukh, 2012). Folate, a methyl donor in the one carbon cycle is required for nucleotide synthesis and methylation reactions (Mason, 2003), while vitamin B_{12} is an important intermediary compound which acts as a cofactor for the conversion of homocysteine to methionine (Fox & Stover, 2008). These micronutrients act as cofactors or molecular donors for epigenetic processes like DNA methylation which modulate gene expression and thereby cellular proliferation and apoptosis during fetal development (Fall, 2009). The growing field of epigenetic research has highlighted the role of folate and vitamin B_{12} in developmental origins of health and disease (Waterland & Michels, 2007).

Epigenetics is defined as "inherited changes in gene expression that are not due to DNA nucleotide sequence but due to some environmental factors" (Godfrey, Lillycrop, Burdge, Gluckman, & Hanson, 2007). Epigenetic gene regulation involves three classes of molecular mechanisms: DNA methylation, histone modifications, and DNA-binding proteins. Among these, DNA methylation is the best characterized epigenetic mechanism in mammals (Waterland & Michels, 2007). Aberrant DNA methylation has been implicated in the pathogenesis of aging, cancer, cardiovascular and neurological diseases (McKay, Williams, & Mathers, 2004). Placental DNA methylation is also reported to influence the placental function (Kim, Friso, et al., 2009; Kim, Hong, et al., 2009). Animal studies clearly demonstrate that maternal undernutrition or overnutrition modify some epigenetic marks in their offspring thereby affecting gene expression (Chavatte-Palmer, Tarrade, & Lévy, 2012). Recent studies indicate that nutrition during the preconceptional period in humans is an important determinant of whole-genome methylation patterns in the newborn which may influence risk for diseases in later life (Khulan et al., 2012). However, the mechanism which links pregnancy-related complications to risk for diseases in adult life is not clearly understood. Although efforts have been made in identifying the mechanisms relating specific exposures in early life to the risk for developing chronic diseases in the adulthood, a complete picture remains obscure (Brenseke, Prater, Bahamonde, & Gutierrez, 2013). Therefore, there is a need to understand the role of maternal nutrition especially micronutrients, in the initiation and progression of the metabolic syndrome and neurodevelopmental disorders in the offspring.

5. CONSEQUENCES OF ALTERED MATERNAL NUTRITION IN PREGNANCY

Maternal nutrition is a modifiable risk factor of public health importance that can be integrated into efforts to prevent adverse birth outcomes, particularly among economically developing/low-income populations (Abu-Saad & Fraser, 2010). Different dietary interventional studies concluded that there is additional high quality randomized controlled trials are required to identify maternal diet intakes that optimize neonatal and infant outcomes (Gresham, Byles, Bisquera, & Hure, 2014). Studies reported that dietary interventions showed some small, but significant differences in pregnancy outcomes including a reduction in the incidence of preterm birth (Gresham et al., 2014). Further high-quality randomized controlled trials are required to identify maternal diet intakes that optimize pregnancy outcomes.

Altered micronutrient levels in the one carbon cycle may induce oxidative stress either by affecting the folate pool or by impairing remethylation of homocysteine (Mohammad et al., 2011). Influences of maternal nutrition on epigenetic programming are most important during prenatal and early postnatal development, when epigenetic mechanisms undergo establishment and maturation. Disruption of normal gene-specific methylation patterns by perturbations in maternal nutrition may affect the pregnancy outcome having long-term implications in the offspring. Thus, studies in pregnancy complications and in first episode psychotic patients suggest that altered metabolism of micronutrients like folic acid, vitamin B_{12}, and DHA through the one carbon cycle may epigenetically regulate neurotrophic factors.

Our study, for the first time, indicates a close association of constituents of one carbon cycle, DHA, and neurotrophins in preterm pregnancy (Dhobale, Wadhwani, Mehendale, Pisal, & Joshi, 2011; Dhobale et al., 2012). The lower levels of cord BDNF and NGF in preterm pregnancies may provide clues to predict cognitive deficits in children. Children born preterm are suggested to be at risk for developing neurodevelopmental disorders in later life. Identification of these genes can be of use in elucidating the molecular basis of the pathology and may further also lead to identification of suitable disease early biomarkers.

These altered levels of neurotrophins (BDNF and NGF) may be a consequence of altered gene expression and promoter methylation due to changes in the one carbon cycle. It is likely that these changes in the one carbon cycle influence epigenetic programming of the placenta. This may have implications for micronutrient mediated fetal programming of diseases in adult life.

REFERENCES

Abu-Saad, K., & Fraser, D. (2010). Maternal nutrition and birth outcomes. *Epidemiologic Reviews*, *32*, 5–25.

Allaj, V., Guo, C., & Nie, D. (2013). Non-steroid anti-inflammatory drugs, prostaglandins, and cancer. *Cell & Bioscience*, *3*, 8.

Allen, K. G., & Harris, M. A. (2001). The role of n-3 fatty acids in gestation and parturition. *Experimental Biology and Medicine (Maywood, NJ)*, *226*, 498–506.

Anderson, O. S., Sant, K. E., & Dolinoy, D. C. (2012). Nutrition and epigenetics: An interplay of dietary methyl donors, one-carbon metabolism and DNA methylation. *The Journal of Nutritional Biochemistry*, *23*, 853–859.

Antony, A. C. (2007). In utero physiology: Role of folic acid in nutrient delivery and fetal development. *The American Journal of Clinical Nutrition*, *85*, 598S–603S.

Barbacid, M. (1995). Structural and functional properties of the TRK family of eurotrophin receptors. *Annals of the New York Academy of Sciences*, *766*, 442–458.

Bártfai, L., Bártfai, Z., Nedeczky, I., Puho, E. H., Bánhidy, F., & Czeizel, A. E. (2012). Rate of preterm birth in pregnant women with vitamin E treatment: A population-based study. *The Journal of Maternal-Fetal & Neonatal Medicine*, *25*, 575–580.

Bartkowska, K., Turlejski, K., & Djavadian, R. L. (2010). Neurotrophins and their receptors in early development of the mammalian nervous system. *Acta Neurobiologiae Experimentalis (Wars)*, *70*, 454–467.

Beaudin, A. E., & Stover, P. J. (2007). Folate-mediated one-carbon metabolism and neural tube defects: Balancing genome synthesis and gene expression. *Birth Defects Research Part C, Embryo Today*, *81*, 183–203.

Bergen, N. E., Jaddoe, V. W., Timmermans, S., Hofman, A., Lindemans, J., Russcher, H., et al. (2012). Homocysteine and folate concentrations in early pregnancy and the risk of adverse pregnancy outcomes: The generation R study. *BJOG*, *119*, 739–751.

Berry, A., Bindocci, E., & Alleva, E. (2012). NGF, brain and behavioral plasticity. *Neural Plasticity*, *2012*. 784040.

Berti, C., Cetin, I., Agostoni, C., Desoye, G., Devlieger, R., Emmett, P. M., et al. (2016). Pregnancy and infants' outcome: Nutritional and metabolic implications. *Critical Reviews in Food Science and Nutrition*, *56*(1), 82–91.

Binder, D. K., & Scharfman, H. E. (2004). Brain-derived neurotrophic factor. *Growth Factors*, *22*, 123–131.

Blasbalg, T. L., Hibbeln, J. R., Ramsden, C. E., Majchrzak, S. F., & Rawlings, R. R. (2011). Changes in consumption of omega-3 and omega-6 fatty acids in the United States during the 20th century. *The American Journal of Clinical Nutrition*, *93*, 950–962.

Blencowe, H., Cousens, S., Modell, B., & Lawn, J. (2010). Folic acid to reduce neonatal mortality from neural tube disorders. *International Journal of Epidemiology*, *39*(Suppl. 1), 110–121.

Brenseke, B., Prater, M. R., Bahamonde, J., & Gutierrez, J. C. (2013). Current thoughts on maternal nutrition and fetal programming of the metabolic syndrome. *Journal of Pregnancy*, *2013*, 368461.

Burdge, G. C., & Calder, P. C. (2006). Dietary alpha-linolenic acid and health-related outcomes: A metabolic perspective. *Nutrition Research Reviews*, *19*, 26–52.

Butte, N. F. (2000). Carbohydarste and lipid metabolism in pregnancy: Normal compared with gestational diabetes mellitus. *The American Journal of Clinical Nutrition*, *71*(5 Suppl.), 1256S–1261S.

Carlson, S. E., Cooke, R. J., Werkman, S. H., & Tolley, E. A. (1992). First year growth of preterm infants fed standard compared to marine oil n-3 supplemented formula. *Lipids*, *27*, 901–907.

Carlson, S. J., Fallon, E. M., Kalish, B. T., Gura, K. M., & Puder, M. (2013). The Rol −3 fatty acid DHA in the human life cycle. *JPEN Journal of Parenteral and Enteral Nutrition*, *37*, 15–22.

Carolan-Olah, M., Duarte-Gardea, M., & Lechuga, J. (2015). A critical review: Early life nutrition and prenatal programming for adult disease. *Journal of Clinical Nursing*, *24*(23–24), 3716–3729.

Casciaro, A., Arcuri, F., Occhini, R., Toti, M. S., De Felice, C., & Toti, P. (2009). Expression of placental neurotrophin-3 (NT-3) in physiological pregnancy, Preeclampsia and Chorioamnionitis. *Clinical Medicine Pathology*, *2*, 9–15.

Cashel, K. M., & Greenfield, H. (1994). Principal sources of dietary fat in Australia: Evidence from apparent consumption data and the national dietary survey of adults. *The British Journal of Nutrition, 71*, 753–773.
Cetin, I., Alvino, G., & Cardellicchio, M. (2009). Long chain fatty acids and dietary fats in fetal nutrition. *Journal of Physiology, 587*, 3441–3451.
Chalazonitis, A. (2004). Neurotrophin-3 in the development of the enteric nervous system. *Progress in Brain Research, 146*, 243–263.
Chang, C. Y., Ke, D. S., & Chen, J. Y. (2009). Essential fatty acids and human brain. *Acta Neurologica Taiwanica, 18*, 231–241.
Chao, M. V., Rajagopal, R., & Lee, F. S. (2006). Neurotrophin signalling in health and disease. *Clinical Science (London, England), 110*, 167–173.
Chavarro, J. E., Halldorsson, T. I., Leth, T., Bysted, A., & Olsen, S. F. (2011). A prospective study of trans fat intake and risk of preeclampsia in Denmark. *European Journal of Clinical Nutrition, 65*, 944–951.
Chavatte-Palmer, P., Tarrade, A., & Lévy, R. (2012). Developmental origins of health and disease in adults: Role of maternal environment. *Gynécologie, Obstétrique & Fertilité, 40*, 517–519.
Chen, Y. F., Zeng, X., Zhang, K., Lai, B. Q., Ling, E. A., & Zeng, Y. S. (2013). Neurotrophin-3 stimulates migration of mesenchymal stem cells overexpressing TrkC. *Current Medicinal Chemistry, 20*, 3022–3033.
Clapp, J. F., 3rd. (2002). Maternal carbohydrate intake and pregnancy outcome. *The Proceedings of the Nutrition Society, 61*, 45–50.
Coletta, J. M., Bell, S. J., & Roman, A. S. (2010). Omega-3 fatty acids and pregnancy. *Reviews in Obstetrics & Gynecology, 3*, 163–171.
Czeizel, A. E., Dudás, I., Paput, L., & Bánhidy, F. (2011). Prevention of neural-tube defects with periconceptional folic acid, methylfolate, or multivitamins? *Annals of Nutrition & Metabolism, 58*, 263–271.
Dhobale, M., Chavan, P., Kulkarni, A., Mehendale, S., Pisal, H., & Joshi, S. (2012). Reduced folate, increased vitamin B(12) and homocysteine concentrations in women delivering preterm. *Annals of Nutrition & Metabolism, 61*, 7–14.
Dhobale, M., & Joshi, S. (2012). Altered maternal micronutrients (folic acid, vitamin B(12)) and omega 3 fatty acids through oxidative stress may reduce neurotrophic factors in preterm pregnancy. *The Journal of Maternal-Fetal & Neonatal Medicine, 25*(4), 317–323.
Dhobale, M. V., Wadhwani, N., Mehendale, S. S., Pisal, H. R., & Joshi, S. R. (2011). Reduced levels of placental long chain polyunsaturated fatty acids in preterm deliveries. *Prostaglandins, Leukotrienes, and Essential Fatty Acids, 85*, 149–153.
Dror, D. K., & Allen, L. H. (2012). Interventions with vitamins B6, B12 and C in pregnancy. *Paediatric and Perinatal Epidemiology, 26*(Suppl. 1), 55–74.
Duggleby, S. L., & Jackson, A. A. (2002). Protein, amino acid and nitrogen metabolism during pregnancy: How might the mother meet the needs of her fetus? *Current Opinion in Clinical Nutrition and Metabolic Care, 5*(5), 503–509.
Ebesunun, M. O., & Obajobi, E. O. (2012). Elevated plasma homocysteine in type 2 diabetes mellitus: A risk factor for cardiovascular diseases. *The Pan African Medical Journal, 12*, 48.
Ecker, J. (2012). Profiling eicosanoids and phospholipids using LC-MS/MS: Principles and recent applications. *Journal of Separation Science, 35*, 1227–1235.
Edwards, R. H., Selby, M. J., Mobley, W. C., Weinrich, S. L., Hruby, D. E., & Rutter, W. J. (1988). Processing and secretion of nerve growth factor: Expression in mammalian cells with a vaccinia virus vector. *Molecular and Cellular Biology, 8*, 2456–2464.
Fall, C. (2009). Maternal nutrition: Effects on health in the next generation. *The Indian Journal of Medical Research, 130*, 593–599.
Finkelstein, J. D. (1998). The metabolism of homocysteine: Pathways and regulation. *European Journal of Pediatrics, 157*(Suppl. 2), S40–S44.

Fox, J. T., & Stover, P. J. (2008). Folate-mediated one-carbon metabolism. *Vitamins and Hormones*, *79*, 1–44.
Gambling, L., Kennedy, C., & McArdle, H. J. (2011). Iron and copper in fetal development. *Seminars in Cell & Developmental Biology*, *22*, 637–644.
Ghio, A., Bertolotto, A., Resi, V., Volpe, L., & Di Cianni, G. (2011). Triglyceride metabolism in pregnancy. *Advances in Clinical Chemistry*, *55*, 133–153.
Gigante, A., Bevilacqua, C., Pagnotta, A., Manzotti, S., Toesca, A., & Greco, F. (2003). Expression of NGF, Trka and p75 in human cartilage. *European Journal of Histochemistry*, *47*, 339–344.
Godfrey, K. M., Lillycrop, K. A., Burdge, G. C., Gluckman, P. D., & Hanson, M. A. (2007). Epigenetic mechanisms and the mismatch concept of the developmental origins of health and disease. *Pediatric Research*, *61*, 5R–10R.
Greenberg, J. A., Bell, S. J., & Ausdal, W. W. (2008). Omega-3 fatty acid supplementation during pregnancy. *Reviews in Obstetrics & Gynecology*, *1*, 162–169.
Gresham, E., Byles, J. E., Bisquera, A., & Hure, A. J. (2014). Effects of dietary interventions on neonatal and infant outcomes: A systematic review and meta-analysis. *The American Journal of Clinical Nutrition*, *100*(5), 1298–1321.
Hanna, S., Lachover, L., & Rajarethinam, R. P. (2009). Vitamin b12 deficiency and depression in the elderly: Review and case report. *Primary Care Companion to the Journal of Clinical Psychiatry*, *11*, 269–270.
Hay, W. W., Jr. (1994). Placental transport of nutrients to the fetus. *Hormone Research*, *42*(4–5), 215–222.
He, T., & Katusic, Z. S. (2012). Brain-derived neurotrophic factor increases expression of MnSOD in human circulating angiogenic cells. *Microvascular Research*, *83*, 366–371.
Herrera, E. (2002). Implications of dietary fatty acids during pregnancy on placental, fetal and postnatal development—A review. *Placenta*, *23*(Suppl. A), S9–S19.
Hovdenak, N., & Haram, K. (2012). Influence of mineral and vitamin supplements on pregnancy outcome. *European Journal of Obstetrics, Gynecology, and Reproductive Biology*, *164*, 127–132.
Huffman, S. L., Harika, R. K., Eilander, A., & Osendarp, S. J. (2011). Essential fats: How do they affect growth and development of infants and young children in developing countries? A literature review. *Maternal & Child Nutrition*, 7(Suppl. 3), 44–65.
Ianatchkova, V., Chaveeva, P., & Shterev, A. (2015). The gestational diabetes mellitus as a specific pregnancy state. *Akush Ginekol (Sofiia)*, *54*(9), 29–33 [Article in Bulgarian].
Imdad, A., & Bhutta, Z. A. (2011). Effect of balanced protein energy supplementation during pregnancy on birth outcomes. *BMC Public Health*, *11*(Suppl. 3), S17.
Imdad, A., & Bhutta, Z. A. (2013). Nutritional management of the low birth weight/preterm infant in community settings: a perspective from the developing world. *J Pediatr*, *162*(Suppl. 3), S107–S114.
Ip, N. Y., Ibáñez, C. F., Nye, S. H., McClain, J., Jones, P. F., Gies, D. R., et al. (1992). Mammalian neurotrophin-4: Structure, chromosomal localization, tissue distribution, and receptor specificity. *Proceedings of the National Academy of Sciences of the United States of America*, *89*, 3060–3064.
Jiang, H., Liu, Y., Zhang, Y., & Chen, Z. Y. (2011). Association of plasma brain-derived neurotrophic factor and cardiovascular risk factors and prognosis in angina pectoris. *Biochemical and Biophysical Research Communications*, *415*, 99–103.
Jo, J., Gavrilova, O., Pack, S., Jou, W., Mullen, S., Sumner, A. E., et al. (2009). Hypertrophy and/or hyperplasia: Dynamics of adipose tissue growth. *PLoS Computational Biology*, *5*, e1000324.
Joshi, S. R., Mehendale, S. S., Dangat, K. D., Kilari, A. S., Yadav, H. R., & Taralekar, V. S. (2008). High maternal plasma antioxidant concentrations associated with preterm delivery. *Annals of Nutrition and Metabolism*, *53*, 276–282.

Kabaran, S., & Besler, H. T. (2015). Do fatty acids affect fetal programming? *Journal of Health, Population, and Nutrition, 33*, 14.

Kalhan, S. C., & Marczewski, S. E. (2012). Methionine, homocysteine, one carbon metabolism and fetal growth. *Reviews in Endocrine & Metabolic Disorders, 13*, 109–119.

Kanai Azuma, M., Kanai, Y., Matsuda, H., Kurohmaru, M., Tachi, C., Yazaki, K., et al. (1997). Nerve growth factor promotes giant-cell transformation of mouse trophoblast cells in vitro. *Biochemical and Biophysical Research Communications, 231*, 309–315.

Kawamura, K., Kawamura, N., Kumazawa, Y., Kumagai, J., Fujimoto, T., & Tanaka, T. (2011). Brain-derived neurotrophic factor/tyrosine kinase B signaling regulates human trophoblast growth in an in vivo animal model of ectopic pregnancy. *Endocrinology, 152*, 1090–1100.

Kawamura, K., Kawamura, N., Sato, W., Fukuda, J., Kumagai, J., & Tanaka, T. (2009). Brain-derived neurotrophic factor promotes implantation and subsequent placental development by stimulating trophoblast cell growth and survival. *Endocrinology, 150*, 3774–3782.

Khulan, B., Cooper, W. N., Skinner, B. M., Bauer, J., Owens, S., Prentice, A. M., et al. (2012). Periconceptional maternal micronutrient supplementation is associated with widespread gender related changes in the epigenome: A study of a unique resource in the Gambia. *Human Molecular Genetics, 21*, 2086–2101.

Kim, K. C., Friso, S., & Choi, S. W. (2009). DNA methylation, an epigenetic mechanism connecting folate to healthy embryonic development and aging. *The Journal of Nutritional Biochemistry, 20*, 917–926.

Kim, J. M., Hong, K., Lee, J. H., Lee, S., & Chang, N. (2009). Effect of folate deficiency on placental DNA methylation in hyperhomocysteinemic rats. *The Journal of Nutritional Biochemistry, 20*, 172–176.

Koletzko, B., Lien, E., Agostoni, C., Böhles, H., Campoy, C., Cetin, I., et al. (2008). World association of perinatal medicine dietary guidelines working group. The roles of long-chain polyunsaturated fatty acids in pregnancy, lactation and infancy: Review of current knowledge and consensus recommendations. *Journal of Perinatal Medicine, 36*, 5–14.

Kulkarni, A., Dangat, K., Kale, A., Sable, P., Chavan-Gautam, P., & Joshi, S. (2011). Effects of altered maternal folic acid, vitamin B12 and docosahexaenoic acid on placental global DNA methylation patterns in Wistar rats. *PloS One, 6*, e17706.

Li, F., Watkins, D., & Rosenblatt, D. S. (2009). Vitamin B(12) and birth defects. *Molecular Genetics and Metabolism, 98*, 166–172.

Lin, G., Wang, X., Wu, G., Feng, C., Zhou, H., Li, D., et al. (2014). Improving amino acid nutrition to prevent intrauterine growth restriction in mammals. *Amino Acids, 46*(7), 1605–1623.

Liu, Q. R., Walther, D., Drgon, T., Polesskaya, O., Lesnick, T. G., Strain, K. J., et al. (2005). Human brain derived neurotrophic factor (BDNF) genes, splicing patterns, and assessments of associations with substance abuse and Parkinson's Disease. *American Journal of Medical Genetics. Part B, Neuropsychiatric Genetics, 134*, 93–103.

Lowe, N. M., Lowe, N. M., Fraser, W. D., & Jackson, M. J. (2002). Is there a potential therapeutic value of copper and zinc for osteoporosis? *The Proceedings of the Nutrition Society, 61*, 181–185.

Mahomed, K., Williams, M. A., King, I. B., Mudzamiri, S., & Woelk, G. B. (2007). Erythrocyte omega-3, omega-6 and trans fatty acids in relation to risk of preeclampsia among women delivering at Harare Maternity Hospital, Zimbabwe. *Physiological Research, 56*, 37–50.

Mason, J. B. (2003). Biomarkers of nutrient exposure and status in one-carbon (methyl) metabolism. *The Journal of Nutrition, 133*(Suppl. 3), 941S–947S.

Mayeur, S., Silhol, M., Moitrot, E., Barbaux, S., Breton, C., Gabory, A., et al. (2010). Placental BDNF/TrkB signaling system is modulated by fetal growth disturbances in rat and human. *Placenta, 31*, 785–791.

McDonald, N. Q., Lapatto, R., Murray-Rust, J., Gunning, J., Wlodawer, A., & Blundell, T. L. (1991). New protein fold revealed by a 2.3-A resolution crystal structure of nerve growth factor. *Nature, 354*, 411–414.

McKay, J. A., Waltham, K. J., Williams, E. A., & Mathers, J. C. (2011). Folate depletion during pregnancy and lactation reduces genomic DNA methylation in murine adult offspring. *Genes & Nutrition, 6*, 189–196.

McKay, J. A., Williams, E. A., & Mathers, J. C. (2004). Folate and DNA methylation during in utero development and aging. *Biochemical Society Transactions, 32*, 1006–1007.

Mennitti, L. V., Oliveira, J. L., Morais, C. A., Estadella, D., Oyama, L. M., Oller do Nascimento, C. M., et al. (2015). Type of fatty acids in maternal diets during pregnancy and/or lactation and metabolic consequences of the offspring. *The Journal of Nutritional Biochemistry, 26*(2), 99–111.

Miralles, F., Philippe, P., Czernichow, P., & Scharfmann, R. (1998). Expression of nerve growth factor and its high-affinity receptor Trk-A in the rat pancreas during embryonic and fetal life. *The Journal of Endocrinology, 156*, 431–439.

Mohammad, N. S., Yedluri, R., Addepalli, P., Gottumukkala, S. R., Digumarti, R. R., & Kutala, V. K. (2011). Aberrations in one carbon metabolism induce oxidative DNA damage in sporadic breast cancer. *Molecular and Cellular Biochemistry, 349*(1–2), 159–167.

Molloy, A. M., Kirke, P. N., Brody, L. C., Scott, J. M., & Mills, J. L. (2008). Effects of folate and vitamin B12 deficiencies during pregnancy on fetal, infant, and child development. *Food and Nutrition Bulletin, 29*, S101–S111. discussion S112–S115.

Molloy, A. M., Kirke, P. N., Troendle, J. F., Burke, H., Sutton, M., Brody, L. C., et al. (2009). Maternal vitamin B12 status and risk of neural tube defects in a population with high neural tube defect prevalence and no folic Acid fortification. *Pediatrics, 123*, 917–923.

Nico, B., Mangieri, D., Benagiano, V., Crivellato, E., & Ribatti, D. (2008). Nerve growth factor as an angiogenic factor. *Microvascular Research, 75*, 135–141.

Numakawa, T., Suzuki, S., Kumamaru, E., Adachi, N., Richards, M., & Kunugi, H. (2010). BDNF function and intracellular signaling in neurons. *Histology and Histopathology, 25*, 237–258.

Ota, E., Tobe-Gai, R., Mori, R., & Farrar, D. (2012). Antenatal dietary advice and supplementation to increase energy and protein intake. *Cochrane Database of Systematic Reviews, 9*CD000032.

Patapoutian, A., & Reichardt, L. F. (2001). Trk receptors: Mediators of neurotrophin action. *Current Opinion in Neurology, 11*, 272–280.

Pathak, P., & Kapil, U. (2004). Role of trace elements zinc, copper and magnesium during pregnancy and its outcome. *Indian Journal of Pediatrics, 71*, 1003–1005.

Pickell, L., Li, D., Brown, K., Mikael, L. G., Wang, X. L., Wu, Q., et al. (2009). Methylenetetrahydrofolate reductase deficiency and low dietary folate increase embryonic delay and placental abnormalities in mice. *Birth Defects Research Part A, Clinical and Molecular Teratology, 85*, 531–541.

Pruunsild, P., Kazantseva, A., Aid, T., Palm, K., & Timmusk, T. (2007). Dissecting the human BDNF locus: Bidirectional transcription, complex splicing, and multiple promoters. *Genomics, 90*, 397–406.

Quadros, E. V., Nakayama, Y., & Sequeira, J. M. (2009). The protein and the gene encoding the receptor for the cellular uptake of transcobalamin-bound cobalamin. *Blood, 113*, 186–192.

Rajasingam, D., Seed, P. T., Briley, A. L., Shennan, A. H., & Poston, L. (2009). A prospective study of pregnancy outcome and biomarkers of oxidative stress in nulliparous obese women. *American Journal of Obstetrics and Gynecology, 200*. 395.e1-9.

Ray, J. G., & Laskin, C. A. (1999). Folic acid and homocyst(e)ine metabolic defects and the risk of placental abruption, pre-eclampsia and spontaneous pregnancy loss: A systematic review. *Placenta, 20*, 519–529.

Rezaei, R., Wu, Z., Hou, Y., Bazer, F. W., & Wu, G. (2016). Amino acids and mammary gland development: Nutritional implications for milk production and neonatal growth. *Journal of Animal Science and Biotechnology*, 7, 20.

Robinson, R. C., Radziejewski, C., Stuart, D. I., & Jones, E. Y. (1995). Structure of the brain-derived neurotrophic factor/neurotrophin 3 heterodimer. *Biochemistry*, *34*, 4139–4146.

Rodriguez, A., Sarda, P., Nessmann, C., Boulot, P., Leger, C. L., & Descomps, B. (1998). Delta6- and delta5-desaturase activities in the human fetal liver: Kinetic aspects. *Journal of Lipid Research*, *39*, 1825–1832.

Rodriguez-Cruz, M., Tovar, A. R., Palacios-González, B., Del Prado, M., & Torres, . N. (2006). Synthesis of long-chain polyunsaturated fatty acids in lactating mammary gland: Role of Delta5 and Delta6 desaturases, SREBP-1, PPARalpha, and PGC-1. *Journal of Lipid Research*, *47*, 553–560.

Rosell, M. S., Lloyd-Wright, Z., Appleby, P. N., Sanders, T. A., Allen, N. E., & Key, T. J. (2005). Long chain n-3 polyunsaturated fatty acids in plasma in British meat-eating, vegetarian, and vegan men. *The American Journal of Clinical Nutrition*, *82*, 327–334.

Sable, P. S., Dangat, K. D., Joshi, A. A., & Joshi, S. R. (2012). Maternal omega 3 fatty acid supplementation during pregnancy to a micronutrient-imbalanced diet protects postnatal reduction of brain neurotrophic factor in the rat offspring. *Neuroscience*, *217*, 46–55.

Sable, P., Dangat, K., Kale, A., & Joshi, S. (2011). Altered brain neurotrophic factor at birth: Consequence of imbalance in maternal folic acid and vitamin B_{12} metabolism. *Neuroscience*, *190*, 127–134.

Segal, R. A. (2003). Selectivity in neurotrophin signaling: Theme and variations. *Annual Review of Neuroscience*, *26*, 299–330.

Selhub, J. (1999). Homocysteine metabolism. *Annual Review of Nutrition*, *19*, 217–246.

Selhub, J. (2002). Folate, vitamin B12 and vitamin B6 and one carbon metabolism. *The Journal of Nutrition, Health & Aging*, *6*, 39–42.

Shapiro, A. L., Kaar, J. L., Crume, T. L., Starling, A. P., Siega-Riz, A. M., Ringham, B. M., et al. (2016). Maternal diet quality in pregnancy and neonatal adiposity: The healthy start study. *International Journal of Obesity*, *40(7)*, 1056–1062.

Siega-Riz, A. M., Promislow, J. H., Savitz, D. A., Thorp, J. M., Jr., & McDonald, T. (2003). Vitamin C intake and the risk of preterm delivery. *American Journal of Obstetrics and Gynecology*, *189*, 519–525.

Simopoulos, A. P. (2011). Evolutionary aspects of diet: The omega-6/omega-3 ratio and the brain. *Molecular Neurobiology*, *44*, 203–215.

Skaper, S. D. (2008). The biology of neurotrophins, signalling pathways, and functional peptide mimetics of neurotrophins and their receptors. *CNS & Neurological Disorders Drug Targets*, 7, 46–62.

Smythies, J. R., Gottfries, C. G., & Regland, B. (1997). Disturbances of one-carbon metabolism in neuropsychiatric disorders: A review. *Biological Psychiatry*, *41*, 230–233.

Sofroniew, M. V., Howe, C. L., & Mobley, W. C. (2001). Nerve growth factor signaling, neuroprotection, and neural repair. *Annual Review of Neuroscience*, *24*, 1217–1281.

Stanger, O. (2002). Physiology of folic acid in health and disease. *Current Drug Metabolism*, *3*, 211–223.

Stover, P. J. (2009). One-carbon metabolism-genome interactions in folate-associated pathologies. *The Journal of Nutrition*, *139*, 2402–2405.

Tapia-Arancibia, L., Rage, F., Givalois, L., & Arancibia, S. (2004). Physiology of BDNF: Focus on hypothalamic function. *Frontiers in Neuroendocrinology*, *25*, 77–107.

Thoenen, H. (1995). Neurotrophic factor and neuronal plasticity. *Science*, *270*, 593–598.

Tometten, M., Blois, S., & Arck, P. C. (2005). Nerve growth factor in reproductive biology: Link between the immune, endocrine and nervous system? *Chemical Immunology and Allergy*, *89*, 135–148.

Toti, P., Ciarmela, P., Florio, P., Volpi, N., Occhini, R., & Petraglia, F. (2006). Human placenta and fetal membranes express nerve growth factor mRNA and protein. *Journal of Endocrinological Investigation, 29*, 337–341.

Tvrzicka, E., Kremmyda, L. S., Stankova, B., & Zak, A. (2011). Fatty acids as biocompounds: Their role in human metabolism, health and disease—A review. Part 1: Classification, dietary sources and biological functions. *Biomedical Papers of the Medical Faculty of the University Palacky, Olomouc, Czech Republic, 155*, 117–130.

Vanderjagt, D. J., Ujah, I. A., Ikeh, E. I., Bryant, J., Pam, V., Hilgart, A., et al. (2011). Assessment of the vitamin B12 status of pregnant women in Nigeria using plasma holo-transcobalamin. *ISRN Obstetrics and Gynecology, 2011*, 365894.

Wall, P. D., Pressman, E. K., & Woods, J. R., Jr. (2002). Preterm premature rupture of the membranes and antioxidants: The free radical connection. *Journal of Perinatal Medicine, 30*, 447–457.

Watanabe, F. (2007). Vitamin B12 sources and bioavailability. *Experimental Biology and Medicine (Maywood, NJ), 232*, 1266–1274.

Waterland, R. A., & Michels, K. B. (2007). Epigenetic epidemiology of the developmental origins hypothesis. *Annual Review of Nutrition, 27*, 363–388.

Weiss, L. A., Chambers, C. D., Gonzalez, V., Hagey, L. R., & Jones, K. L. (2012). The omega-6 fatty acid linoleic acid is associated with risk of gastroschisis: A novel dietary risk factor. *American Journal of Medical Genetics Part A, 158*, 803–807.

Wilk, M., Horodnicka-Józwa, A., Moleda, P., Petriczko, E., Safranow, K., Chojnacka, H., et al. (2015). Assessment of selected carbohydrate parameters in children exposed to gestational diabetes in utero. *Neuro Endocrinology Letters, 36*(5), 504–510.

Williams, P. J., Bulmer, J. N., Innes, B. A., & Broughton Pipkin, F. (2011). Possible roles for folic acid in the regulation of trophoblast invasion and placental development in normal early human pregnancy. *Biology of Reproduction, 84*, 1148–1153.

Yajnik, C. S., & Deshmukh, U. S. (2012). Fetal programming: Maternal nutrition and role of one carbon metabolism. *Reviews in Endocrine & Metabolic Disorders, 13*, 121–127.

Yajnik, C. S., Deshpande, S. S., Lubree, H. G., Naik, S. S., Bhat, D. S., Uradey, B. S., et al. (2006). Vitamin B12 deficiency and hyperhomocysteinemia in rural and urban Indians. *The Journal of the Association of Physicians of India, 54*, 775–782.

Yi, P., Melnyk, S., Pogribna, M., Pogribny, I. P., Hine, R. J., & James, S. J. (2000). Increase in plasma homocysteine associated with parallel increases in plasma *S*-adenosylhomocysteine and lymphocyte DNA hypomethylation. *The Journal of Biological Chemistry, 275*, 29318–29323.

Zhao, R., Diop-Bove, N., Visentin, M., & Goldman, I. D. (2011). Mechanisms of membrane transport of folates into cells and across epithelia. *Annual Review of Nutrition, 31*, 177–201.

Zheng, L. R., & Shao, J. G. (2012). Morphological study on development of nerve growth factorpositive neurons in the cerebellum of human fetus. *Zhongguo Yi Xue Ke Xue Yuan Xue Bao, 34*, 46–50.

CHAPTER FOURTEEN

The Role of Neurotrophin Signaling in Gliomagenesis: A Focus on the p75 Neurotrophin Receptor ($p75^{NTR}$/CD271)

M.M. Alshehri, S.M. Robbins, D.L. Senger[1]
Arnie Charbonneau Cancer Centre, University of Calgary, Calgary, AB, Canada
[1]Corresponding author: e-mail address: senger@ucalgary.ca

Contents

1. General Overview 368
2. Malignant Glioma: A Major Clinical Challenge 368
3. Glioma Disease Reservoirs 370
   3.1 The Brain Tumor Stem Cell Disease Reservoir 370
   3.2 The Invasive Glioma Disease Reservoir 371
4. Molecular Determinants of Glioma Invasion and Progression 372
5. The Role of the Neurotrophins and Their Cognate Receptors in Glioma Progression 374
   5.1 Introduction to the Neurotrophins 374
   5.2 $p75^{NTR}$ (CD271) as a Central Regulator of Glioma Tumorigenesis 376
   5.3 The Role of the Trk Receptors in Gliomagenesis 381
   5.4 Neurotrophins and Their Involvement in the Glioma Microenvironment 382
6. A Role for $p75^{NTR}$ in Cancer: Extending Beyond Glioma 384
7. Conclusions and Future Directions 385
Acknowledgments 389
References 389

Abstract

The p75 neurotrophin receptor ($p75^{NTR}$, a.k.a. CD271), a transmembrane glycoprotein and a member of the tumor necrosis family (TNF) of receptors, was originally identified as a nerve growth factor receptor in the mid-1980s. While $p75^{NTR}$ is recognized to have important roles during neural development, its presence in both neural and nonneural tissues clearly supports the potential to mediate a broad range of functions depending on cellular context. Using an unbiased in vivo selection paradigm for genes underlying the invasive behavior of glioma, a critical characteristic that contributes to poor clinical outcome for glioma patients, we identified $p75^{NTR}$ as a central regulator of glioma

Vitamins and Hormones, Volume 104
ISSN 0083-6729
http://dx.doi.org/10.1016/bs.vh.2016.11.001

invasion. Herein we review the expanding role that $p75^{NTR}$ plays in glioma progression with an emphasis on how $p75^{NTR}$ may contribute to the treatment refractory nature of glioma. Based on the observation that $p75^{NTR}$ is expressed and functional in two critical glioma disease reservoirs, namely, the highly infiltrative cells that evade surgical resection, and the radiation- and chemotherapy-resistant brain tumor-initiating cells (also referred to as brain tumor stem cells), we propose that $p75^{NTR}$ and its myriad of downstream signaling effectors represent rationale therapeutic targets for this devastating disease. Lastly, we provide the provocative hypothesis that, in addition to the well-documented cell autonomous signaling functions, the neurotrophins, and their respective receptors, contribute in a cell nonautonomous manner to drive the complex cellular and molecular composition of the brain tumor microenvironment, an environment that fuels tumorigenesis.

1. GENERAL OVERVIEW

In this chapter we focus on the role of the p75 neurotrophin receptor ($p75^{NTR}$; a.k.a. CD271) signaling axis in regulating glioma progression. Despite large-scale genomic efforts revealing a deeper molecular understanding of the disease, survival rates have not changed dramatically in the past 30 years and glioma remains one of the most lethal human cancers with very few cases of long-term survivors. While there are many reasons that could explain this lack of therapeutic response including difficulties in delivering drugs across the blood–brain barrier (BBB) and the diverse intra- and intertumor heterogeneity, we believe that many of the clinical strategies have not focused on the most relevant therapeutic targets, namely, the highly invasive glioma population and the more recently described brain tumor-initiating cells (a.k.a. glioma stem cells, brain tumor stem cells), both disease "reservoirs" which are refractory to our current therapies and ultimately lead to disease relapse. In this chapter we review the role of $p75^{NTR}$ and other neurotrophin receptors in regulating phenotypic attributes of these glioma populations and highlight the potential of neurotrophin signaling as a rationale target(s) for therapeutic intervention for this devastating disease.

2. MALIGNANT GLIOMA: A MAJOR CLINICAL CHALLENGE

Gliomas are the most common primary brain tumors in adults, representing more than 50% of all primary brain tumors. To date, pathological classification of glioma has been based on histopathologic appearance delineating tumors as astrocytoma, oligodendroglioma, and mixed oligoastrocytoma (Dolecek, Propp, Stroup, & Kruchko, 2012). The World Health Organization (WHO) grade criteria currently classify glioma into four grades (WHO grade I–IV) with astrocytic tumors representing around 82% of the grade IV tumors and other tumor types consisting of histologically unspecified glioma (7%), oligodendroglioma (6%), and mixed oligoastrocytoma (3%), with the latter categories also classified into WHO grade II–III (Dolecek et al., 2012). Several recent studies have further subdivided glioma by molecular classification (Brennan et al., 2013; Cancer Genome Atlas Research Network, 2008, 2013) building on the identification of prognostic and predictive molecular markers for glioma subtypes, indicators increasingly being utilized in the clinical evaluation and treatment outcome of glioma patients (Brennan et al., 2013; Dunn et al., 2016). Of note, the WHO will be announcing a new classification system for brain tumors that will be based on traditional neuropathology together with molecular information that incorporates specific biomarkers such as IDH and TERT mutations (Appin & Brat, 2015; Louis et al., 2014). Implementation of the new classifications will change current clinical trial design by allowing for the establishment of biomarker-stratified trials.

Among gliomas, the aggressive grade IV tumors or glioblastoma (GBM) is typified by high vascular endothelial proliferation, necrosis, cellular atypia, increased mitotic activity, and high cellular density (Westphal & Lamszus, 2011), with the majority of tumors arising de novo and a small percentage developing from low-grade astrocytoma. Despite current therapies including surgical tumor resection followed by radiotherapy and chemotherapy with temozolomide (TMZ), an oral alkylating agent which can cross the BBB, GBM patients' prognosis remains disappointing and survival of these patients is limited on average to 14.6 months with less than 5% of patients surviving long term (Stupp et al., 2005). In addition to overcoming the permeability of drugs across the BBB, other major barriers to effective treatment of glioblastoma include tumor heterogeneity both between patients and within an individual tumor where different populations of tumor cells exist that have invasive and stem-like features, both characteristics that facilitate resistance to current therapies (Bao, Wu, McLendon, et al., 2006; Hemmati et al., 2003; Lee et al., 2006; Singh et al., 2003; Vescovi, Galli, & Reynolds, 2006) and cell populations that have been shown to express the neurotrophin receptor $p75^{NTR}$ (Fig. 1) (Ahn et al., 2015; Forsyth et al., 2014; Johnston et al., 2007; Wang et al., 2008).

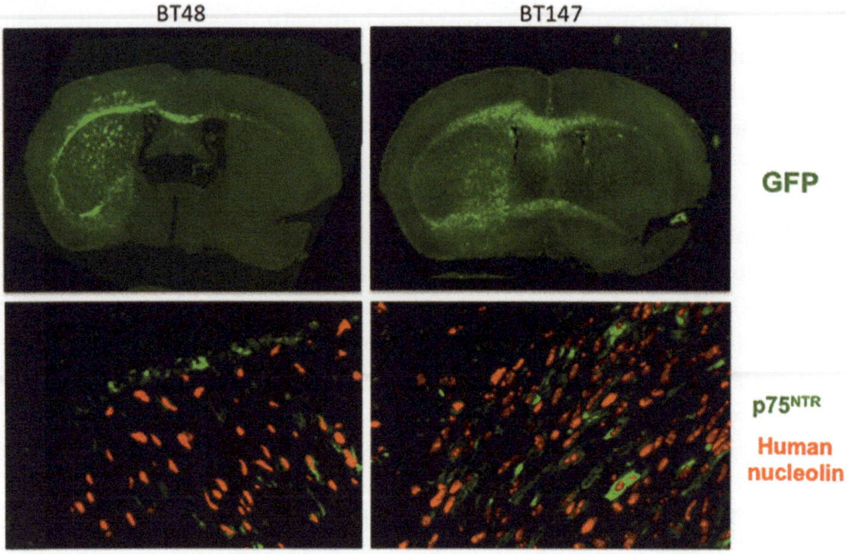

Fig. 1 p75^{NTR}-positive patient-derived brain tumor-initiating cells (BTICs) are highly infiltrative. Whole-brain sections from animals implanted with patient-derived brain tumor-initiating cells (BT48 and BT147) expressing a GFP construct were imaged using the InCell 6000 (*upper panels*). Highly infiltrative tumor cells (GFP; *green*) are seen invading through the brain parenchyma and along the corpus callosum. Whole-brain sections from animals bearing orthotopic patient-derived xenografts (BT48 and BT147) were stained with anti-p75^{NTR} (*green*) and a human-specific nucleolin antibody (*red*) to detect the human tumor cells (*lower panels*).

3. GLIOMA DISEASE RESERVOIRS

3.1 The Brain Tumor Stem Cell Disease Reservoir

Over the decades, a number of models have been established to study glioma biology including many recent studies that have focused on a population of tumor cells with stem-like features referred to as brain tumor stem cells or cells that initiate tumorigenesis (brain tumor-initiating cells; BTICs) (Bao, Wu, McLendon, et al., 2006; Hemmati et al., 2003; Lee et al., 2006; Singh et al., 2003; Vescovi et al., 2006). The existence of this population of cells was first postulated based on the observation that many cancers are organized as a hierarchy that consists of different classes of cells with tumorigenic and differentiation potential. The presence of these stem-like cells has now been identified in several tumor types, including malignant glioma (Bonnet & Dick, 1997; Clarke et al., 2006; Galli et al., 2004; Günther et al., 2008; Nowell, 1986; Shackleton et al., 2006; Sidransky et al., 1992; Singh et al., 2003, 2004) where these cells have been shown to maintain the ability to self-renew, proliferate, and give rise to more differentiated daughter cells that can repopulate the tumor (Clarke et al., 2006; Hemmati et al., 2003; Ignatov et al., 2002; Singh et al., 2003, 2004; Vescovi et al., 2006). Although there is great debate in the field as to the validity and nature of a bona fide cancer stem cell population (Rowan, 2009), their importance remains. These cells have signature characteristics of transformed cells including growth factor independence and tumorigenicity, they harbor the spectrum of molecular genetic alterations that are known to occur in human glioblastoma [e.g., mutations of the p53, PTEN, IDH1, EGFR (Blough, Beauchamp, Westgate, Kelly, & Cairncross, 2011; Kelly et al., 2009; Luchman et al., 2012; Lun et al., 2016)], and they form tumors that look and behave like human glioblastoma, infiltrating the cerebral cortex, spreading across the subependyma, and migrating across the corpus callosum (Fig. 1) (Kelly et al., 2009; Lee et al., 2006; Lun et al., 2016; Singh et al., 2004).

3.2 The Invasive Glioma Disease Reservoir

Infiltration of glioma cells throughout the brain is a hallmark of high-grade glioma and is fundamental to the lack of surgical success. In approximately 95% of cases, recurrent tumors grow within a few centimeters of the resection cavity. This invasive behavior of glioma cells is stimulated by different molecular processes including detachment of glioma cells from the main tumor mass, interaction of the cell with the surrounding extracellular matrix (ECM), degradation of the ECM to provide passage for the cell, and activation of cellular machinery underlying cell movement (Furnari et al., 2007; Nakada et al., 2007). As mentioned, this invasive process has been observed to occur with cells invading within the brain parenchyma through specific routes that include transiting along the basement membrane of blood vessels and axonal white matter tracts. Whether these routes are preferred by glioma cells due to the presence of certain substrates that support adhesion and stimulate invasion or due to the enrichment of oxygen and nutrition sources, or both, is unclear. These cells are genotypically and phenotypically distinct from their noninvasive counterparts and have been shown to activate a number of coordinate cellular programs necessary for migration, invasion (Fortin Ensign, Mathews, Symons, Berens, & Tran, 2013; Xie, Mittal, & Berens, 2014), and survival (e.g., reduced proliferation, marked resistance to apoptosis, upregulation of ATP-binding cassette transporters) (Bleau, Huse, & Holland, 2009; Giese, Bjerkvig, Berens, & Westphal, 2003; Joy et al., 2003; Mariani, Beaudry, McDonough, Hoelzinger, Demuth, et al., 2001). Many individual genes have been implicated in glioma invasion and gene expression profiling identified a subclass of glioma that expresses genes involved in cell migration and invasion that strongly correlates with poor patient survival (Freije et al., 2004; Li et al., 2009; Liang et al., 2005; Phillips et al., 2006; Verhaak et al., 2010). Moreover, and consistent with these phenotypic and genotypic differences, recent studies found striking differences between tumor cells at the invading tumor margin and those residing in the main tumor mass where only 60–80% of mutations were found in common (Kim, Lee, et al., 2015; Kim, Zheng, et al., 2015). In addition, the tumor specimens isolated from the tumor margin were found to have different gene expression profiles, tumor microenvironment, and immune infiltration (Glas et al., 2010; Pistollato et al., 2010), observations that highlight the need for therapeutic approaches that consider all tumor populations. Although we have described the invasive and stem-like glioma disease reservoirs independently, these cell populations are not mutually exclusive and many of the BTICs isolated and characterized to date possess the cardinal features of highly infiltrative glioma (Fig. 1).

4. MOLECULAR DETERMINANTS OF GLIOMA INVASION AND PROGRESSION

Glioma invasion is not random but rather occurs along white matter tracts (optic radiations, corpus callosum, or anterior commissure), adjacent to neurons, and along basement membranes of blood vessels. The brain has its own unique ECM composed of glycosaminoglycans (GAGs) and a few structural glycoproteins, with the main GAGs being hyaluronic acid, chondroitin, keratin, and heparan sulfates (Giese & Westphal, 1996; Gritsenko, Ilina, & Friedl, 2012). In addition, glioma cells themselves synthesize many ECM components including the collagens I, III, IV, and VI, tenascin, fibronectin, vitronectin, laminin, hyaluronan, chondroitin sulfate, and heparan sulfate proteoglycans (Gritsenko et al., 2012). Over the years a large number of individual proteins have been implicated in glioma adhesion and invasion that segregate into functional classes that include proteins that interact with the ECM and surrounding cellular microenvironment, proteins that degrade the ECM, and proteins involved in glioma cell movement. These proteins include several surface receptors such as integrin αVβ3 (D'Abaco & Kaye,

2007), NCAM (Owens et al., 1998; Prag et al., 2002), CD44 (Gunia et al., 1999; Merzak, Koocheckpour, & Pilkington, 1994; Radotra & McCormick, 1997), BEHAB (Nutt, Zerillo, Kelly, & Hockfield, 2001; Zhang, Kelly, Zerillo, Jaworski, & Hockfield, 1998), semaphorins (Li, Law, & Lee, 2012; Neufeld et al., 2005), the cadherins (Maret et al., 2010; Wang et al., 2014), and the ephrin (Eph) receptor–ligand system (Miao et al., 2009; Nakada et al., 2010; Nakada, Drake, Nakada, Niska, & Berens, 2006; Nakada et al., 2004) as well as a number of proteases including the matrix metalloproteinases [MMPs; particularly MMP2 and MMP9 (Kargiotis et al., 2008; Kondraganti et al., 2000; Lakka et al., 2004; Wang, McLeod, & Cassidy, 2003)], the ADAMs (a disintegrin and metalloproteinase), plasmin, and cathepsins (Gondi, Lakka, Dinh, et al., 2004; Gondi, Lakka, Yanamandra, et al., 2004; Lakka et al., 2004; Rao et al., 1994; Sarkar et al., 2014; Wildeboer, Naus, Amy Sang, Bartsch, & Pagenstecher, 2006), to name a few. Identification of the large array of genes and proteins involved in glioma invasion has utilized a number of experimental techniques including the isolation of invasive cells from human cell lines in vitro (Joy et al., 2003; Mariani, Beaudry, McDonough, Hoelzinger, Demuth, et al., 2001), the use of organotypic brain slice cultures (Beadle et al., 2008; Caspani, Echevarria, Rottner, & Small, 2006; Holtkamp et al., 2005; Jung et al., 2001, 1999; Ohnishi, Matsumura, Izumoto, Hiraga, & Hayakawa, 1998), and the collection of tumor and invasive cells from the main tumor mass in frozen glioblastoma patient specimens using laser capture microdissection (Hoelzinger et al., 2005; Kislin, McDonough, Eschbacher, Armstrong, & Berens, 2009; Mariani, Beaudry, McDonough, Hoelzinger, Kaczmarek, et al., 2001; Mariani, McDonough, et al., 2001; Toussaint et al., 2012). Using an alternative method of orthotopic serial in vivo selection (Robbins & Senger, 2013) we isolated highly invasive glioma variants and identified the neurotrophin receptor $p75^{NTR}$ as a driver of the infiltrative nature of glioma (Johnston et al., 2007). Functional, biochemical, and clinical studies established that $p75^{NTR}$ dramatically enhanced migration and invasion of genetically distinct glioma cells in vitro and in vivo and frequently exhibited robust expression in highly invasive glioblastoma patient specimens (Ahn et al., 2015; Johnston et al., 2007; Wang et al., 2008). Furthermore, $p75^{NTR}$-mediated invasion was found to occur via a neurotrophin-dependent regulated intramembrane proteolysis (RIP) that could be inhibited with clinically relevant γ-secretase inhibitors (Wang et al., 2008), and consistent with an increase in invasive ability was found to be associated with changes in the composition and rigidity of the tumor-associated ECM (Fig. 2). In this chapter

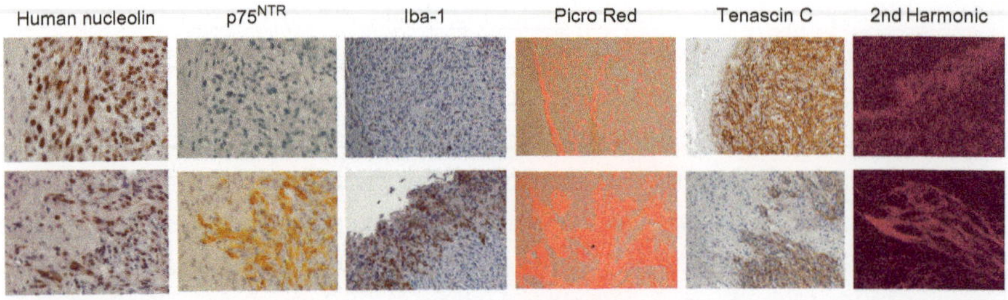

Fig. 2 p75^{NTR} regulates the brain tumor microenvironment. Brain sections from animals implanted with noninvasive human U87 glioma cells (*top row*) and invasive p75^{NTR}-expressing U87 cells (*bottom row*) were stained for human-specific nucleolin antibody to detect the human tumor cells, anti-p75^{NTR}, an Iba-1 antibody to detect tumor-associated microglia/macrophage, or anti-Tenascin C (*brown*) and counterstained with Toluidine Blue (*blue*). In addition, the collagen composition of the extracellular matrix was visualized by Picrosirius Red (Picro Red) and Second Harmonic generation.

we focus on the role of the neurotrophins and their receptors in glioma with specific emphasis on p75^{NTR}, its role in glioma progression, and its ability to mediate distinct biological properties of the highly invasive and stem-like glioma disease reservoirs. While we clearly appreciate the complexity of this receptor system that also includes other p75^{NTR} coreceptors (Nogo, Lingo, and the Sortilin family of receptors), there is currently no evidence in the context of p75^{NTR}-mediated biological processes in glioma for the involvement of these receptors and a more detailed description of these systems in other biological processes is covered elsewhere (Ibanez & Simi, 2012; Reichardt, 2006).

5. THE ROLE OF THE NEUROTROPHINS AND THEIR COGNATE RECEPTORS IN GLIOMA PROGRESSION

5.1 Introduction to the Neurotrophins

The neurotrophins are a family of proteins that control various aspects of development, survival, and function of neurons in both the central and peripheral nervous systems. This family includes nerve growth factor (NGF) (Levi-Montalcini, 1987), brain-derived neurotrophic factor (BDNF) (Leibrock et al., 1989), neurotrophin-3 (NT3) (Ernfors, Ibanez, Ebendal, Olson, & Persson, 1990; Maisonpierre et al., 1990), and neurotrophin-4/5 (NT4/5) (Berkemeier et al., 1991; Hallbook, Ibanez, & Persson, 1991). These proteins are synthesized as 30–35 kilodalton (kDa) precursor polypeptides (proneurotrophins) that can undergo cleavage by a diverse number of enzymes to yield (13–16 kDa) a mature form of the protein. The neurotrophins form homodimers held together by hydrophobic interactions (McDonald et al., 1991) and can bind and activate four transmembrane receptors: the p75 neurotrophin receptor ($p75^{NTR}$) and the tropomyosin-related kinase (Trk) family of tyrosine kinase receptors (TrkA, TrkB, and TrkC). These receptors can function independently or collectively to mediate numerous biological functions (Reichardt, 2006). For instance, neurons express both $p75^{NTR}$ and the Trks where together they interact physically and functionally in ways that alter the signaling of each receptor. In addition, the binding affinity of each neurotrophin varies between the individual receptors providing additional specificity to the receptor–ligand system (Ibanez & Simi, 2012). Several reports, including our own, found that glioma cell lines, highly tumorigenic human malignant glioma patient-derived BTICs, and patient-derived malignant glioma tissues can express the neurotrophins NGF, BDNF, and NT3 and their receptors $p75^{NTR}$, TrkA, TrkB, and TrkC where they have been found to correlate with glioma tumor progression and activate biological processes that include tumor cell proliferation, survival, invasion, migration, differentiation, and angiogenesis (Ahn et al., 2015; Birnbaum et al., 2007; Forsyth et al., 2014; Hamel, Westphal, Szonyi, Escandon, & Nikolics, 1993; Johnston et al., 2007; Lawn et al., 2015; Murphy et al., 1977; Vinores & Koestner, 1980; Wang et al., 2008; Xiong et al., 2015). In contrast, in low-grade glioma, this same receptor system has been shown to inhibit tumor growth and survival (Chiaretti et al., 2004; Kimura, Yoshino, Katayama, Watanabe, & Fukushima, 2002; Vinores & Koestner, 1980), observations that emphasize the contextual signaling differences of this system and caution needed when interpreting experimental outcomes. This ligand–receptor system can act both directly on the tumor cells such as the induction of glioma proliferation by NGF (cell autonomous) (Engebraaten, Bjerkvig, Pedersen, & Laerum, 1993; Oelmann et al., 1995; Singer, Hansen, Martinie, & Karp, 1999) or indirectly (cell nonautonomous) on the tumor microenvironment to mediate tumor progression. For example, one of the hallmarks of high-grade tumors is a significant presence of aberrant blood vessel formation (angiogenesis) generated for nutrient delivery. The involvement of the neurotrophins in the recruitment, proliferation, and alignment of the endothelial blood vessel cells has been shown to occur through a complex interaction between the endothelial cells and the glioma cells (Fischer, Gagner, Law, Newcomb, & Zagzag, 2005). In this process, NGF was found to promote angiogenesis through direct interaction with α9β1 integrin (Walsh et al., 2012) and subsequent activation of the MAPK Erk1/2 pathway (Brown et al., 2008). In addition, neural stem cells have been shown to release BDNF and VEGF to stimulate vascular tubes formation by brain-derived endothelial cells (Li, Ford, Lavik, & Madri, 2006), a mechanism that could occur in the tumor microenvironment where BDNF and VEGF are secreted in high amounts by both glioma cells and endothelial cells resulting in the formation of a perivascular niche, a preferred location of BTICs (Bao, Wu, Sathornsumetee, et al., 2006; Calabrese et al., 2007). NT3 has also been shown to be secreted by glioma cells where it actively recruits TrkC-expressing bone marrow stromal cells to the glioma microenvironment (Birnbaum et al., 2007) and tumor-associated macrophages have been linked with high tumor grade and poor prognosis of many cancers (Bingle, Brown, & Lewis, 2002; He et al., 2006; Hussain, Yang, Suki, Aldape, et al., 2006; Hussain, Yang, Suki, Grimm, & Heimberger, 2006; Komohara, Ohnishi, Kuratsu, & Takeya, 2008; Lorger, 2012).

5.2 p75^{NTR} (CD271) as a Central Regulator of Glioma Tumorigenesis

p75^{NTR}, similar to other tumor necrosis factor receptor family members, is a transmembrane glycosylated receptor that elicits an array of biological functions through its ability to interact with its cognate ligands and coreceptors (Barker, 2004; Ibanez & Simi, 2012; Kaplan & Miller, 2004; Reichardt, 2006). In neurons, where p75^{NTR} is coexpressed with the Trks (A, B, and C), it was initially thought to control a life vs death axis (Frade, Rodriguez-Tebar, & Barde, 1996; Majdan, Walsh, Aloyz, & Miller, 2001). It is now known that p75^{NTR} alone or with other coreceptors can mediate several cellular functions that include cell death, survival, migration, and axonal growth inhibition (see reviews: Chao, 2003; Dechant & Barde, 2002; Ibanez & Simi, 2012; Kaplan & Miller, 2003, 2004; Roux & Barker, 2002). In glioma, p75^{NTR} can modulate hallmarks of glioblastoma including invasion and proliferation (Ahn et al., 2015; Berghoff, Jaisimha, Duggan, MacSharry, & McCarthy, 2015; Forsyth et al., 2014; Johnston et al., 2007; Wang et al., 2008). p75^{NTR} was found to be upregulated in invading glioma cells, and treatment of these cells with either the mature neurotrophins (NGF, BDNF, or NT3) or a proneurotrophin (pro-NGF) that binds to and activates p75^{NTR} but not the Trk receptors resulted in enhanced migration (Johnston et al., 2007). Furthermore, the ectopic expression of p75^{NTR} in genetically diverse glioma cells (U87MG and

U251) was found to induce glioma migration and invasion in vitro and in vivo (Johnston et al., 2007) and is concomitant with changes in cytoskeleton rearrangement and a decrease in RhoA activity (Johnston et al., 2007). Although the glioma cells expressing $p75^{NTR}$ can respond to several neurotrophins, both the invading glioma cells and genetically distinct $p75^{NTR}$-expressing BTICs were found to secrete high levels of BDNF that bound to $p75^{NTR}$ on the surface of the cell resulting in an autocrine loop (Ahn et al., 2015; Johnston et al., 2007; Wang et al., 2008). Consistent with these data, the expression of mutant $p75^{NTR}$ alleles that have been crippled for ligand binding cannot induce glioma invasion in vitro and result in the formation of noninvasive well-circumscribed tumors in orthotopic xenograft models in vivo (Johnston et al., 2007). These data build on the requirement of neurotrophin binding in mediating glioma invasion and, in the context of the glioma models examined thus far, support the concept that $p75^{NTR}$ on the surface of a glioma cell is constitutively associated with its downstream mediators necessary to facilitate glioma invasion. Moreover, and clinically significant, assessment of $p75^{NTR}$ protein expression in a series of glioma patient samples identified a subpopulation of $p75^{NTR}$-expressing glioma cells present in 85% of the high-grade glioma and 22% of mid-grade astrocytoma, cells that were found to be more invasive than the non-$p75^{NTR}$-expressing glioma cells from the same patient (Johnston et al., 2007). Since the initial finding, independent data supporting a strong association with $p75^{NTR}$ and high-grade glioma have been established by the Repository of Molecular Brain Neoplasia Data (REMBRANDT), the TCGA, and the Human Protein Atlas where $p75^{NTR}$ has been shown to be elevated at both the RNA and protein level. Furthermore, accompanying patient data establish that high expression of $p75^{NTR}$ correlates with lower overall patient survival (Fig. 3).

$p75^{NTR}$, similar to many type I transmembrane proteins including Notch and amyloid precursor protein, has been shown to undergo RIP by sequential cleavage, first by an α-secretase and then by γ-secretase (Kanning et al., 2003; Pardossi-Piquard et al., 2005; Podlesniy et al., 2006). The initial cleavage by α-secretase results in the generation of a C-terminal fragment (CTF) that remains membrane bound and is susceptible to a second cleavage by γ-secretase that results in the cytoplasmic release of the intracellular domain (ICD). We demonstrated that neurotrophin-dependent RIP of $p75^{NTR}$ is required for $p75^{NTR}$-mediated glioma invasion, and this proteolytic processing of $p75^{NTR}$ occurs in $p75^{NTR}$-positive patient tumor specimens and BTICs (Wang et al., 2008). Inhibition of this

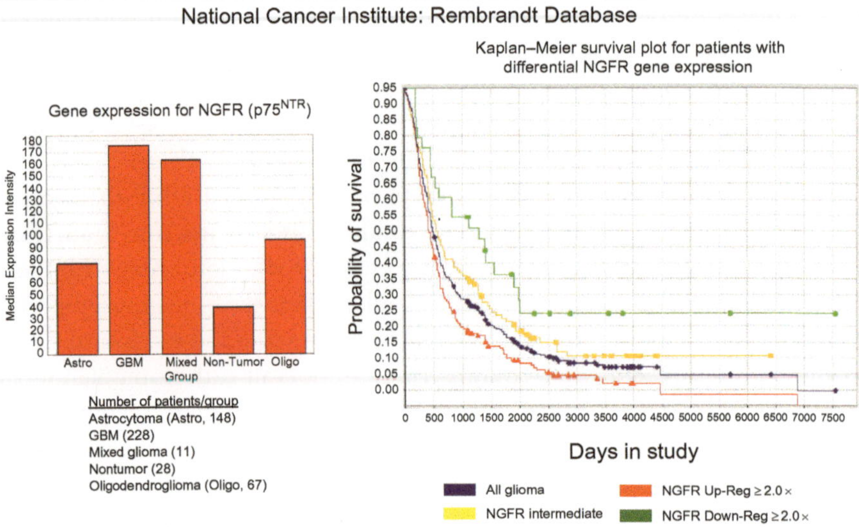

Fig. 3 p75^{NTR} associates with high-grade glioma and poorer overall survival. Graphical data represent the gene expression data for NGFR (p75^{NTR}) from the Rembrandt data portal where data characterizing human brain neoplasias contain subject annotation data from microarray analysis of 482 subjects. Rembrandt is a project of the National Cancer Institute Division of Cancer Treatment and Diagnosis, Cancer Diagnosis Program [National Cancer Institute (https://www.cancer.gov)].

process by clinically relevant γ-secretase inhibitors or expression of cleavage-resistant chimeras of p75^{NTR} significantly impairs the highly invasive nature of genetically distinct glioblastoma and patient-derived BTICs in vitro and in vivo where a dramatic decrease in glioma invasion and prolonged survival in orthotopic xenograft models was observed (Wang et al., 2008). With the generation and release of the p75^{NTR} ICD the possibility exists that RIP might be used as a messenger for the neurotrophins to both activate an intracellular signal and modulate transcriptional events in the nucleus (Chao, 2003; Parkhurst, Zampieri, & Chao, 2010). Although studying the ICD fragment of the Notch receptor has been very informative with respect to the mechanism of Notch signaling (Fortini, 2009), it is not as clear for the ICD of p75^{NTR} where differing observations for its requirement in vitro have been observed (Berghoff, Lassmann, Preusser, & Hoftberger, 2013; Wang et al., 2008). There is certainly supporting evidence that the cleavage and release of the ICD plays a role in some biological processes (Domeniconi et al., 2005; Kanning et al., 2003; Zampieri, Xu, Neubert, & Chao, 2005) possibly by directing or limiting its spatial localization. In this regard p75^{NTR} has been shown to localize within lipid rafts

(Bilderback, Gazula, Lisanti, & Dobrowsky, 1999; Bilderback, Grigsby, & Dobrowsky, 1997; Fujitani et al., 2005; Gil, Cubi, & Aguilera, 2007; Higuchi, Yamashita, Yoshikawa, & Tohyama, 2003; Huang et al., 1999) and PKA activation has been shown to induce the phosphorylation of $p75^{NTR}$, an event required for lipid raft localization and downstream signaling in cerebellar neurons (Higuchi et al., 2003). Consistent with a possible role for lipid raft-mediated and hence compartmentalized signaling in glioma cells we have shown that disruption of these membrane microdomains abrogates $p75^{NTR}$-mediated invasion (Ahn et al., 2015). Similarly, protein palmitoylation, a reversible process, may also regulate the localization of $p75^{NTR}$ through binding to a cysteine residue within the juxtamembrane domain of the receptor (Barker, Barbee, Misko, & Shooter, 1994). Protein palmitoylation has been shown to be regulated by specific stimuli and required for postsynaptic clustering of PSD-95, presumably by targeting PDS-95 via its lipid modification to endosomal vesicles (Craven & Bredt, 1998; El-Husseini, Schnell, Chetkovich, Nicoll, & Bredt, 2000). In neurons palmitoylation of $p75^{NTR}$ on the CTF occurs following α-secretase cleavage and has also been shown to mediate its translocation to the lipid rafts where it undergoes further cleavage by γ-secretase and activation of a neuronal death signal (Underwood, Reid, May, Bartlett, & Coulson, 2008). The exact role of these posttranslational modifications and spatial localization in mediating glioma invasion and progression still needs to be explored.

In contrast to the studies showing a requirement for proteolytic processing of $p75^{NTR}$ for glioma invasion (Wang et al., 2008), $p75^{NTR}$ has also been shown in U87MG glioma cells in vitro to regulate migration in a γ-secretase-independent mechanism through modulation of cadherin-11 (Cdh-11) where knockdown of Cdh-11 resulted in a significant inhibition of $p75^{NTR}$-mediated cell migration (Berghoff et al., 2015). These data suggest that $p75^{NTR}$ may mediate glioblastoma cell migration through both γ-secretase-dependent and -independent mechanisms; however the full complexity of these diverse signaling paradigms will need to be determined.

To identify the $p75^{NTR}$ downstream signaling pathways that mediate glioma tumor progression, a functional analysis using a series of truncation, deletion, and point mutants that disrupt key effector regions of the protein was performed. Such a mutational approach is possible with $p75^{NTR}$ as the receptor itself is devoid of any intrinsic catalytic activity and, instead, requires association with other proteins for signaling. Structurally $p75^{NTR}$ has a cysteine-rich (CRD) extracellular domain (ECD), a single transmembrane domain (TM), and a 152-amino acid cytoplasmic domain (ICD) that contains at least three recognized functional domains: the juxtamembrane region (JM), the death domain (DD), and PDZ-binding domain. Data from a number of cell systems have identified signaling proteins that bind to specific regions of the p75^{NTR} molecule in a cell context-specific manner (Reichardt, 2006). This mutational approach revealed that the extreme C-terminal PDZ-binding motif of p75^{NTR} was required for invasion. Using a peptide-based affinity strategy utilizing the PDZ-binding motif of p75^{NTR}, PDLIM1, a protein involved in cytoskeleton regulation was identified as a novel signaling adaptor for p75^{NTR} (Ahn et al., 2015). PDLIM1 was found to interact with p75^{NTR} in highly invasive p75^{NTR}-expressing human glioma cells and patient-derived BTICs (Fig. 4), and conversely, downregulation of PDLIM1 or the complete loss of the PDZ-binding motif (ΔSPV) resulted in abrogation of invasion (Ahn et al., 2015). Moreover, this study highlighted the regulated nature of this interaction as prevention of phosphorylation on serine 303 located in the cytoplasmic tail of p75^{NTR} by pharmacological inhibition of the cAMP-dependent protein kinase (PKA), or by a mutational strategy (S303G), crippled

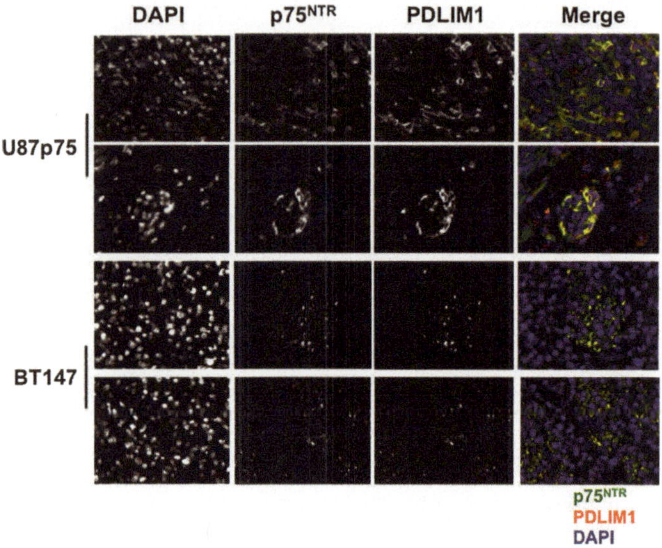

Fig. 4 PDLIM1 and p75^{NTR} colocalize in vivo. Whole-brain sections from animals implanted with human U87 glioma cells expressing p75^{NTR} (U87-p75^{NTR}) or patient-derived BTICs (BT147) were costained with anti-p75^{NTR} (*2nd column*) and anti-PDLIM1 (*3rd column*). Merged image (*right column*) shows strong colocalization of P75^{NTR} and PDLIM1 in vivo (*yellow*). Sections were also stained with DAPI (*left column*) to visualize the cell nuclei (*blue*).

p75^{NTR}-mediated glioma invasion and resulted in phosphorylation of serine 425 within the C-terminal PDZ-binding motif (SPV) (Ahn et al., 2015). As noted, there are a large number of signaling effectors that have been found to lie downstream of p75^{NTR} (Reichardt, 2006) but have not been discussed here based on the current lack of evidence for their role in high-grade glioma.

5.3 The Role of the Trk Receptors in Gliomagenesis

While there is certainly compelling data that p75^{NTR} mediates certain aspects of glioma tumor progression including regulating glioma invasion, a definitive role for the Trk receptors in glioma biology is not as evident. Clearly the Trk receptors play major roles in the developing and mature nervous system (Deinhardt & Chao, 2014), and thus it would not be surprising if they imparted specific functions during the disease process. There is a number of RNA- and protein-based studies that have revealed the differential expression of the Trk family of receptors in different CNS tumors (Assimakopoulou, Kondyli, Gatzounis, Maraziotis, & Varakis, 2007; Wang et al., 1998), but a consensus on whether Trk receptor expression correlates with histological stage or clinical outcome is lacking (Wang et al., 1998). It is noted using publicly available microarray data from the Rembrandt portal [National Cancer Institute (https://www.cancer.gov)] that contains data on 579 central nervous neoplasms that p75^{NTR} associates with high-grade glioblastoma and poorer overall survival of glioma patients (Fig. 3). TrkA, TrkB, and TrkC levels did not associate with patient outcomes based on the data available. There is some data albeit in vitro that suggest that Trk expression may induce differentiation of glioma cells (Kokunai, Iguchi, & Tamaki, 1999) which would be consistent with a role in neuronal differentiation, a function originally attributed to the Trk receptors (Kaplan, Martin-Zanca, & Parada, 1991). Consistent with this, enforced expression of the Trk receptors has been shown to decrease in vivo aggressiveness of C6 rat glioma cells (Lachyankar et al., 1997) and TrkA was found to be expressed by infiltrative cells extending from the main tumor mass and at the tumor margin when compared to the noninvasive cells within the tumor (Edwards et al., 2011). However this correlation between Trk expression and "tumor suppression" maybe context specific since TrkA when in a complex with integrin β1 (ITGB1) causes activation of ZEB-1 transcriptional repressor, resulting in E-cadherin suppression and an increase in the invasiveness of malignant glioma (Edwards et al., 2011).

Recently, Lawn et al. reported that some patient-derived BTICs express TrkB and TrkC and their activation with BDNF and NT3 enhances BTIC viability and neurosphere formation in vitro, effects that were abrogated by pharmacological inhibitors of Trk signaling or TrkB and TrkC knockdown (Lawn et al., 2015). Clearly more in vivo approaches need to be pursued to fully appreciate the role of the Trks in glioma tumorigenesis.

5.4 Neurotrophins and Their Involvement in the Glioma Microenvironment

It is now clear that tumor formation involves the complex interaction of neoplastic cells together with the ECM, tumor vasculature, and normal stromal compartment including immune cells (Hambardzumyan & Bergers, 2015; Hanahan & Coussens, 2012; Quail & Joyce, 2013; Shiao, Ganesan, Rugo, & Coussens, 2011). The presence of diverse cell types with different "activated" states within tumors can dramatically impact therapeutic response and be a contributor to therapeutic resistance (Joyce, 2005; Junttila & de Sauvage, 2013; Ruffell & Coussens, 2015). With respect to brain tumors the brain has a unique composition of cell types including neuronal and glial progenitors, neurons, astrocytes, oligodendrocytes, microglia/macrophages, and brain endothelial cells all of which have been shown in some capacity to impact glioma growth and invasion. In addition these cell types comprise a number of defined niches that have been correlated with brain tumorigenesis including stem cell niches (e.g., subventricular zone) and perivascular niches. Often these nonneoplastic cell compartments can aid and abet by providing factors that promote survival, proliferation, and the invasive behavior of the tumor cells.

Tumor-associated macrophages have been linked with high tumor grade and poor prognosis in many cancers including glioma (Bingle et al., 2002; Hussain, Yang, Suki, Aldape, et al., 2006; Hussain, Yang, Suki, Grimm, & Heimberger, 2006; Komohara et al., 2008). The presence of microglia (brain-resident macrophages) and infiltrated macrophages throughout glioma tissue was first observed by Wilder Penfield in 1925 (Penfield, 1925). While the weighted importance of the resident microglia and the blood-borne macrophages in glioma tumorigenesis is under debate, there is a substantial literature on the interactions of these cells within glioma (da Fonseca & Badie, 2013; Wei, Gabrusiewicz, & Heimberger, 2013) including the ability of the glioma to suppress the immune surveillance functions (Wu et al., 2010; Yang, Han, Kaur, Crane, & Parsa, 2010). Depending on the context, microglia/macrophages have a protumorigenic (Bettinger,

Thanos, & Paulus, 2002; Dziurzynski et al., 2011; Kerber et al., 2008; Markovic et al., 2009; Platten et al., 2003; Simmons et al., 2011; Zhai, Heppner, & Tsirka, 2011; Zhang et al., 2012) or antitumorigenic activity (Brantley et al., 2010; Hoa et al., 2007; Hwang et al., 2009; Mora et al., 2009; Nakagawa et al., 2007). The ability to tailor conditions where microglia/macrophages can inhibit tumor growth is highlighted by a recent study demonstrating that microglia and tissue-infiltrated macrophages can be "programmed" by amphotericin B to suppress tumor growth and improve overall survival in patient-derived intracranial (orthotopic) xenograft models (Sarkar et al., 2014). To further complicate this intricate interplay, macrophages and microglia are functionally defined into two major subsets: classically activated (M1) and alternatively activated (M2) macrophages. The M1 phenotype produces proinflammatory cytokines (such as IL-1, IL-12, iNOS, NO, and TNFα), while the M2 phenotype produces cytokines with antiinflammatory properties (IL-10, TGFβ, and IL-1 receptor antagonist) [reviewed in Wei et al. (2013)]. In situ, many gliomas are characterized by the predominance of tumor-promoting M2 macrophages (based on markers such as arginase, CD163 and CD204), while some in vitro studies suggest M1 macrophages (based on markers such as iNOS) reduce the proliferative capacity of some cancers (Wei et al., 2013). A recent study from Pyonteck et al. (2013) is consistent with the presence of tumor-promoting M2 macrophages and provides evidence that inhibition of the CSF-1R pathway can alter macrophage polarization and convert M2 macrophages to an M1 phenotype that results in suppression of glioma growth.

Inflammation is a major component of the tumor microenvironment (Mantovani, Marchesi, Portal, Allavena, & Sica, 2008), and malignant glioma are characterized by the presence of extensive hypoxic and necrotic areas that trigger recruitment of immune cells and tumor progression by releasing factors, including neurotrophins (Hambardzumyan & Bergers, 2015). Microglia/macrophages have been shown to express neurotrophins in vitro and in the cerebral cortex, globus pallidus, and medulla of the brain (Elkabes, DiCicco-Bloom, & Black, 1996), and stimulation of microglia with lipopolysaccharide induces both BDNF and NGF secretion (Nakajima et al., 2001). Protein expression studies have revealed that neurotrophin receptors TrkA, B, and C as well as $p75^{NTR}$ are expressed by microglia in vitro (Nakajima et al., 1998), and microglia stimulation by neurotrophins results in the phosphorylation of TrkA and TrkB, activation of the NF-$_{k}$B pathway, and increased plasminogen secretion. These effects were abrogated using the protein kinase inhibitor, K252a, suggesting that neurotrophins signaling through Trk receptors regulate microglia function (Nakajima et al., 1998). Furthermore, microglia treatment with neurotrophins BDNF, NGF, and NT3 increased their proliferation and induced their phagocytic activity (Elkabes et al., 1996). Activated microglia also secrete interleukin-6 (IL-6) that induces neurotrophin secretion by astrocytes (Otten et al., 2000), suggesting that neurotrophins can be secreted in the brain as a result of cross talk between immune cells and brain cells. Collectively, these findings suggest that neurotrophins mediate communication between glioma cells and immune cells in order to promote glioma progression. This area of investigation is rapidly evolving and will almost undoubtedly reveal unique roles for the neurotrophin signaling systems in modulating the microenvironment to fuel brain tumor growth and potentially contribute to the lack of a durable therapeutic response.

6. A ROLE FOR $p75^{NTR}$ IN CANCER: EXTENDING BEYOND GLIOMA

Similar to glioma cells, $p75^{NTR}$ has also been implicated in the invasive ability of other cells including Schwann cells, neural crest cells, and melanocytes (Anton, Weskamp, Reichardt, & Matthew, 1994; Carter et al., 1996; Herrmann et al., 1993; Menter, Herrmann, Marchetti, & Nicolson, 1994; Shonukan, Bagayogo, McCrea, Chao, & Hempstead, 2003; Stemple & Anderson, 1992). In addition, a more global role for $p75^{NTR}$ in cancer tumorigenesis is emerging including an association in melanoma, thyroid, and breast cancer (Marchetti, Aucoin, Blust, Murry, & Greiter-Wilke, 2004; Menter et al., 1994; Rocha et al., 2006; Verbeke et al., 2010). The role for $p75^{NTR}$ in cancer was first observed soon after its initial discovery when NGF treatment of human melanoma cells was shown to enhance growth and survival (Fabricant, De Larco, & Todaro, 1977). Subsequently $p75^{NTR}$ was found to aid the metastatic progression of melanoma and specifically the development of brain metastasis (Herrmann et al., 1993; Menter et al., 1994; Shonukan et al., 2003) where NGF, NT3, or pro-NGF was found to be potent chemotactic agents (Fabricant et al., 1977; Shonukan et al., 2003). In addition, similar to the identification of $p75^{NTR}$ in patient-derived BTICs (Ahn et al., 2015; Johnston et al., 2007; Wang et al., 2008) $p75^{NTR}$ (a.k.a. CD271) was detected in melanoma-initiating cells where it was recognized as a cancer stem cell marker (Boiko et al., 2010) where the $p75^{NTR}$-positive cells were found to initiate and maintain tumor growth in vivo when transplanted into immunocompromised mouse models (Boiko et al., 2010; Redmer et al., 2014). These cells were more metastatic and found to retain the capacity to establish similar tumor heterogeneity to the original tumor (Boiko et al., 2010; Civenni et al., 2011). Based on these data and the fact that $p75^{NTR}$ is expressed on a diverse array of stem cells [reviewed in Tomellini, Lagadec, Polakowska, and Le Bourhis (2014)] investigators have now evaluated its presence on different populations of cancer cells with tumor-initiating properties (Chopin, Lagadec, Toillon, & Le Bourhis, 2016). In addition to melanoma this marker has been found on tumor-initiating cells from breast (Kim et al., 2012), esophageal squamous cell carcinomas (Huang, Sarkaria, Cloughesy, & Mischel, 2009; Okumura, Shimada, Imamura, & Yasumoto, 2003), hypopharyngeal carcinoma (Imai et al., 2013), squamous cell carcinoma of the head and neck (Murillo-Sauca et al., 2014), neuroblastoma (Biagiotti et al., 2006), and osteosarcoma (Tian, Li, Si, Liu, & Li, 2014) and in many cases its expression has been correlated with poorer clinical outcome. The discovery however that $p75^{NTR}$-negative populations of melanoma cells have similar tumor-initiating ability suggests that $p75^{NTR}$ is not an exclusive hallmark of cancer stem cells (Cheli et al., 2014; Quintana et al., 2008). Similar to other "cancer stem cell" markers there appears to be some discrepancy in the literature as to their importance, which may be explained by phenotypic plasticity, a concept that is currently being explored in the field (Marjanovic, Weinberg, & Chaffer, 2013). In addition there is some support for a role of $p75^{NTR}$ in tumor suppression. In the setting of gastric, bladder, and prostate cancer, $p75^{NTR}$ has been proposed to act as a tumor suppressor by blocking cell cycle progression and inducing apoptosis (Jin, Pan, He, et al., 2007; Jin, Pan, Zhao, et al., 2007; Khwaja, Tabassum, Allen, & Djakiew, 2006; Tabassum, Khwaja, & Djakiew, 2003). However as in the case of glioma, in the majority of tumors investigated thus far, $p75^{NTR}$ tends to promote tumorigenesis. The ability of $p75^{NTR}$ to suppress or promote tumor growth continues to highlight the differences in neurotrophin signaling based on the cellular context in which it is deciphered.

7. CONCLUSIONS AND FUTURE DIRECTIONS

In this chapter, we summarize the role of neurotrophins and their receptors with detailed focus on the $p75^{NTR}$ receptor signaling system in the progression of malignant glioma and highlight the emerging evidence of neurotrophin signaling in fueling proliferation, mediating invasion, and modulating interaction between glioma cells and immune cells in the tumor microenvironment. Indeed, further studies are required to uncover the downstream signaling pathways of neurotrophin receptors p75^{NTR}, TrkA, TrkB, and TrkC, involved in glioma progression. Accumulating evidence suggests that targeting the neurotrophin signaling pathways may offer a new therapeutic strategy against treatment resistance and tumor recurrence. In general one can consider the cell autonomous effects of signaling downstream of a neurotrophin receptor as one signaling axis and the noncell autonomous effects of neurotrophin signaling as another, concepts that will be discussed here in the context of p75^{NTR} and its role in glioma progression and potential therapeutic opportunity.

A detailed model of p75^{NTR}-mediated signaling with respect to glioma invasion is presented in Fig. 5. Briefly, upon ligand binding there is the regulated phosphorylation of the cytoplasmic tail of p75^{NTR} that facilitates the translocation to specific membrane microdomains and a regulated interaction with the PDZ domain-containing protein PDLIM1. After an initial

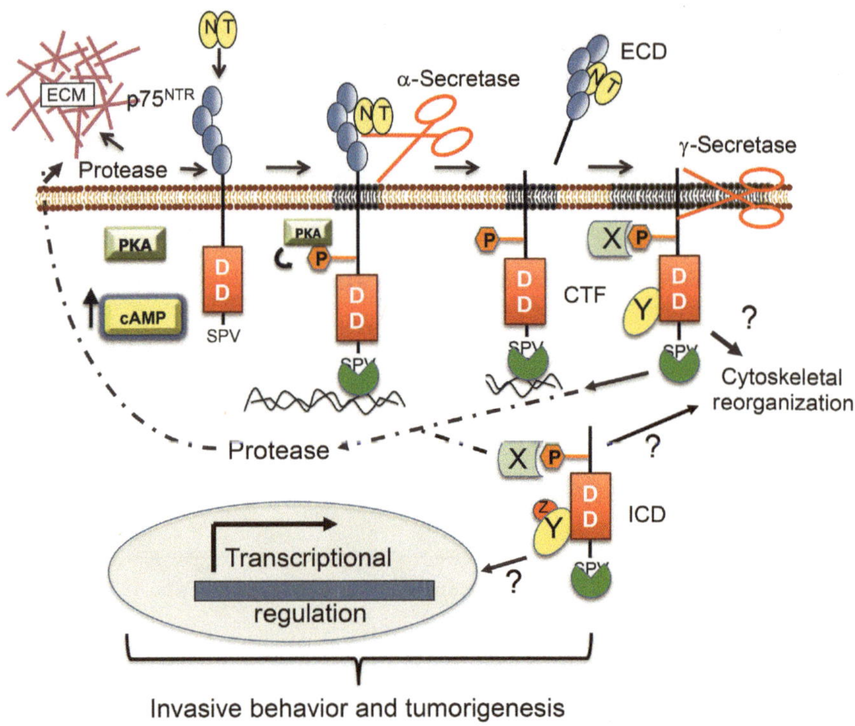

Fig. 5 Model of p75^{NTR}-mediated glioma invasion.

cleavage of the ECD of p75^{NTR} by a member of the α-secretase family, p75^{NTR} is further processed by γ-secretase to release the ICD fragment, a required event to promote glioma invasion. The main outstanding question with respect to this model focuses on the underlying molecular mechanism by which the ICD fragment mediates the increase in migratory phenotype, either by modulating processes within the cytoplasm and/or nucleus and could functionally involve multiple cellular locales. In addition details around the mechanics of p75^{NTR} activation by ligands also remain unanswered which could include conformational changes in the receptor by remodeling of disulfide-linked receptor subunits to elicit ligand-specific signaling events (Vilar, Charalampopoulos, Kenchappa, Reversi, et al., 2009; Vilar, Charalampopoulos, Kenchappa, Simi, et al., 2009). Further detailed mechanistic insight into the downstream signaling from p75^{NTR} via interaction with its respective cognate ligands will be required to fully appreciate the signaling pathways and opportunity for therapeutic intervention.

While we proposed that the γ-secretase-mediated proteolysis of the p75^{NTR} receptor represents a logical therapeutic target for consideration for glioblastoma due to the development of clinically applicable inhibitors (Wang et al., 2008), it is not without its drawbacks. In addition to p75^{NTR}, γ-secretase has over 100 cellular targets (e.g., Notch, Erb4, E-cadherin, CD44, amyloid precursor protein), and to date these inhibitors not surprisingly have shown adverse effects on the intestinal lining of the gut (Henley, Sundell, Sethuraman, Dowsett, & May, 2014). The unwanted side effects are further highlighted by the fact that the Eli Lilly pharmaceutical company halted the development of semagacestat (γ-secretase inhibitor) due to preliminary results from ongoing long-term phase III studies for patients with Alzheimer's disease where they found a decline in cognitive function in patients given the inhibitor (Doody et al., 2013; Henley et al., 2014). In addition, the drug was associated with a greater risk of skin lesions including skin cancers (Doody et al., 2013; Henley et al., 2014; Hsu, Hsu, Lee, Kuo, & Pai, 2013). Newer strategies to define specific inhibitors for γ-secretase substrate specificity are being identified and pursued (Golde, Koo, Felsenstein, Osborne, & Miele, 2013) including in the setting of Alzheimer's disease (He et al., 2010), suggesting a future approach to consider for p75^{NTR}-mediated signaling in the setting of glioma.

Whether the regulated interaction of p75^{NTR} with PDLIM1 represents a therapeutic target remains the topic of current investigations. The development of reversible small-molecule inhibitors that target neuron-specific

PDZ domains has been proposed as a means to specifically target compartmentalized signaling within the CNS (Blazer & Neubig, 2009). We would therefore propose that targeting PDZ domain interactions in glioma with specific small-molecule inhibitors might provide a more subtle tissue- and receptor-specific therapeutic approach. This approach would not be limited to glioma and may have applicability in other metastatic melanoma and breast cancers where $p75^{NTR}$ and/or PDLIM1 have also been implicated (Boiko et al., 2010; Herrmann et al., 1993; Menter et al., 1994; Shonukan et al., 2003; Verbeke et al., 2010, 2013).

There has also been considerable interest surrounding the concept that the proliferating glioma disease reservoir and the slowly dividing invasive glioma disease reservoir may be in very different metabolic states. Whether $p75^{NTR}$ or neurotrophin signaling in general is involved in regulating metabolism in glioma is unknown, but it is interesting to note that $p75^{NTR}$ has been shown to control energy expenditure in the setting of obesity via the regulation of cAMP levels by PKA (Baeza-Raja et al., 2016), an interesting similar association to what we have seen with $p75^{NTR}$ and PDLIM1.

There is certainly an emerging interest in the noncell autonomous functions of the neurotrophins in regulating the complexity of the brain tumor microenvironment that includes the cellular composition such as the presence of Iba-1 positive cells (macrophages and microglia) as well as the noncellular composition that includes changes in the composition and rigidity of the tumor microenvironment (Fig. 2). In this regard it has recently been shown that extracellular vesicles commonly referred to as exosomes can be released within the glioma microenvironment that carry bioactive cargo that could directly modulate tumorigenesis including the transfer of oncogenic proteins such as the EGFRvIII, proteases, and microRNAs (Al-Nedawi et al., 2008; Al-Nedawi, Meehan, & Rak, 2009; D'Asti et al., 2012; Garnier, Jabado, & Rak, 2013; Lee et al., 2014; Montermini et al., 2015). This transfer of bioactive molecules can occur between tumor cells or between tumor cells and the normal cells within the environment. While the literature is scant with respect to the presence of the neurotrophins and/or their receptors in exosomes, there is one study that suggests exosomes isolated from irradiated glioma cells increase the migration and invasion of recipient glioma cells at least in part by the activation of the TrkA receptor (Arscott et al., 2013). While in its infancy we anticipate that investigation in this area will increase, especially considering that $p75^{NTR}$ has been shown to traffic to multivesicular bodies specialized for exosomal release in neuronal cells (Escudero et al., 2014).

We would propose that more research should be conducted to investigate the role of neurotrophin signaling in recruiting and possible reeducating immune cells within the glioma microenvironment. The presence of diverse cell types with different "activated" states can dramatically impact and shape therapeutic response and contribute to therapeutic resistance (Junttila & de Sauvage, 2013). This novel approach to therapeutic targeting of glioma is highlighted by two recent studies. First, Pyonteck et al. targeted glioma-associated macrophages by crippling the CSF-1 signaling axis in the absence of a chemotherapeutic drug and found a decrease in tumor burden with an overall increase in survival using a specific mouse model of glioma (Pyonteck et al., 2013). Second, microglia and tumor-recruited macrophages could be pharmacologically activated to produce factors that suppress the growth of human glioma cells in vitro and significantly improve overall survival in orthotopic human glioma-bearing animals (Sarkar et al., 2014). Glioma is inevitably a fatal disease with very few survivors, a statistic that has not changed dramatically in the past 30 years despite a deeper molecular understanding of the disease. Whether the neurotrophins and their receptors will be an answer to this challenge remains unknown but more mechanistic insight into their specific functions in glioma progression is certainly warranted, an opinion shared by these authors.

ACKNOWLEDGMENTS

The authors wish to thank Dr. Bo-Young Ahn and Dr. Jennifer Rahn for providing some of the images represented in Fig. 2 and to Dr. Astrid de Boeck for analyzing the publicly available databases for clinical outcome data regarding the neurotrophin receptors. M.M.A. acknowledges studentship support from the King Abdullah International Medical Research Center. Research referenced by Robbins' and Senger laboratories was supported by funds from the Canadian Institutes of Health Research, the Terry Fox Research Foundation Pan-Canadian Team Grant, and Alberta Innovates Health Solutions.

REFERENCES

Ahn, B. Y., Saldanha-Gama, R. F., Rahn, J. J., Hao, X., Zhang, J., Dang, N. H., … Senger, D. L. (2015). Glioma invasion mediated by the p75 neurotrophin receptor (p75(NTR)/CD271) requires regulated interaction with PDLIM1. *Oncogene*, *35*(11), 1411–1422. http://dx.doi.org/10.1038/onc.2015.199.

Al-Nedawi, K., Meehan, B., Micallef, J., Lhotak, V., May, L., Guha, A., & Rak, J. (2008). Intercellular transfer of the oncogenic receptor EGFRvIII by microvesicles derived from tumour cells. *Nature Cell Biology*, *10*(5), 619–624. http://dx.doi.org/10.1038/ncb1725.

Al-Nedawi, K., Meehan, B., & Rak, J. (2009). Microvesicles: Messengers and mediators of tumor progression. *Cell Cycle*, *8*(13), 2014–2018.

Anton, E. S., Weskamp, G., Reichardt, L. F., & Matthew, W. D. (1994). Nerve growth factor and its low-affinity receptor promote Schwann cell migration. *Proceedings of the National Academy of Sciences of the United States of America*, *91*(7), 2795–2799.

Appin, C. L., & Brat, D. J. (2015). Molecular pathways in gliomagenesis and their relevance to neuropathologic diagnosis. *Advances in Anatomic Pathology*, *22*(1), 50–58. http://dx.doi.org/10.1097/PAP.0000000000000048.

Arscott, W. T., Tandle, A. T., Zhao, S., Shabason, J. E., Gordon, I. K., Schlaff, C. D., ... Camphausen, K. A. (2013). Ionizing radiation and glioblastoma exosomes: Implications in tumor biology and cell migration. *Translational Oncology*, *6*(6), 638–648.

Assimakopoulou, M., Kondyli, M., Gatzounis, G., Maraziotis, T., & Varakis, J. (2007). Neurotrophin receptors expression and JNK pathway activation in human astrocytomas. *BMC Cancer*, 7, 202. http://dx.doi.org/10.1186/1471-2407-7-202.

Baeza-Raja, B., Sachs, B. D., Li, P., Christian, F., Vagena, E., Davalos, D., ... Akassoglou, K. (2016). p75 neurotrophin receptor regulates energy balance in obesity. *Cell Reports*, *14*(2), 255–268. http://dx.doi.org/10.1016/j.celrep.2015.12.028.

Bao, S., Wu, Q., McLendon, R. E., Hao, Y., Shi, Q., Hjelmeland, A. B., ... Rich, J. N. (2006). Glioma stem cells promote radioresistance by preferential activation of the DNA damage response. *Nature*, *444*(7120), 756–760. http://dx.doi.org/10.1038/nature05236.

Bao, S., Wu, Q., Sathornsumetee, S., Hao, Y., Li, Z., Hjelmeland, A. B., ... Rich, J. N. (2006). Stem cell-like glioma cells promote tumor angiogenesis through vascular endothelial growth factor. *Cancer Research*, *66*(16), 7843–7848. http://dx.doi.org/10.1158/0008-5472.CAN-06-1010. 66/16/7843 [pii].

Barker, P. A. (2004). p75NTR is positively promiscuous: Novel partners and new insights. *Neuron*, *42*(4), 529–533.

Barker, P. A., Barbee, G., Misko, T. P., & Shooter, E. M. (1994). The low affinity neurotrophin receptor, p75LNTR, is palmitoylated by thioester formation through cysteine 279. *The Journal of Biological Chemistry*, *269*(48), 30645–30650.

Beadle, C., Assanah, M. C., Monzo, P., Vallee, R., Rosenfeld, S. S., & Canoll, P. (2008). The role of myosin II in glioma invasion of the brain. *Molecular Biology of the Cell*, *19*(8), 3357–3368. http://dx.doi.org/10.1091/mbc.E08-03-0319.

Berghoff, A. S., Lassmann, H., Preusser, M., & Hoftberger, R. (2013). Characterization of the inflammatory response to solid cancer metastases in the human brain. *Clinical & Experimental Metastasis*, *30*(1), 69–81. http://dx.doi.org/10.1007/s10585-012-9510-4.

Berghoff, J., Jaisimha, A. V., Duggan, S., MacSharry, J., & McCarthy, J. V. (2015). Gamma-secretase-independent role for cadherin-11 in neurotrophin receptor p75 (p75(NTR)) mediated glioblastoma cell migration. *Molecular and Cellular Neurosciences*, *69*, 41–53. http://dx.doi.org/10.1016/j.mcn.2015.10.003.

Berkemeier, L. R., Winslow, J. W., Kaplan, D. R., Nikolics, K., Goeddel, D. V., & Rosenthal, A. (1991). Neurotrophin-5: A novel neurotrophic factor that activates trk and trkB. *Neuron*, 7(5), 857–866.

Bettinger, I., Thanos, S., & Paulus, W. (2002). Microglia promote glioma migration. *Acta Neuropathologica*, *103*(4), 351–355. http://dx.doi.org/10.1007/s00401-001-0472-x.

Biagiotti, T., D'Amico, M., Marzi, I., Di Gennaro, P., Arcangeli, A., Wanke, E., & Olivotto, M. (2006). Cell renewing in neuroblastoma: Electrophysiological and immunocytochemical characterization of stem cells and derivatives. *Stem Cells*, *24*(2), 443–453. http://dx.doi.org/10.1634/stemcells.2004-0264.

Bilderback, T. R., Gazula, V. R., Lisanti, M. P., & Dobrowsky, R. T. (1999). Caveolin interacts with Trk A and p75(NTR) and regulates neurotrophin signaling pathways. *The Journal of Biological Chemistry*, *274*(1), 257–263.

Bilderback, T. R., Grigsby, R. J., & Dobrowsky, R. T. (1997). Association of p75(NTR) with caveolin and localization of neurotrophin-induced sphingomyelin hydrolysis to caveolae. *The Journal of Biological Chemistry*, *272*(16), 10922–10927.

Bingle, L., Brown, N. J., & Lewis, C. E. (2002). The role of tumour-associated macrophages in tumour progression: Implications for new anticancer therapies. *The Journal of Pathology*, *196*(3), 254–265. http://dx.doi.org/10.1002/path.1027.

Birnbaum, T., Roider, J., Schankin, C. J., Padovan, C. S., Schichor, C., Goldbrunner, R., & Straube, A. (2007). Malignant gliomas actively recruit bone marrow stromal cells by secreting angiogenic cytokines. *Journal of Neuro-Oncology*, *83*(3), 241–247. http://dx.doi.org/10.1007/s11060-007-9332-4.

Blazer, L. L., & Neubig, R. R. (2009). Small molecule protein-protein interaction inhibitors as CNS therapeutic agents: Current progress and future hurdles. *Neuropsychopharmacology*, *34*(1), 126–141. http://dx.doi.org/10.1038/npp.2008.151.

Bleau, A. M., Huse, J. T., & Holland, E. C. (2009). The ABCG2 resistance network of glioblastoma. *Cell Cycle*, *8*(18), 2936–2944. doi:9504 [pii].

Blough, M. D., Beauchamp, D. C., Westgate, M. R., Kelly, J. J., & Cairncross, J. G. (2011). Effect of aberrant p53 function on temozolomide sensitivity of glioma cell lines and brain tumor initiating cells from glioblastoma. *Journal of Neuro-Oncology*, *102*(1), 1–7. http://dx.doi.org/10.1007/s11060-010-0283-9.

Boiko, A. D., Razorenova, O. V., van de Rijn, M., Swetter, S. M., Johnson, D. L., Ly, D. P., ... Weissman, I. L. (2010). Human melanoma-initiating cells express neural crest nerve growth factor receptor CD271. *Nature*, *466*(7302), 133–137. http://dx.doi.org/10.1038/nature09161.

Bonnet, D., & Dick, J. E. (1997). Human acute myeloid leukemia is organized as a hierarchy that originates from a primitive hematopoietic cell. *Nature Medicine*, *3*(7), 730–737.

Brantley, E. C., Guo, L., Zhang, C., Lin, Q., Yokoi, K., Langley, R. R., ... Fidler, I. J. (2010). Nitric oxide-mediated tumoricidal activity of murine microglial cells. *Translational Oncology*, *3*(6), 380–388.

Brennan, C. W., Verhaak, R. G., McKenna, A., Campos, B., Noushmehr, H., Salama, S. R., ... Chin, L. (2013). The somatic genomic landscape of glioblastoma. *Cell*, *155*(2), 462–477. http://dx.doi.org/10.1016/j.cell.2013.09.034. S0092-8674(13)01208-7 [pii].

Brown, M. C., Staniszewska, I., Lazarovici, P., Tuszynski, G. P., Del Valle, L., & Marcinkiewicz, C. (2008). Regulatory effect of nerve growth factor in alpha9beta1 integrin-dependent progression of glioblastoma. *Neuro-Oncology*, *10*(6), 968–980. http://dx.doi.org/10.1215/15228517-2008-047.

Calabrese, C., Poppleton, H., Kocak, M., Hogg, T. L., Fuller, C., Hamner, B., ... Gilbertson, R. J. (2007). A perivascular niche for brain tumor stem cells. *Cancer Cell*, *11*(1), 69–82. http://dx.doi.org/10.1016/j.ccr.2006.11.020.

Cancer Genome Atlas Research Network. (2008). Comprehensive genomic characterization defines human glioblastoma genes and core pathways. *Nature*, *455*(7216), 1061–1068. http://dx.doi.org/10.1038/nature07385.

Cancer Genome Atlas Research Network. (2013). Corrigendum: Comprehensive genomic characterization defines human glioblastoma genes and core pathways. *Nature*, *494*(7438), 506. http://dx.doi.org/10.1038/nature11903.

Carter, B. D., Kaltschmidt, C., Kaltschmidt, B., Offenhauser, N., Bohm-Matthaei, R., Baeuerle, P. A., & Barde, Y. A. (1996). Selective activation of NF-kappa B by nerve growth factor through the neurotrophin receptor p75. *Science*, *272*(5261), 542–545.

Caspani, E. M., Echevarria, D., Rottner, K., & Small, J. V. (2006). Live imaging of glioblastoma cells in brain tissue shows requirement of actin bundles for migration. *Neuron Glia Biology*, *2*(2), 105–114. http://dx.doi.org/10.1017/S1740925X06000111.

Chao, M. V. (2003). Neurotrophins and their receptors: A convergence point for many signalling pathways. *Nature Reviews. Neuroscience*, *4*(4), 299–309.

Cheli, Y., Bonnazi, V. F., Jacquel, A., Allegra, M., De Donatis, G. M., Bahadoran, P., ... Ballotti, R. (2014). CD271 is an imperfect marker for melanoma initiating cells. *Oncotarget*, *5*(14), 5272–5283. http://dx.doi.org/10.18632/oncotarget.1967.

Chiaretti, A., Aloe, L., Antonelli, A., Ruggiero, A., Piastra, M., Riccardi, R., ... Di Rocco, C. (2004). Neurotrophic factor expression in childhood low-grade astrocytomas and ependymomas. *Child's Nervous System*, *20*(6), 412–419. http://dx.doi.org/10.1007/s00381-004-0959-6.

Chopin, V., Lagadec, C., Toillon, R. A., & Le Bourhis, X. (2016). Neurotrophin signaling in cancer stem cells. *Cellular and Molecular Life Sciences*, *73*(9), 1859–1870. http://dx.doi.org/10.1007/s00018-016-2156-7.

Civenni, G., Walter, A., Kobert, N., Mihic-Probst, D., Zipser, M., Belloni, B., … Sommer, L. (2011). Human CD271-positive melanoma stem cells associated with metastasis establish tumor heterogeneity and long-term growth. *Cancer Research*, *71*(8), 3098–3109. http://dx.doi.org/10.1158/0008-5472.CAN-10-3997.

Clarke, M. F., Dick, J. E., Dirks, P. B., Eaves, C. J., Jamieson, C. H., Jones, D. L., … Wahl, G. M. (2006). Cancer stem cells—Perspectives on current status and future directions: AACR Workshop on cancer stem cells. *Cancer Research*, *66*(19), 9339–9344.

Craven, S. E., & Bredt, D. S. (1998). PDZ proteins organize synaptic signaling pathways. *Cell*, *93*(4), 495–498. doi:S0092-8674(00)81179-4 [pii].

D'Abaco, G. M., & Kaye, A. H. (2007). Integrins: Molecular determinants of glioma invasion. *Journal of Clinical Neuroscience*, *14*(11), 1041–1048. http://dx.doi.org/10.1016/j.jocn.2007.06.019. S0967-5868(07)00379-7 [pii].

da Fonseca, A. C., & Badie, B. (2013). Microglia and macrophages in malignant gliomas: Recent discoveries and implications for promising therapies. *Clinical & Developmental Immunology*, *2013*, 264124. http://dx.doi.org/10.1155/2013/264124.

D'Asti, E., Garnier, D., Lee, T. H., Montermini, L., Meehan, B., & Rak, J. (2012). Oncogenic extracellular vesicles in brain tumor progression. *Frontiers in Physiology*, *3*, 294. http://dx.doi.org/10.3389/fphys.2012.00294.

Dechant, G., & Barde, Y. A. (2002). The neurotrophin receptor p75(NTR): Novel functions and implications for diseases of the nervous system. *Nature Neuroscience*, *5*(11), 1131–1136.

Deinhardt, K., & Chao, M. V. (2014). Trk receptors. *Handbook of Experimental Pharmacology*, *220*, 103–119. http://dx.doi.org/10.1007/978-3-642-45106-5_5.

Dolecek, T. A., Propp, J. M., Stroup, N. E., & Kruchko, C. (2012). CBTRUS statistical report: Primary brain and central nervous system tumors diagnosed in the United States in 2005–2009. *Neuro-Oncology*, *14*(Suppl. 5), v1–v49. http://dx.doi.org/10.1093/neuonc/nos218.

Domeniconi, M., Zampieri, N., Spencer, T., Hilaire, M., Mellado, W., Chao, M. V., & Filbin, M. T. (2005). MAG induces regulated intramembrane proteolysis of the p75 neurotrophin receptor to inhibit neurite outgrowth. *Neuron*, *46*(6), 849–855. http://dx.doi.org/10.1016/j.neuron.2005.05.029. S0896-6273(05)00476-9 [pii].

Doody, R. S., Raman, R., Farlow, M., Iwatsubo, T., Vellas, B., Joffe, S., … Semagacestat Study, G. (2013). A phase 3 trial of semagacestat for treatment of Alzheimer's disease. *The New England Journal of Medicine*, *369*(4), 341–350. http://dx.doi.org/10.1056/NEJMoa1210951.

Dunn, G. P., Rinne, M. L., Wykosky, J., Genovese, G., Quayle, S. N., Dunn, I. F., … Hahn, W. C. (2016). Emerging insights into the molecular and cellular basis of glioblastoma. *Genes & Development*, *26*(8), 756–784. http://dx.doi.org/10.1101/gad.187922.112.

Dziurzynski, K., Wei, J., Qiao, W., Hatiboglu, M. A., Kong, L. Y., Wu, A., … Heimberger, A. B. (2011). Glioma-associated cytomegalovirus mediates subversion of the monocyte lineage to a tumor propagating phenotype. *Clinical Cancer Research*, *17*(14), 4642–4649. http://dx.doi.org/10.1158/1078-0432.CCR-11-0414.

Edwards, L. A., Woolard, K., Son, M. J., Li, A., Lee, J., Ene, C., … Belova, G. (2011). Effect of brain-and tumor-derived connective tissue growth factor on glioma invasion. *Journal of the National Cancer Institute*, *103*(15), 1162–1178.

El-Husseini, A. E., Schnell, E., Chetkovich, D. M., Nicoll, R. A., & Bredt, D. S. (2000). PSD-95 involvement in maturation of excitatory synapses. *Science*, *290*(5495), 1364–1368. doi:9001 [pii].

Elkabes, S., DiCicco-Bloom, E. M., & Black, I. B. (1996). Brain microglia/macrophages express neurotrophins that selectively regulate microglial proliferation and function. *The Journal of Neuroscience*, *16*(8), 2508–2521.

Engebraaten, O., Bjerkvig, R., Pedersen, P. H., & Laerum, O. D. (1993). Effects of EGF, bFGF, NGF and PDGF(bb) on cell proliferative, migratory and invasive capacities of human brain-tumour biopsies in vitro. *International Journal of Cancer*, *53*(2), 209–214.

Ernfors, P., Ibanez, C. F., Ebendal, T., Olson, L., & Persson, H. (1990). Molecular cloning and neurotrophic activities of a protein with structural similarities to nerve growth factor: Developmental and topographical expression in the brain. *Proceedings of the National Academy of Sciences of the United States of America*, *87*(14), 5454–5458.

Escudero, C. A., Lazo, O. M., Galleguillos, C., Parraguez, J. I., Lopez-Verrilli, M. A., Cabeza, C., … Bronfman, F. C. (2014). The p75 neurotrophin receptor evades the endolysosomal route in neuronal cells, favouring multivesicular bodies specialised for exosomal release. *Journal of Cell Science*, *127*(Pt. 9), 1966–1979. http://dx.doi.org/10.1242/jcs.141754.

Fabricant, R. N., De Larco, J. E., & Todaro, G. J. (1977). Nerve growth factor receptors on human melanoma cells in culture. *Proceedings of the National Academy of Sciences of the United States of America*, *74*(2), 565–569.

Fischer, I., Gagner, J. P., Law, M., Newcomb, E. W., & Zagzag, D. (2005). Angiogenesis in gliomas: Biology and molecular pathophysiology. *Brain Pathology*, *15*(4), 297–310.

Forsyth, P. A., Krishna, N., Lawn, S., Valadez, J. G., Qu, X., Fenstermacher, D. A., … Kenchappa, R. S. (2014). p75 neurotrophin receptor cleavage by alpha- and gamma-secretases is required for neurotrophin-mediated proliferation of brain tumor-initiating cells. *The Journal of Biological Chemistry*, *289*(12), 8067–8085. http://dx.doi.org/10.1074/jbc.M113.513762.

Fortin Ensign, S. P., Mathews, I. T., Symons, M. H., Berens, M. E., & Tran, N. L. (2013). Implications of Rho GTPase signaling in glioma cell invasion and tumor progression. *Frontiers in Oncology*, *3*, 241. http://dx.doi.org/10.3389/fonc.2013.00241.

Fortini, M. E. (2009). Notch signaling: The core pathway and its posttranslational regulation. *Developmental Cell*, *16*(5), 633–647. http://dx.doi.org/10.1016/j.devcel.2009.03.010. S1534-5807(09)00131-2 [pii].

Frade, J. M., Rodriguez-Tebar, A., & Barde, Y. A. (1996). Induction of cell death by endogenous nerve growth factor through its p75 receptor. *Nature*, *383*(6596), 166–168.

Freije, W. A., Castro-Vargas, F. E., Fang, Z., Horvath, S., Cloughesy, T., Liau, L. M., … Nelson, S. F. (2004). Gene expression profiling of gliomas strongly predicts survival. *Cancer Research*, *64*(18), 6503–6510.

Fujitani, M., Kawai, H., Proia, R. L., Kashiwagi, A., Yasuda, H., & Yamashita, T. (2005). Binding of soluble myelin-associated glycoprotein to specific gangliosides induces the association of p75NTR to lipid rafts and signal transduction. *Journal of Neurochemistry*, *94*(1), 15–21.

Furnari, F. B., Fenton, T., Bachoo, R. M., Mukasa, A., Stommel, J. M., Stegh, A., … Cavenee, W. K. (2007). Malignant astrocytic glioma: Genetics, biology, and paths to treatment. *Genes & Development*, *21*(21), 2683–2710. http://dx.doi.org/10.1101/gad.1596707.

Galli, R., Binda, E., Orfanelli, U., Cipelletti, B., Gritti, A., De Vitis, S., … Vescovi, A. (2004). Isolation and characterization of tumorigenic, stem-like neural precursors from human glioblastoma. *Cancer Research*, *64*(19), 7011–7021.

Garnier, D., Jabado, N., & Rak, J. (2013). Extracellular vesicles as prospective carriers of oncogenic protein signatures in adult and paediatric brain tumours. *Proteomics*, *13*(10–11), 1595–1607. http://dx.doi.org/10.1002/pmic.201200360.

Giese, A., Bjerkvig, R., Berens, M. E., & Westphal, M. (2003). Cost of migration: Invasion of malignant gliomas and implications for treatment. *Journal of Clinical Oncology*, *21*(8), 1624–1636.

Giese, A., & Westphal, M. (1996). Glioma invasion in the central nervous system. *Neurosurgery, 39*(2), 235–250. discussion 250–232.
Gil, C., Cubi, R., & Aguilera, J. (2007). Shedding of the p75NTR neurotrophin receptor is modulated by lipid rafts. *FEBS Letters, 581*(9), 1851–1858. http://dx.doi.org/10.1016/j.febslet.2007.03.080. S0014-5793(07)00358-4 [pii].
Glas, M., Rath, B. H., Simon, M., Reinartz, R., Schramme, A., Trageser, D., ... Scheffler, B. (2010). Residual tumor cells are unique cellular targets in glioblastoma. *Annals of Neurology, 68*(2), 264–269. http://dx.doi.org/10.1002/ana.22036.
Golde, T. E., Koo, E. H., Felsenstein, K. M., Osborne, B. A., & Miele, L. (2013). gamma-Secretase inhibitors and modulators. *Biochimica et Biophysica Acta, 1828*(12), 2898–2907. http://dx.doi.org/10.1016/j.bbamem.2013.06.005.
Gondi, C. S., Lakka, S. S., Dinh, D. H., Olivero, W. C., Gujrati, M., & Rao, J. S. (2004). RNAi-mediated inhibition of cathepsin B and uPAR leads to decreased cell invasion, angiogenesis and tumor growth in gliomas. *Oncogene, 23*(52), 8486–8496.
Gondi, C. S., Lakka, S. S., Yanamandra, N., Olivero, W. C., Dinh, D. H., Gujrati, M., ... Rao, J. S. (2004). Adenovirus-mediated expression of antisense urokinase plasminogen activator receptor and antisense cathepsin B inhibits tumor growth, invasion, and angiogenesis in gliomas. *Cancer Research, 64*(12), 4069–4077. http://dx.doi.org/10.1158/0008-5472.CAN-04-1243.
Gritsenko, P. G., Ilina, O., & Friedl, P. (2012). Interstitial guidance of cancer invasion. *The Journal of Pathology, 226*(2), 185–199. http://dx.doi.org/10.1002/path.3031.
Gunia, S., Hussein, S., Radu, D. L., Putz, K. M., Breyer, R., Hecker, H., ... Stan, A. C. (1999). CD44s-targeted treatment with monoclonal antibody blocks intracerebral invasion and growth of 9 L gliosarcoma. *Clinical & Experimental Metastasis, 17*(3), 221–230.
Günther, H. S., Schmidt, N. O., Phillips, H. S., Kemming, D., Kharbanda, S., Soriano, R., ... Lamszus, K. (2008). Glioblastoma-derived stem cell-enriched cultures from distinct subgroups according to molecular and phenotypic criteria. *Oncogene, 27*(20), 2897–2909.
Hallbook, F., Ibanez, C. F., & Persson, H. (1991). Evolutionary studies of the nerve growth factor family reveal a novel member abundantly expressed in Xenopus ovary. *Neuron, 6*(5), 845–858.
Hambardzumyan, D., & Bergers, G. (2015). Glioblastoma: Defining tumor niches. *Trends Cancer, 1*(4), 252–265. http://dx.doi.org/10.1016/j.trecan.2015.10.009.
Hamel, W., Westphal, M., Szonyi, E., Escandon, E., & Nikolics, K. (1993). Neurotrophin gene expression by cell lines derived from human gliomas. *Journal of Neuroscience Research, 34*(2), 147–157. http://dx.doi.org/10.1002/jnr.490340202.
Hanahan, D., & Coussens, L. M. (2012). Accessories to the crime: Functions of cells recruited to the tumor microenvironment. *Cancer Cell, 21*(3), 309–322. http://dx.doi.org/10.1016/j.ccr.2012.02.022.
He, B. P., Wang, J. J., Zhang, X., Wu, Y., Wang, M., Bay, B. H., & Chang, A. Y. (2006). Differential reactions of microglia to brain metastasis of lung cancer. *Molecular Medicine, 12*(7–8), 161–170. http://dx.doi.org/10.2119/2006-00033.He.
He, G., Luo, W., Li, P., Remmers, C., Netzer, W. J., Hendrick, J., ... Greengard, P. (2010). Gamma-secretase activating protein is a therapeutic target for Alzheimer's disease. *Nature, 467*(7311), 95–98. http://dx.doi.org/10.1038/nature09325.
Hemmati, H. D., Nakano, I., Lazareff, J. A., Masterman-Smith, M., Geschwind, D. H., Bronner-Fraser, M., & Kornblum, H. I. (2003). Cancerous stem cells can arise from pediatric brain tumors. *Proceedings of the National Academy of Sciences of the United States of America, 100*(25), 15178–15183. http://dx.doi.org/10.1073/pnas.2036535100.
Henley, D. B., Sundell, K. L., Sethuraman, G., Dowsett, S. A., & May, P. C. (2014). Safety profile of semagacestat, a gamma-secretase inhibitor: IDENTITY trial findings. *Current Medical Research and Opinion, 30*(10), 2021–2032. http://dx.doi.org/10.1185/03007995.2014.939167.

Herrmann, J. L., Menter, D. G., Hamada, J., Marchetti, D., Nakajima, M., & Nicolson, G. L. (1993). Mediation of NGF-stimulated extracellular matrix invasion by the human melanoma low-affinity p75 neurotrophin receptor: Melanoma p75 functions independently of trkA. *Molecular Biology of the Cell*, *4*(11), 1205–1216.

Higuchi, H., Yamashita, T., Yoshikawa, H., & Tohyama, M. (2003). PKA phosphorylates the p75 receptor and regulates its localization to lipid rafts. *The EMBO Journal*, *22*(8), 1790–1800.

Hoa, N. T., Zhang, J. G., Delgado, C. L., Myers, M. P., Callahan, L. L., Vandeusen, G., … Jadus, M. R. (2007). Human monocytes kill M-CSF-expressing glioma cells by BK channel activation. *Laboratory Investigation*, *87*(2), 115–129. http://dx.doi.org/10.1038/labinvest.3700506.

Hoelzinger, D. B., Mariani, L., Weis, J., Woyke, T., Berens, T. J., McDonough, W. S., … Berens, M. E. (2005). Gene expression profile of glioblastoma multiforme invasive phenotype points to new therapeutic targets. *Neoplasia*, *7*(1), 7–16.

Holtkamp, N., Afanasieva, A., Elstner, A., van Landeghem, F. K., Konneker, M., Kuhn, S. A., … von Deimling, A. (2005). Brain slice invasion model reveals genes differentially regulated in glioma invasion. *Biochemical and Biophysical Research Communications*, *336*(4), 1227–1233.

Hsu, C. K., Hsu, C. C., Lee, J. Y., Kuo, Y. M., & Pai, M. C. (2013). Exacerbation of psoriatic skin lesions in a patient with Alzheimer disease receiving gamma-secretase inhibitor. *Journal of the American Academy of Dermatology*, *68*(2), e46–e48. http://dx.doi.org/10.1016/j.jaad.2012.07.012.

Huang, C. S., Zhou, J., Feng, A. K., Lynch, C. C., Klumperman, J., DeArmond, S. J., & Mobley, W. C. (1999). Nerve growth factor signaling in caveolae-like domains at the plasma membrane. *The Journal of Biological Chemistry*, *274*(51), 36707–36714.

Huang, T. T., Sarkaria, S. M., Cloughesy, T. F., & Mischel, P. S. (2009). Targeted therapy for malignant glioma patients: Lessons learned and the road ahead. *Neurotherapeutics*, *6*(3), 500–512. http://dx.doi.org/10.1016/j.nurt.2009.04.008.

Hussain, S. F., Yang, D., Suki, D., Aldape, K., Grimm, E., & Heimberger, A. B. (2006). The role of human glioma-infiltrating microglia/macrophages in mediating antitumor immune responses. *Neuro-Oncology*, *8*(3), 261–279. http://dx.doi.org/10.1215/15228517-2006-008.

Hussain, S. F., Yang, D., Suki, D., Grimm, E., & Heimberger, A. B. (2006). Innate immune functions of microglia isolated from human glioma patients. *Journal of Translational Medicine*, *4*, 15. http://dx.doi.org/10.1186/1479-5876-4-15.

Hwang, S. Y., Yoo, B. C., Jung, J. W., Oh, E. S., Hwang, J. S., Shin, J. A., … Han, I. O. (2009). Induction of glioma apoptosis by microglia-secreted molecules: The role of nitric oxide and cathepsin B. *Biochimica et Biophysica Acta*, *1793*(11), 1656–1668. http://dx.doi.org/10.1016/j.bbamcr.2009.08.011. S0167-4889(09)00225-0 [pii].

Ibanez, C. F., & Simi, A. (2012). p75 neurotrophin receptor signaling in nervous system injury and degeneration: Paradox and opportunity. *Trends in Neurosciences*, *35*(7), 431–440. http://dx.doi.org/10.1016/j.tins.2012.03.007.

Ignatova, T. N., Kukekov, V. G., Laywell, E. D., Suslov, O. N., Vrionis, F. D., & Steindler, D. A. (2002). Human cortical glial tumors contain neural stem-like cells expressing astroglial and neuronal markers in vitro. *Glia*, *39*(3), 193–206.

Imai, T., Tamai, K., Oizumi, S., Oyama, K., Yamaguchi, K., Sato, I., … Tanaka, N. (2013). CD271 defines a stem cell-like population in hypopharyngeal cancer. *PLoS One*, *8*(4). e62002, http://dx.doi.org/10.1371/journal.pone.0062002.

Jin, H., Pan, Y., He, L., Zhai, H., Li, X., Zhao, L., … Fan, D. (2007). p75 neurotrophin receptor inhibits invasion and metastasis of gastric cancer. *Molecular Cancer Research*, *5*(5), 423–433. http://dx.doi.org/10.1158/1541-7786.MCR-06-0407. 5/5/423 [pii].

Jin, H., Pan, Y., Zhao, L., Zhai, H., Li, X., Sun, L., … Fan, D. (2007). p75 neurotrophin receptor suppresses the proliferation of human gastric cancer cells. *Neoplasia*, *9*(6), 471–478.

Johnston, A. L., Lun, X., Rahn, J. J., Liacini, A., Wang, L., Hamilton, M. G., … Forsyth, P. A. (2007). The p75 neurotrophin receptor is a central regulator of glioma invasion. *PLoS Biology*, *5*(8), e212.

Joy, A. M., Beaudry, C. E., Tran, N. L., Ponce, F. A., Holz, D. R., Demuth, T., & Berens, M. E. (2003). Migrating glioma cells activate the PI3-K pathway and display decreased susceptibility to apoptosis. *Journal of Cell Science*, *116*(Pt. 21), 4409–4417.

Joyce, J. A. (2005). Therapeutic targeting of the tumor microenvironment. *Cancer Cell*, 7(6), 513–520. http://dx.doi.org/10.1016/j.ccr.2005.05.024.

Jung, S., Ackerley, C., Ivanchuk, S., Mondal, S., Becker, L. E., & Rutka, J. T. (2001). Tracking the invasiveness of human astrocytoma cells by using green fluorescent protein in an organotypical brain slice model. *Journal of Neurosurgery*, *94*(1), 80–89.

Jung, S., Hinek, A., Tsugu, A., Hubbard, S. L., Ackerley, C., Becker, L. E., & Rutka, J. T. (1999). Astrocytoma cell interaction with elastin substrates: Implications for astrocytoma invasive potential. *Glia*, *25*(2), 179–189. http://dx.doi.org/10.1002/(SICI)1098-1136(19990115)25:2<179::AID-GLIA8>3.0.CO;2-B. [pii].

Junttila, M. R., & de Sauvage, F. J. (2013). Influence of tumour micro-environment heterogeneity on therapeutic response. *Nature*, *501*(7467), 346–354. http://dx.doi.org/10.1038/nature12626.

Kanning, K. C., Hudson, M., Amieux, P. S., Wiley, J. C., Bothwell, M., & Schecterson, L. C. (2003). Proteolytic processing of the p75 neurotrophin receptor and two homologs generates C-terminal fragments with signaling capability. *The Journal of Neuroscience*, *23*(13), 5425–5436.

Kaplan, D. R., Martin-Zanca, D., & Parada, L. F. (1991). Tyrosine phosphorylation and tyrosine kinase activity of the trk proto-oncogene product induced by NGF. *Nature*, *350*(6314), 158–160.

Kaplan, D. R., & Miller, F. D. (2003). Axon growth inhibition: Signals from the p75 neurotrophin receptor. *Nature Neuroscience*, *6*(5), 435–436.

Kaplan, D. R., & Miller, F. D. (2004). Neurobiology: A move to sort life from death. *Nature*, *427*(6977), 798–799.

Kargiotis, O., Chetty, C., Gondi, C. S., Tsung, A. J., Dinh, D. H., Gujrati, M., … Rao, J. S. (2008). Adenovirus-mediated transfer of siRNA against MMP-2 mRNA results in impaired invasion and tumor-induced angiogenesis, induces apoptosis in vitro and inhibits tumor growth in vivo in glioblastoma. *Oncogene*, *27*(35), 4830–4840. http://dx.doi.org/10.1038/onc.2008.122.

Kelly, J. J., Stechishin, O., Chojnacki, A., Lun, X., Sun, B., Senger, D. L., … Weiss, S. (2009). Proliferation of human glioblastoma stem cells occurs independently of exogenous mitogens. *Stem Cells*, *27*(8), 1722–1733. http://dx.doi.org/10.1002/stem.98.

Kerber, M., Reiss, Y., Wickersheim, A., Jugold, M., Kiessling, F., Heil, M., … Machein, M. R. (2008). Flt-1 signaling in macrophages promotes glioma growth in vivo. *Cancer Research*, *68*(18), 7342–7351. http://dx.doi.org/10.1158/0008-5472.CAN-07-6241. 68/18/7342 [pii].

Khwaja, F., Tabassum, A., Allen, J., & Djakiew, D. (2006). The p75(NTR) tumor suppressor induces cell cycle arrest facilitating caspase mediated apoptosis in prostate tumor cells. *Biochemical and Biophysical Research Communications*, *341*(4), 1184–1192.

Kim, J., Lee, I. H., Cho, H. J., Park, C. K., Jung, Y. S., Kim, Y., … Nam, D. H. (2015). Spatiotemporal evolution of the primary glioblastoma genome. *Cancer Cell*, *28*(3), 318–328. http://dx.doi.org/10.1016/j.ccell.2015.07.013.

Kim, J., Villadsen, R., Sorlie, T., Fogh, L., Gronlund, S. Z., Fridriksdottir, A. J., … Petersen, O. W. (2012). Tumor initiating but differentiated luminal-like breast cancer cells are highly invasive in the absence of basal-like activity. *Proceedings of the National*

Academy of Sciences of the United States of America, 109(16), 6124–6129. http://dx.doi.org/10.1073/pnas.1203203109.

Kim, H., Zheng, S., Amini, S. S., Virk, S. M., Mikkelsen, T., Brat, D. J., … Verhaak, R. G. (2015). Whole-genome and multisector exome sequencing of primary and post-treatment glioblastoma reveals patterns of tumor evolution. *Genome Research, 25*(3), 316–327. http://dx.doi.org/10.1101/gr.180612.114.

Kimura, S., Yoshino, A., Katayama, Y., Watanabe, T., & Fukushima, T. (2002). Growth control of C6 glioma in vivo by nerve growth factor. *Journal of Neuro-Oncology, 59*(3), 199–205.

Kislin, K. L., McDonough, W. S., Eschbacher, J. M., Armstrong, B. A., & Berens, M. E. (2009). NHERF-1: Modulator of glioblastoma cell migration and invasion. *Neoplasia, 11*(4), 377–387.

Kokunai, T., Iguchi, H., & Tamaki, N. (1999). Differentiation and growth inhibition of glioma cells induced by transfer of trk A proto-oncogene. *Journal of Neuro-Oncology, 42*(1), 23–34.

Komohara, Y., Ohnishi, K., Kuratsu, J., & Takeya, M. (2008). Possible involvement of the M2 anti-inflammatory macrophage phenotype in growth of human gliomas. *The Journal of Pathology, 216*(1), 15–24. http://dx.doi.org/10.1002/path.2370.

Kondraganti, S., Mohanam, S., Chintala, S. K., Kin, Y., Jasti, S. L., Nirmala, C., … Rao, J. S. (2000). Selective suppression of matrix metalloproteinase-9 in human glioblastoma cells by antisense gene transfer impairs glioblastoma cell invasion. *Cancer Research, 60*(24), 6851–6855.

Lachyankar, M. B., Ross, A. H., Litofsky, N. S., Condon, P. J., Quesenberry, P. J., & Recht, L. D. (1997). TrkA expression decreases the in vivo aggressiveness of C6 glioma cells. *Cancer Research, 57*(3), 532–536.

Lakka, S. S., Gondi, C. S., Yanamandra, N., Olivero, W. C., Dinh, D. H., Gujrati, M., & Rao, J. S. (2004). Inhibition of cathepsin B and MMP-9 gene expression in glioblastoma cell line via RNA interference reduces tumor cell invasion, tumor growth and angiogenesis. *Oncogene, 23*(27), 4681–4689.

Lawn, S., Krishna, N., Pisklakova, A., Qu, X., Fenstermacher, D. A., Fournier, M., … Kenchappa, R. S. (2015). Neurotrophin signaling via TrkB and TrkC receptors promotes the growth of brain tumor-initiating cells. *Journal of Biological Chemistry, 290*(6), 3814–3824.

Lee, J., Kotliarova, S., Kotliarov, Y., Li, A., Su, Q., Donin, N. M., … Fine, H. A. (2006). Tumor stem cells derived from glioblastomas cultured in bFGF and EGF more closely mirror the phenotype and genotype of primary tumors than do serum-cultured cell lines. *Cancer Cell, 9*(5), 391–403. http://dx.doi.org/10.1016/j.ccr.2006.03.030. S1535-6108(06)00117-6 [pii].

Lee, T. H., Chennakrishnaiah, S., Audemard, E., Montermini, L., Meehan, B., & Rak, J. (2014). Oncogenic ras-driven cancer cell vesiculation leads to emission of double-stranded DNA capable of interacting with target cells. *Biochemical and Biophysical Research Communications, 451*(2), 295–301. http://dx.doi.org/10.1016/j.bbrc.2014.07.109.

Leibrock, J., Lottspeich, F., Hohn, A., Hofer, M., Hengerer, B., Masiakowski, P., … Barde, Y. A. (1989). Molecular cloning and expression of brain-derived neurotrophic factor. *Nature, 341*(6238), 149–152.

Levi-Montalcini, R. (1987). The nerve growth factor 35 years later. *Science, 237*(4819), 1154–1162.

Li, Q., Ford, M. C., Lavik, E. B., & Madri, J. A. (2006). Modeling the neurovascular niche: VEGF- and BDNF-mediated cross-talk between neural stem cells and endothelial cells: An in vitro study. *Journal of Neuroscience Research, 84*(8), 1656–1668. http://dx.doi.org/10.1002/jnr.21087.

Li, X., Law, J. W., & Lee, A. Y. (2012). Semaphorin 5A and plexin-B3 regulate human glioma cell motility and morphology through Rac1 and the actin cytoskeleton. *Oncogene, 31*(5), 595–610. http://dx.doi.org/10.1038/onc.2011.256.

Li, A., Walling, J., Ahn, S., Kotliarov, Y., Su, Q., Quezado, M., … Fine, H. A. (2009). Unsupervised analysis of transcriptomic profiles reveals six glioma subtypes. *Cancer Research*, *69*(5), 2091–2099.

Liang, Y., Diehn, M., Watson, N., Bollen, A. W., Aldape, K. D., Nicholas, M. K., … Israel, M. A. (2005). Gene expression profiling reveals molecularly and clinically distinct subtypes of glioblastoma multiforme. *Proceedings of the National Academy of Sciences of the United States of America*, *102*(16), 5814–5819.

Lorger, M. (2012). Tumor microenvironment in the brain. *Cancers (Basel)*, *4*(1), 218–243. http://dx.doi.org/10.3390/cancers4010218.

Louis, D. N., Perry, A., Burger, P., Ellison, D. W., Reifenberger, G., von Deimling, A., … International Society of Neuropathology—Haarlem (2014). International Society of Neuropathology—Haarlem consensus guidelines for nervous system tumor classification and grading. *Brain Pathology*, *24*(5), 429–435. http://dx.doi.org/10.1111/bpa.12171.

Luchman, H. A., Stechishin, O. D., Dang, N. H., Blough, M. D., Chesnelong, C., Kelly, J. J., … Weiss, S. (2012). An in vivo patient-derived model of endogenous IDH1-mutant glioma. *Neuro-Oncology*, *14*(2), 184–191. http://dx.doi.org/10.1093/neuonc/nor207.

Lun, X., Wells, J. C., Grinshtein, N., King, J. C., Hao, X., Dang, N. H., … Senger, D. L. (2016). Disulfiram when combined with copper enhances the therapeutic effects of temozolomide for the treatment of Glioblastoma. *Clinical Cancer Research*, *22*(15), 3860–3875. http://dx.doi.org/10.1158/1078-0432.CCR-15-1798.

Maisonpierre, P. C., Belluscio, L., Squinto, S., Ip, N. Y., Furth, M. E., Lindsay, R. M., & Yancopoulos, G. D. (1990). Neurotrophin-3: A neurotrophic factor related to NGF and BDNF. *Science*, *247*(4949 Pt. 1), 1446–1451.

Majdan, M., Walsh, G. S., Aloyz, R., & Miller, F. D. (2001). TrkA mediates developmental sympathetic neuron survival in vivo by silencing an ongoing p75NTR-mediated death signal. *The Journal of Cell Biology*, *155*(7), 1275–1285. http://dx.doi.org/10.1083/jcb.200110017.

Mantovani, A., Marchesi, F., Portal, C., Allavena, P., & Sica, A. (2008). Linking inflammation reactions to cancer: Novel targets for therapeutic strategies. *Adv Exp Med Biol*, *610*, 112–127.

Marchetti, D., Aucoin, R., Blust, J., Murry, B., & Greiter-Wilke, A. (2004). p75 neurotrophin receptor functions as a survival receptor in brain-metastatic melanoma cells. *Journal of Cellular Biochemistry*, *91*(1), 206–215.

Maret, D., Gruzglin, E., Sadr, M. S., Siu, V., Shan, W., Koch, A. W., … Colman, D. R. (2010). Surface expression of precursor N-cadherin promotes tumor cell invasion. *Neoplasia*, *12*(12), 1066–1080.

Mariani, L., Beaudry, C., McDonough, W. S., Hoelzinger, D. B., Demuth, T., Ross, K. R., … Berens, M. E. (2001). Glioma cell motility is associated with reduced transcription of proapoptotic and proliferation genes: A cDNA microarray analysis. *Journal of Neuro-Oncology*, *53*(2), 161–176.

Mariani, L., Beaudry, C., McDonough, W. S., Hoelzinger, D. B., Kaczmarek, E., Ponce, F., … Berens, M. E. (2001). Death-associated protein 3 (Dap-3) is overexpressed in invasive glioblastoma cells in vivo and in glioma cell lines with induced motility phenotype in vitro. *Clinical Cancer Research*, 7(8), 2480–2489.

Mariani, L., McDonough, W. S., Hoelzinger, D. B., Beaudry, C., Kaczmarek, E., Coons, S. W., … Berens, M. E. (2001). Identification and validation of P311 as a glioblastoma invasion gene using laser capture microdissection. *Cancer Research*, *61*(10), 4190–4196.

Marjanovic, N. D., Weinberg, R. A., & Chaffer, C. L. (2013). Poised with purpose: Cell plasticity enhances tumorigenicity. *Cell Cycle*, *12*(17), 2713–2714. http://dx.doi.org/10.4161/cc.26075.

Markovic, D. S., Vinnakota, K., Chirasani, S., Synowitz, M., Raguet, H., Stock, K., … Kettenmann, H. (2009). Gliomas induce and exploit microglial MT1-MMP expression for tumor expansion. *Proceedings of the National Academy of Sciences of the United States of America*, *106*(30), 12530–12535. http://dx.doi.org/10.1073/pnas.0804273106.

McDonald, N. Q., Lapatto, R., Murray-Rust, J., Gunning, J., Wlodawer, A., & Blundell, T. L. (1991). New protein fold revealed by a 2.3-A resolution crystal structure of nerve growth factor. *Nature*, *354*(6352), 411–414.

Menter, D. G., Herrmann, J. L., Marchetti, D., & Nicolson, G. L. (1994). Involvement of neurotrophins and growth factors in brain metastasis formation. *Invasion & Metastasis*, *14*(1–6), 372–384.

Merzak, A., Koocheckpour, S., & Pilkington, G. J. (1994). CD44 mediates human glioma cell adhesion and invasion in vitro. *Cancer Research*, *54*(15), 3988–3992.

Miao, H., Li, D. Q., Mukherjee, A., Guo, H., Petty, A., Cutter, J., … Wang, B. (2009). EphA2 mediates ligand-dependent inhibition and ligand-independent promotion of cell migration and invasion via a reciprocal regulatory loop with Akt. *Cancer Cell*, *16*(1), 9–20. http://dx.doi.org/10.1016/j.ccr.2009.04.009. S1535-6108(09) 00143-3 [pii].

Montermini, L., Meehan, B., Garnier, D., Lee, W. J., Lee, T. H., Guha, A., … Rak, J. (2015). Inhibition of oncogenic epidermal growth factor receptor kinase triggers release of exosome-like extracellular vesicles and impacts their phosphoprotein and DNA content. *The Journal of Biological Chemistry*, *290*(40), 24534–24546. http://dx.doi.org/10.1074/jbc.M115.679217.

Mora, R., Abschuetz, A., Kees, T., Dokic, I., Joschko, N., Kleber, S., … Regnier-Vigouroux, A. (2009). TNF-alpha- and TRAIL-resistant glioma cells undergo autophagy-dependent cell death induced by activated microglia. *Glia*, *57*(5), 561–581. http://dx.doi.org/10.1002/glia.20785.

Murillo-Sauca, O., Chung, M. K., Shin, J. H., Karamboulas, C., Kwok, S., Jung, Y. H., … Sunwoo, J. B. (2014). CD271 is a functional and targetable marker of tumor-initiating cells in head and neck squamous cell carcinoma. *Oncotarget*, *5*(16), 6854–6866. http://dx.doi.org/10.18632/oncotarget.2269.

Murphy, R. A., Oger, J., Saide, J. D., Blanchard, M. H., Arnason, B. G., Hogan, C., & Pantazis, N. J. (1977). Secretion of nerve growth factor by central nervous system glioma cells in culture. *The Journal of Cell Biology*, *72*(3), 769–773.

Nakada, M., Anderson, E. M., Demuth, T., Nakada, S., Reavie, L. B., Drake, K. L., … Berens, M. E. (2010). The phosphorylation of ephrin-B2 ligand promotes glioma cell migration and invasion. *International Journal of Cancer*, *126*(5), 1155–1165. http://dx.doi.org/10.1002/ijc.24849.

Nakada, M., Drake, K. L., Nakada, S., Niska, J. A., & Berens, M. E. (2006). Ephrin-B3 ligand promotes glioma invasion through activation of Rac1. *Cancer Research*, *66*(17), 8492–8500. http://dx.doi.org/10.1158/0008-5472.CAN-05-4211. 66/17/8492 [pii].

Nakada, M., Nakada, S., Demuth, T., Tran, N. L., Hoelzinger, D. B., & Berens, M. E. (2007). Molecular targets of glioma invasion. *Cellular and Molecular Life Sciences*, *64*(4), 458–478. http://dx.doi.org/10.1007/s00018-007-6342-5.

Nakada, M., Niska, J. A., Miyamori, H., McDonough, W. S., Wu, J., Sato, H., & Berens, M. E. (2004). The phosphorylation of EphB2 receptor regulates migration and invasion of human glioma cells. *Cancer Research*, *64*(9), 3179–3185.

Nakagawa, J., Saio, M., Tamakawa, N., Suwa, T., Frey, A. B., Nonaka, K., … Takami, T. (2007). TNF expressed by tumor-associated macrophages, but not microglia, can eliminate glioma. *International Journal of Oncology*, *30*(4), 803–811.

Nakajima, K., Honda, S., Tohyama, Y., Imai, Y., Kohsaka, S., & Kurihara, T. (2001). Neurotrophin secretion from cultured microglia. *Journal of Neuroscience Research*, *65*(4), 322–331.

Nakajima, K., Kikuchi, Y., Ikoma, E., Honda, S., Ishikawa, M., Liu, Y., & Kohsaka, S. (1998). Neurotrophins regulate the function of cultured microglia. *Glia*, *24*(3), 272–289.

Neufeld, G., Shraga-Heled, N., Lange, T., Guttmann-Raviv, N., Herzog, Y., & Kessler, O. (2005). Semaphorins in cancer. *Frontiers in Bioscience*, *10*, 751–760. doi:1569 [pii].

Nowell, P. C. (1986). Mechanisms of tumor progression. *Cancer Research*, *46*(5), 2203–2207.

Nutt, C. L., Zerillo, C. A., Kelly, G. M., & Hockfield, S. (2001). Brain enriched hyaluronan binding (BEHAB)/brevican increases aggressiveness of CNS-1 gliomas in Lewis rats. *Cancer Research*, *61*(19), 7056–7059.

Oelmann, E., Sreter, L., Schuller, I., Serve, H., Koenigsmann, M., Wiedenmann, B., … Berdel, W. E. (1995). Nerve growth factor stimulates clonal growth of human lung cancer cell lines and a human glioblastoma cell line expressing high-affinity nerve growth factor binding sites involving tyrosine kinase signaling. *Cancer Research*, *55*(10), 2212–2219.

Ohnishi, T., Matsumura, H., Izumoto, S., Hiraga, S., & Hayakawa, T. (1998). A novel model of glioma cell invasion using organotypic brain slice culture. *Cancer Research*, *58*(14), 2935–2940.

Okumura, T., Shimada, Y., Imamura, M., & Yasumoto, S. (2003). Neurotrophin receptor p75(NTR) characterizes human esophageal keratinocyte stem cells in vitro. *Oncogene*, *22*(26), 4017–4026. http://dx.doi.org/10.1038/sj.onc.1206525.

Otten, U., Marz, P., Heese, K., Hock, C., Kunz, D., & Rose-John, S. (2000). Cytokines and neurotrophins interact in normal and diseased states. *Annals of the New York Academy of Sciences*, *917*, 322–330.

Owens, G. C., Orr, E. A., DeMasters, B. K., Muschel, R. J., Berens, M. E., & Kruse, C. A. (1998). Overexpression of a transmembrane isoform of neural cell adhesion molecule alters the invasiveness of rat CNS-1 glioma. *Cancer Research*, *58*(9), 2020–2028.

Pardossi-Piquard, R., Petit, A., Kawarai, T., Sunyach, C., Alves da Costa, C., Vincent, B., … Checler, F. (2005). Presenilin-dependent transcriptional control of the Abeta-degrading enzyme neprilysin by intracellular domains of betaAPP and APLP. *Neuron*, *46*(4), 541–554. http://dx.doi.org/10.1016/j.neuron.2005.04.008.

Parkhurst, C. N., Zampieri, N., & Chao, M. V. (2010). Nuclear localization of the p75 neurotrophin receptor intracellular domain. *The Journal of Biological Chemistry*, *285*(8), 5361–5368. http://dx.doi.org/10.1074/jbc.M109.045054.

Penfield, W. (1925). Microglia and the process of phagocytosis in gliomas. *The American Journal of Pathology*, *1*(1). 77–90.15.

Phillips, H. S., Kharbanda, S., Chen, R., Forrest, W. F., Soriano, R. H., Wu, T. D., … Aldape, K. (2006). Molecular subclasses of high-grade glioma predict prognosis, delineate a pattern of disease progression, and resemble stages in neurogenesis. *Cancer Cell*, *9*(3), 157–173.

Pistollato, F., Abbadi, S., Rampazzo, E., Persano, L., Della Puppa, A., Frasson, C., … Basso, G. (2010). Intratumoral hypoxic gradient drives stem cells distribution and MGMT expression in glioblastoma. *Stem Cells*, *28*(5), 851–862. http://dx.doi.org/10.1002/stem.415.

Platten, M., Kretz, A., Naumann, U., Aulwurm, S., Egashira, K., Isenmann, S., & Weller, M. (2003). Monocyte chemoattractant protein-1 increases microglial infiltration and aggressiveness of gliomas. *Annals of Neurology*, *54*(3), 388–392. http://dx.doi.org/10.1002/ana.10679.

Podlesniy, P., Kichev, A., Pedraza, C., Saurat, J., Encinas, M., Perez, B., … Espinet, C. (2006). Pro-NGF from Alzheimer's disease and normal human brain displays distinctive abilities to induce processing and nuclear translocation of intracellular domain of p75NTR and apoptosis. *The American Journal of Pathology*, *169*(1), 119–131. http://dx.doi.org/10.2353/ajpath.2006.050787.

Prag, S., Lepekhin, E. A., Kolkova, K., Hartmann-Petersen, R., Kawa, A., Walmod, P. S., … Pedersen, N. (2002). NCAM regulates cell motility. *Journal of Cell Science*, *115*(Pt. 2), 283–292.

Pyonteck, S. M., Akkari, L., Schuhmacher, A. J., Bowman, R. L., Sevenich, L., Quail, D. F., … Joyce, J. A. (2013). CSF-1R inhibition alters macrophage polarization and blocks glioma progression. *Nature Medicine*, *19*(10), 1264–1272. http://dx.doi.org/10.1038/nm.3337.

Quail, D. F., & Joyce, J. A. (2013). Microenvironmental regulation of tumor progression and metastasis. *Nature Medicine*, *19*(11), 1423–1437. http://dx.doi.org/10.1038/nm.3394.

Quintana, E., Shackleton, M., Sabel, M. S., Fullen, D. R., Johnson, T. M., & Morrison, S. J. (2008). Efficient tumour formation by single human melanoma cells. *Nature*, *456*(7222), 593–598. http://dx.doi.org/10.1038/nature07567.

Radotra, B., & McCormick, D. (1997). CD44 is involved in migration but not spreading of astrocytoma cells in vitro. *Anticancer Research*, *17*(2A), 945–949.

Rao, J. S., Steck, P. A., Tofilon, P., Boyd, D., Ali-Osman, F., Stetler-Stevenson, W. G., … Sawaya, R. (1994). Role of plasminogen activator and of 92-KDa type IV collagenase in glioblastoma invasion using an in vitro matrigel model. *Journal of Neuro-Oncology*, *18*(2), 129–138.

Redmer, T., Welte, Y., Behrens, D., Fichtner, I., Przybilla, D., Wruck, W., … Regenbrecht, C. R. (2014). The nerve growth factor receptor CD271 is crucial to maintain tumorigenicity and stem-like properties of melanoma cells. *PLoS One*. *9*(5). e92596. http://dx.doi.org/10.1371/journal.pone.0092596.

Reichardt, L. F. (2006). Neurotrophin-regulated signalling pathways. *Philosophical Transactions of the Royal Society of London Series B, Biological Sciences*, *361*(1473), 1545–1564. http://dx.doi.org/10.1098/rstb.2006.1894. G280147415872711 [pii].

Robbins, S. M., & Senger, D. L. (2013). Assessing the mechanisms of glioma invasion. In R. Martinez Murillo & A. Martinez (Eds.), *Neuromethods* (pp. 275–298): *Vol. 77*. New York: Springer Science and Business Media.

Rocha, A. S., Risberg, B., Magalhaes, J., Trovisco, V., de Castro, I. V., Lazarovici, P., … Sobrinho-Simoes, M. (2006). The p75 neurotrophin receptor is widely expressed in conventional papillary thyroid carcinoma. *Human Pathology*, *37*(5), 562–568. http://dx.doi.org/10.1016/j.humpath.2005.12.016. S0046-8177(06)00007-4 [pii].

Rowan, K. (2009). Are cancer stem cells real? After four decades debate still simmers. *Journal of the National Cancer Institute*, *101*(8), 546–547.

Roux, P. P., & Barker, P. A. (2002). Neurotrophin signaling through the p75 neurotrophin receptor. *Progress in Neurobiology*, *67*(3), 203–233.

Ruffell, B., & Coussens, L. M. (2015). Macrophages and therapeutic resistance in cancer. *Cancer Cell*, *27*(4), 462–472. http://dx.doi.org/10.1016/j.ccell.2015.02.015.

Sarkar, S., Doring, A., Zemp, F. J., Silva, C., Lun, X., Wang, X., … Yong, V. W. (2014). Therapeutic activation of macrophages and microglia to suppress brain tumor-initiating cells. *Nature Neuroscience*, *17*(1), 46–55. http://dx.doi.org/10.1038/nn.3597.

Shackleton, M., Vaillant, F., Simpson, K. J., Stingl, J., Smyth, G. K., Asselin-Labat, M. L., … Visvader, J. E. (2006). Generation of a functional mammary gland from a single stem cell. *Nature*, *439*(7072), 84–88.

Shiao, S. L., Ganesan, A. P., Rugo, H. S., & Coussens, L. M. (2011). Immune microenvironments in solid tumors: New targets for therapy. *Genes & Development*, *25*(24), 2559–2572. http://dx.doi.org/10.1101/gad.169029.111.

Shonukan, O., Bagayogo, I., McCrea, P., Chao, M., & Hempstead, B. (2003). Neurotrophin-induced melanoma cell migration is mediated through the actin-bundling protein fascin. *Oncogene*, *22*(23), 3616–3623.

Sidransky, D., Mikkelsen, T., Schwechheimer, K., Rosenblum, M. L., Cavanee, W., & Vogelstein, B. (1992). Clonal expansion of p53 mutant cells is associated with brain tumour progression. *Nature, 355*(6363), 846–847.

Simmons, G. W., Pong, W. W., Emnett, R. J., White, C. R., Gianino, S. M., Rodriguez, F. J., & Gutmann, D. H. (2011). Neurofibromatosis-1 heterozygosity increases microglia in a spatially and temporally restricted pattern relevant to mouse optic glioma formation and growth. *Journal of Neuropathology and Experimental Neurology, 70*(1), 51–62. http://dx.doi.org/10.1097/NEN.0b013e3182032d37.

Singer, H. S., Hansen, B., Martinie, D., & Karp, C. L. (1999). Mitogenesis in glioblastoma multiforme cell lines: A role for NGF and its TrkA receptors. *Journal of Neuro-Oncology, 45*(1), 1–8.

Singh, S. K., Clarke, I. D., Terasaki, M., Bonn, V. E., Hawkins, C., Squire, J., & Dirks, P. B. (2003). Identification of a cancer stem cell in human brain tumors. *Cancer Research, 63*(18), 5821–5828.

Singh, S. K., Hawkins, C., Clarke, I. D., Squire, J. A., Bayani, J., Hide, T., … Dirks, P. B. (2004). Identification of human brain tumour initiating cells. *Nature, 432*(7015), 396–401. http://dx.doi.org/10.1038/nature03128.

Stemple, D. L., & Anderson, D. J. (1992). Isolation of a stem cell for neurons and glia from the mammalian neural crest. *Cell, 71*(6), 973–985.

Stupp, R., Mason, W. P., van den Bent, M. J., Weller, M., Fisher, B., Taphoorn, M. J., … Mirimanoff, R. O. (2005). Radiotherapy plus concomitant and adjuvant temozolomide for glioblastoma. *The New England Journal of Medicine, 352*(10), 987–996. http://dx.doi.org/10.1056/NEJMoa043330. 352/10/987 [pii].

Tabassum, A., Khwaja, F., & Djakiew, D. (2003). The p75(NTR) tumor suppressor induces caspase-mediated apoptosis in bladder tumor cells. *International Journal of Cancer, 105*(1), 47–52.

Tian, J., Li, X., Si, M., Liu, T., & Li, J. (2014). CD271+ osteosarcoma cells display stem-like properties. *PLoS One. 9*(6). e98549http://dx.doi.org/10.1371/journal.pone.0098549.

Tomellini, E., Lagadec, C., Polakowska, R., & Le Bourhis, X. (2014). Role of p75 neurotrophin receptor in stem cell biology: More than just a marker. *Molecular Life Sciences, 71*(13), 2467–2481. http://dx.doi.org/10.1007/s00018-014-1564-9.

Toussaint, L. G., 3rd, Nilson, A. E., Goble, J. M., Ballman, K. V., James, C. D., Lefranc, F., … Uhm, J. H. (2012). Galectin-1, a gene preferentially expressed at the tumor margin, promotes glioblastoma cell invasion. *Molecular Cancer, 11*, 32. http://dx.doi.org/10.1186/1476-4598-11-32.

Underwood, C. K., Reid, K., May, L. M., Bartlett, P. F., & Coulson, E. J. (2008). Palmitoylation of the C-terminal fragment of p75(NTR) regulates death signaling and is required for subsequent cleavage by gamma-secretase. *Molecular and Cellular Neurosciences, 37*(2), 346–358. http://dx.doi.org/10.1016/j.mcn.2007.10.005. S1044-7431(07)00249-7 [pii].

Verbeke, S., Meignan, S., Lagadec, C., Germain, E., Hondermarck, H., Adriaenssens, E., & Le Bourhis, X. (2010). Overexpression of p75(NTR) increases survival of breast cancer cells through p21(waf1). *Cellular Signalling, 22*(12), 1864–1873. http://dx.doi.org/10.1016/j.cellsig.2010.07.014. S0898-6568(10)00207-X [pii].

Verbeke, S., Tomellini, E., Dhamani, F., Meignan, S., Adriaenssens, E., & Xuefen le, B. (2013). Extracellular cleavage of the p75 neurotrophin receptor is implicated in its pro-survival effect in breast cancer cells. *FEBS Letters, 587*(16), 2591–2596. http://dx.doi.org/10.1016/j.febslet.2013.06.039. S0014-5793(13)00497-3 [pii].

Verhaak, R. G., Hoadley, K. A., Purdom, E., Wang, V., Qi, Y., Wilkerson, M. D. … (2010). Cancer Genome Atlas Research Network. Integrated genomic analysis identifies clinically relevant subtypes of glioblastoma characterized by abnormalities in PDGFRA, IDH1, EGFR, and NF1. *Cancer Cell, 17*(1), 98–110.

Vescovi, A. L., Galli, R., & Reynolds, B. A. (2006). Brain tumour stem cells. *Nature Reviews Cancer*, *6*(6), 425–436. http://dx.doi.org/10.1038/nrc1889.

Vilar, M., Charalampopoulos, I., Kenchappa, R. S., Reversi, A., Klos-Applequist, J. M., Karaca, E., … Ibanez, C. F. (2009). Ligand-independent signaling by disulfide-crosslinked dimers of the p75 neurotrophin receptor. *Journal of Cell Science*, *122*(Pt. 18), 3351–3357. http://dx.doi.org/10.1242/jcs.055061.

Vilar, M., Charalampopoulos, I., Kenchappa, R. S., Simi, A., Karaca, E., Reversi, A., … Ibanez, C. F. (2009). Activation of the p75 neurotrophin receptor through conformational rearrangement of disulphide-linked receptor dimers. *Neuron*, *62*(1), 72–83. http://dx.doi.org/10.1016/j.neuron.2009.02.020.

Vinores, S. A., & Koestner, A. (1980). The effect of nerve growth factor on undifferentiated glioma cells. *Cancer Letters*, *10*(4), 309–318.

Walsh, E. M., Kim, R., Del Valle, L., Weaver, M., Sheffield, J., Lazarovici, P., & Marcinkiewicz, C. (2012). Importance of interaction between nerve growth factor and alpha9beta1 integrin in glial tumor angiogenesis. *Neuro-Oncology*, *14*(7), 890–901. http://dx.doi.org/10.1093/neuonc/nos119.

Wang, C., Tao, B., Li, S., Li, B., Wang, X., Hu, G., … Liu, J. (2014). Characterizing the role of PCDH9 in the regulation of glioma cell apoptosis and invasion. *Journal of Molecular Neuroscience*, *52*(2), 250–260. http://dx.doi.org/10.1007/s12031-013-0133-2.

Wang, L., Rahn, J. J., Lun, X., Sun, B., Kelly, J. J., Weiss, S., … Senger, D. L. (2008). Gamma-secretase represents a therapeutic target for the treatment of invasive glioma mediated by the p75 neurotrophin receptor. *PLoS Biology*. *6*(11). e289. http://dx.doi.org/10.1371/journal.pbio.0060289. 08-PLBI-RA-2243 [pii].

Wang, W., McLeod, H. L., & Cassidy, J. (2003). Disulfiram-mediated inhibition of NF-kappaB activity enhances cytotoxicity of 5-fluorouracil in human colorectal cancer cell lines. *International Journal of Cancer*, *104*(4), 504–511. http://dx.doi.org/10.1002/ijc.10972.

Wang, Y., Hagel, C., Hamel, W., Muller, S., Kluwe, L., & Westphal, M. (1998). Trk A, B, and C are commonly expressed in human astrocytes and astrocytic gliomas but not by human oligodendrocytes and oligodendroglioma. *Acta Neuropathologica*, *96*(4), 357–364.

Wei, J., Gabrusiewicz, K., & Heimberger, A. (2013). The controversial role of microglia in malignant gliomas. *Clinical & Developmental Immunology*, *2013*, 285246. http://dx.doi.org/10.1155/2013/285246.

Westphal, M., & Lamszus, K. (2011). The neurobiology of gliomas: From cell biology to the development of therapeutic approaches. *Nature Reviews Neuroscience*, *12*(9), 495–508. http://dx.doi.org/10.1038/nrn3060.

Wildeboer, D., Naus, S., Amy Sang, Q. X., Bartsch, J. W., & Pagenstecher, A. (2006). Metalloproteinase disintegrins ADAM8 and ADAM19 are highly regulated in human primary brain tumors and their expression levels and activities are associated with invasiveness. *Journal of Neuropathology and Experimental Neurology*, *65*(5), 516–527. http://dx.doi.org/10.1097/01.jnen.0000229240.51490.d3.

Wu, A., Wei, J., Kong, L. Y., Wang, Y., Priebe, W., Qiao, W., … Heimberger, A. B. (2010). Glioma cancer stem cells induce immunosuppressive macrophages/microglia. *Neuro-Oncology*, *12*(11), 1113–1125. http://dx.doi.org/10.1093/neuonc/noq082.

Xie, Q., Mittal, S., & Berens, M. E. (2014). Targeting adaptive glioblastoma: An overview of proliferation and invasion. *Neuro-oncology*, *16*(12), 1575–1584. http://dx.doi.org/10.1093/neuonc/nou147.

Xiong, J., Zhou, L. I., Lim, Y., Yang, M., Zhu, Y. H., Li, Z. W., … Zhou, X. F. (2015). Mature brain-derived neurotrophic factor and its receptor TrkB are upregulated in human glioma tissues. *Oncology Letters*, *10*(1), 223–227. http://dx.doi.org/10.3892/ol.2015.3181.

Yang, I., Han, S. J., Kaur, G., Crane, C., & Parsa, A. T. (2010). The role of microglia in central nervous system immunity and glioma immunology. *Journal of Clinical Neuroscience*, *17*(1), 6–10. http://dx.doi.org/10.1016/j.jocn.2009.05.006. S0967-5868(09)00332-4 [pii].

Zampieri, N., Xu, C. F., Neubert, T. A., & Chao, M. V. (2005). Cleavage of p75 neurotrophin receptor by alpha-secretase and gamma-secretase requires specific receptor domains. *The Journal of Biological Chemistry*, *280*(15), 14563–14571.

Zhai, H., Heppner, F. L., & Tsirka, S. E. (2011). Microglia/macrophages promote glioma progression. *Glia*, *59*(3), 472–485. http://dx.doi.org/10.1002/glia.21117.

Zhang, H., Kelly, G., Zerillo, C., Jaworski, D. M., & Hockfield, S. (1998). Expression of a cleaved brain-specific extracellular matrix protein mediates glioma cell invasion In vivo. *The Journal of Neuroscience*, *18*(7), 2370–2376.

Zhang, J., Sarkar, S., Cua, R., Zhou, Y., Hader, W., & Yong, V. W. (2012). A dialog between glioma and microglia that promotes tumor invasiveness through the CCL2/CCR2/interleukin-6 axis. *Carcinogenesis*, *33*(2), 312–319. http://dx.doi.org/10.1093/carcin/bgr289.

CHAPTER FIFTEEN

Neurotrophic Factors Used to Treat Spinal Cord Injury

S.I. Hodgetts*[,†,1], A.R. Harvey*[,†]

*School of Anatomy, Physiology and Human Biology, The University of Western Australia, Perth, WA, Australia

[†]Western Australian Neuroscience Research Institute, Perth, WA, Australia

[1]Corresponding author: e-mail address: stuart.hodgetts@uwa.edu.au

Contents

1. Introduction 406
2. Neurotrophins 428
   2.1 Brain-Derived Neurotrophic Factor 428
   2.2 Neurotrophin-3 430
   2.3 NGF and NT-4/5 433
   2.4 Neuropoietic Cytokines 433
3. Selected Examples of Other Growth Factors 434
   3.1 Fibroblastic Growth Factors 434
   3.2 Glial Cell-Derived Neurotrophic Factor 436
   3.3 IGF-1 and PDGF 437
4. Combinations of Factors 438
5. Location of Neurotrophic Factor Delivery 439
6. Mode of Neurotrophic Factor Delivery 441
7. Timing of Neurotrophic Factor Delivery 443
8. Conclusion 444

Acknowledgments 444

References 444

Abstract

The application of neurotrophic factors as a therapy to improve morphological and behavioral outcomes after experimental spinal cord injury (SCI) has been the focus of many studies. These studies vary markedly in the type of neurotrophic factor that is delivered, the mode of administration, and the location, timing, and duration of the treatment. Generally, the majority of studies have had significant success if neurotrophic factors are applied in or close to the lesion site during the acute or the subacute phase after SCI. Comparatively fewer studies have administered neurotrophic factors in order to directly target the somata of injured neurons. The mode of delivery varies between acute injection of recombinant proteins, subacute or chronic delivery using a variety of strategies including osmotic minipumps, cell-mediated delivery, delivery

Vitamins and Hormones, Volume 104
ISSN 0083-6729
http://dx.doi.org/10.1016/bs.vh.2016.11.007

using polymer release vehicles or supporting bridges of some sort, or the use of gene therapy to modify neurons, glial cells, or precursor/stem cells. In this brief review, we summarize the state of play of many of the therapies using these factors, most of which have been undertaken in rodent models of SCI.

1. INTRODUCTION

A wide range of neurotrophic factors can be delivered to the central nervous system (CNS) in many ways to promote morphological and functional sparing/recovery after spinal cord injury (SCI). These neurotrophic factors can enhance neuronal survival, alter glia phenotype, and/or promote plasticity and axonal regrowth. Neurons contributing axons to different long-distance ascending (sensory) or descending (motor) fiber systems express different types of receptor (Boyce & Mendell, 2014; Hollis & Tuszynski, 2011) and may therefore be preferentially sensitive to a particular factor or combination of factors (Blesch, Fischer, & Tuszynski, 2012). It may also be important to target spinal neurons involved in segmental or short-range intersegmental circuits (Boyce & Mendell, 2014). In addition, targeting of glial cells with different types or combinations of neurotrophic factors may reduce the negative effects of inflammation and scarring, and promote other beneficial actions such as reduced demyelination or enhanced remyelination (Plemel et al., 2014). Importantly, endogenous expression of neurotrophic factors, cytokines, chemokines, and their various receptors in different populations of neurons and glia is altered at different times after SCI (Hougland, Harrison, Magnuson, Rouchka, & Petruska, 2012; King, Bradbury, McMahon, & Priestley, 2000; Liebl, Huang, Young, & Parada, 2001; Nakamura & Bregman, 2001). Knowledge of these spatio/temporal changes in receptor expression as well as sensitivity and reactivity to different species of molecules is therefore of critical value when designing neurotrophic factor-based therapies.

This review of the "state of play" of such growth-promoting factors tested in experimental models and clinical trials of SCI is not designed to be exhaustive, but rather is intended to distil and complement other detailed reviews to which the interested reader is also directed (e.g., Arvanian, 2013 Awad, Carmody, & Steinmetz, 2015; Blesch et al., 2012; Boyce & Mendell,

2014; Bregman et al., 2002; Gerin et al., 2011; Hollis, Lu, Blesch, & Tuszynski, 2009; Kabu, Gao, Kwon, & Labhasetwar, 2015; Lu & Tuszynski, 2008; McCall, Weidner, & Blesch, 2012; Oudega, Bradbury, & Ramer, 2012; Smith, Falone, & Frank, 2012; Tetzlaff et al., 2011; Weishaupt, Blesch, & Fouad, 2012). To date, provision of neurotrophic support after SCI includes some of the most commonly used neurotrophins and neuropoietic cytokines, as well as other growth factors such as glial cell-derived neurotrophic factor (GDNF), the fibroblast growth factors (FGFs), and insulin-like growth factors (IGFs). Combinations of these various factors have also been trialed. A summary of some of the factors that have been tested in SCI, their mode of delivery and their effects on structure and function, is provided in Tables 1–4. This review does not include studies employing avulsion injuries of spinal roots.

As discussed in detail elsewhere (Harvey et al., 2015), it is essential to determine where best to apply such factors and how to administer them. Growth factors can be delivered: (i) in and/or immediately around the spinal cord lesion itself, (ii) to location(s) containing the neurons whose axons are affected by injury, and/or (iii) via intrathecal or systemic application. Injecting neurotrophic factors intrathecally reduces the need for high concentration delivery using conventional systemic injections, but the effectiveness of such therapy is susceptible to rapid dispersion and elimination by cerebrospinal fluid (Pardridge, 2012). The effectiveness of the type of therapeutic intervention and overall mode of delivery, as well as timing of neurotrophic delivery (e.g., acute vs chronic), may also vary markedly depending on the level (e.g., cervical vs thoracic) and type (e.g., incomplete contusion/compression vs complete transection) of SCI. Other issues include whether the neurotrophic factors are to be exogenously administered by acute injection of recombinant proteins, subacute or chronic delivery using osmotic minipumps, cell-mediated delivery, delivery via liposomes, nanoparticles, or polymer scaffolds, or the use of gene therapy to modify neurons, glial cells, or precursor/stem cells. Temporal application of neurotrophic support immediately after SCI, or even after delayed intervals, and the period over which such support must be sustained may also influence outcomes. The ability to modulate support over time is crucial to coordinate promotion of neuronal survival, axonal sprouting, long-distance regeneration of axons, terminal arborization, and reformation of synaptic connections.

Table 1 Summary of Selected Studies Using BDNF and NT-3 (See End of Table 4 for Key to Abbreviations)

BDNF and NT-3

Species	SCI Model	Growth Factor Delivery: Method	Growth Factor Delivery: Time and Duration	Outcomes	References
Rat Lewis	Dorsobilateral transection, lower thoracic spinal cord	NT-3, BDNF, or NGF	NS	*Morphology*: NT-3 √, BDNF X, NGF √ *Function*: NA *Survival*: ~17d	Schnell, Schneider, Kolbeck, Barde, and Schwab (1994)
Rat F344	T8 Transection	SC BDNF/NT-3 (sc, minipump)	Acute (d0)	*Morphology*: BDNF +NT-√ *Function*: NA *Survival*: 4wks	Xu, Guenard, Kleitman, Aebischer, and Bunge (1995)
Rat F344	T7 Dorsal hemisection	NT-3 rFB	Acute (d0)	*Morphology*: √ *Function*: √ *Survival*: 3mo	Grill, Murai, Blesch, Gage, and Tuszynski (1997)
Rat SD (Male)	C3/C4 Hemilaminectomy	BDNF, NT-4/5 PN, minipump	Acute (d7, for 7d)	*Morphology*: BDNF or NT-4/5 √ *Function*: NA *Survival*: 12wks	Kobayashi et al. (1997)
Rat F344	T8 Transection	BDNF SC	Acute (d0)	*Morphology*: √ *Function*: NA *Survival*: 4 h, 24 h, 7d, 30d	Menei, Montero-Menei, Whittemore, Bunge, and Bunge (1998)

Rat F344	T8 Contusion	NT-3, BDNF, NGF, bFGF FB	Acute (d2)	*Morphology*: √ *Function*: NA *Survival*: 10wks	McTigue, Horner, Stokes, and Gage (1998)
Rat Wistar (Male)	T9 Dorsal hemisection	NT-3 Collagen	Acute (d0)	*Morphology*: √ *Function*: √ *Survival*: 4wks	Houweling, Lankhorst, Gispen, Bar, and Joosten (1998)
Rat Wistar	T9 Dorsal hemisection	BDNF Collagen	Acute (d0)	*Morphology*: √ *Function*: √ *Survival*: 4wks	Houweling, van Asseldonk, et al. (1998)
Rat Wistar (Male)	T6 Crush	NT-3, BDNF, GDNF Minipump	Acute (d0 or 4wks)	*Morphology*: NT-3 √, BDNF X, GDNF √ *Function*: NA *Survival*: 4wks	Bradbury et al. (1999)
Rat SD	C3/C4 Dorsolateral transection	BDNF hBF, RV transduction	Acute (d0)	*Morphology*: √ *Function*: √ *Survival*: 2mo	Liu et al. (1999)
Rat, SD	T11 Lateral hemisection	BDNF, CNTF, NT-3, GDNF PN, minipump	Acute (d0 for 15d)	*Morphology*: √ *Function*: NA *Survival*: 15d	Yick, Wu, So, and Yip (1999)
Rat SD	C3 Hemisection	BDNF, NT-3	1–4wks, 1–8wks, 5–8wks	*Morphology*: √ *Function*: NA *Survival*: 8wks	Novikova, Novikov, and Kellerth (2000)

Continued

Table 1 Summary of Selected Studies Using BDNF and NT-3 (See End of Table 4 for Key to Abbreviations)—cont'd

BDNF and NT-3

		Growth Factor Delivery			
Species	**SCI Model**	**Method**	**Time and Duration**	**Outcomes**	**References**
Rat SD	T6–T8 Complete transection	E14 fetal grafts	BDNF, NT-3 Minipump	*Morphology*: √ *Function*: √ *Survival*: 10wks	Coumans et al. (2001)
Rat SD	C4 Dorsolateral aspiration	BDNF rFB (RV, alginate)	Acute (d0)	*Morphology*: √ *Function*: NA *Survival*: 2wks, 4wks	Tobias et al. (2001)
Rat (NS)	T8 Lateral hemisection	BDNF, NT-3 SC, MG	Acute (d0, for 30d)	*Morphology*: √ *Function*: NA *Survival*: ~37d	Bamber et al. (2001)
Rat Fischer	T7 Subcortical lesions	BDNF FB	Acute (d0)	*Morphology*: √ *Function*: NA *Survival*: 2wks or 3mo	Lu, Blesch, and Tuszynski (2001)
Rat SD (Male)	T9–T10 Ablation	BDNF Minipump	Acute (d0, 14d)	*Morphology*: √ *Function*: NA *Survival*: 3wks	Hiebert, Khodarahmi, McGraw, Steeves, and Tetzlaff (2002)
Rat SD	C3 Hemisection	BDNF rFB	Chronic (4wks)	*Morphology*: √ *Function*: √ *Survival*: 2mo	Jin, Fischer, Tessler, and Houle (2002)

Rat SD (Male)	C4 Transection	BDNF Minipump	Chronic (6–12mo)	*Morphology*: √ *Function*: NA *Survival*: >12mo	Kwon et al. (2002)
Rat F344	C4 Transection	BDNF, NT-3 (OEC, AV)	Acute (d0)	*Morphology*: BDNF √, NT-3 X *Function*: √ *Survival*: 4mo	Ruitenberg et al. (2003)
Rat SD (Male)	T13 Hemisection (Unilateral)	BDNF NPC	Chronic (d28)	*Morphology*: NA *Function*: √ *Survival*: 60d	Hains, Johnson, Eaton, Willis, and Hulsebosch (2003)
Rat F344	T9–T10 Transection	SC bridge BDNF, NT-3 AAV	Acute (d0)	*Morphology*: √ *Function*: √ *Survival*: Transgene 16wks	Blits, Oudega, Boer, Bartlett Bunge, and Verhaagen (2003)
Rat SD (Male)	C3/C4 Axotomy	BDNF Gelfoam	Chronic (d56)	*Morphology*: X *Function*: NA *Survival*: ∼10wks	Kwon et al. (2004)
Rat Wistar (Male)	T8 Complete transection	BDNF AAV	Acute (d0)	*Morphology*: √ *Function*: √ *Survival*: 6wks	Koda et al. (2004)
Rat SD	C4 Hemisection	rFB BDNF rFB, alginate	Acute (d0)	*Morphology*: √ *Function*: √ *Survival*: 5wks	Tobias et al. (2005)

Continued

Table 1 Summary of Selected Studies Using BDNF and NT-3 (See End of Table 4 for Key to Abbreviations)—cont'd

BDNF and NT-3

		Growth Factor Delivery			
Species	**SCI Model**	**Method**	**Time and Duration**	**Outcomes**	**References**
Rat F344	C3 Transection	BDNF rMPC, RV	Acute (d0)	*Morphology*: √ *Function*: X *Survival*: 1mo, 3mo	Lu, Jones, and Tuszynski (2005)
Rat F344	C4 CST transection	NT-3 OEC, AV	Acute (d0)	*Morphology*: √ *Function*: X *Survival*: 3mo	Ruitenberg et al. (2005)
Rat SD	C5 or L2 Transection	NT-3 Minipump	Acute (d0, for 2wks)	*Morphology*: √ *Function*: NA *Survival*: 2wks	Hagg, Baker, Emsley, and Tetzlaff (2005)
Rat SD	C5/C6 Lateral hemisection	BDNF, NT-3 Fetal grafts	Subacute (d14)	*Morphology*: X *Function*: √ *Survival*: 6wks	Lynskey et al. (2006)
Rat F344	C4 Transection	rMPC BDNF rMPC, RV	Acute (d0)	*Morphology*: √ *Function*: NA *Survival*: 1mo	Stokols et al. (2006)
Rat SD (Male)	T10 Hemisection	BDNF Agarose	Acute (d0)	*Morphology*: √ *Function*: NA *Survival*: 6wks	Jain, Kim, McKeon, and Bellamkonda (2006)

Rat Lewis	T8 Right dorsal hemisection	BDNF, NT-3 Minipump	Acute (d0 for 2wks)	*Morphology*: X *Function*: X *Survival*: 6wks	Vavrek, Girgis, Tetzlaff, Hiebert, and Fouad (2006)
Rat F344	C3 Transection	NT-3 rMPC, LV	Acute (d0)	*Morphology*: √ *Function*: NA *Survival*: 1mo	Taylor, Jones, Tuszynski, and Blesch (2006)
Rat Long–Evans	T9 Complete ablation	hNT-3 fibrin	Acute (d0)	*Morphology*: √ *Function*: X *Survival*: 12wks	Taylor, Rosenzweig, McDonald, and Sakiyama-Elbert (2006)
Rat SD (Male)	T8 Hemisection	NT- Hydrogel	Acute (d0, for 2wks)	*Morphology*: √ *Function*: √ *Survival*: 6wks	Piantino, Burdick, Goldberg, Langer, and Benowitz (2006)
Rat SD	T8 Complete transection	FGF-1, NT-3 Hydrogel	Acute (d0)	*Morphology*: FGF-1 √, NT-3 X *Function*: √ *Survival*: 8wks	Tsai, Dalton, Shoichet, and Tator (2006)
Rat SD	C4/C5 Lateral hemisection	BDNF, NT-3	Subacute (d14)	*Morphology*: NT-3 √, BDNF X *Function*: NA *Survival*: 6wks	Iarikov et al. (2007)
Cat NS	T11–T12 Complete transection	rFB BDNF NT-3 rFB	Acute (d0)	*Morphology*: √ *Function*: √ *Survival*: 12wks	Boyce, Tumolo, Fischer, Murray, and Lemay (2007)

Continued

Table 1 Summary of Selected Studies Using BDNF and NT-3 (See End of Table 4 for Key to Abbreviations)—cont'd

BDNF and NT-3

		Growth Factor Delivery			
Species	**SCI Model**	**Method**	**Time and Duration**	**Outcomes**	**References**
Rat Wistar (Male)	T8 Transection	BDNF (AV rMPC, MG)	Acute (d0)	*Morphology*: √ *Function*: √ *Survival*: 6wks	Koda et al. (2007)
Rat F344	T7 Dorsal hemisection	Fibroblasts expressing RV-hBDNF FB, RV	Acute (d0, varied 6wks)	*Morphology*: √ *Function*: NNA *Survival*: 12wks	Blesch and Tuszynski (2007)
Rat SD (Male)	C4 Axotomy	BDNF AAV, LV Minipump	Acute (d0)	*Morphology*: √ *Function*: NA *Survival*: 2wks, 4wks	Kwon et al. (2007)
Rat SD	T9 Contusion	NT-3 Minipump	1 Acute (d0 for 1wk)	*Morphology*: X *Function*: NA *Survival*: 9d	Baker, Nakashima, and Hagg (2007)
Rat SD	T8–T9 Transection	NT-3 NSPC, SCs, AV, collagen	Acute (d0)	*Morphology*: √ *Function*: √ *Survival*: ~2mo	Guo et al. (2007)
Rat F344	C3 Transection	NT-3 rMPC, RV	Chronic (6wks)	*Morphology*: √ *Function*: NA *Survival*: 6wk, 3mo	Lu, Jones, and Tuszynski (2007)

Rat SD	T9 Dorsal hemisection	BDNF hMPC, AV	Acute (d0)	*Morphology*: √ *Function*: √ *Survival*: 5wks	Sasaki et al. (2009)
Rat	C1 Transection	NT- MPC, LV	Acute (d0)	*Morphology*: √ *Function*: NA *Survival*: ∼4wks	Alto et al. (2009)
Rat F344	C3 Complete transection	NT-3 MSC, LV	Chronic (6wks and 15mo)	*Morphology*: √ *Function*: NA *Survival*: 12wks or 16.5mo	Kadoya et al. (2009)
Rhesus monkey	C7 Lateral hemisection	BDNF, NT-3 FB, LV, gelfoam	Acute (d0)	*Morphology*: √ *Function*: NA *Survival*: 6wks, 6mo, and 8mo	Brock et al. (2010)
Rat LE	T9 Dorsal hemisection	NT-3, PDGF mNPCs, fibrin	Subacute (d14)	*Morphology*: √ *Function*: NA *Survival*: 4wks	Johnson, Tatara, Shiu, and Sakiyama-Elbert (2010)
Rat SD	T7/T8 Dorsoventral hemisection	NT-3 hNSCs, RV PCL scaffold	Acute (d0)	*Morphology*: √ *Function*: √ *Survival*: 9wks	Hwang et al. (2011)
Rat SD (Male)	T8 Dorsal hemisection	BDNF Hydrogel	Acute (d0)	*Morphology*: √ *Function*: NA *Survival*: 6wks	Jain, McKeon, Brady-Kalnay, and Bellamkonda (2011)

Continued

Table 1 Summary of Selected Studies Using BDNF and NT-3 (See End of Table 4 for Key to Abbreviations)—cont'd

BDNF and NT-3

		Growth Factor Delivery			
Species	**SCI Model**	**Method**	**Time and Duration**	**Outcomes**	**References**
Rat F344	C2/C3 Dorsal columns, CST	NT-3 MPC, LV	Acute (d0 for 4–10wks)	*Morphology*: √ *Function*: NA *Survival*: 10wks	Hou, Nicholson, van Niekerk, Motsch, and Blesch (2012)
Rat SD	T10 Complete transection	BDNF, NT-3 AAV	Acute (d0)	*Morphology*: √ *Function*: BDNF √, NT-3 X *Survival*: 6wks	Boyce, Park, Gage, and Mendell (2012)
Rat SD	T8 Complete transection	BDNF, NT-3, NGF mNSC	Acute (d0)	*Morphology*: X *Function*: √ *Survival*: 1mo	Gu et al. (2012)
Rat Long–Evans	4 mm midline segment removed	BDNF, NT-3 LV, nanoparticles	Acute (d0)	*Morphology*: √ *Function*: NA *Survival*: 1, 2, and 4wks	Tuinstra et al. (2012)
Rat F344	C5–C6 Lateral hemisection	BDNF, NT-3 FB, RV, scAAV	Acute (d7)	*Morphology*: √ *Function*: X *Survival*: 9wks	Fouad, Bennett, Vavrek, and Blesch (2013)
Rat SD	C5 Hemicontusion	BDNF, GDNF Gelfoam, PN graft, ChABC	Chronic (15–17wks)	*Morphology*: ChABC √, BDNF X *Function*: X *Survival*: 25wks	Tom et al. (2013)

Rat SD	T10 Hemisection	rMPCs BDNF rMPCs, AV	Acute (d0)	*Morphology*: √ *Function*: √ *Survival*: 2mo	Zhao et al. (2013)
Rat SD	T10 Complete transection	BDNF mNSC	Acute (d0)	*Morphology*: √ *Function*: √ *Survival*: 4wks	He et al. (2013)
Rat SD	T9 Complete transection	NT-3 Collagen scaffold	Acute (d0)	*Morphology*: √ *Function*: X *Survival*: 4wks	Yao et al. (2013)
Cat	T12 Contusion	BDNF, NT-3 rFB	Acute (d0)	*Morphology*: X *Function*: √ *Survival*: 5wks	Ollivier-Lanvin, Fischer, Tom, Houle, and Lemay (2014)
Rat F344	T7–T8 Contusion	NT-3 rMPC, LV	Acute (d7)	*Morphology*: √ *Function*: √ *Survival*: 5wks	Wang, Zhang, and Li (2014)
Rat Lewis	T8 Hemisection	NT-3 AAV-2	Acute (d0)	*Morphology*: *Function*: NA *Survival*: 54d	Weishaupt et al. (2014)
Rat Wistar (Male)	T9/T10 Complete transection	BDNF AAV	Acute (d0 for 7wks)	*Morphology*: √ *Function*: √ *Survival*: 7wks	Ziemlinska et al. (2014)

Continued

Table 1 Summary of Selected Studies Using BDNF and NT-3 (See End of Table 4 for Key to Abbreviations)—cont'd

BDNF and NT-3

		Growth Factor Delivery			
Species	**SCI Model**	**Method**	**Time and Duration**	**Outcomes**	**References**
Rat F344	C5 Lateral hemisection	BDNF rBMSCs Alginate hydrogel	Acute (d0)	*Morphology*: √ *Function*: NA *Survival*: 4wks	Gunther, Weidner, Muller, and Blesch (2015)
Canine Beagle female	T12 Complete transection	BDNF Collagen scaffold	Acute (d0)	*Morphology*: √ *Function*: √ *Survival*: 38wks	Han et al. (2015)
Mouse C57BL/ 6	T10 Clip compression	BDNF LV	Acute (d3)	*Morphology*: √ *Function*: √ *Survival*: 4wks	Ji et al. (2015)
Rat SD	T9/T10 Complete transection	BDNF, NT-3 cAMP, collagen	Acute (d0)	*Morphology*: √ *Function*: √ *Survival*: 12wks	Li et al. (2016)
Rat SD	T9 Contusion	BDNF BMSC, LV	Acute (d3)	*Morphology*: √ *Function*: X *Survival*: 4wks	Ritfeld et al. (2015)

Rat SD	T9/10 Dorsolateral hemisection	BDNF PLGA/HA scaffolds	Acute (d0)	*Morphology*: √ *Function*: √ *Survival*: 8wks	Wen et al. (2015)
Rat Long–Evans	T8 Dorsolateral hemisection	NT-3, PDGF_AA NPCs, fibrin	Subacute (d14)	*Morphology*: √ *Function*: NA *Survival*: 2wks	Wilems, Pardieck, Iyer, and Sakiyama-Elbert (2015)
Rat Wistar	T7–T8 Complete transection	NT-3 Chitosan	Acute (d0)	*Morphology*: √ *Function*: √ *Survival*: 57wks	Yang et al. (2015)
Rat F344	T9–T10 Hemisection	BDNF, NT-3 AAV	Acute (d0)	*Morphology*: BDNF √, NT-3 X *Function*: NA *Survival*: 3wks	Betz et al. (2016)
Rat SD	T7–T8 Contusion	NT-3 BMSCs, LV	Acute (d7)	*Morphology*: √ *Function*: √ *Survival*: 5wks	Zhang, Wang, He, Xie, and Li (2016)

Table 2 Summary of Selected Studies Using NGF

NGF

		Delivery			
Species	**SCI Model**	**Method**	**Time and Duration**	**Outcomes**	**References**
Rat F344	No injury	rFB, hNGF, bFGF rFB, RV	N/A	*Morphology* NGF √, bFGF X *Function*: NA *Survival*: 1yr	Tuszynski et al. (1994)
Rat	No injury	hNGF SC, RV	N/A	*Morphology*: √ *Function*: Not affected *Survival*: 1yr	Tuszynski et al. (1998)
Rat F344	T7 Dorsal hemisection	NGF SC, FB, Collagen	Acute (d0)	*Morphology*: √ *Function*: NA *Survival*: 6mo	Weidner, Blesch, Grill, and Tuszynski (1999)
Rat SD	T10 Contusion	bFGF, NGF, CNTF, IL1, 4, 6, Minipump	Acute (d0 for 7d)	*Morphology*: bFGF √, CNTF and NGF X *Function*: NA *Survival*: 7d	Lee et al. (1999)
Rat SD	L4/L5 Clip compression	FGF2, NGF AV	Subacute (d16)	*Morphology*: √ *Function*: √ *Survival*: 16d	Romero et al. (2001)

Rat SD	T9 Contusion	NGF rMPCs	Acute (d7)	*Morphology*: √ *Function*: √ *Survival*: 2mo	Huang, Wang, and Chen (2006)
Rat Wistar	T10 Contusion	NGF, BDNF rFSCC, SC	Acute (d7)	*Morphology*: √ *Function*: √ *Survival*: 3mo	Feng et al. (2009)
Rat SD	T9 Clip compression	NGF	Acute (d0)	*Morphology*: √ *Function*: √ *Survival*: Up to d14	Zhang et al. (2014)
Rat SD	T9 Clip compression	NGF Gelatin	Acute (d0)	*Morphology*: √ *Function*: √ *Survival*: 2wks	Zhu et al. (2015)
Rat SD	T9 Clip compression	NGF Hydrogel	Acute (d0)	*Morphology*: √ *Function*: √ *Survival*: 4wks	Zhao et al. (2016)

See also Schnell et al. (1994), McTigue et al. (1998), and Gu et al. (2012) in Table 1.

Table 3 Summary of Selected Studies Using FGF

FGF					
		Delivery			
Species	SCI Model	Method	Time and Duration	Outcomes	References
Rat SD	T10 Contusion	rh-bFGF	Acute (d0 for 1wk)	*Morphology*: √ *Function*: √ *Survival*: 6wks	Rabchevsky et al. (1999)
Rat SD	T1 Clip compression	EGF, bFGF Minipump,	Acute (d0 for 3–14d)	*Morphology*: X *Function*: X *Survival*: 6wks	Kojima and Tator (2002)
Rat F344	T8 Transection	FGF2 Fibrin	Acute (d0)	*Morphology*: X *Function*: X *Survival*: 3–12wks	Meijs et al. (2004)
Rat SD (Male)	T9 Contusion	bFGF rMPC	Acute (d7)	*Morphology*: √ *Function*: √ *Survival*: 8wks	Kim, Oh, Lee, and Yoon (2006)
Rat SD	T8 Complete transection	FGF-1 PN, fibrin glue	Acute (d0)	*Morphology*: √ *Function*: √ *Survival*: 2–4wks	Lee et al. (2008)
Human Trial	Cervical SCI Complete and incomplete	FGF-1 Fibrin glue	5mo–8.5yr	*Morphology*: N/A *Function*: √ *Survival*: N/A	Wu, Huang, Tsai, Chen, and Cheng (2008)

Rat Wistar	T10 Contusion	hbFGF NSCs	Acute (d7)	*Morphology*: √ *Function*: √ *Survival*: 4wks	Meng et al. (2008)
Rat SD	T9 Contusion	FGF1	Acute (d0)	*Morphology*: √ *Function*: √ *Survival*: 28d	Tsai, Shen, Kuo, Cheng, and Chak (2008)
Rat SD	T8 Complete transection	aFGF Fibrin glue, PNG	Acute (d0)	*Morphology*: √ *Function*: √ *Survival*: 6mo	Lee et al. (2010)
Rat SD	T9 Contusion	aFGF AAV	Acute (d0)	*Morphology*: √ *Function*: √ *Survival*: 5wks	Huang et al. (2011)
Zebra-fish	Complete transection	FGF Transgenic	Acute (d0)	*Morphology*: √ *Function*: √ *Survival*: 5wks	Goldshmit et al. (2012)
Rat Wistar (Male)	T11/12 Complete transection	FGF2 Gelfoam	Acute (d0)	*Morphology*: √ *Function*: √ *Survival*: 2mo	Guzen et al. (2012)
Rat SD	T2 Clip compression	FGF2 Nanoparticles, hydrogel	Acute (d0)	*Morphology*: √ *Function*: X *Survival*: 4wks	Kang, Baumann, Tator, and Shoichet (2013)
Rat SD	T10 Contusion	bFGF Gelatin hydrogel	Acute (d7)	*Morphology*: √ (reduced mechanical allodynia)	Furuya et al. (2013)

Continued

Table 3 Summary of Selected Studies Using FGF—cont'd

FGF					
		Delivery			
Species	SCI Model	Method	Time and Duration	Outcomes	References
				Function: X *Survival*: 9wks	
Rat Wistar	T10 Complete transection	FGF2	Acute (d0)	*Morphology*: √ *Function*: √ *Survival*: 6wks	Kasai, Jikoh, Fukumitsu, and Furukawa (2014)
Mice C57BL6 (Male)	T12 Complete left hemisection	hFGF2	Acute (d0, for 5wks)	*Morphology*: √ *Function*: √ *Survival*: 5wks	Goldshmit et al. (2014)
Rat SD	T10 Contusion	FGF-2 hMPCs, PLGA-nanospheres	Acute (d0)	*Morphology*: √ *Function*: √ *Survival*: 9wks	Shin, Pennant, Yoon do, Ha, and Kim (2014)
Rat SD	T9 Complete transection	bFGF-HEMA-MOETACL hydrogel	Acute (d5)	*Morphology*: √ *Function*: √ *Survival*: 8wks	Chen et al. (2015)
Rat Wistar	T8–T9 Balloon compression	bFGF, EGF alginate scaffold	Acute (d7)	*Morphology*: √ *Function*: √ *Survival*: 7wks	Grulova et al. (2015)

See also McTigue et al. (1998) and Tsai et al. (2006) in Table 1; Lee, Green, Dietrich, and Yezierski (1999) and Romero, Rangappa, Garry, and Smith (2001) in Table 2.

Table 4 Summary of Selected Studies Using GDNF

GDNF					
		Delivery			
Species	SCI Model	Method	Time and Duration	Outcomes	References
Rat F344	T7 Complete transection or dorsal hemisection	GDNF FB, MLV	Acute (d0 for 3mo)	*Morphology*: √ *Function*: X *Survival*: 2wks, 1mo, 3mo	Blesch and Tuszynski (2003)
Rat F344	T8 Lateral hemisection	rhGDNF SC, MG	Acute (d0)	*Morphology*: √ *Function*: NA *Survival*: 4–5wks	Iannotti et al. (2003)
Rat SD	C3 Hemisection	GDNF Gelfoam	Subacute (4wks)	*Morphology*: √ *Function*: NA *Survival*: Up to 8wks	Dolbeare and Houle (2003)
Rat SD	T11 Contusion	rhGDNF Minipump	Acute (d0 for 28d)	*Morphology*: √ *Function*: NA *Survival*: 7wks	Iannotti et al. (2004)
Rat SD	T8 Transection	GDNF OEC, RV	Acute (d0)	*Morphology*: √ *Function*: √ *Survival*: 8wks	Cao et al. (2004)

Continued

Table 4 Summary of Selected Studies Using GDNF—cont'd

GDNF

		Delivery			
Species	**SCI Model**	**Method**	**Time and Duration**	**Outcomes**	**References**
Rat SD	T8 Contusion	GDNF mNPC	Subacute (d8)	*Morphology*: X *Function*: √ (Enhanced pain responses) *Survival*: NA	Macias et al. (2006)
Rat SD	T10 Hemisection	rhGDNF Minipump	Acute (d0 for 28d)	*Morphology*: √ *Function*: NA *Survival*: Up to 6wks	Zhang et al. (2009)
Rat SD	T10 Transection	GDNF rSC, LV	Acute (d0)	*Morphology*: √ *Function*: NA *Survival*: 6wks	Deng et al. (2011)
Rat Lewis	T8–T9 Complete transection	GDNF Collagen scaffold	Acute (d0)	*Morphology*: √ *Function*: X *Survival*: Up to 6wks	Cholas, Hsu, and Spector (2012)

Rat SD	T11 Hemisection	GDNF SC, LV	Acute (d0 for 6wks)	*Morphology*: √ *Function*: √ *Survival*: 6–8wks	Deng et al. (2013)
Rat Long–Evans	T9/10 Lateral hemisection	GDNF Alginate fibrogen: hydrogel	Acute (d0 for 6wks or 3mo)	*Morphology*: √ *Function*: √ *Survival*: 6wks or 3mo	Ansorena et al. (2013)

See also Yick et al. (1999), Bradbury et al. (1999), and Tom et al. (2013) in Table 1.
Key: Growth factors: *aFGF*, acidic fibroblast growth factor; *BDNF*, brain-derived neurotrophic factor; *bFGF*, basic fibroblast growth factor; *CNTF*, ciliary neurotrophic factor; *GDNF*, glial cell-derived neurotrophic factor; *HGF*, hepatocyte growth factor; *IGF-1*, insulin-like growth factor; *NGF*, nerve growth factor; *Ngn2*, neurogenin-2; *NT*, neurotrophin; *NT-3*, neurotrophin-3; *PDGF*, platelet-derived growth factor; *VEGF*, vascular endothelial growth factor.
Donor cell types: *FB*, fibroblast; *FSCC*, fetal spinal cord stem cell; *GRP*, glial restricted precursors; *OEC*, olfactory ensheathing cell; *NSPC*, neural stem/progenitor cell; *MPC*, mesenchymal precursor cell from bone marrow ("BMSC" or "MSC"); *SC*, Schwann cell. Note: All cells injected into lesion site unless otherwise stated.
Scaffolds: *MG*, matrigel.
Animal strains: *F344*, Fischer; *LE*, Long–Evans; *SD*, Sprague Dawley.
Viral: *AAV*, adeno-associated virus; *AV*, adenovirus; *LV*, lentivirus; *MMLV*, Moloney murine leukemia retrovirus; *RV*, retrovirus; *scAAV*, self-complementary AAV.
Tracts: *CST*, corticospinal.
General: *C*, cervical; *d*, day; *h*, human; *m*, mouse; *min*, minute; *mo*, month; *N/A*, not applicable; *NA*, not assessed; *NS*, not specified; *PN*, peripheral nerve; *r*, rat; *T*, thoracic; *wk*, week. Note: All host animals female unless stated. √, significant improvement (for one or more factors); X, no significant improvement.

2. NEUROTROPHINS

The family of four structurally related neurotrophin proteins consist of brain-derived neurotrophic factor (BDNF), neurotrophin-3 (NT-3), nerve growth factor (NGF), and neurotrophin-4/5 (NT-4/5) (reviewed in Boyce & Mendell, 2014). All have important roles in peripheral and central nervous system development, cell survival, differentiation, and axonal growth throughout life. Each neurotrophin binds with and signals through specific, high-affinity Trk (tropomyosin receptor kinase) receptors (primarily NGF to TrkA; BDNF, NT-4/5 to TrkB; NT-3 to TrkC), but all members can also bind with lower affinity to the p75(NTR) receptor (Huang & Reichardt, 2001).

2.1 Brain-Derived Neurotrophic Factor

The most highly studied neurotrophin in experimental SCI (see Table 1), BDNF, plays a role in neurogenesis, neuroprotection, axonal sprouting, myelination, adaptive synaptic plasticity, and synaptic transmission (Kovalchuk, Holthoff, & Konnerth, 2004). Evidence also indicates that BDNF has potent antioxidative effects and may be involved in the regulation of the immune response after SCI (Ji et al., 2015). The importance of the location of BDNF administration as well as the timing of delivery has been specifically highlighted by Lu et al. (Lu et al., 2001) and several other groups (Blesch & Tuszynski, 2007; Kwon et al., 2002; Novikova et al., 2000). In most SCI studies, BDNF is applied directly into the spinal cord (via minipump, viral vector injection, or secretion by genetically modified donor cells) either rostral, caudal, or within the lesion, and occasionally in combinations of these sites. Other studies have applied or expressed BDNF in the brainstem region containing the red nucleus (Kobayashi et al., 1997; Kwon et al., 2007, 2002; Ruitenberg et al., 2004; Weishaupt, Li, Di Pardo, Sipione, & Fouad, 2013) or via infusion into sensorimotor cortex (Hiebert et al., 2002; Vavrek et al., 2006). In terms of the timing of BDNF delivery, only partial functional improvements were reported in chronic studies (Hains et al., 2003; Jin et al., 2002), while acute studies showed mixed responses (see later). Some studies using delayed delivery have reported improved survival/growth of axons after SCI (Coumans et al., 2001; Kwon et al., 2002; Novikova et al., 2000). Delayed delivery of BDNF to neurons in the red nucleus is effective in rescuing their lesion-induced atrophy (Ruitenberg et al., 2004).

2.1.1 Neuroprotection, Sparing, and Growth With BDNF

After SCI, BDNF administration can enhance numbers of spared axons and/or induce extensive regrowth or sprouting within the lesion site, although rarely beyond (Blesch & Tuszynski, 2007; Blits et al., 2003; Golden et al., 2007; Jain et al., 2006, 2011; Koda et al., 2004, 2007; Kwon et al., 2002; Liu et al., 1999; Lu et al., 2005, 2012; Lynskey et al., 2006; Ollivier-Lanvin et al., 2014; Ruitenberg et al., 2003; Sharp, Flanagan, Yee, & Steward, 2012; Stokols et al., 2006; Tobias et al., 2005; Tuinstra et al., 2012; Xu et al., 1995). Exogenously applied BDNF reportedly reduces atrophy of cortical (Brock et al., 2010; Sasaki et al., 2009), rubrospinal (Kobayashi et al., 1997; Kwon et al., 2007; Novikova et al., 2000; Ruitenberg et al., 2004), and spinal neurons (Yick et al., 1999). Not all axonal populations are responsive to BDNF, for example, Menei et al. reported sensory and brainstem but no propriospinal regeneration (Menei et al., 1998). The situation with the impact of BDNF on sprouting/regrowth of corticospinal (CST) axons is not settled, with some reporting no effects (Brock et al., 2010; He et al., 2013; Iarikov et al., 2007; Lu et al., 2001; Schnell et al., 1994; Xu et al., 1995), while others described BDNF-induced sprouting of CST axons after SCI in adult rats (Lynskey et al., 2006; Sasaki et al., 2009; Vavrek et al., 2006). In most studies, exogenous BDNF increases myelination after SCI (Bamber et al., 2001; Golden et al., 2007; McTigue et al., 1998; Ollivier-Lanvin et al., 2014; Stokols et al., 2006; Xu et al., 1995; Zhao et al., 2013).

Murine neonatal CST neurons express TrkB, and BDNF has been shown to enhance branching and arborization of these neurons in vitro (Ozdinler & Macklis, 2006). However, adult SCI can alter TrkB expression in and around the lesion site (Liebl et al., 2001) and it has been reported that this receptor is expressed primarily on the soma and dendrites but not on the axons of *adult* CST neurons (Lu et al., 2001). Interestingly, adult SCI is similarly known to alter p75 expression (Ji et al., 2015; King et al., 2000; Liebl et al., 2001), which could potentially affect the action of exogenously applied BDNF. In contrast to these inconsistent effects on CST regeneration, many studies have reported regrowth of 5HT fibers in response to BDNF administration (Boyce et al., 2007; Brock et al., 2010; Coumans et al., 2001; Golden et al., 2007; Koda et al., 2007; Lu et al., 2005;

Tobias et al., 2001), which has the potential to facilitate locomotor recovery (Slawinska, Miazga, & Jordan, 2014).

2.1.2 Functional Recovery With BDNF

Studies using thoracic SCI report either no additional improvements in locomotor function over controls (Koda et al., 2004, 2007; Lu et al., 2005; Sharp et al., 2012; Vavrek et al., 2006) or partial and/or significant improvements (Blits et al., 2003; Boyce et al., 2012; Coumans et al., 2001; Feng et al., 2009; Han et al., 2014; He et al., 2013; Houweling, van Asseldonk, et al., 1998; Lu et al., 2012; Ollivier-Lanvin et al., 2014; Sasaki et al., 2009; Ziemlinska et al., 2014). Hains et al. reported reduced allodynia/mechanical responses that may in part be attributed to BDNF (Hains et al., 2003). With cervical injuries, either no additional improvements in limb function (Lu et al., 2005; Tom et al., 2013) or partial and/or significant improvements (Liu et al., 1999; Ruitenberg et al., 2003; Tobias et al., 2005; Weishaupt et al., 2013) over controls have been reported. Following high cervical hemisection, chronic intrathecal infusion (Mantilla, Gransee, Zhan, & Sieck, 2013) or cell-mediated delivery (Gransee, Zhan, Sieck, & Mantilla, 2014) of BDNF into cervical spinal cord targeting phrenic motoneurons resulted in improved diaphragm function, an effect that is dependent on effective TrkB signaling. Vector-mediated overexpression of TrkB in these neurons is also effective (Gransee et al., 2014).

BDNF delivery using agarose scaffolds enhances neural fiber infiltration of support cells and neural fiber growth into implanted scaffolds (Jain et al., 2006), while BDNF-enriched collagen can increase axonal regeneration of peripheral axons (Han et al., 2014) or CST axons (Houweling, Lankhorst, et al., 1998; Houweling, van Asseldonk, et al., 1998) and thereby enhance functional recovery. Importantly however, from a potential therapeutic perspective, sensitized nociceptive circuits and spasticity have also been reported after BDNF delivery (Boyce et al., 2012; Fouad & Tetzlaff, 2012).

2.2 Neurotrophin-3

This neurotrophin has also been extensively used in experimental SCI studies (Table 1) and occasionally in studies that compare delivery of NT-3 to the effects of BDNF in the same lesion paradigm (Bamber et al., 2001; Boyce et al., 2012; Brock et al., 2010; Iarikov et al., 2007; Li et al., 2016; Lynskey et al., 2006; McTigue et al., 1998; Novikova et al., 2000; Ollivier-Lanvin et al., 2014; Piantino et al., 2006; Ruitenberg et al., 2003; Schnell et al., 1994; Taylor, Jones, et al., 2006; Tuinstra et al., 2012; Wilems et al., 2015; Xu et al., 1995; Yick et al., 1999) (see also Table 1). Most experiments involve mid-to-low thoracic injuries, although several cervical SCI studies have been undertaken (Alto et al., 2009; Brock et al., 2010; Fouad et al., 2013; Gunther et al., 2015; Hagg et al., 2005; Hou et al., 2012; Iarikov et al., 2007; Kadoya et al., 2009; Lu et al., 2007; Lynskey et al., 2006; Novikova et al., 2000; Ruitenberg et al., 2005, 2003; Taylor, Jones, et al., 2006). The majority of studies involve delivery of NT-3 in or close to the SCI lesion site, and most involve acute delivery immediately or soon after injury. Exceptions are Vavrek et al. (2006) who delivered NT-3 intrathecally at the cervical level after a thoracic (T8) hemisection, Zhang et al. (2016) who delivered via lumbar puncture after thoracic (T7–T8) contusion, and two groups who administered NT-3 after 14 days (Johnson, Tatara, Shiu, et al., 2010; Lynskey et al., 2006), and a study that applied NT-3 6 weeks after SCI (Lu et al., 2007).

NT-3 has also been administered in a range of studies in association with scaffolds of various types, composed of collagen (Houweling, Lankhorst, et al., 1998; Houweling, van Asseldonk, et al., 1998; Li et al., 2016; Yao et al., 2013), fibrin (Guo et al., 2007; Johnson, Tatara, Shiu, et al., 2010; Taylor, Jones, et al., 2006; Wilems et al., 2015), matrigel (Bamber et al., 2001), hyaluronic acid/PLGA (Wen et al., 2015), chitosan (Yang et al., 2015), or hydrogels (Piantino et al., 2006; Tsai et al., 2006). Improved outcomes using a combination of NT-3 gradients, peripheral conditioning lesions, and stromal cell grafts up to 15 months after SCI have also been reported (Kadoya et al., 2009). A further study combined delivery of adeno-associated viral (AAV) vectors encoding BDNF to sensorimotor cortex with AAV-NT-3 injection into the ipsilateral reticular formation (Weishaupt et al., 2013) (Table 1). Finally, after C1 dorsal transection, Alto et al. (2009) used sciatic nerve preconditioning lesions and mesenchymal stem cells grafts in the lesion to promote axonal regeneration and then injected lentiviral vectors to overexpress NT-3 in an appropriate brainstem target (nucleus gracilis) or inappropriate target (reticular formation). NT-3 expression in nucleus gracilis promoted reinnervation and axodendritic synapse formation, whereas NT-3 expression in the reticular formation induced ectopic growth of ascending sensory axons, "even bypassing a correct target to grow into an NT-3 source" (Alto et al., 2009).

2.2.1 Neuroprotection, Sparing, and Growth With NT-3

NT-3 applied using minipumps or via cellular or extracellular matrix delivery can enhance the sprouting and local growth of CST axons

(Grill et al., 1997; Houweling, Lankhorst, et al., 1998; Iarikov et al., 2007; Ji et al., 2015; Ruitenberg et al., 2005; Schnell et al., 1994) although others reported no effect (Vavrek et al., 2006) or even reduced sprouting (Hagg et al., 2005) of such axons. Overexpression of NT-3 in lumbar spinal cord using an adenoviral vector 2 weeks but not 4 months after unilateral lesion of the medullary pyramids resulted in enhanced sprouting of intact CST axons (Chen, Zhou, & Shine, 2006). Other axonal populations also respond to exogenous application of NT-3 after SCI (Bradbury et al., 1999; Bradbury, King, Simmons, Priestley, & McMahon, 1998; Hou et al., 2012; Lu et al., 2007; Taylor, Jones, et al., 2006), although in contrast to BDNF, NT-3 has no impact on rubrospinal neurons or their axons (Ruitenberg et al., 2003), and reduced growth of axons from brainstem neurons (Tsai et al., 2006). Positive effects on myelination have been documented (Han et al., 2015; Hwang et al., 2011; Ji et al., 2015; Yang et al., 2015), perhaps associated with infiltrating Schwann cells (Thomas et al., 2014).

2.2.2 Functional Recovery With NT-3

After thoracic injuries, a number of studies that assessed behavior reported an improvement in locomotor function over controls (Blits et al., 2003; Boyce et al., 2012; Coumans et al., 2001; Grill et al., 1997; Guo et al., 2007; Houweling, Lankhorst, et al., 1998; Hwang et al., 2011; Ollivier-Lanvin et al., 2014; Ruitenberg et al., 2003; Wang et al., 2014). Few studies have addressed the impact of NT-3 after cervical SCI; in some studies, no additional improvement in limb function was seen compared to controls (Ruitenberg et al., 2005), while others described an improvement in a forelimb-reaching task (Lynskey et al., 2006; Ruitenberg et al., 2003). In terms of the timing of delivery, delayed application of NT-3 postinjury was associated with greater axonal growth (Coumans et al., 2001), and when used with BDNF, NT-3 administration 5–8 weeks after the SCI prevented the death and atrophy of rubrospinal neurons (Novikova et al., 2000). Later NT-3 administration also was found to improve forelimb function after cervical SCI (14-day delay: Lynskey et al., 2006). NT-3 has also been administered in association with different donor cell types such as fibroblasts (Blits et al., 2003; Boyce et al., 2007; Brock et al., 2010; Fouad et al., 2013; Liu et al., 1999; McTigue et al., 1998; Ollivier-Lanvin et al., 2014), Schwann cells (Bamber et al., 2001: #11794), olfactory ensheathing cells (Ruitenberg et al., 2003, 2005), neuronal precursor/stem cells (Guo et al., 2007; Hwang et al., 2011; Johnson, Tatara, McCreedy, Shiu, &

Sakiyama-Elbert, 2010), and mesenchymal precursor cells (MPCs) (Gu et al., 2012; Hou et al., 2012; Lu et al., 2007; Taylor, Jones, et al., 2006; Wang et al., 2014; Zhang et al., 2016), with generally improved functional outcomes being reported.

2.3 NGF and NT-4/5

The least tested in SCI models, NGF and NT-4/5, have been almost always delivered immediately after the injury (but see Grill et al., 1997) (Table 2). Initial NGF studies reported enhanced growth of sensory, including nociceptive axons (Tuszynski et al., 1994, 1998), while others described improved morphology and greater overall axonal growth (Feng et al., 2009; Huang et al., 2006; Weidner et al., 1999) with improved locomotor outcomes. Delivery of NGF using scaffold-based approaches such as collagen (Weidner et al., 1999), gelatin, nanostructured lipid carriers (Zhu et al., 2015), and heparin poloxamer thermosensitive hydrogels (Zhao et al., 2016) resulted in morphological and functional improvements. NGF released from fibroblasts grafted 1–3 months after SCI enhanced the growth of "putative" cerulospinal axons and primary sensory axons (Grill et al., 1997) and with donor MPCs (Huang et al., 2006) also promoted regeneration and functional improvements. Reportedly, NGF has positive synergistic effects with BDNF (Feng et al., 2009). A recent study reported that NGF reduced endoplasmic reticulum-induced death of spinal cord neurons after T9 contusion, with improved locomotor function (Zhang et al., 2014). NT-4/5 signals primarily through TrkB, as does BDNF, and the effects of NT-4/5 on neuroprotection and axonal growth after SCI appear to be similar and act on similar neuronal populations (Blesch, Yang, Weidner, Hoang, & Otero, 2004; Bregman, McAtee, Dai, & Kuhn, 1997; Kobayashi et al., 1997). Recently, network analysis of the SCI interactome of proteins revealed that NGF disruption is a major potential target for SCI therapies (Alawieh, Sabra, Sabra, Tomlinson, & Zaraket, 2015).

2.4 Neuropoietic Cytokines

Ciliary neurotrophic factor (CNTF) and leukemia inhibitory factor (LIF) are cytokines whose activity and mode of action have a direct impact within the nervous system. CNTF signals through the Jak/STAT pathway and has been extensively and successfully used to promote the regeneration of rodent adult retinal ganglion cell axons (Fischer & Leibinger, 2012; Harvey et al., 2006; Hellstrom & Harvey, 2011), yet the effects of neuropoietic cytokines have rarely been tested after SCI. After thoracic hemisection or contusion injuries, acute application of CNTF to the lesion site has been reported to protect neurons in red nucleus (Ye et al., 2004) and also Clarke's nucleus (Yick et al., 1999). CNTF also protects CST neurons in sensorimotor cortex after intracortical axotomy (Dale, Kuang, Wei, & Varon, 1995), and these neurons, at least in murine neonates, are known to express CNTF receptor α (Ozdinler & Macklis, 2006). LIF also increases the growth of such axons after thoracic SCI (Blesch et al., 1999), again consistent with expression of CNTF receptor α by lamina V pyramidal neurons (Ip et al., 1993). However, this may be complicated by the influence of LIF-secreting cells on increased local expression of NT-3 that may have contributed to enhanced CST plasticity (Blesch et al., 1999). Improved functional outcomes have also been reported (Cao et al., 2010; Wang et al., 2014; Ye et al., 2004).

A study using exogenous LIF delivery found no change in astroglial or microglial/macrophage activation (Blesch et al., 1999), while another CNTF study did report altered astrocytic and microglial activity in close proximity to the injury (Ye et al., 2004). CNTF is expressed in reactive astrocytes after SCI (Lee, Kim, Shin, Moon, & Chun, 1998), and with FGF2, this cytokine likely plays a role in postinjury gliogenesis (Tripathi & McTigue, 2008). Antibody neutralization of CNTF reportedly reduced scar formation and promoted donor cell survival (Ishii et al., 2006). CNTF is also a chemoattractant for macrophages (Cen et al., 2007) with potentially useful or adverse consequences to tissue repair (Benowitz & Popovich, 2011; Yin et al., 2003).

Acute CNTF delivery following contusion SCI using transfected rat MPCs from bone marrow improved axonal regrowth and functional recovery (Abbaszadeh et al., 2015). It remains unresolved whether MPCs (or any donor cell type for that matter) can simply improve the terrain of the injury site to enhance existing endogenous repair processes and promote functional and morphological outcomes after SCI as a result.

3. SELECTED EXAMPLES OF OTHER GROWTH FACTORS

3.1 Fibroblastic Growth Factors

Belonging to a large family of heparin-binding proteins, FGFs interact with membrane-associated proteoglycans and bind to a variety of FGF receptor subtypes. FGF1 (acidic, aFGF) and FGF2 (basic, bFGF) have been widely tested for their potential therapeutic effects in the CNS. Almost all mammalian SCI studies that employed FGF have applied either FGF1 or FGF2 in the vicinity of the lesion, usually immediately or within 7 days after the contusion or transection injury, usually at thoracic levels (Table 3). Intrathecal administration typically involves osmotic pumps to continuously deliver FGF to (or near) the lesion over days or weeks (Kim et al., 2006; Kojima & Tator, 2002; Lee et al., 1999; Rabchevsky et al., 1999). FGF has also been delivered in association with Schwann cells or peripheral nerve (PN) grafts, released by carriers such as gelfoam, gelatin, fibrin glue, HEMA-MOETACL hydrogels alginate scaffolds or from nanoparticles (Chen et al., 2015; Furuya et al., 2013; Grulova et al., 2015; Guzen et al., 2012; Kang et al., 2013; Lee et al., 2008, 2010; Meijs et al., 2004; Meng et al., 2008; Shin et al., 2014; Tsai et al., 2006; Wu et al., 2008). A recent study successfully delivered FGF2 (a known mitogen for stem cell self-renewal) via subcutaneous injection to give improved outcomes (Goldshmit et al., 2014).

Tissue sparing has been improved in most contusion studies with FGF2 infusions (Grulova et al., 2015; Kasai et al., 2014; Kim et al., 2006; Lee et al., 1999; Rabchevsky et al., 1999), one study reporting significant reduction if treated with FGF2 within 1 h post-SCI but not if treatment was delayed for 3 h (Lee et al., 1999). Some studies report βIII-tubulin, MAP2, Gap-43, or 5HT-positive axons within the lesion site of FGF1- or FGF2-treated animals. After more severe or complete transection injuries, FGFs are usually administered in association with fibrin bridges and either Schwann cell or PN grafts, or with hydrogels (Chen et al., 2015). Increased axonal growth has been reported in some of these studies (Chen et al., 2015; Guzen et al., 2012; Kasai et al., 2014; Lee et al., 2008, 2010; Tsai et al., 2006) but not in others (Meijs et al., 2004).

Behaviorally, there have been mixed results, with locomotor improvement reported in several studies after contusion injuries (Grulova et al., 2015; Hwang et al., 2011; Kim et al., 2006; Meng et al., 2008; Rabchevsky et al., 1999; Shin et al., 2014; Tsai et al., 2008), and with some (Chen et al., 2015; Goldshmit et al., 2014; Guzen et al., 2012; Kasai et al., 2014; Lee et al., 2008; Tsai et al., 2006) or no (Meijs et al., 2004) functional improvement after complete transection. A human trial using FGF1 and fibrin glue in nine patients with cervical SCI reported improvement in ASIA grade for some patients, as well as some improvement in bowel/bladder function and reduced neuropathic pain for one patient (Wu et al., 2008).

FGFs are known to affect astrocyte phenotype and proliferation in CNS tissue as well as in nonneuronal cells. Changes in astrocyte phenotype after FGF administration to the injured spinal cord are not always extensive

(Kang et al., 2013; Kim et al., 2006), although FGF2 (Grulova et al., 2015) and FGF1 (Lee et al., 2008) may reduce astrocytic reactivity and macrophage infiltration. In zebrafish SCI studies (Goldshmit et al., 2012), subcutaneous administration of FGF2 in adult mice reduced inflammation, microglia and macrophage activation, and astrocyte reactivity at 2 weeks. Axons were seen entering and crossing the lesion site, closely associated with GFAP-positive astrocytic "bridges" (Goldshmit et al., 2014). Similar types of glial/meningeal bridges that support axonal regrowth have previously been described in other parts of the CNS (Dyson, Harvey, Trapp, & Heath, 1988; Harvey & Tan, 1992), emphasizing the importance of geometry in determining how reactive glia influence regenerative events after injury.

3.2 Glial Cell-Derived Neurotrophic Factor

Most experimental SCI studies that have used GDNF have delivered this protein to the region of the thoracic injury site in the spinal cord (Table 4). Virtually all GDNF administration has been immediate (acute) or shortly after injury (subacute), either as recombinant protein (minipump delivery) (Iannotti et al., 2004; Zhang et al., 2009), via collagen scaffold (Cholas et al., 2012), released from gels (Abdanipour, Tiraihi, & Taheri, 2014; Ansorena et al., 2013; Dolbeare & Houle, 2003; Iannotti et al., 2003; Tom et al., 2013), via viral vector injection (Deng et al., 2013), with nanoparticles (Menon, Muresanu, Sharma, Mossler, & Sharma, 2012), or released from various types of transplanted, genetically modified cells (Blesch & Tuszynski, 2003; Cao et al., 2004; Deng et al., 2013, 2011; Gowing et al., 2014; Macias et al., 2006). Sequential delivery of GDNF in gelfoam immediately after cervical hemisection SCI and again a month later has been described (Dolbeare & Houle, 2003).

Morphologically, studies described reduced lesion volume and greater sparing after GDNF infusion via osmotic minipump (Iannotti et al., 2004) or dehydrothermally cross-linked collagen scaffold (Cholas et al., 2012), and several studies saw reduced astrocytic reactivity around the injury site (Ansorena et al., 2013; Deng et al., 2011; Iannotti et al., 2003; Menon et al., 2012) and reduced macrophage infiltration (Iannotti et al., 2003). Most studies describe increased fiber growth from various supraspinal sources and propriospinal axons (Deng et al., 2011; Dolbeare & Houle, 2003; Iannotti et al., 2003, 2004) as well as some types of sensory axon, for example, CGRP and substance P-positive axons (Blesch & Tuszynski, 2003; Zhang et al., 2009). One study reported no CST tract axons in GDNF-expressing fibroblast grafts (Blesch & Tuszynski, 2003), whereas these axons were seen in GDNF-transduced olfactory glia grafts (Cao et al., 2004). Greater myelination has also been reported, probably by Schwann cells (Blesch & Tuszynski, 2003; Deng et al., 2013, 2011; Iannotti et al., 2003; Zhang et al., 2009). Note here that trapping of axons in GDNF-expressing grafts may occur (Blesch & Tuszynski, 2003).

Behaviorally, improvements in locomotor function have been reported (Abdanipour et al., 2014; Ansorena et al., 2013; Cao et al., 2004; Deng et al., 2013) although others saw little or no change (Blesch & Tuszynski, 2003; Cholas et al., 2012; Iannotti et al., 2003; Tom et al., 2013) and one study (Macias et al., 2006) reported enhanced pain receptivity, possibly related to GDNF stimulation of nociceptive axonal sprouting and regrowth.

3.3 IGF-1 and PDGF

Both IGF-1 and platelet-derived growth factor (PDGF) signaling occur through respective tyrosine kinase receptors and influence various intracellular pathways associated with cell division and growth, including activation via the mTOR pathway. IGF-1 enhances growth of murine neonatal CST neurons in vitro and in vivo via PI3-dependent pathways (Ozdinler & Macklis, 2006). IGF-1 and PDGF delivered with Schwann cells in a Matrigel channel elicited greater ingrowth of brainstem axons and enhanced myelination (Oudega, Xu, Guenard, Kleitman, & Bunge, 1997). After C3 transection, grafting of bone marrow-derived MPCs \transduced with IGF-1 also promoted raphespinal and cerulospinal axons but not CST axons, although death of these neurons was prevented by coadministration of an IGF-binding protein inhibitor (Hollis et al., 2009). Importantly in the context of later discussion about site of neurotrophic factor delivery, adult CST somas express IGF-1 receptors, but their axons do not (Hollis et al., 2009). IGF-1 levels in the injured spinal cord after T11 hemisection increased and locomotor function improved after systemic intravenous administration of an IGF-1 plasmid (Hung et al., 2007). Delayed (14 days' postinjury) transplantation of embryonic stem cells combined with PDGF and NT-3 in fibrin scaffolds enhanced the viability of grafted cells after T9 dorsal hemisection and increased locomotor function (grid-walk) (Johnson, Tatara, McCreedy, et al., 2010; Johnson, Tatara, Shiu, et al., 2010).

4. COMBINATIONS OF FACTORS

Combined neurotrophic factor delivery in therapeutic SCI studies includes NT-3 with BDNF (Bamber et al., 2001; Boyce et al., 2007; Brock et al., 2010; Fouad et al., 2013; Gu et al., 2012; Novikova et al., 2000; Xu et al., 1995), NT-3 and PDGF (Johnson, Tatara, McCreedy, et al., 2010; Johnson, Tatara, Shiu, et al., 2010), a cerebrolysin mixture of BDNF,GDNF, NGF, and CNTF (Menon et al., 2012), and NGF with BDNF (Feng et al., 2009; Gu et al., 2012). PDGF in combination with vascular endothelial growth factor improved morphological and behavioral outcomes after T10 contusion (Chehrehasa, Cobcroft, Young, Mackay-Sim, & Goss, 2014). Subacute or chronic (8 week) grafting of neural precursors combined with chondroitinase ABC (chABC) to degrade chondroitin sulfate proteoglycans (CSPG) with transient intrathecal infusion of epidermal growth factor (EGF), FGF-2, and PDGF elicited greater plasticity of CST axons and brainstem-derived 5HT axons as well as improved functional recovery with no apparent enhancement of nociception (Karimi-Abdolrezaee, Eftekharpour, Wang, Schut, & Fehlings, 2010). Following lateral T8 or T10 hemisection injuries, NT-3 combined with herpes simplex viral vector-induced overexpression of the NMDA receptor 2 subunit and either chABC (Garcia-Alias et al., 2011) or an antibody to Nogo-A (Schnell et al., 2011) increased axonal plasticity with enhanced "functional detours" in spinal cord circuitry. Similarly, after T10 contusion SCI, combination of AAV-NT-3 and chABC near the lesion site resulted in the establishment of a novel detour projection from VLF and dCST with improved locomotor function (Hunanyan, Petrosyan, Alessi, & Arvanian, 2013). PN grafts and chABC treatment were combined with GDNF administration directly into the lesion and lentiviral delivery of BDNF distal to an injury, although the number of regenerating axons did not increase their greater numbers of synaptic connections with spinal cord neurons distally (Tom et al., 2013). Finally, after cervical or thoracic SCI in immunocompromised nude rats, transplantation of human neural stem cells in a fibrin matrix containing a cocktail of nine neurotrophic factors (including BDNF, NT-3, PDGF, IGF-1, GDNF, and FGFs) facilitated long-distance outgrowth of axons from grafted cells, albeit with inconsistent results relating to host axonal growth into the grafts and differing functional outcomes (Lu et al., 2012, 2014; Sharp et al., 2012).

The multitrophin D15 is a modified human NT-3 that binds to and activates TrkB and TrkC receptors and therefore has both BDNF and NT-3 biological activities (Urfer et al., 1994). This modified gene has been incorporated into Schwann cells or various types of neural precursor cell prior to implantation into contused/compressed thoracic spinal cord, either at the time of injury or in the subacute phase (7–9 days) after SCI. A chronic study grafted D15 modified neural precursor cells 6 weeks after the injury (Kusano et al., 2010). D15 therapy consistently increases myelination (Cao et al., 2005; Enomoto, Bunge, & Tsoulfas, 2013; Golden et al., 2007; Kanno et al., 2014; Kusano et al., 2010) and enhances growth of some supraspinal and sensory axons (Enomoto et al., 2013; Golden et al., 2007; Kanno et al., 2014) and some inconsistent effects on CST growth (Golden et al., 2007; Kanno et al., 2014). Significant or partial improvement in motor and sensory function has been reported in some D15 therapies (Cao et al., 2005; Fan et al., 2013; Kanno et al., 2014; Kusano et al., 2010) but not all (Enomoto et al., 2013; Hurtado et al., 2006). D15 transduced Schwann cell grafts combined with enzymatic removal of CSPGs resulted in greater axonal growth (Kanno et al., 2014). Finally, D15 gene mutation reduced binding of the neurotrophic protein to the p75/NTR receptor and enhanced Schwann cell graft survival and produced greater regrowth of supraspinal and sensory axons (Enomoto et al., 2013).

Numerous studies have employed neurotrophic factors in combination with growth-supporting matrices and/or cell or tissue grafts (see McCall et al., 2012; Oudega et al., 2012 for recent reviews). Such combinations may involve the transduction of donor cell types to overexpress the particular neurotrophin(s), or as a combination through codelivery via injection or osmotic minipumps. Other studies that coculture donor cells in the presence of neurotrophins prior to transplantation have been reviewed elsewhere (Tetzlaff et al., 2011). Several examples of combined use of neurotrophic factors with other pharmacotherapeutic agents have already been given, as has the use of chABC to degrade CSPGs. Elevation of cAMP for SCI therapy is comparatively common (Hannila & Filbin, 2008; Pearse et al., 2004), although the combined application of a neurotrophin with cyclic AMP elevation is surprisingly rare (Li et al., 2016; Lu et al., 2007; Lu, Yang, Jones, Filbin, & Tuszynski, 2004).

5. LOCATION OF NEUROTROPHIC FACTOR DELIVERY

In experimental SCI models the most common site of administered neurotrophic support is the spinal cord itself. Often delivered directly into the lesion site, and released from grafted engineered cells or from bridging matrices, or sometimes injected just rostral and/or caudal to the injury, factors can also be chronically delivered intrathecally near the SCI using osmotic minipumps. As summarized earlier, these approaches have generally been found to be effective, usually providing beneficial impacts on tissue sparing, glial reactivity, and axonal plasticity/growth. Although functional improvements are not always observed, a majority of studies have found that most neurotrophic/growth factors can enhance locomotor recovery to some extent, especially when assessing hindlimb function after thoracic SCI. Note that it is the norm to cite and compare mean group values when reporting functional or histological outcomes, but correlative analysis of morphology/connectivity with sensorimotor performance at the individual animal level is rarely undertaken (Harvey et al., 2015). Of course, delivery of factors into the lesion area may affect numerous axons and/or cell populations in different ways, with complex and not always predictable outcomes. The effects of exogenous neurotrophic factor expression on CNS glia, myelination, and infiltrating macrophages/monocytes have mainly been described as beneficial, but this may not always be the case. In addition, when targeting a specific neuronal population and its damaged axons, application of relevant factor(s) proximal to an injury site may induce local sprouting of responsive axons in that location rather than lengthy regrowth of the axons through and beyond a lesion (Harvey, Ooi, & Rodger, 2012). Finally, factors expressed by cells within a lesion cavity may enhance ingrowth of injured axonal populations, but those axons can become trapped, with few able to grow out into distal host spinal cord tissue (Blesch & Tuszynski, 2003; McCall et al., 2012). The use of restorative bridges combined with cellular grafts placed distal to the injury site as a source of neurotrophic attraction and guidance may be helpful under such circumstances.

An alternative, or more likely an additional, mode of delivery of neurotrophic factors as a therapeutic treatment for SCI involves directly targeting the neurons whose axons are compromised or transected by the trauma. Direct targeting of injured neurons is used infrequently in SCI studies, not least because of the large number of potential sites in the brain and spinal cord that project axons into spinal cord tracts, but also the different types of neuronal populations that may express different receptors and respond differentially to exogenous neurotrophic support (Blesch et al., 2012). Application of neurotrophic factors to the spinal cord will obviously directly impact propriospinal neurons and spinal projection neurons, sometimes to excellent effect (Blits et al., 2003; Boyce & Mendell, 2014; Boyce et al., 2012; Bradbury et al., 1998; Deng et al., 2011; Iannotti et al., 2003; Lu et al., 2005; Mantilla et al., 2013; Xu et al., 1995; Yick et al., 1999), but what about other neuronal populations? As mentioned earlier, some studies have focused on increasing BDNF levels in red nucleus (Kobayashi et al., 1997; Kwon et al., 2007, 2002; Ruitenberg et al., 2004) or in sensorimotor cortex (Hiebert et al., 2002; Vavrek et al., 2006). Combined administration of AAV-BDNF to sensorimotor cortex and AAV-NT-3 to the reticular formation has been reported (Weishaupt et al., 2013), and lentiviral vectors have also been used to overexpress trkB in sensorimotor cortex, increasing the regrowth of CST axons after subcortical axotomy (Hollis et al., 2009). The transduction of multiple neuronal populations with a cocktail of different AAV serotypes may be a useful approach to affect a wide range of relevant neuronal populations. Thus after SCI, AAV serotypes 1, 2, 5, 8, and 9 injected just rostral to a spinal cord lesion can be retrogradely transported by severed axons to thousands of neurons within spinal cord and several sites in the brainstem (Klaw, Xu, & Tom, 2013). Importantly, different serotypes displayed different transduction efficiencies and regional specificities.

We believe that, in addition to therapies aimed at repairing the injured spinal cord itself, more attempts should be made to directly target supraspinal neurons that are damaged after such injuries. Previously, we reported preliminary data using bi-*cis*tronic AAV vectors (serotype 2.1) to introduce neurotrophic factors (AAV-CNTF^{mCherry}) into cortical neurons in adult rat sensorimotor cortex, with the eventual aim of combining this approach with cellular bridging grafts in the injured spinal cord (Harvey et al., 2015). We are in the process of carrying out such combinatorial studies using MPCs to stabilize the lesion site to improve the peri-lesion terrain through which CST axons can regenerate. Preliminary data suggest that cell-transplanted animals have reduced injury cavities and more extensive CST fiber regrowth around and beyond the lesion site. Animals transplanted with viable cells have the highest density of transfected AAV-CNTF^{mCherry}-positive fibers and occasional anterogradely labeled CST axons have been observed several mm caudal to the injury site.

6. MODE OF NEUROTROPHIC FACTOR DELIVERY

We have summarized here how neurotrophic factors can be delivered as recombinant proteins (acute injections or infusion using osmotic pumps), by gene therapy using viral vectors or other delivery systems, in biological or polymer bridges/scaffolds, nanoparticles, or via glia or precursor/stem cells that are genetically modified ex vivo prior to transplantation into the SCI site (see Tables 1–4). In all studies, it is important to know the initial dose of a given factor, its bioavailability, and whether sustained or transient delivery of neurotrophic support is the best option. The amount of recombinant protein delivered as a bolus or over a given period can be calibrated when using injection or minipump protocols, although assessing the bioactivity of any remaining neurotrophic factor in minipumps at the end of the delivery period is often not documented. With other modes of delivery system, while the kinetics of neurotrophic factor release from bridging matrices, viral vectors, or transduced cells can readily be measured in vitro, this is much more difficult to assess in vivo although several studies have successfully attempted the task (Cao et al., 2005; Golden et al., 2007; Kanno et al., 2014; Tuszynski, Murai, Blesch, Grill, & Miller, 1997).

Modulation of neurotrophic support over time may be critical to enable sequential regenerative events to occur, such as neuroprotection, followed by local sprouting, or the long-distance regrowth of axons followed by terminal arborization and functional reconnection with appropriate target neurons. The latter in particular may involve more subtle, short-range signals that could be swamped by sustained expression of high levels of exogenous neurotrophic factor. Neurotrophic factor supply may cease abruptly (e.g., from implanted minipumps that need to be removed from the spinal cord to prevent reinjury as animals recover sufficiently and become more active and mobile), or there may be more gradual loss of support from grafted cells that cease neurotrophic factor production or fail to survive transplantation long term. Controlled release of neurotrophic factors such as NT-3, BDNF, GDNF, CNTF, NGF, and FGF has been reported from a wide variety of biopolymers, and natural or synthetic scaffolds. Scaffold degradation can be used to modulate the kinetics of release (Ansorena et al., 2013; Piantino et al., 2006), but it is also possible to control release using affinity-based delivery systems (such as a heparin-based delivery system) that are reliant on noncovalent interactions between the therapeutic agent and the delivery system to sequester proteins and limit the amount of free trophic factor diffusion (Sakiyama-Elbert, Johnson, Hodgetts, Plant, & Harvey, 2012).

Using genetic methods to express neurotrophic factors, maintained expression has been identified in BDNF- or GDNF-transduced olfactory ensheathing glia at 1 and 8 weeks, respectively, after engraftment (Cao et al., 2004; Ruitenberg et al., 2003). NT-3 mRNA expression was seen in transduced fibroblasts 6 months after transplantation to normal non-lesioned spinal cord (Grill et al., 1997). After direct AAV injection, sustained expression of BDNF and NT-3 has been described in cells in the injured spinal cord up to 16 weeks' postinjection (Blits et al., 2003). In contrast, downregulation of the BDNF gene in encapsulated fibroblasts 2 weeks after transplantation into an adult rat cervical SCI was reported (Tobias et al., 2001). Interestingly, expression returned when the cells were harvested and put back into culture, "suggesting that soluble factors originating from the host immune response may contribute to the down-regulation" (Tobias et al., 2001). With viral vectors, temporal regulation of transgene expression using, for example, the tetracycline-activated (tet-ON or tet-OFF) system is now possible, and a modified, immunologically inert variant of the trans-activator protein has been described (Hoyng et al., 2014; Zaldumbide et al., 2010), which may allow the use of this regulatable system in human gene therapy trials. In rats, using fibroblasts that expressed BDNF under the tet-ON promotor, Blesch and Tuszynski (2007) reported that transient BDNF expression was sufficient to promote axonal outgrowth and axons were maintained in the absence of ongoing BDNF support. In contrast, regulation of NT-3 expression using the tet-OFF system reduced axon numbers when NT-3 expression was transient, suggesting a loss of regenerate axons (Hou et al., 2012).

7. TIMING OF NEUROTROPHIC FACTOR DELIVERY

The vast majority of experimental SCI studies in rodents employing neurotrophic factor therapy administer the factors in the acute or, less frequently, subacute (0–14 days) phase, but only rarely in the chronic (~1 year) phase (Tables 1–4). Early delivery would appear to be an optimal strategy in order to minimize atrophy and/or loss of spinal neurons and other more distant neuronal populations, reduce axonal retraction/die-back and demyelination, and modulate the phenotype of astrocytes and microglia and other infiltrating monocytes/macrophages. The sequential delivery of different factors, appropriate for different phases of the repair process, has rarely been tested in experimental SCI (Tom et al., 2013). Nonetheless, a relatively small number of animal SCI studies in which neurotrophic factors have been delivered many weeks or months after the initial injury (Hains et al., 2003; Jin et al., 2002; Kadoya et al., 2009; Kwon et al., 2002, 2004; Lu et al., 2007; Tom et al., 2013) and a human clinical trial involving a 5- to 8.5-year time span (Wu et al., 2008) have reported some improved morphological and behavioral outcomes. Such studies should perhaps be encouraged in order to determine whether there is any consistent benefit in applying (expensive) neurotrophic therapy to patients with chronic SCI.

8. CONCLUSION

Neurotrophic factor delivery strategies in experimental SCI require careful selection of which factor(s) to use depending on the tracts or cells to be targeted, but it is also critically important to consider the location, mode, and timing of application after an injury. These deliberations are further complicated by the likely need to promote the regrowth not only of descending motor pathways but also of ascending sensory tracts after SCI, each tract potentially with its own trophic requirements. Furthermore, it is crucial to know which trophic factor receptors are expressed by specific neuronal (and glial) populations, whether receptors are differentially expressed on cell bodies vs growth cones, and whether expression of individual receptor(s) changes in response to injury and/or application of the neurotrophic factors themselves. Adverse effects such as increased sensitivity to pain have only occasionally been reported. Comparatively fewer SCI studies have used neurotrophic factors to target the cell bodies of injured neurons themselves, an approach that is highly successful in visual system trauma models (Harvey et al., 2012). Most SCI studies using neurotrophic factor therapy have been performed in rat or mouse, but upscaling to larger animals including cats, dogs (Han et al., 2014, 2015), and primates (Brock et al., 2010; Tuszynski, Grill, Jones, McKay, & Blesch, 2002) may be important to confirm the efficacy of such therapies across species. However, even within a given species, it is known that injury responses to SCI as well as behavioral outcomes vary between inbred strains (Basso et al., 2006; Cui, Hodgetts, Hu, Luo, & Harvey, 2007; Kigerl, McGaughy, & Popovich, 2006). Thus, it may also need to be established if the efficacy of a particular neurotrophic factor is consistent across different strains within a species, especially when replicating SCI studies in experimental animals as a precursor for predicting which neurotrophic factor-based therapeutic protocols have the most likely translational potential in clinical trials.

ACKNOWLEDGMENTS

This work was in part funded by grants from the WA Neurotrauma Research Programme.

REFERENCES

Abbaszadeh, H. A., Tiraihi, T., Noori-Zadeh, A., Delshad, A. R., Sadeghizade, M., & Taheri, T. (2015). Human ciliary neurotrophic factor-overexpressing stable bone marrow stromal cells in the treatment of a rat model of traumatic spinal cord injury. *Cytotherapy*, *17*, 912–921.

Abdanipour, A., Tiraihi, T., & Taheri, T. (2014). Intraspinal transplantation of motoneuron-like cell combined with delivery of polymer-based glial cell line-derived neurotrophic factor for repair of spinal cord contusion injury. *Neural Regeneration Research*, *9*, 1003–1013.

Alawieh, A., Sabra, M., Sabra, Z., Tomlinson, S., & Zaraket, F. A. (2015). Molecular architecture of spinal cord injury protein interaction network. *PLoS One*, *10*, e0135024.

Alto, L. T., Havton, L. A., Conner, J. M., Hollis, E. R., 2nd., Blesch, A., & Tuszynski, M. H. (2009). Chemotropic guidance facilitates axonal regeneration and synapse formation after spinal cord injury. *Nature Neuroscience*, *12*, 1106–1113.

Ansorena, E., De Berdt, P., Ucakar, B., Simon-Yarza, T., Jacobs, D., Schakman, O., et al. (2013). Injectable alginate hydrogel loaded with GDNF promotes functional recovery in a hemisection model of spinal cord injury. *International Journal of Pharmaceutics*, *455*, 148–158.

Arvanian, V. (2013). Role of neurotrophins in spinal plasticity and locomotion. *Current Pharmaceutical Design*, *19*, 4509–4516.

Awad, B. I., Carmody, M. A., & Steinmetz, M. P. (2015). Potential role of growth factors in the management of spinal cord injury. *World Neurosurgery*, *83*, 120–131.

Baker, K. A., Nakashima, S., & Hagg, T. (2007). Dorsal column sensory axons lack trkc and are not rescued by local neurotrophin-3 infusions following spinal cord contusion in adult rats. *Experimental Neurology*, *205*, 82–91.

Bamber, N. I., Li, H., Lu, X., Oudega, M., Aebischer, P., & Xu, X. M. (2001). Neurotrophins BDNF and NT-3 promote axonal re-entry into the distal host spinal cord through Schwann cell-seeded mini-channels. *European Journal of Neuroscience*, *13*, 257–268.

Basso, D. M., Fisher, L. C., Anderson, A. J., Jakeman, L. B., McTigue, D. M., & Popovich, P. G. (2006). Basso mouse scale for locomotion detects differences in recovery after spinal cord injury in five common mouse strains. *Journal of Neurotrauma*, *23*, 635–659.

Benowitz, L. I., & Popovich, P. G. (2011). Inflammation and axon regeneration. *Current Opinion in Neurology*, *24*, 577–583.

Betz, V. M., Sitoci-Ficici, K. H., Uckermann, O., Leipnitz, E., Iltzsche, A., Thirion, C., et al. (2016). Gene-activated fat grafts for the repair of spinal cord injury: A pilot study. *Acta Neurochirurgica (Wien)*, *158*, 367–378.

Blesch, A., Fischer, I., & Tuszynski, M. H. (2012). Gene therapy, neurotrophic factors and spinal cord regeneration. *Handbook of Clinical Neurology*, *109*, 563–574.

Blesch, A., & Tuszynski, M. H. (2003). Cellular GDNF delivery promotes growth of motor and dorsal column sensory axons after partial and complete spinal cord transections and induces remyelination. *Journal of Comparative Neurology*, *467*, 403–417.

Blesch, A., & Tuszynski, M. H. (2007). Transient growth factor delivery sustains regenerated axons after spinal cord injury. *Journal of Neuroscience*, *27*, 10535–10545.

Blesch, A., Uy, H. S., Grill, R. J., Cheng, J. G., Patterson, P. H., & Tuszynski, M. H. (1999). Leukemia inhibitory factor augments neurotrophin expression and corticospinal axon growth after adult CNS injury. *Journal of Neuroscience*, *19*, 3556–3566.

Blesch, A., Yang, H., Weidner, N., Hoang, A., & Otero, D. (2004). Axonal responses to cellularly delivered nt-4/5 after spinal cord injury. *Molecular and Cellular Neurosciences*, *27*, 190–201.

Blits, B., Oudega, M., Boer, G. J., Bartlett Bunge, M., & Verhaagen, J. (2003). Adeno-associated viral vector-mediated neurotrophin gene transfer in the injured adult rat spinal cord improves hind-limb function. *Neuroscience*, *118*, 271–281.

Boyce, V. S., & Mendell, L. M. (2014). Neurotrophins and spinal circuit function. *Frontiers in Neural Circuits*, *8*, 1–7. Article 59.

Boyce, V. S., Park, J., Gage, F. H., & Mendell, L. M. (2012). Differential effects of brain-derived neurotrophic factor and neurotrophin-3 on hindlimb function in paraplegic rats. *European Journal of Neuroscience*, *35*, 221–232.

Boyce, V. S., Tumolo, M., Fischer, I., Murray, M., & Lemay, M. A. (2007). Neurotrophic factors promote and enhance locomotor recovery in untrained spinalized cats. *Journal of Neurophysiology*, *98*, 1988–1996.

Bradbury, E. J., Khemani, S., Von, R., King, Priestley, J. V., & McMahon, S. B. (1999). NT-3 promotes growth of lesioned adult rat sensory axons ascending in the dorsal columns of the spinal cord. *European Journal of Neuroscience*, *11*, 3873–3883.

Bradbury, E. J., King, V. R., Simmons, L. J., Priestley, J. V., & McMahon, S. B. (1998). NT-3, but not BDNF, prevents atrophy and death of axotomized spinal cord projection neurons. *European Journal of Neuroscience*, *10*, 3058–3068.

Bregman, B. S., Coumans, J. V., Dai, H. N., Kuhn, P. L., Lynskey, J., McAtee, M., et al. (2002). Transplants and neurotrophic factors increase regeneration and recovery of function after spinal cord injury. *Progress in Brain Research*, *137*, 257–273.

Bregman, B. S., McAtee, M., Dai, H. N., & Kuhn, P. L. (1997). Neurotrophic factors increase axonal growth after spinal cord injury and transplantation in the adult rat. *Experimental Neurology*, *148*, 475–494.

Brock, J. H., Rosenzweig, E. S., Blesch, A., Moseanko, R., Havton, L. A., Edgerton, V. R., et al. (2010). Local and remote growth factor effects after primate spinal cord injury. *Journal of Neuroscience*, *30*, 9728–9737.

Cao, Q., He, Q., Wang, Y., Cheng, X., Howard, R. M., Zhang, Y., et al. (2010). Transplantation of ciliary neurotrophic factor-expressing adult oligodendrocyte precursor cells promotes remyelination and functional recovery after spinal cord injury. *Journal of Neuroscience*, *30*, 2989–3001.

Cao, L., Liu, L., Chen, Z. Y., Wang, L. M., Ye, J. L., Qiu, H. Y., et al. (2004). Olfactory ensheathing cells genetically modified to secrete GDNF to promote spinal cord repair. *Brain*, *127*, 535–549.

Cao, Q., Xu, X. M., Devries, W. H., Enzmann, G. U., Ping, P., Tsoulfas, P., et al. (2005). Functional recovery in traumatic spinal cord injury after transplantation of multineurotrophin-expressing glial-restricted precursor cells. *Journal of Neuroscience*, *25*, 6947–6957.

Cen, L. P., Luo, J. M., Zhang, C. W., Fan, Y. M., Song, Y., So, K. F., et al. (2007). Chemotactic effect of ciliary neurotrophic factor on macrophages in retinal ganglion cell survival and axonal regeneration. *Investigative Ophthalmology & Visual Science*, *48*, 4257–4266.

Chehrehasa, F., Cobcroft, M., Young, Y. W., Mackay-Sim, A., & Goss, B. (2014). An acute growth factor treatment that preserves function after spinal cord contusion injury. *Journal of Neurotrauma*, *31*, 1807–1813.

Chen, B., He, J., Yang, H., Zhang, Q., Zhang, L., Zhang, X., et al. (2015). Repair of spinal cord injury by implantation of bFGF-incorporated HEMA-MOETACL hydrogel in rats. *Scientific Reports*, *5*, 9017.

Chen, Q., Zhou, L., & Shine, H. D. (2006). Expression of neurotrophin-3 promotes axonal plasticity in the acute but not chronic injured spinal cord. *Journal of Neurotrauma*, *23*, 1254–1260.

Cholas, R., Hsu, H. P., & Spector, M. (2012). Collagen scaffolds incorporating select therapeutic agents to facilitate a reparative response in a standardized hemiresection defect in the rat spinal cord. *Tissue Engineering. Part A*, *18*, 2158–2172.

Coumans, J. V., Lin, T. T., Dai, H. N., MacArthur, L., McAtee, M., Nash, C., et al. (2001). Axonal regeneration and functional recovery after complete spinal cord transection in rats by delayed treatment with transplants and neurotrophins. *Journal of Neuroscience*, *21*, 9334–9344.

Cui, Q., Hodgetts, S. I., Hu, Y., Luo, J. M., & Harvey, A. R. (2007). Strain-specific differences in the effects of cyclosporin a and fk506 on the survival and regeneration of axotomized retinal ganglion cells in adult rats. *Neuroscience, 146*, 986–999.

Dale, S. M., Kuang, R. Z., Wei, X., & Varon, S. (1995). Corticospinal motor neurons in the adult rat: Degeneration after intracortical axotomy and protection by ciliary neurotrophic factor (CNTF). *Experimental Neurology, 135*, 67–73.

Deng, L. X., Deng, P., Ruan, Y., Xu, Z. C., Liu, N. K., Wen, X., et al. (2013). A novel growth-promoting pathway formed by GDNF-overexpressing Schwann cells promotes propriospinal axonal regeneration, synapse formation, and partial recovery of function after spinal cord injury. *Journal of Neuroscience, 33*, 5655–5667.

Deng, L. X., Hu, J., Liu, N., Wang, X., Smith, G. M., Wen, X., et al. (2011). GDNF modifies reactive astrogliosis allowing robust axonal regeneration through Schwann cell-seeded guidance channels after spinal cord injury. *Experimental Neurology, 229*, 238–250.

Dolbeare, D., & Houle, J. D. (2003). Restriction of axonal retraction and promotion of axonal regeneration by chronically injured neurons after intraspinal treatment with glial cell line-derived neurotrophic factor (GDNF). *Journal of Neurotrauma, 20*, 1251–1261.

Dyson, S. E., Harvey, A. R., Trapp, B. D., & Heath, J. W. (1988). Ultrastructural and immunohistochemical analysis of axonal regrowth and myelination in membranes which form over lesion sites in the rat visual system. *Journal of Neurocytology, 17*, 797–808.

Enomoto, M., Bunge, M. B., & Tsoulfas, P. (2013). A multifunctional neurotrophin with reduced affinity to p75NTR enhances transplanted Schwann cell survival and axon growth after spinal cord injury. *Experimental Neurology, 248*, 170–182.

Fan, C., Zheng, Y., Cheng, X., Qi, X., Bu, P., Luo, X., et al. (2013). Transplantation of D15a-expressing glial-restricted-precursor-derived astrocytes improves anatomical and locomotor recovery after spinal cord injury. *International Journal of Biological Sciences, 9*, 78–93.

Feng, S. Q., Kong, X. H., Liu, Y., Ban, D. X., Ning, G. Z., Chen, J. T., et al. (2009). Regeneration of spinal cord with cell and gene therapy. *Orthopaedic Surgery, 1*, 153–163.

Fischer, D., & Leibinger, M. (2012). Promoting optic nerve regeneration. *Progress in Retinal and Eye Research, 31*, 688–701.

Fouad, K., Bennett, D. J., Vavrek, R., & Blesch, A. (2013). Long-term viral brain-derived neurotrophic factor delivery promotes spasticity in rats with a cervical spinal cord hemisection. *Frontiers in Neurology, 4*, 187.

Fouad, K., & Tetzlaff, W. (2012). Rehabilitative training and plasticity following spinal cord injury. *Experimental Neurology, 235*, 91–99.

Furuya, T., Hashimoto, M., Koda, M., Murata, A., Okawa, A., Dezawa, M., et al. (2013). Treatment with basic fibroblast growth factor-incorporated gelatin hydrogel does not exacerbate mechanical allodynia after spinal cord contusion injury in rats. *Journal of Spinal Cord Medicine, 36*, 134–139.

Garcia-Alias, G., Petrosyan, H. A., Schnell, L., Horner, P. J., Bowers, W. J., Mendell, L. M., et al. (2011). Chondroitinase ABC combined with neurotrophin NT-3 secretion and NR2D expression promotes axonal plasticity and functional recovery in rats with lateral hemisection of the spinal cord. *Journal of Neuroscience, 31*, 17788–17799.

Gerin, C. G., Madueke, I. C., Perkins, T., Hill, S., Smith, K., Haley, B., et al. (2011). Combination strategies for repair, plasticity, and regeneration using regulation of gene expression during the chronic phase after spinal cord injury. *Synapse, 65*, 1255–1281.

Golden, K. L., Pearse, D. D., Blits, B., Garg, M. S., Oudega, M., Wood, P. M., et al. (2007). Transduced Schwann cells promote axon growth and myelination after spinal cord injury. *Experimental Neurology, 207*, 203–217.

Goldshmit, Y., Frisca, F., Pinto, A. R., Pebay, A., Tang, J. K., Siegel, A. L., et al. (2014). FGF2 improves functional recovery-decreasing gliosis and increasing radial glia and neural progenitor cells after spinal cord injury. *Brain and Behavior, 4*, 187–200.

Goldshmit, Y., Sztal, T. E., Jusuf, P. R., Hall, T. E., Nguyen-Chi, M., & Currie, P. D. (2012). FGF-dependent glial cell bridges facilitate spinal cord regeneration in zebrafish. *Journal of Neuroscience, 32*, 7477–7492.

Gowing, G., Shelley, B., Staggenborg, K., Hurley, A., Avalos, P., Victoroff, J., et al. (2014). Glial cell line-derived neurotrophic factor-secreting human neural progenitors show long-term survival, maturation into astrocytes, and no tumor formation following transplantation into the spinal cord of immunocompromised rats. *Neuroreport, 25*, 367–372.

Gransee, H., Zhan, W. Z., Sieck, G. C., & Mantilla, C. (2014). Localized delivery of BDNF-expressing mesenchymal stem cells enhances functional recovery following cervical spinal cord injury. *Journal of Neurotrauma, 32*, 185–193.

Grill, R., Murai, K., Blesch, A., Gage, F. H., & Tuszynski, M. H. (1997). Cellular delivery of neurotrophin-3 promotes corticospinal axonal growth and partial functional recovery after spinal cord injury. *Journal of Neuroscience, 17*, 5560–5572.

Grulova, I., Slovinska, L., Blasko, J., Devaux, S., Wisztorski, M., Salzet, M., et al. (2015). Delivery of alginate scaffold releasing two trophic factors for spinal cord injury repair. *Scientific Reports, 5*, 13702.

Gu, Y. L., Yin, L. W., Zhang, Z., Liu, J., Liu, S. J., Zhang, L. F., et al. (2012). Neurotrophin expression in neural stem cells grafted acutely to transected spinal cord of adult rats linked to functional improvement. *Cellular and Molecular Neurobiology, 32*, 1089–1097.

Gunther, M. I., Weidner, N., Muller, R., & Blesch, A. (2015). Cell-seeded alginate hydrogel scaffolds promote directed linear axonal regeneration in the injured rat spinal cord. *Acta Biomaterialia, 27*, 140–150.

Guo, J. S., Zeng, Y. S., Li, H. B., Huang, W. L., Liu, R. Y., Li, X. B., et al. (2007). Cotransplant of neural stem cells and NT-3 gene modified Schwann cells promote the recovery of transected spinal cord injury. *Spinal Cord, 45*, 15–24.

Guzen, F. P., Soares, J. G., de Freitas, L. M., Cavalcanti, J. R., Oliveira, F. G., Araujo, J. F., et al. (2012). Sciatic nerve grafting and inoculation of FGF-2 promotes improvement of motor behavior and fiber regrowth in rats with spinal cord transection. *Restorative Neurology and Neuroscience, 30*, 265–275.

Hagg, T., Baker, K. A., Emsley, J. G., & Tetzlaff, W. (2005). Prolonged local neurotrophin-3 infusion reduces ipsilateral collateral sprouting of spared corticospinal axons in adult rats. *Neuroscience, 130*, 875–887.

Hains, B. C., Johnson, K. M., Eaton, M. J., Willis, W. D., & Hulsebosch, C. E. (2003). Serotonergic neural precursor cell grafts attenuate bilateral hyperexcitability of dorsal horn neurons after spinal hemisection in rat. *Neuroscience, 116*, 1097–1110.

Han, S., Wang, B., Jin, W., Xiao, Z., Chen, B., Xiao, H., et al. (2014). The collagen scaffold with collagen binding bdnf enhances functional recovery by facilitating peripheral nerve infiltrating and ingrowth in canine complete spinal cord transection. *Spinal Cord, 52*, 867–873.

Han, S., Wang, B., Jin, W., Xiao, Z., Li, X., Ding, W., et al. (2015). The linear-ordered collagen scaffold-BDNF complex significantly promotes functional recovery after completely transected spinal cord injury in canine. *Biomaterials, 41*, 89–96.

Hannila, S. S., & Filbin, M. T. (2008). The role of cyclic amp signaling in promoting axonal regeneration after spinal cord injury. *Experimental Neurology, 209*, 321–332.

Harvey, A. R., Hu, Y., Leaver, S. G., Mellough, C. B., Park, K., Verhaagen, J., et al. (2006). Gene therapy and transplantation in cns repair: The visual system. *Progress in Retinal and Eye Research, 25*, 449–489.

Harvey, A. R., Lovett, S. J., Majda, B. T., Yoon, J. H., Wheeler, L. P. G., & Hodgetts, S. I. (2015). Neurotrophic factors for spinal cord repair: Which, where, how and when to apply, and for what period of time? *Brain Research, 1619*, 36–71.

Harvey, A. R., Ooi, J. W., & Rodger, J. (2012). Neurotrophic factors and the regeneration of adult retinal ganglion cell axons. *International Review of Neurobiology, 106*, 1–33.

Harvey, A. R., & Tan, M. M. (1992). Spontaneous regeneration of adult rat retinal ganglion cell axons in vivo. *Neuroreport, 3*, 239–242.

Ho, B. L., Ba, Y. C., Wang, X. Y., Liu, S. J., Liu, G. D., Ou, S., et al. (2013). BDNF expression with functional improvement in transected spinal cord treated with neural stem cells in adult rats. *Neuropeptides, 47*, 1–7.

Hellstrom, M., & Harvey, A. R. (2011). Retinal ganglion cell gene therapy and visual system repair. *Current Gene Therapy, 11*, 116–131.

Hiebert, G. W., Khodarahmi, K., McGraw, J., Steeves, J. D., & Tetzlaff, W. (2002). Brain-derived neurotrophic factor applied to the motor cortex promotes sprouting of corticospinal fibers but not regeneration into a peripheral nerve transplant. *Journal of Neuroscience Research, 69*, 160–168.

Hollis, E. R., 2nd., Lu, P., Blesch, A., & Tuszynski, M. H. (2009). IGF-1 gene delivery promotes corticospinal neuronal survival but not regeneration after adult cns injury. *Experimental Neurology, 215*, 53–59.

Hollis, E. R., 2nd., & Tuszynski, M. H. (2011). Neurotrophins: Potential therapeutic tools for the treatment of spinal cord injury. *Neurotherapeutics, 8*, 694–703.

Hou, S., Nicholson, L., van Niekerk, E., Motsch, M., & Blesch, A. (2012). Dependence of regenerated sensory axons on continuous neurotrophin-3 delivery. *Journal of Neuroscience, 32*, 13206–13220.

Hougland, M. T., Harrison, B. J., Magnuson, D. S., Rouchka, E. C., & Petruska, J. C. (2012). The transcriptional response of neurotrophins and their tyrosine kinase receptors in lumbar sensorimotor circuits to spinal cord contusion is affected by injury severity and survival time. *Frontiers in Physiology, 3*, 478.

Houweling, D. A., Lankhorst, A. J., Gispen, W. H., Bar, P. R., & Joosten, E. A. (1998). Collagen containing neurotrophin-3 (nt-3) attracts regrowing injured corticospinal axons in the adult rat spinal cord and promotes partial functional recovery. *Experimental Neurology, 153*, 49–59.

Houweling, D. A., van Asseldonk, J. T., Lankhorst, A. J., Hamers, F. P., Martin, D., Bar, P. R., et al. (1998). Local application of collagen containing brain-derived neurotrophic factor decreases the loss of function after spinal cord injury in the adult rat. *Neuroscience Letters, 251*, 193–196.

Hoyng, S. A., Gnavi, S., de Winter, F., Eggers, R., Ozawa, T., Zaldumbide, A., et al. (2014). Developing a potentially immunologically inert tetracycline-regulatable viral vector for gene therapy in the peripheral nerve. *Gene Therapy, 21*, 549–557.

Huang, W. C., Kuo, H. S., Tsai, M. J., Ma, H., Chiu, C. W., Huang, M. C., et al. (2011). Adeno-associated virus-mediated human acidic fibroblast growth factor expression promotes functional recovery of spinal cord-contused rats. *Journal of Gene Medicine, 13*, 283–289.

Huang, E. J., & Reichardt, L. F. (2001). Neurotrophins: Roles in neuronal development and function. *Annual Review of Neuroscience, 24*, 677–736.

Huang, F., Wang, J., & Chen, A. (2006). Effects of co-grafts mesenchymal stem cells and nerve growth factor suspension in the repair of spinal cord injury. *Journal of Huazhong University of Science and Technology. Medical Sciences, 26*, 206–210.

Hunanyan, A. S., Petrosyan, H. A., Alessi, V., & Arvanian, V. L. (2013). Combination of chondroitinase abc and aav-nt3 promotes neural plasticity at descending spinal pathways after thoracic contusion in rats. *Journal of Neurophysiology, 110*, 1782–1792.

Hung, K. S., Tsai, S. H., Lee, T. C., Lin, J. W., Chang, C. K., & Chiu, W. T. (2007). Gene transfer of insulin-like growth factor-1 providing neuroprotection after spinal cord injury in rats. *Journal of Neurosurgery. Spine. 6*, 35–46.

Hurtado, A., Moon, L. D., Maquet, V., Blits, B., Jerome, R., & Oudega, M. (2006). Poly (D, L-lactic acid) macroporous guidance scaffolds seeded with Schwann cells genetically modified to secrete a bi-functional neurotrophin implanted in the completely transected adult rat thoracic spinal cord. *Biomaterials*, *27*, 430–442.

Hwang, D. H., Kim, H. M., Kang, Y. M., Joo, I. S., Cho, C. S., Yoon, B. W., et al. (2011). Combination of multifaceted strategies to maximize the therapeutic benefits of neural stem cell transplantation for spinal cord repair. *Cell Transplantation*, *20*, 1361–1379.

Iannotti, C., Li, H., Yan, P., Lu, X., Wirthlin, L., & Xu, X. M. (2003). Glial cell line-derived neurotrophic factor-enriched bridging transplants promote propriospinal axonal regeneration and enhance myelination after spinal cord injury. *Experimental Neurology*, *183*, 379–393.

Iannotti, C., Ping Zhang, Y., Shields, C. B., Han, Y., Burke, D. A., & Xu, X. M. (2004). A neuroprotective role of glial cell line-derived neurotrophic factor following moderate spinal cord contusion injury. *Experimental Neurology*, *189*, 317–332.

Iarikov, D. E., Kim, B. G., Dai, H. N., McAtee, M., Kuhn, P. L., & Bregman, B. S. (2007). Delayed transplantation with exogenous neurotrophin administration enhances plasticity of corticofugal projections after spinal cord injury. *Journal of Neurotrauma*, *24*, 690–702.

Ip, N. Y., McClain, J., Barrezueta, N. X., Aldrich, T. H., Pan, L., Li, Y., et al. (1993). The alpha component of the CNTF receptor is required for signaling and defines potential CNTF targets in the adult and during development. *Neuron*, *10*, 89–102.

Ishii, K., Nakamura, M., Dai, H., Finn, T. P., Okano, H., Toyama, Y., et al. (2006). Neutralization of ciliary neurotrophic factor reduces astrocyte production from transplanted neural stem cells and promotes regeneration of corticospinal tract fibers in spinal cord injury. *Journal of Neuroscience Research*, *84*, 1669–1681.

Jain, A., Kim, Y. T., McKeon, R. J., & Bellamkonda, R. V. (2006). In situ gelling hydrogels for conformal repair of spinal cord defects, and local delivery of BDNF after spinal cord injury. *Biomaterials*, *27*, 497–504.

Jain, A., McKeon, R. J., Brady-Kalnay, S. M., & Bellamkonda, R. V. (2011). Sustained delivery of activated rho GTPases and bdnf promotes axon growth in CSPG-rich regions following spinal cord injury. *PLoS One*, *6*e16135.

Ji, X. C., Dang, Y. Y., Gao, H. Y., Wang, Z. T., Gao, M., Yang, Y., et al. (2015). Local injection of lenti-BDNF at the lesion site promotes M2 macrophage polarization and inhibits inflammatory response after spinal cord injury in mice. *Cellular and Molecular Neurobiology*, *35*, 881–890.

Jin, Y., Fischer, I., Tessler, A., & Houle, J. D. (2002). Transplants of fibroblasts genetically modified to express BDNF promote axonal regeneration from supraspinal neurons following chronic spinal cord injury. *Experimental Neurology*, *177*, 265–275.

Johnson, P. J., Tatara, A., McCreedy, D. A., Shiu, A., & Sakiyama-Elbert, S. E. (2010). Tissue-engineered fibrin scaffolds containing neural progenitors enhance functional recovery in a subacute model of SCI. *Soft Matter*, *6*, 5127–5137.

Johnson, P. J., Tatara, A., Shiu, A., & Sakiyama-Elbert, S. E. (2010). Controlled release of neurotrophin-3 and platelet-derived growth factor from fibrin scaffolds containing neural progenitor cells enhances survival and differentiation into neurons in a subacute model of SCI. *Cell Transplantation*, *19*, 89–101.

Kabu, S., Gao, Y., Kwon, B. K., & Labhasetwar, V. (2015). Drug delivery, cell-based therapies, and tissue engineering approaches for spinal cord injury. *Journal of Controlled Release*, *219*, 141–154.

Kadoya, K., Tsukada, S., Lu, P., Coppola, G., Geschwind, D., Filbin, M. T., et al. (2009). Combined intrinsic and extrinsic neuronal mechanisms facilitate bridging axonal regeneration one year after spinal cord injury. *Neuron*, *64*, 165–172.

Kang, C. E., Baumann, M. D., Tator, C. H., & Shoichet, M. S. (2013). Localized and sustained delivery of fibroblast growth factor-2 from a nanoparticle-hydrogel composite for treatment of spinal cord injury. *Cells, Tissues, Organs*, *197*, 55–63.

Kanno, H., Pressman, Y., Moody, A., Berg, R., Muir, E. M., Rogers, J. H., et al. (2014). Combination of engineered Schwann cell grafts to secrete neurotrophin and chondroitinase promotes axonal regeneration and locomotion after spinal cord injury. *Journal of Neuroscience*, *34*, 1838–1855.

Karimi-Abdolrezaee, S., Eftekharpour, E., Wang, J., Schut, D., & Fehlings, M. G. (2010). Synergistic effects of transplanted adult neural stem/progenitor cells, chondroitinase, and growth factors promote functional repair and plasticity of the chronically injured spinal cord. *Journal of Neuroscience*, *30*, 1657–1676.

Kasai, M., Jikoh, T., Fukumitsu, H., & Furukawa, S. (2014). FGF-2-responsive and spinal cord-resident cells improve locomotor function after spinal cord injury. *Journal of Neurotrauma*, *31*, 1584–1598.

Kigerl, K. A., McGaughy, V. M., & Popovich, P. G. (2006). Comparative analysis of lesion development and intraspinal inflammation in four strains of mice following spinal contusion injury. *Journal of Comparative Neurology*, *494*, 578–594.

Kim, K. N., Oh, S. H., Lee, K. H., & Yoon, D. H. (2006). Effect of human mesenchymal stem cell transplantation combined with growth factor infusion in the repair of injured spinal cord. *Acta Neurochirurgica. Supplement*, *99*, 133–136.

King, V. R., Bradbury, E. J., McMahon, S. B., & Priestley, J. V. (2000). Changes in truncated trkb and p75 receptor expression in the rat spinal cord following spinal cord hemisection and spinal cord hemisection plus neurotrophin treatment. *Experimental Neurology*, *165*, 327–341.

Klaw, M. C., Xu, C., & Tom, V. J. (2013). Intraspinal AAV injections immediately rostral to a thoracic spinal cord injury site efficiently transduces neurons in spinal cord and brain. *Molecular Therapy Nucleic Acids*, *2*, e108.

Kobayashi, N. R., Fan, D. P., Giehl, K. M., Bedard, A. M., Wiegand, S. J., & Tetzlaff, W. (1997). BDNF and NT-4/5 prevent atrophy of rat rubrospinal neurons after cervical axotomy, stimulate GAP-43 and Talpha1-tubulin mRNA expression, and promote axonal regeneration. *Journal of Neuroscience*, *17*, 9583–9595.

Koda, M., Hashimoto, M., Murakami, M., Yoshinaga, K., Ikeda, O., Yamazaki, M., et al. (2004). Adenovirus vector-mediated in vivo gene transfer of brain-derived neurotrophic factor (BDNF) promotes rubrospinal axonal regeneration and functional recovery after complete transection of the adult rat spinal cord. *Journal of Neurotrauma*, *21*, 329–337.

Koda, M., Kamada, T., Hashimoto, M., Murakami, M., Shirasawa, H., Sakao, S., et al. (2007). Adenovirus vector-mediated ex vivo gene transfer of brain-derived neurotrophic factor to bone marrow stromal cells promotes axonal regeneration after transplantation in completely transected adult rat spinal cord. *European Spine Journal*, *16*, 2206–2214.

Kojima, A., & Tator, C. H. (2002). Intrathecal administration of epidermal growth factor and fibroblast growth factor 2 promotes ependymal proliferation and functional recovery after spinal cord injury in adult rats. *Journal of Neurotrauma*, *19*, 223–238.

Kovalchuk, Y., Holthoff, K., & Konnerth, A. (2004). Neurotrophin action on a rapid timescale. *Current Opinion in Neurobiology*, *14*, 558–563.

Kusano, K., Enomoto, M., Hirai, T., Tsoulfas, P., Sotome, S., Shinomiya, K., et al. (2010). Transplanted neural progenitor cells expressing mutant NT3 promote myelination and partial hindlimb recovery in the chronic phase after spinal cord injury. *Biochemical and Biophysical Research Communications*, *393*, 812–817.

Kwon, B. K., Liu, J., Lam, C., Plunet, W., Oschipok, L. W., Hauswirth, W., et al. (2007). Brain-derived neurotrophic factor gene transfer with adeno-associated viral and lentiviral vectors prevents rubrospinal neuronal atrophy and stimulates regeneration-associated gene expression after acute cervical spinal cord injury. *Spine*, *32*, 1164–1173.

Kwon, B. K., Liu, J., Messerer, C., Kobayashi, N. R., McGraw, J., Oschipok, L., et al. (2002). Survival and regeneration of rubrospinal neurons 1 year after spinal cord injury. *Proceedings of the National Academy of Sciences of the United States of America*, *99*, 3246–3251.

Kwon, B. K., Liu, J., Oschipok, L., Teh, J., Liu, Z. W., & Tetzlaff, W. (2004). Rubrospinal neurons fail to respond to brain-derived neurotrophic factor applied to the spinal cord injury site 2 months after cervical axotomy. *Experimental Neurology, 189*, 45–57.

Lee, M. J., Chen, C. J., Cheng, C. H., Huang, W. C., Kuo, H. S., Wu, J. C., et al. (2008). Combined treatment using peripheral nerve graft and FGF-1: Changes to the glial environment and differential macrophage reaction in a complete transected spinal cord. *Neuroscience Letters, 433*, 163–169.

Lee, T. T., Green, B. A., Dietrich, W. D., & Yezierski, R. P. (1999). Neuroprotective effects of basic fibroblast growth factor following spinal cord contusion injury in the rat. *Journal of Neurotrauma, 16*, 347–356.

Lee, H., Jin, M. H., Kang, H. J., Hong, C. H., Bang, W. J., Park, K. K., et al. (2010). Timing of prepubertal androgen administration may have different effects on future fertility as well as penile size in normal male rats. *Urology, 75*, 979–984.

Lee, M. Y., Kim, C. J., Shin, S. L., Moon, S. H., & Chun, M. H. (1998). Increased ciliary neurotrophic factor expression in reactive astrocytes following spinal cord injury in the rat. *Neuroscience Letters, 255*, 79–82.

Li, X., Han, J., Zhao, Y., Ding, W., Wei, J., Li, J., et al. (2016). Functionalized collagen scaffold implantation and camp administration collectively facilitate spinal cord regeneration. *Acta Biomaterialia, 30*, 233–245.

Liebl, D. J., Huang, W., Young, W., & Parada, L. F. (2001). Regulation of Trk receptors following contusion of the rat spinal cord. *Experimental Neurology, 167*, 15–26.

Liu, Y., Kim, D., Himes, B. T., Chow, S. Y., Schallert, T., Murray, M., et al. (1999). Transplants of fibroblasts genetically modified to express BDNF promote regeneration of adult rat rubrospinal axons and recovery of forelimb function. *Journal of Neuroscience, 19*, 4370–4387.

Lu, P., Blesch, A., & Tuszynski, M. H. (2001). Neurotrophism without neurotropism: BDNF promotes survival but not growth of lesioned corticospinal neurons. *Journal of Comparative Neurology, 436*, 456–470.

Lu, P., Jones, L. L., & Tuszynski, M. H. (2005). BDNF-expressing marrow stromal cells support extensive axonal growth at sites of spinal cord injury. *Experimental Neurology, 191*, 344–360.

Lu, P., Jones, L. L., & Tuszynski, M. H. (2007). Axon regeneration through scars and into sites of chronic spinal cord injury. *Experimental Neurology, 203*, 8–21.

Lu, P., & Tuszynski, M. H. (2008). Growth factors and combinatorial therapies for CNS regeneration. *Experimental Neurology, 209*, 313–320.

Lu, P., Wang, Y., Graham, L., McHale, K., Gao, M., Wu, D., et al. (2012). Long-distance growth and connectivity of neural stem cells after severe spinal cord injury. *Cell, 150*, 1264–1273.

Lu, P., Woodruff, G., Wang, Y., Graham, L., Hunt, M., Wu, D., et al. (2014). Long-distance axonal growth from human induced pluripotent stem cells after spinal cord injury. *Neuron, 83*, 789–796.

Lu, P., Yang, H., Jones, L. L., Filbin, M. T., & Tuszynski, M. H. (2004). Combinatorial therapy with neurotrophins and camp promotes axonal regeneration beyond sites of spinal cord injury. *Journal of Neuroscience, 24*, 6402–6409.

Lynskey, J. V., Sandhu, F. A., Dai, H. N., McAtee, M., Slotkin, J. R., MacArthur, L., et al. (2006). Delayed intervention with transplants and neurotrophic factors supports recovery of forelimb function after cervical spinal cord injury in adult rats. *Journal of Neurotrauma, 23*, 617–634.

Macias, M. Y., Syring, M. B., Pizzi, M. A., Crowe, M. J., Alexanian, A. R., & Kurpad, S. N. (2006). Pain with no gain: Allodynia following neural stem cell transplantation in spinal cord injury. *Experimental Neurology, 201*, 335–348.

Mantilla, C. B., Gransee, H. M., Zhan, W. Z., & Sieck, G. C. (2013). Motoneuron BDNF/TRKb signaling enhances functional recovery after cervical spinal cord injury. *Experimental Neurology*, *247*, 101–109.

McCall, J., Weidner, N., & Blesch, A. (2012). Neurotrophic factors in combinatorial approaches for spinal cord regeneration. *Cell and Tissue Research*, *349*, 27–37.

McTigue, D. M., Horner, P. J., Stokes, B. T., & Gage, F. H. (1998). Neurotrophin-3 and brain-derived neurotrophic factor induce oligodendrocyte proliferation and myelination of regenerating axons in the contused adult rat spinal cord. *Journal of Neuroscience*, *18*, 5354–5365.

Meijs, M. F., Timmers, L., Pearse, D. D., Tresco, P. A., Bates, M. L., Joosten, E. A., et al. (2004). Basic fibroblast growth factor promotes neuronal survival but not behavioral recovery in the transected and Schwann cell implanted rat thoracic spinal cord. *Journal of Neurotrauma*, *21*, 1415–1430.

Menei, P., Montero-Menei, C., Whittemore, S. R., Bunge, R. P., & Bunge, M. B. (1998). Schwann cells genetically modified to secrete human bdnf promote enhanced axonal regrowth across transected adult rat spinal cord. *European Journal of Neuroscience*, *10*, 607–621.

Meng, X. T., Li, C., Dong, Z. Y., Liu, J. M., Li, W., Liu, Y., et al. (2008). Co-transplantation of bFGF-expressing amniotic epithelial cells and neural stem cells promotes functional recovery in spinal cord-injured rats. *Cell Biology International*, *32*, 1546–1558.

Menon, P. K., Muresanu, D. F., Sharma, A., Mossler, H., & Sharma, H. S. (2012). Cerebrolysin, a mixture of neurotrophic factors induces marked neuroprotection in spinal cord injury following intoxication of engineered nanoparticles from metals. *CNS & Neurological Disorders Drug Targets*, *11*, 40–49.

Nakamura, M., & Bregman, B. S. (2001). Differences in neurotrophic factor gene expression profiles between neonate and adult rat spinal cord after injury. *Experimental Neurology*, *169*, 407–415.

Novikova, L. N., Novikov, L. N., & Kellerth, J. O. (2000). Survival effects of BDNF and NT-3 on axotomized rubrospinal neurons depend on the temporal pattern of neurotrophin administration. *European Journal of Neuroscience*, *12*, 776–780.

Ollivier-Lanvin, K., Fischer, I., Tom, V., Houle, J. D., & Lemay, M. A. (2014). Either brain-derived neurotrophic factor or neurotrophin-3 only neurotrophin-producing grafts promote locomotor recovery in untrained spinalized cats. *Neurorehabilitation and Neural Repair*, *29*, 90–100.

Oudega, M., Bradbury, E. J., & Ramer, M. S. (2012). Combination therapies. *Handbook of Clinical Neurology*, *109*, 617–636.

Oudega, M., Xu, X. M., Guenard, V., Kleitman, N., & Bunge, M. B. (1997). A combination of insulin-like growth factor-1 and platelet-derived growth factor enhances myelination but diminishes axonal regeneration into Schwann cell grafts in the adult rat spinal cord. *Glia*, *19*, 247–258.

Ozdinler, P. H., & Macklis, J. D. (2006). IGF-1 specifically enhances axon outgrowth of corticospinal motor neurons. *Nature Neuroscience*, *9*, 1371–1381.

Pardridge, W. M. (2012). Drug transport across the blood-brain barrier. *Journal of Cerebral Blood Flow and Metabolism*, *32*, 1959–1972.

Pearse, D. D., Pereira, F. C., Marcillo, A. E., Bates, M. L., Berrocal, Y. A., Filbin, M. T., et al. (2004). Camp and Schwann cells promote axonal growth and functional recovery after spinal cord injury. *Nature Medicine*, *10*, 610–616.

Piantino, J., Burdick, J. A., Goldberg, D., Langer, R., & Benowitz, L. I. (2006). An injectable, biodegradable hydrogel for trophic factor delivery enhances axonal rewiring and improves performance after spinal cord injury. *Experimental Neurology*, *201*, 359–367.

Plemel, J. R., Keough, M. B., Duncan, G. J., Sparling, J. S., Yong, V. W., Stys, P. K., et al. (2014). Remyelination after spinal cord injury: Is it a target for repair? *Progress in Neurobiology*, *117*, 54–72.

Rabchevsky, A. G., Fugaccia, I., Fletcher-Turner, A., Blades, D. A., Mattson, M. P., & Scheff, S. W. (1999). Basic fibroblast growth factor (bFGF) enhances tissue sparing and functional recovery following moderate spinal cord injury. *Journal of Neurotrauma, 16*, 817–830.

Ritfeld, G. J., Patel, A., Chou, A., Novosat, T. L., Castillo, D. G., Roos, R. A., et al. (2015). The role of brain-derived neurotrophic factor in bone marrow stromal cell-mediated spinal cord repair. *Cell Transplantation, 24*, 2209–2220.

Romero, M. I., Rangappa, N., Garry, M. G., & Smith, G. M. (2001). Functional regeneration of chronically injured sensory afferents into adult spinal cord after neurotrophin gene therapy. *Journal of Neuroscience, 21*, 8408–8416.

Ruitenberg, M. J., Blits, B., Dijkhuizen, P. A., te Beek, E. T., Bakker, A., van Heerikhuize, J. J., et al. (2004). Adeno-associated viral vector-mediated gene transfer of brain-derived neurotrophic factor reverses atrophy of rubrospinal neurons following both acute and chronic spinal cord injury. *Neurobiology of Disease, 15*, 394–406.

Ruitenberg, M. J., Levison, D. B., Lee, S. V., Verhaagen, J., Harvey, A. R., & Plant, G. W. (2005). Nt-3 expression from engineered olfactory ensheathing glia promotes spinal sparing and regeneration. *Brain, 128*, 839–853.

Ruitenberg, M. J., Plant, G. W., Hamers, F. P., Wortel, J., Blits, B., Dijkhuizen, P. A., et al. (2003). Ex vivo adenoviral vector-mediated neurotrophin gene transfer to olfactory ensheathing glia: Effects on rubrospinal tract regeneration, lesion size, and functional recovery after implantation in the injured rat spinal cord. *Journal of Neuroscience, 23*, 7045–7058.

Sakiyama-Elbert, S., Johnson, P. J., Hodgetts, S. I., Plant, G. W., & Harvey, A. R. (2012). Scaffolds to promote spinal cord regeneration. *Handbook of Clinical Neurology, 109*, 575–594.

Sasaki, M., Radtke, C., Tan, A. M., Zhao, P., Hamada, H., Houkin, K., et al. (2009). BDNF-hypersecreting human mesenchymal stem cells promote functional recovery, axonal sprouting, and protection of corticospinal neurons after spinal cord injury. *Journal of Neuroscience, 29*, 14932–14941.

Schnell, L., Hunanyan, A. S., Bowers, W. J., Horner, P. J., Federoff, H. J., Gullo, M., et al. (2011). Combined delivery of Nogo-A antibody, neurotrophin-3 and the NMDA-NR2d subunit establishes a functional 'detour' in the hemisected spinal cord. *European Journal of Neuroscience, 34*, 1256–1267.

Schnell, L., Schneider, R., Kolbeck, R., Barde, Y. A., & Schwab, M. E. (1994). Neurotrophin-3 enhances sprouting of corticospinal tract during development and after adult spinal cord lesion. *Nature, 367*, 170–173.

Sharp, K. G., Flanagan, L. A., Yee, K. M., & Steward, O. (2012). A re-assessment of a combinatorial treatment involving Schwann cell transplants and elevation of cyclic amp on recovery of motor function following thoracic spinal cord injury in rats. *Experimental Neurology, 233*, 625–644.

Shin, D. A., Pennant, W. A., Yoon do, H., Ha, Y., & Kim, K. N. (2014). Co-transplantation of bone marrow-derived mesenchymal stem cells and nanospheres containing FGF-2 improve cell survival and neurological function in the injured rat spinal cord. *Acta Neurochirurgica, 156*, 297–303.

Slawinska, U., Miazga, K., & Jordan, L. M. (2014). The role of serotonin in the control of locomotor movements and strategies for restoring locomotion after spinal cord injury. *Acta Neurobiologiae Experimentalis (Wars), 74*, 172–187.

Smith, G. M., Falone, A. E., & Frank, E. (2012). Sensory axon regeneration: Rebuilding functional connections in the spinal cord. *Trends in Neurosciences, 35*, 156–163.

Stokols, S., Sakamoto, J., Breckon, C., Holt, T., Weiss, J., & Tuszynski, M. H. (2006). Templated agarose scaffolds support linear axonal regeneration. *Tissue Engineering, 12*, 2777–2787.

Taylor, L., Jones, L., Tuszynski, M. H., & Blesch, A. (2006). Neurotrophin-3 gradients established by lentiviral gene delivery promote short-distance axonal bridging beyond cellular grafts in the injured spinal cord. *Journal of Neuroscience, 26*, 9713–9721.

Taylor, S. J., Rosenzweig, E. S., McDonald, J. W., 3rd., & Sakiyama-Elbert, S. E. (2006). Delivery of neurotrophin-3 from fibrin enhances neuronal fiber sprouting after spinal cord injury. *Journal of Controlled Release, 113*, 226–235.

Tetzlaff, W., Okon, E. B., Karimi-Abdolrezaee, S., Hill, C. E., Sparling, J. S., Plemel, J. R., et al. (2011). A systematic review of cellular transplantation therapies for spinal cord injury. *Journal of Neurotrauma, 28*, 1611–1682.

Thomas, A. M., Seidlits, S. K., Goodman, A. G., Kukushliev, T. V., Hassani, D. M., Cummings, B. J., et al. (2014). Sonic hedgehog and neurotrophin-3 increase oligodendrocyte numbers and myelination after spinal cord injury. *Integrative Biology: Quantitative Biosciences From Nano to Macro, 6*, 694–705.

Tobias, C. A., Dhoot, N. O., Wheatley, M. A., Tessler, A., Murray, M., & Fischer, I. (2001). Grafting of encapsulated BDNF-producing fibroblasts into the injured spinal cord without immune suppression in adult rats. *Journal of Neurotrauma, 18*, 287–301.

Tobias, C. A., Han, S. S., Shumsky, J. S., Kim, D., Tumolo, M., Dhoot, N. O., et al. (2005). Alginate encapsulated BDNF-producing fibroblast grafts permit recovery of function after spinal cord injury in the absence of immune suppression. *Journal of Neurotrauma, 22*, 138–156.

Tom, V. J., Sandrow-Feinberg, H. R., Miller, K., Domitrovich, C., Bouyer, J., Zhukareva, V., et al. (2013). Exogenous BDNF enhances the integration of chronically injured axons that regenerate through a peripheral nerve grafted into a chondroitinase-treated spinal cord injury site. *Experimental Neurology, 239*, 91–100.

Tripathi, R. B., & McTigue, D. M. (2008). Chronically increased ciliary neurotrophic factor and fibroblast growth factor-2 expression after spinal contusion in rats. *Journal of Comparative Neurology, 510*, 129–144.

Tsai, E. C., Dalton, P. D., Shoichet, M. S., & Tator, C. H. (2006). Matrix inclusion within synthetic hydrogel guidance channels improves specific supraspinal and local axonal regeneration after complete spinal cord transection. *Biomaterials, 27*, 519–533.

Tsai, M. C., Shen, L. F., Kuo, H. S., Cheng, H., & Chak, K. F. (2008). Involvement of acidic fibroblast growth factor in spinal cord injury repair processes revealed by a proteomics approach. *Molecular & Cellular Proteomics, 7*, 1668–1687.

Tuinstra, H. M., Aviles, M. O., Shin, S., Holland, S. J., Zelivyanskaya, M. L., Fast, A. G., et al. (2012). Multifunctional, multichannel bridges that deliver neurotrophin encoding lentivirus for regeneration following spinal cord injury. *Biomaterials, 33*, 1618–1626.

Tuszynski, M. H., Grill, R., Jones, L. L., McKay, H. M., & Blesch, A. (2002). Spontaneous and augmented growth of axons in the primate spinal cord: Effects of local injury and nerve growth factor-secreting cell grafts. *Journal of Comparitive Neurology, 449*, 88–101.

Tuszynski, M. H., Murai, K., Blesch, A., Grill, R., & Miller, I. (1997). Functional characterization of ngf-secreting cell grafts to the acutely injured spinal cord. *Cell Transplantation, 6*, 361–368.

Tuszynski, M. H., Peterson, D. A., Ray, J., Baird, A., Nakahara, Y., & Gage, F. H. (1994). Fibroblasts genetically modified to produce nerve growth factor induce robust neuritic ingrowth after grafting to the spinal cord. *Experimental Neurology, 126*, 1–14.

Tuszynski, M. H., Weidner, N., McCormack, M., Miller, I., Powell, H., & Conner, J. (1998). Grafts of genetically modified Schwann cells to the spinal cord: Survival, axon growth, and myelination. *Cell Transplantation, 7*, 187–196.

Urfer, R., Tsoulfas, P., Soppet, D., Escandon, E., Parada, L. F., & Presta, L. G. (1994). The binding epitopes of neurotrophin-3 to its receptors trkc and gp75 and the design of a multifunctional human neurotrophin. *The EMBO Journal, 13*, 5896–5909.

Vavrek, R., Girgis, J., Tetzlaff, W., Hiebert, G. W., & Fouad, K. (2006). BDNF promotes connections of corticospinal neurons onto spared descending interneurons in spinal cord injured rats. *Brain, 129*, 1534–1545.
Wang, L. J., Zhang, R. P., & Li, J. D. (2014). Transplantation of neurotrophin-3-expressing bone mesenchymal stem cells improves recovery in a rat model of spinal cord injury. *Acta Neurochirurgica, 156*, 1409–1418.
Weidner, N., Blesch, A., Grill, R. J., & Tuszynski, M. H. (1999). Nerve growth factor-hypersecreting Schwann cell grafts augment and guide spinal cord axonal growth and remyelinate central nervous system axons in a phenotypically appropriate manner that correlates with expression of l1. *Journal of Comparative Neurology, 413*, 495–506.
Weishaupt, N., Blesch, A., & Fouad, K. (2012). BDNF: The career of a multifaceted neurotrophin in spinal cord injury. *Experimental Neurology, 238*, 254–264.
Weishaupt, N., Li, S., Di Pardo, A., Sipione, S., & Fouad, K. (2013). Synergistic effects of BDNF and rehabilitative training on recovery after cervical spinal cord injury. *Behavioural Brain Research, 239*, 31–42.
Weishaupt, N., Mason, A. L., Hurd, C., May, Z., Zmyslowski, D. C., Galleguillos, D., et al. (2014). Vector-induced NT-3 expression in rats promotes collateral growth of injured corticospinal tract axons far rostral to a spinal cord injury. *Neuroscience, 272*, 65–75.
Wen, Y., Yu, S., Wu, Y., Ju, R., Wang, H., Liu, Y., et al. (2015). Spinal cord injury repair by implantation of structured hyaluronic acid scaffold with PLGA microspheres in the rat. *Cell and Tissue Research, 364*, 17–28.
Wilems, T. S., Pardieck, J., Iyer, N., & Sakiyama-Elbert, S. E. (2015). Combination therapy of stem cell derived neural progenitors and drug delivery of anti-inhibitory molecules for spinal cord injury. *Acta Biomaterialia, 28*, 23–32.
Wu, J. C., Huang, W. C., Tsai, Y. A., Chen, Y. C., & Cheng, H. (2008). Nerve repair using acidic fibroblast growth factor in human cervical spinal cord injury: A preliminary Phase I clinical study. *Journal of Neurosurgery. Spine, 8*, 208–214.
Xu, X. M., Guenard, V., Kleitman, N., Aebischer, P., & Bunge, M. B. (1995). A combination of BDNF and NT-3 promotes supraspinal axonal regeneration into Schwann cell grafts in adult rat thoracic spinal cord. *Experimental Neurology, 134*, 261–272.
Yang, Z., Zhang, A., Duan, H., Zhang, S., Hao, P., Ye, K., et al. (2015). NT3-chitosan elicits robust endogenous neurogenesis to enable functional recovery after spinal cord injury. *Proceedings of the National Academy of Sciences of the United States of America, 112*, 13354–13359.
Yao, L., Daly, W., Newland, B., Yao, S., Wang, W., Chen, B. K., et al. (2013). Improved axonal regeneration of transected spinal cord mediated by multichannel collagen conduits functionalized with neurotrophin-3 gene. *Gene Therapy, 20*, 1149–1157.
Ye, J., Cao, L., Cui, R., Huang, A., Yan, Z., Lu, C., et al. (2004). The effects of ciliary neurotrophic factor on neurological function and glial activity following contusive spinal cord injury in the rats. *Brain Research, 997*, 30–39.
Yick, L. W., Wu, W., So, K. F., & Yip, H. K. (1999). Peripheral nerve graft and neurotrophic factors enhance neuronal survival and expression of nitric oxide synthase in Clarke's nucleus after hemisection of the spinal cord in adult rat. *Experimental Neurology, 159*, 131–138.
Yin, F., Li, P., Zheng, M., Chen, L., Xu, Q., Chen, K., et al. (2003). Interleukin-6 family of cytokines mediates isoproterenol-induced delayed stat3 activation in mouse heart. *Journal of Biological Chemistry, 278*, 21070–21075.
Zaldumbide, A., Weening, S., Cramer, S. J., Rabelink, M. J., Verhaagen, J., & Hoeben, R. C. (2010). A potentially immunologically inert derivative of the reverse tetracycline-controlled transactivator. *Biotechnology Letters, 32*, 749–754.

Zhang, L., Ma, Z., Smith, G. M., Wen, X., Pressman, Y., Wood, P. M., et al. (2009). GDNF-enhanced axonal regeneration and myelination following spinal cord injury is mediated by primary effects on neurons. *Glia*, *57*, 1178–1191.

Zhang, R. P., Wang, L. J., He, S., Xie, J., & Li, J. D. (2016). Effects of magnetically guided, SPIO-labeled, and neurotrophin-3 gene-modified bone mesenchymal stem cells in a rat model of spinal cord injury. *Stem Cells International*, *2016*, 2018474.

Zhang, H., Wu, F., Kong, X., Yang, J., Chen, H., Deng, L., et al. (2014). Nerve growth factor improves functional recovery by inhibiting endoplasmic reticulum stress-induced neuronal apoptosis in rats with spinal cord injury. *Journal of Translational Medicine*, *12*, 130.

Zhao, Y. Z., Jiang, X., Xiao, J., Lin, Q., Yu, W. Z., Tian, F. R., et al. (2016). Using NGF heparin-poloxamer thermosensitive hydrogels to enhance the nerve regeneration for spinal cord injury. *Acta Biomaterialia*, *29*, 71–80.

Zhao, T., Yan, W., Xu, K., Qi, Y., Dai, X., & Shi, Z. (2013). Combined treatment with platelet-rich plasma and brain-derived neurotrophic factor-overexpressing bone marrow stromal cells supports axonal remyelination in a rat spinal cord hemi-section model. *Cytotherapy*, *15*, 792–804.

Zhu, S. P., Wang, Z. G., Zhao, Y. Z., Wu, J., Shi, H. X., Ye, L. B., et al. (2015). Gelatin nanostructured lipid carriers incorporating nerve growth factor inhibit endoplasmic reticulum stress-induced apoptosis and improve recovery in spinal cord injury. *Molecular Neurobiology*, *53*, 4375–4386.

Ziemlinska, E., Kugler, S., Schachner, M., Wewior, I., Czarkowska-Bauch, J., & Skup, M. (2014). Overexpression of bdnf increases excitability of the lumbar spinal network and leads to robust early locomotor recovery in completely spinalized rats. *PLoS One*, *9*, e88833.

CHAPTER SIXTEEN

Neurotrophins and Migraine

L.B. Martins, A.L. Teixeira, R.B. Domingues[1]
Universidade Federal de Minas Gerais, Belo Horizonte, Minas Gerais, Brazil
[1]Corresponding author: e-mail address: contato@renandomingues.med.br

Contents

1. Introduction 459
2. Neurotrophins and Pain 460
   2.1 NGF 461
   2.2 BDNF 462
   2.3 NT-3 and NT-4/5 463
3. Migraine Pathophysiology 463
4. Neurotrophins and Migraine 465
5. Conclusion and Future Directions 469
References 470

Abstract

Neurotrophins (NTs) have been implicated in generation and modulation of nociceptive pathways. Change in NTs levels is associated with painful conditions and neurological diseases such as migraine. Currently, it is generally recognized that migraine headaches result from the activation and sensitization of trigeminal sensory afferent fibers leading to neuropeptides release such as calcitonin gene-related peptide (CGRP) and substance P (SP). This triggers an inflammatory cascade causing a neurogenic inflammation. The agents responsible for trigeminal activation and release of neuropeptides are still unclear. It is known that the transient receptor potential vanilloid receptor-1 (TRPV1) is an important mediator of CGRP and SP release. TRPV1 is closely associated with tyrosine receptors kinases (Trk), which are NTs receptors. NTs can act on TRPV1 increasing its sensitivity to painful stimuli, therefore predisposing to hyperalgesia. Upregulation of ion channels and pain receptors in dorsal root ganglion neurons may be alternative mechanisms by which NTs contribute to pain development. Only a few studies have been performed to investigate the role of NTs in migraine. These studies have reported changes in NTs levels in migraine patients either during the migraine attack or in free-headache periods.

1. INTRODUCTION

Migraine is an incapacitating primary headache characterized by episodes of head pain lasting 4–72 h with at least two of the following

Vitamins and Hormones, Volume 104
ISSN 0083-6729
http://dx.doi.org/10.1016/bs.vh.2016.10.003

characteristics: unilateral location, pulsatile quality, moderate or severe pain intensity, and aggravation with routine physical activity. Migraine attacks are often accompanied by nausea, vomiting, phonophobia, and/or photophobia, and can be preceded by transient focal neurological symptom called aura (IHS, 2004). Affective and cognitive symptoms, such as depression, irritability, attention deficits, transient amnesia, and reduced ability to navigate in familiar environments can also affect migraine patients during attacks (Burstein, Noseda, & Borsook, 2015). Episodic migraine was defined as headache less than 15 days per month. When headache occurs on 15 or more days per month for more than 3 months with migraine features on at least 8 days it is designated as chronic migraine (CM) (IHS, 2004).

The prevalence of migraine in the United States is about 18% in women and 6% in men (Bigal & Lipton, 2009) and CM prevalence rate is 1.3–5.1% of world population (Natoli et al., 2010). The productivity losses caused by migraine lead to an estimated cost of 13 billion dollars per year to American employers (Bigal & Lipton, 2009).

Despite its high prevalence and significant socioeconomic burden, the pathophysiology of migraine is not fully understood. Several studies have implicated neurotrophins (NTs) in painful and neuropsychiatric disorders. For instance, changes in the levels of NTs have been reported in fibromyalgia (Haas, Portela, Bohmer, Oses, & Lara, 2010), Alzheimer's disease (Faria et al., 2014), and mood disorders (Barbosa et al., 2013; Grassi-Oliveira, Stein, Lopes, Teixeira, & Bauer, 2008).

The role of NTs in pain processing includes the upregulation of neuropeptides, ion channels, vanilloid receptors, purinergic receptor of ATP ($P2X_3$), and glutamate receptor in the dorsal root ganglion neurons (Allen & Dawbarn, 2006). Accordingly, it has been proposed that NTs are involved in headache development in migraine.

In this chapter, we review the current knowledge on the role played by NTs in nociceptive pathways, emphasizing its potential participation in migraine pathophysiology.

2. NEUROTROPHINS AND PAIN

The members of neurotrophin family—nerve growth factor (NGF), brain-derived neurotrophic factor (BDNF), neurotrophin-3 (NT-3), and neurotrophin-4/5 (NT-4/5)—have been implicated in the development and modulation of nociceptive pathways.

NTs act by binding to specific tyrosine receptors kinases (Trk). NGF binds preferentially to TrkA, BDNF, and NT-4/5 bind to TrkB, and NT-3 binds to TrkC, while all NTs bind to p75 receptor with a low affinity (Chao, 2003). Several neuronal types, including sensory neurons, express Trks, and respond to NTs stimulus (McMahon, Armanini, Ling, & Phillips, 1994; Merighi et al., 2008).

2.1 NGF

NGF has been associated with peripheral nociception, especially in inflammatory states. In experimental models, there is an increase of NGF and overexpression of TrkA after pain stimuli (Allen & Dawbarn, 2006). Furthermore, local or systemic administration of NGF causes hyperalgesia (Alleva, Aloe, Cirulli, De Acetis, & Padoa Schioppa, 1997). Beyond its role in acute pain, NGF also seems to act in chronic pain maintenance. Allodynia and peripheral hyperalgesia induced by NGF injection may persist for days (Lewin, Rueff, & Mendell, 1994; Svensson, Castrillon, & Cairns, 2008). Accordingly, blocking of NGF signaling has been proposed as an analgesic treatment of chronic pain states (Sarchielli & Gallai, 2004). Anti-NGF strategies have been developed and are currently being tested in clinical trials, such as RN624 (Tanezumab), a humanized anti-NGF monoclonal antibody (Mullard, 2015).

The involvement of NGF in nociception pathways was also evidenced in rare human genetic diseases defined by mutations in NGF or its receptor (TrkA) genes and characterized by decreased pain perception. In a rare autosomal-recessive disorder called "Congenital Insensitivity to Pain with Anhidrosis" (CIPA), which is caused by different mutations in the human TrkA gene (NTRK1) located on chromosome 1 (1q21–q22), there is an hereditary sensory and autonomic neuropathy with severe pain insensitivity, leading to frequent injuries (Allen & Dawbarn, 2006). Einarsdottir et al. (2004) observed that single-nucleotide polymorphisms (rs2490334 and rs2275607) in the NGFB gene cause loss of deep pain perception in a large family from northern Sweden (Einarsdottir et al., 2004).

The following mechanisms were suggested to explain NGF participation in the development of pain: (i) increase in the expression of pain receptors such as vanilloid receptors—including vanilloid receptor 1 (VR-1) and transient receptor potential vanilloid receptor 1 (TRPV1)—and $P2X_3$; (ii) upregulation of ion channels including calcium, potassium, and specific sodium channels, such as tetradodotoxin (TTX)-sensitive and

TTX-resistant sodium; and (iii) upregulation of neuropeptides and neuromodulators release, such as CGRP, SP, and BDNF (Pezet & McMahon, 2006).

Briefly, NGF expression and release at sites of injury seem to be one the very first steps of inflammatory pain. This is followed by increase in neuropeptides release, upregulation of ion channels, and pain-associated receptors. The subsequent increase in BDNF expression in the dorsal horn of the spinal cord contributes to pain perception. Therefore, BDNF appears to act as a modulator of central pain (Allen & Dawbarn, 2006).

2.2 BDNF

BDNF is an important component of synaptic plasticity and survival of several neurons located in regions associated with behavior, mood, and pain modulation, such as hippocampus, somatosensory cortex, and dorsal root ganglion (Angelucci, Brene, & Mathe, 2005; Merighi et al., 2008).

Neurodegenerative and psychiatric diseases, such as Alzheimer's disease, schizophrenia, and depression are associated with impairment in BDNF or its receptor (TrkB) expression. For instance, there are decreased numbers of TrkB and BDNF-positive neurons in the CNS of individuals with schizophrenia and reduction of BDNF mRNA in hippocampus and cortex of patients with Alzheimer's disease (Adachi, Numakawa, Richards, Nakajima, & Kunugi, 2014). In experimental models, BDNF knockout mice exhibit reduction in hippocampus volume when compared with wild-type mice (Adachi et al., 2014). Conversely, previous studies showed a decrease in BDNF or TrkB levels in prefrontal cortex, hippocampus, amygdala, and serum in depressive patients (Adachi et al., 2014). Together these studies reinforce the role played by BDNF in neuronal survival. The same occurs with sensory neurons as BDNF knockout mice lack 85% of neurons in vestibular ganglia, for example (Huang & Reichardt, 2001). BDNF and TrkB receptor are expressed in some sensory ganglia, including dorsal root ganglion, trigeminal, petrosal, and geniculate ganglia.

BDNF is regarded as a pain modulator as it acts both in fast excitatory (glutamatergic) and inhibitory (GABAergic/glycinergic) signals. In experimental models, injection of high doses of BDNF had analgesic effects, while injection of low doses induced pain (Merighi et al., 2008). BDNF action in excitatory signals is potentially mediated by glutamate receptors, especially *N*-methyl-D-asparate (NMDA) receptors. NMDA receptors are heteromeric ion channel complexes strongly recruited in nociceptive pathways.

BDNF binding to TrkB receptor triggers phosphorylation pathways, such as mitogen-activated protein kinases (MAPK)/extracellular signal-regulated kinases (ERK) and phospholipase C (PLC)/diacylglycerol/protein kinase C (PKC) cascades which phosphorylate NMDA increasing the open probability of this receptor (Merighi et al., 2008). Conversely, BDNF appears to regulate both GABA and glycine release, acting also as a modulator of inhibitory synapses in the dorsal horn. The mechanisms involved in modulation of inhibitory transmission by BDNF seem to be more complex and need to be elucidated. TrkB receptors were detected on GABAergic terminals of inhibitory interneurons and this gives subsidy to suppose that BDNF acts in inhibitory synapses (Merighi et al., 2008).

2.3 NT-3 and NT-4/5

Only a few studies have been performed to investigate the role of NT-3 and NT-4/5 in pain (Gratto & Verge, 2003; Wilson-Gerwing, Dmyterko, Zochodne, Johnston, & Verge, 2005).

Unlike other NTs, NT-3 has been shown to have an analgesic effect by downregulation of NGF, BDNF, CGRP, and SP expression. NT-3 injection had no effect on the response to noxious heat (Shu, Llinas, & Mendell, 1999).

NT-4/5 also binds specifically to TrkB as BDNF. Shu et al. (1999) observed that NT-4/5 was even more effective in hyperalgesia induction than BDNF (Shu et al., 1999). This can be partly explained by the fact that, unlike BDNF, NT-4/5 is also able to bind to TrkA (Berkemeier et al., 1991). More studies are needed to establish the mechanisms by which NT-3 and NT-4/5 may participate in nociceptive pathways.

3. MIGRAINE PATHOPHYSIOLOGY

Migraine is a complex and multifactorial disease whose pathophysiology is still being clarified. Migraine attacks are associated with autonomic, affective, cognitive, and sensory symptoms suggesting the involvement of cortical, subcortical, and brainstem areas (Burstein et al., 2015). Several events can trigger migraine attacks, such as stressful periods, insufficient sleep, skipping meals, hormonal fluctuations, certain foods, flickering lights, noise and smell, indicating the interaction of different neural networks (Burstein et al., 2015).

Formerly, it was believed that abnormal dilatation of meningeal or cerebral blood vessels were the primary trigger for migraine attacks. This theory has been debated for decades (Wolff, 1948, 1963). In fact, vasodilator molecules, such as nitric oxide, are able to trigger migraine (Olesen, Thomsen, Lassen, & Olesen, 1995) and vasoconstriction is an effective mechanism for relief of acute migraine (Ferrari & Saxena, 1993). However, studies that used magnetic resonance angiography showed that vasodilatation does not necessarily precede the migraine attack (Schoonman et al., 2008).

Currently, it is generally recognized that the headache phase of migraine results from activation and sensitization of trigeminal sensory afferent fibers which innervate cranial structures, especially meninges and their blood vessels (Vecchia & Pietrobon, 2012). Trigeminal activation leads to the release of vasoactive neuropeptides, such as CGRP and SP, by nonmyelinated (C-fiber) and myelinated (Aδ fibers) axons, that trigger plasma protein extravasation, mast cell degranulation, and release of inflammatory mediators such as bradykinin and prostaglandin causing a neurogenic inflammation (Burstein et al., 2015; Noseda & Burstein, 2013; Waeber & Moskowitz, 2005). Previous studies have shown increases in serum levels of CGRP during migraine attacks and decreases or normalization of its levels after drug treatment and pain relief, supporting the role of this neuropeptide in migraine attack (Durham, 2006).

The triggers of trigeminal activation may arise from brain, blood, or meningeal tissues (Waeber & Moskowitz, 2005). One possible mechanism for trigeminal activation, especially in migraine with aura, is cortical spreading depression (CSD). CSD is a wave of electrophysiological hyperactivity followed by an inhibitory wave that results in the release of several molecules such as potassium and hydrogenous ions, nitric oxide, arachidonic acid, and prostaglandin (Burstein et al., 2015; Moskowitz, 2007).

The TRPV1, expressed by approximately 25% of trigeminal ganglion cells, is an important mediator of CGRP and SP release. It is proposed that 70% of TRPV1 present in trigeminal ganglion cells express CGRP (Shimizu et al., 2007). TRPV1 is a pain receptor cation channel activated by heat, capsaicin, and extracellular protons (Chao, 2003; Shimizu et al., 2007). This channel is expressed also in the nerve fibers of peripheral tissues and plays an important role in conducting nociceptive signals to the CNS (Shimizu et al., 2007). TRPV1 is closely associated with Trk receptors. Therefore, NTs can act on TRPV1 increasing its sensitivity to painful stimuli, predisposing to hyperalgesia. Indeed, it has been proposed that NGF regulates TRPV1 expression and the thermal hypersensitivity induced by this NT is dependent to TRPV1 (Chao, 2003; Shieh et al., 2010). Influence in membrane potential through upregulation of sodium channels, and activation the expression of small GTPase protein Ras, an important modulator

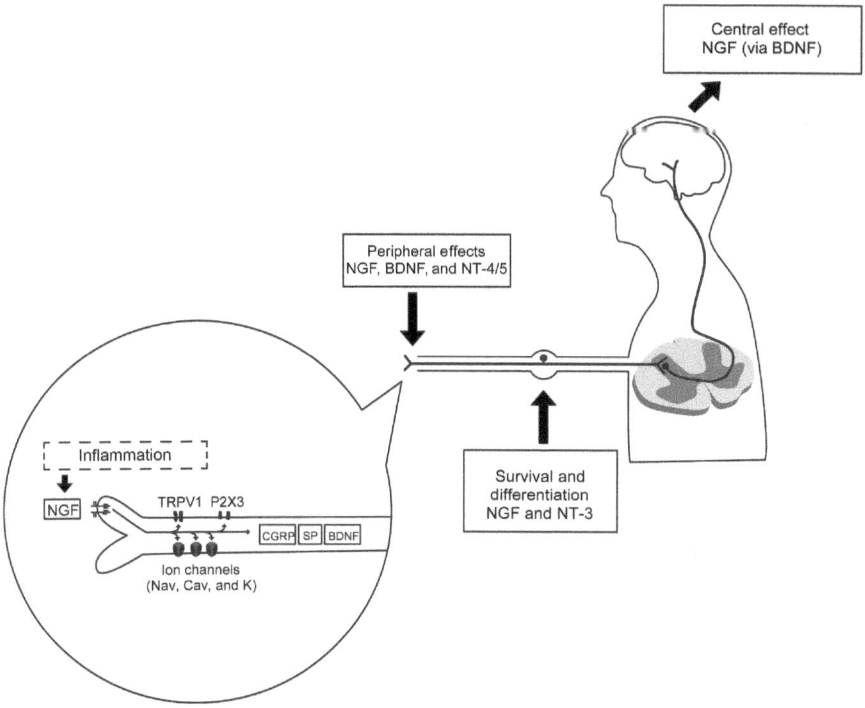

Fig. 1 Role of neurotrophins in nociceptive pathways. NTs are required for neuronal survival and differentiation. NGF, BDNF, and NT-4/5 play a role on peripheral nociceptive pathways, particularly NGF. Inflammatory mediators increase NGF release. This neurotrophin upregulates ion channels and pain-associated receptors (TRPV1 and P2X3) and increases the release of neuropeptides (CGRP and SP) and BDNF. The subsequent increase in BDNF levels contributes to pain development, acting as modulator of central pain. More studies are needed to establish the mechanisms by which NT-3 and NT-4/5 participate in nociceptive pathways.

of TRPV1, are possible mechanisms by which NGF enhances TRPV1 expression (Amaya et al., 2004; Bron, Klesse, Shah, Parada, & Winter, 2003; Shieh et al., 2010).

Despite the known role of NTs in nociceptive process, only a few studies to date have been performed to establish the relationship between NTs and primary headaches, including migraine (Fig. 1).

4. NEUROTROPHINS AND MIGRAINE

Table 1 depicts studies that evaluated NTs levels in biological samples of patients with primary headaches including migraine. Most studies measured NGF and BDNF levels in peripheral (serum, plasma, platelets, or

Table 1 Published Studies on Neurotrophins and Primary Headaches in Humans

References	Patients	Sample	Main Results
NGF			
Sarchielli, Alberti, Floridi, and Gallai (2001)	– Patients with chronic daily headache (CDH): 20 – Control subjects: 20	CSF	(i) ↑ NGF, CGRP, and SP levels in the fluid cerebrospinal in patients with CDH (ii) Positive correlation between NGF and both CGPR and SP levels
Jang, Park, Kho, Chung, and Chung (2011)	– Patients with chronic migraine (CM): 33 – Control subjects: 36	Plasma and saliva	(i) ↑ NGF and neuropeptides (CGRP and SP) in saliva and plasma in patients with CM (ii) Positive correlation between NGF (and neuropeptides in plasma (iii) Association between both NGF (plasma) and neuropeptides (plasma and saliva) and pain intensity measured by Graded Chronic Pain Scale in CM patients
BDNF			
Tanure, Gomez, Hurtado, Teixeira, and Domingues (2010)	– Patients with episodic migraine without aura during migraine attacks and in headache-free periods: 9	Serum	(i) ↑ Serum BDNF during the migraine attack (ii) There was no significant difference in the serum levels of TNF and their receptors (sTNF-R1 and sTNF-R) in migraine attack period and headache-free period
Fischer et al. (2012)	– Patients with migraine with aura: 22 (outside attack) and 13 (during attack) – Patients with migraine without aura: 27 (outside attack) and 8 (during attack) – Cluster headache: 24 (outside bout) and 42 (inside bout) – Frequent episodic tension-type headache: 6 (outside attack) – Control subjects: 22	Serum	(i) ↑ BDNF during migraine attack compared with headache-free periods, tension-type headache, and controls (ii) No difference between patients with migraine with aura compared with those without aura (iii) (iii) ↑ BDNF in cluster headache patients inside and outside cluster bouts compared with control subjects

BDNF and NGF			
Sarchielli, Greco, Stipa, Floridi, and Gallai (2002)	Patients with CDH and previous history of migraine: 25 – Control subjects: 20	CSF	(i) ↑ BDNF, NGF, and glutamate in migraine patients (ii) Positive correlation between BDNF and both NGF and glutamate
Blandini et al. (2006)	– Patients with migraine with aura: 27 – Patients with migraine without aura: 33 – Cluster headache: 14 – Control subjects: 57	Plasma and platelets	(i) ↓ BDNF (platelets) in all primary headaches patients when compared with controls (ii) ↓ NGF (platelets) only in migraine patients when compared with controls (iii) There is no difference in plasma levels (iv) ↑ Platelet levels of 5-HT and 5-HIAA in patients suffering from migraine with aura, migraine without aura and cluster headache when compared with controls
Sarchielli et al. (2007)	– Patients with CM: 20 – Patients with primary fibromyalgia syndrome: 20 – Control subjects: 20	CSF	(i) ↑ BDNF, NGF, and glutamate in patients with fibromyalgia and migraine when compared with controls (ii) Positive correlation between BDNF levels and both glutamate and NGF
BDNF, NGF, NT-3, and NT-4			
Martins et al. (2015)	– Migraine patients: 71 – Control subjects: 70	Plasma	(i) ↑ NT-4 in migraine patients (ii) There was no difference in plasma level of NT-3 and NGF between groups (iii) ↓ BDNF in CM BDNF and that were not influenced by the presence of depressive symptoms

BDNF, brain-derived neurotrophic factor; *CGRP*, calcitonin gene-related peptide; *CM*, chronic migraine; *CSF*, cerebrospinal fluid; *NGF*, nerve growth factor; *NT-3*, neurotrophin-3; *NT-4*, neurotrophin-4; *SNPs*, single-nucleotide polymorphisms; *SP*, substance P; *TNF*, factor de necrose tumoral; *5-HT*, 5-hydroxytryptamine; *5-HIAA*, 5-hydroxyindoleacetic acid.

saliva) and central (cerebrospinal fluid) samples. Only one study measured NT-3 and NT-4 levels in these patients.

As described earlier, NGF-induced pain process seems to involve an upregulation of SP and CGRP expression. Increased levels of NGF, SP, and CGRP were found in peripheral (Sarchielli et al., 2001) and central samples (Jang et al., 2011) of patients with chronic headache and migraine. Moreover, NGF level was positively correlated with neuropeptides levels in both studies, giving support to the evidence that NGF stimulates the SP and CGRP release. These neuropeptides are often colocalized in the trigeminal fibers and they seem to be released during neurogenic inflammation in migraine attacks. It is proposed that CGRP enhances SP action through inhibition of the enzyme involved in SP degradation (Jang et al., 2011). Jang et al. (2011) observed association between both NGF (plasma) and neuropeptides (plasma and saliva) and pain intensity in patients with CM. These results suggest that NGF, SP, and CGRP modulate pain pathways in migraine and these molecules have a regulatory action on each other.

NGF levels were also correlated with BDNF and glutamate levels in patients with chronic daily headache (CDH) (Sarchielli et al., 2002), CM, and fibromyalgia syndrome (Sarchielli et al., 2007) measured in cerebrospinal fluid. These results are in line with experimental studies with the proposed mechanism: NGF → ↑BDNF release → glutamate receptor phosphorylation → excitatory signs regulated, as described earlier.

Previous studies reported changes in BDNF levels during the migraine attack and in headache-free period. Moreover, the SNP rs2049046 in BDNF gene showed a significant association with migraine, especially migraine with aura (Sutherland, Maher, Rodriguez-Acevedo, Haupt, & Griffiths, 2014). Increased BDNF levels during migraine attacks were reported by Tanure et al. (2010) and Fischer et al. (2012). It has been shown that BDNF is coexpressed with CGRP in trigeminal ganglion neurons (Ichikawa et al., 2006).

CGRP and inflammatory stimuli upregulate BDNF expression and release in trigeminal neurons (Balkowiec-Iskra, Vermehren-Schmaedick, & Balkowiec, 2011; Buldyrev et al., 2006). During migraine attack and subsequent neurogenic inflammation, there is the release of inflammatory and neuromodulator molecules which may contribute to BDNF expression (Fischer et al., 2012). It is possible that increased BDNF levels may mediate nociceptive plasticity in trigeminal neurons and contribute to pain control. Simonetti, Giniatullin, and Fabbretti (2008) showed that BDNF is an intermediate to CGRP effect in $P2X_3$ sensitization and synthesis in trigeminal ganglion neurons (Simonetti et al., 2008).

However, Blandini et al. (2006) reported decreased BDNF levels in platelets of migraine and cluster headache patients. BDNF is stored in platelets and it is released upon agonist stimulation (Blandini et al., 2006). Fischer et al. (2012) proposed that platelet activation during migraine attack leads to immediate BDNF release and, hence, decreased levels in platelets of this NT.

BDNF level was also increased in interictal period. Sarchielli et al. (2002, 2007) observed increased in BDNF levels in cerebrospinal fluid of patients with CDH, CM, and fibromyalgia syndrome when compared with control subjects. However, a recent study by our group found lower peripheral BDNF levels in CM patients than in control subjects (Martins et al., 2015). This may be explained by hypothalamic–pituitary–adrenal (HPA) axis hyperactivity associated with chronic stress in these migraine patients. HPA axis hyperactivity is associated with increased pain sensitivity and reduced BDNF production (Pierce, Ryals, Wang, & Christianson, 2014). In summary, these results support the BDNF role as a pain modulator in patients with migraine, contributing to trigeminal nociceptive plasticity and pain control in migraine patients.

Only a one study has been performed to establish NT-3 and NT-4/5 role in migraine. We found that migraine patients showed increased NT-4/5 plasma levels when compared with control subjects (Martins et al., 2015). It has been demonstrated that NT-4/5 facilitates capsaicin-induced currents, suggesting a direct action of this NT on TRPV1 (Shu & Mendell, 1999). As described earlier, TRPV1 is an important mediator of CGRP and substance P release. On the other hand, NT-3 was shown to reduce TRPV1 expression suppressing thermal hyperalgesia (Wilson-Gerwing et al., 2005). These findings suggest the opposite role of NT-3 in nociceptive pathways acting as an analgesic molecule. In fact, NT-3 plasma level was not different in migraine patients in our study. However, more studies are needed to establish NTs role in migraine pathophysiology, especially NT-3 and NT-4/5.

5. CONCLUSION AND FUTURE DIRECTIONS

Migraine pathophysiology is not yet clear. A better knowledge of the roles of specific molecules involved in activation of nociceptive pathways in migraine is important to enable the development of new therapies. NTs play a role in generation and modulation of peripheral and central pain. Previous studies reported changes in NTs levels in patients with primary headache, including migraine.

Anti-NGF therapies are proposed to chronic pain treatment based on at least three major pharmacologic strategies which seemed effective in decreasing hypersensitivity in preclinical models: (i) NGF sequestration or inhibition of NGF-binding of TrkA; (ii) anti-TrkA antibody; and (iii) block of TrkA kinase activity (Mantyh, Koltzenburg, Mendell, Tive, & Shelton, 2011). NTs produce hyperalgesia by upregulating neuropeptides, especially CGRP and TRPV1 expression. Antagonists of CGRP are a promising strategy for migraine attacks treatment. Telcagepant, a CGRP antagonist, was more effective than triptans in pain relief during the migraine attack (Pini, 2009). Another pharmacological strategy is the TRPV1 antagonist (SB-707498) (Pini, 2009). Based on the earlier considerations, it is important to investigate the influence of these new drugs in NTs modulation and production (Tanure et al., 2010). Moreover, future studies are certainly needed to establish the effectiveness and safety of these drugs for migraine treatment.

REFERENCES

Adachi, N., Numakawa, T., Richards, M., Nakajima, S., & Kunugi, H. (2014). New insight in expression, transport, and secretion of brain-derived neurotrophic factor: Implications in brain-related diseases. *World Journal of Biological Chemistry*, *5*, 409–428.

Allen, S. J., & Dawbarn, D. (2006). Clinical relevance of the neurotrophins and their receptors. *Clinical Science (London)*, *110*, 175–191.

Alleva, E., Aloe, L., Cirulli, F., De Acetis, L., & Padoa Schioppa, C. (1997). Postnatal NGF administration causes adult hyperalgesia and overreactivity to social stimuli but does not reverse capsaicin induced hypoalgesia. *Psychoneuroendocrinology*, *22*, 591–602.

Amaya, F., Shimosato, G., Nagano, M., Ueda, M., Hashimoto, S., Tanaka, Y., … Tanaka, M. (2004). NGF and GDNF differentially regulate TRPV1 expression that contributes to development of inflammatory thermal hyperalgesia. *The European Journal of Neuroscience*, *20*, 2303–2310.

Angelucci, F., Brene, S., & Mathe, A. A. (2005). BDNF in schizophrenia, depression and corresponding animal models. *Molecular Psychiatry*, *10*, 345–352.

Balkowiec-Iskra, E., Vermehren-Schmaedick, A., & Balkowiec, A. (2011). Tumor necrosis factor-alpha increases brain-derived neurotrophic factor expression in trigeminal ganglion neurons in an activity-dependent manner. *Neuroscience*, *180*, 322–333.

Barbosa, I. G., Rocha, N. P., Miranda, A. S., Huguet, R. B., Bauer, M. E., Reis, H. J., & Teixeira, A. L. (2013). Increased BDNF levels in long-term bipolar disorder patients. *Revista Brasileira de Psiquiatria*, *35*, 67–69.

Berkemeier, L. R., Winslow, J. W., Kaplan, D. R., Nikolics, K., Goeddel, D. V., & Rosenthal, A. (1991). Neurotrophin-5: A novel neurotrophic factor that activates trk and trkB. *Neuron*, *7*, 857–866.

Bigal, M. E., & Lipton, R. B. (2009). The epidemiology, burden, and comorbidities of migraine. *Neurologic Clinics*, *27*, 321–334.

Blandini, F., Rinaldi, L., Tassorelli, C., Sances, G., Motta, M., Samuele, A., … Leon, A. (2006). Peripheral levels of BDNF and NGF in primary headaches. *Cephalalgia*, *26*, 136–142.

Bron, R., Klesse, L. J., Shah, K., Parada, L. F., & Winter, J. (2003). Activation of Ras is necessary and sufficient for upregulation of vanilloid receptor type 1 in sensory neurons by neurotrophic factors. *Molecular and Cellular Neurosciences*, *22*, 118–132.
Buldyrev, I., Tanner, N. M., Hsieh, H. Y., Dodd, E. G., Nguyen, L. T., & Balkowiec, A. (2006). Calcitonin gene related peptide enhances release of native brain-derived neurotrophic factor from trigeminal ganglion neurons. *Journal of Neurochemistry*, *99*, 1338–1350.
Burstein, R., Noseda, R., & Borsook, D. (2015). Migraine: Multiple processes, complex pathophysiology. *The Journal of Neuroscience*, *35*, 6619–6629.
Chao, M. V. (2003). Neurotrophins and their receptors: A convergence point for many signalling pathways. *Nature Reviews. Neuroscience*, *4*, 299–309.
Durham, P. L. (2006). Calcitonin gene-related peptide (CGRP) and migraine. *Headache*, *46*(Suppl. 1), S3–S8.
Einarsdottir, E., Carlsson, A., Minde, J., Toolanen, G., Svensson, O., Solders, G., … Holmberg, M. (2004). A mutation in the nerve growth factor beta gene (NGFB) causes loss of pain perception. *Human Molecular Genetics*, *13*, 799–805.
Faria, M. C., Goncalves, G. S., Rocha, N. P., Moraes, E. N., Bicalho, M. A., Gualberto Cintra, M. T., … Sousa, L. P. (2014). Increased plasma levels of BDNF and inflammatory markers in Alzheimer's disease. *Journal of Psychiatric Research*, *53*, 166–172.
Ferrari, M. D., & Saxena, P. R. (1993). Clinical and experimental effects of sumatriptan in humans. *Trends in Pharmacological Sciences*, *14*, 129–133.
Fischer, M., Wille, G., Klien, S., Shanib, H., Holle, D., Gaul, C., & Broessner, G. (2012). Brain-derived neurotrophic factor in primary headaches. *The Journal of Headache and Pain*, *13*, 469–475.
Grassi-Oliveira, R., Stein, L. M., Lopes, R. P., Teixeira, A. L., & Bauer, M. E. (2008). Low plasma brain-derived neurotrophic factor and childhood physical neglect are associated with verbal memory impairment in major depression—A preliminary report. *Biological Psychiatry*, *64*, 281–285.
Gratto, K. A., & Verge, V. M. (2003). Neurotrophin-3 down-regulates trkA mRNA, NGF high-affinity binding sites, and associated phenotype in adult DRG neurons. *The European Journal of Neuroscience*, *18*, 1535–1548.
Haas, L., Portela, L. V., Bohmer, A. E., Oses, J. P., & Lara, D. R. (2010). Increased plasma levels of brain derived neurotrophic factor (BDNF) in patients with fibromyalgia. *Neurochemical Research*, *35*, 830–834.
Huang, E. J., & Reichardt, L. F. (2001). Neurotrophins: Roles in neuronal development and function. *Annual Review of Neuroscience*, *24*, 677–736.
Ichikawa, H., Yabuuchi, T., Jin, H. W., Terayama, R., Yamaai, T., Deguchi, T., … Sugimoto, T. (2006). Brain-derived neurotrophic factor-immunoreactive primary sensory neurons in the rat trigeminal ganglion and trigeminal sensory nuclei. *Brain Research*, *1081*, 113–118.
IHS. (2004). The international classification of headache disorders: 2nd edition. *Cephalalgia*, *24*(Suppl. 1), 9–160.
Jang, M. U., Park, J. W., Kho, H. S., Chung, S. C., & Chung, J. W. (2011). Plasma and saliva levels of nerve growth factor and neuropeptides in chronic migraine patients. *Oral Diseases*, *17*, 187–193.
Lewin, G. R., Rueff, A., & Mendell, L. M. (1994). Peripheral and central mechanisms of NGF-induced hyperalgesia. *The European Journal of Neuroscience*, *6*, 1903–1912.
Mantyh, P. W., Koltzenburg, M., Mendell, L. M., Tive, L., & Shelton, D. L. (2011). Antagonism of nerve growth factor-TrkA signaling and the relief of pain. *Anesthesiology*, *115*, 189–204.
Martins, L. B., Duarte, H., Ferreira, A. V., Rocha, N. P., Teixeira, A. L., & Domingues, R. B. (2015). Migraine is associated with altered levels of neurotrophins. *Neuroscience Letters*, *587*, 6–10.

McMahon, S. B., Armanini, M. P., Ling, L. H., & Phillips, H. S. (1994). Expression and coexpression of Trk receptors in subpopulations of adult primary sensory neurons projecting to identified peripheral targets. *Neuron, 12*, 1161–1171.

Merighi, A., Salio, C., Ghirri, A., Lossi, L., Ferrini, F., Betelli, C., & Bardoni, R. (2008). BDNF as a pain modulator. *Progress in Neurobiology, 85*, 297–317.

Moskowitz, M. A. (2007). Pathophysiology of headache—Past and present. *Headache, 47*(Suppl. 1), S58–S63.

Mullard, A. (2015). Drug developers reboot anti-NGF pain programmes. *Nature Reviews. Drug Discovery, 14*, 297–298.

Natoli, J. L., Manack, A., Dean, B., Butler, Q., Turkel, C. C., Stovner, L., & Lipton, R. B. (2010). Global prevalence of chronic migraine: A systematic review. *Cephalalgia, 30*, 599–609.

Noseda, R., & Burstein, R. (2013). Migraine pathophysiology: Anatomy of the trigeminovascular pathway and associated neurological symptoms, cortical spreading depression, sensitization, and modulation of pain. *Pain, 154*(Suppl. 1), S44–S53.

Olesen, J., Thomsen, L. L., Lassen, L. H., & Olesen, I. J. (1995). The nitric oxide hypothesis of migraine and other vascular headaches. *Cephalalgia, 15*, 94–100.

Pezet, S., & McMahon, S. B. (2006). Neurotrophins: Mediators and modulators of pain. *Annual Review of Neuroscience, 29*, 507–538.

Pierce, A. N., Ryals, J. M., Wang, R., & Christianson, J. A. (2014). Vaginal hypersensitivity and hypothalamic-pituitary-adrenal axis dysfunction as a result of neonatal maternal separation in female mice. *Neuroscience, 263*, 216–230.

Pini, L. A. (2009). Future drugs for migraine. *Internal and Emergency Medicine, 4*, 361–362.

Sarchielli, P., Alberti, A., Floridi, A., & Gallai, V. (2001). Levels of nerve growth factor in cerebrospinal fluid of chronic daily headache patients. *Neurology, 57*, 132–134.

Sarchielli, P., & Gallai, V. (2004). Nerve growth factor and chronic daily headache: A potential implication for therapy. *Expert Review of Neurotherapeutics, 4*, 115–127.

Sarchielli, P., Greco, L., Stipa, A., Floridi, A., & Gallai, V. (2002). Brain-derived neurotrophic factor in patients with multiple sclerosis. *Journal of Neuroimmunology, 132*, 180–188.

Sarchielli, P., Mancini, M. L., Floridi, A., Coppola, F., Rossi, C., Nardi, K., … Calabresi, P. (2007). Increased levels of neurotrophins are not specific for chronic migraine: evidence from primary fibromyalgia syndrome. *The Journal of Pain, 8*, 737–745.

Schoonman, G. G., van der Grond, J., Kortmann, C., van der Geest, R. J., Terwindt, G. M., & Ferrari, M. D. (2008). Migraine headache is not associated with cerebral or meningeal vasodilatation—A 3T magnetic resonance angiography study. *Brain, 131*, 2192–2200.

Shieh, K. R., Yi, C. H., Liu, T. T., Tseng, H. L., Ho, H. C., Hsieh, H. T., & Chen, C. L. (2010). Evidence for neurotrophic factors associating with TRPV1 gene expression in the inflamed human esophagus. *Neurogastroenterology and Motility, 22*, 971–977. e252.

Shimizu, T., Toriumi, H., Sato, H., Shibata, M., Nagata, E., Gotoh, K., & Suzuki, N. (2007). Distribution and origin of TRPV1 receptor-containing nerve fibers in the dura mater of rat. *Brain Research, 1173*, 84–91.

Shu, X. Q., Llinas, A., & Mendell, L. M. (1999). Effects of trkB and trkC neurotrophin receptor agonists on thermal nociception: A behavioral and electrophysiological study. *Pain, 80*, 463–470.

Shu, X., & Mendell, L. M. (1999). Nerve growth factor acutely sensitizes the response of adult rat sensory neurons to capsaicin. *Neuroscience Letters, 274*, 159–162.

Simonetti, M., Giniatullin, R., & Fabbretti, E. (2008). Mechanisms mediating the enhanced gene transcription of P2X3 receptor by calcitonin gene-related peptide in trigeminal sensory neurons. *The Journal of Biological Chemistry, 283*, 18743–18752.

Sutherland, H. G., Maher, B. H., Rodriguez-Acevedo, A. J., Haupt, L. M., & Griffiths, L. R. (2014). Investigation of brain-derived neurotrophic factor (BDNF) gene variants in migraine. *Headache, 54*, 1184–1193.

Svensson, P., Castrillon, E., & Cairns, B. E. (2008). Nerve growth factor-evoked masseter muscle sensitization and perturbation of jaw motor function in healthy women. *Journal of Orofacial Pain, 22*, 340–348.

Tanure, M. T., Gomez, R. S., Hurtado, R. C., Teixeira, A. L., & Domingues, R. B. (2010). Increased serum levels of brain derived neurotropic factor during migraine attacks; A pilot study. *The Journal of Headache and Pain, 11*, 427–430.

Vecchia, D., & Pietrobon, D. (2012). Migraine: A disorder of brain excitatory-inhibitory balance? *Trends in Neurosciences, 35*, 507–520.

Waeber, C., & Moskowitz, M. A. (2005). Migraine as an inflammatory disorder. *Neurology, 64*, S9–S15.

Wilson-Gerwing, T. D., Dmyterko, M. V., Zochodne, D. W., Johnston, J. M., & Verge, V. M. (2005). Neurotrophin-3 suppresses thermal hyperalgesia associated with neuropathic pain and attenuates transient receptor potential vanilloid receptor-1 expression in adult sensory neurons. *The Journal of Neuroscience, 25*, 758–767.

Wolff, H. G. (1948). Headache mechanisms. *Bulletin of the U.S. Army Medical Department, 8*, 641–653.

Wolff, H. (1963). *Headache and other head pain*. New York: Oxford.

CHAPTER SEVENTEEN

BDNF During Withdrawal

H. Geoffroy, F. Noble[1]
Université Paris Descartes, Inserm U1124, CNRS ERL3649, Neuroplasticité et Thérapies des addictions, Paris, France
[1]Corresponding author: e-mail address: florence.noble@parisdescartes.fr

Contents

1. Introduction 476
2. BDNF Alterations Following Chronic Administration of Drugs of Abuse 476
   2.1 Juvenile and Young Adult Animals 477
   2.2 Adult Animals 478
   2.3 Dynamic Alterations of BDNF Levels During Withdrawal 481
   2.4 Epigenetics Regulation of BDNF Levels 482
3. Protracted Abstinence, Incubation, and Craving: How Is BDNF Involved? 483
   3.1 Endogenous BDNF 483
   3.2 Pharmacological Intervention on BDNF Signaling 484
4. Peripheral BDNF Levels 486
   4.1 A Potential Biomarker? 486
   4.2 Epigenetics in Clinic 489
5. Polymorphism in BDNF Gene: Predictive Value in Drug Addiction? 489
6. Conclusion and Future Directions 490
Acknowledgments 491
References 491

Abstract

Among the adaptations observed following drugs of abuse consumption, BDNF levels are widely altered in both brain and periphery. In this chapter, we first reviewed these adaptations in preclinical studies, in both juveniles and adult animals. A particular focus was made on protracted withdrawal as incubation is often associated with an increase in central BDNF levels. Then, we reported mixed results regarding the role of BDNF in drug-seeking behavior in animals as BDNF can either enhance reinstatement or have protective properties. Finally, we reviewed recent clinical studies that provide insight into the potential of BDNF to be a good biomarker of vulnerability to relapse.

Vitamins and Hormones, Volume 104
ISSN 0083-6729
http://dx.doi.org/10.1016/bs.vh.2016.10.009

1. INTRODUCTION

Addiction is a complex brain disorder, with a high risk of relapse. Recent estimates from clinical studies suggest that more than two-thirds of individuals relapse within weeks to months of initiating treatment for nicotine, alcohol, and opioid addiction (Hyman et al., 2008; Paliwal, Hyman, & Sinha, 2008; Sinha, Garcia, Paliwal, Kreek, & Rounsaville, 2006). This suggests that the neurobiological adaptations following chronic use of drugs of abuse are long lasting and that the current approved treatments are unable to counteract these neuroadaptations. There is a critical need to understand the mechanisms underlying relapse to drug seeking. It has already been shown that several factors play a role in relapse susceptibility, including for instance the dysregulation of the hypothalamic–pituitary–adrenal axis and an overactivity of the CRF system (Fox, Hong, Siedlarz, & Sinha, 2008; McKee et al., 2011), a high cortisol/ACTH ratios (Sinha et al., 2011), a lower amygdala volume (Wrase et al., 2008), or depressive symptoms (Dodge, Sindelar, & Sinha, 2005; Greenfield et al., 1998). Others studies have also suggested that higher serum brain-derived neurotrophic factor (BDNF) levels are highly predictive of addiction relapse, at least in cocaine abusers (D'Sa, Fox, Hong, Dileone, & Sinha, 2011). In several investigations, correlations have been reported between serum BDNF levels and withdrawal severity (Heberlein, Muschler, et al., 2010; Huang et al., 2008). The aim of this chapter is to summarize data from the literature regarding the role of BDNF during withdrawal, and whether or not this neurotrophic factor may represent a marker with a good potential to be used as predictor of relapse risk and/or outcome measures to assess treatment response.

2. BDNF ALTERATIONS FOLLOWING CHRONIC ADMINISTRATION OF DRUGS OF ABUSE

Adolescence is a critical period of brain development, during which specific synaptic rearrangements and maturation occur. The neurotrophin family, including BDNF, is well known to be involved in the maturation and development of synapses. Generally, adolescence is associated to a high level of novelty/sensation-seeking behaviors that may alter neuroplasticity. Among the behaviors observed, drug consumptions are highly present. Thus, the impact of drugs of abuse on BDNF levels may have different consequences in the different life periods from childhood to adulthood. In this section, we review the data in the literature regarding alterations of BDNF induced by drugs administration during different stages of life in mice and rats.

2.1 Juvenile and Young Adult Animals

2.1.1 Increase in BDNF Levels

Only amphetamine has been reported to enhance BDNF levels in different brain areas. Following a 1-month withdrawal from daily amphetamine injections from postnatal day (PND) 35 to PND 42 in mice, the long 3′ UTR form of BDNF, which is targeted in dendrites rather than in soma (An et al., 2008), was increased by 78% in the hippocampus. In the prefrontal cortex, all the isoforms were altered as increases in total mRNA BDNF, long 3′ UTR, exon IV, and exon VI were reported. These global alterations could be due to phosphorylation of MeCP2 (methyl CpG-binding protein 2), blocking its repressive effect on BDNF transcription. While no alterations were reported at the translational level in the hippocampus, amphetamine treatment in adolescence induces a trend toward increase pro-BDNF and a trend toward decrease mature BDNF (mBDNF) in the prefrontal cortex. This suggests that amphetamine treatment during adolescence would reduce the processing of pro-BDNF in mBDNF while enhancing its transcription (Calabrese et al., 2013). Very interestingly, a recent study has shown BDNF alterations in the brain in young adult male rats, 2 days after amphetamine self-administration, depending on the influence of fatty acid consumption. Thus, hydrogenated vegetable fat-supplemented animals are more prone to self-administer amphetamine compared to control group, and BDNF levels in cortex and hippocampus are increased after 2 days of withdrawal (Kuhn et al., 2015).

2.1.2 Decrease in BDNF Levels

A global BDNF decrease has been reported following chronic exposure to ethanol or cocaine. Cerebellar Purkinje cells seem sensitive to ethanol in an early postnatal period from PND 4 to 6, since intragastric intubation of different doses of ethanol in rats at PND 4 produces a reduction of mRNA expression of BDNF, its receptor TrkB and also the truncated TrkB.T2 isoform from 1 to 8 h postadministration, whereas no change was reported when ethanol was administered later during development, on PND 9. The authors suggest that ethanol alters survival factors in Purkinje neurons during the critical PND 4 period, which are then programmed to cell death, whereas in PND 9, there is a switch from the need of BDNF for neurogenesis and survival to the dependence on NT3, another neurotrophin, for maturation of cells in cerebellum (Ge, Belcher, & Light, 2004). This phenomenon is probably cerebellar specific since the BDNF levels are also decreased later in other brain regions. In adolescent rats trained to oral ethanol self-administration in their home cages during 12 days from PND 25, and withdrawn 7 days, Briones and Woods found both a 21% decrease of BDNF level in hippocampus, and a 32% reduction in the level of pTrkB proteins, without any difference in TrkB levels. These decreases are suspected to mediate the functional consequences (e.g., decrease of hedonic and despair behaviors) of the binge-pattern alcohol exposure by modulating neurogenesis (Briones & Woods, 2013). Finally, after 21 days of cocaine treatment in rats, starting on PND 14, the BDNF level decreases of 35% in the prefrontal cortex 24 h after the end of the treatment (Simchon-Tenenbaum, Weizman, & Rehavi, 2015).

2.2 Adult Animals

As described earlier, in young animals, BDNF alterations after exposure to drug of abuse are complexes. Several parameters such as the drug, the duration and doses of treatments, the withdrawal time analyses, the regions of interest, and the contingency of the administration may play a key role in BDNF regulations. Here we report a summary of the BDNF modifications in adulthood.

2.2.1 Increase in BDNF Levels

In drug-withdrawn rodents, the BDNF–TrkB signaling cascade and the related plasticity are altered in the nucleus *accumbens* (Nac), a brain region known to constitute a target to drugs of abuse to mediate their rewarding and reinforcing properties. Following early (1–3 days) and late (23–29 days) withdrawal from methamphetamine (Ren et al., 2015) or nicotine treatment (Kivinummi, Kaste, Rantamaki, Castren, & Ahtee, 2011), an increase in BDNF protein was found in Nac. Moreover, an increase of pCREB on early withdrawal from nicotine (Kivinummi et al., 2011), and an increase in pTrkB/TrkB ratio was reported on withdrawal day (WD) 23 in methamphetamine withdrawn mice compared to control, accompanied by a higher dendritic spine density specifically in Nac shell (Ren et al., 2015). A subregion-specific BDNF alteration was also shown with cocaine after acute and chronic treatment, as well as during withdrawal. Thus, an increase in BDNF mRNA in the shell was reported, and a decrease of TrkB receptor level in both core and shell 5 days after the end of the drug injection was shown (Filip et al., 2006). The increase BDNF levels observed following methamphetamine treatment seem to be involved in a depression-like phenotype, since the immobility time of the treated animals was higher in the tail suspension test and the forced swim test compared to control, as well as a lower sucrose preference (Ren et al., 2015). Similarly an increase in the duration of the immobility time in the forced swim test during early withdrawal was also observed following cocaine (Filip et al., 2006) and nicotine administration (Roni & Rahman, 2014). With cocaine exposure, the temporal regulation of BDNF was also studied after a single cocaine injection. BDNF mRNA was increased in the Nac shell 60 min after injection and declined after 90 min. Then, BDNF protein levels increased at 120 min postinjection, remained elevated for 4 h, and then decreased to basal levels after 24 h (Graham et al., 2007). A link between BDNF, mGlu5 receptor (a metabotropic glutamate receptor), and long-term depression (LTD) in Nac shell has been described following cocaine treatment. Thus, a progressive increase in BDNF protein levels 14 and 28 days after the end of cocaine treatment was observed, with a decrease in mGlu5 protein levels in Nac shell, leading to a decrease in the induction of mGlu-dependent LTD. This downregulation is thought to be a compensatory response to behavioral sensitization to cocaine (Huang et al., 2011). The hippocampus is also a main target of interest when studying BDNF alterations since it promote learning processes and synaptic plasticity. An increase in c-Fos and BDNF expression in CA3 region was observed after 1 and 3 days of withdrawal from 6% ethanol given in a liquid diet for 14 days, compared to control. Moreover, BDNF level was increased in all regions of the hippocampus, in male and female rats 3 days after the end of the treatment. In this study, c-Fos increase is likely to precede the induction of BDNF (Alele & Devaud, 2013). Some other data provide discrepancies regarding the alterations observed. Kenny and colleagues reported a heterogeneous BDNF alteration depending on hippocampal subregions, with a transient increase in CA1 region of the dorsal hippocampus 2 h after the last nicotine injection, and a return to basal level 24 h after. This acute effect may reflect the synaptic rearrangements in maintenance of nicotine consumption (Kenny, File, & Rattray, 2000). In contrast, 48 h after the end of nicotine treatment, other authors found an increase in BDNF protein levels in hippocampus and a trend toward increased pCREB (Roni & Rahman, 2014). This discrepancy may result from differences into the protocols, as Kenny used passive injections of nicotine for 7 days, whereas in the latter study it was a voluntary consumption of nicotine (mice drank nicotine solution for 21 days) (Roni & Rahman, 2014).

With cocaine, a modest increase of BDNF in hippocampus was reported only after short abstinence and specifically in an enriched environment introduced during abstinence (Thiel et al., 2012).

Very interestingly, when studying protracted ethanol abstinence, Schunck and colleagues found an increase in BDNF level in prefrontal cortex of rats after 12 days of withdrawal from ethanol administration by oral gavage. This was accompanied by an increase in interleukin-10 (IL-10), an antiinflammatory cytokine, as well as an increase in the nociceptive threshold, measured in the tail flick and hot plate tests (Schunck et al., 2015). This is consistent with the observation that BDNF infusions in different brain regions or in spinal cord may produce analgesia. The increase in IL-10 could reverse damages caused by ethanol intoxication and constitute an adaptive response to restore physiological homeostasis.

Finally, following chronic morphine treatment an increase in mRNA BDNF level has been reported 24, 48, and 72 h after the last morphine injection in the nucleus *paragigantocellularis*, one of the major afferents to the locus *coeruleus* (LC). This increase is not limited to BDNF as the other neurotrophin, NT3 was also upregulated compared to control group (Hatami et al., 2007).

2.2.2 Decrease in BDNF Levels

Following 1 and 4 days after injections of escalating doses of D-amphetamine, BDNF was found to be decreased in the dentate gyrus of the hippocampus, potentially underlying the memory deficit that is observed following withdrawal from psychostimulant (Fuller, Murray, & Horner, 2015). With morphine, a downregulation of BDNF mRNA was also observed, both with morphine pellets and subcutaneous injections. Seven days after removal of morphine pellets, both BDNF and TrkB mRNA, were downregulated in the dorsal raphe nucleus in rats, raising the possible contribution of BDNF in the hypoactivity of the serotoninergic system observed during protracted withdrawal from morphine (Lunden & Kirby, 2013). Similarly 4 weeks of abstinence following escalating doses of morphine for 5 days induced a decrease of BDNF in the hippocampus and an altered social interaction. Interestingly, these molecular and behavioral alterations were prevented by acupuncture stimulation at the SP6 acupoint on rats (Lee, Sur, Shim, Hahm, & Lee, 2014).

In the medial and ventral amygdala a reduced BDNF signaling was also observed 24 h after a chronic ethanol treatment in rats, as well as a reduced Arc expression (an immediate early gene implicated in synaptic plasticity), associated with a decrease in dendritic spine density, accompanied by anxiogenic behavior in animals. This BDNF-Arc signaling in amygdala is suggested to be a common target for anxiety and alcoholism (Pandey et al., 2008).

2.3 Dynamic Alterations of BDNF Levels During Withdrawal

After a drug treatment in animals, the BDNF level is sometimes reported to have opposed trends in the same animal regarding the brain regions studied or the time of withdrawal. These discrepancies are difficult to interpret and raise the question of the reliability of BDNF to be a good marker of psychiatric diseases (see Section 4).

An increase in BDNF protein is observed in prelimbic cortex, while a decrease is observed in dorsal hippocampus following a subchronic exposure to either cocaine (15 mg/kg for 10 days, once a day, i.p.) or heroin (1 mg/kg for 5 days, once a day, i.p.). No changes were seen neither in Nac nor in ventral hippocampus. This is consistent with BDNF alterations observed in animal models of depression and fit well with the neurotrophic hypothesis of depression that associate this pathology with a reduction of neurogenesis in hippocampus (Zilkha, Feigin, Barnea-Ygael, & Zangen, 2014). Furthermore, during precipitated opiate withdrawal with naloxone, a gradual BDNF mRNA increase was observed from 2 to 20 h of withdrawal, with a progressive partial return toward control levels at 70 h in the LC. This increase could be the consequence of an increase in neuronal firing. However, in the same study, no change in ventral tegmental area (VTA), another catecholaminergic nucleus was reported, highlighting complex adaptations following drug exposure (Numan et al., 1998).

Using a microarray approach increases of BDNF and TrkB proteins in dorsal striatum 2 and 24 h after methamphetamine self-administration compared to yoked saline controls were observed, but decreases at 1-month abstinence that may contribute to relapse vulnerability. With this experimental approach, other gene expression changes were also detected. Thus, 2 h after methamphetamine cessation an upregulation in c-fos was observed, suggesting a role in transcriptional regulation, as well as in several other neuropeptides including cholecystokinin and prodynorphin, known to modulate dopamine neurotransmission. Finally, an increase in pCREB levels was detected 2 and 24 h after methamphetamine self-administration, with a decrease after 1 month of abstinence. The authors also showed enrichment of pCREB on BDNF promoters only 2 h after cessation of methamphetamine. Interestingly, this study reported dissociation between mRNA and protein expression alterations in particular with TrkB and BDNF, suggesting other mechanisms than increase mRNA levels to upregulate the protein (e.g., increase stabilization of the protein, reduction in the protein degradation, enhanced processing of pro-BDNF to mBDNF, or a more efficient anterograde transport from cortical neurons to striatum) (Krasnova et al., 2013).

The contingency of drug administration adds another level of complexity as Fumagalli and colleagues reported BDNF alterations depending on both the brain region studied and the contingency of cocaine self-administration. Briefly, immediately after the last self-administration session, while no change was observed in medial prefrontal cortex (mPFC), they reported an increase in mBDNF protein in Nac in self-administering and yoked-cocaine animals. However, an increase of the mature protein was only observed in the crude synaptosomal fraction obtained from brains of yoked animals. On the contrary, after 24 h of withdrawal, no change was evident in Nac, whereas an increase of BDNF was reported in mPFC in the crude synaptosomal fraction for all cocaine-treated animals (Fumagalli et al., 2013). Thus, the early abstinence appears as a dynamic phenomenon and depends on several parameters such as the presence of the drug, the brain location, and how the drug is administered.

2.4 Epigenetics Regulation of BDNF Levels

Generally, the interaction between genomic DNA elements (specific sequences with regulatory function), epigenetic modifiers, and transcription factors determines the expression state of genes. The study of epigenetics in relation to drugs of abuse has been a rapidly emerging field during the past several years. Interestingly BDNF region promoters are known to be targets for epigenetic alterations, and several studies have reported that exposures to drugs of abuse modulate epigenetic regulation of BDNF gene expression.

After cocaine self-administration, increased BDNF protein in VTA after 7 days of forced abstinence was correlated to a selective increase in BDNF exon I-containing transcript. The molecular mechanisms leading to these regulations involved an increase in CREB-binding protein (CBP) with BDNF promoters. As CBP has a histone acetyltransferase activity, these associations increased acetylation of histone 3 (H3), opening chromatin configuration that enhanced the transcription of BDNF target genes (Schmidt et al., 2012). With morphine, a lower histone methylation at promoter II and III of BDNF was reported in both VTA and LC 7 days after the last passive injection in rats. As the trimethylation of H3 and K9 is associated with repression, it would have caused the increase of mRNA BDNF during withdrawal (Mashayekhi et al., 2012). Finally, as ethanol withdrawal is associated with increased histone deacetylase (HDAC) activity and a decrease in BDNF in amygdala, You and colleagues demonstrated that a HDAC inhibitor treatment blocked the BDNF alterations and anxiety-like behaviors in rats (You, Zhang, Sakharkar, Teppen, & Pandey, 2014).

3. PROTRACTED ABSTINENCE, INCUBATION, AND CRAVING: HOW IS BDNF INVOLVED?

3.1 Endogenous BDNF

It has been reported that the response for drug-associated stimuli progressively increases over time during withdrawal from opiates and psychostimulants in animal models of reinstatement of reward seeking. This phenomenon is called "incubation of craving." In the past 15 years, researchers demonstrated that this incubation was paralleled by an increase in BDNF levels in some brain regions.

After a history of cocaine use in rats, an increase of BDNF levels in Nac and amygdala is observed on WD30 and WD90, while no difference was reported compared to control 1 day after self-administration. Interestingly, this was drug specific since no BDNF alterations were reported for sucrose, although an increase in the responsiveness to sucrose cues during withdrawal was shown. Moreover, these alterations seem specifics to the neurotrophin BDNF, as no NGF alterations were reported (Grimm et al., 2003). Focusing on WD45 after cocaine self-administration, an increase of BDNF protein in whole Nac was also observed, due in fact to a specific increase in Nac core, without modification in Nac shell (Li et al., 2013). Interestingly, on WD90 a BDNF increase was observed in both Nac shell and core, and in parallel a significant increase in surface expression of pTrkB was observed in whole Nac, suggesting an activation of TrkB receptor. Furthermore, using viral vectors to attenuate BDNF–TrkB signaling, these authors demonstrated opposite role of core and shell in the incubation of cocaine seeking. Injection in the core enhanced cocaine seeking on early withdrawal in rats (WD1), whereas no effect was reported on protracted withdrawal. On the other hand, viral injections in the shell attenuated cocaine seeking on WD90, suggesting that the Nac shell is only involved in incubation in protracted withdrawal. These results highlight the complex regulations of BDNF, with cocaine differently affecting BDNF levels in core and shell subregions and with different time-course elevation during withdrawal. Mechanistically, a BDNF-dependent synaptic sensitization of excitatory synapses of dopamine neurons in the VTA after cocaine withdrawal (10–15 days after the last cocaine injection) has been described and this phenomenon required both NMDA receptor activation and postsynaptic TrkB signaling. This stimulation of synapses by endogenous BDNF may contribute to relapse and compulsive drug seeking by potentiating cue-associated activity in VTA dopamine neurons (Pu, Liu, & Poo, 2006).

In rats trained to self-administer methamphetamine and after prolonged withdrawal, increases in BDNF and TrkB mRNA, glutamate receptors subunits, and several epigenetics enzymes (e.g., Hdac3, Hdac4) were reported in dorsal striatum in Fos-positive neurons only, without anatomical specificity of the subregions (dorsolateral, center, and dorsomedial) and dopamine type neurons (D1- and D2-type). When blocking cue- and drug-induced Fos induction by a D1 antagonist administration, a decrease of cue-induced methamphetamine seeking was observed at 1 month of withdrawal, with no effect on WD2 (Li et al., 2015).

With heroin, whole-genome analysis of mPFC reveals an increase of BDNF mRNA only after 14 days of abstinence (with no change after 1 day), and analyses highlight that BDNF alterations were correlated with goal-directed behaviors, suggesting that BDNF may be a good target for drug interventions to prevent relapse (Kuntz-Melcavage, Brucklacher, Grigson, Freeman, & Vrana, 2009).

3.2 Pharmacological Intervention on BDNF Signaling

There is a growing literature showing that BDNF is implicated in long-term modulation of drug-induced behaviors. However, the consequences of pharmacological manipulations with BDNF are inconsistent and depend on the region of interest. We report here a summary of data that either suggests protective mechanisms against relapse or enhancement of drug seeking.

3.2.1 When BDNF Is a Promising Therapeutic Target …

Infusion of BDNF into PFC immediately after a last cocaine self-administration session in rats but not after 6 days attenuated cocaine-seeking behavior both after 22 h and 6 days of abstinence. This suggests that during early withdrawal, BDNF may interfere with the neuroadaptations induced by the drug, to prevent relapse on cocaine seeking (Berglind et al., 2007). The neural mechanisms involved seem to be dependent on phosphorylation of synapsin—a protein controlling neurotransmitter exocytosis when phosphorylated—in Nac during early withdrawal from cocaine. Indeed, it was shown that p-synapsin Ser9 protein expression in Nac was increased by cocaine 2 and 22 h after the last cocaine session of drug self administration, which was normalized when infusing BDNF in PFC. This phenomenon may restore neuronal activity in the PFC–Nac pathway, potentially recovering the pool of synaptic vesicles during early withdrawal from cocaine (Sun, Eisenstein, Zelek-Molik, & McGinty, 2014). In rats trained to self-administer ethanol, binge consumption causes depressive-like symptoms and decreases neuronal survival in the dentate gyrus in the hippocampal region. Interestingly, when enhancing the BDNF signaling with a TrkB agonist during drug consumption, the neurogenesis is restored, and the anhedonia and despair behavior abolished (Briones & Woods, 2013).

3.2.2 … Or When BDNF Enhances Drug Seeking

As previously described, cocaine enhances BDNF signaling in Nac (Fumagalli et al., 2013; Grimm et al., 2003; Huang et al., 2011; Li et al., 2013). When animals received BDNF infusions in Nac shell during cocaine self-administration and were tested after 10 days of withdrawal in extinction conditions, an increase in cocaine infusion was observed, suggesting an enhancement of cocaine-seeking behavior by BDNF infusions. Furthermore, when animals were submitted to reinstatement testing triggered by exposure to cues previously associated with cocaine, BDNF-treated animals also showed enhanced cocaine seeking (Graham et al., 2007). Consistent with this later finding, CREB activity in the Nac shell produced an enhancement of cocaine reinforcement in self-administering rats measured in a progressive ratio procedure, correlated with BDNF increase. Increasing CREB expression in the shell during withdrawal facilitates relapse to cocaine seeking (Larson et al., 2011), a pattern opposite to the one observed in dorsal striatum, where the upregulation of CREB produces a tolerance-like escalation. In the VTA, BDNF infusion after cocaine self-administration enhanced cocaine seeking after 3 and 30 days of withdrawal, a phenomenon mediated by the MAPK pathway. This effect is anatomically and temporally selective, since BDNF infusion in *substantia nigra*, and VTA 2 h before the test for cocaine seeking, was both ineffective (Lu, Dempsey, Liu, Bossert, & Shaham, 2004). With methamphetamine, as previously mentioned Ren et al. demonstrated mBDNF and pTrkB/TrkB ratio increase in Nac shell after withdrawal from repeated methamphetamine administration. Subchronic administration of a TrkB antagonist, ANA-12, after 1 week of withdrawal from methamphetamine improved the behavioral performances on the forced swimming test, the tail-suspension test, and the preference for sucrose. Finally, the increase of spine density observed in methamphetamine-treated mice in Nac shell was attenuated by repeated administration of ANA-12. Thus, ANA-12 appears to mediate the therapeutic effect on depression after methamphetamine exposure and may be a promising approach to counteract the behavioral alterations following repeated drug exposure (Ren et al., 2015).

4. PERIPHERAL BDNF LEVELS

Previous studies have shown that blood level regulations of BDNF are related to brain BDNF expression and are not a simple epiphenomenon (Karege, Schwald, & Cisse, 2002; Klein et al., 2011) owing to the existence of a high-capacity, saturable transport system, allowing BDNF to cross the blood–brain barrier (Pan, Banks, Fasold, Bluth, & Kastin, 1998). In blood, BDNF is essentially stored in circulating platelets (Fujimura et al., 2002). Human platelets circulate for about 10 days in blood, while the half-life of BDNF in plasma is less than 1 h (Kishino et al., 2001). Thus, while plasma BDNF might reflect free circulating concentration, the serum content might be related to neurotrophin quantities contained in platelets and released in vitro during clotting, reflecting the whole blood content.

4.1 A Potential Biomarker?

The results of clinical studies regarding the association between BDNF serum levels and drug withdrawal are relatively limited and often inconsistent with mixed results. The discrepancy observed may be partly explained by factors including depression, poly-consumption of drugs of abuse, stress, or type of treatment prescribed to patients (such as methadone). All have been reported to influence serum BDNF levels (Jamal, Van der Does, & Penninx, 2015; Lee et al., 2015; Matrisciano et al., 2009).

One of the good examples illustrating the discrepancy in the results can be found in the papers reporting the serum BDNF levels during methamphetamine withdrawal. Chen, Qin, Su, Liu, and Yang (2012) reported that the serum BDNF levels were significantly lower than those of healthy controls after 3 weeks of withdrawal (Chen et al., 2012). On the contrary, two studies (Kim et al., 2005; Ren et al., 2016) reported a peripheral increase of BDNF levels in methamphetamine addicts who had been abstinent for at least 1 month compared to healthy subjects. However, while both studies from Chen et al. (2012) and Kim et al. (2005) only measured BDNF levels at one time point, Ren et al. (2016) evaluated the BDNF levels in methamphetamine users and healthy controls at baseline (abstinent for 1–7 days), and at 1-month endpoint in drug addicts. Very interestingly, they observed time-dependent variations of BDNF, with a higher serum BDNF level during early withdrawal in methamphetamine addicts than healthy subjects, and a decrease during the first month of abstinence. At this time, no more differences were observed compared to the controls (Ren et al., 2016). These results suggest that serum BDNF regulations are dynamic phenomena.

At this time in heroin-dependent patients, there have been four studies measuring levels of serum BDNF (Angelucci et al., 2007; Heberlein et al., 2011; Zhang et al., 2016, 2014). As with methamphetamine literature, the results are inconsistent. Thus, while Angelucci et al. (2007) and Zhang et al. (2016) observed a significant decrease in serum BDNF levels in patients with history of chronic heroin abuse, in contrast Zhang et al. (2014) observed increased levels in heroin-dependent patients during both early withdrawal (1–7 days) and 1 month after heroin cessation compared to healthy controls, a result consistent with those of Heberlein et al. (2011) (Angelucci et al., 2007; Zhang et al., 2016, but see also Heberlein et al., 2011; Zhang et al., 2014). They found a positive correlation between peripheral BDNF levels and opiate craving in opiate-dependent patients. Interestingly, as previously observed with methamphetamine, regulation of BDNF may follow a dynamic process. Thus, in their study, Zhang et al. (2016) found a lower serum BDNF level in heroin-dependent patients than in healthy patients in the early withdrawal, which increased after 6 months abstinence, but remained still below the levels of healthy controls.

In alcohol-associated dependence, here again, mixed results regarding the association between BDNF levels and withdrawal severity have been reported. The first studies have investigated BDNF levels in alcohol-dependent patients in early withdrawal (1–7 days). While several different studies reported no difference in baseline serum BDNF levels in alcoholic patients and controls (Costa, Girard, Dalmay, & Malauzat, 2011; Heberlein, Muschler, et al., 2010; Huang et al., 2008; Zanardini et al., 2011), other authors observed either a modest increase in serum BDNF concentration in the first days of withdrawal (Huang et al., 2008; Reynolds, Mueller, & MacLaren, 2015) or a decrease (Kohler, Klimke, Hellweg, & Lang, 2013). However, the use of different medications in these studies (benzodiazepines, α2-agonists, carbamazepine) may explain these discrepancies, as it is well known that these medications have an effect on peripheral blood levels of neurotrophic factors (Chang, Rapoport, & Rao, 2009; Heberlein, Lenz, et al., 2010). Few studies have also investigated regulation of serum BDNF levels in patients who remained abstinent several weeks. Two of them observed a continuing tendency of BDNF level to increase 2 weeks and 6 months after withdrawal (Costa et al., 2011; Kohler et al., 2013), while another one observed decreased serum BDNF levels in patients abstinent for at least 30 days (Joe et al., 2007).

Several authors investigated the BDNF levels in men or women with cocaine or crack dependence compared with controls. BDNF levels increased during the early abstinence in patients prone to relapse. This raised the question of whether peripheral BDNF could be a good marker of cocaine relapse (Corominas-Roso et al., 2015; D'Sa et al., 2011; McGinty & Mendelson, 2011). While these results seem consistent in the literature, in contrast, discrepancies have been reported considering basal levels of BDNF in cocaine-dependent patients at the beginning of the detoxification treatment compared to healthy subjects. Thus, while some studies reported lower serum BDNF levels (Corominas-Roso, Roncero, Eiroa-Orosa, Gonzalvo, et al., 2013; Corominas-Roso, Roncero, Eiroa-Orosa, Ribases, et al., 2013; von Diemen et al., 2014), other studies did not observe any difference (Angelucci et al., 2007).

One study compared plasma BDNF levels of smokers and nonsmokers prior to smoking cessation, and 4 and 12 weeks after quitting smoking, in presence or in absence of treatments (varenicline or nicotine patch) (Bhang, Choi, & Ahn, 2010). BDNF levels in smokers at baseline were lower than in nonsmokers, and whatever the subgroup of patients under treatments, the rate of BDNF increased 4 weeks after smoking cessation compared to the basal state, followed by a decreasing trend between 4 and 12 weeks. Consistent with these results, Kim, Kim, Lee, and Kim (2007) also reported an increase in plasma BDNF levels in smokers following smoking cessation without treatments (Kim et al., 2007). However, in another study, while the serum BDNF levels of smokers prior to varenicline treatment were lower than in healthy control, the authors did not report an increase in BDNF after treatment with varenicline for 8 weeks (Umene-Nakano et al., 2010).

To summarize, whatever the drug of abuse investigated, the results reported in the literature on serum or plasma BDNF levels are not consistent. The reasons for the discrepancies remain unknown. However, age, duration of drug use, frequency of consumption, time of withdrawal, course of different drugs of abuse, small size of the cohorts, and heterogeneity of the populations could be factors associated with such discrepancies. On the other hand, in a recent animal study, Geoffroy, Puig, Benturquia, and Noble (2014) have observed specific variations of plasma BDNF levels at the usual hour of drug injection, suggesting that the moment at which the blood samples are collected may also be an essential factor (Geoffroy et al., 2014).

4.2 Epigenetics in Clinic

As previously mentioned BDNF gene is under epigenetic regulations (see Section 2.4). Although epigenetic studies have usually been performed on postmortem brain tissue with identification of specific changes, similar alterations at specific genomic regions have also been detected in peripheral samples (Hayashi-Takagi, Vawter, & Iwamoto, 2014). Only a few studies have investigated correlations between peripheral epigenetic marks, and addictive behaviors. In opioid users, no significant promoter methylation variations of BDNF gene were demonstrated compared to control group, without any history of drug abuse (Kordi-Tamandani, Tajoddini, & Salimi, 2015). In alcohol addiction, a significant decrease in BDNF promoter methylation in exon IV was observed during withdrawal, without modifications in serum BDNF levels (Heberlein et al., 2015). The difficulty with these studies can certainly come from subtle changes commonly observed in DNA methylation levels (typically <5%).

5. POLYMORPHISM IN BDNF GENE: PREDICTIVE VALUE IN DRUG ADDICTION?

A common single-nucleotide polymorphism in the human BDNF gene has been identified, with an amino-acid substitution of valine to methionine at amino-acid residue 66 (Val66Met). This genetic variation is associated with alteration of intracellular trafficking and packaging of pro-BDNF, and consequently in the secretion of mBDNF (Egan et al., 2003). The Val66Met BDNF gene polymorphism has been linked with decreased BDNF serum concentrations (Lang, Hellweg, Sander, & Gallinat, 2009).

Very interestingly, while a study in medication-free subjects with alcohol dependence did not show significant association between this BDNF polymorphism and alcohol dependence compared to control group (Nedic et al., 2013), another group reported, in patients admitted to addiction treatment programs, that those with the BDNF Val66Met polymorphism presented a higher risk of relapse (Wojnar et al., 2009). All together, these results may suggest that this genetic variation may be a marker of treatment outcomes in alcohol dependence. This could be correlated to the neuroimaging study performed in alcohol-dependent subjects, showing that the BDNF Val66Met polymorphism influences the brain gray and white matter tissue volumes recovery during abstinence from alcohol (Mon et al., 2013).

Similar results are described in methamphetamine-dependent subjects, with worse treatment outcomes in patients with the Val66Met polymorphism (Heinzerling, McCracken, Swanson, Ray, & Shoptaw, 2012), while no differences in allele distributions between methamphetamine abusers and controls were observed (Su et al., 2014). However, an interaction between sex and BDNF polymorphism has also been reported, with a high frequency of methamphetamine use among females with Val/Val genotype (Heinzerling & Shoptaw, 2012). Furthermore, in the Thai population, the BDNF genetic polymorphism is associated with methamphetamine dependence (Iamjan, Thanoi, Watiktinkorn, Nudmamud-Thanoi, & Reynolds, 2015).

It is difficult to find a consensus regarding BDNF polymorphism and vulnerability to addiction, whatever the drug of abuse. This may be explained by different factors, including the study samples which are generally relatively small, and also the effects that can be gender and ethnicity specifics.

6. CONCLUSION AND FUTURE DIRECTIONS

In this chapter, we discussed recent findings which enhanced the comprehension of the neuroadaptations during withdrawal that follow the consumption of drugs.

These studies indicate that drugs of abuse extensively altered peripheral and central BDNF levels during withdrawal, but the variations depend on several parameters such as the drug itself, the pattern of administration, the gender, or even the developmental period. Even with the same drug of abuse, in both clinical and preclinical studies, there are still discrepancies regarding the BDNF levels. Those mixed results highlight that to date BDNF alone is not suitable to be a good biomarker for relapse vulnerability. Further studies should investigate whether BDNF in association with other(s) protein(s) may be good markers of either drug intoxication, treatment outcomes, or vulnerability to relapse.

ACKNOWLEDGMENTS

The authors declare that there is no conflict of interest regarding the publication of this chapter.

REFERENCES

Alele, P. E., & Devaud, L. L. (2013). Expression of cFos and brain-derived neurotrophic factor in cortex and hippocampus of ethanol-withdrawn male and female rats. *Journal of Pharmacology and Pharmacotherapeutics*, *4*(4), 265–274.

An, J. J., Gharami, K., Liao, G. Y., Woo, N. H., Lau, A. G., Vanevski, F., et al. (2008). Distinct role of long 3′ UTR BDNF mRNA in spine morphology and synaptic plasticity in hippocampal neurons. *Cell*, *134*(1), 175–187.

Angelucci, F., Ricci, V., Pomponi, M., Conte, G., Mathe, A. A., Attilio Tonali, P., et al. (2007). Chronic heroin and cocaine abuse is associated with decreased serum concentrations of the nerve growth factor and brain-derived neurotrophic factor. *Journal of Psychopharmacology*, *21*(8), 820–825.

Berglind, W. J., See, R. E., Fuchs, R. A., Ghee, S. M., Whitfield, T. W., Jr., Miller, S. W., et al. (2007). A BDNF infusion into the medial prefrontal cortex suppresses cocaine seeking in rats. *The European Journal of Neuroscience*, *26*(3), 757–766.

Bhang, S. Y., Choi, S. W., & Ahn, J. H. (2010). Changes in plasma brain-derived neurotrophic factor levels in smokers after smoking cessation. *Neuroscience Letters*, *468*(1), 7–11.

Briones, T. L., & Woods, J. (2013). Chronic binge-like alcohol consumption in adolescence causes depression-like symptoms possibly mediated by the effects of BDNF on neurogenesis. *Neuroscience*, *254*, 324–334.

Calabrese, F., Richetto, J., Racagni, G., Feldon, J., Meyer, U., & Riva, M. A. (2013). Effects of withdrawal from repeated amphetamine exposure in peri-puberty on neuroplasticity-related genes in mice. *Neuroscience*, *250*, 222–231.

Chang, Y. C., Rapoport, S. I., & Rao, J. S. (2009). Chronic administration of mood stabilizers upregulates BDNF and bcl-2 expression levels in rat frontal cortex. *Neurochemical Research*, *34*(3), 536–541.

Chen, J., Qin, J., Su, Q., Liu, Z., & Yang, J. (2012). Treadmill rehabilitation treatment enhanced BDNF-TrkB but not NGF-TrkA signaling in a mouse intracerebral hemorrhage model. *Neuroscience Letters*, *529*(1), 28–32.

Corominas-Roso, M., Roncero, C., Daigre, C., Grau-Lopez, L., Ros-Cucurull, E., Rodriguez-Cintas, L., et al. (2015). Changes in brain-derived neurotrophic factor (BDNF) during abstinence could be associated with relapse in cocaine-dependent patients. *Psychiatry Research*, *225*(3), 309–314.

Corominas-Roso, M., Roncero, C., Eiroa-Orosa, F. J., Gonzalvo, B., Grau-Lopez, L., Ribases, M., et al. (2013a). Brain-derived neurotrophic factor serum levels in cocaine-dependent patients during early abstinence. *European Neuropsychopharmacology*, *23*(9), 1078–1084.

Corominas-Roso, M., Roncero, C., Eiroa-Orosa, F. J., Ribases, M., Barral, C., Daigre, C., et al. (2013b). Serum brain-derived neurotrophic factor levels and cocaine-induced transient psychotic symptoms. *Neuropsychobiology*, *68*(3), 146–155.

Costa, M. A., Girard, M., Dalmay, F., & Malauzat, D. (2011). Brain-derived neurotrophic factor serum levels in alcohol-dependent subjects 6 months after alcohol withdrawal. *Alcoholism, Clinical and Experimental Research*, *35*(11), 1966–1973.

Dodge, R., Sindelar, J., & Sinha, R. (2005). The role of depression symptoms in predicting drug abstinence in outpatient substance abuse treatment. *Journal of Substance Abuse Treatment*, *28*(2), 189–196.

D'Sa, C., Fox, H. C., Hong, A. K., Dileone, R. J., & Sinha, R. (2011). Increased serum brain-derived neurotrophic factor is predictive of cocaine relapse outcomes: A prospective study. *Biological Psychiatry*, *70*(8), 706–711.

Egan, M. F., Kojima, M., Callicott, J. H., Goldberg, T. E., Kolachana, B. S., Bertolino, A., et al. (2003). The BDNF val66met polymorphism affects activity-dependent secretion of BDNF and human memory and hippocampal function. *Cell*, *112*(2), 257–269.

Filip, M., Faron-Gorecka, A., Kusmider, M., Golda, A., Frankowska, M., & Dziedzicka-Wasylewska, M. (2006). Alterations in BDNF and trkB mRNAs following acute or sensitizing cocaine treatments and withdrawal. *Brain Research*, *1071*(1), 218–225.

Fox, H. C., Hong, K. I., Siedlarz, K., & Sinha, R. (2008). Enhanced sensitivity to stress and drug/alcohol craving in abstinent cocaine-dependent individuals compared to social drinkers. *Neuropsychopharmacology*, *33*(4), 796–805.

Fujimura, H., Altar, C. A., Chen, R., Nakamura, T., Nakahashi, T., Kambayashi, J., et al. (2002). Brain-derived neurotrophic factor is stored in human platelets and released by agonist stimulation. *Thrombosis and Haemostasis*, *87*(4), 728–734.

Fuller, J. J., Murray, R. C., & Horner, K. A. (2015). D-Amphetamine withdrawal-induced decreases in brain-derived neurotrophic factor in Sprague-Dawley rats are reversed by treatment with ketamine. *Neuropharmacology*, *97*, 7–17.

Fumagalli, F., Moro, F., Caffino, L., Orru, A., Cassina, C., Giannotti, G., et al. (2013). Region-specific effects on BDNF expression after contingent or non-contingent cocaine i.v. self-administration in rats. *The International Journal of Neuropsychopharmacology*, *16*(4), 913–918.

Ge, Y., Belcher, S. M., & Light, K. E. (2004). Alterations of cerebellar mRNA specific for BDNF, p75NTR, and TrkB receptor isoforms occur within hours of ethanol administration to 4-day-old rat pups. *Brain Research. Developmental Brain Research*, *151*(1–2), 99–109.

Geoffroy, H. A., Puig, S., Benturquia, N., & Noble, F. (2014). Temporal regulation of peripheral BDNF levels during cocaine and morphine withdrawal: Comparison with a natural reward. *The International Journal of Neuropsychopharmacology*, *18*(5), pyu088.

Graham, D. L., Edwards, S., Bachtell, R. K., DiLeone, R. J., Rios, M., & Self, D. W. (2007). Dynamic BDNF activity in nucleus accumbens with cocaine use increases self-administration and relapse. *Nature Neuroscience*, *10*(8), 1029–1037.

Greenfield, S. F., Weiss, R. D., Muenz, L. R., Vagge, L. M., Kelly, J. F., Bello, L. R., et al. (1998). The effect of depression on return to drinking: A prospective study. *Archives of General Psychiatry*, *55*(3), 259–265.

Grimm, J. W., Lu, L., Hayashi, T., Hope, B. T., Su, T. P., & Shaham, Y. (2003). Time-dependent increases in brain-derived neurotrophic factor protein levels within the mesolimbic dopamine system after withdrawal from cocaine: Implications for incubation of cocaine craving. *The Journal of Neuroscience*, *23*(3), 742–747.

Hatami, H., Oryan, S., Semnanian, S., Kazemi, B., Bandepour, M., & Ahmadiani, A. (2007). Alterations of BDNF and NT-3 genes expression in the nucleus paragigantocellularis during morphine dependency and withdrawal. *Neuropeptides*, *41*(5), 321–328.

Hayashi-Takagi, A., Vawter, M. P., & Iwamoto, K. (2014). Peripheral biomarkers revisited: Integrative profiling of peripheral samples for psychiatric research. *Biological Psychiatry*, *75*(12), 920–928.

Heberlein, A., Buscher, P., Schuster, R., Kleimann, A., Lichtinghagen, R., Rhein, M., et al. (2015). Do changes in the BDNF promoter methylation indicate the risk of alcohol relapse? *European Neuropsychopharmacology*, *25*(11), 1892–1897.

Heberlein, A., Dursteler-MacFarland, K. M., Lenz, B., Frieling, H., Grosch, M., Bonsch, D., et al. (2011). Serum levels of BDNF are associated with craving in opiate-dependent patients. *Journal of Psychopharmacology*, *25*(11), 1480–1484.

Heberlein, A., Lenz, B., Muschler, M., Frieling, H., Buechl, R., Groschl, M., et al. (2010). BDNF plasma levels decrease during benzodiazepine withdrawal in patients suffering from comorbidity of depressive disorder and benzodiazepine dependence. *Psychopharmacology*, *209*(2), 213–215.

Heberlein, A., Muschler, M., Wilhelm, J., Frieling, H., Lenz, B., Groschl, M., et al. (2010). BDNF and GDNF serum levels in alcohol-dependent patients during withdrawal. *Progress in Neuro-Psychopharmacology & Biological Psychiatry*, *34*(6), 1060–1064.

Heinzerling, K. G., McCracken, J. T., Swanson, A. N., Ray, L. A., & Shoptaw, S. J. (2012). COMT Val158Met, BDNF Val66Met, and OPRM1 Asn40Asp and methamphetamine dependence treatment response: Preliminary investigation. *Journal of Clinical Psychopharmacology*, *32*(1), 135–137.

Heinzerling, K. G., & Shoptaw, S. (2012). Gender, brain-derived neurotrophic factor Val66Met, and frequency of methamphetamine use. *Gender Medicine*, *9*(2), 112–120.

Huang, M. C., Chen, C. H., Chen, C. H., Liu, S. C., Ho, C. J., Shen, W. W., et al. (2008). Alterations of serum brain-derived neurotrophic factor levels in early alcohol withdrawal. *Alcohol and Alcoholism*, *43*(3), 241–245.

Huang, M. C., Chen, C. H., Liu, H. C., Chen, C. C., Ho, C. C., & Leu, S. J. (2011). Differential patterns of serum brain-derived neurotrophic factor levels in alcoholic patients with and without delirium tremens during acute withdrawal. *Alcoholism, Clinical and Experimental Research*, *35*(1), 126–131.

Hyman, S. M., Paliwal, P., Chaplin, T. M., Mazure, C. M., Rounsaville, B. J., & Sinha, R. (2008). Severity of childhood trauma is predictive of cocaine relapse outcomes in women but not men. *Drug and Alcohol Dependence*, *92*(1–3), 208–216.

Iamjan, S. A., Thanoi, S., Watiktinkorn, P., Nudmamud-Thanoi, S., & Reynolds, G. P. (2015). BDNF (Val66Met) genetic polymorphism is associated with vulnerability for methamphetamine dependence. *Pharmacogenomics*, *16*(14), 1541–1545.

Jamal, M., Van der Does, W., & Penninx, B. W. (2015). Effect of variation in BDNF Val(66) Met polymorphism, smoking, and nicotine dependence on symptom severity of depressive and anxiety disorders. *Drug and Alcohol Dependence*, *148*, 150–157.

Joe, K. H., Kim, Y. K., Kim, T. S., Roh, S. W., Choi, S. W., Kim, Y. B., et al. (2007). Decreased plasma brain-derived neurotrophic factor levels in patients with alcohol dependence. *Alcoholism, Clinical and Experimental Research*, *31*(11), 1833–1838.

Karege, F., Schwald, M., & Cisse, M. (2002). Postnatal developmental profile of brain-derived neurotrophic factor in rat brain and platelets. *Neuroscience Letters*, *328*(3), 261–264.

Kenny, P. J., File, S. E., & Rattray, M. (2000). Acute nicotine decreases, and chronic nicotine increases the expression of brain-derived neurotrophic factor mRNA in rat hippocampus. *Brain Research. Molecular Brain Research*, *85*(1–2), 234–238.

Kim, T. S., Kim, D. J., Lee, H., & Kim, Y. K. (2007). Increased plasma brain-derived neurotrophic factor levels in chronic smokers following unaided smoking cessation. *Neuroscience Letters*, *423*(1), 53–57.

Kim, D. J., Roh, S., Kim, Y., Yoon, S. J., Lee, H. K., Han, C. S., et al. (2005). High concentrations of plasma brain-derived neurotrophic factor in methamphetamine users. *Neuroscience Letters*, *388*(2), 112–115.

Kishino, A., Katayama, N., Ishige, Y., Yamamoto, Y., Ogo, H., Tatsuno, T., et al. (2001). Analysis of effects and pharmacokinetics of subcutaneously administered BDNF. *Neuroreport*, *12*(5), 1067–1072.

Kivinummi, T., Kaste, K., Rantamaki, T., Castren, E., & Ahtee, L. (2011). Alterations in BDNF and phospho-CREB levels following chronic oral nicotine treatment and its withdrawal in dopaminergic brain areas of mice. *Neuroscience Letters*, *491*(2), 108–112.

Klein, A. B., Williamson, R., Santini, M. A., Clemmensen, C., Ettrup, A., Rios, M., et al. (2011). Blood BDNF concentrations reflect brain-tissue BDNF levels across species. *The International Journal of Neuropsychopharmacology, 14*(3), 347–353.

Kohler, S., Klimke, S., Hellweg, R., & Lang, U. E. (2013). Serum brain-derived neurotrophic factor and nerve growth factor concentrations change after alcohol withdrawal: Preliminary data of a case-control comparison. *European Addiction Research, 19*(2), 98–104.

Kordi-Tamandani, D. M., Tajoddini, S., & Salimi, F. (2015). Promoter methylation and BDNF and DAT1 gene expression profiles in patients with drug addiction. *Pathobiology, 82*(2), 94–99.

Krasnova, I. N., Chiflikyan, M., Justinova, Z., McCoy, M. T., Ladenheim, B., Jayanthi, S., et al. (2013). CREB phosphorylation regulates striatal transcriptional responses in the self-administration model of methamphetamine addiction in the rat. *Neurobiology of Disease, 58*, 132–143.

Kuhn, F. T., Dias, V. T., Roversi, K., Vey, L. T., de Freitas, D. L., Pase, C. S., et al. (2015). Cross-generational trans fat consumption favors self-administration of amphetamine and changes molecular expressions of BDNF, DAT, and D1/D2 receptors in the cortex and hippocampus of rats. *Neurotoxicity Research, 28*(4), 319–331.

Kuntz-Melcavage, K. L., Brucklacher, R. M., Grigson, P. S., Freeman, W. M., & Vrana, K. E. (2009). Gene expression changes following extinction testing in a heroin behavioral incubation model. *BMC Neuroscience, 10*, 95.

Lang, U. E., Hellweg, R., Sander, T., & Gallinat, J. (2009). The Met allele of the BDNF Val66Met polymorphism is associated with increased BDNF serum concentrations. *Molecular Psychiatry, 14*(2), 120–122.

Larson, E. B., Graham, D. L., Arzaga, R. R., Buzin, N., Webb, J., Green, T. A., et al. (2011). Overexpression of CREB in the nucleus accumbens shell increases cocaine reinforcement in self-administering rats. *The Journal of Neuroscience, 31*(45), 16447–16457.

Lee, S. Y., Chen, S. L., Chang, Y. H., Chu, C. H., Chen, S. H., Chen, P. S., et al. (2015). A placebo-controlled trial of dextromethorphan as an adjunct in opioid-dependent patients undergoing methadone maintenance treatment. *The International Journal of Neuropsychopharmacology, 18*(7), pyv008.

Lee, B., Sur, B. J., Shim, I., Hahm, D. H., & Lee, H. (2014). Acupuncture stimulation attenuates impaired emotional-like behaviors and activation of the noradrenergic system during protracted abstinence following chronic morphine exposure in rats. *Evidence-Based Complementary and Alternative Medicine, 2014*, 216503.

Li, X., DeJoseph, M. R., Urban, J. H., Bahi, A., Dreyer, J. L., Meredith, G. E., et al. (2013). Different roles of BDNF in nucleus accumbens core versus shell during the incubation of cue-induced cocaine craving and its long-term maintenance. *The Journal of Neuroscience, 33*(3), 1130–1142.

Li, X., Rubio, F. J., Zeric, T., Bossert, J. M., Kambhampati, S., Cates, H. M., et al. (2015). Incubation of methamphetamine craving is associated with selective increases in expression of Bdnf and trkb, glutamate receptors, and epigenetic enzymes in cue-activated fos-expressing dorsal striatal neurons. *The Journal of Neuroscience, 35*(21), 8232–8244.

Lu, L., Dempsey, J., Liu, S. Y., Bossert, J. M., & Shaham, Y. (2004). A single infusion of brain-derived neurotrophic factor into the ventral tegmental area induces long-lasting potentiation of cocaine seeking after withdrawal. *The Journal of Neuroscience, 24*(7), 1604–1611.

Lunden, J. W., & Kirby, L. G. (2013). Opiate exposure and withdrawal dynamically regulate mRNA expression in the serotonergic dorsal raphe nucleus. *Neuroscience, 254*, 160–172.

Mashayekhi, F. J., Rasti, M., Rahvar, M., Mokarram, P., Namavar, M. R., & Owji, A. A. (2012). Expression levels of the BDNF gene and histone modifications around its promoters in the ventral tegmental area and locus ceruleus of rats during forced abstinence from morphine. *Neurochemical Research, 37*(7), 1517–1523.

Matrisciano, F., Bonaccorso, S., Ricciardi, A., Scaccianoce, S., Panaccione, I., Wang, L., et al. (2009). Changes in BDNF serum levels in patients with major depression disorder (MDD) after 6 months treatment with sertraline, escitalopram, or venlafaxine. *Journal of Psychiatric Research*, *43*(3), 247–254.

McGinty, J. F., & Mendelson, J. E. (2011). Is brain-derived neurotrophic factor a selective biomarker that predicts cocaine relapse outcomes? *Biological Psychiatry*, *70*(8), 700–701.

McKee, S. A., Sinha, R., Weinberger, A. H., Sofuoglu, M., Harrison, E. L., Lavery, M., et al. (2011). Stress decreases the ability to resist smoking and potentiates smoking intensity and reward. *Journal of Psychopharmacology*, *25*(4), 490–502.

Mon, A., Durazzo, T. C., Gazdzinski, S., Hutchison, K. E., Pennington, D., & Meyerhoff, D. J. (2013). Brain-derived neurotrophic factor genotype is associated with brain gray and white matter tissue volumes recovery in abstinent alcohol-dependent individuals. *Genes, Brain, and Behavior*, *12*(1), 98–107.

Nedic, G., Perkovic, M. N., Sviglin, K. N., Muck-Seler, D., Borovecki, F., & Pivac, N. (2013). Brain-derived neurotrophic factor Val66Met polymorphism and alcohol-related phenotypes. *Progress in Neuro-Psychopharmacology & Biological Psychiatry*, *40*, 193–198.

Numan, S., Lane-Ladd, S. B., Zhang, L., Lundgren, K. H., Russell, D. S., Seroogy, K. B., et al. (1998). Differential regulation of neurotrophin and trk receptor mRNAs in catecholaminergic nuclei during chronic opiate treatment and withdrawal. *The Journal of Neuroscience*, *18*(24), 10700–10708.

Paliwal, P., Hyman, S. M., & Sinha, R. (2008). Craving predicts time to cocaine relapse: Further validation of the Now and Brief versions of the cocaine craving questionnaire. *Drug and Alcohol Dependence*, *93*(3), 252–259.

Pan, W., Banks, W. A., Fasold, M. B., Bluth, J., & Kastin, A. J. (1998). Transport of brain-derived neurotrophic factor across the blood-brain barrier. *Neuropharmacology*, *37*(12), 1553–1561.

Pandey, S. C., Zhang, H., Ugale, R., Prakash, A., Xu, T., & Misra, K. (2008). Effector immediate-early gene arc in the amygdala plays a critical role in alcoholism. *The Journal of Neuroscience*, *28*(10), 2589–2600.

Pu, L., Liu, Q. S., & Poo, M. M. (2006). BDNF-dependent synaptic sensitization in midbrain dopamine neurons after cocaine withdrawal. *Nature Neuroscience*, *9*(5), 605–607.

Ren, Q., Ma, M., Yang, C., Zhang, J. C., Yao, W., & Hashimoto, K. (2015). BDNF-TrkB signaling in the nucleus accumbens shell of mice has key role in methamphetamine withdrawal symptoms. *Translational Psychiatry*, *5*. e666.

Ren, W., Tao, J., Wei, Y., Su, H., Zhang, J., Xie, Y., et al. (2016). Time-dependent serum brain-derived neurotrophic factor decline during methamphetamine withdrawal. *Medicine (Baltimore)*, *95*(5). e2604.

Reynolds, P. M., Mueller, S. W., & MacLaren, R. (2015). A comparison of dexmedetomidine and placebo on the plasma concentrations of NGF, BDNF, GDNF, and epinephrine during severe alcohol withdrawal. *Alcohol*, *49*(1), 15–19.

Roni, M. A., & Rahman, S. (2014). The effects of lobeline on nicotine withdrawal-induced depression-like behavior in mice. *Psychopharmacology*, *231*(15), 2989–2998.

Schmidt, H. D., Sangrey, G. R., Darnell, S. B., Schassburger, R. L., Cha, J. H., Pierce, R. C., et al. (2012). Increased brain-derived neurotrophic factor (BDNF) expression in the ventral tegmental area during cocaine abstinence is associated with increased histone acetylation at BDNF exon I-containing promoters. *Journal of Neurochemistry*, *120*(2), 202–209.

Schunck, R. V., Torres, I. L., Laste, G., de Souza, A., Macedo, I. C., Valle, M. T., et al. (2015). Protracted alcohol abstinence induces analgesia in rats: Possible relationships with BDNF and interleukin-10. *Pharmacology, Biochemistry, and Behavior*, *135*, 64–69.

Simchon-Tenenbaum, Y., Weizman, A., & Rehavi, M. (2015). Alterations in brain neurotrophic and glial factors following early age chronic methylphenidate and cocaine administration. *Behavioural Brain Research*, *282*, 125–132.

Sinha, R., Fox, H. C., Hong, K. I., Hansen, J., Tuit, K., & Kreek, M. J. (2011). Effects of adrenal sensitivity, stress- and cue-induced craving, and anxiety on subsequent alcohol relapse and treatment outcomes. *Archives of General Psychiatry*, *68*(9), 942–952.

Sinha, R., Garcia, M., Paliwal, P., Kreek, M. J., & Rounsaville, B. J. (2006). Stress-induced cocaine craving and hypothalamic-pituitary-adrenal responses are predictive of cocaine relapse outcomes. *Archives of General Psychiatry*, *63*(3), 324–331.

Su, H., Tao, J., Zhang, J., Xie, Y., Sun, Y., Li, L., et al. (2014). An association between BDNF Val66Met polymorphism and impulsivity in methamphetamine abusers. *Neuroscience Letters*, *582*, 16–20.

Sun, W. L., Eisenstein, S. A., Zelek-Molik, A., & McGinty, J. F. (2014). A single brain-derived neurotrophic factor infusion into the dorsomedial prefrontal cortex attenuates cocaine self-administration-induced phosphorylation of synapsin in the nucleus accumbens during early withdrawal. *The International Journal of Neuropsychopharmacology*, *18*(1), pyu049.

Thiel, K. J., Painter, M. R., Pentkowski, N. S., Mitroi, D., Crawford, C. A., & Neisewander, J. L. (2012). Environmental enrichment counters cocaine abstinence-induced stress and brain reactivity to cocaine cues but fails to prevent the incubation effect. *Addiction Biology*, *17*(2), 365–377.

Umene-Nakano, W., Yoshimura, R., Yoshii, C., Hoshuyama, T., Hayashi, K., Hori, H., et al. (2010). Varenicline does not increase serum BDNF levels in patients with nicotine dependence. *Human Psychopharmacology*, *25*(3), 276–279.

von Diemen, L., Kapczinski, F., Sordi, A. O., de Magalhaes Narvaez, J. C., Guimaraes, L. S., Kessler, F. H., et al. (2014). Increase in brain-derived neurotrophic factor expression in early crack cocaine withdrawal. *The International Journal of Neuropsychopharmacology*, *17*(1), 33–40.

Wojnar, M., Brower, K. J., Strobbe, S., Ilgen, M., Matsumoto, H., Nowosad, I., et al. (2009). Association between Val66Met brain-derived neurotrophic factor (BDNF) gene polymorphism and post-treatment relapse in alcohol dependence. *Alcoholism, Clinical and Experimental Research*, *33*(4), 693–702.

Wrase, J., Makris, N., Braus, D. F., Mann, K., Smolka, M. N., Kennedy, D. N., et al. (2008). Amygdala volume associated with alcohol abuse relapse and craving. *The American Journal of Psychiatry*, *165*(9), 1179–1184.

You, C., Zhang, H., Sakharkar, A. J., Teppen, T., & Pandey, S. C. (2014). Reversal of deficits in dendritic spines, BDNF and Arc expression in the amygdala during alcohol dependence by HDAC inhibitor treatment. *The International Journal of Neuropsychopharmacology*, *17*(2), 313–322.

Zanardini, R., Fontana, A., Pagano, R., Mazzaro, E., Bergamasco, F., Romagnosi, G., et al. (2011). Alterations of brain-derived neurotrophic factor serum levels in patients with alcohol dependence. *Alcoholism, Clinical and Experimental Research*, *35*(8), 1529–1533.

Zhang, K., Jiang, H., Zhang, Q., Du, J., Wang, Y., & Zhao, M. (2016). Brain-derived neurotrophic factor serum levels in heroin-dependent patients after 26 weeks of withdrawal. *Comprehensive Psychiatry*, *65*, 150–155.

Zhang, J., Zhang, X., Su, H., Tao, J., Xie, Y., Han, B., et al. (2014). Increased serum brain-derived neurotrophic factor levels during opiate withdrawal. *Neuroscience Letters*, *571*, 61–65.

Zilkha, N., Feigin, E., Barnea-Ygael, N., & Zangen, A. (2014). Induction of depressive-like effects by subchronic exposure to cocaine or heroin in laboratory rats. *Journal of Neurochemistry*, *130*(4), 575–582.

CHAPTER EIGHTEEN

Neurotrophin Receptors and Perineural Invasion: Analyses in Select Lineage-Unrelated Cutaneous Malignancies With a Propensity for Perineural Invasion

N.F. Frydenlund*, M. Mahalingam†,1
*University of Iowa Carver College of Medicine, Iowa City, IA, United States
†VA Consolidated Laboratories, West Roxbury, MA, United States
[1]Corresponding author: e-mail address: me@meeramahalingam.com

Contents

1. Introduction 498
2. PNI in Cancer 498
3. Definitions 499
4. Incidence of PNI in Cutaneous Malignancies 500
5. Prognostic Impact of PNI in cSCC and DM 502
6. The Biological Underpinnings of PNI: The Role of Neurotrophins 506
   6.1 The NGF–TRKA Axis and PNI 507
   6.2 The GDNF–RET Axis and PNI 508
   6.3 P75NGFR and PNI 512
   6.4 Other Potential Players: NCAM and CXCR4 516
7. Clinical Potential of Neurotrophins as Therapeutic Targets 518
8. Personal Observations 519
9. Conclusions 523
References 524

Abstract

In this chapter, we parse the literature on neurotrophins that have been implicated in the pathogenesis of perineural invasion (PNI) in select lineage-unrelated malignancies. We also detail evidence linking neurotrophins and their receptors (TrkA, RET, p75NGFR, and NCAM) to the pathogenesis of PNI in desmoplastic melanoma and cutaneous squamous cell carcinoma—both malignancies with an established propensity for PNI. Lastly, the clinical potential of neurotrophins as receptors for targeted therapies is explored.

Vitamins and Hormones, Volume 104
ISSN 0083-6729
http://dx.doi.org/10.1016/bs.vh.2016.11.003

ABBREVIATIONS

AJCC American Joint Committee on Cancer
cSCC cutaneous squamous cell carcinoma
DM desmoplastic melanoma
DNM desmoplastic neurotropic melanoma
GDNF glial cell line-derived neurotrophic factor
H&E hematoxylin and eosin
H&N head and neck
IHC immunohistochemistry
NM neurotropic melanoma
PNI perineural invasion
RETp functional RET polymorphism

1. INTRODUCTION

Perineural invasion (PNI) in cutaneous squamous cell carcinoma (cSCC) and desmoplastic melanoma (DM) may be a negative prognostic finding, and likely contributes to increased rates of local recurrence. The biological mechanisms underlying PNI remain unclear, although several lines of evidence implicate neurotrophins, a family of signaling peptides important to the development and maintenance of the peripheral nervous system, and their receptors. Expression of the high-affinity nerve growth factor (NGF) receptor TrkA has been shown to be associated with PNI in numerous malignancies, although literature in cutaneous neoplasms is sparse.

2. PNI IN CANCER

PNI refers to infiltration of the perineural sheath by tumor cells with subsequent tumor proliferation and spread, either proximally or distally along the length of the nerve (Feasel, Brown, Bogle, Tschen, & Nelson, 2001). Best studied in carcinomas of the pancreas and prostate, PNI has long been understood be an important route of tumor spread (Liebig, Ayala, Wilks, Berger, & Albo, 2009). However PNI has been reported in a multitude of lineage-unrelated malignancies. The first recorded observation of PNI in a cutaneous malignancy was reported by Neumann in 1862. Since then, PNI has been reported in multiple cutaneous malignancies including DM, basal cell carcinoma, microcystic adnexal carcinoma, and cSCC

(Busam et al., 2004; Chen-Tsai, Colome-Grimmer, & Wagner, 2004; Feasel et al., 2001; Leibovitch et al., 2005). Of these, PNI is best studied in cSCC and DM both of which are known to have a marked propensity for this invasive feature.

3. DEFINITIONS

While the definition of PNI as tumor cells present within the perineural sheath appears straightforward at first glance, in practice there are inconsistencies in the reporting of PNI. In cSCC, the main point of contention centers on differentiating between true PNI and focal abutment of tumor on nerve (when the nerve is surrounded by tumor but not invaded) (Dunn, Morgan, & Beer, 2009). While originally defined in SCC by Batsakis (1985) as "Tumor cell invasion in, around and through the nerves," which would include focal abutment, recent definitions have become more stringent. In her 2009 review, Liebig suggested that PNI should be defined as "finding tumor cells within any of the three layers of the nerve sheath or at least 33% of the circumferences of the nerve surrounded by tumor cells." Even more stringently, some authors propose that only observing tumor cells within the perineural sheath should be classified as PNI (Scanlon et al., 2014). For our purposes, we have chosen to use this definition when reporting PNI in our cases.

In melanoma, a wholly separate problem presents itself when defining PNI. Namely, the conflation of the terms "Neurotropism or Neurotropic" with PNI. The term neurotropism was initially used by Reed and Leonard (1979) to describe a subset of DMs that displayed prominent "Schwannian differentiation" and/or extensive PNI, which they termed neurotropic melanoma (NM). However, it should be noted that "neurotropic melanoma" may be used to describe any melanomas that exhibit the aforementioned characteristics, and there are reports of both epithelioid and spindle cell melanoma (SCM) without desmoplasia being classified as NM (Baer, Schultz, Synnestvedt, & Elder, 1995; Scanlon et al., 2014). As a result, the label of desmoplastic neurotropic melanoma (DNM) has been used in the case of DMs that present with Schwannian differentiation and/or PNI while NM is reserved for those cases without desmoplasia. Further, because neurotropism can be used to describe both PNI and the distinct cytomorphologic feature of Schwannian differentiation (defined as melanoma cells taking on the cytology of nerve), confusion may result when attempting to determine its prognostic significance (Baer et al., 1995; Su et al., 2014).

In an effort to reduce the equivocality of the nomenclature, we suggest the use of the term PNI as opposed to neurotropism when describing the histopathological observation invasion of peripheral nerve by melanoma cells. This usage is consistent with the terminology employed in describing other cutaneous malignancies (Feasel et al., 2001). Hence, melanomas that display PNI are termed NM, and more specifically DMs that display the same are termed DNMs. In the rare instances where PNI is not present but the melanoma displays a clear neural cytomorphology, we agree with the recent suggestions of Su et al. in their case report of such a lesion, that these tumors should be classified as distinct from NM/DNMs, the authors suggested the use of the term "malignant neurotization" (Su et al., 2014). Alternatively, the term neural transformation has been used to describe such lesions as well (Smithers, McLeod, & Little, 1992). The importance of properly distinguishing between NM and DNM so-named for prominent PNI vs those named for neural transformation is more than just semantic. Neurotropism as PNI is likely to have a different prognostic implication than neurotropism as neural transformation, and at least one study has demonstrated this (Smithers et al., 1992). A clearer understanding of the prognostic relevance of neurotropism would result from separating PNI from neural transformation in diagnosing NM and DNM.

4. INCIDENCE OF PNI IN CUTANEOUS MALIGNANCIES

The incidence of PNI varies substantially across cutaneous malignancies. Within a given malignancy there may be further variations in incidence, dependent on histopathologic subtype. As a result, reported rates of PNI in the literature span a broad range. In cSCC the reported incidence of PNI varies substantially, with a range of 2–14%, though most studies report rates between 2% and 5% (Campoli, Brodland, & Zitelli, 2014; Goepfert, Dichtel, Medina, Lindberg, & Luna, 1984; Leibovitch et al., 2005). Furthermore, it has been suggested that rates of PNI are higher in cSCCs of the head and neck (H&N) compared to other anatomic sites which may confound past reports (Chang et al., 2004). Comparisons of

PNI incidence are complicated by multiple factors that include the challenges associated with detection of PNI using hematoxylin and eosin (H&E) staining (obscured nerve and morphologically imperceptible tumor cells), lack of a standardized histological definition of PNI (delineating true PNI from focal abutment secondary to impingement of nerve by tumor), differences in detection rates between cryostat and formalin-fixed paraffin-embedded tissue sections (Mohs micrographic surgery vs traditional histopathologic examination), and histological mimics such as Renault bodies, perineural fibrosis, reactive neuroepithelial aggregates, and reparative perineural hyperplasia (Campoli et al., 2014; Dunn, Morgan, & Beer, 2009; Dunn, Morgan, Beer, Chen, & Acker, 2009; Hassanein et al., 2005; Kurtz, Hoffman, Zimmerman, & Robinson, 2005; Ronaghy, Yaar, Goldberg, Mahalingam, & Bhawan, 2010; Zhou, Xu, Zhang, Zhao, & Wu, 2014). Lastly, in select cutaneous and noncutaneous malignancies it has been shown that the use of immunohistochemistry (IHC) can aid in the detection of PNI, especially in the tumor bulk where nerve can be obscured by tumor (intratumoral PNI) (Berlingeri-Ramos, Detweiler, Wagner, & Kelly, 2015; Kurtz et al., 2005; Scanlon et al., 2014; Zhou et al., 2014). Such observations indicate that the true incidence of PNI in cSCC may be greater than that reported using H&E alone.

While it is well understood that the greatest rates of PNI are seen in the DM variant, the conflation of PNI with neural transformation in the reporting of both NM and DNM makes it difficult to get a precise figure. In their 1998 study, on 280 cases of DM, Quinn et al. (1998) identified 30% of cases as the DNM variant. Although, the proportion of cases diagnosed as DNM for PNI as compared to exclusively neural transformation was not specified. Busam et al. reported that of 98 cases of DM, 53% displayed nerve involvement (Busam et al., 2004). The highest reported rate of 82%, occurred in a study of 28 DMs of the H&N, suggesting that anatomic location may be of relevance in determining incidence (Kay et al., 2004). There may be further variation in incidence within DM subtypes (pure vs mixed), with one study reporting that rates of neurotropism varied significantly, with a higher incidence in the pure subtype (43.9%) vs the mixed subtype (27.1%) (Murali et al., 2010). The incidence of PNI in non-DM is substantially less. Scanlon et al. report that a query of 1984 primary melanoma cases from the Interdisciplinary Melanoma Cooperative Group at NYU found an overall incidence of 3% (Scanlon et al., 2014). However, this would include melanomas of all subtypes (both DMs and non-DMs). It was noted that cases of DM and acral lentiginous melanoma (ALM) were overrepresented in the PNI-positive group.

5. PROGNOSTIC IMPACT OF PNI IN cSCC AND DM

The prognostic relevance of PNI in both cSCC and melanoma remains equivocal. In most malignancies in which it is observed (both cutaneous and noncutaneous) PNI is associated with a more aggressive phenotype, and increased risk of recurrence (Liebig et al., 2009). In select malignancies (pancreatic and prostate carcinomas), PNI may be associated with increased mortality (Ayala et al., 2004; Fouquet, Germain, Brunaud, Bresler, & Ayav, 2014).

In cutaneous malignancies, the prognostic impact of PNI has best been studied in cSCC. In 2010, based on a confluence of evidence from numerous prospective cohort studies, the American Joint Committee on Cancer (AJCC) added PNI as a high-risk tumor characteristic in tumor staging (Farasat et al., 2011). The rationale for inclusion was drawn from Geopfert et al.'s findings in a retrospective review of 967 cSCCs that PNI was linked to metastasis and reduced survival (Goepfert et al., 1984) as well as a study from Leibovitch et al. of 1177 cSCCs demonstrating an association with recurrence, decreased degree of differentiation, and larger tumor diameter (Leibovitch et al., 2005). Further support for the AJCC's use of PNI in tumor staging comes from Cherpelis et al.'s investigation of 200 cSCCs in which a correlation with metastasis was found (Cherpelis, Marcusen, & Lang, 2002), Clayman et al.'s (2005) analysis of 277 cases linking PNI to decreased disease-specific survival and Moore et al.'s finding in 193 patients with cSCC of the H&N area that PNI is associated with lymph node metastases (Moore et al., 2005). The above, combined with the prevalence of cSCC as the second most common cutaneous malignancy underscore the clinical relevance of PNI identification (Alam & Ratner, 2001).

Recently, subtleties that have emerged from investing PNI's utility as a prognosticator in cSCC. Namely, the importance of nerve diameter and the presence of multiple foci of nerve involvement in determining outcomes of patients with PNI-positive cSCC (Ross et al., 2009). At least two studies have reported that in cSCC, PNI in nerves with a diameter >0.1 mm is associated with worse outcomes (metastasis and recurrence) while involvement of nerves <0.1 mm was not associated with negative outcomes (Carter, Johnson, Chua, Karia, & Schmults, 2013; Ross et al., 2009). However, Lin et al. report that the diameter of the involved nerve had no impact on outcomes in cSCC (Lin et al., 2012). Additionally, Carter et al. also report that the involvement of multiple nerves is associated with risk factors that portend worse outcomes when compared to solitary foci of PNI (Carter et al., 2013). While, Lin et al. similarly found that patients with focal PNI had better outcomes than those who had diffuse PNI with multiple foci of invasion (Lin et al., 2012).

A definitive understanding of PNI's prognostic relevance in melanoma remains to be established. As mentioned previously, discordant definitions of neurotropism serve as a barrier to developing a comprehensive understanding of PNI as a potential prognostic indicator in its own right. Initially neurotropism was understood to be a negative prognostic finding, with Baer et al. finding a significant decrease in 8 year overall survival when comparing both NM and DNM to their nonneurotropic counterparts (Baer et al., 1995). Soon after, in the largest study to date (280 cases), neurotropism was correlated with a twofold increase in local recurrences but was not associated with survival (Quinn et al., 1998). In both of these studies, the potential for "skip lesions" in which the tumor extension along the nerves results in falsely negative margins in biopsies was hypothesized to be the cause for the increase in local recurrence. Since then, the majority of studies have reported that neurotropism is not of significant prognostic relevance (summarized in chronologic order in Table 1). Recently, Scanlon et al. reported that PNI is associated with increased rates of local recurrence in both DM and non-DM cases (Scanlon et al., 2014). Future studies that attempt to determine the prognostic relevance of PNI would benefit greatly by laying out an explicit definition of neurotropism as PNI. Additionally, to the best of our knowledge, the prognostic effects of nerve diameter and multiple foci of nerve involvement have not been investigated in melanoma. Future research into these variables would significantly enhance our understanding of the role PNI plays in affecting outcomes in melanoma patients.

While the prognostic value of PNI remains equivocal, special care should be taken with cases of DNM/NM from the H&N as anatomic location may significantly impact the prognostic influence of PNI. The involvement of cranial nerves may result in significant malignant neuropathies (Smithers et al., 1992), and potentially increase the risk of metastasis to the central nervous system (Chang et al., 2004). In melanomas, perineural spread has been observed to occur beyond the skull base as far as the brain (Chang et al., 2004).

Table 1 Chronological Review of Published Studies Regarding the Prognostic Relevance of Neurotropism

Author/Year	Design	Definition of Neurotropism[a]	Samples Studied (*n*)	Findings	Conclusion
Smithers et al. (1992)	Retrospective	Defined as PNI, while neural transforming melanoma was considered separately	DM (58) (Neurotropism reported in 23 cases, neural transformation reported in 20 cases)	Visceral recurrence in tumors with neurotropism was less frequent than in those neural transformation	Neural transformation is more predictive on local recurrence than neurotropism
				Local recurrence rate appeared to be higher in tumors with neural transformation than in tumors displaying neurotropism	
				Neurotropism in tumors of the head and neck was associated with malignant neuropathies	
Baer et al. (1995)	Prospective	Defined as PNI; authors explicitly state that no cases of Schwannian differentiation in the absence of PNI were observed	NM (5) DNM (11)	The 8-year survival rate was 90% in patients without neurotropism and 60% in those with neurotropism	Neurotropism is a negative prognostic indicator
				Local recurrence occurred in 36% of DNMs and 20% of NMs	
Quinn et al. (1998)	Retrospective	Defined as either PNI or "neural transformation"	DM (190) DNM (90)	Local recurrence occurred in 45% of DNMs compared to 21% of DMs	Neurotropism portends local recurrence, but not decreased survival in DM
				There was no difference in survival between DM and DNM patients	
Su et al. (2014)	Retrospective	Not explicitly defined	DNM (14) DM (19)	There was no significant association between sentinel lymph node metastasis and neurotropism	Neurotropism does not increase the risk for lymph node metastasis in DM

Busam et al. (2004)	Retrospective	Not explicitly defined	DM (92) ("Nerve involvement" reported in 52 of the cases)	Neurotropism was not significantly associated with reduced disease-free survival ($p=0.21$)	Neurotropism was only marginally associated with reduced disease-free survival
Kay et al. (2004)	Retrospective	Defined as PNI.	DNM (23) DM (5)	Neurotropism was not significantly associated with overall survival, local recurrence, or distant metastasis	Neurotropism is not of prognostic relevance in DM
Livestro et al. (2005)	Case–control retrospective	Not explicitly defined	DM (39) DNM (50)	Neurotropism was not significantly associated with survival or local recurrence Of cases with lymph node metastasis 75% were neurotropic	Neurotropism is not of prognostic relevance in DM
Chen et al. (2008)	Retrospective	Defined as either PNI, or "Neural differentiation"	DNM (128)	Local recurrence occurred in 6% of DNM cases	Neurotropism may not predict local recurrence to the extent previously thought
Murali et al. (2010)	Retrospective	Not explicitly defined	DM (163) DNM (89)	Neurotropism was not associated with local regional recurrence	Neurotropism is not a prognostic indicator of increased risk for local regional recurrence
Scanlon et al. (2014)	Prospective	Defined as PNI	DNM (19) NM (23)	PNI detected associated with local recurrence PNI detected by H&E was associated reduced rates of ulceration On multivariate analysis PNI detected no impact on disease-free or overall survival	Neurotropism portends local recurrence but not decreased survival in multiple melanoma variants

[a]Definition of neurotropism was considered to be PNI if any mention of neural invasion, neural involvement, or neural infiltration was seen.
Abbreviations: *DM*=desmoplastic melanoma, *NM*=neurotropic melanoma, *DNM*=desmoplastic neurotropic melanoma.

6. THE BIOLOGICAL UNDERPINNINGS OF PNI: THE ROLE OF NEUROTROPHINS

A comprehensive understanding of the biological processes that promote PNI has yet to be established in any malignancy with a propensity for PNI. Two potential frameworks to explain the pathogenesis of PNI have been proposed. In one, tumor cells adopt a phenotype that enables them to respond to the distinctive microenvironment of nerves and thus acquire properties of enhanced invasiveness and proliferation. On the other hand, tumors themselves may develop a microenvironment that promotes neurite outgrowth, resulting in the proliferation of nerve twigs into and around the tumor with resulting PNI. These two mechanisms are not mutually exclusive and it is likely that some level of both occur at the same time through reciprocal signaling between tumor cells and nerve (Liebig et al., 2009).

To date, attempts to illuminate the molecular players involved in the pathogenesis of PNI have focused primarily on a family of peptides known as neurotrophins and their respective receptors (Fig. 1). Neurotrophins are principally known for their role in controlling the development and maintenance of the peripheral nervous system but are increasingly understood to play significant roles in behavior of multiple lineage-unrelated cancers (Liebig et al., 2009). The most well-characterized neurotrophin is NGF, but there are several other members of the neurotrophin family including

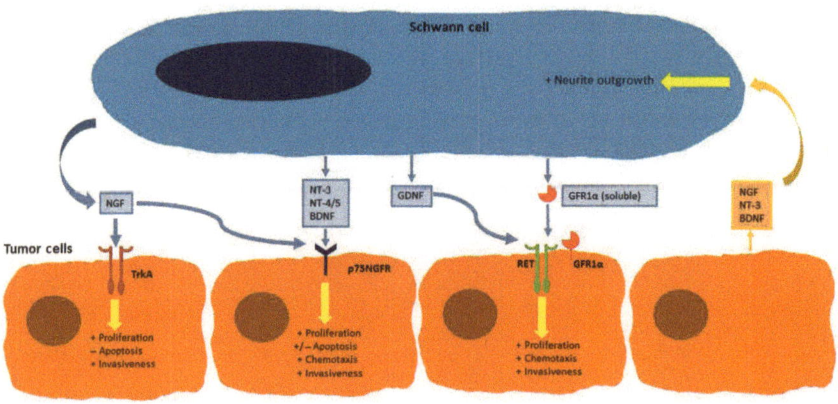

100 pixel

Fig. 1 Crosstalk between Schwann cells and tumor cells involving neurotrophins is thought to influence perineural invasion. Shown here is an overview of the neurotrophin/receptor pairs most studied.

brain-derived neurotrophic factor (BDNF), neurotrophin-3 (NT-3), and neurotrophin 4/5 (NT-4/5) (Boyd & Gordon, 2003; Chen et al., 2008). Glial cell line-derived neurotrophic factor (GDNF) has a similar functional role to the aforementioned neurotrophins and is often grouped together with the NGF-related neutrophins. However, structural differences between GDNF and the NGF family warrant that it be considered in its own separate class of neurotrophins (the GDNF family of neurotrophic factors) (Boyd & Gordon, 2003). There are several receptors that interact with varying degrees of specificity with neurotrophins. The most studied are the Trk family of receptors (which includes TrkA, TrkB, and TrkC), p75NGFR, and RET. The Trk family and RET are comprised of receptor tyrosine kinases, while p75NGFR is a member of the tumor necrosis factor receptor superfamily (He & Garcia, 2004; Truzzi, Marconi, & Pincelli, 2011). The Trk family and RET are known to bind their target neurotrophins with relatively high-affinity and selectivity (TrkA binds NGF and NT-3, TrkB binds BDNF, NT-3, and NT 4/5, TrkC binds NT-3, RET binds GDNF) while p75NGFR is more promiscuous, binding a multitude of neurotrophins with different affinities. Within the context of PNI in cutaneous malignancies, the TrkA–NGF axis, RET, and p75NGFR have emerged as the most promising neurotrophin/neurotrophin receptor candidates for having a mechanistic role.

6.1 The NGF–TRKA Axis and PNI

TrkA serves as the membrane-bound high-affinity receptor for NGF and is expressed in multitude of lineage-unrelated cells. In normal epidermis, TrkA expression is found to be the greatest in basal keratinocytes with expression dissipating in the superficial layers of the epidermis. There are conflicting reports of TrkA expression in normal melanocytes (Marco et al., 1993; Yaar et al., 1994). In epidermal keratinocytes as well as keratinocytes of the follicular epithelia, TrkA is thought to responsible for mediating NGF-induced stimulation of proliferation and inhibition of apoptosis (Marco et al., 1993; Marconi, Vaschieri, Zanoli, Giannetti, & Pincelli, 1999; Paus, Lüftl, & Czarnetzki, 1994; Pincelli et al., 1994). The proproliferative effects of NGF on cells expressing TrkA combined with the NGF-rich microenvironment surrounding nerves as a result of its secretion by Schwann cells (Johnson, Taniuchi, & DiStefano, 1988), has led many to hypothesize a connection between TrkA and PNI.

Studies in several noncutaneous malignancies have shown TrkA expression to be associated with PNI (Dang, Zhang, Ma, & Shimahara, 2006;

Descamps et al., 2001; Kolokythas, Cox, Dekker, & Schmidt, 2010; Wang et al., 2006; Zhu et al., 1999, 2002). In both salivary adenoid cystic carcinoma and pancreatic ductal adenocarcinoma, TrkA was found to mediate enhanced invasiveness in cell cultures in response to NGF treatment (Wang et al., 2006; Zhu et al., 2002). In pancreatic cancer cell lines, this proinvasive response has been linked to an increased production of select matrix metalloproteinases (MMP-2) in response to NGF (Okada et al., 2004). It is hypothesized that the expression of TrkA receptors on tumor cells allows them to respond to the NGF-rich microenvironment surrounding nerves, resulting in increased production of proteolytic enzymes, and subsequent invasion through the perineural sheath. In cutaneous malignancies, evidence for TrkA's association with PNI was first reported in a study by Chen-Tsai et al. in which four cases of squamous cell carcinoma with PNI and three cases without PNI were analyzed immunohistochemically (Chen-Tsai et al., 2004). Notably, the authors used a promiscuous antibody that binds TrkA, B, and C. The authors found staining was the strongest in the PNI-positive cases suggestive of a relationship between TrkA and PNI but limited sample size and nonspecificity of the antibody prevented any conclusive conclusions from being made.

Evidence for TrkA expression in melanomas is conflicting. Data from in vitro studies have shown TrkA expression to be absent (Herrmann et al., 1993; Innominato, Libbrecht, & van den Oord, 2001; Walch, Albino, & Marchetti, 1999). While an immunohistochemical study in 152 primary and 64 metastatic melanoma biopsies detected significant levels of expression of the activated form of the TrkA receptor (p-TrkA), with expression being observed more frequently in biopsies of primary tumors (55%) compared to metastases (30%) (Flørenes et al., 2004). In the same study, the authors reported that membranous p-TrkA expression was significantly associated with decreased overall survival, and elevated markers of proliferation (cyclin-A and Ki-67). Although, p-TrkA expression was seen more often in nodular melanomas than the superficial spreading variant, spindle cell, and DMs were not included in the study.

6.2 The GDNF–RET Axis and PNI

RET is a prooncogene encoding a membrane-bound receptor tyrosine kinase for GDNF. RET acts in concert with the GFRα1 coreceptor to bind GDNF, and like NGF–TrkA signaling, the RET–GDNF axis serves to support the survival and development of the peripheral nervous system (Takahashi,

2001). Activating mutations in the RET gene have been associated with numerous cancers including neuroblastoma, papillary thyroid carcinoma, pancreatic cancers, and colon cancer (Mulligan, 2014; Takahashi, 2001; Tallini & Asa, 2001). Interestingly, inactivating mutations in RET have been associated with deficiencies in neural crest cell migration during development, most notably in Hirschsprung's disease which is caused by insufficient development of the enteric nervous system (Puri & Shinkai, 2004).

Most of the research into RET's role in malignancies has been done in pancreatic cancers where RET–GDNF signaling has emerged as an important mediator of PNI (Cavel et al., 2012; He et al., 2014; Zeng et al., 2008). Using IHC, Zeng et al. have shown strong positive staining for GDNF in 88% of pancreatic ductal adenocarcinomas with PNI compared to 60% of those without PNI (Zeng et al., 2008). Furthermore, in vitro experiments have shown that treatment of pancreatic cancer cell lines with GDNF results in an increased production of matrix metalopreinase-9 (MMP-9) with a corresponding increase in invasive capacity—an effect that is lost after treatment with an inhibitor of RET (Okada, Eibl, Duffy, Reber, & Hines, 2003). Secretion of proteolytic enzymes like MMP-9 may facilitate degradation of the perineural sheath, essential for PNI. Taken together, the proinvasive effects of GDNF–RET signaling, combined with the increased expression of GDNF in PNI-positive pancreatic cancers, lend credence to the hypothesis that RET–GDNF signaling plays a role in the development of PNI.

Adding another dimension to the RET–PNI connection, a study conducted by Cavel et al. showed nerves invaded by pancreatic cancer had significantly elevated populations of endoneurial macrophages in excisions from both patients and mice (Cavel et al., 2012). These macrophages were found to secrete high levels of GDNF which when applied to pancreatic cells in culture promoted chemotactic migration of the cancer cells. Furthermore, the migration enhancing effect was lost after the pancreatic cancer cells were transfected with either siRNA against GFRα1 or RET inhibitor treatment, suggesting paracrine signaling between cancer cells expressing GFRα1/RET and GDNF secreting endoneurial macrophages. Most recently, it has been shown in a mouse model for PNI that peripheral nerves are capable of releasing soluble GFRα1 receptor, which enables GDNF binding in pancreatic cancer cells that lack GFRα1 but express RET, resulting in PNI (He et al., 2014). Conversely, in this same study it was shown that tumor cells that lack RET did not undergo PNI, even in the presence of soluble GFRα1. Taken together, these results suggest that

RET's causative relationship in PNI may be dependent on both paracrine signaling involving endoneurial macrophages, and additionally requires the presence of the GFRα1 coreceptor either in its soluble form or expressed on the tumor cells themselves.

While the literature is lacking regarding the role of RET in cSCC, numerous lines of evidence suggest that RET is of importance in the pathogenesis of melanoma. Transgenic mouse lines with either overactivated RET mutations or overexpression of wild-type RET develop melanosis and benign melanocytic lesions that eventually progress to metastatic melanoma (Abschuetz et al., 2012; Kato et al., 1998, 2000; Ohshima et al., 2010; Umansky & Sevko, 2013). It has been shown that the progression from benign lesions to melanomas corresponds with increasing levels of RET expression and its activation in tumors from these mouse models (Kato et al., 1998). It is also interesting to note that in RET transgenic mice, repeated UV irradiation results in the activation of RET leading to melanocytic tumor development at an accelerated rate compared to sham-irradiated controls (Kato et al., 2000). Evidence from human melanoma cell cultures further underscores the potential role of RET–GDNF signaling in melanomagenisis. In select human melanoma cell lines, mRNA transcripts for RET, GFRα1, and GDNF were detected, while human melanocytes were found to only express GFRα1 (Ohshima et al., 2010). A functional polymorphism (G691S) in the RET gene (functional RET polymorphism—RETp, Fig. 2) has been shown to amplify RET's responsiveness to GDNF treatment in melanoma cell culture (Barr, Amato, Robinson, Kounalakis, & Robinson, 2012; Narita et al., 2009). This mutation is found as a germline mutation in 15–20% of the general population but may also occur as somatic mutation in some instances (Barr et al., 2012). Compared to RETwt melanoma cells, those with RETp displayed amplified proliferation, migration, and invasion after the addition of GDNF to culture medium, suggesting that RETp sensitizes the RET receptor to GDNF stimulation (Narita et al., 2009). Furthermore, when Narita et al. analyzed 71 non-DM biopsies and 70 DM biopsies, the incidence of RETp was greater in the DM group than in the non-DM group (61% and 31%, respectively) (Narita et al., 2009). Our own experience indicates that within DM, RETp is more frequently found in the pure subtype as compared to the mixed subtype (Miller et al., 2012). As DMs are found predominantly in chronically sun exposed areas, the known role of UV light in modifying RET signaling may be of importance. Although RET–GDNF signaling appears to play a role in the development of PNI in pancreatic cancers,

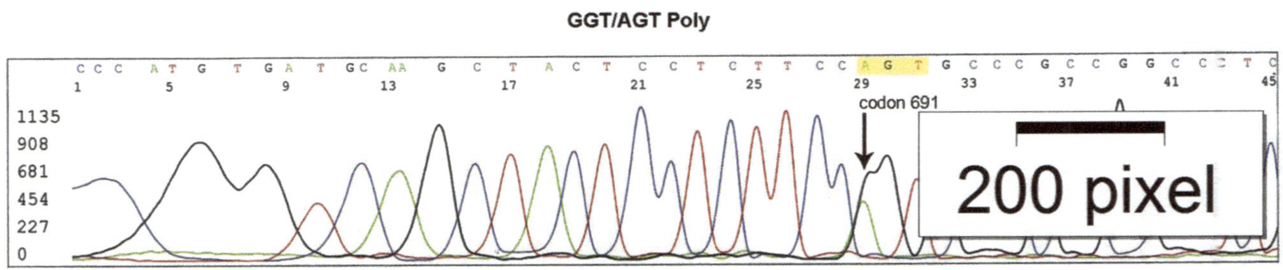

Fig. 2 Direct DNA sequencing results from a case of desmoplastic melanoma displaying the GGT/AGT functional RET polymorphism (*RETp*) at codon 691 at exon 11.

and additionally the activating RETp mutation is found to be more frequently expressed in DM, a firm connection between RET and PNI in melanoma remains to be established.

6.3 P75NGFR and PNI

The most studied of all the neurotrophin receptors in melanoma, especially in the spindle cell and desmoplastic variants, is p75NGFR (alternatively called p75NTR, CD271, and the low affinity NGF receptor). The p75NGFR receptor is also the most promiscuous of neurotrophin receptors and is known to bind NGF, NT-3, NT-4/5, and BDNF with somewhat lower affinity than the Trk family of neurotrophin receptors. Unlike the Trk family and RET, p75NGFR is not a receptor tyrosine kinase, belonging instead to the tumor necrosis factor receptor superfamily (He & Garcia, 2004). In neurons, the role of p75NGFR is complex and poorly understood. It has been suggested that in the presence of TrkA, p75NGFR acts to increase TrkA's binding affinity for NGF and promote survival while, in the absence of TrkA, NGF binding to p75NGFR induces apoptosis (He & Garcia, 2004). Additionally, the migration of Schwann cells during embryonic development is thought to be regulated by p75NGFR (Anton, Weskamp, Reichardt, & Matthew, 1994). In cancers, strong expression of p75NGFR is often seen in schwannoma, neuroblastoma, and other tumors of neural crest origin, although it's functional role is still unclear (Perosio & Brooks, 1988).

Conflicting reports exist regarding the expression of p75NGFR in melanocytes. Numerous in vitro studies have found expression of p75NGFR in human melanocyte cell culture (Peacocke, Yaar, Mansur, Chao, & Gilchrest, 1988). In addition, it has been shown that the treatment of melanocyte cell cultures with phorbol 12-tetradecanoate 13-acetate, a potent activator of protein kinase C, induces upregulation of p75NGFR expression and causes the melanocytes to adopt a dendritic cytomorphology (Peacocke et al., 1988). However, ex vivo studies using IHC have reported negative p75NGFR staining in melanocytes but positive staining in nevic corpuscles within neurotized nevi (Argenyi, Rodgers, & Wick, 1996). Most immunohistochemical studies on formalin-fixed paraffin-embedded tissue samples report negative p7NGFR staining in epidermal melanocytes.

In 1996, two studies reported that p75NGFR was strongly expressed in melanomas with a spindled phonotype, including DM and SCM, but not in epithelioid melanomas (Iwamoto, Odland, Piepkorn, & Bothwell, 1996;

Kanik, Yaar, & Bhawan, 1996). Iwamoto et al. observed that p75NGFR was strongly expressed in all DMs and SCMs studied while only expressed in 1/11 (9%) of epithelioid melanomas (Iwamoto et al., 1996). At the same time, Kanik et al. emphasized the diagnostic utility of p75NGFR as an immunohistochemical adjunct for diagnosing DM in S100-negative spindle celled lesions, finding that p75NGFR had superior staining characteristics compared to S100 in SCM, DM, and NM (Kanik et al., 1996). They also noted that p75NGFR staining had an inverse relationship with HMB-45 staining, as tumors that stained strongly for p75NGFR did not express HMB-45 and vice versa. Since these early reports from 1996, several studies have used p75NGFR immunohistochemical staining in an attempt to enhance diagnostic accuracy when dealing with spindled cell melanomas. Usually, p75NGR staining is compared to that of S100, a well-established marker of neural crest-derived cells. S100 is nearly ubiquitously expressed across all melanoma variants, however in rare instances S100-negative SCMs and DMs have been reported, presenting a diagnostic pitfall (Aisner, Maker, Rosenberg, & Berman, 2005). In these instances p75NGFR has been shown to stain positively, and, based on this, many have recommended inclusion of p75NGFR in an initial immunohistochemical screening panel when dealing with equivocal cutaneous spindle celled lesions (Kanik et al., 1996; Lazova, Tantcheva-Poor, & Sigal, 2010; Sigal, Keenan, & Lazova, 2012).

Although p75NGFR may indeed be useful in the identification of SCM and DM in S100-negative cutaneous spindle cell lesions, discrepancies in the literature regarding its sensitivity (Chan & Tahan, 2010; Huttenbach, Prieto, & Reed, 2002), as well as staining in scars (Otaibi, Jukic, Drogowski, Bhawan, & Radfar, 2011; Trejo, Reed, & Prieto, 2002) indicate p75NGFR-specific diagnostic pitfalls. Immunohistochemical studies have shown variable expression of p75NGFR in SCM and DM, ranging from 17% to 100% (Iwamoto et al., 1996; Lazova et al., 2010; Ohshima et al., 2010; Sigal et al., 2012). The reasons for these discrepancies are not clear, and may be the result of a combination of factors. Briefly, these may include variability in antibodies employed, concentrations used, and epitope retrieval methods. Our own analysis of studies utilizing p75NGFR staining indicates no unifying reason for the discrepancy in results. Notably, the studies by Kanik et al. (1996) and Huttenbach et al. (2002) each used the same antibody (p75NGF-R from Boehringer, Manheim) with markedly different results (Kanik et al. reported 100% positivity in SCM/DM while Huttebach et al. reported 33%). A second source of variability may be a result of variable scoring rubrics used to establish criteria for p75NGFR positivity.

Our own experience underscores the need for establishing rigid, clear, and scientifically grounded cutoffs for positivity when attempting to correlate immunohistochemical staining with histopathological parameters. This is more complicated in the realm of DMs where it has been shown that in the mixed subtype, p75NGFR staining is mostly present in the spindle cell component but not in cells with epithelioid cytomorphology (Lazova et al., 2010).

Iwamota et al. first hypothesized that p75NGFR may play a role in mediating PNI in melanomas after observing that p75NGFR was expressed in a significantly greater proportion of DMs than epithelioid melanomas (Iwamoto et al., 2001, 1996). The authors suggested that the expression of p75NGFR in melanoma cells may allow them to respond with increased invasiveness and proliferation to the NGF-rich microenvironment surrounding nerves. It was known at the time that p75NGFR expressed in melanoma cell lines mediated NGF-induced extracellular matrix invasion, most likely through increased production of heparanase (Flørenes et al., 2004; Herrmann et al., 1993; Walch et al., 1999). Additionally, Iwamota et al. reported that p75NGFR-mediated NGF's survival promoting effects in melanoma cell cultures. Taken together, these observations supported Iwamoto et al.'s hypothesis. However, since they did not report directly on rates of PNI, they could only conclude that p75NGFR was associated with the SCM and DM variants and not functionally involved in PNI. The first study to attempt to directly correlate p75NGFR expression and PNI came from Chan and Tahan (2010), in which the authors used IHC to compare p75NGFR expression in a cohort of PNI-positive vs PNI-negative melanomas of diverse histopathologic subtypes (DM, SCM, and epithelioid melanomas). They observed p75NGFR expression in 8/10 PNI-positive melanomas compared to 1/9 PNI-negative melanomas, and that "all invasive melanomas with PNI showed some degree of spindle cell differentiation in the vertical growth phase" (Chan & Tahan, 2010).

Outside of its connection to PNI, additional perspectives on the role of p75NGFR in melanomagenesis have emerged in recent years. In 2010, Boiko et al. published a report of a subpopulation of "melanoma-initiating cells" that behaved as cancer stem cells (Boiko et al., 2010). When transplanted into mice with engrafted human skin, these cells consistently resulted in the development of melanoma tumors and in some cases metastasis to lungs and liver. Using fluorescence-activated cell sorting in freshly resected melanomas, it was determined that CD271 (p75NGFR) was a reliable marker for these melanoma-initiating cells. Notably, when taken from the same tumors CD271+ but not CD271− cells resulted in melanoma development after transplantation into mice. Of the resected tumors expressing CD271+, expression was found in between 2.5% and 41% (average of 16.7%) of the tumor cell population, and similar proportions were observed in the melanomas that developed in mice after transplantation of CD271+ cells. These results suggest that those CD271+ cells are capable of regenerating the more differentiated CD271− cells in vivo. The authors also report that CD271+ cells lack expression of traditional melanocyte (tyrosinase/MART/HMB-45). These findings were replicated by Civenni et al. with the additional finding that transplanted CD271− cells were at times capable of forming tumors but only in natural killer cell depleted and SCID mice (Civenni et al., 2011). Interestingly, in the same study, the authors report that staining a microarray of 200 primary and metastatic melanomas taken from patients showed that coexpression of CD271 and SOX10 was associated with metastasis and worse prognosis. The inverse correlation between p75NGFR expression and melanocyte markers in vitro was further elucidated in two recent studies (Furuta, Inozume, Harada, & Shimada, 2014; Redmer et al., 2014). Furuta et al. observed that cytotoxic T-lymphocytes secrete NGF which, upon binding to p75NGFR on melanoma cells, results in the downregulation of melanocyte markers (Furuta et al., 2014). Additionally, p75NGFR was found to be frequently coexpressed with programmed cell death ligand 1 (PDL-1), a potent inducer of cytotoxic T-lymphocyte apoptosis and that both p75NGFR and PDL-1 may be upregulated in response to interferon gamma. These results have important implications in explaining the frequent failures of targeted T-cell therapy toward traditional melanoma antigens. In the second study, Redmer et al. showed that knockdown of CD271 in melanoma cell lines results in a loss of other neural crest stem cell markers such as SOX10 and furthermore that expression of melanocyte markers and CD271 was mutually exclusive (Redmer et al., 2014). Of potential relevance to DM, this change was associated with an alteration in cytomorphology from spindled to an epithelioid variant.

From a review of the literature, it appears that pathologists and basic scientists have distinct schools of thought regarding the role of p75NGFR. Pathologists have been focused on employing staining for p75NGFR to improve diagnostic accuracy in spindle cell lesions as well as understanding its potential role in mediating PNI, while investigators seeking to understand the underlying fundamental biology of melanomas have focused on p75NGFRs role as a potential marker for melanoma stem cells. It is difficult to reconcile the observations that p75NGFR is associated with tumor-initiating cells and suppression of T-lymphocytes while at the same time pathologists know it to be associated with DM which as discussed previously, has a relatively good prognosis. To reconcile these perspectives, significant gains in our understanding of the complex behavior of the p75NGFR receptor are required. The potentially varying effects of its numerous ligands, the consequences of its interactions with the Trk family of neurotrophin receptors, and its newly appreciated role as a marker for tumor-initiating cells are all areas ripe for future investigation.

6.4 Other Potential Players: NCAM and CXCR4

The previously discussed group of receptors and ligands implicated in the pathogenesis of PNI represents only the most well studied. However, there are two other receptors worth mentioning. Both neural cell adhesion molecule (NCAM) and the chemokine receptor CXCR4 have been suggested to play a role in the development of PNI in select cutaneous and non-cutaneous malignancies. Since the expression of each has been detected in melanomas it is reasonable to hypothesize that they may be of relevance in the pathogenesis of PNI.

The NCAM (CD56) protein is an important homophillic cell adhesion molecule known to play a significant role in the development of the nervous system (Paratcha, Ledda, & Ibáñez, 2003). During embryogenesis, it is expressed early on in neural crest cells, with its expression diminishing during migration and reappearing after migration is complete (Crossin, Chuong, & Edelman, 1985). It has been shown that in nervous tissue, NCAM can bind GDNF family ligands and serves as a signaling receptor, resulting in RET-independent activation of GDNF signaling pathways (Paratcha et al., 2003). Conflicting reports exist regarding NCAM's association with PNI. In colorectal adenocarcinoma, expression of NCAM is associated with an increased risk for PNI but a decreased risk of metastasis (Kenmotsu et al., 1994). Select studies on SCC of the H&N (McLaughlin et al., 1999; Vural, Hutcheson, Korourian, Kechelava, & Hanna, 2000), pancreatic cancers (Aloysius et al., 2010), and adenoid cystic carcinoma (Gandour-Edwards, Kapadia, Barnes, Donald, & Janecka, 1997) have found associations between NCAM and PNI. In contrast, there are at least two other studies in pancreatic cancers and SCCs of the H&N which have failed to demonstrate such an association (Kameda et al., 1999; Solares, Brown, Boyle, Parsons, & Panizza, 2009).

In cells of melanocytic lineage, Reed et al. observed NCAM in 92% of nevi, with its expression seen exclusively in Type C nevus cells and nevic corpuscles (Reed, Finnerty, & Albino, 1999). In the same study, expression was detected in 17% of primary invasive melanomas which were histopathologically all of the non-DM variants (Reed et al., 1999). In DM, Huttenbach et al. reported NCAM in 6/12 (50%) of cases—results that when compared to the findings of Reed et al., suggest it may be more frequently expressed in the DM variant when compared to epithelioid melanomas (Huttenbach et al., 2002). Additionally, in vitro studies in melanoma cell lines have found expression of NCAM to be linked with secretion of MMP-2 and therefore enhanced invasiveness (Shi et al., 2011). Furthermore, in melanoma cell lines derived from central nervous system metastases, NCAM is constitutively expressed (Geertsen, Zenklusen, Kamarashev, Burg, & Dummer, 1999). Despite these observations and the potential connection to PNI in other malignancies, to the best of our knowledge, we are not aware of any study that has investigated NCAM's relationship to PNI in melanoma. Enhanced expression of NCAM in DMs, in conjunction with studies demonstrating correlations with PNI in SCCs and pancreatic cancers suggest that such studies may be fruitful.

CXCR4 is a G protein-coupled chemokine receptor that binds the CXCL12 ligand, and together they are believed to play a role in cell migration and proliferation (Mitchell & Mahalingam, 2014). Aberrant functioning of the CXCR4–CXCL12 axis has been implicated in the development of multiple lineage-unrelated malignancies (Balkwill, 2004). Cutaneous malignancies in which CXCR4 is thought to play a functional role include but are not limited to cSCC, dermatofibrosarcoma, merkel cell carcinoma, and melanoma (Basile, Thiers, Maize, & Lathers, 2008; Knapp et al., 2012; Mitchell et al., 2014; Toyozawa et al., 2010). Of these, the prognostic implications of CXCR4 are best understood in melanoma, where it's expression was first observed in primary melanomas and pulmonary metastasis in mouse models by Murakami et al. (2002). Since then, multiple studies have found CXCR4 expression to be associated with ulceration, tumor thickness, lymph node metastasis, and increasing mortality, suggesting that CXCR4 expression may serve as a prognosticator of poor outcome (Longo-Imedio, Longo, Treviño, Lázaro, & Sánchez-Mateos, 2005; Scala et al., 2005; Toyozawa et al., 2012). In contrast, other studies, including our own have failed to find associations between CXCR4 expression and poor outcomes or negative histopathological prognosticators (Kim et al., 2006; Kühnelt-Leddihn,

Müller, Eisendle, Zelger, & Weinlich, 2012; Mitchell et al., 2014). Thus the status of CXCR4 as a prognostic indicator remains equivocal.

CXCR4 has just recently been implicated in the pathogenesis of PNI in select noncutaneous malignancies. The connection between the CXCR4–CXCL12 axis and PNI was first made by Zhang et al. in their study of prostate cancers (Zhang et al., 2008). In this study, the authors reported that when mice with prostate tumors were treated with a CXCR4 antagonist (AMD3100), it resulted in a reduction in the level of PNI seen on histologic examination. Similar findings were reported by Tan et al. in their study of cholangiocarcinoma, in which they reported a correlation between CXCR4 expression and PNI (Tan, Chang, Liu, & Tang, 2014). The authors also investigated the effects of siRNA targeted to CXCR4 on tumor cell invasion in vitro and found that silencing of the CXCR4 gene diminished invasive capacity. The association between CXCR4 and PNI has also been noted in pancreatic cancer (Jiang, Li, Sun, Zhao, & Zhou, 2014). Lastly, Jeong et al. showed that CXCR4 was expressed in the regions surrounding the areas of PNI in adenoid cystic carcinoma samples (Jeong et al., 2014). In the same study, it was reported that administering AMD31000 in a murine PNI model reduced the incidence of PNI from 71.4% to 18.8%. Although CXCR4 is not a neurotrophin receptor, these first few studies provide a solid basis for implicating the CXCR4–CXCL12 axis in the pathogenesis of PNI. To date, no studies have directly attempted to correlate CXCR4 expression with PNI in melanomas but its well-established expression in melanomas of various subtypes and its connection to PNI in other malignancies warrant such investigations.

7. CLINICAL POTENTIAL OF NEUROTROPHINS AS THERAPEUTIC TARGETS

Owing to their association with proliferation, migration, and survival, neurotrophin receptors and CXCR4 make attractive targets for therapeutic intervention. Inhibitors of TrkA, RET, and CXCR4 have been tested in cell culture, in in vivo animal models as well as in select clinical trials. A number of in vitro experiments have used K252a and related molecules (CEP-701) which function as tyrosine kinase inhibitors to target the Trk family of receptors (Osipov et al., 2014). K252a has been shown induce apoptosis and slow tumor growth in pancreatic and prostate cancer animal models, while CEP-701 (Lestaurtinib) is currently undergoing phase III trials in acute myeloid leukemia (Shabbir & Stuart, 2010). Although no clinical trials have been performed in melanoma patients, in vitro studies treating melanoma cell lines with K252a have been shown to inhibit neurotrophin-induced proliferation and migration (Truzzi et al., 2008). These results suggest that animal studies may validate the effects of k252a and CEP-701 on melanoma in vivo. RET inhibitors have also shown promise in clinical trials of noncutaneous malignancies, notably the multitargeted inhibitor vandetanib (which includes RET as one of its targets) has been FDA approved for use in medullary carcinoma of the thyroid after being shown to increase progression-free survival in phase III trials (Wells et al., 2012). In melanoma cell lines, the RET/Flk-1 inhibitor SU5416 was shown suppress GDNF-stimulated proliferation (Ohshima et al., 2010). At least two phase II clinical trials using SU5416 have been completed in melanoma patients (Mita et al., 2007; Peterson et al., 2004). In one study of 31 patients with advanced melanoma, SU5416, although well tolerated showed only “modest clinical activity” (Peterson et al., 2004). The second phase II trial investigated SU5416 in combination with thalidomide in 12 patients with metastatic melanoma, and found the combination to have clinical efficacy in patients who had failed prior therapies (Mita et al., 2007). Of note, in each of these trials it not possible to determine whether the observed effect was due to inhibition of RET or alternatively Flk-1 (the VEGF receptor) as SU5416 targets both. Lastly, as previously mentioned, AMD3100 (plerixafor) a CXCR4 inhibitor, has been shown inhibit PNI in mouse models of adenoid cystic carcinoma (Jeong et al., 2014). Plerixafor is better known for its use in mobilizing hematopoietic stem cells into the peripheral circulation in cancer patients (Uy, Rettig, & Cashen, 2008) although CXCR4’s potential association with poor prognosis in melanomas suggests targeting this receptor may be worth investigating in melanomas.

8. PERSONAL OBSERVATIONS

We recently investigated the relationship between PNI and TrKA in both of the entities elaborated earlier, i.e., cSCC and DM, cutaneous malignancies with a known propensity for PNI. Furthermore, p75NGFR and RETp were additionally investigated in DM.

In cSCC, we observed a trend toward greater rates of PNI in the H&N areas vs non-H&N areas (Frydenlund et al., 2015). Also in cSCC, despite its lack of correlation with PNI, we observed other noteworthy correlations between TrkA and histopathologic prognosticators. Of note, our finding of more frequent TrkA expression in cSCCs from the H&N may relate to previous observations that ultraviolet radiation modulates TrkA expression in human keratinocytes. Ultraviolet radiation is a known potent inducer of apoptosis and there is in vitro evidence to indicate that TrkA and NGF expression are reduced in UV-irradiated human keratinocytes in culture (Marconi et al., 1999). In the same study, transfection of normal human keratinocytes with NGF resulted in resistance to UV-induced apoptosis, suggesting that shut down of the TrkA–NGF axis promotes apoptosis. In areas chronically sun exposed, a failure to downregulate TrkA expression may cause a loss of this protective apoptotic response leading to an increased mutational burden and tumor formation.

Additionally, we found TrkA was more commonly expressed in high-risk morphologic variants and less well-differentiated cSCCs, suggesting an association between TrkA expression and changes in cytomorphology.

Also in cSCC, we noted TrkA expression to be the strongest in keratinocytes in the basal layer of the epidermis in uninvolved skin, a finding which is in keeping with the in vitro observation of a role for TrkA in proliferation and growth of keratinocytes. Of note, this gradient appeared to be lost in epidermis over cSCC tumors with cytoplasmic staining typically observed pan-epidermally suggesting aberrant TrkA expression in the neoplastic area (Fig. 3).

Using the stringent criterion of TrkA positivity being restricted to cases displaying membranous staining, we found that TrkA expression was not associated with PNI in cSCC, and not expressed at all in DM. From a scientific perspective, our findings suggest that in cSCC PNI may be mediated by a different Trk receptor (TrkB or TrkC). Our finding that TrkA was not expressed in any cases of DM, indicates that TrkA is unlikely to play a role in its pathogenesis (Fig. 4).

We observed a nonsignificant threefold increase in the proportion of RETp among PNI-positive samples compared to PNI-negative samples (39% vs 13%). While not significant, our findings suggest an association between RETp and PNI, and a larger study may be better powered to detect a significant correlation. Additionally, for cases in which RETp data were available, five displayed PNI but were p75NGFR negative. Of these, three (60%) had RETp, suggesting that in the absence of p75NGFR expression RETp may be of relevance in PNI. Our observation that only 7% of PNI-positive cases lacked both RETp and p75NGFR, as compared to 57% of PNI-negative cases ($p = 0.0008$), underscores the potential role played by each.

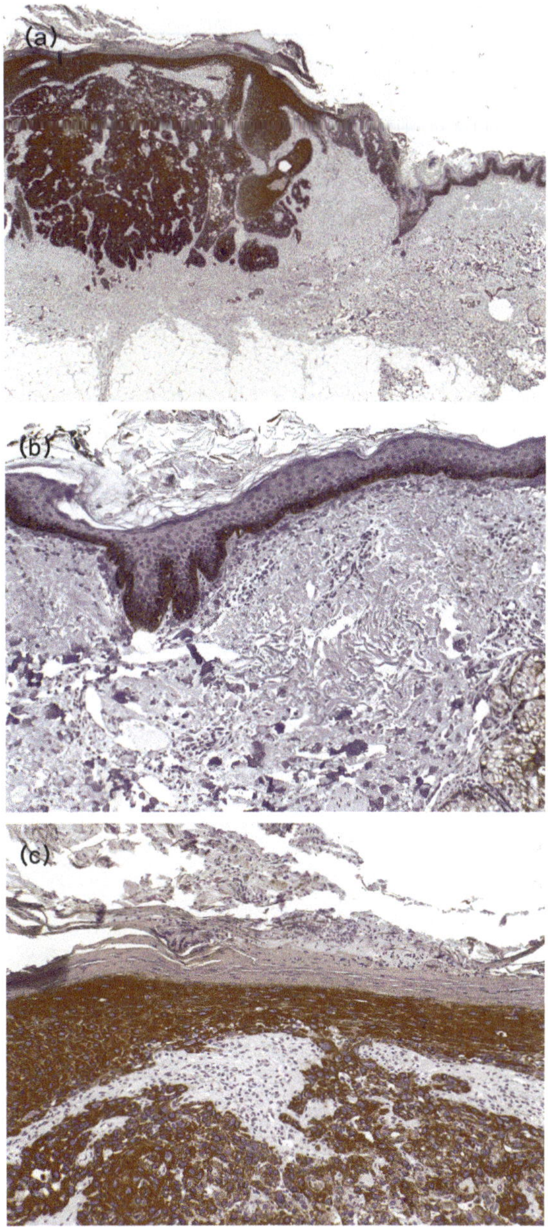

Fig. 3 Immunohistochemical staining for TrkA (*brown*) in a case of cSCC comparing tumor (A and C) and unaffected skin (B). Note the change of TrkA expression from confinement to the basal layer in normal skin to panepidermal staining in the neoplastic region. (A) Scanning magnification. (B) Basal epidermal staining in uninvolved areas (×10). (C) Panepidermal staining over tumor (×10).

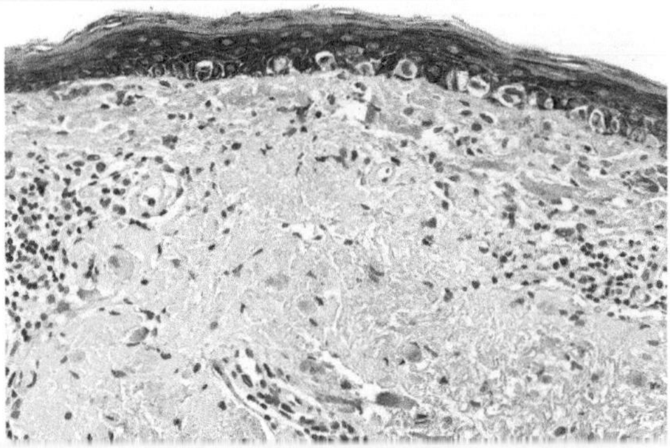

Fig. 4 Case of DM staining negative for TrkA (*brown*) as was observed in all cases we studied. Note the sparing of atypical epidermal melanocytes (x20).

While expression of p75NGFR has long been associated with the spindled cytomorphology in melanomas, much of the previous research has focused on the utility of staining for p75NGFR as an adjunct to help identify S100-negative SCMs and, to differentiate SCMs from other lineage-unrelated spindle cell lesions of the skin (Kanik et al., 1996; Lazova et al., 2010; Sigal et al., 2012). As mentioned earlier, some have demonstrated staining for p75NGFR in 100% of cases of DM (Iwamoto et al., 2001; Kanik et al., 1996; Lazova et al., 2010). We observed positive staining in 65% of our cases in our own analysis (Frydenlund et al., 2015). The fact that p75NGFR expression is found more often in DM than in other subtypes, when taken together with DM's significant propensity for PNI has led some to hypothesize that p75NGFR may play a functional role in PNI (Chan & Tahan, 2010; Iwamoto et al., 2001). Our recent findings have demonstrated that beyond p75NGFR's previously known association with this variant, it is further associated with PNI, even after controlling for depth (Fig. 5).

In vitro studies on melanoma cell lines have shown that p75NGFR mediates the induction of protease expression in response to NGF treatment, resulting in an enhanced invasive capacity (Walch et al., 1999). Furthermore, NGF has been shown to exert chemotactic effects on melanoma cell cultures, with p75NGFR mediating the response (Shonukan, Bagayogo, McCrea, Chao, & Hempstead, 2003). Taken together, from a scientific perspective it is reasonable to hypothesize that p75NGFR is associated with a greater propensity for PNI by allowing for both an enhanced invasiveness in response to NGF (permitting proteolytic digestion of the perineurium) and a chemotactic attraction to the NGF-rich microenvironment of nerve as NGF

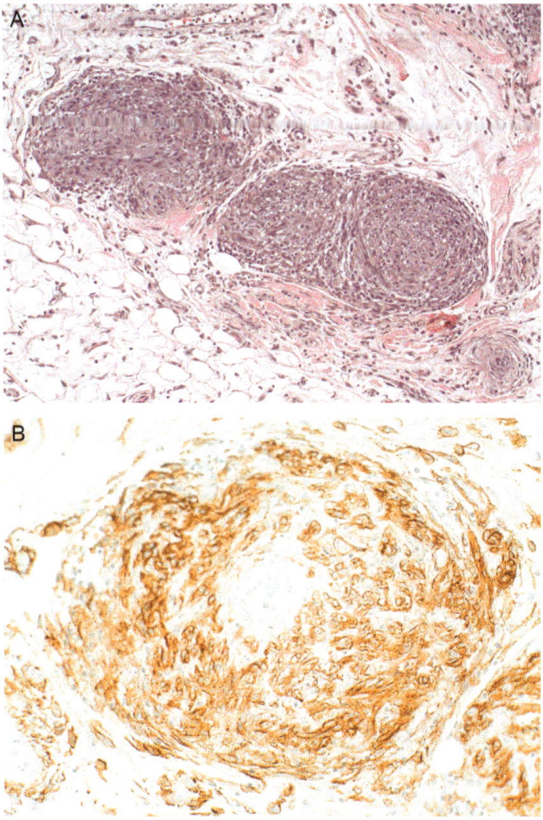

Fig. 5 PNI displayed in a case of DM showing strong expression of p75NGFR in the perineural foci of invasion (*brown*). (A) DM displaying two foci of PNI stained with hematoxylin and eosin (x20). (B) High power view of a focus of PNI in DM positive for p75NGFR expression (x40).

is known to be secreted by Schwann cells (Johnson et al., 1988). Lastly, p75NGFR is a promiscuous receptor and binds other neurotrophins besides NGF (such as NT-3 and BDNF) each of which may have an additional role to play in affecting the behavior of melanomas expressing p75NGFR.

9. CONCLUSIONS

In summary, PNI is most frequently associated with DM (often termed neurotropism) and cSCC. While PNI has been associated with increased risk of local recurrence and decreased disease-free survival, these associations remain equivocal. The correlation of PNI with poor outcomes in several cutaneous as well as noncutaneous malignancies suggests that considering PNI as a histopathological variable in its own right (separate from other definitions of neurotropism), would help clarify its precise prognostic relevance. Additionally, it is still unknown whether diameter of the involved nerve or the presence of multiple foci of PNI modifies prognosis. Biologically, PNI is not well understood, but multiple lines of evidence suggest neurotrophin receptor expression in tumor cells allows for a proproliferative and proinvasive response to the neural microenvironment. While many of the neurotrophin receptors have been considered here in isolation it is possible that there is a significant amount of crosstalk between pathways. Specifically, p75NGFR has been shown to modulate the behavior of Trk receptors and examinations of patterns of coexpression could be revealing. Several of the neurotrophin receptors play additional functional roles in tumor development and may thus be putative targets for future therapies. It is likely that there are multiple pathways that lead to PNI, and what we have presented here are only the best studied.

REFERENCES

Abschuetz, O., Osen, W., Frank, K., Kato, M., Schadendorf, D., & Umansky, V. (2012). T-cell mediated immune responses induced in ret transgenic mouse model of malignant melanoma. *Cancers*, *4*(2), 490–503.

Aisner, D. L., Maker, A., Rosenberg, S. A., & Berman, D. M. (2005). Loss of S100 antigenicity in metastatic melanoma. *Human Pathology*, *36*(9), 1016–1019.

Alam, M., & Ratner, D. (2001). Cutaneous squamous-cell carcinoma. *New England Journal of Medicine*, *344*(13), 975–983.

Aloysius, M. M., Zaitoun, A. M., Awad, S., Ilyas, M., Rowlands, B. J., & Lobo, D. N. (2010). Mucins and CD56 as markers of tumour invasion and prognosis in periampullary cancer. *British Journal of Surgery*, *97*(8), 1269–1278.

Anton, E. S., Weskamp, G., Reichardt, L. F., & Matthew, W. D. (1994). Nerve growth factor and its low-affinity receptor promote Schwann cell migration. *Proceedings of the National Academy of Sciences*, *91*(7), 2795–2799.

Argenyi, Z. B., Rodgers, J., & Wick, M. (1996). Expression of nerve growth factor and epidermal growth factor receptors in neural nevi with nevic corpuscles. *The American Journal of Dermatopathology*, *18*(5), 460–464.

Ayala, G. E., Dai, H., Ittmann, M., Li, R., Powell, M., Frolov, A., … Rowley, D. (2004). Growth and survival mechanisms associated with perineural invasion in prostate cancer. *Cancer Research*, *64*(17), 6082–6090.

Baer, S. C., Schultz, D., Synnestvedt, M., & Elder, D. E. (1995). Desmoplasia and neurotropism. Prognostic variables in patients with stage I melanoma. *Cancer*, *76*(11), 2242–2247.

Balkwill, F. (2004). The significance of cancer cell expression of the chemokine receptor CXCR4. *Seminars in Cancer Biology*, *14*(3), 171–179.

Barr, J., Amato, C. M., Robinson, S. E., Kounalakis, N., & Robinson, W. A. (2012). The RET G691S polymorphism is a germline variant in desmoplastic malignant melanoma. *Melanoma Research*, *22*(1), 92–95.

Basile, J., Thiers, B., Maize, J., & Lathers, D. M. R. (2008). Chemokine receptor expression in non-melanoma skin cancer. *Journal of Cutaneous Pathology*, *35*(7), 623–629.

Batsakis, J. G. (1985). Nerves and neurotropic carcinomas. *The Annals of Otology, Rhinology, and Laryngology*, *94*(4 Pt. 1), 426–427.

Berlingeri-Ramos, A. C., Detweiler, C. J., Wagner, R. F., & Kelly, B. C. (2015). Dual S-100-AE1/3 immunohistochemistry to detect perineural invasion in nonmelanoma skin cancers. *Journal of Skin Cancer*, *2015*, 620235.

Boiko, A. D., Razorenova, O. V., van de Rijn, M., Swetter, S. M., Johnson, D. L., Ly, D. P., … Weissman, I. L. (2010). Human melanoma initiating cells express neural crest nerve growth factor receptor CD271. *Nature*, *466*(7302), 133–137.

Boyd, J. G., & Gordon, T. (2003). Neurotrophic factors and their receptors in axonal regeneration and functional recovery after peripheral nerve injury. *Molecular Neurobiology*, *27*(3), 277–324.

Busam, K. J., Mujumdar, U., Hummer, A. J. M., Nobrega, J., Hawkins, W. G., Coit, D. G., & Brady, M. S. (2004). Cutaneous desmoplastic melanoma: Reappraisal of morphologic heterogeneity and prognostic factors. *The American Journal of Surgical Pathology*, *28*(11), 1518–1525.

Campoli, M., Brodland, D. G., & Zitelli, J. (2014). A prospective evaluation of the clinical, histologic, and therapeutic variables associated with incidental perineural invasion in cutaneous squamous cell carcinoma. *Journal of the American Academy of Dermatology*, *70*(4), 630–636.

Carter, J. B., Johnson, M. M., Chua, T. L., Karia, P. S., & Schmults, C. D. (2013). Outcomes of primary cutaneous squamous cell carcinoma with perineural invasion: An 11-year cohort study. *JAMA Dermatology*, *149*(1), 35–42.

Cavel, O., Shomron, O., Shabtay, A., Vital, J., Trejo-Leider, L., Weizman, N., … Gil, Z. (2012). Endoneurial macrophages induce perineural invasion of pancreatic cancer cells by secretion of GDNF and activation of RET tyrosine kinase receptor. *Cancer Research*, *72*(22), 5733–5743.

Chan, M. M., & Tahan, S. R. (2010). Low-affinity nerve growth factor receptor (P75 NGFR) as a marker of perineural invasion in malignant melanomas. *Journal of Cutaneous Pathology*, *37*, 336–343.

Chang, P. C., Fischbein, N. J., McCalmont, T. H., Kashani-Sabet, M., Zettersten, E. M., Liu, A. Y., & Weissman, J. L. (2004). Perineural spread of malignant melanoma of the head and neck: Clinical and imaging features. *American Journal of Neuroradiology*, *25*(1), 5–11.

Chen, J. Y., Hruby, G., Scolyer, R. A., Murali, R., Hong, A., Fitzgerald, P., et al. (2008). Desmoplastic neurotropic melanoma: A clinicopathologic analysis of 128 cases. *Cancer*, *113*(10), 2770–2778.

Chen-Tsai, C. P., Colome-Grimmer, M., & Wagner, R. F. (2004). Correlations among neural cell adhesion molecule, nerve growth factor, and its receptors, TrkA, TrkB, TrkC, and p75NGFR, in perineural invasion by basal cell and cutaneous squamous cell carcinomas. *Dermatologic Surgery*, *30*(7), 1009–1016.

Cherpelis, B. S., Marcusen, C., & Lang, P. G. (2002). Prognostic factors for metastasis in squamous cell carcinoma of the skin. *Dermatologic Surgery*, *28*(3), 268–273.

Civenni, G., Walter, A., Kobert, N., Mihic-Probst, D., Zipser, M., Belloni, B., … Sommer, L. (2011). Human CD271-positive melanoma stem cells associated with metastasis establish tumor heterogeneity and long-term growth. *Cancer Research*, *71*(8), 3098–3109.

Clayman, G. L., Lee, J. J., Holsinger, F. C., Zhou, X., Duvic, M., El-Naggar, A. K., et al. (2005). Mortality risk from squamous cell skin cancer. *Journal of Clinical Oncology*, *23*(4), 759–765.

Crossin, K. L., Chuong, C. M., & Edelman, G. M. (1985). Expression sequences of cell adhesion molecules. *Proceedings of the National Academy of Sciences*, *82*(20), 6942–6946.

Dang, C., Zhang, Y., Ma, Q., & Shimahara, Y. (2006). Expression of nerve growth factor receptors is correlated with progression and prognosis of human pancreatic cancer. *Journal of Gastroenterology and Hepatology, 21*(5), 850–858.

Descamps, S., Pawlowski, V., Révillion, F., Hornez, L., Hebbar, M., Boilly, B., … Peyrat, J.-P. (2001). Expression of nerve growth factor receptors and their prognostic value in human breast cancer. *Cancer Research, 61*(11), 4337–4340.

Dunn, M., Morgan, M. B., & Beer, T. W. (2009). Perineural invasion: Identification, significance, and a standardized definition. *Dermatologic Surgery, 35*(2), 214–221.

Dunn, M., Morgan, M. B., Beer, T. W., Chen, K. T. K., & Acker, S. M. (2009). Histologic mimics of perineural invasion. *Journal of Cutaneous Pathology, 36*(9), 937–942.

Farasat, S., Yu, S. S., Neel, V. A., Nehal, K. S., Lardaro, T., Mihm, M. C., … Liégeois, N. J. (2011). A new American Joint Committee on Cancer staging system for cutaneous squamous cell carcinoma: Creation and rationale for inclusion of tumor (T) characteristics. *Journal of the American Academy of Dermatology, 64*(6), 1051–1059.

Feasel, A. M., Brown, T. J., Bogle, M. A., Tschen, J. A., & Nelson, B. R. (2001). Perineural invasion of cutaneous malignancies. *Dermatologic Surgery, 27*(6), 531–542.

Flørenes, V. A., Mælandsmo, G. M., Holm, R., Reich, R., Lazarovici, P., & Davidson, B. (2004). Expression of activated TrkA protein in melanocytic tumors relationship to cell proliferation and clinical outcome. *American Journal of Clinical Pathology, 122*(3), 412–420.

Fouquet, T., Germain, A., Brunaud, L., Bresler, L., & Ayav, A. (2014). Is perineural invasion more accurate than other factors to predict early recurrence after pancreatoduodenectomy for pancreatic head adenocarcinoma? *World Journal of Surgery, 38*(8), 2132–2137.

Frydenlund, N., Leone, D. A., Mitchell, B., Abbas, O., Dhingra, J., & Mahalingam, M. (2015a). Perineural invasion in cutaneous squamous cell carcinoma: Role of immunohistochemistry, anatomical site, and the high-affinity nerve growth factor receptor TrkA. *Human Pathology, 46*(8), 1209–1216.

Frydenlund, N., Leone, D., Mitchell, B., Yang, S., Deng, A., Hoang, M. P., & Meera, M. (2015b). Neurotrophin receptors and perineural nvasion in desmoplastic melanoma. *Journal of the American Academy of Dermatology, 72*(5), 851–858.

Furuta, J., Inozume, T., Harada, K., & Shimada, S. (2014). CD271 on melanoma cell is an IFN-γ-inducible immunosuppressive factor that mediates downregulation of melanoma antigens. *Journal of Investigative Dermatology, 134*(5), 1369–1377.

Gandour-Edwards, R., Kapadia, S. B., Barnes, L., Donald, P. J., & Janecka, I. P. (1997). Neural cell adhesion molecule in adenoid cystic carcinoma invading the skull base. *Otolaryngology—Head and Neck Surgery: Official Journal of American Academy of Otolaryngology-Head and Neck Surgery, 117*(5), 453–458.

Geertsen, R., Zenklusen, R., Kamarashev, J., Burg, G., & Dummer, R. (1999). Inverse regulation of neuronal cellular adhesion molecule (NCAM) by IFN-γ in melanoma cell cultures established from CNS lesions. *International Journal of Cancer, 83*(1), 135–140.

Goepfert, H., Dichtel, W. J., Medina, J. E., Lindberg, R. D., & Luna, M. D. (1984). Perineural invasion in squamous cell skin carcinoma of the head and neck. *American Journal of Surgery, 148*(4), 542–547.

Hassanein, A. M., Proper, S. A., Depcik-Smith, N. D., & Flowers, F. P. (2005). Peritumoral fibrosis in basal cell and squamous cell carcinoma mimicking perineural invasion: Potential pitfall in mohs micrographic surgery. *Dermatologic Surgery, 31*(9), 1101–1106.

He, S., Chen, C.-H., Chernichenko, N., He, S., Bakst, R. L., Barajas, F., … Wong, R. J. (2014). GFRα1 released by nerves enhances cancer cell perineural invasion through GDNF-RET signaling. *Proceedings of the National Academy of Sciences of the United States of America, 111*(19), E2008–E2017.

He, X.-L., & Garcia, K. C. (2004). Structure of nerve growth factor complexed with the shared neurotrophin receptor p75. *Science (New York, N.Y.), 304*(5672), 870–875.
Herrmann, J. L., Menter, D. G., Hamada, J., Marchetti, D., Nakajima, M., & Nicolson, G. L. (1993). Mediation of NGF-stimulated extracellular matrix invasion by the human melanoma low affinity p75 neurotrophin receptor: Melanoma p75 functions independently of trkA. *Molecular Biology of the Cell, 4*(11), 1205.
Huttenbach, Y., Prieto, V. G., & Reed, J. A. (2002). Desmoplastic and spindle cell melanomas express protein markers of the neural crest but not of later committed stages of Schwann cell differentiation. *Journal of Cutaneous Pathology, 29*(9), 562–568.
Innominato, P. F., Libbrecht, L., & van den Oord, J. J. (2001). Expression of neurotrophins and their receptors in pigment cell lesions of the skin. *The Journal of Pathology, 194*(1), 95–100.
Iwamoto, S., Burrows, R. C., Agoff, S. N., Piepkorn, M., Bothwell, M., & Schmidt, R. (2001). The p75 neurotrophin receptor, relative to other Schwann cell and melanoma markers, is abundantly expressed in spindled melanomas. *The American Journal of Dermatopathology, 23*(4), 288–294.
Iwamoto, S., Odland, P. B., Piepkorn, M., & Bothwell, M. (1996). Evidence that the p75 neurotrophin receptor mediates perineural spread of desmoplastic melanoma. *Journal of the American Academy of Dermatology, 35*(5 Pt. 1), 725–731.
Jeong, W.-J., Choi, I. J., Park, M.-W., An, S.-Y., Jeon, E.-H., Paik, J. H., … Ahn, S.-H. (2014). CXCR4 antagonist inhibits perineural invasion of adenoid cystic carcinoma. *Journal of Clinical Pathology, 67*, 992–998.
Jiang, Y.-M., Li, G., Sun, B.-C., Zhao, X.-L., & Zhou, Z.-K. (2014). Study on the relationship between CXCR4 expression and perineural invasion in pancreatic cancer. *Asian Pacific Journal of Cancer Prevention: APJCP, 15*(12), 4893–4896.
Johnson, E. M., Taniuchi, M., & DiStefano, P. S. (1988). Expression and possible function of nerve growth factor receptors on Schwann cells. *Trends in Neurosciences, 11*(7), 299–304.
Kameda, K., Shimada, H., Ishikawa, T., Takimoto, A., Momiyama, N., Hasegawa, S., … Ichikawa, Y. (1999). Expression of highly polysialylated neural cell adhesion molecule in pancreatic cancer neural invasive lesion. *Cancer Letters, 137*(2), 201–207.
Kanik, A. B., Yaar, M., & Bhawan, J. (1996). p75 nerve growth factor receptor staining helps identify desmoplastic and neurotropic melanoma. *Journal of Cutaneous Pathology, 23*(3), 205–210.
Kato, M., Liu, W., Akhand, A. A., Hossain, K., Takeda, K., Takahashi, M., & Nakashima, I. (2000). Ultraviolet radiation induces both full activation of Ret kinase and malignant melanocytic tumor promotion in RFP-RET-transgenic mice. *Journal of Investigative Dermatology, 115*(6), 1157–1158.
Kato, M., Takahashi, M., Akhand, A. A., Liu, W., Dai, Y., Shimizu, S., … Nakashima, I. (1998). Transgenic mouse model for skin malignant melanoma. *Oncogene, 17*(14), 1885–1888.
Kay, P. A., Pinheiro, A. D., Lohse, C. M., Pankratz, V. S., Olsen, K. D., Lewis, J. E., & Nascimento, A. G. (2004). Desmoplastic melanoma of the head and neck: Histopathologic and immunohistochemical study of 28 cases. *International Journal of Surgical Pathology, 12*(1), 17–24.
Kenmotsu, M., Gouchi, A., Maruo, Y., Murashima, N., Hiramoto, Y., Iwagaki, H., & Orita, K. (1994). The expression of neural cell adhesion molecule (NCAM), neural invasion and recurrence patterns in rectal cancer—A study using anti-NACM (neural cell adhesion molecule) antibody. *Nihon Geka Gakkai Zasshi, 95*(2), 66–70.
Kim, J., Mori, T., Chen, S. L., Amersi, F. F., Martinez, S. R., Kuo, C., … Hoon, D. S. B. (2006). Chemokine receptor CXCR4 expression in patients with melanoma and colorectal cancer liver metastases and the association with disease outcome. *Annals of Surgery, 244*(1), 113–120.
Knapp, C. F., Sayegh, Z., Schell, M. J., Rawal, B., Ochoa, T., Sondak, V. K., & Messina, J. L. (2012). Expression of CXCR4, E-cadherin, Bcl-2, and survivin in Merkel cell

carcinoma: An immunohistochemical study using a tissue microarray. *The American Journal of Dermatopathology, 34*(6), 592–596.

Kolokythas, A., Cox, D. P., Dekker, N., & Schmidt, B. L. (2010). Nerve growth factor and tyrosine kinase a receptor in oral squamous cell carcinoma: Is there an association with perineural invasion? *Journal of Oral and Maxillofacial Surgery, 68*(6), 1290–1295.

Kühnelt-Leddihn, L., Müller, H., Eisendle, K., Zelger, B., & Weinlich, G. (2012). Overexpression of the chemokine receptors CXCR4, CCR7, CCR9, and CCR10 in human primary cutaneous melanoma: A potential prognostic value for CCR7 and CCR10? *Archives of Dermatological Research, 304*(3), 185–193.

Kurtz, K. A., Hoffman, H. T., Zimmerman, M. B., & Robinson, R. A. (2005). Perineural and vascular invasion in oral cavity squamous carcinoma: Increased incidence on re-review of slides and by using immunohistochemical enhancement. *Archives of Pathology & Laboratory Medicine, 129*(3), 354–359.

Lazova, R., Tantcheva-Poor, I., & Sigal, A. C. (2010). P75 nerve growth factor receptor staining is superior to S100 in identifying spindle cell and desmoplastic melanoma. *Journal of the American Academy of Dermatology, 63*(5), 852–858.

Leibovitch, I., Huilgol, S. C., Selva, D., Hill, D., Richards, S., & Paver, R. (2005). Cutaneous squamous cell carcinoma treated with Mohs micrographic surgery in Australia II. Perineural invasion. *Journal of the American Academy of Dermatology, 53*(2), 261–266.

Liebig, C., Ayala, G., Wilks, J. A., Berger, D. H., & Albo, D. (2009). Perineural invasion in cancer. *Cancer, 115*(15), 3379–3391.

Lin, C., Tripcony, L., Keller, J., Poulsen, M., Martin, J., Jackson, J., & Dickie, G. (2012). Perineural infiltration of cutaneous squamous cell carcinoma and basal cell carcinoma without clinical features. *International Journal of Radiation Oncology, Biology, Physics, 82*(1), 334–340.

Livestro, D. P., Muzikansky, A., Kaine, E. M., Flotte, T. J., Sober, A. J., Mihm, M. C., Jr., et al. (2005). Biology of Desmoplastic melanoma: A case control comparison with other melanomas. *Journal of Clinical Oncology, 23*(27), 6739–6746.

Longo-Imedio, M. I., Longo, N., Treviño, I., Lázaro, P., & Sánchez-Mateos, P. (2005). Clinical significance of CXCR3 and CXCR4 expression in primary melanoma. *International Journal of Cancer, 117*(5), 861–865.

Marco, E. D., Mathor, M., Bondanza, S., Cutuli, N., Marchisio, P. C., Cancedda, R., & Luca, M. D. (1993). Nerve growth factor binds to normal human keratinocytes through high and low affinity receptors and stimulates their growth by a novel autocrine loop. *Journal of Biological Chemistry, 268*(30), 22838–22846.

Marconi, A., Vaschieri, C., Zanoli, S., Giannetti, A., & Pincelli, C. (1999). Nerve growth factor protects human keratinocytes from ultraviolet-B-induced apoptosis. *Journal of Investigative Dermatology, 113*(6), 920–927.

McLaughlin, R. B., Montone, K. T., Wall, S. J., Chalian, A. A., Weinstein, G. S., Roberts, S. A., … Weber, R. S. (1999). Nerve cell adhesion molecule expression in squamous cell carcinoma of the head and neck: A predictor of propensity toward perineural spread. *The Laryngoscope, 109*(5), 821–826.

Miller, D. D., Emley, A., Yang, S., Richards, J. E., Lee, J. E., Deng, A., … Mahalingam, M. (2012). Mixed versus pure variants of desmoplastic melanoma: A genetic and immunohistochemical appraisal. *Modern Pathology, 25*(4), 505–515.

Mita, M. M., Rowinsky, E. K., Forero, L., Eckhart, S. G., Izbicka, E., Weiss, G. R., … Patnaik, A. (2007). A phase II, pharmacokinetic, and biologic study of semaxanib and thalidomide in patients with metastatic melanoma. *Cancer Chemotherapy and Pharmacology, 59*(2), 165–174.

Mitchell, B., Leone, D., Feller, K., Menon, S., Bondzie, P., Yang, S., … Mahalingam, M. (2014). Protein expression of the chemokine receptor CXCR4 and its ligand CXCL12 in primary cutaneous melanoma—Biomarkers of potential utility? *Human Pathology, 45*(10), 2094–2100.

Mitchell, B., & Mahalingam, M. (2014). The CXCR4/CXCL12 axis in cutaneous malignancies with an emphasis on melanoma. *Histology and Histopathology*, *29*, 1539–1546.
Moore, B. A., Weber, R. S., Prieto, V., El-Naggar, A., Holsinger, F. C., Zhou, X., … Clayman, G. L. (2005). Lymph node metastases from cutaneous squamous cell carcinoma of the head and neck. *The Laryngoscope*, *115*(9), 1561–1567.
Mulligan, L. M. (2014). RET revisited: Expanding the oncogenic portfolio. *Nature Reviews. Cancer*, *14*(3), 173–186.
Murakami, T., Maki, W., Cardones, A. R., Fang, H., Tun Kyi, A., Nestle, F. O., et al. (2002). Expression of CXCR-4 enhances the pulmonary metastatic potential of murine B16 melanoma cells. *Cancer Research*, *62*(24), 7328–7334.
Murali, R., Shaw, H. M., Lai, K., McCarthy, S. W., Quinn, M. J., Stretch, J. R., … Scolyer, R. A. (2010). Prognostic factors in cutaneous desmoplastic melanoma: A study of 252 patients. *Cancer*, *116*(17), 4130–4138.
Narita, N., Tanemura, A., Murali, R., Scolyer, R. A., Huang, S., Arigami, T., … Hoon, D. S. (2009). Functional RET G691S polymorphism in cutaneous malignant melanoma. *Oncogene*, *28*(34), 3058–3068.
Ohshima, Y., Yajima, I., Takeda, K., Iida, M., Kumasaka, M., Matsumoto, Y., & Kato, M. (2010). c-RET molecule in malignant melanoma from oncogenic RET-carrying transgenic mice and human cell lines. *PLoS One*, *5*(4), e10279.
Okada, Y., Eibl, G., Duffy, J. P., Reber, H. A., & Hines, O. J. (2003). Glial cell-derived neurotrophic factor upregulates the expression and activation of matrix metalloproteinase-9 in human pancreatic cancer. *Surgery*, *134*(2), 293–299.
Okada, Y., Eibl, G., Guha, S., Duffy, J. P., Reber, H. A., & Hines, O. J. (2004). Nerve growth factor stimulates MMP-2 expression and activity and increases invasion by human pancreatic cancer cells. *Clinical & Experimental Metastasis*, *21*(4), 285–292.
Osipov, A. V., Terpinskaya, T. I., Kryukova, E. V., Ulaschik, V. S., Paulovets, L. V., Petrova, E. A., … Utkin, Y. N. (2014). Nerve growth factor from cobra venom inhibits the growth of Ehrlich tumor in mice. *Toxins*, *6*(3), 784–795.
Otaibi, S., Jukic, D. M., Drogowski, L., Bhawan, J., & Radfar, A. (2011). NGFR (p75) expression in cutaneous scars; Further evidence for a potential pitfall in evaluation of reexcision scars of cutaneous neoplasms, in particular desmoplastic melanoma. *The American Journal of Dermatopathology*, *33*(1), 65–71.
Paratcha, G., Ledda, F., & Ibáñez, C. F. (2003). The neural cell adhesion molecule NCAM is an alternative signaling receptor for GDNF family ligands. *Cell*, *113*(7), 867–879.
Paus, R., Lüftl, M., & Czarnetzki, B. M. (1994). Nerve growth factor modulates keratinocyte proliferation in murine skin organ culture. *British Journal of Dermatology*, *130*(2), 174–180.
Peacocke, M., Yaar, M., Mansur, C. P., Chao, M. V., & Gilchrest, B. A. (1988). Induction of nerve growth factor receptors on cultured human melanocytes. *Proceedings of the National Academy of Sciences of the United States of America*, *85*(14), 5282.
Perosio, P. M., & Brooks, J. J. (1988). Expression of nerve growth factor receptor in paraffin-embedded soft tissue tumors. *The American Journal of Pathology*, *132*(1), 152.
Peterson, A. C., Swiger, S., Stadler, W. M., Medved, M., Karczmar, G., & Gajewski, T. F. (2004). Phase II study of the Flk-1 tyrosine kinase inhibitor SU5416 in advanced melanoma. *Clinical Cancer Research: An Official Journal of the American Association for Cancer Research*, *10*(12 Pt. 1), 4048–4054.
Pincelli, C., Sevignani, C., Manfredini, R., Grande, A., Fantini, F., Bracci-Laudiero, L., … Giannetti, A. (1994). Expression and function of nerve growth factor and nerve growth factor receptor on cultured keratinocytes. *Journal of Investigative Dermatology*, *103*(1), 13–18.
Puri, P., & Shinkai, T. (2004). Pathogenesis of Hirschsprung's disease and its variants: Recent progress. *Seminars in Pediatric Surgery*, *13*(1), 18–24.
Quinn, M. J., Crotty, K. A., Thompson, J. F., Coates, A. S., O'Brien, C. J., & McCarthy, W. H. (1998). Desmoplastic and desmoplastic neurotropic melanoma. *Cancer*, *83*(6), 1128–1135.

Redmer, T., Welte, Y., Behrens, D., Fichtner, I., Przybilla, D., Wruck, W., … Regenbrecht, C. R. A. (2014). The nerve growth factor receptor CD271 is crucial to maintain tumorigenicity and stem-like properties of melanoma cells. *PLoS One*, *9*(5), e92596.

Reed, J. A., Finnerty, B., & Albino, A. P. (1999). Divergent cellular differentiation pathways during the invasive stage of cutaneous malignant melanoma progression. *The American Journal of Pathology*, *155*(2), 549–555.

Reed, R. J., & Leonard, D. D. (1979). Neurotropic melanoma. A variant of desmoplastic melanoma. *The American Journal of Surgical Pathology*, *3*(4), 301–311.

Ronaghy, A., Yaar, R., Goldberg, L. J., Mahalingam, M., & Bhawan, J. (2010). Perineural involvement: What does it mean? *The American Journal of Dermatopathology*, *32*(5), 469–476.

Ross, A. S., Miller Whalen, F., Elenitsas, R., Xu, X., Troxel, A. B., & Schmults, C. D. (2009). Diameter of involved nerves predicts outcomes in cutaneous squamous cell carcinoma with perineural invasion: An investigator-blinded retrospective cohort study. *Dermatologic Surgery*, *35*(12), 1859–1866.

Scala, S., Ottaiano, A., Ascierto, P. A., Cavalli, M., Simeone, E., Giuliano, P., … Castello, G. (2005). Expression of CXCR4 predicts poor prognosis in patients with malignant melanoma. *Clinical Cancer Research: An Official Journal of the American Association for Cancer Research*, *11*(5), 1835–1841.

Scanlon, P., Tian, J., Zhong, J., Silva, I., Shapiro, R., Pavlick, A., … Darvishian, F. (2014). Enhanced immunohistochemical detection of neural infiltration in primary melanoma: Is there a clinical value? *Human Pathology*, *45*(8), 1656–1663.

Shabbir, M., & Stuart, R. (2010). Lestaurtinib, a multitargeted tyrosine kinase inhibitor: From bench to bedside. *Expert Opinion on Investigational Drugs*, *19*(3), 427–436.

Shi, Y., Liu, R., Zhang, S., Xia, Y.-Y., Yang, H.-J., Guo, K., … Feng, Z.-W. (2011). Neural cell adhesion molecule potentiates invasion and metastasis of melanoma cells through CAMP-dependent protein kinase and phosphatidylinositol 3-kinase pathways. *The International Journal of Biochemistry & Cell Biology*, *43*(4), 682–690.

Shonukan, T., Bagayogo, I., McCrea, P. D., Chao, M., & Hempstead, B. (2003). Neurotrophin-induced melanoma cell migration is mediated through the actin-bundling protein fascin. *Oncogene*, *22*(23), 3616–3623.

Sigal, A. C., Keenan, M., & Lazova, R. (2012). P75 nerve growth factor receptor as a useful marker to distinguish spindle cell melanoma from other spindle cell neoplasms of sun-damaged skin. *The American Journal of Dermatopathology*, *34*(2), 145–150.

Smithers, B. M., McLeod, G. R., & Little, J. H. (1992). Desmoplastic melanoma: Patterns of recurrence. *World Journal of Surgery*, *16*(2), 186–190.

Solares, C. A., Brown, I., Boyle, G. M., Parsons, P. G., & Panizza, B. (2009). Neural cell adhesion molecule expression: No correlation with perineural invasion in cutaneous squamous cell carcinoma of the head and neck. *Head & Neck*, *31*(6), 802–806.

Su, A., Dry, S. M., Binder, S. W., Said, J., Shintaku, P., & Sarantopoulos, G. P. (2014). Malignant melanoma with neural differentiation: An exceptional case report and brief review of the pertinent literature. *The American Journal of Dermatopathology*, *36*(1), e5–e9.

Takahashi, M. (2001). The GDNF/RET signaling pathway and human diseases. *Cytokine & Growth Factor Reviews*, *12*(4), 361–373.

Tallini, G., & Asa, S. L. (2001). RET oncogene activation in papillary thyroid carcinoma. *Advances in Anatomic Pathology*, *8*(6), 345–354.

Tan, X.-Y., Chang, S., Liu, W., & Tang, H.-H. (2014). Silencing of CXCR4 inhibits tumor cell proliferation and neural invasion in human hilar cholangiocarcinoma. *Gut and Liver*, *8*(2), 196–204.

Toyozawa, S., Kaminaka, C., Furukawa, F., Nakamura, Y., Matsunaka, H., & Yamamoto, Y. (2012). Chemokine receptor CXCR4 is a novel marker for the progression of cutaneous malignant melanomas. *Acta Histochemica et Cytochemica*, *45*(5), 293–299.

Toyozawa, S., Yamamoto, Y., Ishida, Y., Kondo, T., Nakamura, Y., & Furukawa, F. (2010). Immunohistochemical analysis of CXCR4 expression in fibrohistiocytic tumors. *Acta Histochemica et Cytochemica*, *43*(2), 45–50.

Trejo, O., Reed, J. A., & Prieto, V. G. (2002). Atypical cells in human cutaneous re-excision scars for melanoma express p75NGFR, C56/N-CAM and GAP-43: Evidence of early Schwann cell differentiation. *Journal of Cutaneous Pathology*, *29*(7), 397–406.

Truzzi, F., Marconi, A., Lotti, R., Dallaglio, K., French, L. E., Hempstead, B. L., & Pincelli, C. (2008). Neurotrophins and their receptors stimulate melanoma cell proliferation and migration. *Journal of Investigative Dermatology*, *128*(8), 2031–2040.

Truzzi, F., Marconi, A., & Pincelli, C. (2011). Neurotrophins in healthy and diseased skin. *Dermato-Endocrinology*, *3*(1), 31–35.

Umansky, V., & Sevko, A. (2013). Ret transgenic mouse model of spontaneous skin melanoma: Focus on regulatory T cells. *Pigment Cell & Melanoma Research*, *26*(4), 457–463.

Uy, G. L., Rettig, M. P., & Cashen, A. F. (2008). Plerixafor, a CXCR4 antagonist for the mobilization of hematopoietic stem cells. *Expert Opinion on Biological Therapy*, *8*(11), 1797–1804.

Vural, E., Hutcheson, J., Korourian, S., Kechelava, S., & Hanna, E. (2000). Correlation of neural cell adhesion molecules with perineural spread of squamous cell carcinoma of the head and neck. *Otolaryngology – Head and Neck Surgery*, *122*(5), 717–720.

Walch, T., Albino, A. P., & Marchetti, D. (1999). Correlation of overexpression of the low-affinity p75 neurotrophin receptor with augmented invasion and heparanase production in human malignant melanoma cells. *International Journal of Cancer*, *82*(1), 112–120.

Wang, L., Sun, M., Jiang, Y., Yang, L., Lei, D., Lu, C., … Li, J. (2006). Nerve growth factor and tyrosine kinase a in human salivary adenoid cystic carcinoma: Expression patterns and effects on in vitro invasive behavior. *Journal of Oral and Maxillofacial Surgery*, *64*(4), 636–641.

Wells, S. A., Robinson, B. G., Gagel, R. F., Dralle, H., Fagin, J. A., Santoro, M., … Schlumberger, M. J. (2012). Vandetanib in patients with locally advanced or metastatic medullary thyroid cancer: A randomized, double-blind phase III trial. *Journal of Clinical Oncology: Official Journal of the American Society of Clinical Oncology*, *30*(2), 134–141.

Yaar, M., Eller, M. S., DiBenedetto, P., Reenstra, W. R., Zhai, S., McQuaid, T., … Gilchrest, B. A. (1994). The trk family of receptors mediates nerve growth factor and neurotrophin-3 effects in melanocytes. *Journal of Clinical Investigation*, *94*(4), 1550–1562.

Zeng, Q., Cheng, Y., Zhu, Q., Yu, Z., Wu, X., Huang, K., … Zhang, Q. (2008). The relationship between over-expression of glial cell-derived neurotrophic factor and its RET receptor with progression and prognosis of human pancreatic cancer. *Journal of International Medical Research*, *36*(4), 656–664.

Zhang, S., Qi, L., Li, M., Zhang, D., Xu, S., Wang, N., & Sun, B. (2008). Chemokine CXCL12 and its receptor CXCR4 expression are associated with perineural invasion of prostate cancer. *Journal of Experimental & Clinical Cancer Research: CR*, *27*, 62.

Zhou, Z.-H., Xu, G.-F., Zhang, W.-J., Zhao, H.-B., & Wu, Y.-Y. (2014). Reevaluating significance of perineural invasion in gastric cancer based on double immunohistochemical staining. *Archives of Pathology & Laboratory Medicine*, *138*(2), 229–234.

Zhu, Z., Friess, H., diMola, F. F., Zimmermann, A., Graber, H. U., Korc, M., & Büchler, M. W. (1999). Nerve growth factor expression correlates with perineural invasion and pain in human pancreatic cancer. *Journal of Clinical Oncology*, *17*(8), 2419–2428.

Zhu, Z., Kleeff, J., Kayed, H., Wang, L., Korc, M., Büchler, M. W., & Friess, H. (2002). Nerve growth factor and enhancement of proliferation, invasion, and tumorigenicity of pancreatic cancer cells. *Molecular Carcinogenesis*, *35*(3), 138–147.

FSC
www.fsc.org
MIX
Papier | Fördert gute Waldnutzung
FSC® C083411

Kolibri 360 GmbH
Ferdinand-Jühlke-Straße 7
99095 Erfurt, Deutschland
produktsicherheit@kolibri360.de